Functional Neuroanatomy and Clinical Neuroscience

Functional Neuroanatomy and Clinical Neuroscience

Foundations for Understanding Disorders of Cognition and Behavior

Suzan Uysal

OXFORD
UNIVERSITY PRESS

Oxford University Press is a department of the University of Oxford. It furthers
the University's objective of excellence in research, scholarship, and education
by publishing worldwide. Oxford is a registered trade mark of Oxford University
Press in the UK and certain other countries.

Published in the United States of America by Oxford University Press
198 Madison Avenue, New York, NY 10016, United States of America.

© Oxford University Press 2023

All rights reserved. No part of this publication may be reproduced, stored in
a retrieval system, or transmitted, in any form or by any means, without the
prior permission in writing of Oxford University Press, or as expressly permitted
by law, by license, or under terms agreed with the appropriate reproduction
rights organization. Inquiries concerning reproduction outside the scope of the
above should be sent to the Rights Department, Oxford University Press, at the
address above.

You must not circulate this work in any other form
and you must impose this same condition on any acquirer.

Library of Congress Cataloging-in-Publication Data
Names: Uysal, Suzan, 1960– author.
Title: Functional neuroanatomy and clinical neuroscience : foundations for
understanding disorders of cognition and behavior / Suzan Uysal.
Description: New York, NY : Oxford University Press, [2023] |
Includes bibliographical references and index.
Identifiers: LCCN 2022032441 (print) | LCCN 2022032442 (ebook) |
ISBN 9780190943608 (hardback) | ISBN 9780190943622 (epub) |
ISBN 9780190943615 (updf) | ISBN 9780190943639 (digital-online)
Subjects: MESH: Nervous System—anatomy & histology |
Cognitive Neuroscience—methods | Neurocognitive Disorders—physiopathology |
Neurobehavioral Manifestations—physiology
Classification: LCC QP360.5 (print) | LCC QP360.5 (ebook) | NLM WL 101 |
DDC 612.8/233—dc23/eng/20220907
LC record available at https://lccn.loc.gov/2022032441
LC ebook record available at https://lccn.loc.gov/2022032442

DOI: 10.1093/oso/9780190943608.001.0001

9 8 7 6 5 4 3 2 1
Printed by Sheridan Books, Inc., United States of America

This book is gratefully dedicated to my husband, Aries Arditi, whose love, friendship, and support have paved my path.

Brief Contents

Preface xxiii
Acknowledgments xxv

1. Neurons, Glia, and Basic Neuroanatomy   1
2. Electrical Signaling in Neurons   14
3. Chemical Neurotransmission and Neuropsychopharmacology   24
4. Cranium, Spine, Meninges, Ventricles, and Cerebrospinal Fluid   35
5. Blood Supply of the Brain   49
6. The Peripheral Nervous System   61
7. The Spinal Cord   75
8. The Brainstem   84
9. The Cerebellum   94
10. The Diencephalon   108
11. The Basal Ganglia   123
12. Limbic Structures   138
13. The Cerebral Cortex   150
14. The Occipital Lobes and Visual Processing   168
15. The Parietal Lobes and Associated Disorders   184
16. The Temporal Lobes and Associated Disorders   199
17. The Frontal Lobes and Associated Disorders   212
18. Stroke and Vascular Cognitive Impairment   226
19. Epilepsy   239
20. Traumatic Brain Injury   256
21. Brain Neoplasms   270
22. Brain Infections   279

23. White Matter Disease   289
24. The Motor System and Motor Disorders   301
25. Disorders of Consciousness   319
26. Memory and Amnesia   329
27. Language and the Aphasias   340
28. Alexia, Agraphia, and Acalculia   353
29. Brain Imaging   361
30. The Neurological Examination   378
31. Neuropsychological Assessment   397

Index   411

Contents

Preface   xxiii
Acknowledgments   xxv

1. Neurons, Glia, and Basic Neuroanatomy   1
   Introduction   1
   Neurons   1
      Structural and Functional Components of the Neuron   2
      The Synapse   3
      Neuron Classification   3
   Glia   3
      Oligodendrocytes   4
      Schwann Cells   4
      Astrocytes   4
      Microglia   4
      Ependymal Cells   4
      Gliomas   4
   Basic Neuroanatomical Terminology   5
      Central vs. Peripheral Nervous System   5
      Afferents vs. Efferents   5
      Gray Matter vs. White Matter   5
   Anatomical Terms of Direction and Planes of Section   6
      Directional Terms   6
      Planes of Section   7
   Basic Brain Anatomy   8
      Brainstem   8
      Cerebellum   9
      Cerebrum   9
      Limbic System   10
   Developmental Basis of the Major Brain Subdivisions   10
   Neurodegenerative Disease   11
   Summary   12

2. Electrical Signaling in Neurons   14
   Introduction   14
   Basic Chemistry Concepts Governing Diffusion   15
      Concentration Gradients   15
      Electrostatic Gradients   15
      Balance of Concentration and Electrostatic Gradients   15
   Membrane Structure and Permeability   16
      Lipid Solubility   16
      Membrane Transport Proteins   17
   Membrane Potentials   17
      The Resting Membrane Potential   18
      The Action Potential   19
      Postsynaptic Potentials   21

Clinical Considerations   22
   Summary   22

3. **Chemical Neurotransmission and Neuropsychopharmacology   24**
   Introduction   24
   Synaptic Transmission   24
      Neurotransmitters   24
      Transmitter Release   25
      Neurotransmitter Receptors   25
      Removal of Neurotransmitter from the Synapse   27
      Mechanisms of Drug Action   27
   Neurotransmitter Systems   28
      Acetylcholine   28
      The Monoamines   29
      The Amino Acid Neurotransmitters   30
      The Neuropeptides   30
   Neuropsychopharmacology   31
      Drug Classification   31
      Antipsychotics   31
      Antidepressants   32
      Anxiolytics   33
      Anti-Manic Agents (Mood Stabilizers)   33
      Psychostimulants   33
      Sedative-Hypnotics   33
   Summary   34

4. **Cranium, Spine, Meninges, Ventricles, and Cerebrospinal Fluid   35**
   Introduction   35
   Cranium and Spine   35
      Basic Anatomy of the Cranium   35
      Basic Anatomy of the Spine   37
   Meninges   37
      Dura Mater   37
      Arachnoid Membrane   39
      Pia Mater   40
   Ventricular System   40
   Cerebrospinal Fluid   41
      Production, Circulation, and Reabsorption   41
   Clinical Considerations   42
      Intracranial Pressure, Mass Effect, and Brain Herniation   42
      Meningitis   43
      Meningeal Headaches   43
      Meningiomas   43
      Arachnoid Cysts   43
      Subarachnoid Hemorrhage   43
      Epidural and Subdural Hematomas   43
      Lumbar Puncture   45
      Hydrocephalus   45
   Summary   48

5. **Blood Supply of the Brain   49**
   Introduction   49
   Circulation Overview   49
      The Systemic and Pulmonary Circulations   50
      Extracranial Origin of the Arteries That Supply the Brain   50
   Arterial Supply of the Brain   51
      The Internal Carotid Arterial System/Anterior Circulation   51

     The Vertebrobasilar Arterial System/Posterior Circulation   52
     Blood Supply of the Deep Structures   53
     Watershed Zones   54
     The Cerebral Arterial Circle (Circle of Willis)   54
     Segmentation of the Major Cerebral Arteries   54
   Brain Venous Blood Outflow   55
   The Blood-Brain Barrier   55
   Cerebral Blood Flow and Metabolism   56
   Imaging of the Cerebrovasculature   57
   Cerebrovascular Abnormalities   58
     Arterial Stenosis   58
     Brain Aneurysm   58
     Arteriovenous Malformation   59
     Moyamoya Disease   59
   Summary   60

6. The Peripheral Nervous System   61
   Introduction   61
   Functional Subdivisions of the PNS   61
     The Autonomic Nervous System   62
   The Spinal Nerves   64
     Dermatomes   64
     Myotomes   66
     Peripheral Nerves   66
     Clinical Considerations   66
   The Cranial Nerves   66
     CN1: Olfactory Nerves   69
     CN2: Optic Nerves   70
     CN3: Oculomotor Nerves   70
     CN4: Trochlear Nerves   71
     CN5: Trigeminal Nerves   71
     CN6: Abducens Nerves   71
     CN7: Facial Nerves   71
     CN8: Vestibulocochlear Nerves   72
     CN9: Glossopharyngeal Nerves   72
     CN10: Vagus Nerves   73
     CN11: Accessory Nerves   73
     CN12: Hypoglossal Nerves   74
   Summary   74

7. The Spinal Cord   75
   Introduction   75
   Anatomy   75
     External Gross Anatomy   75
     Internal Gross Anatomy   76
     Blood Supply   77
     Segmental Organization   77
     Regional Differences in Spinal Cord Anatomy   78
   Functional Organization of the Spinal Cord   78
     Spinal Cord Gray Matter   78
     Spinal Cord White Matter   79
   Spinal Reflexes   80
   Spinal Cord Injury Syndromes   81
   Epidural and Spinal Anesthesia   82
   Summary   82

8. The Brainstem   84
   Introduction   84
   Basic Brainstem Anatomy   84

Ascending Somatosensory Tracts 85
  Descending Motor Tracts 85
  Brainstem Blood Supply 86
 The Medulla 86
  Somatosensory Nuclei 86
  Cranial Nerve Nuclei 87
  Nuclei Regulating Vital Life Functions 87
 The Pons 87
  The Ventral Pons 87
  The Dorsal Pons 88
 The Midbrain 88
  The Midbrain Tectum 88
  The Midbrain Tegmentum 89
  The Cerebral Peduncles 89
 The Reticular Formation 89
  Consciousness and Sleep-Wake Cycle 90
  Motor Function 90
  Autonomic Functions 90
  Sensory Modulation 90
 Clinical Considerations 91
 Summary 91

9. The Cerebellum 94
 Introduction 94
 Anatomy of the Cerebellum 94
  Lobes and Lobules 94
  Deep Cerebellar Nuclei 95
  Zonal Organization 96
  Cerebellar Peduncles 96
  Cerebellar Circuitry 97
  Blood Supply 98
 Functional Divisions of the Cerebellum 98
  The Vestibulocerebellum 98
  The Spinocerebellum 100
  The Cerebrocerebellum 100
 Signs and Symptoms of Cerebellar Damage 101
  Truncal Ataxia 102
  Appendicular Ataxia 102
  Ataxic Dysarthria 103
  Nystagmus 103
  Cerebellar Cognitive Affective Syndrome 104
 Cerebellar Pathology 104
  Cerebellar Infarction 104
  Tumors 105
  Acute Alcohol Intoxication and Chronic Alcoholism 105
  The Spinocerebellar Ataxias 105
  Congenital Malformations 105
 Summary 106

10. The Diencephalon 108
 Introduction 108
 Thalamus 108
  Anatomy 108
  Blood Supply 110
  Thalamic Nuclei and Thalamic Function 111
  Thalamic Pathology 112
 Hypothalamus 113
  Anatomy 113
  Function 115

        Autonomic Function   115
        Endocrine Function and the Pituitary Gland   115
        Motivated Behaviors   117
        Thermoregulation   120
        Circadian Rhythms   121
        Hypothalamic Syndromes   121
    Summary   121

11. The Basal Ganglia   123
    Introduction   123
    Anatomy   123
        Blood Supply   124
        Basal Ganglia Circuits   125
        The Sensorimotor Basal Ganglia Circuit   125
        The Associative Basal Ganglia Circuit   125
    Clinical Syndromes Resulting from Basal Ganglia Pathology   126
    Parkinson's Disease   126
        Resting Tremor   127
        Rigidity   127
        Hypokinesia and Bradykinesia   127
        Postural Instability and Gait Impairment   127
        Myerson's Sign   127
        Non-Motor Features   128
        Pathology   128
        Genetics of Parkinson's Disease   129
        Treatment   129
        Other Causes of Parkinsonism   130
    Parkinson Plus Syndromes   131
        Diffuse Lewy Body Disease   131
        Progressive Supranuclear Palsy   132
        Corticobasal Syndrome   134
        Multisystem Atrophy   134
    Huntington's Disease   135
    Wilson's Disease   136
    Sydenham's Chorea   136
    Basal Ganglia Calcification   136
    Neurodegeneration with Brain Iron Accumulation   137
    Summary   137

12. Limbic Structures   138
    Introduction   138
    History   138
    The Cingulate Cortex   139
        Blood Supply   140
        Cingulotomy   140
    The Amygdala: Fear, Anxiety, and Aggression   140
        Anatomy   140
        Klüver-Bucy Syndrome   141
        Aggression   141
        Fear and Anxiety   142
    The Nucleus Accumbens: Reward and Pleasure   143
        Anatomy   143
        Reward   143
    The Septal Region   145
        Anatomy   145
        Function   145
    Limbic Encephalitis   146
        Herpes Simplex Virus Limbic Encephalitis   146
        Autoimmune Limbic Encephalitis   146

Deep Brain Stimulation for Psychiatric Disorders   147
   Addiction   147
     Major Depressive Disorder   147
Summary   147

13. The Cerebral Cortex   150

Introduction   150
Comparative Neuroanatomy   150
Cortical Anatomy   151
   Major Landmarks and Lobes   151
     Lateral Surface   152
     Medial Surface   153
     Inferior Surface   154
Cortical Connections   154
   Projection Fibers   154
   Commissural Fibers   155
   Association Fibers   156
Cortical Localization   157
   History   157
Functional Maps of the Cerebral Cortex   158
Primary, Secondary, and Tertiary Cortical Zones   159
Cytoarchitecture of the Cerebral Cortex   161
   Cortical Cells   161
   Cortical Layers   161
   Types of Cerebral Cortex Based on Microstructure of Cortical Layers   162
   Cytoarchitectural Maps   162
   Columnar Organization   162
Cortical Disconnections   163
   Corpus Callosotomy   163
Agenesis of the Corpus Callosum   165
Summary   166

14. The Occipital Lobes and Visual Processing   168

Introduction   168
Basic Anatomy of the Occipital Lobes   168
The Retina   169
   The Receptive Field   170
   Pattern Processing within the Retina   170
   Color Coding   171
The Visual Field   171
The Visual Pathways   172
   From the Retina to the Lateral Geniculate Nucleus   172
   From the Lateral Geniculate Nucleus to the Primary Visual Cortex   173
   The Retinotopic Map   173
   Secondary Visual Cortex   173
   The Extrageniculate Pathways   174
Visual Field Defects and Cerebral Blindness   174
Disorders of Higher-Order Visual Processing: Visual Agnosia   176
   The Concept of Visual Agnosia   176
   Apperceptive Visual Agnosia   176
Disorders of Higher-Order Visual Processing: Lesions of the Ventral Stream   177
   Associative Visual Agnosia/Visual Object Agnosia   177
   Prosopagnosia   178
   Acquired Achromatopsia   178
Disorders of Higher-Order Visual Processing: Lesions of the Dorsal Stream   179
   Astereopsis   179

    Akinetopsia 179
    Optic Ataxia 179
    Oculomotor Apraxia 180
  Visual Disconnection Syndromes 180
    Pure Alexia 180
    Color Anomia 180
  Visual Hallucinations and Illusions 181
    Visual Release Hallucinations (Charles Bonnet Syndrome) 182
  Summary 182

15. The Parietal Lobes and Associated Disorders 184
  Introduction 184
  Basic Anatomy of the Parietal Lobes 184
  Somatosensation 185
    The Somatosensory Pathways 185
    Somatosensory Cortex 186
  Disorders of Cortical Somatosensory Processing 187
    Lesions of Primary Somatosensory Cortex 187
    Disorders of Higher-Order Cortical Somatosensory Processing 188
  Somatosensory Hallucinations and Illusions 188
    Positive Somesthetic Symptoms 188
    Parietal Lobe Epilepsy 189
    Phantom Limb 189
  Disorders of Spatial Cognition 189
    Visuomotor Disorders 190
    Visuospatial Disorders 190
    Topographic Disorientation 190
    Body Schema Disorders 190
  Disorders of Spatial Attention 191
    Unilateral Spatial Neglect 192
    Simultanagnosia 193
  Bálint's Syndrome 194
  Left Angular Gyrus Syndrome and Gerstmann's Syndrome 194
  Anosognosia 194
    Anosognosia for Hemiplegia (Babinski's Syndrome) 194
    Visual Anosognosia (Anton's Syndrome) 195
  Posterior Cortical Atrophy (Benson's Syndrome) 195
  Summary 197

16. The Temporal Lobes and Associated Disorders 199
  Introduction 199
  Basic Anatomy of the Temporal Lobes 199
    The Hippocampal Formation 200
    Blood Supply to the Temporal Lobes 201
  Audition 202
    The Auditory Stimulus 202
    The Ear 202
    The Auditory Pathways 204
    Auditory Deficits 204
    Cortical Auditory Disorders 205
    Auditory Illusions and Hallucinations 206
  Olfaction 206
    Anosmia 207
    Olfactory Hallucinations 207
  Alzheimer's Disease 207
    Clinical Presentation 207
    Epidemiology 208

Pathology 208
Etiology 209
Diagnosis 210
Treatment 210
Summary 210

17. The Frontal Lobes and Associated Disorders 212
Introduction 212
Basic Anatomy of the Frontal Lobes 212
Motor Cortex 213
History 213
Primary Motor Cortex 214
Secondary Motor Cortex 215
Prefrontal Cortex 217
Anatomy 217
The Frontal Lobe Controversy 217
Phineas Gage 218
Prefrontal Lobotomy 219
Prefrontal Injury and Disease Syndromes 220
The Dorsolateral Prefrontal Syndrome (Dysexecutive Syndrome) 221
The Orbitofrontal Syndrome (Disinhibited Syndrome) 221
The Medial Frontal/Anterior Cingulate Syndrome (Akinetic/Apathetic Syndrome) 222
Signs and Symptoms Due to Diffuse Lesions 222
Behavioral Variant Frontotemporal Dementia 223
Summary 224

18. Stroke and Vascular Cognitive Impairment 226
Introduction 226
Ischemic Stroke 226
Pathophysiology 226
Anatomic Classification of Ischemic Stroke 227
Transient Ischemic Attacks 231
Hemorrhagic Stroke 232
Pathophysiology 232
Cerebral Amyloid Angiopathy 233
Stroke Diagnosis 233
Stroke Treatment 233
Ischemic Stroke 233
Subarachnoid Hemorrhage 234
Intracerebral Hemorrhage 234
Stroke Risk Factors and Stroke Prevention 234
Vascular Cognitive Impairment (VCI) 235
Definition 235
Clinical Presentation of Vascular Cognitive Impairment 235
Neuropsychological Assessment and Differential Diagnosis 237
Treatment Recommendations for Patients with Vascular Cognitive Impairment 237
Summary 237

19. Epilepsy 239
Introduction 239
Definitions 239
The EEG and Epilepsy Diagnosis 240
The International 10–20 System 240
Physiological Basis of the EEG 241
EEG Interpretation 242
Classification of Seizures 244
Focal vs. Generalized Seizures 245

Seizures with Preserved Awareness vs. Impaired Awareness   245
Motor vs. Non-Motor Seizures   246
Focal Onset Seizures   246
Focal Onset Seizures with Preserved Awareness   246
Focal Onset Seizures with Impaired Awareness   248
Generalized Seizures   248
Generalized Motor Seizures   249
Generalized Non-Motor Seizures   249
Status Epilepticus   249
Etiology: Idiopathic (Primary) vs. Symptomatic (Secondary) Epilepsy   249
Non-Epileptic Seizures   250
Epilepsy Syndromes   250
Lennox-Gastaut Syndrome   250
Rasmussen's Syndrome   251
Landau-Kleffner Syndrome   251
Reflex Epilepsy   251
Differential Diagnosis   251
Medical Treatment   251
Epilepsy Surgery   252
Seizure Focus Localization   252
Intracarotid Amobarbital Procedure and Functional Mapping   252
Surgical Procedures   253
Neuromodulation   254
Summary   254

20. Traumatic Brain Injury   256
Introduction   256
Definition   256
Epidemiology and Etiology   257
Pathophysiology of TBI   257
Primary vs. Secondary Injury   257
Primary Injuries   257
Secondary Injuries   259
Classification of Injuries   261
Penetrating vs. Non-Penetrating Head Injuries   261
Focal vs. Diffuse Injuries   261
Coup vs. Contrecoup Injuries   261
Clinical Classification of Acute TBI and Measures of TBI Severity   262
The Glasgow Coma Scale   262
Duration of Loss of Consciousness   263
Post-Traumatic Amnesia   263
Moderate-Severe TBI   263
Mild TBI   265
Concussion   265
TBI Management   266
Chronic Traumatic Encephalopathy   267
Summary   268

21. Brain Neoplasms   270
Introduction   270
Tumor Classification   270
Gliomas   271
Meningiomas   272
Pituitary Tumors   273
Schwannomas   274
Metastatic Tumors   275
Clinical Signs and Symptoms   275
Focal Neurological Signs and Symptoms   275

Seizures 275
Headache and Increased Intracranial Pressure 276
Endocrine Abnormalities 276
Diagnosis and Prognosis 276
Treatment 276
Surgical Intervention 276
Radiation Therapy 277
Chemotherapy 277
Neurological Effects 277
Summary 278

22. Brain Infections 279
Introduction 279
Classification 279
Portal of Entry 280
Meningitis 280
Viral Meningitis 280
Bacterial Meningitis 281
Encephalitis 281
Herpes Simplex Virus Encephalitis 281
Rabies Encephalitis 282
Brain Abscess 282
Fungal Infections of the CNS 283
Parasitic Infections of the CNS 283
Neurocysticercosis 283
Prion Diseases 284
Kuru 285
Creutzfeldt-Jakob Disease 285
Fatal Familial Insomnia 286
Gerstmann-Sträussler-Scheinker Syndrome 286
Human Immunodeficiency Virus 286
Lyme Disease 287
Summary 287

23. White Matter Disease 289
Introduction 289
White Matter 289
The Leukodystrophies 291
Autoimmune Inflammatory Demyelinating Diseases 292
Multiple Sclerosis 292
Guillain-Barré Syndrome 295
Acute Disseminated Encephalomyelitis 295
Toxic Leukoencephalopathy 295
Drugs of Abuse and Environmental Toxins 295
Radiation-Induced Toxic Leukoencephalopathy 296
Chemotherapy-Induced Toxic Leukoencephalopathy 297
Infectious Demyelinating Disorders 297
Progressive Multifocal Leukoencephalopathy 297
Other Viral Infections Associated with Demyelination 298
Acquired Metabolic Leukoencephalopathy 298
Central Pontine Myelinolysis 298
Cobalamin Deficiency 299
Hypoxic-Ischemic Leukoencephalopathy 299
Hypoxic Leukoencephalopathy 299
Ischemic Leukoencephalopathy 299
Summary 300

24. The Motor System and Motor Disorders 301
   Introduction 301
   Muscle 302
      Basic Mechanics of Movement 302
      Muscle Fibers 303
      Muscle Contraction 303
      The Monosynaptic Stretch Reflex 304
   Lower Motor Neurons 304
      Spinal Cord 304
      Cranial Nerve Nuclei 304
   Upper Motor Neurons: The Descending Motor Pathways 305
      Motor Cortex 305
      Pathways from the Cerebral Cortex 306
      Indirect Cortical-Brainstem-Spinal Cord Pathways 307
      Two Descending Motor Systems 308
      Basal Ganglia and Cerebellum 309
   Lower Motor Neuron versus Upper Motor Neuron Lesions 309
      Bulbar versus Pseudobulbar Palsy 309
   Diagnostic Testing 309
   Myasthenia Gravis 310
   Cerebral Palsy 310
   Motor Neuron Disease 311
      Amyotrophic Lateral Sclerosis 311
      Progressive Bulbar Palsy 311
      Primary Lateral Sclerosis 311
   Apraxia 312
      Ideomotor Apraxia 313
      Ideational Apraxia 314
      Apraxia of Speech 314
      Apraxic Agraphia 314
      Disconnection Apraxias 315
      Conduction Apraxia 315
      Apraxia-Like Syndromes 315
   Alien Hand Syndrome 317
   Summary 317

25. Disorders of Consciousness 319
   Introduction 319
   Neurological Assessment: The Glasgow Coma Scale 320
      Motor Response 320
      Verbal Response 322
      Eye-Opening Response 322
   Coma 322
      Pathophysiology of Coma 322
      Etiology 323
      Examination 323
      Prognosis 323
   The Vegetative State 324
   The Minimally Conscious State 324
   Delirium 324
   Syndromes That Mimic Disorders of Consciousness 326
      Locked-In Syndrome 326
      Brain Death 327
   Summary 327

26. Memory and Amnesia 329
   Introduction 329
   The Case of H.M. 329
      Anatomy of Medial Temporal Amnesia 331
         Pathologies Causing Medial Temporal Amnesia 332
         Material-Specific Anterograde Amnesia 332
      Remote Memory and Retrograde Amnesia 332
      Amnesia Due to Lesions in Non-Medial Temporal Lobe Structures 333
         Fornix 333
         Diencephalic Amnesia and Korsakoff's Syndrome 333
         Basal Forebrain Amnesia 334
         Frontal Amnesia 334
      Transient Amnesia Syndromes 334
         Transient Global Amnesia 334
         Other Transient Amnesia Syndromes 335
      Dissociative Amnesia 335
      Forms of Memory 335
         Declarative Memory 336
         Nondeclarative Memory 336
      Forms of Memory Based on Time Span 337
         Immediate Memory 337
         Recent Memory 337
         Remote Memory 337
      Summary 338

27. Language and the Aphasias 340
   Introduction 340
   Oral Language 341
      Speech 341
   Aphasia 342
      Language Assessment 343
   The Classic Aphasia Syndromes 344
      Broca's Aphasia 344
      Wernicke's Aphasia 345
      Global Aphasia 346
      Conduction Aphasia 346
      The Extrasylvian Aphasias 346
      Nominal (Anomic) Aphasia 347
   Other Aphasias 347
      Subcortical Aphasia 347
      Dynamic Aphasia 347
   The Primary Progressive Aphasias 348
      Semantic Variant PPA 348
      Nonfluent/Agrammatic Variant PPA 350
      Logopenic Variant PPA 350
   Summary 351

28. Alexia, Agraphia, and Acalculia 353
   Introduction 353
   Alexia 354
      Pure Alexia 354
      Alexia with Agraphia 354
      Surface Alexia 354
      Deep Alexia 355
      Spatial Alexia 356
   Agraphia 356
      Pure Agraphia 356
      Surface Agraphia 356

Spatial Agraphia   356
Apraxic Agraphia   357
Agraphia Due to Non-Apraxic Motor Disturbances   357
Acalculia   357
Primary Acalculia   357
Aphasic, Alexic, and Agraphic Acalculias   358
Spatial Acalculia   358
Dysexecutive Acalculia   358
Summary   358

29. Brain Imaging   361
Jacqueline C. Junn and Suzan Uysal
Introduction   361
Plain Film X-Ray   362
Angiography   362
Structural Brain Imaging   363
Imaging Planes   363
Contrast Enhancement   363
Interpretation Basics   363
Computed Tomography   368
Magnetic Resonance Imaging   369
Ultrasound   373
Functional Neuroimaging   374
Single Photon Emission Computed Tomography   374
Positron Emission Tomography   374
Functional MRI   376
Summary   376

30. The Neurological Examination   378
Suzan Uysal and Stephen Krieger
Introduction   378
The Clinical History   379
The Mental Status Exam   379
The Cranial Nerve Exam   379
CN1 Testing   380
CN2 Testing   380
CN2 and CN3 Testing   381
CN3, CN4, and CN6 (Eye Movement) Testing   383
CN5 Testing   385
CN7 Testing   386
CN5 and CN7 Testing   387
CN8 Testing   387
CN9 and CN10 Testing   387
CN11 Testing   388
CN12 Testing   388
The Motor Exam   388
Appearance   388
Muscle Tone   389
Strength   389
Abnormal Movements   390
The Sensory Exam   390
Discriminative Touch   391
Proprioception   391
Vibration   391
Pain   391
Cortical Somatosensory Function   391
The Reflex Exam   392
Deep Tendon Reflexes   392
Primitive Reflexes   393

    Coordination/Cerebellar Exam   394
    Station and Gait   395
    Summary   396

31. Neuropsychological Assessment   397
    Introduction   397
    The Referral Question   398
    History   398
        History of the Presenting Problem/Present Illness   398
        Previous Examinations and Studies   398
        Medical History   399
        Family Medical History   399
        Developmental History   399
        Education, Work, and Social History   399
    Behavioral Observations   399
        Appearance   399
        Sensory and Motor Function   399
        Speech and Language   399
        Cognitive Process   400
        Behavioral Regulation   400
        Affect   400
        Comportment, Tact, and Interpersonal Relatedness   400
    Testing Procedures and Results   400
        Attention   401
        Visuoperceptual Function and Spatial Cognition   402
        Praxis and Constructional Ability   403
        Language   404
        Number Processing and Calculation   405
        Learning and Memory   405
        Abstract Thinking and Reasoning   406
        Executive Functions   406
        Performance Validity Testing   409
        Mood and Personality   409
    Diagnostic/Descriptive Conclusions   409
    Recommendations   409
    Summary   410

Index   411

Preface

The human nervous system is an exquisite piece of biological machinery. It underlies the greatest achievements of humankind, such as the sciences, the arts, and technology. But it also underlies the most mundane activities that most of us perform effortlessly every day, such as seeing, hearing, walking, talking, reading, writing, remembering, dressing, navigating, and planning one's day. These, too, are feats of biological engineering, as becomes readily apparent when one has contact with someone who has lost one or more of these fundamental abilities from brain disease or injury.

This volume is intended for a broad range of non-physician healthcare professionals who work with patients with brain disease or injury, and particularly for clinical neuropsychologists. It works well as either a textbook or a reference. Its purpose is to present functional neuroanatomy and clinical neuroscience (i.e., the "*neuro*" in *neuropsychology*) in a simple but not simplistic way that is accessible to an audience that has had exposure to brain and behavioral science but does not have a medical school education. It interleaves discussion of functional neuroanatomy, clinical neuroscience, and disorders of the human central nervous system with discussion of neurocognitive and neurobehavioral syndromes. It provides a comprehensive overview of key neuroanatomic concepts without being overly detailed, and clearly links those concepts to cognitive and behavioral disorders that are relevant for clinical practice.

The book reflects my many years of experience teaching basic behavioral neuroscience and clinical neuropsychology. I believe that neuropsychologists and other non-physician healthcare professionals benefit greatly from foundational education in functional neuroanatomy and clinical neuroscience. Armed with such knowledge, they can better understand the nature of their patients' conditions and medical records, better communicate with other professionals, and better contribute to multidisciplinary research.

The chapters follow an order that I have found best for teaching: (1) functional neuroanatomy that is clinically relevant, with subsections discussing clinical syndromes that are localized to specific brain areas; (2) common neuropathologies that may affect many different parts of the brain; (3) domain-specific syndromes that often involve more than one area of the brain; and (4) clinical assessment methods. But while the book can be read (or taught) in chapter order, there is ample cross-referencing, so it is optional to follow that order. That is, the volume functions as well as a reference as it does as a textbook. For readers who have limited familiarity with brain imaging, it may be helpful to read that chapter (Chapter 29) to better understand neuroimaging findings discussed in earlier chapters.

A word about stylistic choices I have made is in order. First, I have attempted to be vigilant in defining all commonly used technical terms, with synonyms in parentheses. While this makes certain sentences denser and less easy to vocalize in the mind's ear, it has the advantage of presenting nearly all equivalent terminology that might be encountered in other books and articles. Second, I have referred to *people* when describing those who experience the effects of brain disease or injury on function in daily life, but to *patients* when discussing findings on clinical examinations, such as brain imaging, the neurological exam, and neuropsychological assessment. Finally, I have adopted a contemporary but not universally accepted gender-neutral yet compact writing style, and therefore I employ "they" and "their" as both singular and plural definite pronouns, to avoid both gender-biased language and clumsier constructions such as "she or he," "s/he," and alternating "he's" and "she's." I hope this choice offends no one's grammatic sensibilities!

Acknowledgments

I extend my gratitude toward many people who have helped make this book possible. First and foremost, I thank Aries Arditi, Ph.D., for carefully editing the entire manuscript, offering insightful suggestions about the content, and improving the language throughout. The final product is infinitely better thanks to his input. I also thank Lisa Glukhovsky, Ph.D., for her reading of the text from the viewpoint of a postdoctoral fellow in clinical neuropsychology and for her valuable editorial input. Words alone, however, do not tell the whole story—I thank Jill K. Gregory for her beautiful illustrations and Jacqueline C. Junn, M.D., for providing neuroimages.

I thank my late parents Turhan Uysal, M.D., and Gloria Uysal for their strong belief in the importance of education, and for their belief in me. I also thank my teachers and mentors, all of whom have shaped my thinking; my students for their enthusiasm for learning; and my patients for their critical role in my clinical education and for sharing with me their humanity. Finally, I thank my family—my husband, Aries Arditi, and our daughters Anika and Zoë. Aries is my soul mate and life partner; he has been supportive of all my endeavors, since the earliest years of my graduate education and throughout my career, and in all things beyond work. Zoë and Anika, the greatest gifts in our lives, have made life rich with love and purpose. To all, my deepest gratitude.

Suzan Uysal
Icahn School of Medicine at Mount Sinai

Neurons, Glia, and Basic Neuroanatomy

Introduction

The human nervous system forms the underlying physiological basis for our thoughts, feelings, and behaviors (including even our interest in and ability to analyze and understand that system). A basic knowledge of structural neuroanatomy is the starting point for developing an understanding of nervous system function, the mechanisms of brain diseases and injuries, and their resulting effects on cognition and behavior. This chapter acquaints the reader with the cells of the nervous system, basic neuroanatomical terminology, anatomical terms of direction, and basic brain anatomy. It concludes with a general description of one class of brain disorder, the neurodegenerative diseases, which selectively target specific anatomical regions or functional systems of the nervous system and are therefore discussed throughout this book.

Neurons

The nervous system as a whole: (1) continuously receives information about the world and the state of the organism, monitoring both the external and internal environments; (2) integrates incoming information with information previously learned through experience; and (3) guides physical movements and behaviors. These three main roles are often referred to as sensory, integrative, and motor, and their functions are accomplished by an extensive network of **neurons** (also known as **nerve cells**), cells that have evolved special features for carrying out the main information-processing tasks of the nervous system and which connect to each other to form neural circuits. **Glial cells**, a second major cell type in the nervous system, perform essential support functions for neurons. Neurons and glia together constitute the nervous system **parenchyma**, the essential tissue that makes up the bulk of an organ and conducts its specific function.

The neuron is the basic anatomical and physiological unit of the nervous system, an assertion that is known as the **neuron doctrine**. Early support for the doctrine was based on the work of neuroanatomist **Santiago Ramón y Cajal** (1852–1934), who applied the Golgi method of silver staining to visualize the histology of nervous tissue under a light microscope. This method stains a small number of cells (1%) in their entirety, including the fine dendrites and axons, allowing whole cells, which would not be visible using other staining techniques, to stand out against the background of densely packed neurons,. At the time the young field of neuroscience was dominated by a debate as to whether the nervous system was composed of a diffuse network of continuous tissue fibers (**reticular theory**) or of distinct individual cells (**neuron theory**). Cajal's work established the latter, and in 1906 he and **Camillo Golgi** (1843–1926) were awarded the Nobel Prize in Physiology or Medicine for this discovery.

Neurons are concentrated within the central nervous system (CNS), which consists of the brain and spinal cord. Neurons are also present throughout most of the body, where they make up the peripheral nervous system (PNS), passing information from the periphery to the CNS and from the CNS to the periphery, as well as processing information. Individual neurons mirror the functions of the nervous system as a whole: they receive incoming signals, integrate signals from multiple sources, and transmit signals to other neurons. Neurons are electrically excitable; they receive, integrate, and transmit information through changes in electrical state (see Chapter 2, "Electrical Signaling in Neurons").

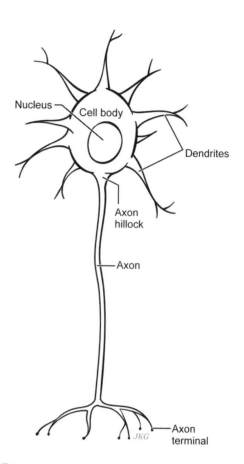

FIGURE 1.1. Standard neuron, with dendrites, cell body, axon, and axon terminals. The axon hillock is a region between the cell body and initial segment of the axon, where action potentials are generated.

Structural and Functional Components of the Neuron

All neurons have four fundamental components, each having a distinct morphology and function: dendrites, a cell body, an axon, and axon terminals (Figure 1.1). The distribution of organelles and macromolecular components varies among these regions and plays a role in determining function.

The **cell body** (**soma**, **perikaryon**) is the metabolic center of the cell. It contains organelles that are common to all animal cells, such as the nucleus, mitochondria, Golgi apparatus, ribosomes, and endoplasmic reticulum. Neuron cell bodies vary in diameter, ranging from 5 to 135 μm. Two kinds of fibers arise from the neuron cell body: a single, non-branching process known as the **axon**, and numerous branching extensions known as **dendrites** (Greek *dendro*, "tree").

The **axon** (**nerve fiber**) is a single tubular process that extends from the soma, originating from a specialized region at the point of transition between the cell body and axon, known as the **axon hillock**. The axon functions as the conducting unit of the neuron, carrying electrical signals known as **action potentials** that propagate like an electrochemical wave along the length of the axon and carry information from one end of the neuron to the other. Axons are longer and thinner than dendrites, have a uniform diameter, and range from 1 μm to over 1 meter in length. Examples of neurons with especially long axons are those that carry motor signals from the motor cortex to the lower spinal cord to control the lower extremities, and those that carry somatosensory signals from the lower extremities up to the lower brain. Despite the fact that some neurons have very long axons, the neurons of most animals are generally too small to be visible to the unaided eye; a notable exception is the squid giant axon, which can have a diameter as large as 1 mm! Axons may give rise to one or more large collateral branches, which generally branch off the main axon at a right angle. The axon diameter remains uniform despite this collateral branching. Axons commonly branch profusely just before they terminate, often thousands of times.

The distal ends of the terminal branches of the axon, which are enlarged, are called **terminal buttons** (**axon**

terminals). These contain vesicles that are filled with a chemical **neurotransmitter** substance. When an action potential arrives at the axon terminals, neurotransmitter is released from the terminal buttons into the **synapse** (**synaptic cleft**), a small (approximately 20 nanometers wide) space between the signaling neuron and the receiving cell.

Dendrites constitute the chief information-receiving apparatus of the neuron. They often branch profusely, and their diameter tapers as they extend a short distance from the cell body with a maximum distance of approximately 1 µm. The dendritic branches may have **dendritic spines**, small protrusions that increase the surface area of dendrites and thereby allow for a greater number of synaptic contacts. Not surprisingly, the number and extent of dendritic processes emanating from a given cell correlate with the number of synaptic contacts that other neurons make with that cell.

To summarize, the four basic structural components of neurons serve four distinct functions in signal transmission: (1) dendrites serve as the chief receptive (input) component; (2) the cell body serves as the principal integrative (summing) component; (3) the axon serves as the conductive (signaling) component; and (4) axon terminal buttons serve as the neurotransmitter-secreting (output) component. Thus, information signals are conducted *within* neurons electrically, while information signals are transmitted *between* neurons chemically.

The Synapse

Within the CNS, axon terminals most often synapse with juxtaposed dendrites of the neurons with which they communicate. The axon terminal of the signaling neuron is the **presynaptic** element, and the dendrite of the receiving neuron is the **postsynaptic** element; this contact is known as an **axodendritic synapse**. Synaptic contacts also occur between axon terminals and neuronal cell bodies in **axosomatic** synapses, as well as between axon terminals and the axons of other neurons in **axoaxonic** synapses. In the periphery, efferent neurons synapse on the cells of effector organs such as a muscle or gland; this type of synapse is known as a **neuroeffector junction**. The synapse between efferent neurons and muscle cells is known as the **neuromuscular junction**.

Neuron Classification

Neurons come in many shapes and sizes and are more varied than any other type of cell in the body. They are classified along several structural dimensions, including the type of tissue to which they connect, the number of **neurites** (axons and dendrites) that extend from the soma, cell body shape, dendritic tree branching pattern, and axon length.

Classification based on the type of tissue to which the neuron connects relates to neuron function. Neurons that connect to sensory receptors (e.g., skin, retina) are known as **primary sensory neurons**. Neurons that synapse with muscle are called **primary motor neurons** (**motoneurons**). Most neurons connect to other neurons and are classified broadly as **interneurons**.

Neurons are also classified based on the number of neurites that extend from the soma. **Multipolar** neurons have several dendrites and one axon emanating from the cell body. Most of the neurons in mammalian nervous systems are multipolar. **Bipolar** neurons have one dendrite emanating from one end of their cell body and an axon emanating from the other end. They are found in peripheral components of the visual, olfactory, auditory, and vestibular sensory systems. **Unipolar** neurons have a single process that extends from the cell body and bifurcates into a dendrite and an axon. They are found in peripheral components of the somatosensory system.

Cell body shape is another structural feature by which neurons are classified. For example, pyramidal cells have pyramid-shaped cell bodies, stellate cells have star-shaped cell bodies, and basket cells have a dense plexus of terminals that surround the cell body of target cells. Dendritic arborization (branching) patterns vary widely and account for much of the morphological variability between neurons. For example, cerebellar Purkinje cells have a characteristic planar fan-shaped dendritic tree. One classification based on dendritic morphology is based on whether cells are **spiny** (i.e., have dendritic spines) or **aspinous** (i.e., without spines).

Finally, neurons may be classified according to the length of the axon, which reflects the signaling function of the neuron. Neurons with long axons that extend from one brain region to another are called **projection neurons**; they are also referred to as **relay neurons** or Golgi Type I neurons. These neurons make up the long fiber tracts of the brain, spinal cord, and peripheral nerves (e.g., corticospinal neurons). Neurons that have short axons terminating in the neighborhood of the soma are called **local circuit neurons**; they are also referred to as **association neurons** or Golgi Type II neurons. These neurons play an important role in local information processing within a brain region.

Glia

Glial cells, also known as **glia** or **neuroglia**, are found between nerve cell bodies and axons. The term *glia* is derived from the Greek word for "glue." Initially it was thought that glial cells simply hold neurons in place, but it is now known that glia provide much more than structural support. Glia account for approximately half the volume of the brain and spinal cord. They are typically small and extremely numerous. Unlike neurons, glial cells are not electrically excitable and do not generate electrical signals.

One of the most important roles of glia is axon myelination. Some types of glial cells manufacture and wrap axons with **myelin**, a fatty substance, in a series of segments that are about 1 mm in length and separated by small gaps known as **nodes of Ranvier** that are approximately 1 μm long and evenly spaced along the length of the axon. Myelin acts as an insulating sheath, ensuring that electrical signals from one axon do not spread to adjacent neurons. Myelin also vastly increases the speed of action potential conduction down the length of the axon (by as much as 100 times!). This is especially important for cells that conduct signals between structures that are separated by long distances and that play a role in time-critical functions such as motor and sensory pathways, where muscles and sense organs lie outside the brain. The mechanism by which myelination increases the speed of action potential conduction down the axon is described in Chapter 2.

Myelination is an important aspect of nervous system maturation. In species in which birth occurs prior to myelination (such as humans), the young are quite helpless motorically until myelination is well under way. Myelination also plays an important role in human cognitive and behavioral development from infancy to young adulthood. Neuroimaging studies have shown that with normal aging, there is a loss of myelin in late life that follows an anterior-posterior gradient, with frontal-temporal regions and anterior corpus callosum preferentially affected. The loss of myelin is associated with age-related cognitive changes. There is evidence that myelin also enhances the speed of mentation and that loss of myelin and slower conduction decrease the rate of mentation.

There are five types of glial cells: oligodendrocytes, Schwann cells, astrocytes, microglia, and ependymal cells.

Oligodendrocytes

Oligodendrocytes (**oligodendroglia**) provide myelin to neurons of the brain and spinal cord. Each oligodendrocyte has numerous processes that extend from its cell body to the axons of multiple neurons, with each process providing approximately 1 mm of myelin sheath by wrapping around each axon in concentric layers.

Schwann Cells

Schwann cells provide myelin to the neurons of the PNS that make up the peripheral nerves, having a similar function to the oligodendrocytes of the CNS. Each Schwann cell provides one segment of myelin sheath for one axon, as the entire Schwann cell concentrically wraps around the axon. Thus, multiple Schwann cells ensheathe a single axon.

Astrocytes

Astrocytes (**astroglia**) are the most numerous cell type in the CNS. They have star-shaped cell bodies and many irregularly shaped processes that extend from the cell body. Astrocytes provide physical and nutritive support to neurons and surround and isolate synapses minimizing transmitter dispersion. Many astrocytes have processes that surround capillaries and play a role in forming the **blood-brain barrier**, a chemical barrier that can prevent harmful substances from entering the brain (see Chapter 5, "Blood Supply of the Brain"). Astrocytes regulate the chemical composition of the extracellular fluid by maintaining the appropriate ionic composition and removing chemical transmitters that have been released into the extracellular space. Astrocytes also play a role in CNS scar tissue formation.

Microglia

Microglia act as scavengers in response to injury, infection, and disease within the CNS. Resting microglial cells are evenly spaced throughout the CNS. When the brain or spinal cord is injured, resting microglial cells in the region transform into reactive microglial cells; they retract their processes, actively proliferate, exhibit ameboid movement (a crawling type of movement similar to that of amoeba and accomplished by foot-like extensions of the cytoplasm), acquire phagocytic properties (they engulf and digest debris after neuronal death), and form scar tissue by a process known as **gliosis**.

Ependymal Cells

Ependymal cells line the brain ventricles (a network of hollow spaces within the brain that are filled with cerebrospinal fluid) and the spinal cord central canal. Ependymal cells play a role in the production and circulation of cerebrospinal fluid, a clear fluid that surrounds the brain and spinal cord (see Chapter 4, "Cranium, Spine, Meninges, Ventricles, and Cerebrospinal Fluid").

Gliomas

Especially relevant to clinical brain science, all glia types can give rise to **neoplasms** (tumors) known as **gliomas**, which may be benign or cancerous (see Chapter 21, "Brain Neoplasms"). The naming scheme is logical. Astrocytes give rise to **astrocytomas**, oligodendrocytes give rise to **oligodendrogliomas**, Schwann cells give rise to **schwannomas**, and ependymal cells give rise to **ependymomas**.

Basic Neuroanatomical Terminology

Central vs. Peripheral Nervous System

As discussed earlier, the nervous system is conventionally divided into two regions: central and peripheral. The CNS consists of brain and spinal cord, the structures of the nervous system that are encased in bone (Figure 1.2). The brain lies within the **cranium**; it consists of the brainstem, cerebellum, and cerebrum. The spinal cord lies within the **vertebral column (spinal column)**.

The PNS lies outside the cranium and vertebral column (i.e., outside the CNS). It consists of 12 pairs of **cranial nerves** that enter and exit the cranium, and 31 pairs of **spinal nerves** that enter and exit the spinal column, as well as their associated branches and ganglia. The nerves of the PNS carry sensory signals from sensory organs and other tissues in the periphery to the CNS, and motor signals from the CNS to muscle cells. The PNS is divided into somatic and autonomic subdivisions. The somatic nervous system consists of nerves innervating the sensory organs and skeletal muscles. The autonomic nervous system consists of nerves innervating the visceral organs.

Afferents vs. Efferents

Nerve fibers that carry sensory information from the body and special sense organs to the CNS are called **afferents** (mnemonic: *afferents arrive*). Nerve fibers that carry motor information from the CNS to the effector organs of the body (muscles and glands) are called **efferents** (mnemonic: *efferents exit*). The terms *afferent* and *efferent*, when used in describing nerve fibers of the PNS, are defined with respect to the CNS. Within the CNS the terms *afferent* and *efferent* apply to the specific brain region under discussion. For example, fibers traveling to cortex (i.e., cortical inputs) are afferent to cortex, while fibers leaving cortex (i.e., cortical outputs) are efferent to cortex.

Spinal nerves are mixed; they carry both sensory afferents and motor efferents. Cranial nerves may be purely afferent, purely efferent, or mixed.

Gray Matter vs. White Matter

The nervous system may be described as composed of gray matter and white matter (Figure 1.3). This is because in an exsanguinated brain that is preserved in formaldehyde, collections of myelin-coated axons appear white, and collections of cell bodies appear relatively pale ("gray").

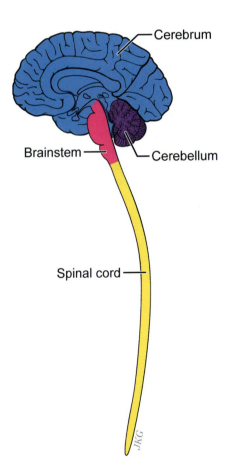

FIGURE 1.2. The central nervous system consists of the brain and spinal cord. The three major divisions of the brain are the cerebrum, cerebellum, and brainstem which is structurally continuous with the spinal cord.

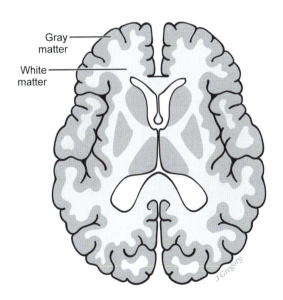

FIGURE 1.3. Gray matter and white matter.

GRAY MATTER

Gray matter consists mainly of nerve cell bodies and dendrites. Within the CNS, gray matter is organized into ball-like clusters of cell bodies called **nuclei** and sheet-like layers of cell bodies called **cortex** (e.g., cerebral cortex, cerebellar cortex). Within the PNS, collections of nerve cell bodies are called **ganglia**, ball-like structures that function as relay stations for nerve signals.

WHITE MATTER

White matter consists of collections of myelinated axons. While gray matter carries out the information-processing functions of the nervous system, white matter allows for the rapid and metabolically efficient transfer of information. In the human brain, white matter makes up about 50% of the total brain volume.

Within the PNS, bundles of myelinated axons are called **nerves**. Nerves conduct information from the CNS to the body by efferent fibers, and from the body to the CNS by afferent fibers.

Within the CNS, bundles of myelinated axons are called fiber **tracts**. Tracts are often named after the two CNS regions that they connect; the first part of the tract name indicates the site of origin, and the second part of the name reflects the site of termination. For example, the spinothalamic tract travels from the spinal cord to the thalamus, and the corticospinal tract travels from the cerebral cortex to the spinal cord. Thus, if the tract name begins with "spino" it carries afferent sensory signals from the spinal cord to the brain, and if the tract name ends with "spinal" it carries efferent motor signals to the spinal cord.

Within the nervous system, the terms *fasciculus, lemniscus, peduncle, funiculus,* and *bundle* refer to fiber tracts or collections of fiber tracts. A **fasciculus** (**fascicle**) is a bundle of fibers that are functionally related (e.g., gracile fasciculus, cuneate fasciculus, medial longitudinal fasciculus, arcuate fasciculus, superior longitudinal fasciculus, uncinate fasciculus). A **lemniscus** is a bundle of neural fibers in the form of a ribbon or band (e.g., medial lemniscus, lateral lemniscus). A **peduncle** is a stalk-like aggregate of tracts connecting different parts of the brain (e.g., cerebral peduncles, cerebellar peduncles). A **funiculus** is an aggregate of fiber tracts running up and down the spinal cord; it is also referred to as a *column* (e.g., dorsal, ventral, and lateral columns of the spinal cord).

The white matter connections of the brain are classified into three main categories: projection fibers, commissural fibers, and association fibers. **Projection fibers** connect different structures of the CNS (e.g., the corticospinal tract). **Commissures** interconnect symmetrical structures between the two halves of the brain (e.g., the largest commissure is the corpus callosum, which connects the cerebral cortex of the two hemispheres). **Association fibers** connect cortical regions within the same hemisphere. White matter also is found within the cerebral cortex and subcortical nuclei.

Anatomical Terms of Direction and Planes of Section

Directional Terms

Directional terms describe the positions of structures relative to other structures or locations in the body. The **neuraxis** (**neuroaxis**) is the central axis of the CNS that runs from the front of the brain to the end of the spinal cord. The three major axes of the CNS, defined relative to the neuraxis, are **rostral-caudal**, **dorsal-ventral**, and **medial-lateral**. **Rostral** (Latin *rostrum*) means toward the beak/nose, **caudal** (Latin *cauda*) means toward the tail, **dorsal** (Latin *dorsum*) means toward the back, and **ventral** (Latin *venter*) means toward the abdomen. **Medial** means toward the midline and **lateral** means away from the midline.

These terms are used especially when describing the relative positions of structures in animals that have a straight neuraxis, such as fish and quadrupeds. The human being, however, is bipedal and has an upright posture and bent neuraxis, with the bend (**cephalic flexure**) occurring at the junction between the brainstem and the cerebrum. Therefore, the rostral-caudal and dorsal-ventral axes, which are defined relative to the neuraxis, are not constant with respect to the environment (Figure 1.4). In human anatomy, another system is often used that designates relative positions of structures with respect to the environment in the **standard anatomic position**, standing erect with head facing forward (it does not depend on the whether the patient is standing, sitting, supine, or prone); these are the **anterior-posterior** and **superior-inferior axes**. **Anterior** means toward the front of the body, **posterior** means toward the back of the body, **superior** means toward the top of the body (or ceiling), and **inferior** means toward the bottom of the body (or floor).

In humans, structures located above the cephalic flexure (i.e., the cerebrum) have the same orientation with respect to the ground as in four-legged animals: anterior = rostral, posterior = caudal, superior = dorsal, and inferior = ventral. At the cephalic flexure there is a 90-degree rotation of the neuraxis with respect to the environment. Consequently, for structures located below the cephalic flexure (i.e., brainstem, cerebellum, and spinal cord): superior = rostral, inferior = caudal, anterior = ventral, and posterior = dorsal.

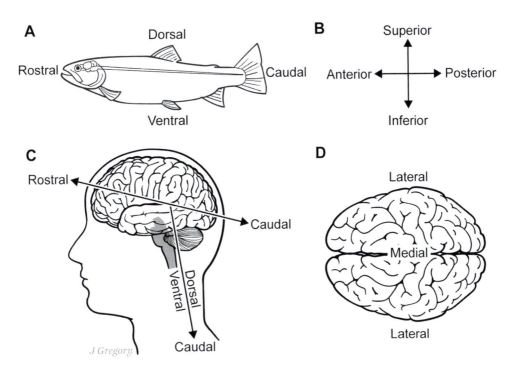

FIGURE 1.4. Directional terms. (A) The three major axes of the CNS are rostral-caudal, dorsal-ventral, and medial-lateral; these terms are used to describe the position of structures relative to the neuraxis. (B) The anterior-posterior and superior-inferior axes describe the relative positions of structures with respect to the environment. (C) Humans have an upright posture and a bent neuraxis, with the bend occurring at the junction between the brainstem and the cerebrum. Therefore, the rostral-caudal and dorsal-ventral axes, which are defined relative to the neuraxis, are not constant with respect to the environment. (D) Medial means toward the midline and lateral means away from midline.

Other positional terms that are frequently used are **unilateral** (on one side or within one hemisphere), **bilateral** (on both sides or within both hemispheres), **ipsilateral** (same side relative to midline), and **contralateral** (on the opposite side relative to midline). These terms are used to describe the location of lesions (e.g., unilateral or bilateral frontal lobe contusion), the location of neurological deficits (e.g., unilateral weakness, bilateral tremor), and the relationship between lesion location and resulting neurological deficit (e.g., cerebellar hemispheric infarct with ipsilateral ataxia, occipital lobe infarct with contralateral visual field defect). A related term is **decussation**, a crossing of nerve fibers or tracts that forms an X. A unilateral lesion in motor cortex results in contralateral motor deficits because the descending motor pathway from the motor cortex to the motor neurons in the spinal cord decussates (crosses the midline). The neurological deficit is on the opposite side of the body as the lesion (it is **contralesional**). A unilateral lesion in the cerebellum (an important component of the motor system) results in ipsilateral motor deficits because there are two decussations in the motor pathway between the cerebellum and the spinal cord. The neurological deficit is on the same side of the body as the lesion (it is **ipsilesional**).

Additional directional terms that are used to describe location for structures that have a point of origin, such as limbs and blood vessels, are **proximal** (near the site of origin) and **distal** (away from the point of origin). With reference to the extremities (i.e., the arms and legs), proximal refers to closest to the trunk, and distal refers to farthest from the trunk. For example, stroke due to occlusion of a blood vessel may selectively affect the proximal musculature of the extremities (e.g., upper arm, upper leg) or the distal musculature (e.g., hand, foot). With reference to arteries, proximal refers to closest to the heart, and distal refers to farthest from the heart. For example, a proximal cerebral artery occlusion occurs near the site of origin of the vessel, while a distal cerebral artery occlusion occurs away from the site of origin of the vessel.

Planes of Section

There are three standard planes of orientation in anatomy that can be thought of as two-dimensional slices (sections) through three-dimensional space, and are used in gross structural examination (e.g., brain dissection), histological analysis, or medical imaging. These planes are coronal, axial, and sagittal (Figure 1.5).

The **coronal** sectioning plane, also known as the **frontal** plane, is parallel to a vertical plane through both ears, approximating the same plane as a tiara

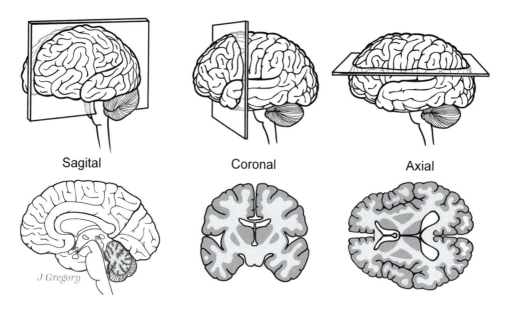

FIGURE 1.5. The three standard planes of section: sagittal, coronal, and axial.

crown (coronal) worn on the head. Because sectioning in this plane results in anterior and posterior portions, this plane is **orthogonal** to the anterior-posterior axis (i.e., it divides the brain into anterior and posterior portions). The **axial** sectioning plane, also known as the **horizontal** plane or the **transverse** plane, is parallel to the floor. Because sectioning in this plane results in inferior and superior portions, this plane is orthogonal to the superior-inferior axis (i.e., it divides the brain into superior and inferior portions). The **sagittal** sectioning plane, also known as the **longitudinal**, **median**, or **anteroposterior** plane, divides the body into left and right portions. This plane is called the sagittal plane because it is in the same plane as a bow and arrow held by an archer (as in the constellation Sagittarius). The **midsagittal** plane divides the brain into two halves. **Parasagittal** planes are sagittal planes that are off midline. Because sectioning in the sagittal plane results in right and left portions, this plane is orthogonal to the right-left axis. An **oblique plane** of section is intermediate between two of these three principal planes.

Basic Brain Anatomy

The adult human brain weighs about 1,400 grams (3 pounds). It is estimated that there are about 100 billion neurons within the human brain, and that each neuron makes synaptic contact with at least 10,000 other neurons. The brain has three major divisions: the brainstem, cerebellum, and cerebrum. Each of these has further subdivisions (Figure 1.6).

Brainstem

The **brainstem** is the lowermost part of the brain. It consists of three structures: the medulla, pons, and midbrain (Figure 1.7). The **medulla** (Latin, "oblong core") is the caudal-most portion of the brainstem. It was once referred to as the "bulb" because it is a rounded mass of tissue similar in shape to a tulip bulb; the term *bulbar* refers to the medulla, although in some cases it refers to the brainstem more generally (e.g., corticobulbar fibers run from the cerebral cortex to the cranial nerve nuclei of the brainstem, pons and midbrain included). The **pons** (Latin, "bridge") lies between the medulla and the midbrain. The **midbrain** is the rostral-most portion of the brainstem; it is only visible from the inferior surface of the brain.

The brainstem is structurally continuous with the spinal cord, and has a similar organization as the spinal cord, with gray matter located centrally and surrounded by a zone of white matter. It controls the basic vital life functions necessary for physiologic survival, such as heart rate, blood pressure, breathing, and digestion, as well as sleep-wakefulness and level of arousal in higher animals with a cerebral cortex (mammals). The brainstem is the phylogenetically oldest part of the human brain. It is similar in structure and function to the brains of reptiles and fish, and therefore it is also referred to as the **reptilian brain**.

Neurons, Glia, and Basic Neuroanatomy | 9

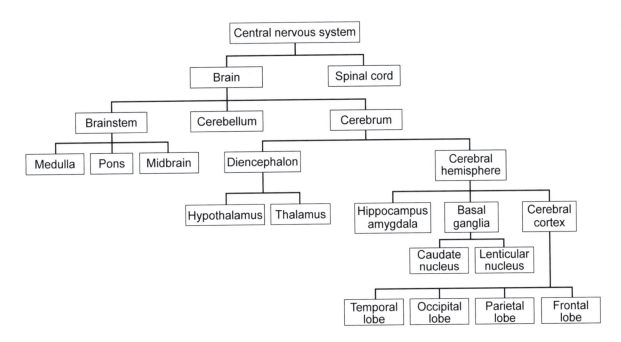

FIGURE 1.6. Subdivisions of the central nervous system.

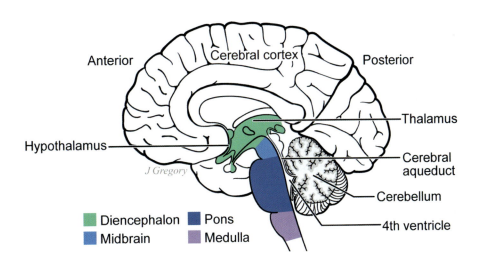

FIGURE 1.7. Midsagittal section of the brain. The brainstem, the lowermost part of the brain, consists of the medulla, pons, and midbrain. The cerebellum lies dorsal to the brainstem. The cerebrum is located rostral to the brainstem (in upright humans it is located superior to the brainstem and cerebellum); it consists of the diencephalon and cerebral hemispheres.

Cerebellum

The **cerebellum** (Latin, "little brain") lies dorsal to the brainstem (see Figure 1.7). It is an important part of the motor system that controls balance and coordination. It consists of two hemispheres, each composed of three lobes. It communicates with the rest of the brain by three pairs of cerebellar peduncles: the superior cerebellar peduncles connect to the midbrain, the middle cerebellar peduncles connect to the pons, and the inferior cerebellar peduncles connect to the medulla. The cerebellum evolved more recently than the brainstem.

Cerebrum

The **cerebrum**, also known as the forebrain, is located rostral to the brainstem. In upright humans it is located superior to the brainstem and cerebellum (see Figure 1.7), above the sharp bend of the neuraxis at the midbrain-diencephalic junction. It is the most recently evolved major component of the central nervous system. Following the developmental scheme of brain subdivision, the cerebrum consists of the diencephalon and cerebral hemispheres (see "Developmental Basis of the Major Brain Subdivisions," below).

DIENCEPHALON

The **diencephalon** (Latin, "in-between brain, interbrain") is defined as those structures surrounding the slit-shaped third ventricle. It lies between the brainstem and cerebral hemispheres. The most prominent structures of the diencephalon are the thalamus and hypothalamus, both of which are composed of multiple nuclei that perform many different functions.

The **thalamus** (Latin, "inner chamber") is a paired ellipsoid-shaped mass of gray matter that makes up about 80% of the diencephalon. Each thalamus is divided into a number of distinct nuclei, named according to their positions using the six terms of orientation. Virtually every nucleus of the thalamus sends axons to the cerebral cortex, and every part of the cortex receives afferent fibers from the thalamus. Thus, nearly all information received by the cortex is first relayed through the thalamus. For this reason, the thalamus is often called the "gateway to the cerebral cortex." Since the thalamus provides so much of the input to the cerebral cortex, it drives cortical activity, and consequently thalamic lesions may disrupt cortical function.

The **hypothalamus** lies immediately ventral to the thalamus, at the base of the brain. It contains many nuclei and fiber tracts. The overall aim of hypothalamic function is to preserve homeostasis. The hypothalamus coordinates autonomic, endocrine, and somatic motor responses; plays a role in thermoregulation and circadian rhythms; and mediates drives (hunger, thirst, sex, and aggression) for individual and species survival behaviors (feeding, drinking, sex, and defense reactions). It is regarded as the master control of autonomic function by virtue of its connections to lower autonomic centers in the brainstem and spinal cord. It also plays a role in endocrine gland function (pancreas, ovaries, testes, thyroid, parathyroid, and adrenals) by virtue of its control over the superior endocrine organ, the pituitary gland.

CEREBRAL HEMISPHERES

The **cerebral hemispheres** are composed of the **cerebral cortex**, the underlying white matter, and the subcortical structures of the **basal ganglia** (caudate nucleus, putamen, globus pallidus), **hippocampus**, and **amygdala**. The basal ganglia are a group of subcortical gray matter masses that lie deep (i.e., basal) within the cerebral hemispheres; the term is an accepted misnomer since a ganglion is a group of nerve cells located outside of the CNS. There are two cerebral hemispheres, right and left, separated by the interhemispheric fissure (longitudinal fissure). Each hemisphere is composed of four lobes: frontal, temporal, parietal, and occipital.

"Higher-order" (i.e., cognitive and behavioral) functions were long thought to be exclusively the domain of the cerebral cortex. It is important to recognize, however, that nervous system functions are subserved by neural circuits consisting of distinct nodes that are geographically distributed throughout the brain. Subcortical structures are involved in not only sensorimotor circuits and behaviors, but also complex behaviors and cognition, based on their links with the cerebral cortex.

Limbic System

The term **limbic** (Latin, "border") **system** is sometimes used to refer to the nuclear structures of the cerebrum (i.e., thalamus, hypothalamus, amygdala, hippocampus, and basal ganglia) along with the cingulate cortex, a ring of cerebral cortex that surrounds the lateral ventricles and lies buried beneath the cortical surface. The limbic system is associated with motivation, emotion, learning, and memory. Despite its name, the concept of the limbic system as a specific neural system with a unified functional role has not held up as neuroscience has advanced. Consequently, the preferred term is **limbic structures**.

Developmental Basis of the Major Brain Subdivisions

In order to understand some of the nomenclature used in describing adult brain anatomy, it is important to know some basic information about its development. Early in the embryological development of all vertebrates, the nervous system begins as a **neural tube**. At the rostral end, the neural tube balloons into three primary vesicles that are the precursor of the brain. The vesicles are named using the Greek root word *enkephalon* (meaning "brain") and a prefix describing its position along the length of the developing nervous system: (1) the **prosencephalon (forebrain)**, located most rostrally; (2) the **mesencephalon (midbrain)**; and (3) the **rhombencephalon (hindbrain)**, located most caudally (Figure 1.8). In human embryogenesis, the three primary vesicles form at the fourth week of development.

In the next stage, the prosencephalon (forebrain) and rhombencephalon (hindbrain) each divide into two vesicles; the mesencephalon does not subdivide. This results in a five-vesicle structure: (1) **telencephalon** (endbrain), (2) **diencephalon** (interbrain), (3) **mesencephalon** (midbrain), (4) **metencephalon** (afterbrain), and (5) **myelencephalon** (marrow brain). This occurs at the fifth week of embryonic development.

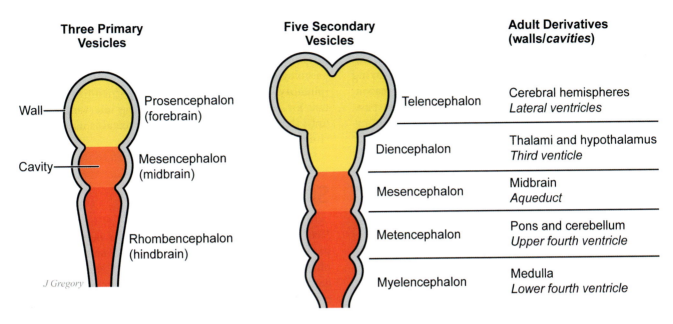

FIGURE 1.8. Divisions of the brain based on development: prosencephalon (forebrain), mesencephalon (midbrain), and rhombencephalon (hindbrain); telencephalon (endbrain), diencephalon (interbrain), mesencephalon (midbrain), metencephalon (afterbrain), and myelencephalon (marrow brain).

With further development, the telencephalon differentiates into the structures of the cerebral hemispheres (cerebral cortex, cerebral white matter, and basal ganglia); the diencephalon differentiates into the thalamus and hypothalamus; and the metencephalon differentiates into the cerebellum and pons. The mesencephalon (midbrain) again does not further differentiate.

Familiarity with this developmental scheme and terminology is useful to clinicians for several reasons. It explains why the midbrain, rather than being a large structure occupying the middle third of the brain, as its name might suggest, is a relatively small structure (i.e., because the mesencephalon begins as one of three primary vesicles but does not differentiate in the same way that the prosencephalon and rhombencephalon do). It also clarifies the misconception that the terms *cerebrum* and *cerebral hemispheres* are synonymous. The term *cerebrum* refers to the forebrain (prosencephalon) structures, which encompass the cerebral hemispheres and diencephalon. It also clarifies that the diencephalon is *not* a component of the brainstem, although it was considered so in classic neuroanatomy. Familiarity with this terminology is also useful because some of these terms are used clinically (e.g., diencephalic amnesia, mesencephalic syndrome).

Neurodegenerative Disease

Neurodegenerative diseases are characterized by a progressive death of neurons and a progressive loss of neurological function. These diseases have two important characteristics: (1) they have an insidious onset and a gradually progressive course, and (2) they affect specific parts of the nervous system. This group of disorders is introduced here in a general way because they are discussed more specifically throughout this book in the context of the specific region or functional system affected, rather than in a dedicated chapter.

Regarding onset and course, neurodegenerative diseases manifest as a change in neurological function that begins insidiously after a long period of normal function. It usually is not possible to assign a date of onset, although the clinical history usually reveals that subtle symptoms were present for some time, which attracted little attention at the time but in retrospect are understood to be early signs or symptoms. Signs and symptoms develop **de novo**; there are no antecedent events. The neurodegenerative diseases also are characterized by steady progression over years. Clinical **signs** (i.e., objective evidence of injury or disease) and **symptoms** (i.e., subjective experience of injury or disease) are late expressions of the pathologic process; they manifest only when the degree of neuronal loss exceeds a certain threshold. Most manifest later in life, and in the early stages the clinical challenge is in differentiating changes due to early stage neurodegenerative disease from those related to normal aging. At the cellular and subcellular level, the neurodegenerative diseases are distinctively different from the programmed cell loss of normal aging. They are irreversible; currently there are no cures.

The neurodegenerative diseases selectively target specific anatomical regions or functional systems of the nervous system (e.g., cerebral cortex, cerebellum, the motor system), at least early in their course, leaving other regions and systems unaffected. For this reason, they were referred to as **system atrophies** in the past. They differ with respect to the degree of neuroanatomical specificity; some affect very specific brain nuclei or neocortical regions throughout their course, while others affect specific neuronal regions but progress to have more diffuse effects. The neuroanatomical specificity of these diseases gives rise to distinct clinical **syndromes**, a combination of signs and symptoms that are characteristic of a disease or disorder. This anatomical-syndromal specificity forms the basis of clinical classification of neurodegenerative diseases.

In general, the neurodegenerative diseases result in a loss of cognitive function, motor function, or both, and their resulting syndromes are classified accordingly. Those diseases that primarily involve the cerebral cortex result in a loss of cognitive function that eventually progresses to dementia. **Dementia** is defined as a decline in cognitive function that affects at least two domains of cognition and is associated with a decline in occupational and or social functioning. The terms *neurodegenerative disease* and *dementia* are often used interchangeably, but this is not accurate. In the strictest sense, the term *dementia* does not imply any single substrate or pathology, and therefore can be applied to non-neurodegenerative conditions (e.g., post-stroke dementia). While all neurodegenerative diseases that diffusely affect the cerebrum will result in dementia, the presence of dementia does not necessarily connote neurodegenerative disease. The syndromes characterized by progressive dementia without other neurologic signs include Alzheimer's disease (Chapter 16, "The Temporal Lobes and Associated Disorders"), the frontotemporal dementias (Chapter 17, "The Frontal Lobes and Associated Disorders," and Chapter 27, Language and the Aphasias), and posterior cortical atrophy (Chapter 15, "The Parietal Lobes and Associated Disorders"). Neurodegenerative diseases that involve subcortical brain structures often give rise to movement disorders, such as Parkinson's disease; some movement disorders are accompanied by progressive dementia, such as Huntington's disease (chorea), Lewy-body dementia, and corticobasal ganglionic degeneration (Chapter 11, "The Basal Ganglia").

While neurodegenerative diseases can be classified according to primary clinical features (e.g., dementia, parkinsonism, motor neuron disease) and anatomic distribution of neurodegeneration (e.g., frontotemporal, extrapyramidal, spinocerebellar), the diagnostic gold standard is neuropathological evaluation at autopsy. The neurodegenerative diseases all involve misfolded proteins and abnormal protein deposits in the brain, and they are classified based on the protein molecule affected (e.g., amyloidoses, tauopathies, alpha-synucleinopathies). Our knowledge of the molecular pathologies underlying these disorders, however, is incomplete, and multiple molecular pathologies may give rise to a single clinical syndrome (**phenotype**). Understanding the abnormalities is important for developing disease-modifying interventions, but until then, syndromal classification remains useful to the clinician.

Summary

Neurons and glia together constitute the nervous system parenchyma. The neuron is the basic anatomical and physiological unit of the nervous system. Neurons are organized into circuits and are electrically excitable; they receive, integrate, and transmit information through changes in electrical state. Glial cells do not generate electrical signals; they provide myelin and a variety of support functions for neurons.

The brain has three major divisions: the brainstem, cerebellum, and cerebrum. The brainstem consists of the medulla, pons, and midbrain. The cerebrum consists of the diencephalon and cerebral hemispheres; the diencephalon consists of the thalamus and hypothalamus; the cerebral hemispheres consist of the cerebral cortex, the underlying white matter, and the subcortical structures of the basal ganglia (caudate nucleus, putamen, globus pallidus), hippocampus, and amygdala.

When examining and describing the nervous system, we use directional terms to describe the positions of structures relative to other structures or locations in the body. The three major axes defined relative to the neuraxis are rostral-caudal, dorsal-ventral, and medial-lateral. In human anatomy, another system is often used that designates relative positions of structures with respect to the environment; these are the anterior-posterior and superior-inferior axes. We also examine CNS structures by making two-dimensional sections, whether it be in gross structural examination by brain dissection, histological analysis, or medical imaging. These sections are made in three standard planes: coronal, axial, and sagittal.

Additional Reading

1. Blumenfeld H. Neuroanatomy overview and basic definitions. In: Blumenfeld H, *Neuroanatomy through clinical cases*, 3rd ed., pp. 13–46. Sinauer Associates; 2021.
2. Brodal P. Structure of the neuron and organization of nervous tissue. In: Brodal P., *The central nervous system: structure and function*. 5th ed. Oxford University Press; 2016. https://oxfordmedicine.com/view/10.1093/med/9780190228958.001.0001/med-9780190228958-chapter-1.

3. Brodal P. Glia. In: Brodal P., *The central nervous system: structure and function*. 5th ed. Oxford University Press; 2016. https://oxfordmedicine.com/view/10.1093/med/9780190228958.001.0001/med-9780190228958-chapter-2.
4. Brodal P. Parts of the nervous system. In: Brodal P, *The central nervous system: structure and function*. 5th ed. Oxford University Press; 2016. https://oxfordmedicine.com/view/10.1093/med/9780190228958.001.0001/med-9780190228958-chapter-6.
5. DeArmond SJ, Fusco MM, Dewey MM. *Structure of the human brain: a photographic atlas*. 3rd ed. Oxford University Press; 1989.
6. Diamond MC, Scheibel AB, Elson LM. *The human brain coloring book*. Harper & Row; 1985.
7. Vanderah TW, Gould D. Introduction to the nervous system. In: Vanderah TW, Gould DJ, *Nolte's the human brain: an introduction to its functional anatomy*. 8th ed., pp. 1–35. Elsevier; 2021.
8. Vanderah TW, Gould D. Gross anatomy and general organization of the central nervous system. In: Vanderah TW, Gould DJ, *Nolte's the human brain: an introduction to its functional anatomy*. 8th ed., pp. 55–79. Elsevier; 2021.
9. Waxman SG. Fundamentals of the nervous system. In: Waxman SG, *Clinical neuroanatomy*. 29th ed. McGraw-Hill; 2020. https://accessmedicine-mhmedical-com.proxy.library.nyu.edu/content.aspx?bookid=2850&sectionid=242762730.

2

Electrical Signaling in Neurons

Introduction

Neurons, the basic functional unit of the nervous system, are electrically excitable; they receive, integrate, and transmit information via electric currents generated by the flow of ions. This chapter provides an overview of the process by which electrical signals (voltage changes) are generated and propagated within neurons. A basic understanding of these processes is fundamental to understanding the mechanisms underlying seizures, electroencephalography, and brain stimulation, as well as some aspects of neuropharmacology and demyelinating disorders.

In the nervous system, information is conveyed from one place to another by large electrical signals known as **action potentials** (AP). These voltage changes are generated at the axon hillock and travel the length of the axon to the axon terminals. The arrival of the AP at the neuron's axon terminals triggers the release of chemical neurotransmitter from vesicles into the synaptic cleft, a small space between the presynaptic axon terminal and the postsynaptic cell. Neurotransmitter molecules diffuse across the synaptic cleft and bind with transmitter receptor molecules located on the membrane of the postsynaptic cell (Figure 2.1).

The binding of transmitter molecule to receptor molecule in the postsynaptic membrane causes a small electrical signal known as the **postsynaptic potential** (PSP) within the receiving neuron. These voltage changes are either excitatory or inhibitory, and either promote or inhibit, respectively, the firing of APs in the receiving neuron. The receiving neuron integrates (sums) all incoming PSPs from multiple neurons and multiple synaptic contacts. If the net level exceeds a threshold value, the neuron will generate an AP and thereby relay the message to subsequent neurons in the circuit.

Communication *within* neurons is accomplished by electrical signaling, the topic of this chapter. Communication *between* neurons is accomplished (in large part) by chemical signaling via neurotransmitters, the topic of Chapter 3. Within a cell, APs flow in one direction down the axon, from the axon hillock to the axon terminals. Communication at the synapse also generally proceeds in only one direction, from terminal button to the postsynaptic site (usually a dendrite).

To understand how electrical signals are generated and conducted within neurons, it is necessary to understand some basic principles of chemistry that govern the diffusion

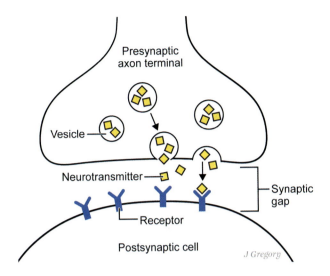

FIGURE 2.1. Neurotransmitter molecules diffuse across the synaptic cleft, a small space between the presynaptic axon terminal and the postsynaptic cell, and bind with transmitter receptors located on the membrane of the postsynaptic cell.

(movement) of molecules and ions, some basic principles of membrane structure and permeability, and the ionic basis of membrane potentials.

Basic Chemistry Concepts Governing Diffusion

In this section, we introduce a few basic chemistry concepts that are essential in understanding electrical signaling within neurons.

Concentration Gradients

When a substance is in solution (dissolved in a medium), its molecules are in constant random motion. This **Brownian motion** drives the **diffusion** (movement) of molecules from areas of high concentration to areas of low concentration.

A difference in concentration between one region and an adjacent region is called a **concentration gradient**. Substances in solution diffuse along their concentration gradient until the concentration is uniform throughout the solution (until there is no longer a gradient). If two such regions are separated by a membrane, the rate of diffusion will depend on how **permeable** (permissive) the membrane is in allowing a specific type of molecule or ion to diffuse through it. For instance, suppose that we have a vessel of water that we divide into two compartments, separated by a membrane that is semipermeable to sugar molecules, and we dissolve ordinary table sugar into the water on one side of the membrane. Since the concentration of sugar is higher on one side of the membrane than the other, and the membrane is semipermeable to sugar molecules, diffusion will eventually eliminate the concentration gradient across the membrane. The more permeable the membrane is to sugar, the faster the sugar molecules will diffuse across the membrane and achieve equal concentrations on the two sides of the membrane.

Electrostatic Gradients

In addition to concentration gradients, there are other important forces at work affecting the movement of *charged* molecules or atoms across membranes. These charged chemical elements, called **ions**, are particularly important in neuron chemistry. Positively charged ions are called **cations**, negatively charged ions are called **anions**, and solutions containing ions are called **electrolytes**. An example that is familiar to everyone is table salt, or sodium chloride (NaCl), which when dissolved in water dissociates into sodium (Na^+) and chloride (Cl^-) ions. Now suppose that in our water-filled vessel (divided into two compartments by a membrane) we add NaCl to one compartment; the NaCl, when dissolved, dissociates into charged Na^+ and Cl^- ions. As with non-ionic molecules (such as sugar), ions diffuse across the membrane from a region of high concentration to a region of low concentration (down the concentration gradient) until the concentrations of Na^+ and Cl^- are equal on both sides of the membrane. But now, there is also an electrical force at work. Since unlike charges are attracted to each other, and like charges are repelled by each other, this **electrostatic force** also influences the movement of ions.

To further extend our example, if all the Na^+ cations were on one side of the membrane, and all the Cl^- anions were on the other side, there would be a difference of charge across the membrane, an **electrostatic gradient**. Since the Na^+ ions are repelled by other Na^+ ions and attracted to Cl^- ions, and Cl^- ions are repelled by other Cl^- ions and attracted to Na^+ ions, Na^+ ions will diffuse away from the positively charged compartment into the negatively charged compartment, and Cl^- ions will diffuse away from the negatively charged compartment into the positively charged compartment, until the positive and negative charges within each compartment and across the membrane cancel each other out (until the electrostatic gradient no longer exists.) Thus, when cations and anions are separated by a membrane, electrostatic pressure promotes movement of ions across the membrane to neutralize the electrostatic gradient.

Balance of Concentration and Electrostatic Gradients

Charged particles in solution distribute themselves to achieve two goals: equal concentrations *and* electrical neutrality. In the above examples, both goals were achieved completely since (1) the membrane was permeable to both Na⁺ and Cl⁻; (2) there were no other ions in solution that could influence the movement of Na⁺ and Cl⁻; and (3) there were equal amounts of both ions. However, if we had equal Na⁺ and Cl⁻ concentrations on both sides of the membrane (no concentration or electrostatic gradient), and then we added to one compartment a substance that dissolves into large anions to which the membrane was not permeable, the distribution of Na+ and Cl⁻ would change. Na⁺ would be attracted into the compartment containing the large anion, resulting in an uneven distribution (concentration gradient) for Na⁺ across the membrane. On the other hand, Cl⁻ would be repelled by the large anion and would become more concentrated on the other side of the membrane. Ultimately a steady state would develop for both Na⁺ and Cl⁻ in which neither electrical neutrality nor concentration equivalence are achieved. In other words, for each ion to which the membrane is permeable, the two opposing forces balance each other out, and an equilibrium is achieved in which there is no net flux of those ions across the membrane.

This phenomenon is especially important in neurons because changes in the separation of charges across the membrane form the basis of electrical signaling within neurons.

The separation of charge across the neuronal membrane consists of a thin cloud of positive ions spread over the outer surface of the membrane, and a thin cloud of negative ions spread over the inner surface of the membrane. This separation of charges is maintained by the neuronal membrane, which acts as a barrier to the diffusion of ions and other molecules. But what determines membrane permeability? Transport of any particular molecule type across a neuronal membrane depends upon two factors: (1) the degree to which the molecule can pass through the lipid portion of membrane, and (2) whether the membrane contains special transport mechanisms for the molecule.

Lipid Solubility

The lipids are a group of organic compounds that are insoluble in water (e.g., fatty acids, neutral fats, waxes, steroids). Like other cell membranes, the neuronal membrane is composed of phospholipid molecules that have a **hydrophilic** (having an affinity for water molecules through forming hydrogen or ionic bonds) phosphate "head" and a **hydrophobic** (having the property of repelling water molecules) fatty acid "tail." The phospholipid molecules are arranged in a double layer (bilayer), with the hydrophilic phosphate heads oriented so that they are in contact with the extra- and intracellular fluids, and the hydrophobic tails lying within the interior of the cell membrane (Figure 2.2). The phospholipid cell membrane is permeable to **lipid-soluble** molecules but impermeable to **water-soluble** molecules.

Membrane Structure and Permeability

All cells operate on principles similar to those just discussed, where intracellular and extracellular fluid compartments are separated by the cell membrane. Since intracellular and extracellular fluids differ in their chemical composition due to the permeability characteristics of the cell membrane, there is a separation of charge (an electrostatic gradient) across the membranes of all types of cells.

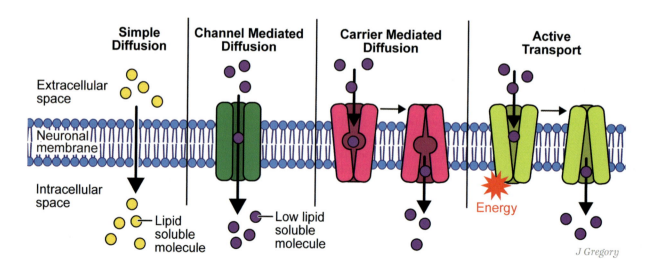

FIGURE 2.2. The neuronal membrane is composed of a phospholipid bilayer with the hydrophilic phosphate heads oriented so that they are in contact with the extra- and intracellular fluids and the hydrophobic tails lying within the interior of the cell membrane. Lipid-soluble molecules diffuse freely across the phospholipid membrane. Molecules with low lipid solubility cross the neuronal membrane by passive transport by channels or carriers, or active transport by pumps.

Lipid solubility depends upon the charge and size of the molecule. Electrically neutral (**nonpolar**) molecules are more lipid soluble than charged (**polar**) molecules, and small molecules are more lipid soluble than large molecules. Small nonpolar molecules therefore have the highest lipid solubility, while large polar molecules have the lowest lipid solubility.

Lipid soluble molecules move freely through the phospholipid bilayer by **passive transport**; movement across the membrane is driven by the electrochemical gradient and does not require expenditure of the cell's metabolic energy. Molecules that diffuse freely across the neuronal membrane include oxygen, carbon dioxide, alcohol, nicotine, and most psychoactive drugs.

Membrane Transport Proteins

Molecules with low lipid solubility do not pass freely through the phospholipid bilayer of the neuronal membrane. Rather, they cross the membrane via **membrane transport proteins**, of which there are three main types: channels, carriers, and pumps. Protein **channels** form pores that traverse the neuronal membrane. Protein **carriers** carry molecules from one side of a membrane to the other; the molecule binds to the carrier on one side of the membrane, the carrier changes shape and flips within the membrane, then releases the molecule on the other side of the membrane. Movement across the membrane through channels and carriers is known as **facilitated diffusion** because it occurs via a membrane protein, unlike **simple diffusion** where molecules pass through the membrane without the help of membrane proteins. Like simple diffusion, however, facilitated diffusion occurs by passive transport (it is driven by the electrochemical gradient and does not require cellular energy). Passive transport neutralizes electrochemical gradients. By contrast, protein **pumps** move substances across the membrane against an electrochemical gradient by **active transport**, which requires expenditure of **adenosine triphosphate (ATP)**, the main energy carrying molecule within cells. Active transport creates electrochemical gradients.

Ions, being charged molecules or atoms, have low lipid solubility and are unable to traverse the phospholipid portion of the neuronal membrane. They do, however, traverse the membrane via ion channels and pumps. Ion channels and ion pumps therefore are particularly important in neuronal signaling, which is based on electric currents generated by the flow of ions across the neuronal membrane.

ION CHANNELS

Ion channels are specialized proteins that form pores in the neuronal membrane that ions can travel through. There are many types of ion channels, differing in their selectivity and whether they are gated or not.

Selectivity refers to the kind of ion that the channel lets through; selectivity results from characteristics of the channel such as the diameter, shape, and charge along the inside surface. **Gating** refers to the opening or closing of ion channels. **Non-gated channels** are always opened. **Gated ion channels** can be opened or closed; the change in state from closed to opened (or opened to closed) results from a change in the conformation (shape) of the protein molecule and occurs very rapidly. Channels are gated in two basic ways. **Voltage-gated ion channels** open and close depending on the membrane potential. **Chemically gated (ligand-gated) ion channels** open and close depending on whether a specific molecule (e.g., a neurotransmitter) is present; these channels are components of larger molecules that function as neurotransmitter receptors (see Chapter 3, "Chemical Neurotransmission and Neuropsychopharmacology").

There may be multiple types of ion channels for any individual ion type. For example, potassium ions (K^+) can traverse the neuronal membrane through non-gated channels that are highly selective for K^+, relatively nonselective cation channels that allow several types of small cation to pass (e.g., Na^+ and K^+), voltage-gated K^+ channels, and ligand-gated K^+ channels.

The permeability of a membrane to a particular ion type depends on the number of open channels that allow the ion type to pass through. The movement and direction of flow of the ion type across the cell membrane is the product of permeability and the net electrochemical driving force (the combined effects of the concentration and electrostatic gradients). Ion channel gating provides a means of controlling the permeability of the membrane.

ION PUMPS

Ion pumps (**ion transporters**) are specialized transmembrane proteins that move ions across a cell membrane against electrochemical gradients. Ion pumps require ATP to function, and they generate electrochemical gradients. A particularly important ion pump in neuronal function is the sodium-potassium pump (see "The Sodium-Potassium Pump," below).

Membrane Potentials

The details of electrical signaling within neurons were first elucidated in the 1940s by **Alan Hodgkin** (1914–1998) and **Andrew Huxley** (1917–2012), for which they won the Nobel Prize in Physiology or Medicine in 1963. They performed **intracellular recording** (**single-unit recording**) in the **squid giant axon**. These axons are part of the squid's water jet propulsion system and are up to 1.5 mm in diameter, hundreds of times larger than the largest mammalian axon. To measure the difference in electrical charge (**electrical potential**) across the **axolemma** (the cell membrane of the axon), an isolated squid giant axon is placed in a dish filled with a fluid that is similar in chemical composition to the extracellular fluid within which the axon normally exists. An **extracellular electrode** (a wire which conducts current) is placed into the extracellular medium, and an **intracellular microelectrode** (composed of fine wire or an electrolyte-filled micropipette that conducts current) is placed in the cell's **axoplasm** (the cytoplasm within the axon); the microelectrode has a very fine tip that does not damage the cell membrane. The difference in voltage across the membrane is measured with a **voltmeter**, amplified, and displayed on an **oscilloscope** which graphically depicts voltage (amplitude of the charge difference) over time. This technique may also be used **in vivo** (in a living organism) to record from single cells, either acutely in anesthetized animals or chronically in awake behaving animals.

The term **membrane potential** (V_m) refers to the voltage across the membrane at any given moment. By convention, the potential outside of the cell is defined as zero, and the direction of current flow is defined as the direction of net movement of positive charge. Thus, across the cell membrane, cations move in the same direction as the current, and anions move in the opposite direction of the current flow.

The membrane potential at any given time is determined by the relative numbers of ions on both sides of the membrane. This apportioning is determined by the permeability characteristics of the membrane and the makeup of the intra- and extracellular media. Changes in the permeability of the cell membrane to ionic flow, which is controlled by ion channels embedded in the cell membrane, result in a movement of ions across the membrane.

The electrical potential across the membrane that will exactly prevent a net diffusion of a particular ion type across the membrane in either direction is called the **equilibrium potential** (**Nernst potential**) for that ion (e.g., V_{Na}, V_K). When a membrane becomes permeable to an ion, those ions flow across the membrane in the direction that aims to achieve the equilibrium potential of that ion.

The Resting Membrane Potential

The **resting membrane potential** (V_r) reflects the magnitude of electrical charge across the cell membrane when the neuron is quiescent (when it is not receiving or signaling any information). With respect to the outside of the cell (the extracellular fluid), the inside of the cell (the intracellular fluid) is negatively charged. This charge varies between neurons, ranging from –40 to –90 millivolts (mV), but –70 mV is conventionally used as the standard. Electrical signaling in neurons (APs and PSPs) is accomplished by deviations from the resting potential.

Four ion types play a major role in membrane potentials: Na^+, K^+, Cl^-, and organic anions (A^-). These four are the most abundant types of ions found on either side of the membrane. The ionic basis of the resting membrane potential can be understood in terms of (1) how these ions are distributed across the membrane in a neuron at rest, and (2) how membrane permeability and electrostatic and concentration forces result in this distribution (Figure 2.3).

Organic anions are large, negatively charged proteins, amino acids, and intermediate metabolites. They are concentrated intracellularly and cannot cross the neuronal membrane, thereby imparting a negative charge to the inside of the cell.

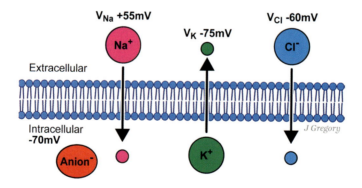

FIGURE 2.3. Four major ion types play a role in the membrane potential: sodium (Na^+), potassium (K^+), chloride (Cl^-), and organic anions (A^-). Organic anions are concentrated intracellularly and cannot cross the neuronal membrane, thereby imparting a negative charge to the inside of the cell. Na^+ and Cl^- ions are concentrated in the extracellular fluid, and K^+ is concentrated intracellularly. At rest, the cell membrane is permeable to K^+ and Cl^-; they are distributed across the membrane, nearly achieving their equilibrium potentials (E_K and E_{Cl}). At rest, the cell membrane is highly impermeable to Na^+, which is positively charged and concentrated extracellularly; thus both concentration and electrostatic gradients produce a great driving force for Na^+ to enter the cell and achieve its equilibrium potential (E_{Na}).

Na+, K+, and Cl- are present on both sides of the membrane. Their distributions across the membrane and equilibrium potentials are:

- Na+: 10 × more concentrated extracellularly, $V_{Na} = +55$ mV
- Cl-: 30 × more concentrated extracellularly, $V_{Cl} = -60$ mV
- K+: 35 × more concentrated intracellularly, $V_K = -75$ mV.

It is easy to remember that Na+ and Cl- are concentrated in the extracellular fluid, since extracellular fluid is similar to seawater (and we evolved from sea creatures).

Non-gated K+ and Cl- channels are primarily responsible for the membrane potential when the cell is not signaling. At rest, the cell membrane is very permeable to K+. Because K+ is concentrated within the cell, the K+ concentration gradient promotes K+ outflow from the cell. However, the negative charge of the intracellular fluid arising from the trapped A- has the opposite effect, resisting outflow of K+ due to electrostatic attraction between A- and K+ ions. Therefore K+ distributes itself across the membrane so that it nearly achieves its equilibrium potential. At rest, the neuronal membrane is also permeable to Cl-. Because Cl- is concentrated outside the cell, the Cl- concentration gradient promotes Cl- flow into the cell. However, the negative charge of the intracellular fluid arising from the trapped A- has the opposite effect, resisting Cl- inflow due to electrostatic repulsion between A- and Cl- ions. Therefore Cl- distributes itself across the membrane so that it nearly achieves its equilibrium potential (but to a lesser degree than K+).

At rest, the cell membrane is highly impermeable to Na+. This ion is positively charged and concentrated outside the cell, contributing to the net negative charge within the cell relative to the outside of the cell. But unlike K+ and Cl-, both concentration and electrostatic gradients produce a great driving force for Na+ to enter the cell, but this does not occur because the cell membrane is highly impermeable to Na+. Should the cell membrane suddenly become permeable to Na+, however, Na+ would rush into the cell.

The Action Potential

The primary characteristic that distinguishes neurons from most other cells is that they are excitable: they generate electrical signals. This excitability allows them to respond to incoming information and communicate this information to other neurons via the AP.

The AP is a transient alteration in the membrane potential. During an AP the polarity of the charge distribution across the neuronal membrane rapidly reverses from its resting potential of -70 mV, and the inside of the axon becomes positively charged to about +50 mV (Figure 2.4). This reversal of polarity is very brief, and within 3 msec the membrane potential is restored to its resting state of

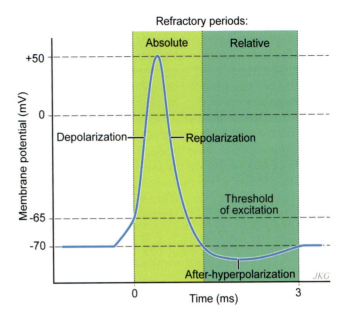

FIGURE 2.4. The action potential (AP) is a transient alteration in the membrane potential.

-70 mV. Because of its brevity, the AP is also called the **nerve impulse**.

The AP can be explained as a series of changes in the movement of Na+ and K+ across the neuronal membrane. At rest, the neuronal membrane is highly impermeable to Na+. If the cell becomes **depolarized** (the difference between intracellular and extracellular charge distribution is reduced or brought closer to 0) by about 5 mV at the axon hillock, reaching the **threshold of excitation** (bringing the V_m from -70 mV to about -65 mV), an AP is triggered. This normally occurs with the summation of PSPs reaching the threshold of excitation at the axon hillock, but APs can also be artificially initiated. Indeed, Hodgkin and Huxley elucidated the ionic basis of APs by recording the membrane potential (difference in electrical charge across the axolemma) in the isolated squid giant axon, while simultaneously applying an intracellular depolarizing current by a stimulating electrode and triggering APs.

During the rising phase of the AP, there is an abrupt and enormous change in the membrane's permeability to Na+ by a factor of 5,000, due to the opening of voltage-gated Na+ channels in the axonal membrane. Sodium ions rush into the cell down electrostatic and concentration gradients, driving the membrane potential toward the equilibrium potential for sodium (V_{Na}). Within about 1 msec, approximately the time that the peak of the AP is achieved (+50 mV), Na+ channels rapidly close and the membrane again becomes impermeable to Na+, terminating the rising phase of the AP. The rising phase of the AP is said to be a depolarization because it moves in the direction of 0 mV, even though there is an overshoot past 0 mV.

The rising phase of the AP is followed by **repolarization**, which occurs by K+ efflux. As the inside of the cell becomes positively charged at the peak of the AP, voltage-gated K+ channels open, increasing the membrane's permeability to K+. Potassium ions are driven out of the cell by both electrostatic and concentration forces. This efflux of positive charge from the cell restores the cell's intracellular negativity.

The third phase of the AP is the **after-hyperpolarization**, an overshoot in the restoration of the cell's internal negativity. This occurs because the electrical potential at which K+ is in equilibrium is –75 mV and just slightly more negative than V_r, and at the moment that V_r is attained some K+ channels are still open. When all the voltage-gated K+ channels are finally closed, the resting membrane potential is restored.

During the rising phase of the AP, voltage-gated Na+ channels open rapidly and briefly. Channel opening is followed by a period of channel inactivation, during which the Na+ channels cannot open again. Consequently, the cell enters a **refractory period**. During the first phase of the refractory period, it is impossible for the cell to fire another AP, no matter how strong the stimulus. This is the **absolute refractory period**, and it sets the maximum possible firing frequency of the cell (in large fibers the absolute refractory period is approximately 0.4 msec, setting the maximal firing frequency at 2,500 APs per second). The absolute refractory period is followed by a **relative refractory period**, during which it is possible to stimulate an AP, but only with a stimulus that is stronger than what is ordinarily sufficient to trigger an AP (a depolarization greater than +5 mV).

The amplitude of the AP is independent of the magnitude of the stimulus that triggered it, a principle that is known as the **all-or-none law**. All stimuli above the neuron's threshold will trigger a full AP, while all stimuli below the neuron's threshold will not. Thus, the magnitude of the stimulus is not coded by the magnitude of the AP, because that is fixed. Rather, the magnitude of the stimulus is coded by the frequency with which the neuron "fires" APs (the firing frequency).

ACTION POTENTIAL PROPAGATION

The speed of AP conduction varies among nerve fibers, depending on the axon diameter and whether the axon is myelinated or not. Small unmyelinated fibers conduct APs at about 1 meter per second, while very large, myelinated fibers can conduct APs up to 120 meters per second (roughly the length of a football field in one second). The thicker the axon, the more rapid the AP conduction because there is less resistance to longitudinal current spread within the axoplasm. Myelination increases AP conduction by a process known as saltatory conduction.

Once generated at the axon hillock, the AP propagates along the length of the axon toward the axon terminals as an electrochemical wave. In unmyelinated axons, action potential propagation involves activation of voltage-gated sodium channels along the entire length of the axon. An AP is triggered at the axon hillock, voltage-gated ion channels at this site open, and there is a local influx of Na+ into the cell. Once inside the axoplasm, Na+ diffuses toward the adjacent patch of neuronal membrane and causes depolarization that triggers opening of the voltage-gated Na+ channels in that region. In this way, the AP is regenerated along the entire length of the axon.

Myelinated axons are insulated by myelin segments that are separated by **nodes of Ranvier**. The nodes are 1–2 μm long and are evenly spaced, whereas the **internodes** (myelinated segments) can be up to 1,500 μm long. In myelinated axons, AP propagation involves activation of voltage-gated sodium channels only at the nodes of Ranvier, because it is only at the nodes where the cell membrane is in contact with the extracellular fluid (Figure 2.5). As Na+ diffuses within the axoplasm from one node to the next, the depolarization triggers the opening of voltage-gated Na+ channels. In this way, the AP is regenerated at each node along the axon. Passive conduction of the current between nodes is very rapid, whereas active regeneration at the nodes is time-consuming. Because transmission of the AP is rapid internodally and is slowed down at the nodes, it gives the appearance of jumping from node to node; thus this mode of transmission is called **saltatory conduction** (Latin *saltare*, "dance"). Myelination increases the AP conduction velocity because it incorporates the passive spread of electric current.

THE SODIUM-POTASSIUM PUMP

Although the resting membrane potential of –70 mV is restored following an AP, the resting ionic distribution is not. There is now more Na+ and less K+ inside the cell than there was prior to the AP. There is, however, a mechanism for restoring the original ionic distribution, the **sodium-potassium pump** (Na+/K+ pump). The Na+/K+ pump is a membrane protein that extrudes Na+ from the intracellular fluid and recovers K+. This pump protein is present in all cells of the body, but it is especially abundant in the membranes of excitable cells such as neurons.

The Na+/K+ pump has three binding sites for Na+ on its intracellular surface and two binding sites for K+ on its extracellular surface. When all five sites have ions bound to them, the membrane protein changes conformation, and Na+ ions are moved from the inside of the cell to the outside of the cell, while K+ ions are moved from the outside of the cell to the inside of the cell. The pump is said to be **electrogenic** because it produces a net flux of charge across the membrane, extruding three Na+ for every two K+ that it includes.

The Na+/K+ pump restores and maintains the resting membrane potential by reversing the ionic movements

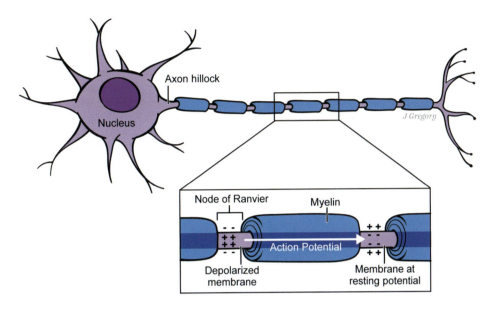

FIGURE 2.5. Myelinated axons are insulated by myelin segments that are separated by nodes of Ranvier. The action potential travels down myelinated axons by saltatory conduction. Axon potentials are generated at the axon hillock by the opening of voltage-gated Na$^+$ channels. Na$^+$ enters the cell, diffuses within the axoplasm to the adjacent node, and the depolarization triggers opening of voltage-gated Na$^+$ channels at that node. Thus, the AP is regenerated at each node.

that occur during the AP, as well as correcting for the minor "leaking" of Na$^+$ and K$^+$ ions across the membrane through non-gated channels that normally occurs in neurons at rest. Without the pump, ionic gradients would gradually run down due to APs and passive fluxes. The Na$^+$/K$^+$ pump requires ATP to move Na$^+$ and K$^+$ against their electrochemical gradients, and approximately 40% of the neuron's energy is expended on the pump. This is one reason why neurons need a nearly constant supply of energy. Insufficient supply of ATP, as occurs with disorders of blood or oxygen delivery to neurons, results in the breakdown of ionic gradients and failure of neuronal function.

Postsynaptic Potentials

Arrival of an AP to axon terminals causes the release of neurotransmitter from the presynaptic cell into the synaptic cleft. Neurotransmitter diffuses across the cleft and binds to transmitter receptors on the postsynaptic cell. The neurotransmitter-receptor binding triggers PSPs, due to local alterations in permeability and ionic distribution across the membrane. Postsynaptic potentials differ from APs in that they are much smaller and are graded in size; their strength varies depending on the amount of neurotransmitter substance secreted into the synaptic cleft.

Postsynaptic potentials may be excitatory or inhibitory. **Excitatory postsynaptic potentials** (EPSPs) are depolarizing; they bring the membrane potential closer to threshold and increase the probability of an AP being triggered.

Inhibitory postsynaptic potentials (IPSPs) are hyperpolarizing; they increase the potential across the membrane, bringing it further from threshold and decreasing the likelihood of an AP being triggered.

Some neurotransmitters that cause EPSPs do so by increasing Na$^+$ conductance through ligand-gated Na$^+$ channels. The cell membrane becomes more permeable to Na$^+$ and Na$^+$ moves toward its equilibrium potential (V_{Na} = +55 mV), causing a small, local depolarization at the dendritic or somatic synaptic site. The influx of sodium is insufficient to reach sodium's equilibrium potential of +55 mV, however, because too few of these Na$^+$ channels open. Furthermore, this influx of Na$^+$ does not initiate an AP at the postsynaptic site because the voltage-gated Na$^+$ channels responsible for generating the AP are present only at the axon hillock and within the axon. Some neurotransmitters cause IPSPs by increasing K+ conductance. The cell membrane becomes even more permeable to K$^+$, and K$^+$ leaves the cell, moving the membrane toward its equilibrium potential (V_K = –75 mV). EPSPs and IPSPs can be generated by increases in the membrane permeability of other ion types. They can also be caused by decreases in membrane permeability to ions. For example, EPSPs can be generated by a decrease in K$^+$ conductance; such a synapse is called a **conductance decrease synapse**.

NEURAL INTEGRATION

An individual neuron may have tens of thousands of synapses on it and may receive numerous PSPs along its dendritic and somatic surfaces at any given time. These signals are summated in a process known as **neural integration**. Because PSPs are based on ionic currents, they spread as the ions diffuse within the intracellular fluid, and their magnitude diminishes over space (distance from the site of origin) and time (since onset of the PSP). Neural integration is the summation of PSPs over space and time; successive PSPs summate by **temporal integration**, while adjacent PSPs summate by **spatial integration**.

If the summation of PSPs reaches the threshold of excitation at the axon hillock, an AP is fired. The rate of firing of any individual axon at a particular time is determined by the degree to which the excitatory state is above threshold (the relative number of EPSPs and IPSPs), although the upper limit is constrained by the absolute refractory period. The axon hillock is the site of AP initiation because it has the lowest threshold for firing an AP due to the high concentration of voltage-sensitive Na+ channels there. Therefore, PSPs that occur proximal (closer) to the axon hillock have more weight in determining whether the neuron will fire an AP or not, because there is less spatiotemporal decay than for PSPs that occur further away from the hillock.

Clinical Considerations

A basic understanding of electrical signaling in neurons is fundamental to understanding clinical neurophysiology methods, some aspects of neuropharmacology, and some neurological disorders.

Clinical neurophysiology is a diagnostic discipline concerned with measuring electrical activity in the central nervous system, peripheral nervous system, and muscle. The four main techniques of clinical neurophysiology are electroencephalography (EEG), evoked potentials (EP), nerve conduction studies (NCS), and electromyography (EMG). The EEG measures electrical activity arising from the cerebral cortex by electrodes placed on the scalp surface (see Chapter 19, "Epilepsy"). Specialized invasive EEG methods involve recording from electrodes that have been surgically implanted on the surface of the cerebral cortex or within the deep structures of the brain. The principal use of EEG is in epilepsy, but it is also used in assessing disorders of consciousness and other brain disorders. Evoked potentials are electrical signals produced by the nervous system in response to an external stimulus. They can be measured to assess the integrity of neural pathways. Sensory EPs can be recorded following stimulation in any sensory modality, but visual EPs (VEPs), auditory EPs (AEPs), and somatosensory EPs (SEPs) are most often used for clinical diagnostic testing and intraoperative monitoring. Nerve conduction and EMG studies measure the electrical activities of peripheral nerves and muscles, respectively. They are used in diagnosing **neuromuscular disorders**, a category encompassing disorders affecting muscle, the nerves that control muscle, or communication between nerve and muscle.

With respect to neuropharmacology, **local anesthetics** such as lidocaine and procaine block the transmission of nerve impulses in peripheral nerves by reversibly blocking voltage-gated sodium channels in the neuronal membrane, thereby inducing **analgesia** (loss of pain sensation) and **anesthesia** (numbness, a loss of somatosensation) in a restricted location of the body (such as a tooth or an area of skin).

Over the last few decades, it has been recognized that an increasing number of neurological disorders are associated with dysfunctional ion channels. These disorders are classified as **neurological channelopathies** because they share ion channel dysfunction as a common pathogenesis. Neurological channelopathies are frequently due to genetic mutations, but they may also be acquired through autoimmune mechanisms. Some forms of epilepsy are associated with genetic brain sodium channelopathies. **Autoantibodies** are **antibodies** (immune proteins) that mistakenly target and react with a person's own tissues or organs In autoimmune neurological channelopathies, autoantibodies are directed against voltage-gated or ligand-gated ion channels and receptors (e.g., voltage-gated calcium channels, voltage-gated potassium channels, NMDA glutamate receptors, AMPA glutamate receptors, $GABA_B$ receptors). These disorders are diagnosed by laboratory tests and are treated by immunotherapies that reduce the levels of the pathogenic autoantibodies.

Summary

Neurons are electrically excitable; they receive, integrate, and transmit information by changes in their electrical state. The intracellular and extracellular fluid compartments of the neuron differ in chemical composition and consequently there is an electrostatic gradient across the cell membrane. Changes in membrane permeability result in ionic current flow across the neuronal membrane; the resulting deviations in membrane potential are the basis of electrical signaling. Neurons receive information in the form of PSPs and communicate information in the form of APs. A basic knowledge of the process by which electrical signals are generated and propagated within neurons is fundamental to understanding mechanisms underlying seizures, electroencephalography, brain stimulation, and some aspects of neuropharmacology.

Additional Reading

1. Brodal P. Neuronal excitability. In: Brodal P, *The central nervous system: structure and function*. 5th ed. Oxford University Press; 2016. https://oxfordmedicine.com/view/10.1093/med/9780190228958.001.0001/med-9780190228958-chapter-3.
2. Shepherd GM. *The synaptic organization of the brain*. 5th ed. Oxford University Press, 2003.
3. Vanderah TW, Gould D. Electrical signaling by neurons. In: Vanderah TW, Gould D, *Nolte's the human brain: an introduction to its functional anatomy*. 8th ed., pp. 147–171. Elsevier; 2021.

Chemical Neurotransmission and Neuropsychopharmacology

Introduction

Abnormalities in the function of neurotransmitter systems contribute to a wide range of brain disorders, and many neuropharmacological agents work by altering neurotransmitter release, neurotransmitter receptor binding, and/or neurotransmitter removal from the synapse. A basic knowledge of the neurotransmission processes therefore serves as a foundation for understanding the mechanisms of many neurological and psychiatric disorders, as well as the mechanisms of pharmacological intervention. This chapter presents on overview of chemical neurotransmission at the synapse, the major neurotransmitters, mechanisms of drug action, and neuropsychopharmacology. Common psychotropic drug classifications discussed include the antipsychotics, antidepressants, anxiolytics, antimanics, psychostimulants, and sedative-hypnotics.

Synaptic Transmission

The process of chemical **neurotransmission**, also called **synaptic transmission**, involves multiple steps: (1) neurotransmitter synthesis and packaging into vesicles in the presynaptic cell; (2) neurotransmitter release into the synaptic cleft in response to arrival of an action potential (AP) at the axon terminal; (3) neurotransmitter binding to receptors on postsynaptic cells and generation of **postsynaptic potentials** (PSP); and (4) rapid removal of the transmitter from the synaptic cleft by reuptake and/or enzymatic degradation.

Neurotransmitters

Acetylcholine was the first neurotransmitter to be identified by **Otto Loewi** (1873–1961) in 1921. Prior to its discovery, it was unknown whether neuron-to-neuron signaling was electrical or chemical. Norepinephrine was the second neurotransmitter to be identified, in 1946. Early in the study of neurotransmitters within the brain, the idea prevailed that the brain used only one excitatory neurotransmitter (norepinephrine) and one inhibitory neurotransmitter (acetylcholine). More than 100 substances are now recognized to act as neurotransmitters.

Early knowledge of neurotransmitters came from experiments on peripheral nerve synapses at the neuromuscular junction (the site of chemical communication between neurons and muscle cells) and autonomic ganglia (the site of chemical communication between autonomic neurons originating centrally and autonomic neurons innervating target organs in the periphery). Analysis at these structures is relatively easy because there is only one neurotransmitter. The identification and study of neurotransmitters within the central nervous system (CNS) is much more difficult than in the peripheral nervous system (PNS) because a given region may contain numerous neurotransmitters, and a given cell may be acted upon by numerous neurotransmitters.

Neurons can be differentiated according to the type of neurotransmitter that they use for signaling. Any single neuron releases the same neurotransmitter at all of its terminals (**Dale's principle**). Neurotransmitters, then, differ in the pattern of their distribution among neurons in the CNS, but serve as the singular vehicle for synaptic signaling of any particular neuron. The study of the **chemoarchitecture** of the brain involves identifying the cell groups, pathways, and terminals containing specific neurotransmitters.

Transmitter Release

Neurotransmitter is stored in synaptic vesicles within the axon terminals (terminal buttons). Arrival of an AP at the axon terminal triggers the opening of voltage-dependent calcium channels in the membrane of the axon terminal, and an influx of calcium ions (Ca^{2+}) into the axon terminal (Figure 3.1). The increase in calcium ion concentration in the cytoplasm of the axon terminal causes the vesicles to migrate to the presynaptic membrane. The vesicular membrane and presynaptic membrane fuse and form a pore that grows larger until the vesicle membrane collapses into the presynaptic membrane, releasing neurotransmitter into the synaptic cleft. Thus, neurotransmitter release from the presynaptic axon terminal into the synaptic cleft occurs by the process of **exocytosis**.

Neurotransmitter Receptors

Neurotransmitter receptors are membrane-bound proteins that interact with extracellular physiological signals and convert them into intracellular effects. When neurotransmitter is released into the synaptic cleft, it diffuses across the cleft and binds to transmitter receptors in the postsynaptic membrane, usually on dendrites.

Neurotransmitters bind to specific receptors. For example, acetylcholine acts at cholinergic receptors, and dopamine acts at dopaminergic receptors. Transmitter receptors have a specific molecular configuration that determines which molecules can bind with them. The binding of neurotransmitters to their receptors is often described with a **lock and key** metaphor; each receptor binds only with chemicals of a particular structure, similar to how locks accept only specifically shaped keys. Molecules that bind with receptors are referred to as **ligands**.

The binding of a neurotransmitter to a receptor in the postsynaptic membrane causes a conformational (shape)

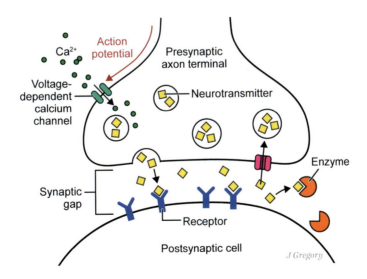

FIGURE 3.1. Synaptic transmission. Arrival of an action potential at the axon terminal triggers opening of voltage-gated calcium channels, which causes an influx of calcium ions, which causes neurotransmitter-containing vesicles to migrate to the presynaptic membrane. The vesicular membrane and presynaptic membrane fuse and form a pore, releasing neurotransmitter into the synaptic cleft. Neurotransmitters are cleared from the synaptic cleft by reuptake or enzymatic degradation.

change in the receptor protein. This conformational change alters the postsynaptic membrane permeability to a specific ion type. These events result in a local change in the postsynaptic membrane potential that is either depolarizing or hyperpolarizing. Depolarization moves the membrane potential toward the threshold of excitation for triggering an AP (e.g., from –70 mV to –65 mV); this is known as an **excitatory postsynaptic potential** (EPSP). Hyperpolarization moves the membrane potential away from the threshold of excitation for triggering an AP (e.g., from –70 mV to –75 mV); this is known as an **inhibitory postsynaptic potential** (IPSP). Postsynaptic potentials are **graded** in magnitude; the amplitude of the PSP depends on the amount of neurotransmitter at the synapse. This stands in contrast to the AP which follows the all-or-none law; the amplitude of the AP is independent of the magnitude of the triggering stimulus. Most EPSPs and IPSPs occur by opening of ion channels for a specific ion type, which increases conductance of that ion type across the membrane. Some PSPs, however, occur by closing ion channels that are normally open, which decreases conductance of the ion type across the membrane. For example, opening of sodium (Na$^+$) channels results in EPSPs (V_{Na} = +55 mV), opening of potassium (K$^+$) channels results in IPSPs (V_K = –75 mV), and decreased conductance of K$^+$ by channel closing results in EPSPs.

There are multiple receptors for any particular neurotransmitter, and these are called **receptor subtypes**. Thus, a single neurotransmitter can have different physiological effects on cells, depending on the nature of the receptor at the particular synapse. This realization led to the concept of multi-action neurons that synthesize and release a single neurotransmitter but produce different responses at different target cells that have different receptors.

Whether a transmitter has an excitatory or inhibitory effect depends on the nature of the receptor to which it binds and the ion channel to which the receptor is linked. Some neurotransmitters have exclusively excitatory effects, some have exclusively inhibitory effects, and some are excitatory at some synapses but inhibitory at other synapses, depending on the receptors at the synapse.

There are two broad categories of postsynaptic neurotransmitter receptors, ionotropic and metabotropic.

IONOTROPIC RECEPTORS

Ionotropic receptors control ion channels directly (Greek *tropos*, "to move in response to a stimulus"). The ion channel is part of the receptor protein complex (Figure 3.2). When transmitter binds to the receptor, the receptor undergoes a change in conformation that alters the permeability of the membrane to an ion type (Na$^+$, K$^+$, Cl$^-$, or Ca^{2+}). Ionotropic receptors are therefore also referred to as **ligand-gated ion channels**. The change in ionotropic receptor conformation causes either (1) opening of an ion

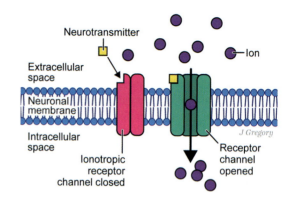

FIGURE 3.2. Ionotropic receptors control ion channels directly; when transmitter binds to the receptor, the receptor undergoes a change in conformation that alters the permeability of the membrane to an ion species.

channel, allowing for influx or efflux of an ion type, or (2) closing of an ion channel that is normally open, reducing influx or efflux of an ion type. Ionotropic receptors mediate rapid signaling. The neurotransmitter binds for just several msec; when it is released, the ion channel returns to its resting configuration and the neurotransmitter is broken down enzymatically, ending the signal. Thus, ionotropic receptors produce fast postsynaptic responses lasting just several msec.

METABOTROPIC RECEPTORS

Metabotropic receptors do not have ion channels as part of their structure. Rather, the neurotransmitter-receptor interaction results in one or more metabolic steps, eventually forming an intracellular messenger that acts on ion channels.

Metabotropic receptors are linked to guanine nucleotide binding protein (**G-protein**) and therefore also are referred to as **G protein-coupled receptors** (Figure 3.3). When neurotransmitter binds to a metabotropic receptor, the G-protein is activated by conversion of guanyl nucleotide diphosphate (GDP) into guanyl nucleotide

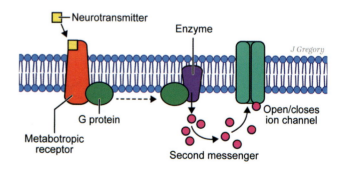

FIGURE 3.3. Metabotropic receptors activate a second messenger through a linked G-protein. The second messenger mediates the cellular response to the transmitter.

triphosphate (GTP). This in turn activates a **second messenger**, which mediates the cellular response to the transmitter (first messenger). For example, G-protein activation may convert adenosine triphosphate (ATP) into cyclic adenosine monophosphate (c-AMP), one type of second messenger. Cyclic-AMP in turn activates a protein kinase that phosphorylates the ion channel, causing a change in its configuration (either opening or closing) and a PSP. The signal transduction mechanism from transmitter-receptor binding to ion channel conformational change involves multiple steps and requires cellular energy (metabolism). Thus, second messengers are intracellular signaling molecules that form in response to extracellular signaling molecules and trigger an intracellular biochemical cascade, ultimately leading to the formation of an effector molecule that results in a cellular response (in this case, a change in ion channel state).

The effects of metabotropic receptors on the membrane potential are slower than those of ionotropic receptors, as the change in configuration of the ion channel lasts longer, sometimes several minutes. Furthermore, metabotropic receptors may have more widespread effects through a cell, as compared to ionotropic receptors that have effects only in the immediate vicinity of the receptor.

AUTORECEPTORS

Autoreceptors are receptors that are located in the presynaptic nerve cell membrane and are sensitive to the neurotransmitter released by the neuron. Autoreceptors provide feedback to the releasing cell on its level of activity, serving as part of a negative feedback loop in regulating the activity (signal transduction) of the cell.

RECEPTOR REGULATION

A cell's ability to respond to a chemical message depends on the presence of receptors. Receptors are proteins that are manufactured by cells; the number of receptors created is regulated. When neural tissue is exposed chronically to a chemical substance that simulates a specific neurotransmitter (i.e., binds to a receptor and causes similar postsynaptic effects), it leads to a decrease in the number of receptors for that transmitter, a process known as **receptor down-regulation**. In contrast, when neural tissue is exposed chronically to a chemical substance that interferes with neurotransmission of a specific neurotransmitter, it leads to an increase in the number of receptors for that transmitter, a process known as **receptor up-regulation**.

Similarly, when the nerve supply to an effector organ (e.g., skeletal muscle) has been severed, the effector is deprived of normal transmitter substance. The lack of postsynaptic transmitter input due to denervation results in **denervation supersensitivity**, an enhanced response to exogenously applied neurotransmitter or drugs that simulate the transmitter. Denervation supersensitivity is a compensatory response, and in the case of the neuromuscular junction of skeletal muscle, it occurs by receptor up-regulation, as evidenced by elevated binding of radiolabeled ligand (ligand tagged with a radioactive tracer for quantification and localization). Denervation supersensitivity also occurs when central targets are deprived of their normal levels of transmitter input, in some cases due to receptor up-regulation.

Removal of Neurotransmitter from the Synapse

In order for signals to remain temporally distinct, neurotransmitters must be cleared from the synaptic cleft. This occurs by reuptake into the presynaptic cell or enzymatic degradation (see Figure 3.1). Reuptake involves actively pumping the neurotransmitter back into the presynaptic axon terminals or nearby glia via a specialized protein **reuptake transporter**. When neurotransmitters are broken down enzymatically within the synapse, the products of transmitter metabolism are taken up by the terminal buttons and are recycled in the synthesis of new transmitter molecules.

Mechanisms of Drug Action

Neuropharmacological and psychopharmacological agents mostly produce their effects by altering synaptic transmission. They may do so at any stage of the neurotransmission process, by blocking or enhancing neurotransmitter synthesis, interfering with neurotransmitter storage, blocking or enhancing neurotransmitter release, acting at receptors, augmenting or reducing postsynaptic effects, or altering neurotransmitter deactivation by blocking reuptake or interfering with the catabolic enzyme that breaks down the transmitter.

Any drug that has the net effect of increasing the efficiency or enhancing transmission of a particular neurotransmitter is an **agonist** of that neurotransmitter. Similarly, any drug that has the net effect of decreasing the efficiency or inhibiting transmission of a particular neurotransmitter is an **antagonist** of that neurotransmitter. Drugs that mimic a neurotransmitter by binding to its receptor, causing a conformational change in the receptor that leads to similar postsynaptic effects, are called **receptor agonists**. Drugs that bind to a receptor but block it, preventing the neurotransmitter from acting at the receptor, are **receptor antagonists**.

Drugs that act directly on receptors may be further classified. **Full agonists** bind to a receptor and elicit the maximum possible response. **Partial agonists** bind to a receptor but do not elicit the maximum possible response, even at full receptor occupancy. A partial agonist acts as an antagonist in the presence of a full agonist because it competes with the full agonist for the same receptors and reduces the ability of the full agonist to produce its maximum effect. Thus, partial agonists have both agonist and antagonist effects. **Inverse agonists**, also known as **negative antagonists**, bind to a receptor but elicit the opposite response of an agonist. This contrasts with **neutral receptor antagonists** described above, which bind to a receptor but simply block it, preventing the neurotransmitter from acting at the receptor.

Neurotransmitter Systems

In order for a chemical substance to qualify as a **neurotransmitter**, several criteria must be met: (1) the substance is synthesized by the neuron; (2) the substance is present in axonal terminals and released in sufficient amounts to exert an effect on the postsynaptic neuron; (3) exogenous administration of the transmitter mimics the action of the endogenously released substance at the postsynaptic cell; and (4) a mechanism exists for terminating the action of the substance at the synapse.

Most neurotransmitters are classified into two broad classes: the small molecule neurotransmitters and the neuropeptides. The **small molecule neurotransmitters** are fast-acting, affecting the postsynaptic membrane within 1 msec of being released. They are responsible for the most acute responses of the nervous system, such as motor and sensory signals. The small molecule transmitters can be classified into three groups: acetylcholine, the monoamines, and the amino acids. The neuropeptides constitute a large class of signaling molecules in the nervous system that are responsible for slow-onset, long-lasting effects (see "The Neuropeptides," below).

Acetylcholine

Acetylcholine (ACh) is the neurotransmitter at the neuromuscular junction of skeletal muscle. It is also the transmitter at the neuroeffector junctions of the parasympathetic division of the autonomic system (i.e., at the postganglionic neurons that synapse on glands, smooth muscles, and cardiac muscle), and the principal neurotransmitter released by preganglionic neurons at the autonomic ganglia in both sympathetic and parasympathetic divisions of the autonomic system.

In the CNS, ACh is located in several cell groups that give rise to long axon projection neurons (e.g., the nucleus basalis of Meynert, which projects to the amygdala and the entire neocortex). Additionally, cholinergic local circuit neurons are found throughout the neostriatum (caudate, putamen, nucleus accumbens). The role of ACh in the CNS, however, is not well understood.

There are two families of ACh receptors, nicotinic and muscarinic, each with several subtypes. **Nicotinic receptors**, so named because historically they were found to bind the tobacco plant alkaloid nicotine, are ionotropic. They are present in the PNS at the neuromuscular junction of skeletal muscles, where they are linked to calcium channels. Stimulation of nicotinic receptors at the neuromuscular junction causes Ca^{2+} influx into muscle cells and muscle contraction (see Chapter 24, "The Motor System and Motor Disorders"). Nicotinic receptors are also present at the autonomic ganglia synapses. Nicotinic receptors at non-neuromuscular junction synapses are linked directly to Na^+ channels and produce EPSPs.

Muscarinic receptors are so named because they were found to bind with muscarine, a compound produced by the mushroom *Amanita muscaria*. They are metabotropic and produce IPSPs via channels that increase the conductance of K^+ across the cell membrane. In the parasympathetic division of the PNS, ACh is released from postganglionic fibers, where it binds to muscarinic receptors on the end organs (e.g., cardiac muscle, smooth muscle, glands). Acetylcholine receptors in the brain are primarily of the muscarinic type.

Deactivation of ACh occurs by enzymatic breakdown via **acetylcholinesterase** (AChE). This enzyme is one of the fastest-acting enzymes in the body. It rapidly deactivates ACh by breaking it down into acetate and choline. The liberated choline is then transported from the synapse into cholinergic terminals by reuptake and is recycled into new ACh.

There are multiple mechanisms of blocking transmission at the neuromuscular junction of skeletal muscles and causing paralysis, including blocking nicotinic receptors, preventing release of ACh from motor neurons, and altering the enzymatic breakdown of ACh by AChE. Many poisonous snake venoms, such as cobra venom, contain toxins that block neuromuscular transmission by irreversibly binding to nicotinic ACh receptors. This leads to paralysis of the snake's prey because motor neuron activity cannot elicit skeletal muscle contraction. **Curare** is a plant extract that acts as a nicotinic receptor antagonist and produces paralysis. It was used by South American indigenous people, who dipped their arrows and blowgun darts in curare, leading to paralysis in their prey. Curare-like drugs are used in medicine to produce muscular relaxation during surgery and procedures to reset dislocated joints or align broken bones, and to prevent gagging during examination of the esophagus and trachea. **Botulinus toxin** is a neurotoxin produced by the bacterium *Clostridium botulinum*. It inhibits the release of ACh from motor neurons at the neuromuscular junction, causing muscle paralysis. It is used in the management of conditions with muscular overactivity, such as focal dystonia, strabismus, hemifacial spasm, spastic movement disorders, and tension headaches caused by contraction of the neck and scalp muscles. Botulinus toxin is also used in cosmetic dermatology to reduce face wrinkles. The **organophosphates** inhibit AChE, causing ACh to accumulate at cholinergic synapses. Overstimulation of nicotinic receptors leads to depolarizing block of postsynaptic cells and causes neuromuscular paralysis. Organophosphates are highly toxic to insects, but less so to humans and domestic animals; thus, they are commonly used as insecticides. They are also used as chemical weapon nerve gases (e.g., sarin nerve gas). Accumulation of ACh at nicotinic synapses results in death from respiratory failure.

Acetylcholine also plays an important role in learning and memory. **Scopolamine**, a muscarinic receptor antagonist, is well known for its ability to produce amnesia. Alzheimer's disease is a cortical neurodegenerative disease that is characterized by amnesia as the earliest and most prominent symptom. **Acetylcholinesterase inhibitors** increase the amount of ACh available at cholinergic synapses; donepezil is an AChE inhibitor that is used in the treatment of Alzheimer's disease.

The Monoamines

The **monoamine** transmitters, also known as the **biogenic amines**, contain a single amine group. They include **dopamine** (DA), **norepinephrine** (NE, known as noradrenalin in the periphery), **epinephrine** (Epi, known as adrenalin in the periphery), and **serotonin** (5-hydroxytryptamine, 5-HT); DA, NE, and Epi are also classified as **catecholamines** (CA) because they contain a catechol nucleus.

Monoamine synthesizing neurons are located mainly in the brainstem in discrete nuclei. They have fine-fibered, profusely ramifying ascending and descending projections to many regions of the brain and spinal cord. The monoamines are removed from the synaptic cleft by reuptake into the presynaptic terminal, where they are broken down intracellularly by the enzyme **monoamine oxidase** (MAO), which has two forms. MAO-A degrades DA, NE, and 5-HT; MAO-B degrades DA only. Many psychotropic drugs affect the synthesis, receptor binding, or catabolism (breakdown) of the monoamines (see "Neuropsychopharmacology," below).

DOPAMINE

Dopamine-synthesizing neurons are located in discrete cell groups in the diencephalon and mesencephalon, giving rise to three distinct brain dopaminergic projections: the nigrostriatal, mesolimbic, and mesocortical pathways (Figure 3.4).

The **nigrostriatal pathway** originates in the substantia nigra of the midbrain and terminates in the striatum (caudate and putamen). This pathway undergoes

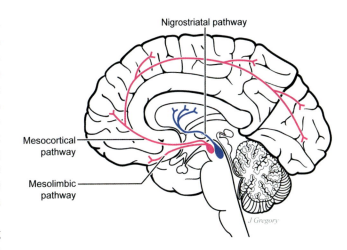

FIGURE 3.4. Dopamine-synthesizing cells are located in discrete cell groups in the diencephalon and mesencephalon and give rise to three distinct brain dopaminergic projections: the nigrostriatal, mesolimbic, and mesocortical pathways.

selective degeneration in Parkinson's disease (see Chapter 11, "The Basal Ganglia"). Providing L-DOPA (L-3,4-dihydroxyphenylalanine), an intermediate molecule in the DA biosynthesis pathway, increases the rate of DA synthesis in surviving nigrostriatal cells. This strategy forms the basis for the pharmacological treatment of Parkinson's disease.

The **mesolimbic pathway** originates in the ventral tegmental area of the midbrain and projects to limbic structures (nucleus accumbens, amygdala, olfactory tubercle). This pathway plays a key role in reward, motivation, and incentive drive; therefore, it is also known as the **reward pathway** (see Chapter 12, "Limbic Structures"). It signals natural rewards such as food, sex, and social interaction, and it mediates the rewarding properties of many drugs of abuse.

The **mesocortical pathway** also originates in the ventral tegmental area and projects to limbic cortex (entorhinal, suprarhinal, and anterior cingulate cortex), hippocampus, and widespread neocortical regions (i.e., the entire frontal lobe and all regions of association cortex in the parietal and temporal lobes). This pathway is important in a wide range of functions, including executive functions (e.g., working memory, planning, decision-making).

Five subtypes of DA receptor, named D_1 through D_5, all of which are metabotropic, have been identified; they differ in various ways, including molecular structure, binding properties, and distribution throughout the brain. Dopamine plays a role in many functions, including motivation, pleasure, and cognition. Abnormal DA receptor signaling is implicated in several neuropsychiatric disorders, thus DA receptors are a common target in psychopharmacology (e.g., antipsychotics, psychostimulants).

NOREPINEPHRINE

Norepinephrine is the neurotransmitter used at sympathetic postganglionic synapses. Within the brain, NE-synthesizing neurons are located in discrete cell groups in the pons and medulla. Their ascending and descending projections branch extensively, so that the entire cortex and spinal cord receive noradrenergic input. The **locus coeruleus** is the major noradrenergic nucleus of the brain, containing more than 50% of all noradrenergic brain cells. There are two norepinephrine receptor subtype families, α and β, each with further subtypes.

SEROTONIN

Serotonin (5-HT) is produced by several distinct cell groups known as the **raphe nuclei**. These nuclei have widespread ascending and descending projections. There are seven general serotonin receptor classes, 5-HT_1 through 5HT_7, some with several subtypes (e.g., 5-HT_{1A}, 5-HT_{1B}). Serotonin usually has inhibitory effects on its targets.

The Amino Acid Neurotransmitters

Several amino acids act as neurotransmitters. As a group, they are the most abundant neurotransmitters in the CNS; however, few have been implicated in human disease. They are classified further as excitatory amino acid transmitters and inhibitory amino acid transmitters. **Glutamate** is the main excitatory transmitter in the CNS. **Gamma-aminobutyric acid** (GABA) is the main inhibitory neurotransmitter in the brain, and **glycine** is the main inhibitory neurotransmitter in the spinal cord.

GLUTAMATE

Glutamate is the principal excitatory amino acid transmitter in the CNS and the most abundant neurotransmitter in the vertebrate nervous system. There are several ionotropic glutamate receptors: **AMPA receptors**, **NMDA receptors**, and **kianate receptors**. There are also several metabotropic glutamate receptor subtypes. Glutamate signaling within the synapse is terminated by reuptake.

GABA

GABA is exclusively an inhibitory neurotransmitter. It is the main inhibitory neurotransmitter in the vertebrate nervous system and is present throughout the brain. GABA is synthesized and released primarily by local circuit neurons, but a number of long axon projection systems use this neurotransmitter as well. The two major GABA receptor subtypes are GABA_A ionotropic receptors and GABA_B metabotropic receptors. Both are linked to Cl^- channels. GABA is removed from the synapse by reuptake into presynaptic terminals and glia.

GABA receptors have modulatory sites for alcohol, benzodiazepines, and barbiturates, which all facilitate GABAergic transmission (i.e., increase neural inhibition mediated by GABA). Benzodiazepines increase the frequency of GABA receptor ion channel opening, and barbiturates increase the duration of GABA receptor ion channel opening. Alcohol, benzodiazepines, and barbiturates all have anxiolytic and sedative effects. The effects of these substances are additive; mixing small doses of each can sum to be fatal.

Many antiepileptic drugs, including benzodiazepines and barbiturates, are GABA agonists. Gabapentin works by increasing the production of GABA, and sodium valproate and vigabatrin work by decreasing the breakdown of GABA. Combining antiepileptic drugs with alcohol may increase the sedative effects of the antiepileptic, reduce tolerance to alcohol lowering the dose required for intoxication, and alter the therapeutic efficacy of antiepileptic drugs.

GLYCINE

Glycine is the major inhibitory neurotransmitter in the lower brainstem and spinal cord of vertebrates. It is released mainly by local circuit neurons. Very few drugs act specifically at glycine receptors or affect glycine synthesis or reuptake.

The Neuropeptides

Since the 1970s, it has been recognized that peptides play a role in chemical communication between neurons. The **neuropeptides** (peptides produced and released by discrete neuron populations that function in neural signaling) are typically slow acting and produce effects in the range of seconds or longer, a time span considerably longer than that of the small molecule neurotransmitters. The neuropeptides tend to act as **neuromodulators** (i.e., they modulate the actions of other neurotransmitters).

Many neuropeptides were identified initially in the periphery, and their names reflect the organ in which they were first found or the physiological action that was first studied. Some neuropeptides are found in both the CNS and the gastrointestinal tract and therefore are called **brain-gut peptides**; these include vasoactive intestinal peptide (VIP), cholecystokinin (CCK), substance P, and neuropeptide Y. Another class of neuropeptides is the **opioid peptides.**

OPIOID PEPTIDES

In 1973 it was discovered that brain tissue contained molecules that bind selectively with opiate drugs, through studies in which opiate drugs were radioactively labeled (radioligand binding studies). Following the discovery of **opiate receptors**, the search for the endogenous ligands of these receptors revealed the **endogenous opioids**, also known as **opioid peptides**.

It is now known that there are three families of endogenous opioid peptides: the **endorphins** (for *endogenous morphine*), **enkephalins**, and **dynorphins**. The major opioid receptor subtypes are mu (endorphin receptor), delta (enkephalin receptor) and kappa (dynorphin receptor). They are all metabotropic receptors that are inhibitory (i.e., produce IPSPs). Different opioid receptor subtypes have different distributions in the brain and different behavioral effects. The opioid peptides are broken down enzymatically by **peptidases**.

The brain opioid peptide systems play a role in stress and pain, motivation, reward, and attachment behaviors.

Neuropsychopharmacology

A **drug** is a chemical substance that causes a physiological effect when ingested or otherwise introduced into the body. **Pharmacology** is a branch of medical science concerned with the study of drug action. **Neuropharmacology** is a branch of pharmacology concerned with the study of drug action on the nervous system. **Neuropsychopharmacology** (**behavioral neuropharmacology**) is a branch of neuropharmacology concerned with the study of how drugs influence behavior, mood, and cognition.

Drug Classification

Drugs can be classified in multiple ways. One classification system involves grouping by chemical structure; however, this does not necessarily correspond to drugs having similar effects. Minor changes in molecular structure can lead to considerable changes in a drug's effects, so drugs with similar structures can have quite different effects. Slight alterations in molecular structure can change a molecule from one that is active at a receptor into one that is inactive, a mixed agonist-antagonist, or a pure antagonist. Conversely, drugs with very different chemical structures can have very similar effects (e.g., MAO inhibitors, tricyclics, and selective serotonin reuptake inhibitors all have antidepressant effects).

Drugs therefore may also be classified according to their main therapeutic effect. This is not a perfect scheme because drugs may be used for different purposes in different doses and in different individuals. For example, carbamazepine and lamotrigine are used to treat seizure disorders, but they are also used to treat bipolar disorder. Common psychotropic drug classifications include the antipsychotics, antidepressants, anxiolytics, antimanics, psychostimulants, and sedative-hypnotics.

Antipsychotics

TYPICAL ANTIPSYCHOTICS

The first drug discovered to have antipsychotic effects was **chlorpromazine** (Thorazine). It was introduced to the market in the 1950s, dramatically changing the field of psychiatry and ushering in the psychopharmacological revolution. It largely replaced biological therapies of the day, such as prefrontal lobotomy and insulin coma, and to some extent electroconvulsive therapy (although there has been a recent resurgence in psychosurgery and deep brain stimulation for patients with medication-refractory psychiatric disorders). The psychopharmacological revolution also played a prominent role in the deinstitutionalization movement during the 1960s and 1970s, a federal policy responsible for the release of thousands of patients with psychiatric disorders from U.S. asylums.

Chlorpromazine is a phenothiazine, and it was soon discovered that other phenothiazines also had antipsychotic effects. In the 1960s it was discovered that a variety of non-phenothiazine drugs such as haloperidol also had antipsychotic properties, and these were introduced to the market during the 1960s and 1970s. The antipsychotics of the time were all found to cause a syndrome referred to as "**neurolepsis**," characterized by psychomotor slowing, apathy, lack of initiative, limited range of emotion, reduced agitation, and reduced confusion; thus, these drugs also came to be known as **neuroleptics**.

By the 1970s, it was recognized that despite their different chemical structures, all antipsychotic medications are dopamine receptor (specifically D_2) antagonists. Furthermore, their potency as antipsychotics correlates highly with their ability to block D_2 receptors. This observation led to the formulation of the **dopamine hypothesis of schizophrenia**, which in its original formulation stated that schizophrenia is based in excess DA activity. This hypothesis is supported by the fact that L-DOPA, which enhances DA synthesis and is used to treat Parkinson's disease, can elicit paranoid psychosis. Stimulant drugs with DA agonist properties, such as amphetamine (which retards CA reuptake and stimulates DA release) and cocaine (which retards CA reuptake, especially of DA), exacerbate psychotic symptoms in people with schizophrenia; they can also elicit symptoms resembling schizophrenia, particularly paranoid delusions, in people without schizophrenia. The antipsychotics reverse stimulant-induced psychotic episodes.

The main side effects of the antipsychotic drugs are parkinsonism with acute use, and tardive dyskinesia with chronic use. The parkinsonism consists of bradykinesia, rigidity, and tremor, and it is caused by blockade of D_2

receptors within the basal ganglia. **Tardive dyskinesia** is a movement disorder characterized by late (tardive) onset of involuntary, repetitive movements that particularly affect the face, lips, and tongue (e.g., grimacing, lip-smacking, sticking out the tongue), but also may affect the trunk and extremities. Tardive dyskinesia occurs with long-term use of dopaminergic antagonists; it is due to an induced hypersensitivity of D_2 receptors within the striatum.

The etiology and pathophysiology of schizophrenia are unknown. However, there is evidence that excessive dopaminergic activity within the mesolimbic pathway underlies the positive symptoms (i.e., presence of abnormal behaviors, beliefs, or perceptual experiences) of schizophrenia, such as hallucinations and delusions, which are alleviated by D_2 receptor antagonists. The D_2 receptor antagonists also act within the mesocortical and nigrostriatal pathways. There is evidence that reduced dopaminergic activity within the mesocortical pathway underlies the negative symptoms (i.e., deficits) of schizophrenia, such as restricted emotional expression, avolition (lack of motivated self-initiated purposeful activities), anhedonia (inability to feel pleasure), and asociality (little or no desire to socialize with others). D_2 receptor antagonists do not alleviate the negative symptoms and may exacerbate them. D_2 receptor antagonists acting within the nigrostriatal pathway produce parkinsonism and tardive dyskinesia side effects.

In the 1990s, a group of new antipsychotics emerged that worked by mechanisms other than D_2 receptor blockade, leading to a distinction between **first-generation/typical antipsychotics** (also referred to as conventional antipsychotics, classical antipsychotics, or neuroleptics), and **second-generation/atypical antipsychotics**.

ATYPICAL ANTIPSYCHOTICS

Second-generation antipsychotics have largely supplanted the first-generation antipsychotics due to their decreased risk for parkinsonism and tardive dyskinesia as side effects, and their efficacy in treating the negative symptoms of schizophrenia. These drugs include clozapine, risperidone, quetiapine, olanzapine, and ziprasidone.

It is not known precisely how the atypical antipsychotics work. They bind to D_2 receptors, but much more loosely than the typical antipsychotics. The transient D_2 receptor antagonism of atypical antipsychotics is sufficient to obtain an antipsychotic effect, while the longer-lasting D_2 receptor antagonism of typical antipsychotics increases the risk of adverse effects. However, many atypical antipsychotics have combined D_2 and serotonin 5-HT_{2A} receptor antagonist effects, with 5-HT_{2A} effects modulating dopaminergic activity. Some atypical antipsychotics act as D_2 partial agonists, some act as 5-HT_{1A} partial agonists, and some interact with D_3 or D_4 receptors.

Antidepressants

There are three classes of antidepressants: MAO inhibitors, reuptake inhibitors, and atypical antidepressants.

MONOAMINE OXIDASE INHIBITORS

The first molecules discovered to have antidepressant effects were the monoamine oxidase inhibitors (MAOIs). They were introduced to the market in the 1950s, and some are still in use. The MAOIs inhibit the breakdown of monoamine transmitters within the presynaptic terminal and increase the availability of monoamines for release into the synapse. This discovery led to the **monoamine theory of depression**, which postulates that depression is due to deficient neurotransmission in serotonergic and noradrenergic pathways. The MAOIs carry a risk of tyramine toxicity and require dietary restriction of foods rich in the amino acid tyramine (e.g., wine, cheese, yeast products, chocolate, milk, beer). Tyramine toxicity produces sympathomimetic effects, including a marked rise in blood pressure, headaches, intracranial bleeding, and potentially, death. For this reason, use of MAOIs for treating depression has largely been supplanted by reuptake inhibitors and atypical antidepressants. Given the dietary restrictions and side effects, the MAOIs are usually considered as medications of last resort for patients who have not responded to other antidepressant medications.

REUPTAKE INHIBITORS

The **tricyclic antidepressants** were introduced in the 1960s. They block the reuptake of both 5-HT and NE by acting on the monoamine transporters, which increases the availability of 5-HT and NE in the synapse. The tricyclics differ in their relative effects on 5-HT and NE systems. Abrupt discontinuation of tricyclic antidepressants produces withdrawal symptoms; therefore, the dose should be reduced gradually when treatment is discontinued. The tricyclic antidepressants include clomipramine, desipramine, imipramine, amitriptyline, and nortriptyline.

Selective serotonin reuptake inhibitors (SSRIs), also known as serotonin-specific reuptake inhibitors, inhibit the binding of 5-HT to the serotonin transporter, thereby increasing the level of 5-HT in the synapse and availability for binding to postsynaptic receptors. The first SSRI, fluoxetine, was introduced in the late 1980s, and this class has grown since then to include sertraline, paroxetine, citalopram, escitalopram, and fluvoxamine. The SSRIs are regarded as first-line treatment for depression because they have high clinical effectiveness and fewer side effects than other classes of antidepressants. The SSRIs are the most frequently prescribed antidepressants.

Serotonin-norepinephrine reuptake inhibitors (SNRIs), also referred to as dual reuptake inhibitors or dual-acting antidepressants, block reuptake of both 5-HT and NE. These include venlafaxine, desvenlafaxine, and duloxetine.

ATYPICAL ANTIDEPRESSANTS

The **atypical antidepressants** are a group of medications in which the mechanism of antidepressant action differs from the other classes of antidepressants. These medications are frequently used in patients with major depression who have inadequate responses or intolerable side effects to first-line treatment with SSRIs. The atypical antidepressants include bupropion (a norepinephrine-dopamine reuptake inhibitor/NDRI), mirtazapine (a tetracyclic), and trazodone (a $5-HT_{2A}$ and $5-HT_{2C}$ antagonist).

Anxiolytics

The **anxiolytics** (antianxiety drugs) are used for the treatment of anxiety disorders, including panic disorder, posttraumatic stress disorder, obsessive-compulsive disorder, generalized anxiety disorder, social anxiety disorder, and specific phobias.

One major class of anxiolytics is the **benzodiazepines**. The first benzodiazepine introduced to the market in 1960 was chlordiazepoxide, followed by diazepam in 1963. Other benzodiazepines include alprazolam, clonazepam, lorazepam, and triazolam.

There is evidence that the anxiety disorders are based on decreased inhibition of the brain circuitry mediating fear due to GABAergic dysfunction. Benzodiazepines enhance the effects of GABA at the $GABA_A$ receptor by potentiating chloride ion conductance across the postsynaptic neuronal cell membrane, causing greater postsynaptic hyperpolarization and decreased probability of firing.

Benzodiazepines are associated with tolerance and dependence. They also have synergistic effects when used with other CNS depressants such as alcohol, and thereby can lead to coma or fatal overdose. They are therefore best used for short-range treatment. Non-benzodiazepine drugs with anxiolytic properties are preferred for long-term use, such as some SSRIs (e.g., fluvoxamine, paroxetine, sertraline), tricyclic antidepressants, bupropion, and buspirone.

Anti-Manic Agents (Mood Stabilizers)

The **antimanics** are used to treat episodes of mania in patients with bipolar disorder. They are also effective in preventing or attenuating the frequency and intensity of recurrent manic and depressive episodes; thus, they are also referred to as **mood-stabilizers**.

The most commonly used antimanic drug is **lithium**. The induced mood stability is unusually strong in some patients, to the point where patients report feeling emotionless in situations when mood shifts are expected and appropriate. Lithium's mechanism of action is poorly understood. Lithium has a very low therapeutic index (i.e., there is a very small difference between the blood concentration at which the drug becomes toxic and the blood concentration at which the drug is effective). It is therefore important to monitor blood concentrations regularly. Toxic signs include diarrhea, vomiting, drowsiness, confusion, and muscular weakness. At higher levels, ataxia, tinnitus, and impaired kidney function occur. At even higher levels, coma, respiratory depression, and death can occur.

Some medications first used as antiepileptic drugs have also been found to be effective mood stabilizers for bipolar disorder, such as carbamazepine, lamotrigine, and valproate. Some atypical antipsychotics are also effective mood stabilizers, such as quetiapine, olanzapine, and aripiprazole.

Psychostimulants

The psychostimulants are the most commonly used medications in the treatment of attention-deficit hyperactivity disorder (ADHD) and narcolepsy. The main mechanism of action of stimulants such as methylphenidate, dextroamphetamine, and amphetamine/dextroamphetamine is as CA agonists that act by blocking DA and NE reuptake.

Low-dose psychostimulants are the first-line treatment for ADHD. At low doses, psychostimulants have calming behavioral effects and improve cognitive function in those with ADHD, in contrast to high doses that have activating behavioral effects and impair cognitive function. The precise therapeutic mechanism of action is yet to be elucidated. Non-stimulant medications are also used to treat ADHD, such as atomoxetine, a NE reuptake inhibitor.

Sedative-Hypnotics

The sedative-hypnotics induce and maintain sleep and are used to treat insomnia. Most sedative-hypnotics exert their effects by enhancing transmission at $GABA_A$ receptors. The sedative-hypnotics include benzodiazepines such as temazepam and triazolam, and non-benzodiazepine $GABA_A$ receptor agonists such as zolpidem, zaleplon, and eszopiclone (also known as **Z-drugs**). Trazodone is an antidepressant with sedative effects and is often used when the underlying cause of insomnia is depression.

Summary

The process of synaptic transmission involves: (1) neurotransmitter synthesis and packaging into vesicles in the presynaptic cell; (2) neurotransmitter release into the synaptic cleft in response to arrival of an AP at the axon terminal; (3) neurotransmitter binding to receptors on postsynaptic cells and generation of PSPs; and (4) rapid removal of the transmitter from the synaptic cleft by reuptake and/or enzymatic degradation. There are two broad classes of neurotransmitters: (1) the small molecule neurotransmitters, which include acetylcholine, the monoamines, and the amino acids; and (2) the neuropeptides. Neuropharmacological and psychopharmacological agents mostly produce their effects by altering synaptic transmission. Drugs may be classified according to their main therapeutic effect; common psychotropic drug classifications include the antipsychotics, antidepressants, anxiolytics, antimanics, psychostimulants, and sedative-hypnotics.

Additional Reading

1. Brodal P. Neurotransmitters and their receptors. In: Brodal P, *The central nervous system: structure and function*. 5th ed. Oxford; 2016. https://oxfordmedicine.com/view/10.1093/med/9780190228958.001.0001/med-9780190228958-chapter-5.
2. Ebenezer, IS. *Neuropsychopharmacology and therapeutics*. Wiley-Blackwell; 2015.
3. Iversen L, Iversen S, Bloom FE, Roth RH. *Introduction to neuropsychopharmacology*. Oxford University Press; 2008.
4. Vanderah TW, Gould D. Synaptic transmission between neurons. In: Vanderah TW, Gould D, *Nolte's the human brain: an introduction to its functional anatomy*. 8th ed., pp. 173–194. Elsevier; 2021.

4

Cranium, Spine, Meninges, Ventricles, and Cerebrospinal Fluid

Introduction

The central nervous system (CNS) is covered with several layers of protective tissues: the bony coverings, consisting of the cranium and the spine, and the meninges. The **cerebrospinal fluid** (CSF), which surrounds the brain and spinal cord, also serves a protective function, and fills a series of interconnected ventricles that run within the core of the brain. This chapter introduces the reader to these external and internal structures, as well as related clinical considerations such as intracranial pressure and mass effect, meningitis, meningiomas, arachnoid cysts, subarachnoid hemorrhage, epidural and subdural hematomas, brain herniation, and hydrocephalus, as well as the neurodiagnostic procedure of lumbar puncture.

Cranium and Spine

The brain and spinal cord are encased in bone. The brain lies within the cranium, while the spinal cord lies within the vertebral column. The cranium and the facial bones together make up the **skull**.

Basic Anatomy of the Cranium

The **cranium** is also known as the **braincase** or **calvarium**. It is made up of eight bones: two parietal bones, two temporal bones, one occipital bone, one frontal bone, one ethmoid bone, and one sphenoid bone (Figure 4.1). Thus, four of the cranial bones are paired (two temporal and two parietal) and four are unpaired and midline (frontal, occipital, ethmoid, and sphenoid). The lobes of the cerebrum are named after the overlying cranial bones.

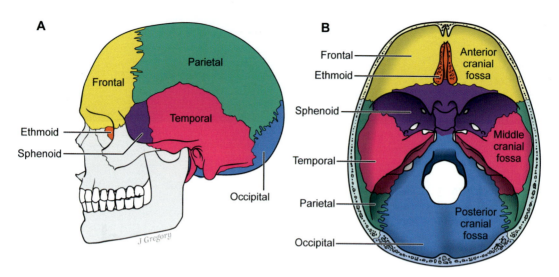

FIGURE 4.1. The cranium is composed of eight bones: two parietal, two temporal, one occipital, one frontal, one ethmoid, and one sphenoid. (A) Lateral (side) view. (B) The skull base, superior view. The inner surface of the cranial cavity has several ridges of bone that divide the skull base into the anterior fossa, middle fossa, and posterior fossa.

A **foramen** (pl. **foramina**) is an aperture that allows the passage of structures from one region to another. Cranial nerves and blood vessels enter and exit the cranial compartment through the **cranial foramina**.

Anatomically, the cranium can be subdivided into a roof and a floor. The roof is the rounded top of the cranium and is known as **skullcap** or **calvaria** (not to be confused with the calvarium). The calvaria is made up of the superior portions of the frontal bone, the occipital bone, and the two parietal bones. The floor of the cranial cavity is referred to as the **skull base**. The ethmoid and sphenoid bones are located within the base of the skull, as are the inferior portions of the frontal bone, the occipital bone, and the two temporal bones. The skull base has numerous **foramina**. The largest of these is the **foramen magnum** within the occipital bone; it is a landmark that is used to identify the point of transition from upper cervical spinal cord to the lower medulla (i.e., the cervico-medullary junction).

The **ethmoid** bone (Greek *ethmos*, "sieve") is a small, unpaired midline bone that lies at the anterior base of the cranium above the nasal cavity. It is one of the most complex bones of the head; components of the ethmoid bone include the cribriform plate and the crista galli. The **cribriform plate** (Latin *cribrum*, sieve) is a sieve-like region of the ethmoid bone that is perforated by small holes that allow bundles of **olfactory nerve** fibers to pass from the nasal cavity to the cranial cavity; the perforations are called the **olfactory foramina**. The **olfactory fossae (olfactory grooves)** of the cribriform plate are shallow depressions, within which lie the **olfactory bulbs**. The **crista galli** (Latin, "crest of the rooster") is a bony midline projection that lies in the sagittal plane between the two olfactory fossae; it serves for the attachment of the cerebral falx, a membranous sheet of tissue (see the section "Dura Mater," below).

The **sphenoid** (Greek, "wedge-like") bone is a butterfly-shaped, unpaired midline bone that lies at the middle base of the cranium. It is the most complex bone in the human body and has several components: a midline body; two pairs of wings extending laterally from the body (the smaller pair referred to as the lesser wings, and the larger pair referred to as the greater wings); and two **pterygoid** (Greek, "wing-like") **processes** that project from each greater wing. The **sella turcica** (Latin, "Turkish saddle") is a saddle-shaped depression in the body of the sphenoid bone within which the lies the pituitary gland; it is also known as the **pituitary fossa**. The sella turcica is surrounded by the **anterior and posterior clinoid processes**, like the four corners of a four-poster bed (clinoid meaning "resembling a bed"); these processes serve as attachment points for the tentorium, a membranous sheet of tissue that separates the cerebellum from the overlying occipital lobes (see the section "Dura Mater," below).

The inner surface of the skull base has several ridges of bone that divide the skull base into three **cranial fossae** (depressions): the anterior fossa, middle fossa, and posterior fossa (see Figure 4.1). The **anterior fossa** contains the frontal lobes. The **middle fossa** houses the temporal lobes. The **posterior fossa** houses the cerebellum and the brainstem. The anterior fossa is separated from each middle fossa by the **sphenoid ridge** of the lesser wing of the sphenoid bone, and each middle fossa is separated from the posterior fossa by the **petrous ridge** of each temporal bone. Because the inner surface of the skull base has many bony edges, in contrast to the smooth inner surface of the calvarium, the inferior surface of the cerebrum is much more prone to bruising with head trauma than is the superior surface of the cerebrum.

Cranium, Spine, Meninges, Ventricles, and Cerebrospinal Fluid | 37

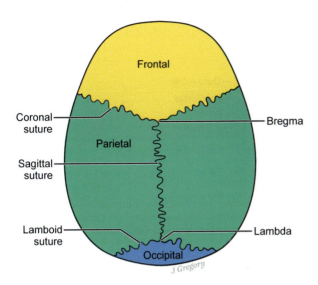

FIGURE 4.2. The calvaria (skullcap). The coronal suture connects the frontal and two parietal bones, the lambdoid suture connects the two parietal bones to the occipital bone, and the sagittal suture connects the two parietal bones. Bregma is a midline landmark where the coronal and sagittal sutures meet, and lambda is a midline landmark where the lambdoid and sagittal sutures meet.

The bones of the cranium are joined by fibrous joints known as **skull sutures** (Figure 4.2). The **coronal suture** runs in the coronal plane and joins the frontal bone and the two parietal bones. The **lambdoid suture** connects the two parietal bones to the occipital bone. The **sagittal suture** runs in the midsagittal plane, from the coronal suture to the lambdoid suture, connecting the two parietal bones. The points where skull sutures join serve as anatomic landmarks for surgery and radiology. **Bregma** is a midline landmark where the coronal and sagittal sutures meet between the frontal bone and the two parietal bones, and **lambda** is a midline landmark where the lambdoid and sagittal sutures meet between the occipital bone and the two parietal bones. Other skull landmarks are the **nasion**, a depression between the two eyes just superior to the bridge of the nose and inferior to the forehead, and the **inion**, a small bony protuberance of the occipital bone at the base of the skull.

Before the age of 18 months, the cranial bones are not fused, and there are spaces between the bones known as **fontanelles** (soft spots). The **anterior fontanelle** is a diamond-shaped space located between the frontal bone and the two parietal bones, at the junction between the coronal suture and the sagittal suture (i.e., at bregma). It closes by approximately 18 months of age. The **posterior fontanelle** is a triangle-shaped space located between the occipital bone and the two parietal bones, at the junction between the sagittal and lambdoid sutures (i.e., at lambda). It closes within 6–8 months of age.

Basic Anatomy of the Spine

The spinal cord is protected by the spine, also known as the **spinal column** or the **vertebral column**. The vertebral column has 5 subdivisions: cervical, thoracic, lumbar, sacral, and coccygeal. There are 7 cervical **vertebrae**, located at the neck level (C1–C7); 12 thoracic vertebrae, located at the thorax (chest) level (T1–T12); 5 lumbar vertebrae, located at the level of the lower back between the ribs and the pelvis (L1–L5); 5 sacral vertebrae (S1–S5) that are fused and form the sacrum, a large flat triangular bone located at the level of the small of the back; and 4 fused coccygeal vertebrae (Co1–Co4) that form the coccyx, also known as the tailbone. The 24 vertebrae of the cervical, thoracic, and lumbar spine (i.e., above the sacrum) are unfused and give the vertebral column its flexibility, allowing for movement. Fusion of the sacral and coccygeal vertebrae begins at about age 20 and is not complete until middle age (which explains, in part, why teenagers make the best gymnasts).

Adjacent vertebrae are separated from each other by an **intervertebral disk** that lies within the **intervertebral** space, and which may herniate, resulting in the condition known as "slipped disk." The disks are composed of cartilage and act as shock absorbers. The spinal cord lies within the **vertebral canal**, passing through the **spinal (vertebral) foramen** of each vertebra. The spinal nerves exit the vertebral canal via the **intervertebral foramens**.

Meninges

Several layers of connective tissue, known collectively as the **meninges**, cover the central and peripheral components of the nervous system. The brain and spinal cord are covered by three meningeal layers: the dura mater, arachnoid mater, and pia mater (Figure 4.3). The spinal nerves and cranial nerves are covered by the dura and the pia, which fuse to form a **nerve sheath**. Within the CNS, the dura is also referred to as the **pachymeninx** because it is thick (Greek, *pachy*, "thick"); the pia and arachnoid together are called the **leptomeninges** (Greek, *lepto*, "thin") because they are thin.

Dura Mater

The dura mater ("hard," as in durable; "mother") is the outermost meninx. It is a tough, strong, flexible, non-stretchable fibrous tissue that covers the brain and the spinal cord.

The spinal dura, which surrounds the spinal cord, is a single-layered membrane. The cranial dura, which surrounds the brain, consists of an outer, **periosteal**

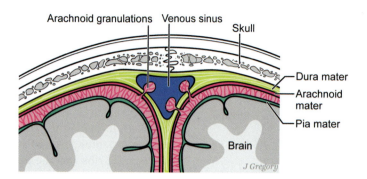

FIGURE 4.3. The brain and spinal cord are covered by three meningeal layers: the dura mater, arachnoid mater, and pia mater. The cranial dura consists of two layers. These layers split away from each other at certain points to form infoldings and venous sinuses. Arachnoid granulations extend from the arachnoid membrane and protrude through the dural wall into the venous sinuses.

(endosteal) layer that adheres tightly to the inner surface of the cranium, and an inner **meningeal layer**. These layers split away from each other at certain points to form infoldings and venous sinuses (see Figure 4.3).

The dural infoldings are known as **dural septa**, **dural reflections**, and **dural folds** (Figure 4.4). There are four main dural septa: the cerebral falx (*falx cerebri*), the cerebellar falx (*falx cerebelli*), the tentorium (*tentorium cerebelli*), and the sellar diaphragm (*diaphragma sellae*). The **cerebral falx** (Latin *falx*, "sickle") is the largest of these septa. It is a midline, sickle-shaped reflection that dips down into the longitudinal fissure dividing the two hemispheres. The **cerebellar falx** is a small midline dural reflection that lies between the cerebellar hemispheres. The **tentorium** is a tent-shaped reflection that lies within the transverse fissure, between the overlying occipital lobes and the underlying cerebellum. The midbrain passes through a space in the tentorium known as the **tentorial notch** (**tentorial incisure**). The **sellar diaphragm** is a circular-shaped reflection that forms a roof over the sella turcica and lies between the pituitary gland and the overlying hypothalamus. It has a small opening through which the **pituitary stalk** (**infundibulum**), which connects the pituitary and hypothalamus, passes.

The dural reflections are firmly attached to the cranium and are stretched rather tautly, which allows them to provide support and helps prevent excessive movement of the brain. They also act as structural dividers and partition the cranial cavity into compartments. The tentorium divides the intracranial vault into supratentorial and infratentorial compartments. The **supratentorial** compartment contains the cerebrum. The **infratentorial** compartment, also known as the posterior fossa, contains the brainstem and the cerebellum. The cerebral falx divides the supratentorial compartment into halves and separates the right and left hemispheres of the cerebrum.

The **dural venous sinuses** are an interconnected system of channels that receive venous blood from veins of the brain, and CSF from the subarachnoid space (Figure 4.5). This system ultimately empties into the two internal jugular veins, which drain blood from the brain. The main venous sinuses are the **superior sagittal sinus**, the **inferior sagittal sinus**, the **straight sinus**, the **transverse sinuses**, and the **sigmoid (S-shaped) sinuses**. The cerebral falx houses the superior and inferior sagittal sinuses. The line of attachment between the cerebral falx and the tentorium houses the straight sinus, which is continuous with the inferior sagittal sinus. The tentorium houses the left and right transverse sinuses. The superior sagittal sinus, straight sinus, and transverse sinuses all meet at the **confluence of sinuses**. Venous blood flows posteriorly from the superior sagittal, inferior sagittal,

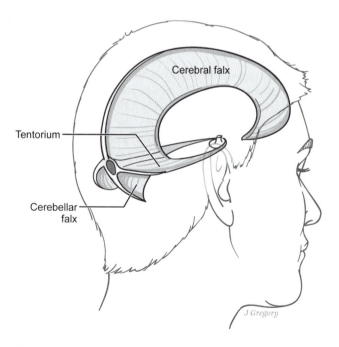

FIGURE 4.4. The major dural reflections: the cerebral falx, the cerebellar falx, and the tentorium.

Cranium, Spine, Meninges, Ventricles, and Cerebrospinal Fluid | 39

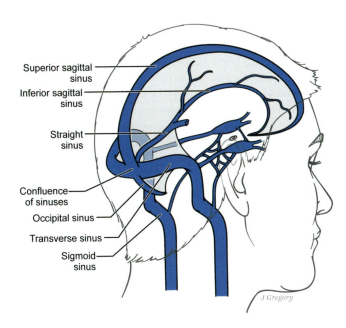

FIGURE 4.5. The venous sinuses are an interconnected system of channels that receive venous blood from veins of the brain and CSF from the subarachnoid space and drain into the internal jugular veins.

Arachnoid Membrane

The **arachnoid** membrane is a delicate, soft, spongy membrane that lies beneath the dura of the brain and spinal cord. It is connected to the underlying pia mater by web-like filaments called **arachnoid trabeculae** (Greek *arachne*, "spider"; *trabecula*, "track") that traverse the **subarachnoid space**. The subarachnoid space is filled with CSF.

The subarachnoid space is continuous across the cerebral and cerebellar convexities and along the spinal cord. There are several areas where the subarachnoid space is enlarged and devoid of trabeculae, forming **subarachnoid cisterns** (Latin *cisterna*, "box") that are also filled with CSF (Figure 4.6). The cisterns serve as important landmarks for detecting certain pathologies in brain-imaging studies; thus it is useful to be familiar with their names and locations. Effacement of one or more of the cisterns may indicate brain swelling or the presence of a space-occupying lesion that displaces the soft tissues of the brain; blood within the cisterns indicates that there has been a subarachnoid hemorrhage. The largest cistern, the **cisterna magna (cerebellomedullary cistern)**, is located between the cerebellum and the dorsal surface of the medulla. Other cisterns and their locations include the following: the **Sylvian cistern (insular cistern)**, lying over the insula deep within the Sylvian fissure; the **suprasellar cistern**

and straight sinuses into the confluence, then into the transverse sinuses and sigmoid sinuses, exiting the head by the internal jugular veins.

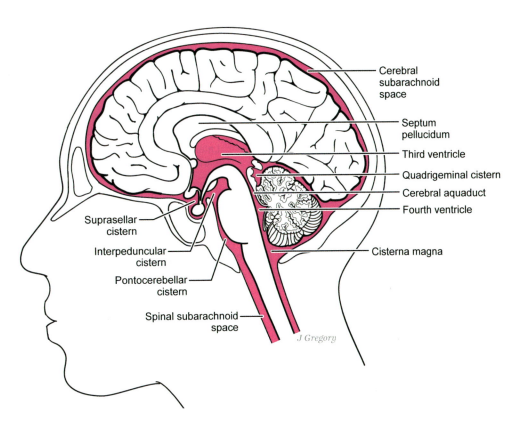

FIGURE 4.6. The subarachnoid space is continuous across the cerebral and cerebellar convexities and along the spinal cord. The subarachnoid cisterns are areas where the subarachnoid space is enlarged and devoid of trabeculae.

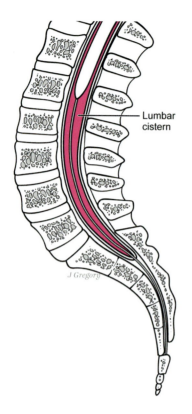

FIGURE 4.7. The lumbar cistern is a large subarachnoid cistern lying below the end of the spinal cord, beginning at the L1/L2 level of the spinal column and extending to the lower border of S2.

(**chiasmatic cistern**), lying superior to the sella turcica surrounding the optic chiasm; the **interpeduncular cistern** (**basal cistern**), lying between the cerebral peduncles of the ventral midbrain; the **superior cistern** (**quadrigeminal cistern**), lying dorsal to the midbrain; the **ambient cistern**, lying lateral to the midbrain; and the **pontine cistern**, lying on the ventral surface of the brainstem between the pons and the medulla.

There is also a large subarachnoid cistern lying below the end of the spinal cord, known as the **lumbar cistern**, beginning at the L1/L2 level of the spinal column and extending to the lower border of S2 (Figure 4.7).

Pia Mater

The pia mater ("soft mother") is a very fine, translucent vascular membrane that firmly adheres to the surface of the brain and spinal cord. It is the only meningeal layer that follows all the brain's contours (the gyri and sulci), and it also follows the nutrient vessels (arteries, arterioles) that penetrate the brain substance all the way down to the capillary level.

Ventricular System

The brain surrounds an interconnected system of cavities called ventricles that are filled with CSF (Figure 4.8). There are **two lateral ventricles**, each connecting via a

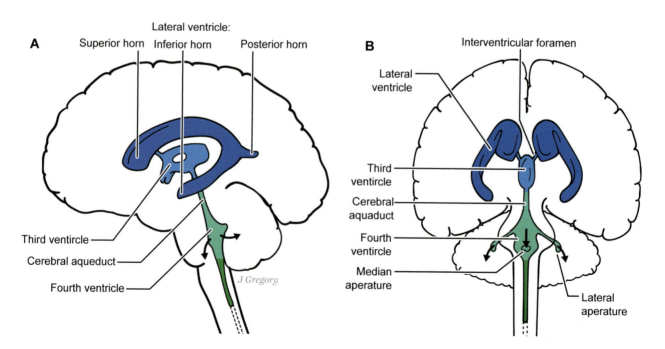

FIGURE 4.8. Anatomy of the ventricular system. Two lateral ventricles each connect via a foramen of Monro (interventricular foramen) to the third ventricle, which connects via the cerebral aqueduct (aqueduct of Sylvius) to the fourth ventricle. The fourth ventricle is continuous with the subarachnoid space by one median aperture and two lateral apertures. (A) Lateral view. (B) Anterior (front) view.

foramen of Monro (**interventricular foramen**) to the midline **third ventricle**, which connects via the **cerebral aqueduct (aqueduct of Sylvius)** to the midline **fourth ventricle**.

The lateral ventricles lie within the telencephalon/cerebral hemispheres (i.e., basal ganglia, cerebral cortex, and cerebral white matter). They are C-shaped. Several brain structures follow the curve of the lateral ventricle and are therefore also C-shaped: the caudate nucleus (a component of the basal ganglia); the corpus callosum (a thick band of white matter interconnecting the cerebral cortex of the right and left hemispheres); the **fornix** (the major output tract of the hippocampus); and the **stria terminalis** (a major output tract of the amygdala).

The **septum pellucidum**, a thin translucent double membrane made of glial cells (ependyma), separates the lateral ventricles. The lateral ventricles have extensions, or horns, named according to the surrounding lobe; these are the **frontal horn (anterior horn)**, the **temporal horn (inferior horn)**, and the **occipital horn (posterior horn)**. Two other components of the lateral ventricles are the **body**, located within the parietal lobe, and the **atrium (collateral trigone)**, which is the most expanded part of the lateral ventricle, where the occipital/posterior and temporal/inferior horns meet.

The third ventricle runs through the diencephalon at midline. An outline of the third ventricle reveals four **recesses** that are named according to their location and related diencephalic structures: the optic recess, infundibular recess, suprapineal recess, and pineal recess. The **massa intermedia (interthalamic adhesion)** crosses the midline of the third ventricle. This structure is a bridge of tissue that crosses the midline and connects the two thalami, but it is not a commissure (a band of nerve fibers that interconnect symmetrical structures). Each thalamus functions separately within the ipsilateral cerebral hemisphere with little communication from its contralateral partner. The massa intermedia is absent in approximately 30% of people without apparent consequence.

The third ventricle is connected to the fourth ventricle via the cerebral aqueduct, which runs through the midbrain. It divides the midbrain into a **tectum** (roof) and a **tegmentum** (floor). Because of its constricted lumen (2 mm in diameter), the aqueduct is subject to obstruction from infection, trauma, or space-occupying lesions.

The fourth ventricle runs within the hindbrain (cerebellum, pons, and medulla). The floor of the fourth ventricle is formed by the pons and the medulla. The roof of the fourth ventricle is formed by two vela (veil-like membranous partitions) underlying the cerebellum: the **superior medullary velum (anterior medullary velum)** and the **inferior medullary velum (posterior medullary velum)**. The fourth ventricle is also referred to as the **rhomboid fossa** (depression) because it is diamond-shaped due to its narrowing rostrally into the cerebral aqueduct and caudally into the **central canal** of the spinal cord. The point at which the fourth ventricle narrows into the central canal is known as the **obex**. The obex occurs at the same level as the foramen magnum, and it serves as an additional anatomic landmark distinguishing the intracranial medulla and the extracranial spinal cord.

The fourth ventricle is also continuous with the subarachnoid space due to three foramina in its roof: one **median aperture** known as the **foramen of Magendie**, and two **lateral apertures** known as the **foramina of Luschka** (see Figure 4.8). These allow CSF to escape from the ventricular system and enter the subarachnoid space. The foramen of Magendie opens into the cisterna magna, while the foramina of Luschka open into the pontine cistern.

Cerebrospinal Fluid

CSF is a clear, colorless fluid that circulates through the ventricular system and subarachnoid space and bathes the external and internal surfaces of the brain. By supporting the brain through flotation, it decreases the effective weight of the brain from 1,500 grams in air to less than 50 grams, thereby reducing pressure on the base of the brain. The brain itself is soft and jelly-like and cannot support its own weight. CSF also serves as a cushion between the brain and the cranium, protecting against mechanical trauma by absorbing shock due to sudden head movements. CSF also plays a role in nourishment of the CNS and the removal of metabolites.

The total volume of CSF is 150 milliliters (ml). Approximately 10% of the intracranial and intraspinal space is filled with CSF. Only 25 ml occupy the ventricles; the rest occupies the subarachnoid space. The third and fourth ventricles together contain only about 2 ml of CSF, so the lateral ventricles contain nearly all the ventricular CSF.

Production, Circulation, and Reabsorption

CSF is manufactured by the **choroid plexus**, a vascular ruffle of tissue that lines the lateral, third, and fourth ventricles. CSF is produced and reabsorbed at a rate of about 20 ml/hour, or 500 ml/day, so the total CSF volume is renewed about four times per day. The choroid plexus is composed of capillaries, pia mater, and **ependymal cells** (a type of glial cell that lines the ventricles). CSF is essentially extracted from blood and is similar in composition to both blood plasma and the extracellular fluid surrounding the cells of the brain. The most extensive choroid plexuses line the inferior surfaces of the lateral ventricles, where most of the CSF is produced and located. The frontal and occipital horns are devoid of choroid plexus (making them a choice location for surgeons to place pressure-relieving shunts). The arterial supply of the cerebral choroid plexus is provided mainly by the **anterior and posterior choroidal arteries**.

CSF circulates in a characteristic pattern, flowing from the lateral ventricles, through the third ventricle, to the fourth ventricle. CSF exits the fourth ventricle by the foramina of Luschka and Magendie, enters the subarachnoid space, and flows upward over the surface of the cerebral hemispheres and downward around the spinal cord.

CSF is reabsorbed into the venous blood pool through **arachnoid granulations**, finger-like extensions that extend from the arachnoid membrane and protrude through the dural wall into the venous sinuses (see Figure 4.3). Each granulation is composed of numerous **arachnoid villi** (elongated projections). The granulations function as pressure-dependent, one-way valves; the hydrostatic pressure of CSF is greater than the venous blood in the dural sinuses, so fluid moves from the subarachnoid space into the venous system. Most CSF reabsorption occurs at the superior sagittal venous sinus.

Clinical Considerations

Intracranial Pressure, Mass Effect, and Brain Herniation

The cranium, being rigid, has a fixed volume. The pressure within the cranium is termed the **intracranial pressure**. Increases in the volume of the intracranial contents, as occurs with space-occupying lesions (e.g., hematoma, tumor), diffuse brain swelling (**cerebral edema**), and hydrocephalus, result in increased intracranial pressure (**intracranial hypertension**). They may also result in **mass effect**—compression, distortion, and/or displacement of intracranial contents.

Signs and symptoms of increased intracranial pressure include (1) headache, which is often worse in the morning and on coughing, sneezing, or bending; (2) vomiting, which is often projectile; (3) blurred double vision (due to cranial nerve defects); and (4) altered mental status characterized by drowsiness, depressed level of alertness and attention, or coma.

Space-occupying lesions, also known as **mass lesions** (e.g., hematoma, tumor), result in increased intracranial pressure in the compartment that they occupy. They can also cause mass effect, a distortion of the normal brain geometry (Figure 4.9). Types of mass effect include compression of structures, **midline shift**, **effacement** (compression of a space) of the cortical sulci and ventricles, and brain **herniation** (shift of tissue from its normal location into an adjacent space).

Brain herniation is a displacement of brain tissue that occurs with increased intracranial pressure. There are

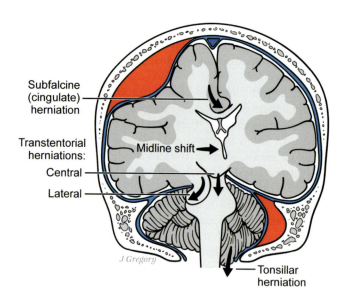

FIGURE 4.9. An intracranial mass may exert pressure on surrounding tissue (mass effect) and result in increased intracranial pressure, midline shift, effacement of ventricles and sulci, and/or herniation beneath the cerebral falx, across the tentorium, or through the foramen magnum.

three categories of brain herniation: subfalcine, transtentorial, and foramen magnum. In **subfalcine herniation** (**cingulate herniation**), the cingulate gyrus of one hemisphere is displaced under the cerebral falx. This results in a shift of the septum pellucidum from the midline. **Transtentorial herniations** occur when brain tissue is displaced across the tentorium through the tentorial notch; these may be descending or ascending. Descending transtentorial herniations are caused by mass effect in the forebrain, which pushes supratentorial brain structures into the infratentorial compartment. There are two types of descending transtentorial herniations, uncal and central. In **uncal herniation**, the **uncus**, a hook-shaped region that is located at the anterior and medial extreme of the inferior temporal lobe, is displaced across the tentorium and compresses the midbrain. In **central herniation**, the diencephalon is displaced across the tentorium. Ascending transtentorial herniation is caused by mass effect in the posterior cranial fossa, usually due to a cerebellar or brainstem tumor, which leads to displacement of brain tissue from the infratentorial to the supratentorial compartment. **Foramen magnum herniation** (**tonsillar herniation, transforaminal herniation**) occurs when the cerebellar tonsils (a rounded lobule on the undersurface of each cerebellar hemisphere) and brainstem are displaced through the foramen magnum. This often leads to loss of consciousness, cardiovascular dysfunction, respiratory failure, and death, since the brainstem controls sleep-wakefulness and level of arousal, as well as basic vital life functions such as heart rate, blood pressure, and breathing.

Meningitis

Meningitis is an inflammation of the meninges of the brain and spinal cord. It usually affects the arachnoid and pia. It is caused either by bacteria (i.e., bacterial meningitis) or viruses (viral meningitis, **aseptic meningitis**). Bacterial meningitis is less common but much more serious than viral meningitis; it can be life threatening if not treated promptly. Symptoms of meningitis include severe headache, severe neck stiffness (nuchal rigidity), high fever, nausea, vomiting, altered mental state, **photophobia** (inability to tolerate light), and **phonophobia** (inability to tolerate loud sounds). The neck stiffness is caused by the muscles of the neck contracting strongly (guarding) to prevent bending of the neck and subsequent painful stretching of the inflamed meninges.

Imaging signs of meningitis include effacement of the CSF spaces (sulci, fissures, basal cisterns) due to inflammation of the meninges.

Meningeal Headaches

The brain itself is insensitive to pain because it lacks sensory receptors. However, the dura mater and arteries do contain pain receptors, and therefore are the site of origin of headaches. Alcohol-related headaches are due to the direct toxic effects of alcohol on the meninges, and therefore are a type of meningeal headache.

Meningiomas

Meningiomas are a type of brain tumor that arises from the arachnoid cells. They are the most common type of brain tumor and occur twice as often in women than in men. Meningiomas are benign in that they do not spread (metastasize) to other organs, but because they are space-occupying lesions within the cranial vault, they cannot be considered benign in the sense of being harmless. They may disturb brain function by putting pressure on the brain, but because they are very slow growing, they can become immense before giving rise to symptoms. Meningiomas often produce seizures as an initial symptom.

Meningiomas are classified according to their location. Examples include cerebral convexity meningiomas, which grow on the surface of the brain directly under the skull; parasagittal (parafalcine) meningiomas located in or adjacent to the cerebral falx; sphenoid wing meningiomas located near the sphenoid bone behind the eyes; foramen magnum meningiomas; posterior fossa meningiomas; tentorium meningiomas; and spinal meningiomas. Parasagittal meningiomas are the most common, and sphenoid wing meningiomas are the second most common.

Meningiomas are most often treated by surgical resection. Treatment outcome is good if the tumor is in an accessible area; meningiomas located infratentorially present considerable risks of morbidity (see Chapter 21, "Brain Neoplasms").

Arachnoid Cysts

A **cyst** is an abnormal, noncancerous sac-like structure composed of an outer membrane that is filled with fluid or other material. Cysts can grow in many of the body tissues, and they are classified according to the type of tissue that they originate from or the material they contain. Several types of cysts occur within the CNS, such as arachnoid cysts, ependymal cysts, and colloid cysts.

Arachnoid cysts (**leptomeningeal cysts**) are formed from arachnoid membrane and are filled with CSF. They may be intracranial or intraspinal. Primary (developmental) arachnoid cysts arise during gestation and are present at birth. Secondary arachnoid cysts are less common and often develop after various pathologies (e.g., traumatic brain injury, infection, intracranial hemorrhage).

Arachnoid cysts are space-occupying lesions, and the specific signs and symptoms depend on whether the cyst is large enough to affect CNS function, and the specific location. Small cysts are usually asymptomatic. Large arachnoid cysts often compress surrounding structures, including blood vessels, leading to neurological symptoms such as headaches, seizures, and mental status changes, at which point they require neurosurgical treatment. Most arachnoid cysts (approximately 50%–65%) form within the middle cranial fossa around the temporal lobe. Arachnoid cysts are a common incidental finding on brain imaging studies.

Subarachnoid Hemorrhage

The major cerebral arteries lying on the surface of the brain travel within the subarachnoid space (although they are covered by pia). Rupture of these arteries results in **subarachnoid hemorrhage**, spontaneous bleeding into the subarachnoid space due to nontraumatic causes, usually rupture of aneurysms or arteriovenous malformations (see Chapter 18, "Stroke and Vascular Cognitive Impairment"). The most common cause of subarachnoid hemorrhage, however, is head trauma.

Epidural and Subdural Hematomas

The cranial dura is very tightly bound to the inner surface of the skull and less tightly bound to the underlying arachnoid layer. There are two potential spaces associated with the cranial dura: the **epidural (extradural) space** and the **subdural space**. The spinal dura is bound loosely via connective tissue to the inner surface of the vertebral column, forming an actual epidural space; the subdural space, however, is potential.

The potential epidural and subdural spaces around the brain may become fluid-filled under certain pathological conditions, most often the result of bleeding. This leads to a **hematoma**, a collection of blood outside of blood vessels. The blood is usually clotted or partially clotted. Epidural and subdural hematomas are diagnosed by neuroimaging. Most are treated by neurosurgical evacuation.

Epidural hematoma is a collection of blood between the dura and the cranium. Epidural hematomas most often result from rupture of one of the meningeal arteries (which supply the dura and bone) due to trauma. The middle meningeal artery is the largest of these. It runs along the outer surface of the dura, beneath the temporal bone (grooves formed by this artery and its branches can be seen on the inner surface of the skull). The middle meningeal artery and its branches are subject to rupture with temporal bone fractures. Less commonly, tearing of a dural venous sinus can cause an epidural hematoma.

An acute epidural hematoma is biconvex (lens-shaped, elliptical, football-shaped) in appearance on neuroimaging (Figure 4.10). This is because the dura tightly adheres to the inner surface of the skull; thus, as the bleeding continues and the hematoma expands, the dura bulges inward. Epidural hematomas may cross the midline because they are above the dura and are not blocked by the cerebral falx.

A **subdural hematoma** is a collection of blood on the surface of the brain, beneath the dura and above the arachnoid membrane. Subdural hematomas usually occur with tearing of superficial **bridging veins** that drain the cerebral cortex, at the point where they traverse the subdural space en route to the dural venous sinuses. Rapid accelerations and decelerations of the head, as commonly occur in motor vehicle accidents, are the most common cause of subdural hematomas. Because the blood is not constrained by a tight dura, subdural hematomas extend freely along the convexity of the brain and are crescent-shaped on neuroimaging (Figure 4.11). Subdural hematomas do not cross the midline because they cannot pass the cerebral falx.

Subdural hematomas are classified as acute, subacute, or chronic, depending on the onset of symptoms after the trauma. Acute subdural hematomas grow rapidly; neurological signs and symptoms develop immediately to within 72 hours. Acute subdural hematomas occur with high-impact head injuries, have a high risk of death, and

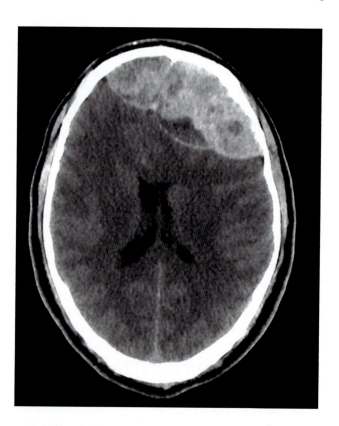

FIGURE 4.10. An axial CT image of the head demonstrates a bifrontal epidural hematoma. Epidural hematomas are extra-axial collections that can cross the midline, but do not cross the suture lines.

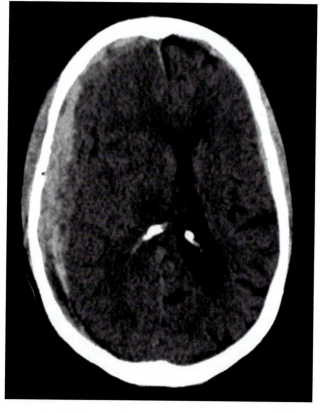

FIGURE 4.11. An axial CT image of the head demonstrates mixed-density right frontoparietal subdural hematoma. Subdural hematomas are extra-axial collections that can cross the suture lines, but do not cross the midline.

require emergency neurosurgery. Subacute subdural hematomas grow more slowly, and neurological signs and symptoms develop within 4–21 days after trauma. Chronic subdural hematomas grow slowly; signs and symptoms emerge > 21 days after trauma. Chronic subdural hematomas may become surprisingly large before producing symptoms. These hematomas can develop after minor head traumas and falls, where there is no immediate loss or alteration of consciousness. The precipitating trauma may be so minor that the patient does not recall it upon clinical interview. Chronic subdural hematomas commonly occur in the elderly due to falls. The cerebral atrophy that occurs with aging increases the distance that bridging veins must traverse between the two meningeal layers, making them more vulnerable to rupture.

Lumbar Puncture

Lumbar puncture, also known as a **spinal tap**, is a technique for sampling and analyzing CSF to help diagnose CNS pathology. This procedure involves inserting a needle into the lumbar cistern (lumbar sac) between L3/L4 or L4/L5, at a level safely below the spinal cord (see Figure 4.6). A sample of CSF is withdrawn for analysis, and the opening and closing pressures (i.e., at the beginning and end of fluid collection) are measured.

The opening pressure of the CSF within the lumbar cistern is used as an index of the intracranial pressure, based on the assumption that pressures are equal throughout the neuraxis as the cranium and spinal column together are considered to be a closed system. Decreased pressure within the lumbar cistern occurs when CSF flow is obstructed above the puncture site; increased pressure within the lumbar cistern reflects elevated intracranial pressure.

Because the composition of CSF is so constant in healthy individuals, CSF chemical composition analysis plays an important role in neurological diagnosis. Basic CSF analysis includes appearance (color, clarity), glucose content, protein content, and cell content. Normally, CSF is clear and colorless. Pink, orange, or yellow CSF may indicate that there has been bleeding within the CNS or subarachnoid space, while green CSF may be seen with infection. Cloudy CSF may indicate the presence of microorganisms or white blood cells. Glucose level may be decreased when abnormal cells within the CSF metabolize glucose (e.g., bacteria, fungi, malignancy). Protein levels are normally very low; elevated levels are observed in a variety of pathologies (e.g., meningitis, multiple sclerosis). The white blood cell count in CSF is normally very low, but increases in a variety of pathologies (e.g., CNS infection, multiple sclerosis, leukemia). Routine CSF tests may be supplemented by additional tests based on initial findings or the presumptive diagnosis. With infection of the CNS, pathogens may be cultured to identify the offending microorganism. With primary CNS tumors, the type of tumor may be identified based on tumor cell examination. Demyelinating diseases are associated with increased gamma globulin content. CSF analysis is becoming increasingly useful in the differential diagnosis of neurodegenerative diseases.

Hydrocephalus

The cranium, vertebral canal, and inelastic dura together form a rigid container that surrounds the CNS. The total volume of the parenchyma, CSF, and blood remains constant; an increase in the volume of any of these tissue compartments must be accompanied by a decrease in the volume of the other tissue compartments, otherwise there will be an increase in intracranial pressure. This principle of homeostatic intracerebral volume regulation is known as the **Monro-Kellie doctrine**.

The CSF pressure is regulated at all levels of CSF hydrodynamics: secretion, circulation, and absorption. **Hydrocephalus** is a broad term that refers to any disorder of CSF hydrodynamics due to obstruction in the pathway of CSF circulation, CSF over-secretion, or diminished CSF reabsorption. In hydrocephalus, the CSF within the ventricles is under increased pressure relative to the surrounding brain, resulting in an increase in CSF volume and **ventricular dilation**. While cerebral atrophy, focal destructive lesions (e.g., lobectomy), and hypoplasia (failure in development) also lead to an abnormal increase of CSF volume and ventricular dilation, the additional CSF replaces rather than displaces brain tissue. In these disorders, the increase in CSF volume is *not* the result of a hydrodynamic disorder. There is no increase in intracranial pressure, and the increase in CSF volume has no impact on brain function. These conditions are distinct from hydrocephalus, although the term *hydrocephalus ex vacuo* (i.e., arising out of a vacuum) is an archaic misnomer that was previously used to describe such conditions.

Hydrocephalus is divided into two main types, communicating and noncommunicating (obstructive), depending on whether there is a blockage of CSF flow within the ventricular system or between the ventricular system and the subarachnoid space. This terminology derives from a diagnostic procedure from an early twentieth-century procedure that is no longer used. In this procedure, a small burr hole was drilled into the skull, a catheter was passed through the brain substance into a lateral ventricle, a dye was injected, and a spinal tap was performed 30 minutes later. If the dye was present in the lumbar cistern, there was communication between the ventricular system and the subarachnoid space, hence the term *communicating hydrocephalus*. If the dye was not present in the lumbar cistern, then there was no communication between the lateral ventricles and subarachnoid space,

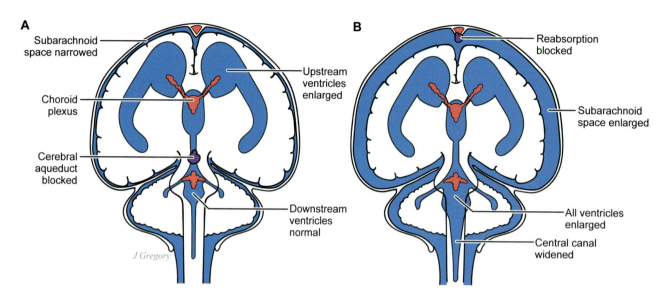

FIGURE 4.12. Hydrocephalus. (A) Non-communicating hydrocephalus is due to obstruction in the pathway of cerebrospinal fluid circulation by a space-occupying lesion. Obstructions most often occur at bottlenecks of the ventricular pathway, such as the cerebral aqueduct or interventricular foramens, leading to rapid dilation of the ventricles upstream to the obstruction. (B) Communicating hydrocephalus occurs with hypersecretion or diminished reabsorption of cerebrospinal fluid; it is characterized by dilation of the entire ventricular system as well as the subarachnoid space.

indicating an obstruction, hence the term *noncommunicating hydrocephalus*.

Noncommunicating hydrocephalus (obstructive hydrocephalus) is more common than communicating hydrocephalus. Obstruction in the pathway of CSF circulation occurs when pressure exerted by a space-occupying lesion partially or completely obstructs the ventricular pathway. In this case, CSF of the ventricles is prevented from reaching the subarachnoid space. Obstructions most often occur at bottlenecks of the ventricular pathway (Figure 4.12). The cerebral aqueduct is particularly vulnerable and is the most common site of obstruction (e.g., aqueductal stenosis), but obstructions also occur at one or both the interventricular foramens, or the median and lateral apertures of the fourth ventricle. These lead to rapid dilation of the ventricles upstream to the obstruction. Common causes of obstructive hydrocephalus are tumors and **colloid cysts** which are filled with a gelatinous material (colloid) and usually form in the third ventricle or a foramen of Monroe.

Communicating hydrocephalus occurs with hypersecretion or diminished reabsorption of CSF, leading to dilation of the entire ventricular system and the subarachnoid space (see Figure 4.12). Over-secretion occurs with choroid plexus tumor. Diminished reabsorption of CSF occurs with impaired function of the arachnoid granulations due to obstruction at the arachnoid villi that may be caused by protein and detritus, or obstruction within the subarachnoid space due to thickening of the arachnoid membrane, both of which may occur with subarachnoid hemorrhage or meningitis.

The symptoms of hydrocephalus vary with age and progression of the disorder. In infants, the skull expands to accommodate the buildup of CSF because the skull sutures have not yet closed. This is reflected in unusually large or rapidly increasing head size and is the reason why pediatricians measure head circumference in young children. Skull sutures fuse by about age 6, after attainment of full brain size. Older children and adults experience neurological symptoms of increased intracranial pressure because their skulls cannot expand to accommodate the buildup of CSF.

Hydrocephalus is diagnosed through computed tomography (CT) and magnetic resonance imaging (MRI) scans. It is often treated with a **shunt** that is surgically implanted into a ventricle and drains excess CSF. The shunt has three components: a proximal (inflow) catheter that is placed into a CSF space; a distal (outflow) catheter that is placed into a distal site of the body where CSF can be reabsorbed; and a valve mechanism that regulates the rate of CSF flow through the catheter tubing. Most commonly, CSF diversion is accomplished by **ventriculoperitoneal (VP) shunt**, in which the proximal catheter is implanted intraventricularly and the distal catheter is placed into the peritoneal cavity. Less commonly, **lumboperitoneal shunting** is employed, in which the proximal catheter is placed in the lumbar subarachnoid space and the distal catheter is placed into the peritoneal cavity. An alternative treatment is **endoscopic third ventriculostomy**, in which a small perforation is made in the floor of the third ventricle to allow CSF to flow out of the third ventricle and into the subarachnoid space. It is used in cases of congenital and acquired aqueductal stenosis and other causes of hydrocephalus.

NORMAL PRESSURE HYDROCEPHALUS

Normal pressure hydrocephalus (**NPH**) is a form of communicating hydrocephalus in which there is ventricular dilation without symptoms of increased intracranial pressure and without significant cortical atrophy. There is a mismatch between CSF production and CSF resorption. Normal pressure hydrocephalus is classified as either primary/idiopathic (of unknown cause), or secondary/symptomatic due to other CNS conditions such as **intracranial hemorrhage** (epidural, subdural, subarachnoid, intracerebral), trauma, and infection. Most cases are idiopathic.

Computed tomography and MRI studies show ventricular enlargement (**ventriculomegaly**) that is disproportionate to the degree of cerebral atrophy. Lumbar puncture, however, shows normal or near normal CSF pressure in NPH, in contrast to other forms of communicating hydrocephalus. Other radiographic features include rounded frontal horns, ballooning of the third ventricle, widening of the temporal horns without evidence of hippocampal atrophy, and a periventricular halo reflecting increased tissue fluid content due to compensatory CSF flow from the ventricles into the periventricular white matter (Figure 4.13).

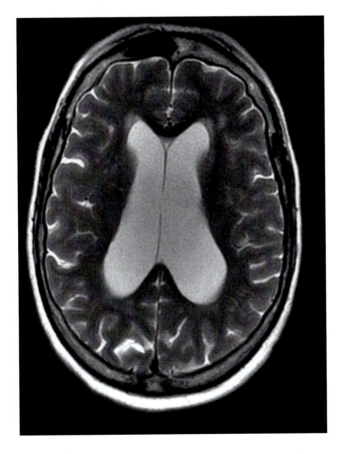

FIGURE 4.13. Normal pressure hydrocephalus. MRI T2 axial image shows ventricular enlargement disproportionate to cerebral atrophy, indicated by the width of the cortical sulci.

Idiopathic NPH is a disease of the elderly (i.e., age ≥ 65 years). It is characterized by a classic triad of symptoms known as **Hakim's triad** or the **Adams triad**, consisting of gait disturbance, urinary incontinence, and progressive dementia. A mnemonic for this triad is "wet, wobbly, and wacky." There is considerable variation in the symptom presentation, course, and severity. The entire triad need not be present. Cognitive impairment does not occur in all persons with NPH, and its severity does not correlate with the severity of gait disturbance. Incontinence also does not occur in all, or it may occur rarely or intermittently. Differential diagnosis of idiopathic NPH can be challenging because of this variability in presentation, and because gait difficulties, dementia, and urinary incontinence due to other causes are common in the elderly.

Gait disturbance is the cardinal sign of NPH, as it is usually the earliest and most prominent symptom. The gait is slow, broad-based (i.e., the horizontal stride width when both feet are in contact with the ground > 4 inches), and shuffling or magnetic (i.e., step height/foot-floor clearance is diminished so that the feet seem stuck to the walking surface as if by a magnet). The gait disorder is progressive. In the early stage, it presents as cautious gait, unsteadiness, reduced balance (especially when encountering uneven surfaces such as stairs and curbs), and difficulty with tandem gait. At this point, those with NPH may begin using a mobility aid such as a cane or walker for added stability. The gait disturbance progresses to obvious difficulty walking and considerable instability. In its most severe form, unaided gait is not possible. Because of the gait disturbance, those with NPH often have a history of falls.

Urinary symptoms range in severity from urinary frequency, urgency, to frank incontinence, and can also be progressive. Frank incontinence is often associated with indifference and lack of concern.

The dementia of NPH is often characterized by slowed thinking (**bradyphrenia**), inattention, and deficits in executive functions (e.g., initiation, planning, problem-solving) most prominently, and as such conforms to the so-called **frontal-subcortical dementia syndrome**. In many cases, however, there is a global cognitive impairment, indicating diffuse brain dysfunction. Idiopathic NPH accounts for approximately 6% of all dementias. It is one of the few reversible causes of dementia.

The pathophysiology of idiopathic NPH is not understood, but intraventricular pressure monitoring has shown that there are modest intermittent elevations of CSF pressure (usually occurring at night), unlike the sustained high intraventricular pressures observed in patients with noncommunicating hydrocephalus. It has been hypothesized that in idiopathic NPH, the intermittent intraventricular elevations cause ventricular enlargement, and the distortion is maintained despite a return to normal CSF pressure levels. Many of the symptoms of NPH have been attributed to frontal involvement. Gait and urinary symptoms are thought to be related to the stretching of descending motor fibers from the mesial surface of the motor cortex that project through the vicinity of the frontal horns and innervate the legs and sphincters. The indifference and lack of concern associated with frank incontinence is consistent with frontal involvement. Compression of the frontal white matter also accounts for the cognitive profile of NPH.

NPH is treated by diversion of CSF by shunt. Gait disturbance is most responsive to shunting; urinary symptoms respond to a lesser extent; and cognitive impairment is least responsive. Shunt function should be monitored, and the valve settings can be adjusted to limit the risk of over- or under-drainage.

The outcomes of shunt surgery are variable. In general, advanced stage NPH (i.e., with more severe symptoms or longer duration of symptoms) is associated with poorer outcomes. Thus, early diagnosis and treatment are key to achieving optimal outcomes. In cases where there is no clinical improvement and the shunt is functioning properly, the patient's symptoms may not be due to NPH, or there may be comorbid conditions that contribute significantly to the clinical symptoms. Several tests can be used to predict patients who will be **shunt-responsive**, including the CSF tap test and external lumbar drainage. The **CSF tap test** (**lumbar tap test**) involves lumbar drainage of 30–50 mL of CSF to temporarily simulate the effect of shunt drainage; gait and cognition are assessed before and after CSF drainage. **External lumbar drainage** involves implanting a catheter within the lumbar cistern and removing some CSF over several days into an external bag; gait and cognition are assessed before and after removal of CSF. In these tests, gait disturbance is more responsive than cognitive impairment, similar to shunting.

Summary

The CNS is protected by bony coverings (cranium and spine) and three layers of meninges (dura, arachnoid, and pia mater). The CSF, which surrounds the brain and spinal cord and fills a series of interconnected ventricles that run within the core of the brain, is also protective. A basic knowledge of these external and internal structures is fundamental to understanding a wide variety of neurological conditions, such as increased intracranial pressure and mass effect, meningitis, meningiomas, arachnoid cysts, subarachnoid hemorrhage, epidural and subdural hematomas, brain herniation, and hydrocephalus, as well as the neurodiagnostic procedure of lumbar puncture.

Additional Reading

1. Blumenfeld H. Brain and environs: cranium, ventricles, and meninges. In: Blumenfeld H, *Neuroanatomy through clinical cases*. 3rd ed., pp. 125–219. Sinauer Associates; 2021.
2. Brodal P. The coverings of the brain and the ventricular system. In: Brodal P, *The central nervous system: structure and function*. 5th ed. Oxford University Press; 2016. https://oxfordmedicine.com/view/10.1093/med/9780190228958.001.0001/med-9780190228958-chapter-7.
3. Vanderah TW, Gould DJ. Meningeal coverings of the brain and spinal cord. In: Vanderah TW, Gould DJ, *Nolte's the human brain: an introduction to its functional anatomy*. 8th ed., pp. 81–97. Elsevier; 2021.
4. Vanderah TW, Gould DJ. Ventricles and cerebrospinal fluid. In: Vanderah TW, Gould DJ, *Nolte's the human brain: an introduction to its functional anatomy*. 8th ed., pp. 99–116. Elsevier; 2021.
5. Waxman SG. Ventricles and coverings of the brain. In: Waxman SG, *Clinical neuroanatomy*. 29th ed. McGraw-Hill; 2020. https://accessmedicine-mhmedical-com.proxy.library.nyu.edu/content.aspx?bookid=2850&sectionid=242763936.

Blood Supply of the Brain

Introduction

The central nervous system (CNS) is one of the most metabolically active tissues in the body, but because neurons are unable to store glucose as glycogen or perform anaerobic metabolism, they are vulnerable to even brief periods of ischemia. This chapter provides a general overview of circulation to the brain, arterial cerebrovascular anatomy, and the blood-brain barrier, as well as cerebral angiography which permits visualization of the cerebral blood vessels and identification of cerebrovascular abnormalities such as stenosis, aneurysm, arteriovenous malformation, and moyamoya disease. A basic knowledge of cerebrovascular anatomy, cerebrovascular imaging, and cerebrovascular pathologies is essential to understanding mechanisms of stroke and neurovascular interventions.

Circulation Overview

Blood transports oxygen, glucose, and other substances necessary for normal physiological function to tissue and carries metabolic waste products away. It is at the capillaries where there is an exchange of nutrients from blood to tissue, and waste products from tissue to blood. The brain is one of the most highly perfused organs in the body. The human brain makes up only about 2% of the total body mass, yet it accounts for 15%–20% of the body's total oxygen and glucose consumption. Because of the brain's high energy needs, blood vessels in the CNS are arranged in a dense meshwork of capillaries. Since the neuronal cell body is responsible for meeting the metabolic needs of the entire cell, including dendrites and the axon, the density of blood capillaries is two to four times greater in gray matter than white matter. The average distance from a gray matter neuronal cell body to the nearest capillary segment is 11 μm.

Although the CNS is one of the most metabolically active tissues in the body, neurons—unlike other body cells—are unable to store glucose as glycogen, or to generate adenosine triphosphate (ATP, the energy currency of cells) in the absence of oxygen (i.e., they are unable to perform **anaerobic metabolism**). Therefore, even brief failures of blood delivery to brain tissue adversely affect neuronal function. In humans, total loss of blood flow to the brain for about 10 seconds results in loss of consciousness, for about 20 seconds results in cessation of electrical activity, and for 4–8 minutes results in irreversible neuronal death.

Ischemia is a loss of blood supply to an organ, including no blood flow or inadequate blood flow. In ischemia, blood supply is inadequate to meet the organ's metabolic needs, and there is a mismatch between oxygen/fuel delivery and metabolic demand. Ischemia results in localized tissue death, known as **infarction**. **Brain ischemia** is also referred to as **cerebral ischemia**—a slight misnomer since the ischemia does not selectively affect the cerebrum. Brain ischemia may be global or focal. **Global cerebral ischemia** refers to a loss of blood flow to the entire brain. This occurs when there is a general circulatory failure, as in cases of cardiac arrest or severe hypotension. Global cerebral ischemia results in global cerebral infarction and **brain death** (a complete and irreversible loss of brain function characterized by coma, loss of brainstem reflexes, and cessation of breathing). **Focal cerebral ischemia** occurs with occlusion of arteries that deliver blood to the brain. Such occlusions result in focal infarcts. This is the most common cause of **ischemic stroke** (see Chapter 18, "Stroke and Vascular Cognitive Impairment"). The resulting clinical signs and symptoms depend on the location of the infarct.

The Systemic and Pulmonary Circulations

The heart has four chambers: right atrium, right ventricle, left atrium, and left ventricle. Blood enters the heart at each atrium and leaves the heart from the ventricle on the same side. Thus, under normal circumstances, the left and right sides of the heart operate as separate circulatory channels.

The left side of the heart pumps oxygenated blood to the body tissues via the **systemic circulation**. The right side of the heart pumps deoxygenated blood to the lungs to be oxygenated via the **pulmonary circulation**. **Arteries** carry blood away from the heart and **veins** carry blood to the heart. Thus, in the systemic circulation arteries carry oxygenated blood away from the heart and veins carry deoxygenated blood to the heart, while in the pulmonary circulation arteries carry deoxygenated blood away from the heart and veins carry oxygenated blood to the heart.

Circulation of oxygenated blood through the body proceeds from the left ventricle to the **aorta**, to large arteries, to small arteries, to arterioles, and then to capillaries. Deoxygenated blood flows from capillaries to venules, to small veins, to large veins, and then returns to the right atrium via the **vena cava**. The deoxygenated blood in the right atrium is then pumped into the right ventricle, which in turns pumps the blood to the lungs via the **pulmonary arteries**. At the lungs, carbon dioxide is removed from the blood and oxygen is replaced. Oxygenated blood travels from the lungs to the heart via the **pulmonary veins** and enters the left atrium. Blood in the left atrium is pumped into the left ventricle, and the whole process begins again.

Extracranial Origin of the Arteries That Supply the Brain

The aorta ascends from the left ventricle, curves laterally toward the left side of the body as the **aortic arch**, and then descends through the thorax (chest) and abdomen. The first three vessels branching off the aorta originate from the aortic arch and ultimately give rise to the arteries that deliver blood to the brain (Figure 5.1). The aortic arch branches are, in order of their branching off the aorta (from right to left), the **brachiocephalic (innominate) artery**, the **left common carotid artery**, and the **left subclavian artery**. The brachiocephalic artery bifurcates into the **right subclavian artery** and the **right common carotid artery**. Thus, ultimately there are two common carotid arteries and two subclavian arteries.

The common carotid arteries ascend in the neck. Just below jaw level, each common carotid artery divides into an **external carotid artery** and an **internal carotid artery** (ICA) at a branch point known as the **carotid bifurcation**. The external carotid arteries branch to supply the jaw, face, neck, and meninges (mostly by the middle meningeal artery). The ICAs ascend in the neck without giving rise to branches and enter the cranial cavity via the **carotid canals** of the temporal bones. The proximal portion of the ICAs (i.e., at the base of the ICA, just superior to the carotid bifurcation) is dilated. This dilation, known as the **carotid sinus (carotid bulb)**, contains **baroreceptors** that monitor the pressure in the arterial blood. Baroreceptors detect mechanical stretch that occurs with increased arterial blood pressure and play a role in reflexes aimed at maintaining blood pressure within a normal range (i.e., blood pressure homeostasis).

The subclavian arteries, which run beneath the clavicles, give rise to the **vertebral arteries** that supply blood to the brain, as well as arterial branches that supply the upper extremities and portions of the thorax and throat. The vertebral arteries ascend along the sides of the neck through holes (**intertransverse foramina**) in the transverse processes of the cervical vertebra, hence their naming (see Figure 5.1). The vertebral arteries enter the cranial cavity via the foramen magnum of the occipital bone.

It is important to understand the anatomy of the extracranial blood vessels that give rise to the arteries supplying the brain because vascular pathology at this level

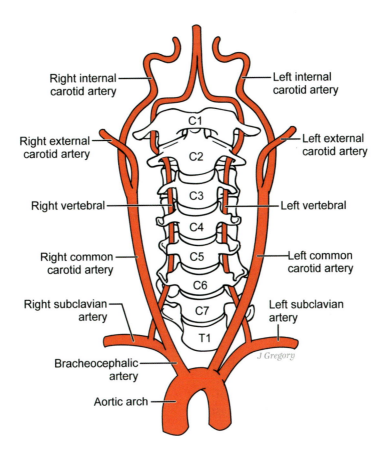

FIGURE 5.1. The first three vessels that branch off the aorta are the brachiocephalic (innominate) artery, the left common carotid artery, and the left subclavian artery. The brachiocephalic artery bifurcates into the right subclavian artery and the right common carotid artery. Each common carotid artery branches into an external carotid artery and an internal carotid artery. Each subclavian artery gives rise to a vertebral artery which ascends through foramina in the cervical vertebra. Two pairs of arteries supply blood to the brain: the internal carotid arteries and the vertebral arteries. The diagram depicts the vessels in standard anatomical position, oriented as if the subject is standing upright and facing the viewer, so that the right side of the body (anatomical right) is depicted on the left side of the image, and the left side of the body (anatomical left) is depicted on the right side of the image.

can affect brain perfusion (i.e., the delivery of oxygenated blood to tissue).

Arterial Supply of the Brain

Two pairs of arteries supply blood to the brain: the ICAs and the vertebral arteries (see Figure 5.1). The **internal carotid arterial system** supplies the anterior portion of the brain and is therefore also referred to as the **anterior circulation**. The anterior circulation provides 80% of the entire circulation to the brain and most of the blood flow to the cerebrum. The vertebral arteries merge to form a single basilar artery and are therefore known as the **vertebrobasilar system**. This system supplies the posterior portion of the brain and is therefore also referred to as the **posterior circulation**. The posterior circulation provides 20% of the entire circulation to the brain and most of the blood flow to the brainstem, cerebellum, and occipital lobes.

The Internal Carotid Arterial System/Anterior Circulation

THE INTERNAL CAROTID ARTERIES

As the ICAs enter the cranial cavity they continue their ascent, taking a curving course before bifurcating into a large **middle cerebral artery** (MCA) and smaller **anterior cerebral artery** (ACA). Prior to that bifurcation, each ICA gives rise to three branches: an **ophthalmic artery** that supplies the eye, a **posterior communicating artery**, and an **anterior choroidal artery** (which supplies parts of the choroid plexus that lines the ventricles and several deep brain structures).

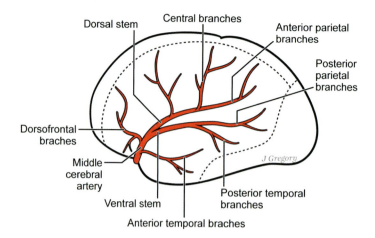

FIGURE 5.2. Each middle cerebral artery emerges from the lateral sulcus and branches into a dorsal stem that supplies the cortical area above the lateral fissure and a ventral stem that supplies the cortical area below the lateral fissure. Supply to the parietal lobe is variable, from either the dorsal stem or ventral stem.

THE MIDDLE CEREBRAL ARTERIES

The MCAs are large, direct extensions of the ICAs. They receive 75%–80% of the blood passing through the anterior arterial system. The MCAs are the most frequently affected part of the cerebral vasculature when debris is cast off from the heart or heart vessel walls.

The MCAs are situated on the external lateral surface of the cerebral hemispheres (Figure 5.2). Each MCA emerges from the lateral sulcus and gives rise to cortical branches that spread over most of the lateral surface of the cerebral hemisphere. Each MCA branches into a **dorsal stem** (also known as the **superior division**) that supplies the cortical area above the lateral fissure (i.e., the frontal convexity), and a **ventral stem** (also known as the **inferior division**) that supplies the cortical area below the lateral fissure (i.e., the lateral temporal lobe). Supply to the parietal lobe is variable from individual to individual, from either the dorsal stem or ventral stem.

The MCA cortical branch vessels, named for the cortical territory they supply, are the orbitofrontal artery, prefrontal artery, precentral (pre-Rolandic) artery, central (Rolandic) artery, parietal artery, angular gyrus artery, temporo-occipital artery, posterior temporal artery, middle temporal artery, anterior temporal artery, and temporopolar artery.

THE ANTERIOR CEREBRAL ARTERIES

The ACAs branch off the ICAs at nearly a right angle, course medially, and run within the interhemispheric fissure (longitudinal fissure) along the medial surface of the brain. They supply the medial surface of the frontal and parietal lobes, parts of the inferior surface of the frontal lobe, and the anterior four-fifths of the corpus callosum (Figure 5.3).

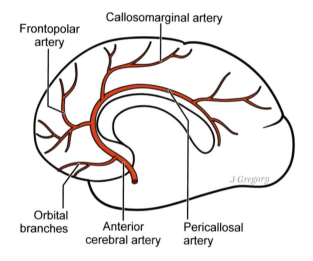

FIGURE 5.3. Each anterior cerebral artery branches off an internal carotid artery, courses medially, and runs within the interhemispheric fissure along the medial surface of the brain. They supply the medial surface of the frontal and parietal lobes, parts of the inferior surface of the frontal lobe, and the anterior four-fifths of the corpus callosum.

The ACA cortical branch vessels, named for the cortical territory they supply, are the pericallosal artery, orbitofrontal artery, frontopolar artery, anterior internal frontal artery, middle internal frontal artery, posterior internal frontal artery, paracentral artery, superior internal parietal artery, and inferior internal parietal artery.

The Vertebrobasilar Arterial System/Posterior Circulation

The vertebral arteries pass along the ventral surface of the medulla (Figure 5.4). They merge at the caudal end of the pons to form a single **basilar artery**, which ascends

Blood Supply of the Brain | 53

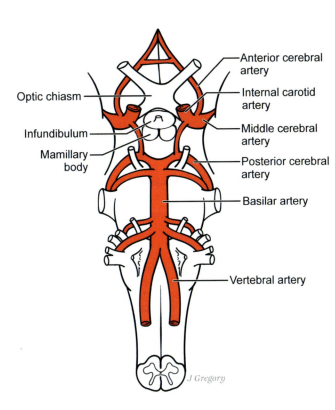

FIGURE 5.4. The vertebral arteries pass along the ventral surface of the medulla and merge at the caudal end of the pons to form a single basilar artery. The basilar artery ascends along the inferior aspect of the pons and bifurcates into two posterior cerebral arteries at the rostral end of the pons.

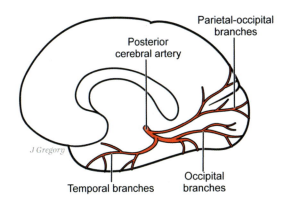

FIGURE 5.5. The posterior cerebral arteries supply blood to the occipital lobes, the splenium of the corpus callosum, and the medial and inferior surfaces of the temporal lobes, including the hippocampi.

along the inferior aspect of the pons and bifurcates into two **posterior cerebral arteries** (PCA) at the rostral end of the pons.

Prior to this bifurcation, the vertebral and basilar arteries send branches to the spinal cord, cerebellum, and brainstem. These branches include the anterior spinal arteries (ASA), posterior spinal arteries (PSA), and posterior inferior cerebellar arteries (PICA), originating from the vertebral arteries; and the anterior inferior cerebellar arteries (AICA), superior cerebellar arteries (SCA), and pontine arteries, originating from the basilar artery.

THE POSTERIOR CEREBRAL ARTERIES

The PCAs supply blood to the occipital lobes, the posterior corpus callosum (splenium), and the medial and inferior surfaces of the temporal lobes, including the hippocampi (Figure 5.5). The PCA cortical branches are the anterior temporal artery, posterior temporal artery, and occipital artery.

Blood Supply of the Deep Structures

The deep structures of the cerebral hemispheres (e.g., basal ganglia, internal capsule, diencephalon) are supplied by small branches of the anterior, middle, and posterior cerebral arteries known as **perforating (ganglionic) arteries**. These branches arise from their parent arteries around the base of the brain at an approximate right angle and penetrate the brain near their origin. These penetrations form the anterior and posterior perforated substances, areas of gray matter lying within the **interpeduncular fossa** (i.e., a deep depression on the inferior surface of the midbrain between the two cerebral peduncles) that are perforated by these numerous small blood vessels as they enter the brain tissue on their way to deep structures. The perforating arteries of the anterior circulation enter the brain through the **anterior perforated substance**; those of the posterior circulation enter the brain through the **posterior perforated substance**.

The **striate arteries** are a collection of small perforating arteries arising from the anterior circulation that supply the basal ganglia (Figure 5.6). The ACAs give rise

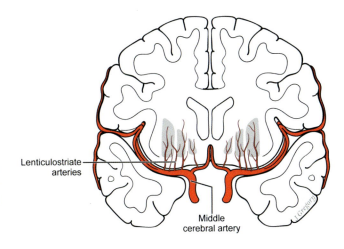

FIGURE 5.6. An example of deep penetrating arteries. The middle cerebral artery gives rise to the lenticulostriate arteries, which provide the main blood supply to the basal ganglia and the internal capsule.

to the **medial striate arteries**, and the MCAs give rise to the **lateral striate** (lenticulostriate) **arteries**. The **thalamogeniculate arteries** are the main blood supply to the thalamus; they arise from the PCAs.

The perforating arteries are very small in diameter and the regions that they supply do not have significant collateral circulation (i.e., an alternate pathway for blood flow). Consequently, they are vulnerable to blockage and are susceptible to lacunar infarcts. In addition, their thin walls make them vulnerable to rupture. Damage to these small vessels can result in neurologic deficits with surprisingly devastating impact. For example, a small lesion to the **internal capsule** (a compact bundle of white matter projection fibers located deep in the brain) can result in a corticospinal tract lesion that produces **hemiplegia** or **hemiparesis** (paralysis or weakness on the opposite side of the body) similar in severity to damage to a large expanse of motor cortex.

Watershed Zones

Regions of the body that receive dual blood supply from the most distal branches of two large arteries are known as **watershed zones**. Within the CNS, the watershed areas are those regions of the nervous system that receive their blood supply from the terminal branches of major vascular trees; these lie between the territories of the major cerebral arteries (i.e., ACA-MCA, MCA-PCA), cerebellar arteries, and spinal arteries. Because the watershed zones receive the lowest blood flow, they are particularly vulnerable to injury due to insufficient blood flow (ischemia) or insufficient blood oxygen levels (hypoxia).

The Cerebral Arterial Circle (Circle of Willis)

The **cerebral arterial circle** is a ring of interconnected blood vessels located at the inferior surface of the brain (Figure 5.7). It lies within the interpeduncular fossa and circles the optic chiasm, infundibulum, and mammillary bodies. It is also known as the **circle of Willis**, named after the English neuroanatomist **Sir Thomas Willis** (1621–1675). The circle of Willis connects the internal carotid (anterior) and vertebrobasilar (posterior) arterial systems via two **posterior communicating arteries** (Pcom) that join the PCAs to the ICAs, and a single **anterior communicating artery** (Acom) that connects the ACAs near their entrance into the longitudinal fissure.

Due to equality of blood pressures, there is normally little mixing of blood between the anterior and posterior circulation or between the right and left hemisphere circulation. If a major vessel within or proximal to the circle of Willis becomes occluded, the obstruction blocks blood flow and reduces blood pressure in that region. Because

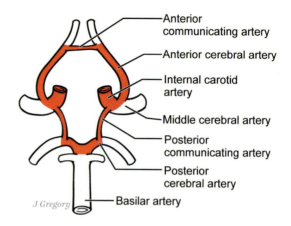

FIGURE 5.7. The circle of Willis connects the internal carotid and vertebrobasilar arterial systems via two posterior communicating arteries that join the posterior cerebral arteries to the internal carotid arteries, and a single anterior communicating artery that connects the anterior cerebral arteries near their entrance into the longitudinal fissure.

normal blood flow in a neighboring region is now at a relatively higher pressure, blood will cross over and help nourish the tissue in the region of the occlusion. The communicating arteries provide alternate routes for blood to flow between the posterior and anterior circulation, and between the right and left anterior circulation. The alternative, collateral circulatory routes created by the communicating arteries thereby decrease the probability of local ischemia and neurological damage due to occlusion of a vessel within or proximal to the circle.

The health of the vessels within the circle determines whether an alternative blood flow route develops. For example, a complete ICA occlusion can produce no symptoms if the vessels around the circle of Willis are healthy and allow for collateral blood flow. Alternately, it can produce a massive ischemic stroke in the entire ICA territory if the vessels around the circle of Willis are unhealthy and do not allow for collateral blood flow. The rapidity of occlusion also determines whether an alternative route of blood flow develops. Slow onset occlusions, like those caused by **atherosclerosis** (a buildup of plaque on the inner walls of the arteries), may allow for alternative routes to develop; rapid occlusions, like those caused by clots that circulate and then become lodged in a vessel, may not.

Segmentation of the Major Cerebral Arteries

The major cerebral arteries are divided into segments by classifications that are used clinically by neuroradiologists, neurosurgeons, and neurologists for greater specificity in describing the location of vascular pathologies affecting these vessels. Familiarity with this concept and terminology is useful because neuroimaging descriptions of pathology often refer to the specific segment affected. The segmentation uses a numerical scale in the direction of blood flow (from proximal to distal).

Segmentation of the ICA and vertebral artery is based on surrounding anatomy. The ICA is divided into seven anatomical segments: C1 (cervical segment); C2 (petrous/horizontal segment); C3 (lacerum segment); C4 (cavernous segment); C5 (clinoid segment); C6 (ophthalmic/supraclinoid segment); and C7 (communicating segment). The vertebral artery is divided into four anatomical segments: V1 (pre-foraminal segment); V2 (foraminal segment); V3 (extradural segment); and V4 (intradural/intracranial segment).

The anterior, middle, and posterior cerebral arteries are subdivided into segments based on major branch points. The MCA is divided into four segments: M1 (sphenoidal, horizontal segment); M2 (insular segment); M3 (opercular segment); and M4 (cortical segment). The ACA is divided into five segments: A1 (horizontal, pre-communicating segment); A2 (vertical, post-communicating, infra-callosal segment); A3 (pre-callosal segment); A4 (supra-callosal segment); and A5 (post-callosal segment). The PCA is divided into four segments: P1 (pre-communicating segment); P2 (post-communicating segment); P3 (quadrigeminal segment); and P4 (calcarine segment).

Brain Venous Blood Outflow

Three sets of vessels play a role in venous blood outflow of the brain: superficial veins, deep veins, and venous sinuses. **Superficial veins** lie on the surface of the cerebral hemispheres and drain the lateral and inferior surfaces of the hemispheres. **Deep veins** drain the internal structures. The large veins pool their oxygen-depleted blood into a system of interconnected **venous sinuses** (see Figure 4.5 in Chapter 4), including the superior sagittal sinus, inferior sagittal sinus, straight sinus, occipital sinus, transverse sinuses, and sigmoid (S-shaped) sinuses. The superior sagittal sinus, straight sinus, and occipital sinus meet at the confluence of sinuses, where blood then flows into the transverse sinuses and sigmoid sinuses. The sigmoid sinuses then drain into the **internal jugular veins**, which exit the skull via the **jugular foramina**.

The Blood-Brain Barrier

More so than any other organ, the CNS requires a constant and stable environment to function normally. Other organs of the body can tolerate the wide and frequent fluctuations in the extracellular concentrations of hormones, amino acids, and electrolytes that occur after meals, or with physical exertion and stress. In the CNS, similar changes in the **interstitial** (between cells) fluid would disrupt the ability of neurons to generate and transmit electrical signals.

The **blood-brain barrier** (BBB) regulates the composition of the extracellular fluid within the CNS and provides a stable chemical environment for neurons by preventing substances that are circulating in the blood from nonselectively entering the brain and cerebrospinal fluid (CSF). The site of the barrier is at the interfaces between the blood and brain, and between the blood and CSF. In addition to protecting the brain from variations in blood composition, the BBB protects the brain from many toxins and pathogens.

The BBB was first observed by **Paul Ehrlich** (1854–1915) in the 1880s when he found that an intravenous injection of the dye trypan blue (which binds with albumin protein) into a laboratory animal stains the tissues of the body, but not the brain or spinal cord. When this dye was injected directly into the brain or ventricles, the CNS was stained but not the peripheral (i.e., non-CNS) tissues. These findings demonstrated the presence of a barrier between the blood and the fluid that surrounds the cells of the brain.

The cellular basis of the BBB was later revealed by electron microscopy examination of the brain and body microvasculature. In the brain, the walls of the capillaries are composed of endothelial cells that are fused together by **tight junctions**, whereas in other organs the capillary walls have **fenestrations** (small gaps) between the endothelial cells, permitting free passage of many small molecules from blood plasma to extracellular fluid. In addition, brain capillaries rest on a continuous basement membrane (basal lamina), which also plays a role in creating the BBB. A barrier between the blood and CSF is predominantly produced by the tight junctions between the epithelial cells of the choroid plexuses that line the ventricles. There is no barrier, however, between the CSF and the brain. The pia mater and ependymal cells (a type of glial cell) lining the ventricles are both extremely permeable. Consequently, the CSF space is continuous with the extracellular space. All substances that enter the CSF can also diffuse into the interstitial fluid of the brain, and vice versa.

The BBB has selective permeability, allowing only certain substances to pass through. Transport across the barrier depends on whether the molecule is lipid soluble, or whether the membranes of the barrier cells contain special transport mechanisms (i.e., channels or pumps) for the molecule. The BBB is efficient at keeping microorganisms out of the brain. It is, however, also efficient at keeping antibiotics and other drugs out of the brain, posing an obstacle for treating intracranial infections and other CNS disorders.

There are a few brain regions that are unique in that they lack a blood-brain barrier because their capillaries are fenestrated, allowing them to sample the contents of the systemic circulation or to secrete substances directly into the circulation. These regions are known as **circumventricular organs** because they are located close to the midline (i.e., 3rd and 4th) ventricles. These structures are isolated from adjacent neural structures and the ventricular CSF by specialized ependymal cells known as **tanycytes**. The circumventricular organs play a role in monitoring the chemical composition of the blood for maintaining homeostasis and protecting the organism. For example, the **area postrema** of the medulla, also known as the **chemoreceptor trigger zone**, detects bloodborne toxins (e.g., general anesthetics, cancer chemotherapy) and initiates vomiting (**emesis**). Another example is the posterior pituitary, where the lack of BBB allows for feedback control of hormone release. Because the circumventricular organs lack a BBB, they are vulnerable to circulating pathogens and can be portals for their entry into the brain.

Cerebral Blood Flow and Metabolism

Adequate blood flow (**perfusion**) is critically important to supply the brain with necessary oxygen and energy substrates to support normal function. **Cerebral blood flow** (CBF) is defined as the volume of blood that passes through a specific quantity of brain tissue during a specific time interval. Both **hypoperfusion** (insufficient CBF) and **hyperperfusion** (excessive CBF) can damage brain tissue. The standard unit of measurement for CBF is milliliters (ml) of blood per 100 grams (g) of tissue per minute. Under normal conditions, the average rate of blood flow in the human brain is typically about 50 ml per 100 g per minute, with lower values in the white matter (about 20 ml per 100 g per minute) and greater values in the gray matter (about 80 ml per 100 g per minute).

Cerebral blood flow is influenced and regulated by multiple factors, including systemic arterial blood pressure, intracranial pressure, the levels of oxygen and carbon dioxide in the arterial blood, and vascular reactivity (the ability of blood vessel to respond by vasoconstriction or vasodilation). The force driving blood into the brain is known as the **cerebral perfusion pressure**; it is a pressure gradient defined as the difference between the mean systemic arterial pressure driving blood into the brain and backpressure (resistance) due to the intracranial pressure. Normally, blood flow to the brain is remarkably constant, despite the normal fluctuations in systemic blood pressure that occur throughout the course of a day with activities such as sleep, changes in posture, and exercise. **Cerebral autoregulation** is the homeostatic process that maintains CBF at a constant level despite changes in systemic blood pressure and cerebral perfusion pressure. The cerebral arteries maintain constant flow by constricting in response to increased blood pressure and relaxing (dilating) in response to decreased pressure. In healthy adults, autoregulation fails when the cerebral perfusion pressure falls below a certain level.

Cerebral metabolism is the major determinant of **regional cerebral blood flow**. Increases in regional cellular metabolism result in the release of vasodilator substances that cause local vasodilation and increased blood flow to the region, while decreases in metabolism lead to decreases in the release of these vasodilator substances. This ensures that oxygen supply to tissue is adequate to meet the metabolic demands of that region and that products of metabolism are removed.

The measures of blood flow and pressure described above have important clinical correlates. Increases in intracranial pressure that occur with a variety of intracranial pathologies (e.g., traumatic brain injury), as well as severe hypotension (e.g., due to blood loss or inadequate pumping action of the heart), result in decreased cerebral perfusion pressure and decreased CBF. Critical care management of patients with elevated intracranial pressure or severe hypotension therefore requires maintaining an adequate cerebral perfusion pressure to decrease the risk of ischemic brain injury.

Hypothermia drastically reduces the cerebral metabolic demand for oxygen and glucose, due to the simple fact that the rate of all biochemical reactions is dependent on temperature. Hypothermia preserves ATP levels and enables the brain to endure longer periods of interrupted oxygen and glucose delivery before cell death and gross neurological injury due to ischemia, as well as anoxia (no oxygen supply with normal blood flow) or hypoxia (low oxygen supply with normal blood flow). This fact was first appreciated when it was observed that people who drown in very cold water have much better neurological recovery than those who drown in warm water, even after prolonged periods of hypoxia. These observations paved the way for research exploring the potential of induced hypothermia as a neuroprotective strategy for various forms of **hypoxic-ischemic injury** (i.e., due to oxygen deprivation or lack of perfusion to the brain), and studies performed in animals showed that hypothermia improves neurological recovery and reduces mortality. **Therapeutic hypothermia**, now known as **targeted temperature management**, was first introduced in the 1950s to prevent ischemic brain injury during cardiac surgery. It is currently used as a neuroprotective strategy during coronary artery bypass graft surgery, surgical repair of thoracoabdominal and intracranial aneurysms, and **carotid endarterectomy** (a surgical procedure that removes plaque buildup inside the carotid artery, with the goal of stroke prevention). Hypothermia is induced by surface cooling and/or endovascular cooling (by a catheter attached to a cooling unit or infusion of cooled blood or saline). Targeted temperature management is also used as a neuroprotective strategy in patients with cardiac arrest and infants with birth asphyxia. It is being studied in other forms of neurologic injury, including ischemic stroke, subarachnoid hemorrhage, intracerebral hemorrhage, traumatic brain injury, and spinal cord injury.

Imaging of the Cerebrovasculature

Angiography refers to a group of imaging techniques that allow visualization of blood vessels. There are several modalities for imaging the brain vasculature: arterial catheter angiography, computerized tomographic angiography, and magnetic resonance angiography. There are two types of angiograms: **arteriograms** image arteries, and **venograms** image veins.

Arterial catheter angiography involves injecting a radio-opaque dye into the vascular system to make blood vessels stand out against brain tissue when imaged by X-ray. The angiography technique was developed in 1927 by **Egas Moniz** (1874–1955), a Portuguese neurologist from the University of Lisbon (who also developed the surgical procedure of lobotomy). A catheter is placed into the femoral artery of the thigh and threaded under fluoroscopic guidance through the heart, up the aorta, into the aortic arch, and into the cerebral artery of interest (i.e., internal carotid artery or vertebral artery). Radio-opaque contrast medium that does not cross the BBB is injected into the artery, and a series of X-rays (angiograms) are taken at 2-second intervals in both antero-posterior and lateral projections. The flow of blood as the dye circulates through the vasculature is observed, first through the arteries, then into capillaries, and finally into veins and venous sinuses (Figure 5.8). This technique is used to image the structural anatomy of the cerebral vasculature and can reveal structural cerebrovascular abnormalities. Prior to the advent of computerized tomography (CT) and magnetic resonance imaging (MRI), cerebral angiography also was used to detect the presence of space-occupying lesions inferred from distortion in the shape and/or position of the vessels.

Arterial catheter angiography, however, is invasive and carries a risk of stroke, thus **computerized tomographic angiography** (CTA) and **magnetic resonance angiography** (MRA) are more commonly used to image the brain vasculature (see Chapter 29, "Brain Imaging"). These techniques are less invasive because they do not involve catheterization, and they also have the advantage of being able to simultaneously image the CNS itself. Arterial catheter angiography, however, produces the most detailed images of the cerebrovasculature (i.e., it has the highest degree of spatial and temporal resolution); it is used when other imaging studies do not provide a definitive diagnosis, and for further characterization of and therapy for vascular intracranial lesions.

The quantity of blood that flows through a particular volume of brain tissue can be measured by CT and MR perfusion techniques. Regional cerebral blood flow varies in a pattern that is correlated with neural activity and can be measured by **single-photon emission computerized tomography** (SPECT), a **functional imaging modality**

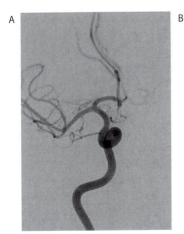
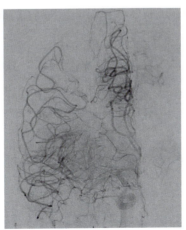
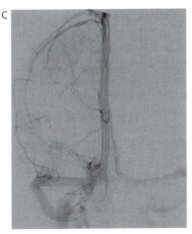

FIGURE 5.8. Conventional catheter cerebral angiography is used to evaluate the intracranial vasculature. (A) Arterial phase. (B) Capillary phase. (C) Venous phase.

that indirectly reveals patterns of brain metabolic activity (see Chapter 29, "Brain Imaging").

Cerebrovascular Abnormalities

Angiography permits visualization of the cerebral blood vessels and identification of cerebrovascular abnormalities such as stenosis, aneurysm, arteriovenous malformation, and moyamoya disease. Arterial catheter angiography is also used during neurovascular interventions such as balloon angioplasty, aneurysm coil embolization, arteriovenous malformation glue embolization, and neurovascular revascularization (see Chapter 18, "Stroke and Vascular Cognitive Impairment").

Arterial Stenosis

Arterial stenosis is a narrowing of the inner surface (lumen) of an artery that blocks or inhibits blood flow. It is usually caused by atherosclerosis. Arterial stenosis often affects the carotid artery and intracranial arteries; it carries a risk of ischemic stroke.

High-grade carotid stenosis, as revealed by carotid Doppler scan, is treated by carotid endarterectomy, stenting, or balloon angioplasty (Figure 5.9). **Endarterectomy** is a surgical procedure in which an incision is made in the neck and carotid artery, and the artery is cleared of plaque. **Stenting** is a minimally invasive endovascular procedure in which a tiny mesh tube (i.e., a stent) is inserted into the diseased artery, deployed, and left in place to widen the vessel and increase blood flow. **Balloon angioplasty** is a minimally invasive, endovascular procedure to widen narrowed or obstructed arteries or veins, typically to treat arterial atherosclerosis. A deflated balloon attached to a catheter (a balloon catheter) is passed over a guidewire into the narrowed vessel and then inflated to a fixed size. The balloon forces expansion of the blood vessel and the surrounding muscular wall; the balloon is then deflated and withdrawn. Stenting and balloon angioplasty are also used to treat **intracranial atherosclerotic disease**.

Brain Aneurysm

An aneurysm is a balloon-like dilatation defect in the wall of a blood vessel. Aneurysms can occur anywhere in the body, but one of the most common locations is in the brain. Brain aneurysms carry a high risk of intracranial hemorrhage. They may be congenital (present at birth) or may develop with risk factors such as age greater than 40, hypertension (high blood pressure), atherosclerosis, and cigarette smoking. They usually arise at the bifurcation of cerebral vessels, particularly on the large vessels of the internal carotid system that are part of the circle of Willis. The apexes of vessel bifurcations are prone to developing aneurysms because that is where there is maximum hemodynamic stress in a vascular network (Figure 5.10). The anterior communicating artery is the

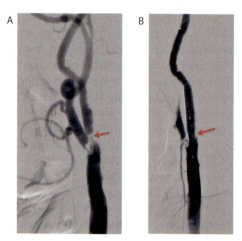

FIGURE 5.9. (A) Common carotid angiogram showing high-grade stenosis of the left ICA and the external carotid artery. (B) The patient underwent angioplasty and stenting with improved flow in the left ICA.

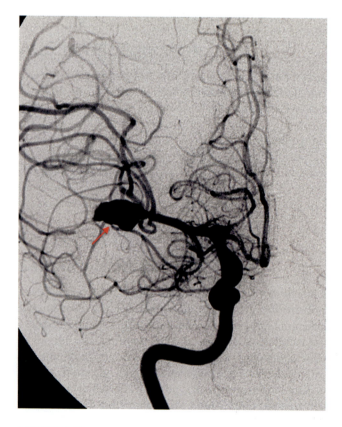

FIGURE 5.10. Conventional angiogram (AP) view demonstrates a right MCA bifurcation saccular aneurysm.

most frequent site of aneurysms (40%). Unruptured aneurysms typically are asymptomatic; they are usually discovered by chance as an **incidental imaging finding** (an imaging abnormality found in a symptomatic patient that is unrelated to the patient's symptoms, or an imaging abnormality found in a healthy, asymptomatic person).

Intracranial aneurysms are treated by radiologically guided **endovascular coiling** (see Chapter 18, "Stroke and Vascular Cognitive Impairment").

Arteriovenous Malformation

An **arteriovenous malformation** (AVM) is an abnormal tangle of arteries and veins (Figure 5.11). They may be found in any part of the brain, but the majority (80%) are within the cerebrum, and they especially occur in the parietal lobe. They may be very small or so large that they cover the greater part of a cerebral hemisphere. The pathogenesis of brain AVMs is not well understood; they are believed to be static (unchanging) lesions that are congenital. The vast majority occur sporadically, but some are genetically based.

AVMs have one or more large feeding arteries, several draining veins, and abnormal direct connections between arteries and veins without the normal intervening capillary bed. Blood flow is directly diverted from the arteries to the veins, interfering with circulation in the region. An abnormally high blood flow goes through AVMs due to the low resistance of the vascular bed. Thus, AVMs function as **arteriovenous shunts**, resulting in the pathological process of **vascular steal** in which blood is shunted away from the area. The draining veins of an AVM have abnormally thin walls that are exposed to a much greater blood pressure than normal veins and are prone to rupture; thus AVMs carry a high risk of cerebral hemorrhage (subarachnoid, intracerebral, or intraventricular). Most AVMs are discovered after hemorrhage, but they also present with other neurological symptoms and signs that may lead to their discovery, especially seizures and migraine-like headaches. Unruptured AVMs may even be associated with findings of cognitive weakness on neuropsychological examination.

Brain AVMs are treated by three methods: surgical removal of the AVM network, stereotactic radiosurgery, and intra-arterial embolization (see Chapter 18, "Stroke and Vascular Cognitive Impairment"). The primary rationale for treating brain AVMs is to prevent new or recurrent hemorrhage.

Moyamoya Disease

Moyamoya disease is a progressive narrowing and obstruction of the arteries within the circle of Willis. The brain attempts to compensate by forming collateral pathways of blood flow adjacent to the stenotic vessels at the base of the brain. The term *moyamoya* means "puff of smoke" in Japanese and refers to the characteristic angiographic appearance of the abnormal vascular collateral networks (Figure 5.12). Left untreated, moyamoya disease can lead to progressive ischemic strokes.

Moyamoya is treated by **neurosurgical revascularization** to improve the blood flow to the brain, by either direct or indirect bypass. Direct revascularization involves using a branch of a scalp artery for direct **anastomosis** (connection) to a branch of a brain artery in a procedure known as superficial temporal artery to middle cerebral artery **bypass** (a surgical procedure to create new pathways for blood flow). This procedure provides improved blood supply to the brain immediately. Indirect bypass methods provide more blood flow to the brain, but they do not involve making a direct connection to a brain blood vessel. The **encephalo-duro-arterio-synangiosis** procedure involves dissecting a branch of the temporal artery and implanting it directly on the surface of the brain. The **encephalo-myo-synangiosis** procedure involves

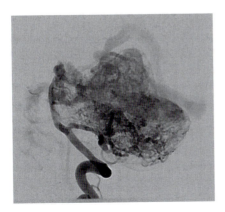

FIGURE 5.11. A conventional angiogram with left vertebral artery injection shows a left cerebellar AVM.

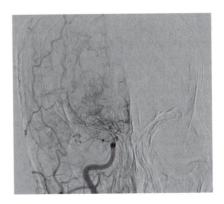

FIGURE 5.12. Moyamoya disease. There is stenosis of the terminal ICA, and the characteristic "puff of smoke" appearance of the abnormal vascular collateral networks on angiography.

dissecting the temporalis muscle and implanting it on the surface of the brain. Both procedures result in the formation of new blood supply.

Summary

The CNS is one of the most metabolically active tissues in the body; in humans, the brain makes up only about 2% of the total body mass, yet it accounts for 15%–20% of the body's total oxygen and glucose consumption. Unlike other body cells, however, neurons are unable to store glucose as glycogen, or to perform anaerobic metabolism; thus the CNS is vulnerable to even brief periods of ischemia. Two pairs of arteries supply blood to the brain: the ICAs and the vertebral arteries. The ICAs provide the anterior circulation and give rise to the anterior and middle cerebral arteries. The vertebral arteries provide the posterior circulation and give rise to the posterior cerebral arteries. The anterior and posterior arterial systems are connected through the circle of Willis, a ring of interconnected blood vessels located at the inferior surface of the brain. Cerebral angiography permits visualization of the cerebral blood vessels and identification of cerebrovascular abnormalities such as stenosis, aneurysm, arteriovenous malformation, and moyamoya disease. A basic knowledge of cerebrovascular anatomy, cerebrovascular imaging, and cerebrovascular pathologies is essential to understanding mechanisms of stroke and neurovascular interventions.

Additional Reading

1. Blumenfeld H. Cerebral hemispheres and vascular supply. In: Blumenfeld H, *Neuroanatomy through clinical cases*, 3rd ed., pp. 389–457. Sinauer Associates; 2021.
2. Brodal P. The blood supply of the CNS. In: Brodal P, *The central nervous system: structure and function*. 5th ed. Oxford University Press; 2016. https://oxfordmedicine.com/view/10.1093/med/9780190228958.001.0001/med-9780190228958-chapter-8. Accessed July 4, 2022.
3. Vanderah TW, Gould D. Blood supply of the brain. In: Vanderah TW, Gould D, *Nolte's the human brain: an introduction to its functional anatomy*. 8th ed., pp. 119-145. Elsevier; 2021.
4. Waxman SG. Vascular supply of the brain. In: Waxman SG, *Clinical neuroanatomy*. 29th ed. McGraw-Hill; 2020. https://accessmedicine-mhmedical-com.proxy.library.nyu.edu/content.aspx?bookid=2850&sectionid=242764058. Accessed July 4, 2022.

6

The Peripheral Nervous System

Introduction

A basic knowledge of the functional anatomy of the peripheral nervous system (PNS) helps form a foundation for understanding the neurocircuitry of sensory and motor functions, the methods of the neurological examination, and interpretation of neurological exam findings. The PNS consists of 31 pairs of spinal nerves and 12 pairs of cranial nerves, as well as their associated branches and ganglia. These nerves carry afferent fibers transmitting sensory information from the body and head to the central nervous system (CNS), and efferent fibers from the CNS to the effector organs (i.e., muscles and glands) of the body and head. The spinal nerves emanate from the spine and serve the sensory and motor functions of the body. The cranial nerves emanate from the cranium and serve the sensory and motor functions of the head and neck, the special senses, and parasympathetic innervation. Each cranial nerve is associated with one or more cranial nerve nuclei, mostly located in the brainstem.

Functional Subdivisions of the PNS

Besides the functional division of afferent and efferent components, the PNS has two additional functional divisions, the somatic system and the visceral system, each with afferent and efferent components.

The efferent component of the **somatic system** innervates **skeletal muscle**, which is attached to bone via tendons and is responsible for movements of the body and for posture. Skeletal muscles allow for voluntary movements (i.e., movements that are under conscious control); the pathways of this system originate in the cerebral cortex and descend through the CNS as the corticospinal and corticobulbar pathways, which in turn innervate motor neurons that directly synapse on skeletal muscles. Some skeletal muscle movements, however, are not under voluntary control, such as shivering, twitching, tremor, spasm, cramps, and reflexes. The efferent component of the **visceral system** innervates cardiac muscle, smooth muscle, and exocrine glands. Unlike the somatic efferents, the visceral efferents are generally not under voluntary control. The visceral efferent component of the PNS is therefore also known as the **autonomic nervous system**.

The descending signals of this system originate in the hypothalamus and the brainstem.

The afferent component of the somatic system originates from the body wall (i.e., not the internal organs) and carries signals from skin, skeletal muscles, and joints. The afferent component of the visceral system originates from the viscera (i.e., the soft internal organs contained within the abdominal and thoracic cavities). In addition, there are PNS afferents that originate from the specialized sense organs (i.e., the eye, ear, nose, and tongue) that carry information from the **special senses** (vision, audition, vestibular senses, olfaction, gustation); the special senses are distinguished from **somatosensation**, the more general somatic sense from all over the body (skin, muscles, joints, viscera).

Peripheral nerve fibers are characterized into six types: (1) **general somatic efferents** innervating skeletal muscle; (2) **general visceral efferents** innervating smooth muscle, cardiac muscle, and exocrine glands; (3) **general somatic afferents** carrying sensory impulses from the skin, muscles and joints; (4) **general visceral afferents** carrying sensory impulses from the internal organs; (5) **special somatic afferents** carrying the special senses of vision, audition, and the vestibular sense (position and movement of the head); and (6) **special visceral afferents** carrying the special senses of taste and smell (i.e., the **chemosenses**).

The Autonomic Nervous System

The **autonomic nervous system** (ANS) consists of the general visceral efferent component of the PNS and regulates the functions of the internal organs. This system functions largely involuntarily (i.e., independent of conscious control), a feature that is reflected by the name *autonomic*, which means "self-governing." It innervates smooth muscle (e.g., muscles of the blood vessels, internal organs, walls and sphincters of gut, gall bladder, urinary bladder, pupillary muscles, lens muscles, and muscles associated with hair follicles), cardiac muscle, and exocrine (ducted) glands (e.g., salivary, lacrimal, and sweat glands). It regulates the **vegetative** processes that are concerned with maintenance of life, such as the heart rate, respiratory rate, and digestion. The general function of the ANS is to maintain stability of the internal environment of the body required for optimal organ function.

The ANS is regulated by the hypothalamus, and while it operates primarily in the absence of volition, it is influenced by emotions by way of this hypothalamic control (see Chapter 10, "The Diencephalon").

SYMPATHETIC AND PARASYMPATHETIC DIVISIONS OF THE ANS

The ANS has two divisions: sympathetic and parasympathetic. In general, the **sympathetic system** is responsible for promoting "fight or flight" functions, and the **parasympathetic system** is responsible for promoting "rest and digest" functions.

The sympathetic system regulates **catabolic** physiological processes, which involve the expenditure of energy from reserves that are stored in the body. It prepares the body for emergency by accelerating the heart rate, increasing blood pressure, and redistributing blood away from skin and intestine, to the brain, heart, and skeletal muscle. It relaxes smooth muscles of the bronchi, producing bronchodilation and thereby increasing oxygen availability. It relaxes smooth muscles of the intestinal tract, inhibiting peristalsis. It closes sphincter muscles, inhibiting elimination processes. It dilates pupils. It causes sweating and piloerection (goosebumps, hair to stand on end). It also causes the release of adrenaline and noradrenaline from the adrenal glands into the circulation, where they act as hormones on internal organs. This causes breakdown of liver glycogen into glucose for immediate use and provides a relatively rapid and widespread reinforcement of sympathetic activities during emergency situations and stress. Sympathetic activation underlies the **fight or flight response** (see Chapter 10, "The Diencephalon").

The parasympathetic system regulates **anabolic** physiological processes, which are aimed at conserving and restoring energy. It slows the heart rate, increases peristalsis of the intestine, increases glandular activity of the digestive system, contracts the bladder wall, opens sphincters, and constricts pupils. It inhibits the use of liver glycogen.

In many organs, the sympathetic and parasympathetic systems have opposing effects. For example, in the heart, sympathetic activity increases the heart rate and force of contraction, while parasympathetic activity reduces the heart rate. In the lungs, sympathetic activity produces bronchodilation, while parasympathetic activity produces constriction of bronchioles and increased glandular secretions. In the gastrointestinal tract, sympathetic activity reduces peristalsis, decreases secretions, and contracts sphincter muscles, while parasympathetic activity increases peristalsis, increases secretions of gastric and intestinal juices, and relaxes sphincter muscles. Autonomic activity, however, is seldom solely either sympathetic or parasympathetic. Rather, homeostasis is maintained by the interplay between sympathetic and parasympathetic activity. In other instances, one subdivision of the ANS is unopposed (e.g., the sweat glands and limb vasculature receive only sympathetic innervation), or both divisions act cooperatively (e.g., in male sexual function, erection is maintained primarily by parasympathetic activity and ejaculation occurs by sympathetic activity).

The efferent pathways of the ANS traveling from the CNS to the visceral organs consist of a two-neuron chain with an intervening synapse at a ganglion. The **preganglionic neurons** originate in the CNS cranial nerve nuclei and terminate in **autonomic ganglia**; the **postganglionic**

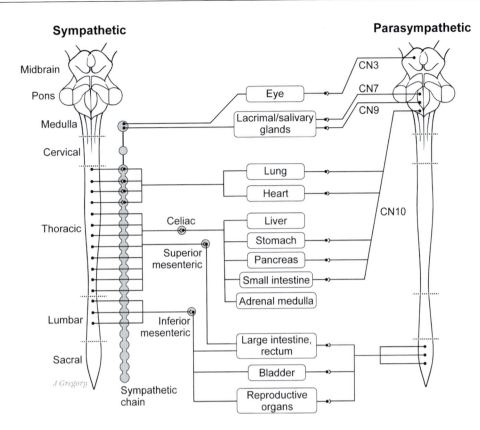

FIGURE 6.1. The autonomic system consists of sympathetic and parasympathetic divisions. The efferent pathways consist of a two-neuron chain: preganglionic neurons originate in the CNS and terminate in an autonomic ganglion, and postganglionic neurons originate in the autonomic ganglion and terminate in the target effector organ. In the parasympathetic division, the preganglionic efferents originate in cranial nerve nuclei and the sacral level of the spinal cord. In the sympathetic division, the preganglionic efferents emerge from the thoracic and lumbar levels of the spinal cord.

neurons originate in autonomic ganglia and terminate in their target effector organs (Figure 6.1).

In the sympathetic division, the preganglionic neuron efferents emerge from the thoracic and lumbar levels of the spinal cord. The sympathetic division of the ANS is therefore also known as the **thoracolumbar division**. The preganglionic axons are relatively short, terminating in a chain of ganglia flanking the vertebral column known as the **paravertebral ganglia (sympathetic chain)**, or in **prevertebral ganglia** (celiac, superior mesenteric, and inferior mesenteric) that innervate the gastrointestinal system and reproductive organs. The postganglionic axons of the ANS sympathetic division are relatively long. The preganglionic sympathetic neurons release acetylcholine, and the postganglionic neurons release norepinephrine. The exception to this arrangement is the adrenal medulla, which is connected directly to the preganglionic neuron and releases adrenalin directly into the circulation.

In the parasympathetic division, the preganglionic efferents originate in cranial nerve nuclei and the sacral level of the spinal cord. The parasympathetic division of the ANS is therefore also known as the **craniosacral division**. The preganglionic neuron axons are relatively long and synapse in **intramural ganglia** located in the walls of the target organ, and the postganglionic axons are relatively short. Both pre- and postganglionic neurons of the parasympathetic division release acetylcholine.

DYSAUTONOMIA

Dysautonomia is a general term used to describe any disorder of the ANS. It generally involves failure of the sympathetic or parasympathetic components of the ANS, but may involve ANS overactivity. There are many forms of dysautonomia, and they are classified as primary or secondary. The primary dysautonomias are due to neurological disease affecting the ANS, while the secondary dysautonomias are due to non-neurologic systemic illnesses and drug side effects. Primary dysautonomia may occur with some neurodegenerative diseases (e.g., Parkinson's disease).

The most striking symptom of dysautonomia is **orthostatic hypotension (postural hypotension)**, which is characterized by lightheadedness, dizziness, or **syncope** (fainting) that occurs when a person stands up from sitting, or sits up from lying down, caused by a drop in blood pressure. Many conditions and medications may cause orthostatic hypotension, including those that decrease the volume of circulating blood (e.g., cardiac pump failure, dehydration, diuretics) and medications that widen blood vessels (vasodilators). Neurogenic orthostatic hypotension is caused by neurologic disorders that directly affect the ANS and cause failure of the autonomic cardiovascular reflexes that compensate for the normal drop in blood pressure that occurs upon standing or sitting up. Normally, these reflexes cause blood vessels to constrict and the heart to beat faster and harder so that more blood is pushed upward in the body to preserve blood flow to the brain. In dysautonomia, the reflex-mediated compensatory responses do not occur; when the patient stands, the blood pressure falls beyond tolerable limits, redistribution of perfusion does not occur, and blood flow to the brain is reduced, resulting in postural lightheadedness, dizziness, or syncope.

Other symptoms of dysautonomia are often subtler than those of orthostatic hypotension and include abnormalities of sweating (lack of sweating, excessive sweating), urination (incontinence, incomplete bladder emptying), bowel function (constipation, diarrhea), and sexual function (e.g., erectile dysfunction, vaginal dryness).

The Spinal Nerves

There are 31 pairs of spinal nerves that connect to the spinal cord, one on each side of the vertebral column, named according to the vertebrae from which they emerge (Figure 6.2). There are 8 cervical nerves (C1–C8) emanating from the cervical spine; 12 thoracic nerves (T1–T12) emanating from the thoracic spine; 5 lumbar nerves (L1–L5) emanating from the lumbar spine; 5 sacral nerves (S1–S5) emanating from the sacrum; and one coccygeal nerve (Co1) emanating from the coccyx. The nerves emerge from the spinal column through openings between adjacent vertebrae known as **intervertebral foramen** (except for C1, which emerges between the occipital bone and the C1 vertebra). There are 7 cervical vertebrae, named C1–C7. Cervical nerves C1–C7 emerge above their corresponding vertebrae, while nerve C8 emerges below the C7 vertebra. The thoracic and lumbar nerves emerge below their corresponding vertebrae.

The spinal nerves are **mixed nerves**; they carry both afferent and efferent fibers. Each spinal nerve splits into a **dorsal root** (**posterior root**) by which sensory information enters the spinal cord, and a **ventral root** (**anterior root**) by which motor information exits the spinal cord (Figure 6.3). The cell bodies of the unipolar sensory afferents are located in the **dorsal root ganglia**. The discovery that the sensory and motor nerve fibers are segregated, with its implication that nerve fibers carry information in only one direction, was a landmark discovery in the history of neuroscience and is known as the **Bell-Magendie law**.

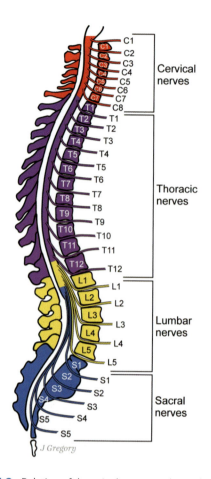

FIGURE 6.2. Relation of the spinal nerves and spinal cord segments to the vertebrae. Note the termination of the spinal cord at the level of the L1 or L2 vertebra.

Dermatomes

The skin is our largest sensory organ. It is innervated by afferent fibers that carry somatosensory information to the CNS. Cutaneous signals from the skin enter the spinal cord via the dorsal roots of the spinal nerves. Consequently, the skin surface is divided into discrete regions called dermatomes (*derma*, "skin," and *tome*, "a thin slice") that reflect the segmental innervation of the skin. Thus, a **dermatome** is a restricted area of skin that is innervated by a single dorsal root of a spinal nerve (Figure 6.4).

The skin surface of each half of the body is divided into 30 dermatomes, named after their corresponding spinal nerves. Spinal nerve C1, however, does not have an associated dermatome because it predominantly carries motor fibers innervating the neck and does not have a dorsal root. The dermatomes of the thorax and abdomen are arranged in a pattern of horizontal stripes (referenced to an upright human), but the dermatomes of the arms and the legs run longitudinally along the limbs. This pattern has its origin in the way that the limbs bud and rotate during early embryonic development.

The Peripheral Nervous System | 65

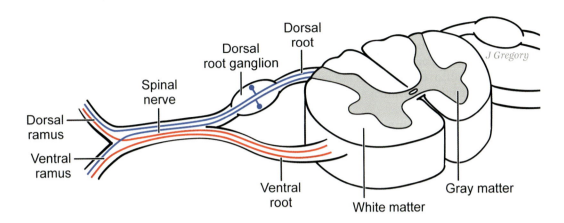

FIGURE 6.3. The spinal nerves carry both afferent and efferent fibers. Afferents (depicted in blue) enter the spinal cord by the dorsal roots, and efferents (depicted in red) leave the spinal cord by the ventral roots.

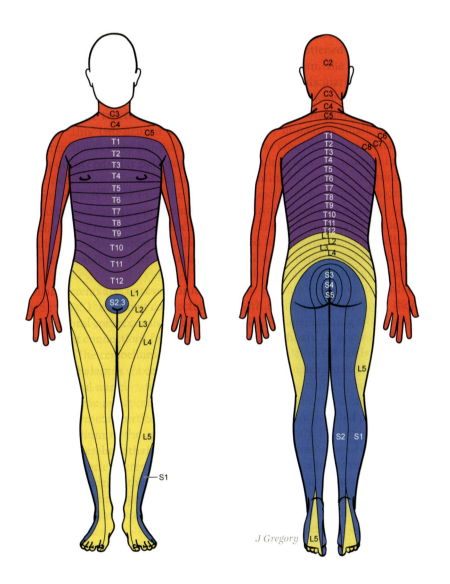

FIGURE 6.4. The dermatomes. Dermatome landmarks include the nipples (T4), xiphoid process (T7), umbilicus (T10), and perineum (S4 and S5). The first cervical spinal nerve (C1) carries motor fibers predominantly, thus C1 is not associated with a dermatome.

There is some variability between individuals in the **dermatomal map**, and there may be some overlap between adjacent dermatomes. In general, the neck is served by spinal nerves C2–C4, the upper limbs are served by spinal nerves C5–T2, the thorax and abdomen are served by spinal nerves T3–T12, and the lower limbs and genitalia are served by spinal nerves L1–S5.

Sensory symptoms (loss, pain) that follow a dermatomal distribution are indicative of pathology at the nerve root level. Shingles, which is caused by reactivation of varicella zoster (chickenpox) virus that was lying dormant in the dorsal root ganglia, typically involves a rash in a single dermatome.

Myotomes

A **myotome** is the set of muscles innervated by a single spinal nerve on one side of the body. Muscle weakness that has a myotomal distribution is indicative of pathology at the nerve root level. Individual muscles, however, can be innervated by more than one nerve and by nerves that originate from different spinal cord levels. Myotome testing involves checking the ability to perform specific actions.

Peripheral Nerves

As each spinal nerve emanates from the spine, it divides into a ventral (anterior) **ramus** (branch) carrying fibers that give rise to peripheral nerves that supply the anterolateral parts of the trunk and the limbs, and a dorsal (posterior) ramus carrying fibers that give rise peripheral nerves that supply the posterior parts of the trunk and the limbs (see Figure 6.3).

The **anterior rami** of most spinal nerves contribute to axonal networks called **plexuses** in which fibers from multiple spinal nerves merge; these are the **cervical**, **brachial**, **lumbar**, and **sacral** plexuses. The plexuses give rise to **peripheral nerves** (e.g., the median, ulnar, and radial nerves supplying the upper limbs), which are distal to the spinal cord and nerve roots and are composed of fibers from two or more spinal nerves. For example, the radial nerve originates from the brachial plexus and carries motor efferents (from the ventral roots) of spinal nerves C5, C6, C7, C8 and T1. The specific area of skin from which the afferents are carried by a specific peripheral nerve is known as the **peripheral nerve field** (**cutaneous nerve distribution**). Thus, peripheral nerve pathology gives rise to sensory symptoms that follow a pattern that differs from the dermatomal pattern arising from spinal nerve pathology, and similarly gives rise to motor symptoms that follow a different pattern than the myotomes.

Clinical Considerations

Diagnosis of PNS disorders usually begins with localization. **Neuropathy**, also known as **peripheral neuropathy**, refers to disease or dysfunction of the peripheral nerves. Peripheral neuropathies may affect one nerve (**mononeuropathy**), several discrete nerves (**multiple mononeuropathy**), multiple nerves diffusely (**polyneuropathy**), a plexus (**plexopathy**), or a nerve root (**radiculopathy**). Disorders affecting spinal nerves that do not affect the spinal cord cause sensory abnormalities, motor abnormalities, or both, in the area innervated by the affected spinal nerve(s). Common causes of radiculopathy and mononeuropathy are trauma and nerve compression (**entrapment**). Polyneuropathies are most often caused by a systemic process, such as metabolic diseases (e.g., diabetes) and toxins (e.g., chronic alcohol abuse).

Neuropathies are distinguished as either demyelinating or denervating. In **demyelinating neuropathies**, only the myelin sheath is affected. In **denervating neuropathies**, both the myelin sheath and the axon itself are damaged. The **hereditary motor and sensory neuropathies** (**Charcot-Marie-Tooth disease**) are a group of genetic disorders affecting the structure and function of the peripheral nerves. These disorders affect either the myelin sheath or the peripheral nerve axon, but they affect both the afferent and efferent pathways.

The Cranial Nerves

There are 12 pairs of cranial nerves (Table 6.1). They are identified by both a name and number; Roman numerals have been used traditionally, but Arabic numerals are also used. The cranial nerves are: the olfactory nerves (CN1), the optic nerves (CN2), the oculomotor nerves (CN3), the trochlear nerves (CN4), the trigeminal nerves (CN5), the abducens nerves (CN6), the facial nerves (CN7), the vestibulocochlear nerves (CN8), the glossopharyngeal nerves (CN9), the vagus nerves (CN10), the accessory nerves (CN11), and the hypoglossal nerves (CN12). A classic mnemonic for remembering the cranial nerves is "**O**h **O**h **O**h **T**o **T**ouch **A**nd **F**eel **V**ery **G**ood **V**elvet, **AH**."

The cranial nerves enter and exit the skull through foramina. They are numbered according to the rostrocaudal sequence in which they pierce the dura mater during their course between brain and periphery (Figure 6.5). Eleven of the cranial nerves enter/exit the brain from its ventral surface; the trochlear (CN4) exits the brain from its dorsal surface. Knowledge of the levels of the cranial nerves and their associated nuclei allows neurologists to localize lesions.

The cranial nerves are primarily distributed in the head and neck. The exception is the vagus nerve, which provides parasympathetic innervation to the structures in the thorax and abdomen. The cranial nerves serve three general functions: (1) motor and sensory innervation of the head and neck; (2) innervation of the special sense

The Peripheral Nervous System | 67

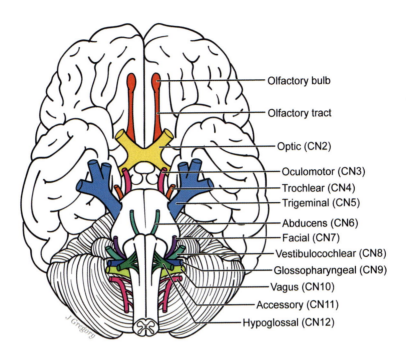

FIGURE 6.5. The cranial nerves. The olfactory nerves (CN1) are not depicted; they are composed of multiple small nerve bundles which enter the cranium through olfactory foramina and synapse in the olfactory bulbs.

organs (olfaction, vision, audition, vestibular, gustation); and (3) parasympathetic innervation.

The cranial nerves carry afferents to and efferents from the cranial nerve nuclei. Each cranial nerve is associated with one or more cranial nerve nuclei that lie within the CNS, most of which are located in the brainstem (Figure. 6.6). Cranial nerve efferents originate from cranial nerve motor nuclei. Cranial nerve afferents terminate in cranial nerve sensory nuclei; the cell bodies of those neurons lie in sensory ganglia in the periphery.

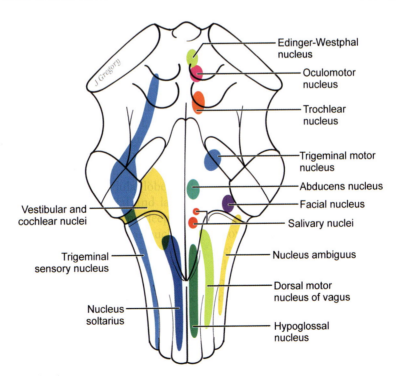

FIGURE 6.6. The brainstem viewed from the dorsal aspect, with the afferent cranial nerve nuclei shown on the left and the efferent cranial nerve nuclei shown on the right. The nuclei of first four cranial nerves (CN1–CN4) are located in the midbrain or higher subcortical structures, the nuclei of the second four cranial nerves (CN5–CN8) are located in the pons, and the nuclei of the last four cranial nerves (CN9–CN12) are located in the medulla or upper cervical cord.

TABLE 6.1 Summary of the cranial nerves, their nuclei and type [sensory (S), motor (M), parasympathetic (P)], and CNS level of origin/termination

CN1: Olfactory (Sensory)		
smell	olfactory bulb (S)	telencephalon
CN2: Optic (Sensory)		
vision	Lateral geniculate nucleus (S)	diencephalon
pupillary light reflex	pretectal area (S)	midbrain
circadian rhythm	suprachiasmatic nuclei (S)	diencephalon
reflexive orienting eye and head movements	superior colliculi (S)	midbrain
CN3: Oculomotor (Motor)		
elevation, depression, and adduction of the eye (superior rectus, inferior rectus, medial rectus, inferior oblique), eyelid opening (levator palpebrae)	oculomotor nucleus (M)	midbrain
Pupillary constriction (iris sphincter), lens accommodation (ciliary muscle)	Edinger-Westphal nucleus (P)	midbrain
CN4: Trochlear (Motor)		
depression of the adducted eye, intorsion of the abducted eye (superior oblique)	trochlear nucleus (M)	midbrain
CN5: Trigeminal (Mixed)		
proprioception of the mandible	mesencephalic trigeminal nucleus (S)	midbrain
sensation of the face	main sensory trigeminal nucleus (S)	pons
sensation from ear	spinal trigeminal nucleus (S)	mid-pons to spinal cord
Motor control of the muscles of mastication	trigeminal motor nucleus (M)	Pons
CN6: Abducens (Motor)		
abduction of the eye (lateral rectus)	abducens nucleus (M)	pons
CN7: Facial (Mixed)		
muscles of facial expression	facial motor nucleus (M)	pons
submandibular and sublingual salivary glands lacrimal glands	superior salivatory nucleus (P)	pons
Taste to the anterior 2/3 of the tongue	rostral solitary nucleus (S)	medulla
Sensation from ear	spinal trigeminal nucleus (S)	mid-pons to spinal cord
CN8: Vestibulocochlear (Sensory)		
hearing	2 cochlear nuclei (dorsal, ventral) (S)	medulla
balance	4 vestibular nuclei (superior, lateral, medial, ascending) (S)	pons and medulla
CN9: Glossopharyngeal (Mixed)		
swallowing (stylopharyngeous)	nucleus ambiguous (M)	medulla
parotid salivary gland	inferior salivatory nucleus (P)	pons

TABLE 6.1 Continued		
taste of posterior 1/3 of the tongue	rostral solitary nucleus (S)	medulla
carotid sinus baroreceptors	caudal solitary nucleus (S)	medulla
pharynx and ear sensation	spinal trigeminal nucleus (S)	mid-pons to spinal cord
CN10: Vagus (Mixed)		
speech, swallowing, coughing and vomiting reflexes (motor control of soft palate, pharynx, larynx)	nucleus ambiguus (M)	medulla
autonomic fibers to esophagus, stomach, small intestine, heart, trachea	dorsal motor nucleus of vagus (P)	medulla
taste from epiglottis and root of tongue	rostral solitary nucleus (S)	medulla
visceral sensation	caudal solitary nucleus (S)	medulla
sensation from ear	spinal trigeminal nucleus (S)	mid-pons to spinal cord
CN11: Accessory (Motor)		
Head and shoulder movement (sternocleidomastoid and trapezius muscles)	spinal accessory nucleus (M)	C1–C5
CN12: Hypoglossal (Motor)		
tongue movement	hypoglossal nucleus (M)	medulla

Some cranial nerves are composed entirely of sensory afferent fibers (CN1, CN2, CN8), some are composed entirely of motor efferent fibers (CN3, CN4, CN6, CN11, CN12), and some are mixed, possessing both afferent and efferent fibers (CN5, CN7, CN9, CN10); a mnemonic is "**S**ome **S**ay **M**oney **M**atters **B**ut **M**y **B**rother **S**ays **B**ig **B**rains **M**atter **M**ore," with the B meaning both motor and sensory components. Some cranial nerves carry parasympathetic fibers of the autonomic system (CN3, CN7, CN9, CN10).

Each cranial nerve is associated with one or more cranial nerve nuclei (see Table 6.1). In most cases the cranial nerve nuclei are located at the same rostrocaudal level at which their nerves leave the brainstem. For efferent pathways, the cranial nerve nuclei are the sites of pathway origination, while for afferent pathways the cranial nerve nuclei are the sites of pathway termination. In contrast to cranial nerves, which may carry both sensory afferents and motor/parasympathetic efferents, cranial nerve nuclei are either sensory or motor, but never both. The cranial nerve nuclei are generally located within the brainstem. The exceptions to this are the olfactory nerve (CN1) sensory afferents that project to the olfactory bulb (telencephalon), the optic nerve (CN2) sensory afferents that project to the lateral geniculate nucleus of the diencephalon, and the accessory nerve (CN11) efferents which originate from the spinal accessory nucleus within the C1–C5 levels of the spinal cord.

The cranial nerve afferents are sensory neurons whose cell bodies reside outside of the CNS in ganglia (just as the cell bodies of the sensory afferents to the spinal cord reside in the dorsal root ganglia) and axons terminate in cranial nerve nuclei. Sensory cranial nerve nuclei in turn conduct signals to the thalamus and the cerebral cortex. Many cranial nerve nuclei receive afferents from only one cranial nerve, but some receive afferents from more than one cranial nerve. For example, the rostral solitary nucleus receives inputs from CN7, CN9, and CN10, conveying taste information from the anterior two-thirds of the tongue (CN7), posterior one-third of the tongue (CN9), and epiglottis and root of tongue (CN10).

The cranial nerve efferents are motor neurons that innervate skeletal muscle or parasympathetic preganglionic motor neurons. Their cell bodies lie within the brainstem in one or more cranial nerve nuclei. Efferent cranial nerve nuclei controlling skeletal muscles of the head and neck generally have ipsilateral control (the exception being the trochlear nucleus); they receive direction from higher brain centers that is most often contralateral.

CNI: Olfactory Nerves

The **olfactory nerves** (CN1) are pure sensory nerves that are responsible for the sense of smell (olfaction). They carry sensory afferents that project to the **olfactory bulbs** of the telencephalon, located on the inferior surface of the frontal lobes (see Figure 6.5). Each olfactory nerve is actually composed of multiple small nerve bundles (**olfactory filaments/fila**). These bundles of olfactory nerve fibers enter the cranium through perforations in the cribriform plate known as the **olfactory foramina**. From each olfactory bulb, second-order neurons project via an **olfactory tract** to olfactory cortex. The olfactory system is notable for being the only sensory system that reaches the cerebral cortex without first synapsing in the thalamus.

Damage to the olfactory nerves, bulbs, or tracts results in **anosmia**, a loss of the sense of smell, or **hyposmia**, a reduced sense of smell. Anosmia usually is due to nasal congestion, but it can also be due to neurologic injury to the olfactory nerves, bulbs, or tracts, as often occurs with head trauma. Its presence can serve as a marker of additional structural damage and dysfunction of the orbitofrontal cortex, an area of the inferior surface of the frontal lobe that sits just above the orbits (eye sockets).

CN2: Optic Nerves

The **optic nerves** (CN2) are pure sensory nerves that are responsible for vision. They carry sensory afferents from the retina, the light-sensitive layer of tissue that lines the interior of the eye. Each optic nerve enters the cranium via an **optic foramen** located in the lesser wing of the sphenoid bone. The majority (90%) of optic nerve fibers project to the **lateral geniculate nuclei** (LGN) of the thalami within the diencephalon. From the LGN, second-order neurons project via the optic radiations to primary visual cortex. This two-neuron retinogeniculostriate pathway is necessary for conscious visual awareness (see Chapter 14, "The Occipital Lobes and Visual Processing").

The remaining 10% of the optic nerve fibers project to other subcortical nuclei. Projections to the pretectal nuclei play a role in regulating pupil diameter with regard to light levels. Projections to the suprachiasmatic nuclei of the hypothalamus via the retinohypothalamic pathway play a role in circadian rhythm (day/night cycle) regulation. Projections to the superior colliculi of the midbrain tectum play a role in reflexive orienting eye and head movements toward visual stimuli (e.g., bright light, movement).

Pathology of the optic nerve results in reduced visual acuity or blindness of the eye. **Optic neuropathy** is a general term applied to optic nerve dysfunction. A wide variety of conditions may result in optic neuropathy, including glaucoma, inflammation (**optic neuritis**), demyelinating disease, ischemia, compression due to space-occupying lesions, toxins, and metabolic disorders.

CN3: Oculomotor Nerves

The **oculomotor nerves** (CN3) are pure motor nerves with a parasympathetic component; they play a role in eye movements, upper eyelid opening, pupil constriction, and accommodation. Each nerve exits the cranium via the **superior orbital fissure**, a slit-like foramen in the sphenoid bone, along with the trochlear and abducens nerves, which also play a role in eye movements.

Motor efferents from the **oculomotor nuclei** of the midbrain innervate (1) four of the six **extrinsic/extraocular muscles** mediating eye movements; and (2) the levator palpebrae muscle, which raises the upper eyelid. Eye movements are controlled by three pairs of antagonistic muscles (i.e., that move the eye in opposite directions such that as one muscle contracts, the other relaxes): the **lateral** and **medial rectus muscles**, the **superior** and **inferior rectus muscles**, and the **superior** and **inferior oblique muscles**. Horizontal movements toward the midline/nose (adduction) or away from the midline/nose (abduction) are controlled by the medial and lateral rectus muscles. Vertical movements up (elevation) or down (depression) require the coordinated action of the superior and inferior rectus muscles as well as the oblique muscles. Torsional movements that bring the top of the eye toward the nose (intorsion) or away from the nose (extorsion) are controlled primarily by the oblique muscles. The oculomotor nerve controls the contralateral superior rectus (which elevates the eye), ipsilateral inferior rectus (which depresses the eye), ipsilateral medial rectus (which adducts the eye), and ipsilateral inferior oblique (which elevates, abducts, and laterally rotates the eye).

Parasympathetic efferents originate from the **Edinger-Westphal nuclei**, which innervate the pupillary sphincter muscles of the irises mediating pupil constriction, and the ciliary muscles of the lenses mediating accommodation for near vision.

Clinical signs of oculomotor nerve dysfunction consist of oculomotor nerve palsy, diplopia, ptosis, and abnormal pupil dilation. **Oculomotor nerve palsy (third-nerve palsy)** is characterized by a misalignment of the affected eye such that it is in a "down and out" position. This is due to the unopposed actions of the lateral rectus muscle, which is responsible for eye abduction and under the control of CN6, and the superior oblique muscle which causes a combination of adduction and downward movement of the eye and is under the control of CN4. Third-nerve palsy is a cardinal early sign of uncal herniation in coma; as the uncus of the temporal lobe herniates across the tentorium, it compresses the midbrain and oculomotor nerve.

Diplopia is double vision when fixating (looking directly at) an object; it is due to misalignment of the eyes. **Ptosis** is a droopy eyelid; it can be due to paralysis of the levator palpebrae muscle. Abnormal pupil dilation (**mydriasis**) is due to lack of parasympathetic innervation of the pupillary muscles mediating pupil constriction; this also produces a loss of the pupillary light reflex (in which light triggers pupillary constriction). Lack of parasympathetic innervation of the eye also produces a loss of accommodation (the ability of the eye to adjust focus at close distance), due to ciliary muscle paralysis.

CN4: Trochlear Nerves

The **trochlear** (pulley) **nerves** (CN4) are pure motor nerves. Each nerve innervates the superior oblique muscle on the opposite side of the body, which mediates depression of the adducted eye and inward rotation (intorsion) of the abducted eye. Each nerve exits the cranium via the superior orbital fissure. The trochlear nerve is unique in that (1) it is the smallest cranial nerve in terms of number of axons; (2) it exits the brain dorsally; (3) it is the only cranial nerve to originate entirely from a contralateral nucleus; and (4) it takes the longest intracranial path of any cranial nerve.

Injury or disease of the trochlear nerve causes paralysis of the superior oblique muscle and **trochlear nerve palsy** (**fourth nerve palsy**) in which the affected eye drifts upward relative to the normal eye. This results in a vertical diplopia, double vision in which the images from the two eyes are separated vertically. The diplopia is most pronounced when the person looks downward, such as when reading and walking downstairs, so rather than making down-gaze movements, the patient will tilt the head forward. The long course of the trochlear nerve renders it particularly prone to injury from blunt head trauma or compression from increased intracranial pressure, space-occupying lesions, or edema anywhere along its course.

CN5: Trigeminal Nerves

The **trigeminal nerves** (CN5) are mixed nerves; they carry both afferent and efferent fibers. They carry sensory information from the face and mouth, and motor efferents mediating mastication (chewing) movements. Each nerve is associated with four cranial nerve nuclei.

The trigeminal (Latin, "born three at a time") has three extracranial branches, each of which enters the cranium by a separate route. The **ophthalmic branch** carries sensation from the skin around the eyes, forehead, and anterior half of the scalp, as well as sensation from the eyeball, conjunctiva, cornea, and lacrimal gland. It enters the cranium through the superior orbital fissure of the sphenoid bone. The **maxillary branch** carries sensation from the skin of the cheek; mucous membrane of the nose, upper jaw, upper teeth, and gums; and dura of the anterior and middle cranial fossae. It enters the cranium through the foramen rotundum of the sphenoid bone. The **mandibular branch** carries sensation from the skin of the face over the mandible and the side of the head in front of the ear; the tongue, lower jaw, lower teeth, and gums; and dura of anterior and middle cranial fossae. The mandibular branch also carries motor efferents from the trigeminal motor nucleus to the muscles of mastication. It traverses the cranium through the foramen ovale of the sphenoid bone.

Trigeminal sensory afferents project to three cranial nerve nuclei: the **chief** (**main**) **trigeminal sensory nucleus**, the **mesencephalic trigeminal nucleus**, and the **spinal trigeminal nucleus**. These nuclei form a long column of cells that extends from the rostral midbrain to the upper cervical spinal cord. Second-order neurons from the trigeminal sensory nuclei project to the ventral posterior medial nucleus of the thalamus via the trigeminothalamic tract. Third-order neurons ascend in the posterior limb of the internal capsule and project to the face area of primary somatosensory cortex. It is notable that sensory input from the external ear is conducted by four cranial nerves (CN5, CN7, CN9, and CN10), all terminating in the spinal trigeminal nucleus.

Motor efferents originating from the **trigeminal motor nucleus** travel exclusively in the mandibular division of the trigeminal nerve and innervate the muscles of mastication used for chewing. The trigeminal motor nucleus receives corticobulbar projections from the face area of the primary motor cortex of the precentral gyrus, providing for voluntary control of chewing.

Trigeminal nerve injuries cause sensory deficits affecting the face, eye, and oral cavity, facial pain, and muscular denervation of the masticator muscles. If a specific branch is affected, it results in reduced sensation (**hypesthesia**) or loss of sensation (**anesthesia**) of the area supplied. When the ophthalmic branch is affected, there is no blink reflex (involuntary blinking elicited by stimulation of the cornea). Trigeminal nerve injury is often associated with facial bone fractures. **Trigeminal neuralgia** (**tic douloureux**) is a neuropathic pain syndrome that originates from CN5 and produces sudden sharp pain in the face and jaw.

CN6: Abducens Nerves

The **abducens nerves** (CN6) are pure motor nerves that innervate the lateral rectus muscle of the ipsilateral eye and mediate eye abduction (outward/lateral eye movement). Each nerve exits the cranium via the superior orbital fissure. The efferents originate from the **abducens nucleus** of the pons.

Injury or disease of the abducens nerve causes paralysis of the lateral rectus muscle and **abducens nerve palsy** (**sixth nerve palsy**), characterized by an ipsilateral cross-eye and an inability to abduct the eye beyond the midline of gaze. This produces a horizontal diplopia. Because the abducens runs a long course on the ventral surface of the pons, it is particularly vulnerable to increased intracranial pressure.

CN7: Facial Nerves

The **facial nerves** (CN7) mediate facial movements, taste from the anterior two-thirds of the tongue, and parasympathetic responses of salivation and tearing. They are mixed nerves and have a parasympathetic component.

Each facial nerve traverses the cranium through the ipsilateral internal acoustic meatus of the temporal bone, along with the vestibulocochlear nerve. Each nerve is associated with four cranial nerve nuclei.

Efferents from the **facial motor nucleus** innervate the **muscles of facial expression** through five main extracranial branches: **temporal, zygomatic, buccal, mandibular,** and **cervical** (an old mnemonic is "**T**o **Z**anzibar **B**y **M**otor **C**ar"). Impulses for voluntary movement of the facial muscles are conveyed through corticobulbar fibers arising from the face region of primary motor cortex projecting to the facial motor nucleus.

Sensory afferents from the anterior two-thirds of the tongue mediating taste terminate in the **rostral solitary nucleus**. Second-order neurons project ipsilaterally to the most medial part of the ventral posterior medial nucleus of the thalamus; third-order neurons project to gustatory cortex located in the parietal operculum and adjacent insular cortex (located adjacent to the somesthetic representation for the tongue). This nerve also carries sensory input from the external ear (along with CN5, CN9, and CN10) to the spinal trigeminal nucleus.

Parasympathetic efferents from the **superior salivatory nucleus** innervate two of the three major salivary glands (submandibular and sublingual), as well as parasympathetic efferents from the **lacrimal nucleus**, a component of the superior salivary nucleus that innervates the lacrimal glands and mediates tearing. The secretion of tears and saliva are examples of how visceral functions can be influenced from higher levels of the brain. The secretion of saliva is initiated by the stimulation of taste receptors, as well as by mentation (e.g., the thought of eating a lemon). The lacrimal nucleus stimulates the production of three types of tears: basal, reflex, and emotional. Basal tears are the thin layer of tear that lubricates the eye. Reflex tears are produced reflexively in response to irritation of the eye by foreign particles or other irritants (e.g., onion vapors, tear gas, pepper spray). Emotional tears (psychic tears) occur when one is overcome with emotion (sadness, happiness, laughing, pain) via inputs from the hypothalamus, which is under the influence of cortical and limbic structures.

A lesion involving the facial nerve results in a **facial nerve palsy** in which there is a complete paralysis of the facial muscles ipsilaterally, manifesting with mouth droop, flattening of the nasolabial fold, inability to close the eye, and smoothing of the brow. **Bell's palsy** is an idiopathic form of facial nerve palsy that is believed to be caused by viral infection. It has a rapid onset over several hours, and there usually is recovery within several weeks to 6 months after onset.

CN8: Vestibulocochlear Nerves

The **vestibulocochlear nerves** (CN8) consist of a bundle of two different sensory nerves, the **cochlear nerve** and the **vestibular nerve**. Both nerves originate from the labyrinth of the inner ear, which is made up of the cochlea and the vestibular apparatus. CN8 enters the cranium through the internal acoustic meatus of the temporal bone, along with the facial nerve.

Auditory signals originate in the **cochlea** and are carried by the cochlear nerve. The cochlear afferents terminate bilaterally in two pairs of **cochlear nuclei**, the dorsal cochlear nuclei and the ventral cochlear nuclei. The cochlear nuclei in turn project to higher regions of the auditory brainstem (superior olivary nucleus, nucleus of the lateral lemniscus, inferior colliculus), which in turn project to the medial geniculate nuclei of the thalamus, which in turn project to auditory cortex (see Chapter 16, "The Temporal Lobes and Associated Disorders").

Vestibular signals originate in the **vestibular apparatus** and are carried by the vestibular nerve. The vestibular apparatus has two main components: two otoliths (utricle and saccule) and three semicircular canals. The otoliths detect linear movement (motion in a straight line) of the head in the horizontal plane (forward or backward, left or right), as well as in the vertical plane relative to gravity (i.e., whether we are right-side up or upside down). The semicircular canals detect rotational movement of the head. These signals mediate perception of movement of the head in space and are used for maintaining posture and equilibrium. The vestibular afferents terminate in four **vestibular nuclei** (superior, lateral, medial, and descending). Second-order neurons project to the cerebellum, brainstem, and spinal cord to influence motor neurons controlling posture. The lateral vestibular nucleus gives rise to the lateral vestibulospinal tract, which facilitates the action of antigravity muscles.

Damage to the vestibulocochlear nerve causes hearing loss, **tinnitus** (an illusory auditory perception of ringing, buzzing, roaring, or hissing), **vertigo** (a sensation of spinning or swaying movement), reduced balance, and **nystagmus** (abnormal involuntary rapid oscillatory eye movements). The most common pathology affecting this nerve is **acoustic neuroma (vestibular schwannoma)**, a noncancerous tumor that arises from the Schwann cells myelinating the nerve fibers. Head trauma is another common cause of CN8 pathology.

CN9: Glossopharyngeal Nerves

The **glossopharyngeal** (*glosso*, "relating to the tongue"; *pharyngeal*, "relating to the pharynx/throat") nerves (CN9) are involved in motor control of swallowing, salivation, intra-oral sensations, and taste from the posterior one-third of the tongue. This nerve is mixed and includes a parasympathetic component. It traverses the cranium through the jugular foramen, along with the vagus (CN10) and accessory (CN11) nerves. It is associated with five cranial nerve nuclei, all located in the medulla.

Motor efferents originating in the **nucleus ambiguus** mediate swallowing (by innervating the stylopharyngeus muscle). Each nucleus receives corticobulbar projections from both cerebral hemispheres. Parasympathetic efferents innervating the parotid salivary gland originate in the **inferior salivatory nucleus**, which receives inputs from the hypothalamus, the olfactory system, and the gustatory system. Afferents signaling sensation (pain, temperature, and nondiscriminative touch) from the pharynx and external ear synapse in the **spinal trigeminal nucleus**. Taste afferents from the posterior one-third of the tongue synapse in the **rostral solitary nucleus**, converging with afferents from the anterior two-thirds of the tongue carried by the facial nerve; second-order neurons from the rostral solitary nucleus project ipsilaterally to the most medial part of the ventral posterior medial of the thalamus; and third-order neurons project to gustatory cortex. This nerve also carries visceral afferents from carotid sinus baroreceptors that synapse in the **caudal solitary nucleus**. The information about arterial blood pressure at the carotid bulb plays a role in reflex responses that ensure that blood pressure is maintained within certain limits and that cerebral blood flow is sufficient at all times.

Glossopharyngeal nerve lesions produce difficulty swallowing; impaired taste over the posterior one-third of the tongue and palate, impaired sensation over the posterior one-third of the tongue, palate, and pharynx; loss of the gag reflex; and unilateral decrease in saliva production. Glossopharyngeal nerve lesions rarely occur in isolation; injury may be a complication of carotid endarterectomy, a surgical procedure that removes plaque buildup inside the carotid artery with the goal of stroke prevention (see Chapter 18, "Stroke and Vascular Cognitive Impairment").

CN10: Vagus Nerves

The **vagus nerves** (CN10) send branches to widespread regions of the body, in the neck, thorax, and abdomen. The word *vagus* is Latin for "wandering" (as in "vagabond," "vague"). The vagus nerve is a mixed nerve and carries a parasympathetic component. It traverses the cranium through the jugular foramen, along with the glossopharyngeal (CN9) and accessory (CN11) nerves. It is associated with five cranial nerve nuclei.

Motor efferents from the **nucleus ambiguus** innervate the pharyngeal constrictor muscles and the intrinsic muscles of the larynx that mediate swallowing and lift the palate for speech. The nucleus ambiguus receives corticobulbar fibers that descend from premotor and motor cortex for voluntary control for speech and eating. The nucleus ambiguus also plays a role in coughing and vomiting (**emesis**) reflexes. Nausea and vomiting can be elicited by various sensory (e.g., taste, vestibular) stimuli, as well as cognitive stimuli giving rise to the emotion of disgust, which is under the influence of cortical and limbic structures.

Parasympathetic efferents originate from the **dorsal motor nucleus of vagus** and innervate the viscera to control gastrointestinal motility. The visceral afferent component of the vagus nerve is massive and relays sensory information from the viscera (thorax, abdomen, pharynx, and larynx) and receptors of the carotid body (which monitors oxygen tension) to the **caudal solitary nucleus**. These visceral signals are important for maintaining homeostasis. Visceral afferent signals generally do not reach consciousness, except for feelings of fullness and emptiness of the stomach (hunger and satiety), urinary bladder pressure, and bowel pressure.

Taste afferents from the epiglottis and root of the tongue synapse in the **rostral solitary nucleus**, converging with inputs from CN7 and CN9. Sensory afferents from the external ear project to the **spinal trigeminal nucleus**, converging with inputs from CN5, CN7, and CN9.

Complete interruption of the vagus nerve results in a characteristic syndrome consisting of ipsilateral soft palate drooping, ipsilateral lost gag reflex, **dysphonia** (hoarse, nasal voice), and **dysphagia** (difficulty swallowing). One of the leading causes of vagus nerve injury is complications of carotid endarterectomy and thoracic surgery.

Vagotomy is a surgical operation in which one or more branches of the vagus nerve are cut. It was used in the past to reduce the rate of gastric secretion in treating peptic ulcers; however, with the advent of acid secretion control medications, the need for surgical management of peptic ulcer disease has greatly decreased.

Vagus nerve stimulation therapy is used to treat medication-resistant epilepsy. Activation of vagal afferents through electrical stimulation leads to stabilization of seizure-related circuitry in the brain, although the specific therapeutic mechanism of action remains unknown. Vagus nerve stimulation is also used for treatment-resistant depression; the stimulation influences activity within limbic and higher cortical brain regions implicated in mood disorders.

CN11: Accessory Nerves

The **accessory nerves** (CN11), also referred to as the **spinal accessory nerves**, mediate head and shoulder movements. This nerve is a pure motor nerve. It exits the cranium through the jugular foramen, along with the glossopharyngeal (CN9) and vagus (CN10) nerves. It innervates two muscles of the neck ipsilaterally, (1) the **sternocleidomastoid** muscle, which tilts and rotates the head; and (2) the **trapezius** muscles, which moves the scapula and mediates shoulder elevation and arm adduction.

This cranial nerve is unique in that it originates from a nucleus located in the upper cervical cord (C1–C5) rather than the brain, and is the only cranial nerve to both enter and exit the cranium. Fibers from the **spinal**

accessory nucleus emerge between the dorsal and ventral roots of the spinal cord, ascend and enter the cranium through the foramen magnum, and then exit the cranium through the jugular foramen. This nerve innervates the ipsilateral sternocleidomastoid and trapezius muscles.

Lesions of CN11 produce weakness of the ipsilateral trapezius and sternocleidomastoid muscles, resulting in reduced shoulder shrug on the side of the lesion and reduced strength to rotate the head in the direction opposite the lesion. This nerve is susceptible to traumatic injury due to its long and superficial course.

CN12: Hypoglossal Nerves

The **hypoglossal nerves** (CN12) mediate tongue movements. CN12 is a pure motor nerve. It exits the cranium through the hypoglossal canal, a foramen in the occipital bone. Motor efferents originate from the **hypoglossal nucleus** and innervate the intrinsic and extrinsic muscles of the tongue ipsilaterally. The hypoglossal nucleus is under the influence of the pyramidal tract coming from the face region of the motor cortex of the contralateral hemisphere.

Hypoglossal nerve injury causes paralysis of one side of the tongue; when the person sticks out their tongue, it deviates toward the side of the lesion. Because of this, it also results in dysarthria (unclear articulation) and dysphagia.

Summary

The PNS comprises a vast network of nerves that carry afferents from the periphery to the CNS, and efferents from the CNS to the periphery; 31 pairs of spinal nerves emanate from the spine and 12 pairs of cranial nerves emanate from the cranium. PNS efferents innervating skeletal muscle make up the somatic motor system. PNS efferents innervating cardiac muscle, smooth muscle, and exocrine glands make up the autonomic system.

The spinal nerves provide motor and sensory innervation of the body and autonomic innervation of viscera. The spinal nerves are mixed, possessing both afferent and efferent fibers. The spinal nerves divide and recombine into peripheral nerves.

The cranial nerves provide motor and sensory innervation of the head and neck, innervation of the special sense organs, and parasympathetic innervation. Some cranial nerves are composed entirely of sensory afferent fibers, some are composed entirely of motor efferent fibers, and some are mixed, possessing both afferent and efferent fibers. Some cranial nerves carry parasympathetic fibers of the autonomic system. Each cranial nerve is associated with one or more cranial nerve nuclei.

A basic knowledge of the functional anatomy of the PNS helps form a foundation for understanding the neurocircuitry of sensory and motor functions and the methods of the neurological examination.

Additional Reading

1. Brodal P. Visceral efferent neurons: the sympathetic and parasympathetic divisions. In: Brodal P, *The central nervous system: structure and function*. 5th ed. Oxford University Press; 2016. https://oxfordmedicine.com/view/10.1093/med/9780190228958.001.0001/med-9780190228958-chapter-28. Accessed July 4, 2022.
2. Brodal P. The cranial nerves. In: Brodal P, *The central nervous system: structure and function*. 5th ed. Oxford University Press; 2016. https://oxfordmedicine.com/view/10.1093/med/9780190228958.001.0001/med-9780190228958-chapter-27. Accessed July 4, 2022.
3. Vanderah TW, Gould D. Sensory receptors and the peripheral nervous system. In: Vanderah TW, Gould D, *Nolte's the human brain: an introduction to its functional anatomy*. 8th ed., pp. 196–218. Elsevier; 2021.

The Spinal Cord

Introduction

The spinal cord serves as a conduit for efferent pathways leaving the brain en route to the effector organs (i.e., muscles and exocrine glands) of the body (below the head), and somatosensory pathways from the body en route to the brain; thus, it functions as the neural link between the brain and the body. Additionally, the spinal cord serves integrative and signal-processing functions and coordinates subconscious activities through spinal reflexes (such as immediate withdrawal of a limb from a potentially harmful stimulus). This chapter introduces the reader to spinal cord gross anatomy, segmental organization, and functional organization, as well as spinal reflexes and spinal cord injury syndromes.

Anatomy

External Gross Anatomy

The spinal cord is a long conical structure, approximately 1–1.5 centimeters (cm) in diameter in cross section (smaller than a dime) and 42–45 cm in length. It begins at the level of the foramen magnum at the base of the skull, where it is continuous with the medulla. Inferiorly, the spinal cord tapers off, terminating at the level of the intervertebral disc between the first and second lumbar vertebrae (i.e., L1 and L2 vertebrae). The tapered, terminal end of the spinal cord is the **conus medullaris** (**conus terminalis**). In adults, the spinal cord occupies only the upper two-thirds of the vertebral canal (see Figure 6.2 in Chapter 6).

The nerve roots of the lumbar and sacral segments run obliquely downward to reach their respective entry/exit points from the vertebral column through the intervertebral foramens, resulting in a collection of nerve roots called the **cauda equina** (horse's tail). This arrangement emerges over the course of prenatal development. In the first 3 months of fetal life, the spinal cord occupies the entire length of the vertebral canal, so

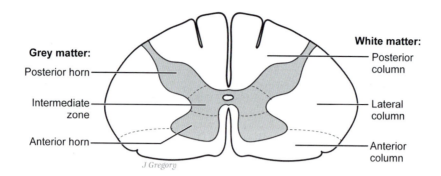

FIGURE 7.1. Cross section of the spinal cord showing fissures and sulci, an outer ring of white matter divided into posterior (dorsal), lateral, and anterior (ventral) columns on each side, and an inner core of gray matter that is divided into a posterior horn (dorsal horn), an anterior horn (ventral horn), and an intermediate zone on each side.

that the spinal cord segments and exiting roots match up alongside the corresponding intervertebral foramen. After that time, growth of the vertebral column exceeds that of the spinal cord, so that the nerves exiting the cord at more caudal levels take an oblique course before exiting the vertebral column.

The spinal cord has two major landmark surface indentations: the ventral median fissure and the dorsal median sulcus. The **ventral (anterior) median fissure** is a wide midline groove on the ventral side of the cord. The **dorsal (posterior) median sulcus** is a narrow midline furrow on the dorsal side of the cord. In addition, there is an **anterolateral sulcus** on each side of the cord from which the **dorsal roots** emerge, and a **posterolateral sulcus** on each side of the cord from which the **ventral roots** emerge. These surface fissures and sulci can be seen in a spinal cord cross section (Figure 7.1).

The cord is covered by three layers of meninges: dura, arachnoid, and pia. The spinal dura is bound loosely via connective tissue to the inner surface of the vertebral column, forming an epidural space; by contrast, the cranial dura is bound tightly to the inner surface of the skull and there is no actual epidural space.

Internal Gross Anatomy

Viewing the spinal cord in transverse section (i.e., 90 degrees to the longitudinal axis), we see that the spinal cord is made of an inner core of gray matter, which is surrounded by a ring of white matter that is composed of myelinated axons ascending and descending longitudinally along the length of the cord (see Figure 7.1). At the very center of the cord is the **central canal**, a remnant of the neural tube that is no longer patent. The central canal of the spinal cord is continuous with the central canal of the lower half of the medulla, which opens up into the fourth ventricle within the upper half of the medulla.

SPINAL CORD GRAY MATTER

In transverse cross section, the spinal cord gray matter is shaped like a butterfly. Each symmetrical half of the gray matter is divided into three main sections: a **dorsal horn (posterior horn)**, a **ventral horn (anterior horn)**, and an **intermediate zone**. The intermediate zone expands into a triangular extension known as the **lateral horn** at the thoracic and upper lumbar levels.

The gray matter of the spinal cord is divided into 10 regions that are histologically distinct. They are analogous to the cortical Brodmann areas (see Chapter 13, "The Cerebral Cortex") in that they are defined by their cellular structure and share a common function (Figure 7.2). These zones are the **Rexed laminae**. They are designated by Roman numerals (I–X); the numbering of the laminae progresses from dorsal to ventral (laminae I–IX). Lamina X, also known as the **gray commissure**, is a thin bridge of gray matter that surrounds the central canal and connects the gray matter on each side of the cord.

SPINAL CORD WHITE MATTER

In cross section, each symmetrical half of the white matter is divided into three zones: a **dorsal column (posterior**

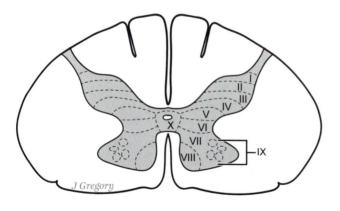

FIGURE 7.2. Laminae of Rexed.

column), a **ventral column (anterior column)**, and a **lateral column** (see Figure 7.1). Each dorsal column lies between the dorsal midline at the dorsal median sulcus and the emergence of the spinal nerve dorsal roots at the anterolateral sulcus. Each ventral column lies between the midline at the ventral median fissure and the emergence of ventral spinal nerve root at the posterolateral sulcus. Each lateral column lies between the dorsal and ventral roots (i.e., between the anterolateral and posterolateral sulci).

The columns, also known as **funiculi**, consist of an aggregate of **fasciculi** (tracts) that run longitudinally up or down the spinal cord (Figure 7.3). Some of the fasciculi carry afferents to the CNS and ascend the spinal cord toward the brain; some of the fasciculi carry efferents from the CNS and descend the spinal cord from the brain; and some fasciculi are mixed (i.e., bi-directional). Many of the ascending tracts bear a name with the prefix *spino-* (e.g., spinothalamic, spinocerebellar) indicating their site of origin, while many of the descending tracts bear a name with the suffix *-spinal* (e.g., corticospinal, vestibulospinal) indicating their site of termination.

The dorsal columns carry ascending (afferent) tracts exclusively, the descending columns carry predominantly descending (efferent) tracts, and the lateral columns carry both ascending and descending tracts. Thus, ascending tracts are present in all three columns, and descending tracts are present in the lateral and ventral columns.

The dorsal columns carry the **gracile** (Latin, "slender, graceful") **fasciculus**, located medially, and the **cuneate** (Latin, "wedge-shaped") **fasciculus**, located laterally. The ventral columns primarily carry descending tracts: these are the ventral (anterior) corticospinal tract, the ventral (anterior) spinothalamic tract, the tectospinal tract, the vestibulospinal tract, and the reticulospinal tract. The lateral columns carry the lateral corticospinal tract, lateral spinothalamic tract, rubrospinal tract, and the spinocerebellar tracts (i.e., dorsal and ventral).

Immediately adjacent to the spinal cord gray matter is a thin zone of **intersegmental tracts** containing both ascending and descending fibers that run from one segment of the cord to another, forming **spino-spinal** connections. Within each half of the cord there is a **dorsal intersegmental tract**, a **lateral intersegmental tract**, and a **ventral intersegmental tract**. These tracts establish the circuitry for intersegmental spinal reflexes (see "Spinal Reflexes," below).

Blood Supply

The blood supply to the spinal cord is provided by a single **anterior spinal artery** and two **posterior spinal arteries**, all of which arise from the vertebral arteries. The anterior spinal artery descends the length of the anterior spinal cord within the anterior median fissure. The posterior spinal arteries descend along the dorsolateral spinal cord. The anterior spinal artery supplies the anterior two-thirds of the spinal cord, and the two posterior spinal arteries supply the posterior one-third of the spinal cord.

Segmental Organization

The spinal cord is segmentally organized according to the region of the body that it monitors and regulates by way of the spinal nerves. There are 31 spinal cord segments, corresponding to the 31 spinal nerve pairs that enter/exit the cord at the intervertebral foramens. Thus, each spinal cord segment is a "slice" of spinal cord that is associated with a single pair of spinal nerves (see Figure 6.2 in Chapter 6). The spinal cord segments are named and numbered according to attachment of the spinal nerves; they are C1–C8, T1–T12, L1–L5, S1–S5, and Co1. Like the vertebral

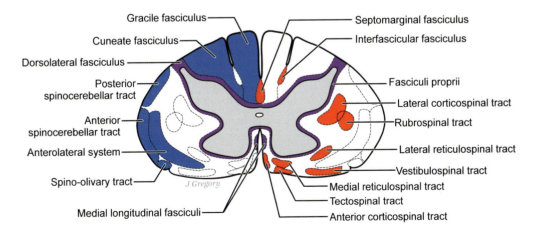

FIGURE 7.3. Fasciculi (tracts) that run longitudinally along the length of the cord. Ascending tracts carrying sensory afferents to the brain are indicated in blue on the left side of the figure, descending tracts carrying efferents from the brain are indicated in red on the right side of the figure, and mixed tracts are indicated in purple.

column and spinal nerves, the spinal cord is divided into four different regions: cervical, thoracic, lumbar, and sacral/coccygeal.

Regional Differences in Spinal Cord Anatomy

Spinal cord anatomy varies at different levels with respect to the relative proportion of gray and white matter, and the size and shape of the gray matter (Figure 7.4).

From lower to progressively higher levels of the spinal cord, there is a sequential addition of afferent axons as well as a loss of descending efferent axons, as more afferents are brought into the cord and more efferents leave the cord via the spinal nerves. This results in a greater volume of white matter relative to gray matter moving from lower sacral to upper cervical spinal segments.

The amount of gray matter present at any given level of the spinal cord is related to the amount of tissue innervated (both motor and sensory) at that level. Thus, its size is greatest in the regions of the spinal cord that innervate both trunk and limbs, as compared to those regions that innervate only the trunk. These are the **cervical and lumbar enlargements**, present at the C3–T1 and L1–S2 segments, respectively. In humans, the cervical segments are characterized by bulging ventral horns, indicative of the large number of motor neurons dedicated to the upper limbs. By contrast, the thoracic segments are characterized by thin dorsal and ventral horns. They have the least amount of gray matter due to the lesser innervation density (i.e., number of receptors and effectors per unit area) of the thorax, abdominal wall, and mid-back. The dorsal horns of the sacral segments are also particularly large, due to the large number of sensory fibers from the genitals.

The lateral horns are present at the T1–L2 levels; they provide preganglionic sympathetic innervation.

Functional Organization of the Spinal Cord

A general principle of spinal cord organization is that the dorsal portion of the spinal cord serves sensory functions, while the ventral portion serves somatic motor functions. The intermediate zone and lateral horns serve autonomic functions.

Spinal Cord Gray Matter

Afferents conveying somatosensory information enter the spinal cord via the spinal nerve dorsal roots and arrive at the dorsal horns. The dorsal horn contains laminae I to VI. Laminae I to IV are concerned with tactile (i.e., skin) sensation. Laminae V and VI are concerned with proprioceptive (i.e., movement and position) sensation.

Motor neuron cell bodies lie within laminae VIII–IX of the ventral horn. These neurons innervate striate muscle (i.e., they are general somatic efferents). They exit the ventral horn via the spinal nerve ventral roots and innervate skeletal muscle on the ipsilateral side of the body. Motor neurons that directly innervate muscle are called **lower motor neurons**. These lower motor neurons receive inputs from **upper motor neurons** which make up the descending motor pathways that originate in the brain and project to the spinal cord (e.g., the corticospinal tract).

Lamina VII, also known as the **intermediate zone**, is located between the dorsal and ventral horns. It contains the visceral motor neuron cell bodies of the autonomic nervous system; these are the preganglionic neurons that project to autonomic ganglia, which in turn give rise to

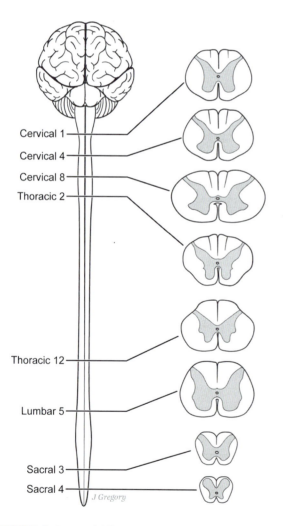

FIGURE 7.4. Regional differences in gross spinal cord anatomy. Moving from lower to higher levels there is a sequential addition of afferent axons, and from higher to lower levels there is a loss of descending efferent axons, resulting in a greater volume of white matter relative to gray matter at higher levels of the cord.

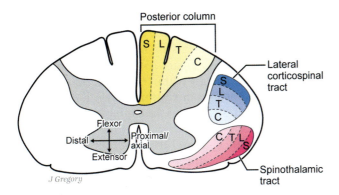

FIGURE 7.5. Somatotopic organization of the ventral horn gray matter and the ascending sensory and descending motor white matter pathways of the spinal cord. Cervical (C); thoracic (T); lumbar (L); sacral (S).

postganglionic neurons that project to the target organ (see Figure 6.1 in Chapter 6). At the T1–L2 spinal segments, the intermediate zone expands to form the lateral horns, small triangular extensions that give rise to preganglionic sympathetic fibers. The intermediate zone of the S2–S4 spinal segments gives rise to preganglionic parasympathetic fibers, but these do not form distinct lateral horns. The intermediate zone also receives visceral afferents and contains interneurons that receive sensory inputs from lamina II to VI.

The gray matter of the spinal cord also has a somatotopic organization; **somatotopy** is the point-to-point mapping of the body's somatosensory surface or musculature within the CNS (Figure 7.5). In the ventral horn gray matter, motor neuron cell bodies innervating the **axial muscles** (i.e., the trunk) are located medially and are present at all spinal segments, while those innervating the **appendicular muscles** (i.e., the limbs) are located laterally and are only present within the segments of the cervical and lumbosacral enlargements. Motor neuron cell bodies innervating the **flexor muscles** (i.e., that decrease the angle between bones on two sides of a joint with contraction, such as the biceps) are located dorsally, while those innervating the **extensor muscles** (i.e., that increase the angle between bones on two sides of a joint with contraction, such as the triceps) are located ventrally.

Spinal Cord White Matter

ASCENDING PATHWAYS

The ascending tracts of the spinal cord convey conscious and subconscious somatosensory signals from skin, muscles, joints, and viscera of the body to the brain. They also play a role in reflexes and regulation of muscle tone. Somatosensation comprises several submodalities: touch, proprioception (position and movement sense), vibratory sensation, temperature, and pain. The sense of touch (**tactile sensation**) originates from the skin and informs us of pressure at the surface of the body; it is an **exteroceptive** sense (i.e., it informs about external stimuli). Tactile sensation consists of **fine touch** (**discriminative touch**) that enables one to sense the stimulus, and **crude touch** (**nondiscriminative touch**) that enables one to sense the stimulus but not localize it. The difference between the two follows from differences in the tactile receptors with respect to sensitivity, the size of the area that they gather information from (i.e., receptive field size), and how deeply they lie within the tissue they innervate. **Proprioception** is the sense of stationary position and movement (**kinesthesia**); it originates from receptors in the muscle, tendons, and joints and is an **interoceptive** sense (i.e., it informs about internal stimuli) that we are largely unaware of. Vibratory sensation (**pallesthesia**) is the ability to perceive mechanical oscillations on or in the body. This is the sense that allows us to feel a mobile phone vibrating in our pocket, but it also allows us to detect vibrations that propagate through solid surfaces that the body is in contact with, such as the floor, as may be caused by an approaching train, approaching herd of elephants, approaching footsteps, or an earthquake, even if inaudible. This sense has an even greater significance in marine animals such as sharks that can detect successive fluid compressions and rarefactions in the water surrounding them due to movement of potential predators and prey in the immediate vicinity. In humans, vibratory sensation arises from receptors located both superficially at the level of the skin and in the deeper tissues of muscles, ligaments, joints, and bone; it is tested clinically with a vibrating tuning fork. Temperature sensation (**thermosensation, thermesthesia**) is our ability to detect warmth and cold. Pain sensation (**nociception**) is triggered by physical tissue damage due to mechanical stimuli, extreme heat and cold, and chemicals released by damaged tissue.

Two ascending systems of particular clinical importance are the dorsal column-medial lemniscal system and the anterolateral system. These somatosensory systems differ with respect to their location within the spinal cord funiculi and the level of the CNS at which they decussate.

The **dorsal column–medial lemniscal system** carries the somatosensory modalities of discriminative touch, proprioception, and vibratory sensation. The primary afferent neurons of this system carry information from the body into the spinal cord, and ascend ipsilaterally within the dorsal columns as the gracile and cuneate fasciculi. These tracts synapse in the brainstem (although they give rise to axon collaterals that synapse with the spinal cord dorsal horn gray matter). Second-order neurons originating from brainstem nuclei decussate en route to the thalamus, which in turn projects via third-order neurons to the primary somatosensory cortex in the parietal lobe. The decussation results in a contralateral representation of the somatosensory surface within the parietal lobe. Although discriminative touch, proprioception, and vibration senses share a spinal cord and brainstem pathway, they arise in different receptors and terminate in divergent pathways within the thalamus and cerebral cortex.

The **anterolateral system** consists of two separate tracts: the **lateral spinothalamic tract** which carries the somatosensory modalities of pain and temperature, and the **anterior spinothalamic tract** which carries the sensory modality of crude touch. In the **lateral spinothalamic pathway**, the primary afferent neurons carry information from the body into the spinal cord and synapse within the spinal cord in the dorsal horn gray matter. Second-order neurons decussate within the anterior commissure of the spinal cord (which lies immediately anterior to the gray commissure, Rexed lamina X), and then ascend in the lateral columns toward the brain. Thus, the lateral spinothalamic tract carries information from the contralateral side of the body.

DESCENDING PATHWAYS

The descending tracts of the spinal cord carry efferent signals to the effector skeletal muscles of the somatic system, and to the effectors of the autonomic somatic system (cardiac muscle, smooth muscle, and exocrine glands). Descending tracts influencing the activity of the skeletal muscles of the body control voluntary movement, reflexes, muscle tone, posture, and balance. These descending pathways include the corticospinal, rubrospinal, tectospinal, vestibulospinal, and reticulospinal pathways. Descending autonomic pathways innervating cardiac muscle, smooth muscle, and glands originate from the hypothalamus and the brainstem reticular formation.

A descending tract of particular clinical importance is the **lateral corticospinal tract**, which carries 75%–90% of all corticospinal fibers. These fibers originate from motor cortex, decussate in the internal capsule, descend in the lateral column contralateral to their origin, and synapse either directly or indirectly (via interneurons) on motor neurons within the ventral horn ipsilateral to their side of descent in the spinal cord. This pathway mediates voluntary movement.

SOMATOTOPY

The dorsal columns, lateral corticospinal tract, and lateral spinothalamic tract all have a somatotopic organization (see Figure 7.5). Primary motor cortex and primary somatosensory cortex have somatotopic maps, and it of course makes sense that lower portions of the pathway also would have somatotopy (because somatotopy at the cortical level requires somatotopy at the subcortical level).

Regarding the dorsal columns, as one moves from the caudal to the rostral end of the spinal cord, nerve fibers are added to the dorsal columns laterally, resulting in a somatotopic representation such that sacral, lumbar, thoracic, and cervical segments are represented in a medial-to-lateral sequence (see Figure 7.5). The gracile fasciculus, which is located medially within the dorsal columns, carries ascending fibers from vertebral levels T7 and below (thus it is present throughout the cord), serving the lower limbs. The cuneate fasciculus, which is located laterally within the dorsal columns, carries ascending fibers from vertebral levels T6 and above, serving the upper limbs. The lateral corticospinal tracts also have a somatotopic organization, such that cervical, thoracic, lumbar, and sacral segments are represented in a medial-to-lateral sequence.

Spinal Reflexes

A **reflex** is an automatic response to a stimulus that is quick, stereotyped, and involuntary. Reflexes are mediated by neural pathways called **reflex arcs**, consisting of a sensory neuron, a motor neuron, and few (if any) interneurons. Reflex responses are quick because they involve few synapses, so there is minimum synaptic delay. They are stereotyped in that they are predictable and repeatable under the same stimulus conditions. They are involuntary; they occur without conscious intention, often without awareness, and are difficult to suppress.

A **spinal cord reflex** is a reflex that is carried out by a neural circuit within the spinal cord and occurs completely independently of the brain (Figure 7.6). The spinal reflex arc consists of sensory afferents that enter the spinal cord and activate spinal motor neurons, either directly or indirectly through one or more spinal interneurons. The reflex is evoked (triggered) by stimulation of the afferent (sensory) limb of the reflex arc, which elicits activity in the efferent limb of the reflex arc and causes muscle contraction.

When a reflex arc consists of only two neuronal components, with sensory afferents synapsing directly on motor efferents, it is known as a **monosynaptic** reflex arc. Reflexes that are mediated by monosynaptic reflex arcs are known as **monosynaptic reflexes**. The **knee-jerk reflex** (**patellar reflex**, **myotactic stretch reflex**) is a monosynaptic spinal reflex. When the leg is bent (as when sitting), a sharp tap on the patellar tendon (just below the kneecap) results in a sudden extension of the lower leg (see Chapter 24, "The Motor System and Motor Disorders").

Most reflex arcs, however, are **polysynaptic**. An example of a polysynaptic spinal reflex is the **withdrawal reflex**, which serves the purpose of protecting the body from damaging stimuli. A **noxious** stimulus that is potentially tissue-damaging (e.g., a hot stove) stimulates **nociceptors** (pain receptors) in the skin and triggers a withdrawal flexion movement. Reflexes and reflex testing

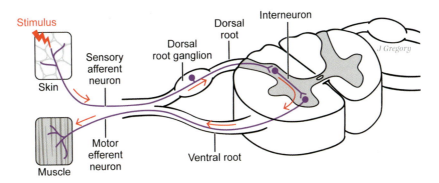

FIGURE 7.6. Reflexes are mediated by neural pathways called reflex arcs, consisting of sensory neurons, motor neurons, and in some cases interneurons. A spinal reflex arc consists of sensory afferents that enter the spinal cord and activate spinal motor neurons, either directly or indirectly through one or more spinal interneurons.

are discussed further in Chapter 30, "The Neurological Examination."

Spinal reflexes can also be classified as segmental or intersegmental. **Segmental spinal reflexes** are mediated by circuits in which the input and output occur at the same level of the spinal cord. **Intersegmental spinal reflexes** are mediated by circuits in which the input and output occur at different levels of the spinal cord (e.g., stepping barefoot on a piece of glass causes withdrawal of the foot by contraction of abdominal and hip muscles). Intersegmental reflexes involve the intersegmental tracts that interconnect different segmental levels of the cord; intersegmental spinal circuitry plays a particularly important role in maintaining muscle tone and posture.

While the circuitry of spinal reflexes is confined to the cord, supraspinal (i.e., brain) centers can modulate the activity of spinal reflex circuits through descending pathways.

Spinal Cord Injury Syndromes

Peripheral nerve, spinal nerve, and spinal cord pathologies all cause sensory and/or motor abnormalities consisting of motor paralysis or paresis (weakness), loss of or diminished sensation, changes in reflexes, and/or autonomic dysfunction. The pattern of motor and sensory loss varies, depending on whether the pathology is located at the level of peripheral nerve, spinal nerve, or spinal cord.

With peripheral nerve lesions, the distribution of signs and symptoms follows the areas innervated by the specific nerves. Spinal nerve pathology produces neurologic dysfunction only in the areas supplied by the affected spinal nerve(s), which each arise from one segment of the spinal cord, providing sensation for a dermatome (the area of skin innervated by a single spinal nerve pair or spinal cord segment) and motor innervation to a myotome (a muscle group innervated by a single spinal nerve pair or spinal cord segment) (see Chapter 6, "The Peripheral Nervous System").

In contrast, spinal cord pathology produces a loss of neurologic function at and below the level of the lesion, as transmission of ascending sensory signals and descending motor signals beyond the affected spinal cord segment is disrupted. Spinal cord pathology arises from a variety of causes, including trauma, ischemia, tumors, and neurodegenerative disease. **Spinal cord injury** (SCI) produces a variety of patterns of deficit, depending on the level of the injury, whether the injury is complete or incomplete, and the nerve tracts that are damaged.

In a **complete SCI** the spinal cord is completely transected and there is a complete loss of sensation and movement below the injury level, affecting both sides of the body equally. The higher the level of the injury, the greater the extent of the neurologic loss. **Paraplegia** affects the lower half of the body, including the lower extremities. Paraplegia occurs with injuries at the T1 spinal segment level or below. **Quadriplegia** (**tetraplegia**) affects all four limbs plus the torso. Quadriplegia occurs with SCIs above the T1 spinal segment (i.e., at the cervical level). Injuries at or above the C4 spinal segment will also affect innervation of the diaphragm; those with such injuries require mechanical breathing assistance with a ventilator.

With **incomplete SCI**, there is some remaining neurological function below the injury level. There are several incomplete spinal cord syndromes: anterior cord syndrome, posterior cord syndrome, Brown-Séquard syndrome, and central cord syndrome (Figure 7.7).

In the **anterior cord syndrome**, the damage disproportionately affects the anterior (ventral) spinal cord. This usually occurs with infarction resulting from occlusion of the anterior spinal artery, and involves the anterior two-thirds of the cord. Anterior cord syndrome is characterized by a complete loss of movement below the level of the lesion due to involvement of the ventral horn, loss of pain

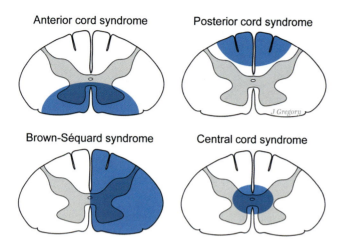

FIGURE 7.7. There are several incomplete spinal cord injury syndromes: anterior cord syndrome, posterior cord syndrome, Brown-Séquard syndrome, and central cord syndrome.

and temperature sensation below the level of the lesion due to involvement of the anterior commissure (which carries the decussating fibers) and possibly the lateral spinothalamic tract, and autonomic dysfunction below the level of the lesion. As the posterior columns are spared, so are discriminative touch, proprioception, and vibratory sensation.

In the **posterior cord syndrome**, the damage disproportionately affects the posterior (dorsal) spinal cord. There is a loss of discriminative touch, proprioception, and vibratory sensation due to involvement of the dorsal horns and dorsal columns, with preserved movement, and preserved pain and temperature sensation.

In the **Brown-Séquard syndrome**, the damage involves one side of the spinal cord only (i.e., **hemisection, hemitransection**). This is a rare form of incomplete spinal cord injury that is most often caused by penetrating traumatic SCI, such as bullet or knife wounds, but it is also caused by spinal cord tumors and other conditions. This syndrome is characterized by (1) ipsilesional weakness or paralysis below the level of injury due to disruption of the lateral corticospinal tract (**hemiparaplegia**); (2) ipsilesional loss of discriminative touch, proprioception, and vibration sensation below the level of injury due to disruption of the dorsal columns; and (3) contralesional loss of pain and temperature sensation below the level of injury due to disruption of the lateral spinothalamic tract (which has already decussated within the cord below the level of the lesion). In practice, "pure" Brown–Séquard syndrome is rare because most lesions of the spinal cord are irregular.

In the **central cord syndrome**, the damage involves the central portion of the spinal cord. The pattern of neurological deficit depends on the size of the lesion. With small lesions, the anterior white commissure of the spinal cord carrying decussating spinothalamic tract fibers is selectively involved, resulting in loss of pain and temperature sensation bilaterally at the affected levels. Below these levels, cord function is intact. Larger lesions result in a loss of movement that is greater in the upper extremities than the lower extremities due to the anatomical organization within the corticospinal tract, where fibers innervating the arms are located medially and fibers innervating the legs are located laterally. Central cord syndrome usually occurs with trauma that affects the cervical cord (e.g., forceful hyperextension of the neck that occurs when one falls forward and hits their chin on the ground).

Epidural and Spinal Anesthesia

Injection of local anesthetic into the epidural space around the spinal cord blocks signal conduction in the nerve roots and ascending and descending cord tracts, resulting in anesthesia (loss of sensation) and paralysis at the level of the injection and below. This procedure is known as **epidural anesthesia** and is commonly used for childbirth. Local anesthetic agents can also be injected into the lumbar cistern to produce a **spinal anesthesia** (see Figure 4.7 in Chapter 4).

Summary

The spinal cord functions as the link between the brain and the body, carrying descending motor and other efferent pathways from the brain to the body and ascending somatosensory pathways from the body to the brain. It also serves integrative and signal processing functions. In transverse section the spinal cord is composed of an inner core of gray matter that is surrounded by a ring of white matter carrying ascending and descending pathways. Thirty-one spinal nerve pairs enter/exit the cord, and each pair defines a "slice" of spinal cord known as a spinal cord segment. A general principle of spinal cord organization is that the dorsal portion serves sensory functions, the ventral portion serves somatic motor functions, and the intermediate zone and lateral horns serve autonomic functions. Additionally, there is somatotopic organization of the gray and white matter.

Spinal cord injury syndromes follow in a straightforward manner from the functional anatomy of the spinal cord. Full transection of the cord (complete SCI) results in a complete loss of sensation and movement below the injury level, affecting both sides of the body equally; paraplegia affects the lower half of the body and occurs with injuries at the T1 spinal segment level or below, while quadriplegia affects all four limbs and the torso and occurs with injuries above the T1 spinal segment. With partial transection of the cord (incomplete SCI) there is some remaining neurological function below the level of the injury; incomplete SCI syndromes include the anterior cord syndrome, posterior cord syndrome, Brown-Séquard syndrome, and central cord syndrome.

Additional Reading

1. Diaz E, Morales H. Spinal cord anatomy and clinical syndromes. *Semin Ultrasound CT MR.* 2016;37(5):360–371. doi:10.1053/j.sult.2016.05.002
2. Novy J. Spinal cord syndromes. *Front Neurol Neurosci.* 2012;30:195–198. doi:10.1159/000333682
3. Vanderah TW, Gould D. Spinal cord. In: Vanderah TW, Gould D, *Nolte's the human brain: an introduction to its functional anatomy.* 8th ed., pp. 221–256. Elsevier; 2021.
4. Waxman SG. The spinal cord. In: Waxman SG, *Clinical neuroanatomy.* 29th ed. McGraw-Hill; 2020. https://accessmedicine-mhmedical-com.proxy.library.nyu.edu/content.aspx?bookid=2850&sectionid=242763051. Accessed July 4, 2022.

8

The Brainstem

Introduction

The brainstem is the lowermost, stalk-like part of the brain. It is said to have received its name from German anatomists in the nineteenth century who used it as a handle to lift the fixed brain from a jar (fixation is the chemical preservation of tissue by fixatives such as formaldehyde for subsequent histological analyses). Phylogenetically, the brainstem is the oldest part of the human brain, similar in structure and function to the brains of reptiles and fish. The brainstem controls basic functions necessary for survival, such as heart rate, breathing, and digestion, as well as sleep-wakefulness and level of arousal in higher animals with cerebral cortex (i.e., mammals). Extensive damage to the brainstem usually results in death, while lesser degrees of damage result in coma and persistent vegetative state. Focal lesions involve various combinations of impaired function of the cranial nerve nuclei and long ascending and descending tracts, forming the basis of brainstem syndromes. This chapter describes basic brainstem anatomy and function, and clinical considerations related to brainstem pathology.

Basic Brainstem Anatomy

The brainstem is located within the posterior cranial fossa, alongside and ventral to the cerebellum. When examining an intact human brain specimen, only the inferior surface of the brainstem can be seen; the other surfaces are covered by the cerebrum and cerebellum.

Moving in a rostro-caudal direction, the brainstem consists of the midbrain, pons, and medulla (see Figure 1.7 in Chapter 1). The caudal end of the medulla is continuous with the spinal cord; the transition between the medulla and the spinal cord is marked by the foramen magnum of the occipital bone. The cerebellum is connected to the brainstem by three pairs of cerebellar peduncles: the superior cerebellar peduncles connect to the midbrain; the middle cerebellar peduncles connect to the pons; and the inferior cerebellar peduncles connect to the medulla.

The cerebral aqueduct and fourth ventricle run through the brainstem, with the cerebral aqueduct at the level of midbrain and the fourth ventricle at the level of the pons and medulla (Figure 8.1). The floor of the fourth ventricle is formed by the pons and

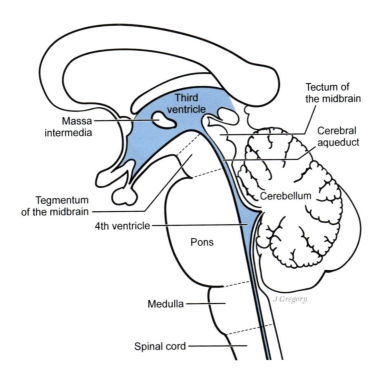

FIGURE 8.1. Midsagittal section showing that the cerebral aqueduct and fourth ventricle run through the brainstem. The cerebral aqueduct runs through the midbrain, separating the midbrain tectum (roof) from the tegmentum (floor).

medulla, while the roof of the fourth ventricle is formed by two **vela** (veil-like membranous partitions) underlying the cerebellum: the **superior medullary velum (anterior medullary velum)** and the **inferior medullary velum (posterior medullary velum)**. Cranial nerves 3–12 emerge from the brainstem (see Figure 6.5 in Chapter 6).

In the brainstem, gray matter is organized into nuclear groups that are subdivided by bands of white matter. The brainstem contains many different functional areas that are tightly packed and can be classified into four main types of neural elements: (1) long somatosensory tracts ascending from the spinal cord up to brain, long motor tracts descending from the higher brain centers to lower brain centers and from brain motor centers to the spinal cord, and nuclei associated with these sensory and motor pathways; (2) nuclei and tracts of CN3–CN12; (3) nuclei regulating the sleep-wake cycle and level of consciousness; and (4) nuclei regulating vital life functions such as cardiovascular and respiratory reflexes. Nuclei regulating level of consciousness, the sleep-wake cycle, and vital life functions are components of the reticular formation, which runs throughout the brainstem (see "The Reticular Formation," below).

Ascending Somatosensory Tracts

The major ascending tracts of the brainstem are: (1) the gracile and cuneate fasciculi and medial lemnisci, (2) the lateral and anterior spinothalamic tracts, and (3) the posterior and anterior spinocerebellar tracts. The **gracile and cuneate fasciculi** and **medial lemnisci** carry discriminative touch and proprioceptive signals. The lateral and anterior **spinothalamic tracts** carry pain and temperature signals. The posterior (dorsal) and anterior (ventral) **spinocerebellar tracts** carry somatosensory inputs to the cerebellum.

Descending Motor Tracts

Major descending tracts of the brainstem include: (1) the corticospinal tracts (lateral and ventral), (2) the corticobulbar tracts, and (3) the corticopontine projections. The **corticospinal tracts** carry motor signals from primary motor cortex to the ventral horn of the spinal cord, synapsing on motor neurons controlling voluntary movements by the skeletal muscles of the trunk and limbs. The **corticobulbar tracts** carry motor signals from primary motor cortex to the cranial nerve motor nuclei for CN5 (trigeminal), CN7 (facial), CN9 (glossopharyngeal), CN10 (vagus), and CN12 (hypoglossal), located in the pons and medulla; CN11 (accessory) originates from the cervical spinal cord but exits the cranium at the level of the medulla. These nuclei mediate control of the muscles of the face and neck, and are used in chewing, facial expression, speech, and swallowing. The corticobulbar fibers accompany the corticospinal axons in their descent from primary motor cortex through the internal capsule and cerebral peduncles, and then gradually diverge as they enter the pons and medulla to terminate in their target cranial nerve nuclei.

Other descending motor pathways that travel through the brainstem and influence the activity of motor neurons in the spinal cord include the corticopontine, rubrospinal, vestibulospinal, tectospinal, and reticulospinal tracts. The corticopontine projections go from motor cortex to deep pontine nuclei, which in turn project to the cerebellum via pontocerebellar fibers traveling in the middle cerebellar peduncle; this cortico-ponto-cerebellar pathway provides a substrate for cerebro-cerebellar communication and is critical to the cerebellum's function in movement coordination (see "Ventral Pons," below). The rubrospinal tracts originate from the red nuclei of the midbrain and terminate at the level of the cervical segments of the spinal cord; these tracts are functionally specialized for motor control of the upper limbs (see "The Midbrain Tegmentum," below). The vestibulospinal tracts originate from the vestibular nuclei, a group of nuclei in the lower pons and upper medulla that receive vestibular signals from the vestibular nerve component of the vestibulocochlear nerve; the vestibulospinal pathways increase antigravity extensor muscle tone in response to vestibular signals resulting from head movement. The tectospinal tracts originate from the midbrain tectum; these tracts mediate postural reflexes in response to visual and auditory stimuli (see "The Midbrain Tectum," below). The reticulospinal tracts originate from the pontine and medullary reticular formation; the pontine reticulospinal tracts innervate extensor muscles and drive muscle tone, while the medullary reticulospinal tracts inhibit extensor muscle tone and activate flexor muscles.

Brainstem Blood Supply

The main supply of blood to the brainstem is provided by small penetrating arteries that arise from the basilar and vertebral arteries (i.e., posterior circulation). Because the small penetrating vessels have narrow lumens and there is little or no collateral circulation, the brainstem is vulnerable to ischemic stroke.

The Medulla

The **medulla oblongata** (*oblong marrow*) is the caudal-most portion of the brainstem. Situated between the spinal cord and the pons, it is connected to the cerebellum by the inferior cerebellar peduncles.

The ventral surface of the medulla has two raised triangular bumps, the **medullary pyramids** (Figure 8.2). The pyramids contain descending motor fibers, the majority of which are part of the corticospinal tract. These fibers originate from motor cortex, descend through the brain, and on their way to the spinal cord decussate (cross) at the base of the pyramids in the **pyramidal**

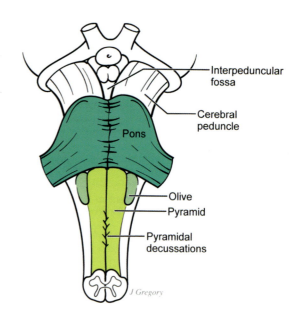

FIGURE 8.2. Ventral surface of the brainstem. The medullary pyramids carry the corticospinal tracts. The pyramidal decussations mark the transition from the medulla to the spinal cord. The ventral surface of the pons has a ridged appearance due to transverse pontine fibers that give rise to the middle cerebellar peduncles. The cerebral peduncles carry large motor tracts descending from the cerebrum (corticospinal, corticopontine, and corticobulbar).

decussation; this anatomical landmark can be seen on the ventral surface of the medulla, and it marks the transition from the medulla to the spinal cord. Pathology affecting the corticospinal tract above the level of the decussation produces paralysis or **paresis** (weakness) on the contralateral side of the body, while pathology affecting the corticospinal tract below the level of the decussation produces paralysis or paresis on the ipsilateral side of the body.

When the cerebellum is dissected away from the brainstem and the dorsal surface of the brainstem is viewed, the floor of the fourth ventricle becomes visible (Figure 8.3). The floor of the fourth ventricle is also referred to as the **rhomboid fossa** because it is a diamond-shaped depression, due to its narrowing rostrally into the cerebral aqueduct and caudally into the central canal of the spinal cord. The most caudal end of the fourth ventricle is the **obex**, an internal landmark that marks the transition from the medulla to the spinal cord.

Somatosensory Nuclei

The dorsal column–medial lemniscal pathways, carrying touch and proprioceptive signals from the periphery to the brain, synapse in the **dorsal column nuclei** (the **cuneate nucleus** and the **gracile nucleus**) located at the caudal end of the medulla.

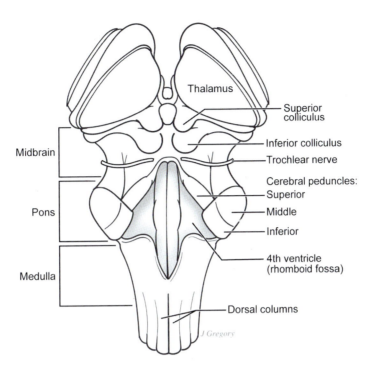

FIGURE 8.3. Dorsal surface of the brainstem. The floor of the fourth ventricle, also known as the rhomboid fossa, narrows rostrally into the cerebral aqueduct and caudally into the central canal. The dorsal surface of the midbrain contains the superior and inferior colliculi. The trochlear nerve (CN4) exits from the dorsal surface of the midbrain.

Cranial Nerve Nuclei

CN9 (glossopharyngeal), CN10 (vagus), and CN12 (hypoglossal) exit from the medulla. The medulla contains cranial nerve motor nuclei that control the muscles of the tongue, pharynx, and larynx (voice box), as well as reflex centers for vomiting, coughing, sneezing, and swallowing (see Table 6.1 in Chapter 6).

The medulla was once referred to as the bulb. In modern clinical usage, however, the word **bulbar** is used to refer to the brainstem in general (e.g., corticobulbar), or specifically to the medulla or the lower cranial nerves connected to the medulla (CN9–CN12). **Bulbar palsy** is a clinical syndrome caused by dysfunction of the cranial nerve efferents originating from the medulla (i.e., CN9–CN12), which is characterized by weakness of the tongue, pharynx, and/or larynx muscles. The most prominent symptoms are **dysphagia** (difficulty swallowing) and **dysarthria** (unclear articulation of speech).

Nuclei Regulating Vital Life Functions

The medulla contains control centers for cardiac, respiratory, and vasomotor functions, and plays a critical role in regulating autonomic (involuntary) functions such as heart rate, breathing, and blood pressure. Pathology of the medulla, therefore, is often fatal. Some cases of sudden unexpected death in healthy persons have been found to be due to a medullary brain lesion on postmortem examination.

The Pons

The **pons** (bridge) lies between the medulla and the midbrain, and ventral to the cerebellum. When viewed from the ventral surface, the pons appears as a broad bulge with transverse ridges (see Figure 8.2). A shallow sulcus runs along the midline of the ventral surface of the pons; this is the **basilar sulcus**, and the basilar artery runs within this sulcus. CN5–CN8 emerge from the pons. The dorsal surface of the pons forms part of the floor of the fourth ventricle (see Figure 8.3). The pons is connected to the overlying cerebellum by the middle cerebellar peduncles. The **cerebellopontine angle cistern** (also referred to simply as the *cerebellopontine angle*) is a triangular cistern located lateral to the pons and ventral to the cerebellum. The pons is broadly divided into two regions: the ventral pons and the dorsal pons.

Most of the blood supply to the pons comes from the **pontine arteries**, which arise from the basilar artery.

The Ventral Pons

The **ventral pons** (**basilar pons**, **basis pontis**) occupies the ventral two-thirds of the pons. It contains **deep pontine**

nuclei that give rise to axons that form the **transverse pontine fibers**; these fibers give the external ventral surface of the pons its transversely ridged appearance. The transverse pontine fibers cross the midline and project to the contralateral cerebellar hemisphere via the **middle cerebellar peduncles**, which consist exclusively of these **pontocerebellar** projections. The pontine nuclei and transverse pontine fibers therefore play an important role in movement. The pontine nuclei receive their input via **corticopontine** projections from the ipsilateral primary motor cortex. Thus, there is a **cortico-ponto-cerebellar** pathway that originates from primary motor cortex and projects via corticopontine projections to ipsilateral pontine nuclei, which in turn project via pontocerebellar projections through the transverse pontine fibers and the middle cerebellar peduncle to the contralateral cerebellum. These inputs provide the cerebellum with information about volitional movements that are anticipated or are in progress.

The basilar pons also contains descending corticospinal and corticobulbar tracts that originate from the cerebral cortex. These projections run longitudinally with respect to the neuraxis (i.e., cranio-caudally), and perpendicularly to the orientation of the transverse pontine fibers.

The Dorsal Pons

The **dorsal pons** (**pontine tegmentum**) houses the cranial nerve nuclei for CN5–CN8 and their associated fiber tracts (see Table 6.1 in Chapter 6). Functions associated with these cranial nerve nuclei include facial sensations (CN5), chewing (CN5), eye movement (CN6), facial expressions (CN7), secretion of saliva and tears (CN7), hearing (CN8), and balance (CN8).

The **locus coeruleus** (Latin, "blue spot") is a small bluish-pigmented, paired nucleus located in the dorsal pons near the floor of the fourth ventricle; its pigmentation is due to melanin granules within its neurons. Despite its small size, it has widespread projections to the entire forebrain, brainstem, and cerebellum. It is the primary site of norepinephrine synthesis within the brain, and is the major source of norepinephrine throughout the brain. The locus coeruleus plays an important role in the sleep-wake cycle, arousal, and attention.

The Midbrain

The **midbrain**, also known as the **mesencephalon**, is the smallest region of the brainstem. It is the most rostral of the brainstem structures, lying between the pons and the diencephalon of the forebrain. It is traversed by the cerebral aqueduct, which connects the third and fourth ventricles.

Two cranial nerves exit from the midbrain, the oculomotor nerve (CN3) and the trochlear nerve (CN4), both of which control eye movements. The oculomotor nerve (CN3) exits from the **interpeduncular fossa**, a deep depression on the inferior surface of the midbrain between the two cerebral peduncles (see Figure 8.2). The trochlear nerve (CN4) exits from the dorsal surface of the midbrain; it is notable for being the only cranial nerve to emerge from the dorsal surface of the brain (see Figure 8.3).

The midbrain is stratified along the dorsoventral dimension into three components: the **tectum** (roof), the **tegmentum** (floor), and the **cerebral peduncles**. As seen in midsagittal section, the tectum lies dorsal to the cerebral aqueduct, the tegmentum lies ventral to the cerebral aqueduct, and the cerebral peduncles lie ventral to the tegmentum (see Figure 8.1). As seen in horizontal sections of the midbrain, the tectum extends dorsally from the cerebral aqueduct, and the tegmentum extends ventrally from the cerebral aqueduct to the cerebral peduncles, two thick masses of fiber tracts (Figure 8.4). Following the plan of the spinal cord, the dorsal midbrain (tectum) serves sensory functions, while the ventral midbrain (tegmentum and cerebral peduncles) serves motor functions.

The Midbrain Tectum

The midbrain tectum is composed of two pairs of rounded surface swellings that are visible from the outside of the midbrain, known as **corpora quadrigemina** (quadruple bodies), comprising the superior and inferior colliculi. The **superior colliculi** receive and integrate visual, auditory, and somatosensory spatial information, and they generate and control orienting movements of the eyes and head toward salient objects in space (e.g., the automatic scanning

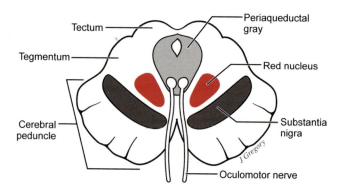

FIGURE 8.4. Horizontal section of the midbrain at the level of the superior colliculus. The tectum extends dorsally from the cerebral aqueduct, and the tegmentum extends ventrally from the cerebral aqueduct to the cerebral peduncles. At this level, the oculomotor nerves emerge from the interpeduncular fossa and the midbrain tegmentum contains three nuclear regions that are distinguished by their color: the substantia nigra, the red nucleus, and the periaqueductal gray.

movements of the eyes and head when reading, automatic orienting movements of the eyes and head toward sudden visual or auditory stimuli). Afferents come from the optic nerve, while efferents project through the tectospinal and tectobulbar tracts to influence motor neurons. The **inferior colliculi** are part of lower auditory centers, but they also play a role in reflexes to sudden loud auditory stimuli.

The Midbrain Tegmentum

The midbrain tegmentum contains three nuclear regions that are distinguished by their color: the substantia nigra, the red nucleus, and the periaqueductal gray (see Figure 8.4). It also contains the ventral tegmental area and the cranial nerve nuclei associated with CN3 and CN4 (i.e., the oculomotor complexes and the trochlear nuclei).

The **substantia nigra** (Latin, "black substance") is a large cluster of neurons that are darkly pigmented by melanin. These cells synthesize dopamine and give rise to the **nigrostrial tract**; this pathway undergoes selective degeneration in Parkinson's disease (see Chapter 11, "The Basal Ganglia").

The **red nucleus** is pale pink in color due to iron. It is a component of the extrapyramidal motor system and is important for motor coordination. It receives cerebellar inputs via the **superior cerebellar peduncle** and gives rise to the **rubrospinal tract**. Little is known about the function of the red nucleus and rubrospinal tract in the human nervous system, other than the fact that it is functionally specialized for motor control of the upper limbs and upper limb flexion. Clinically, this nucleus and descending motor pathway are best known for the difference between decorticate and decerebrate posturing, pathological posturing responses to noxious stimuli that occur with significant brain injury (see Chapter 25, "Disorders of Consciousness"). In experimental studies, brain transections rostral to the superior colliculus result in **decorticate posturing**, which is characterized by a flexor response of the upper limbs. By contrast, brainstem transections at or below the level of superior colliculus result in **decerebrate posturing**, which is characterized by extension of the lower and upper limbs. The difference between decorticate and decerebrate posturing is attributed to activity within the descending rubrospinal tract, which has a facilitatory effect on upper limb flexor muscles. Lesions above the red nucleus that damage inhibitory pathways from the cerebral cortex to the red nucleus result in disinhibition of the red nuclei and rubrospinal pathways, accounting for the flexor response in decorticate posturing. With lesions at or below the level of the red nucleus which disrupt the rubrospinal pathways, this effect is lost and there is extension of the upper limbs in decerebrate posturing. Clinically, decerebrate posturing occurs with lesions within the brainstem, while decorticate posturing occurs with lesions in motor pathways above the brainstem.

The **periaqueductal gray** (PAG, midbrain central gray) is a region of gray matter within the midbrain that surrounds the cerebral aqueduct. It is rich in opioid receptors and plays an important role in natural pain suppression; therefore it is known as an endogenous analgesia center. The PAG receives input from ascending sensory pathways that carry pain signals related to tissue damage, and gives rise to pathways that inhibit the transmission of pain signals from the periphery to the brain. Its role in pain suppression (antinociception) is supported by studies showing lower activation of the PAG in chronic pain states. Deep brain stimulation of the PAG is an effective treatment for severe refractory neuropathic pain.

The **ventral tegmental area** is the origin of the dopaminergic **mesocorticolimbic system**. It is a central component of reward circuitry of the brain and plays a role in motivated behaviors and drug addiction (see Chapter 12, "Limbic Structures").

The Cerebral Peduncles

The **cerebral peduncles** (feet of the brain) are two thick masses of white matter that are visible on the ventral surface of the midbrain (see Figure 8.2). They carry large motor tracts descending from the cerebrum: these are the corticospinal, corticopontine, and corticobulbar tracts. As these fibers course caudally, they converge and enter the pons. The depression between the peduncles is the **interpeduncular fossa**.

The Reticular Formation

The **reticular formation** is a net-like mixture of gray and white matter that occupies the central core of the brainstem, from the lower medulla to the upper midbrain. The cells of the reticular formation are not aggregated into well-defined nuclei and tracts. The reticular formation plays a crucial role in attention, arousal, and sleep, as well as many other functions.

The reticular formation receives many afferent projections from widely varied brain regions. Each reticular neuron may receive input from over 40,000 other neurons, many of which lie at great distances from each other. The reticular formation also has numerous efferent pathways and communicates directly and indirectly with all levels of the nervous system. Some axons bifurcate and project both upstream and downstream, influencing many neurons along the way by collateral branches. By virtue of its extensive efferent connections, the reticular formation can act on virtually all other parts of the CNS. Thus, the reticular formation receives input from, and influences the activity of, most other regions of the CNS. It is a neural system that seems designed for the overall integration of brain function.

Topographically, the nuclei of the reticular formation are arranged into five columns that run along the length of the brainstem: a median group, two paramedian groups, and two lateral groups. The **median group** nuclei lie midline and are unpaired. These nuclei, also known as the **raphe** (*seam*) **nuclei**, are characterized by high serotonin content. The **paramedian groups** lie on either side of the median group. The paired nuclei of this group are made up of neurons with large cell bodies, and thus are referred to as **gigantocellular reticular nuclei**. The **lateral groups** lie most laterally. The paired nuclei of this group are made up of cells with small cell bodies, and thus are referred to as **parvocellular** (*parvo* meaning small) **reticular nuclei**.

The components of the reticular formation are also often described according to the level of the brainstem; thus, we speak of the **medullary reticular formation**, the **pontine reticular formation**, and the **midbrain (mesencephalic) reticular formation**, which lies within the midbrain tegmentum.

The projections of the reticular formation are either ascending or descending. The ascending reticular formation projects to the midline and intralaminar thalamic nuclei that give rise to diffuse projections to the cerebral cortex, forming the **reticulothalamocortical pathway**.

Consciousness and Sleep-Wake Cycle

The ascending reticular formation, also known as the **reticular activating system**, is crucial for the maintenance of wakefulness, awareness, and attention (see Chapter 25, "Disorders of Consciousness"). Selective lesions of the mesencephalic reticular formation in humans and laboratory animals are associated with prolonged periods of unconsciousness. Electrical stimulation of the mesencephalic reticular formation in anesthetized animals results in a change in the electroencephalogram (EEG) from a synchronized pattern characteristic of slow-wave sleep to a desynchronized pattern characteristic of wakefulness. In animals that are awake and lying still with no obvious interest in their surroundings, electrical stimulation of the mesencephalic reticular formation produces both EEG desynchronization and behavioral signs of arousal, alertness, and attentiveness. The EEG activation and behavioral arousal effects are mediated by the reticulothalamocortical pathway. Selective lesions of the mesencephalic area are associated with prolonged periods of unconsciousness.

Since the core of the brainstem is responsible for maintaining a normal level of consciousness and the sleep/wake cycle, a blow to the head which is strong enough to result in sudden compression and shearing of brainstem neuronal elements results in a loss of consciousness. If the trauma causes no serious physical damage to the brainstem, consciousness returns. With more extensive injury, the period of unconsciousness is longer. After consciousness returns, there may be attentional difficulties, or the victim may seem dazed.

Motor Function

Portions of the descending reticular formation are involved in motor function. The reticulospinal pathways play a role in regulating muscle tone, and hence posture (i.e., maintaining an upright position) and equilibrium, by influencing the activity of spinal motor neurons. The **pontine reticulospinal tracts** innervate extensor muscles and drive muscle tone; the **medullary reticulospinal tracts** inhibit extensor muscle tone and activate flexor muscles. Higher brain regions, such as the cerebral cortex and basal ganglia, influence muscle tone via their connections to the reticular formation and the reticulospinal pathways.

Reticulospinal tracts also play a role in crude stereotyped movements through **premotor centers**, networks within the reticular formation that influence the activity of large groups of muscles. Walking movements are mediated by the **mesencephalic locomotor region**. Orienting movements of the head and body elicited by visual and auditory stimuli are mediated by the **tectoreticulospinal pathway**.

Autonomic Functions

The reticular formation also plays a role in autonomic function via networks of afferents from the hypothalamus to the reticular formation, and reticulospinal efferents that act on preganglionic sympathetic neurons and the dorsal motor nucleus of the vagus. The **pontine reticulospinal tract** facilitates sympathetic outflow and facilitates cardiovascular responses, such as increased pulse and blood pressure. The **medullary reticulospinal tract** facilitates parasympathetic outflow and inhibits cardiovascular responses.

Sensory Modulation

The descending reticular formation plays a role in modulating the central transmission of sensory information. Many secondary sensory neurons send collateral branches to the reticular formation. These include collaterals from the spinothalamic tract carrying pain and temperature signals from the body, collaterals from the ascending pathway originating from the sensory trigeminal nucleus carrying pain and temperature signals from the face, collaterals from ascending auditory and vestibular pathways, and collaterals from ascending fibers originating from the solitary nucleus carrying visceral sensory impulses. Descending projections from the reticular formation suppress the central conduction of pain signals.

Clinical Considerations

When a neurologist examines a patient with a neurological disorder, one of their first objectives is to determine where (i.e., at what the level of the nervous system) the pathology is located. Knowledge of the functional anatomy of the nervous system makes it possible to interpret signs and symptoms to localize lesions and to make diagnoses based on functional anatomy. This is particularly clear with brainstem lesions, which are typically focal and involve various combinations of cranial nerve nuclei, pigmented nuclei (i.e., red nucleus, substantia nigra, locus coeruleus), and the long ascending (e.g., spinothalamic, spinocerebellar) and descending (e.g., corticospinal, corticopontine) tracts passing through the brainstem. Cranial nerve-related signs provide rostral-caudal localization, and long tract-related signs provide medial-lateral localization, akin to the lines of latitude and longitude, respectively, where the point of intersection indicates the site of the lesion.

Focal brainstem lesions give rise to specific constellations of neurologic signs and symptoms known as **brainstem syndromes**. These are most commonly caused by infarcts due to occlusion of small perforating arteries of the posterior circulation; however, they may also occur with hemorrhage, neoplasm, or demyelination. They generally involve an ipsilateral cranial nerve palsy and contralateral hemiplegia/hemiparesis and/or hemisensory loss. The **rule of 4** is a shorthand method to remember the essential brainstem anatomy, clinical signs and symptoms, and syndromes resulting from focal brainstem lesions (Table 8.1). **Lateral medullary syndrome** (**Wallenberg syndrome**) is the most common posterior circulation ischemic stroke syndrome.

Locked-in syndrome is a rare neurological condition in which there is complete paralysis of all voluntary muscles except for those controlling eye movements, without alteration of conscious, awareness, or cognitive function (see Chapter 25, "Disorders of Consciousness"). Locked-in syndrome is usually caused by damage to the ventral pons, which interrupts nerve fibers that travel from the cerebrum to the spinal cord and the brainstem cranial nerve nuclei that play a role in speaking; the most common cause is brainstem stroke. Locked-in syndrome may also be caused by pathologies outside of the pons that affect motor neurons (e.g., amyotrophic lateral sclerosis).

Evaluation of **brainstem reflexes** is a critical component of the neurological examination, especially in neurocritical care patients; these include pupillary, corneal, gag, cough, oculocephalic, and vestibulo-ocular reflexes (see Chapter 30, "The Neurological Examination").

The most common causes of focal brainstem lesions are stroke, neoplasm, and multiple sclerosis. Less common etiologies for focal brainstem lesions and brainstem syndromes include central pontine myelinolysis and neurodegenerative disease. **Central pontine myelinolysis** is characterized by demyelination of the white matter fibers in the central part of the basal pons, which houses the corticospinal and corticobulbar tracts. Central pontine myelinolysis occurs as a complication of severe and prolonged sodium depletion (hyponatremia) that is corrected too rapidly (see Chapter 23, "White Matter Disease"). **Progressive supranuclear palsy** is a neurodegenerative movement disorder that is associated with atrophy of the midbrain that may be visible on neuroimaging in the "hummingbird" or "penguin" sign (Figure 8.5; see Chapter 11, "The Basal Ganglia").

Pressure in the intracranial cavity from a blood clot, fluid buildup, or brain herniation may cause secondary pressure on the brainstem, which in turn may produce unconsciousness, cardiorespiratory failure, and death. Extensive damage to the brainstem usually results in death. This is the reason why a gunshot into the oral cavity or to the lower back of the head is lethal; the gun barrel at jaw level directs the bullet through the brainstem. In contrast, directing a gun barrel to the temple or forehead produces damage to higher brain centers that mediate cognitive and behavioral functions; damage at the level of the cerebrum therefore may produce devastating cognitive and behavioral changes without being lethal.

Summary

The brainstem lies within the posterior cranial fossa and below the tentorium, along with the cerebellum. It consists of the midbrain, pons, and medulla. It contains four main types of neural elements: (1) long ascending somatosensory tracts and long descending motor tracts, and nuclei associated with these pathways; (2) nuclei and tracts of CN3–CN12; (3) nuclei regulating the sleep-wake cycle and level of consciousness; and (4) nuclei regulating vital life functions such as heart rate, breathing, and digestion. Extensive damage to the brainstem usually results in death. Lesser degrees of damage result in coma and persistent vegetative state. Focal brainstem lesions produce various combinations of impaired function of the cranial nerve nuclei, long ascending tracts, and long descending tracts, forming the basis of brainstem syndromes. On neurological examination, cranial nerve–related signs provide rostral-caudal localization, and long tract–related signs provide medial-lateral localization.

TABLE 8.1. The Rule of 4 is a Simplified Method for Understanding Brainstem Anatomy and Syndromes

RULE 1: THERE ARE 4 STRUCTURES IN THE "MIDLINE" (MEDIAL), ALL BEGINNING WITH M	
Motor pathway (corticospinal tract)	Contralateral weakness of the arm and leg
Medial Lemniscus	Contralateral loss of vibration and proprioception in the arm and leg
Medial longitudinal fasciculus	Ipsilateral internuclear ophthalmoplegia (failure of adduction of the ipsilateral eye toward the nose and nystagmus in the opposite eye as it looks laterally)
Motor nucleus and nerve	Ipsilateral loss of the cranial nerve that is affected (CN3, CN4, CN6 or CN12)
RULE 2: THERE ARE 4 STRUCTURES TO THE SIDE (LATERAL), ALL BEGINNING WITH S	
Spinocerebellar pathways	Ipsilateral ataxia of the arm and leg
Spinothalamic pathway	Contralateral alteration of pain and temperature affecting the arm, leg, and rarely the trunk
Sensory nucleus of the 5th	Ipsilateral alteration of pain and temperature on the face in the distribution of CN5 (this nucleus is a long vertical structure that extends in the lateral aspect of the pons down into the medulla)
Sympathetic pathway	Ipsilateral Horner's syndrome: partial ptosis and a small pupil (miosis)
RULE 3: THERE ARE 4 CRANIAL NERVES *ABOVE* THE PONS (2 IN THE MIDBRAIN [CN3–CN4], AND 2 ABOVE THE MIDBRAIN (CN1–CN2]), 4 IN THE PONS (CN5–CN8), AND 4 IN THE MEDULLA (CN9–CN12)	
Midbrain	
3 Oculomotor	Misalignment of the affected eye in a downward and outward position ("down and out"); the pupil may be dilated
4 Trochlear	Deviation of the eye upward, inability to look down when adducting the eye (looking in toward the nose)
Pons	
5 Trigeminal	Ipsilateral alteration of pain, temperature, and light touch on the face back as far as the anterior two-thirds of the scalp and sparing the angle of the jaw
6 Abducens	Ipsilateral weakness of abduction (lateral movement) of the eye
7 Facial	Ipsilateral facial weakness
8 Auditory	Ipsilateral deafness
Medulla	
9 Glossopharyngeal	Ipsilateral loss of pharyngeal sensation
10 Vagus	Ipsilateral palatal weakness
11 Accessory	Ipsilateral weakness of the trapezius and sternocleidomastoid muscles
12 Hypoglossal	Ipsilateral weakness of the tongue
RULE 4: THERE ARE 4 PURE MOTOR CRANIAL NERVES ORIGINATING FROM MOTOR NUCLEI LOCATED MEDIALLY WITHIN THE BRAINSTEM AND THEIR NUMBERS DIVIDE INTO 12 (I.E., THE MOTOR NUCLEI FOR CN3, CN4, CN6 AND CN12 ARE LOCATED MEDIALLY AND THE MOTOR NUCLEI FOR CN5, CN7, CN9 AND CN11 ARE LOCATED LATERALLY WITHIN THE BRAINSTEM)	
Brainstem Syndromes	
Medial midbrain syndrome	The 4 M's plus cranial nerves CN3 and CN4
Lateral midbrain syndrome	The 4 S's
Medial pons syndrome	The 4 M's plus cranial nerve CN6
Lateral pons syndrome	The 4 S's plus cranial nerves CN5, CN7, and CN8
Medial medulla syndrome	The 4 M's plus CN 12
Lateral medulla syndrome	The 4 S's plus CN 9–CN11

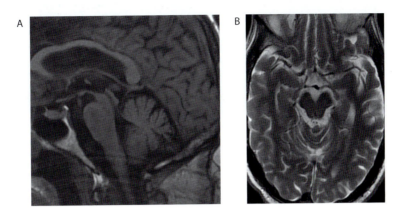

FIGURE 8.5. Progressive supranuclear palsy is a neurodegenerative movement disorder that is associated with atrophy of the midbrain that may be visible on neuroimaging as (A) the "hummingbird" sign on mid-sagittal images, or (B) the "penguin" sign ("Mickey Mouse" sign) on axial images.

Additional Reading

1. Brodal P. The reticular formation: premotor networks, consciousness, and sleep. In: Brodal P, *The central nervous system: structure and function*. 5th ed. Oxford University Press; 2016. https://oxfordmedicine.com/view/10.1093/med/9780190228958.001.0001/med-9780190228958-chapter-26. Accessed July 4, 2022.
2. Blumenfeld H. Brainstem I: surface anatomy and cranial nerves. In: Blumenfeld H, *Neuroanatomy through clinical cases*. 3rd ed., pp. 495–565. Sinauer Associates; 2021.
3. Blumenfeld H. Brainstem III: internal structures and vascular supply. In: Blumenfeld H, *Neuroanatomy through clinical cases*. 3rd ed., pp. 615–697. Sinauer Associates; 2021.
4. D'aes T, Mariën P. Cognitive and affective disturbances following focal brainstem lesions: a review and report of three cases. *Cerebellum*. 2015;14(3):317–340. doi:10.1007/s12311-014-0626-8
5. Gates P. The rule of 4 of the brainstem: a simplified method for understanding brainstem anatomy and brainstem vascular syndromes for the non-neurologist. *Intern Med J*. 2005;35(4):263–266. doi:10.1111/j.1445-5994.2004.00732.x
6. Gates P. Work out where the problem is in the brainstem using "the rule of 4." *Pract Neurol*. 2011;11(3):167–172. doi:10.1136/practneurol-2011-000014
7. Hurley RA, Flashman LA, Chow TW, Taber KH. The brainstem: anatomy, assessment, and clinical syndromes. *J Neuropsychiatry Clin Neurosci*. 2010;22(1):iv–7. doi:10.1176/jnp.2010.22.1.iv
8. Kochar PS, Kumar Y, Sharma P, Kumar V, Gupta N, Goyal P. Isolated medial longitudinal fasciculus syndrome: review of imaging, anatomy, pathophysiology and differential diagnosis. *Neuroradiol J*. 2018;31(1):95–99. doi:10.1177/1971400917700671
9. Moruzzi G, Magoun HW. Brain stem reticular formation and activation of the EEG. 1949. *J Neuropsychiatry Clin Neurosci*. 1995;7(2):251–267. doi:10.1176/jnp.7.2.251
10. Silverman IE, Liu GT, Volpe NJ, Galetta SL. The crossed paralyses: the original brain-stem syndromes of Millard-Gubler, Foville, Weber, and Raymond-Cestan. *Arch Neurol*. 1995;52(6):635–638. doi:10.1001/archneur.1995.00540300117021
11. Tacik P, Alfieri A, Kornhuber M, Dressler D. Gasperini's syndrome: its neuroanatomical basis now and then. *J Hist Neurosci*. 2012;21(1):17–30. doi:10.1080/0964704X.2011.568045
12. Vanderah TW, Gould D. Organization of the brainstem. In: Vanderah TW, Gould D, *Nolte's the human brain: an introduction to its functional anatomy*. 8th ed., pp. 258–284. Elsevier; 2021.
13. Vanderah TW, Gould D. Cranial nerves and their nuclei. In: Vanderah TW, Gould D, *Nolte's the human brain: an introduction to its functional anatomy*. 8th ed., pp. 286–308. Elsevier; 2021.

9

The Cerebellum

Introduction

As the largest motor structure of the central nervous system (CNS), the cerebellum plays a critical role in movement, while damage to it leads to disorders of motor function. The cerebellum regulates the rate, rhythm, force, and accuracy of movements, and cerebellar damage produces movement incoordination. The cerebellum also plays a role in non-motor functions, although in ways that are poorly understood. This chapter presents the anatomy of the cerebellum, functional divisions of the cerebellum, and signs and symptoms of cerebellar damage, and gives an overview of cerebellar pathologies.

Anatomy of the Cerebellum

The cerebellum (Latin, "little brain") lies within the **posterior cranial fossa**, dorsal to the brainstem and separated from the overlying cerebrum by the **tentorium**. In humans, the cerebellum is about the size of a softball and accounts for approximately 10% of the total brain volume; however, because many of its neurons are small it contains more than 50% of all neurons in the entire brain! The cerebellum consists of a highly convoluted gray matter cortex, a medullary core of white matter, four pairs of deep cerebellar nuclei, and three pairs of cerebellar peduncles that carry afferents to and efferents from the cerebellum.

Lobes and Lobules

The cerebellum is divided into three lobes: the **anterior lobe**, **posterior lobe**, and **flocculonodular lobe** (Figure 9.1). In the human brain, the posterior lobe is the largest of the three cerebellar lobes and the flocculonodular lobe is the smallest. The anterior and posterior lobes account for the main mass of the human cerebellum. The three lobes of the cerebellum are divided by two fissures. The **primary fissure** separates the anterior and posterior lobes. It is the deepest fissure of the cerebellum; this is apparent when viewing midsagittal sections, but not when examining the external surface of the cerebellum. On the ventral surface of the cerebellum, the **posterolateral fissure** separates the posterior

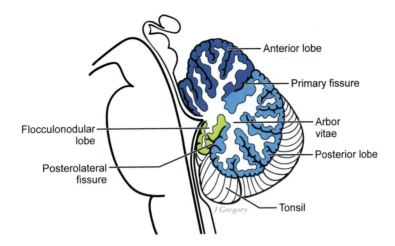

FIGURE 9.1. Sagittal view of the cerebellum. The cerebellum consists of a highly convoluted gray matter cortex and an inner core of white matter (arbor vitae) organized into two hemispheres, each consisting of three lobes: the anterior, posterior, and flocculonodular lobes. The cerebellar tonsils are part of the posterior lobes. The primary fissure separates the anterior and posterior lobes, and the posterolateral fissure separates the posterior and flocculonodular lobes.

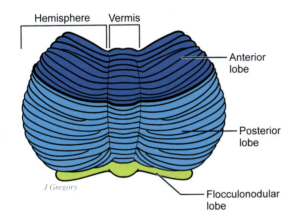

FIGURE 9.2. Superior view of the cerebellum. The anterior and posterior lobes consist of a midline vermis that is flanked by two cerebellar hemispheres. The flocculonodular lobe consists of a midline nodulus (part of the vermis) flanked by two flocculi.

and flocculonodular lobes. The largest and most conspicuous fissure is the **horizontal fissure**, which divides the posterior lobe into superior and inferior portions.

The flocculonodular lobe is visible only from the inferior surface of the cerebellum. The cerebellum curls under itself, and if it were unrolled and flattened out, the flocculonodular lobe would be most caudal; in other words, the rostral-caudal order of the three lobes would be anterior, posterior, and flocculonodular (Figure 9.2). The anterior and posterior lobes consist of a midline **vermis** (Latin, "worm") that is flanked on each side by two **cerebellar hemispheres**. The flocculonodular lobe consists of a midline **nodulus** which is part of the vermis, flanked by two **flocculi**.

The **cerebellar cortex** consists of a very tightly folded layer of gray matter that forms narrow gyri-like ridges called **folia** (Latin, "leaves"; singular, **folium**). The folia are oriented transversely to the longitudinal axis of the organism. A set of deeper folds subdivide the cerebellum into 10 transversely oriented **lobules**, each spanning the vermis and hemispheres, identified by Roman numerals (Figure 9.3). Lobules I–V are located within the anterior lobe, lobules VI–IX are located within the posterior lobe, and lobule X consists of the flocculonodular lobe. The lobules of the anterior and posterior hemispheres each contain multiple folia.

The **cerebellar tonsils** are the hemispheric portion of the most caudal lobule of the posterior lobe. They are ovoid-shaped structures that lie on the undersurface of the cerebellum, just above the foramen magnum. **Tonsilar herniation** through the foramen magnum occurs in a variety of congenital and acquired conditions. When due to increased intracranial pressure, tonsilar herniation may cause secondary compression of the medulla, leading to respiratory and cardiac arrest.

Deep Cerebellar Nuclei

The white matter underlying the cerebellar cortex is composed of myelinated axons entering and exiting the cerebellar cortex. It has a branched, tree-like appearance and is therefore known as the **arbor vitae**, meaning "tree of life" (see Figure 9.1).

Buried within the white matter of each cerebellar hemisphere are four **deep (intrinsic) cerebellar nuclei** (Figure 9.4). From lateral to medial, they are the **dentate** ("toothed") **nucleus**, the **emboliform nucleus**, the **globose nucleus**, and the **fastigial nucleus**. A mnemonic to remember their names and positions, moving from the lateral-most and toward the midline, is the phrase "**D**on't **E**at **G**reasy **F**ood." In lower animals, the emboliform and

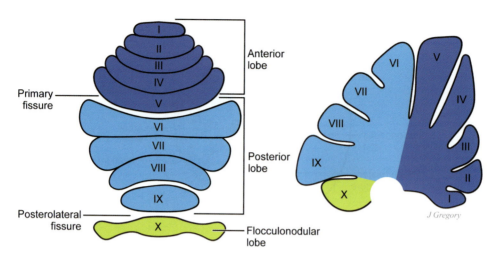

FIGURE 9.3. The ten lobules of the cerebellum.

globose nuclei are not differentiated, but rather form a single **interposed nucleus** (**nucleus interpositus**). In humans and other animals that have distinct emboliform and globose nuclei, the two are often referred to collectively as the **interposed nuclei**. The dentate nucleus is also referred to as the **lateral cerebellar nucleus**, and the fastigial nucleus is also referred to as the **medial cerebellar nucleus**.

Zonal Organization

The anterior and posterior lobes of the cerebellum are organized into three longitudinal zones that run perpendicular to the folia and fissures, based on anatomical connections to the deep cerebellar nuclei (see Figure 9.4). These are the **vermis**, the **intermediate (paravermal, paramedial) zones** which are the medial-most portion of the hemispheres bordering the vermis, and the **lateral zones (lateral hemispheres)** which are the lateral-most portions of the hemispheres. There are no clear gross anatomic borders between the intermediate and lateral zones. The corticonuclear connections of the anterior and posterior lobes are topographically organized in a medial-to-lateral fashion (see Figure 9.4). The vermis projects to the fastigial nuclei, the intermediate zones project to the interposed nuclei, and the lateral zones project to the dentate nuclei.

This zonal organization is relevant to the functional organization of the cerebellum (see "Functional Divisions of the Cerebellum," below). All inputs to the cerebellum terminate in the cerebellar cortex. Outputs from the cerebellar cortex of the anterior and posterior lobes project to the deep cerebellar nuclei, which in turn give rise to most of the cerebellar output fibers. The vermis and intermediate zones of the anterior and posterior lobes and their connections to the fastigial and interposed nuclei have reciprocal connections with the spinal cord. The lateral hemispheres and the dentate nuclei have reciprocal

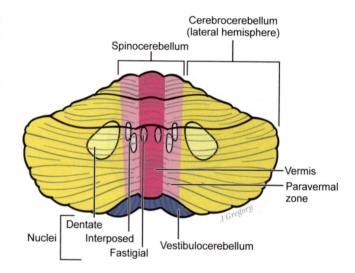

FIGURE 9.4. The anterior and posterior lobes of the cerebellum are organized into three longitudinal zones that run perpendicular to the folia and fissures, based on anatomical connections to the deep cerebellar nuclei: the vermis (dark pink), the intermediate zones (light pink), and the lateral zones (yellow). The cerebellum is also divided into three functional zones based on afferent inputs. The flocculonodular lobes makes up the vestibulocerebellum, the vermis and intermediate zones make up the spinocerebellum, and the lateral hemispheres make up the cerebrocerebellum.

connections with the cerebral cortex. The flocculonodular lobe has a different pattern of connectivity; outputs from the cerebellar cortex of the flocculonodular lobe project directly to the medial and lateral vestibular nuclei of the brainstem.

Cerebellar Peduncles

The cerebellum is connected to and communicates with the rest of the brain by three pairs of fiber bundles known as **peduncles** (see Figure 8.3 in Chapter 8). The **superior cerebellar peduncles** (**brachium conjunctivum**, meaning "joined together arm") connect to the midbrain. The **middle cerebellar peduncles** (**brachium pontis**, meaning "arm of the pons"), connect to the pons. The **inferior cerebellar peduncles** connect to the medulla; they are composed of a large component known as the **restiform** ("ropelike") **body**, and a smaller component known as the **juxtarestiform body**.

Cerebellar afferents are conveyed primarily through the middle and inferior cerebellar peduncles (Table 9.1). These afferents originate from three main sources: the cerebral cortex/pontine nuclei, brainstem nuclei (e.g., vestibular nuclei, reticular formation, olivary nuclei), and the spinal cord. Cerebellar efferents are conveyed primarily through the superior cerebellar peduncles. These efferents project to the cerebral cortex and brainstem nuclei (e.g., vestibular nuclei, reticular formation, red nucleus).

The middle cerebellar peduncles carry cerebellar afferents exclusively, all originating from the pontine nuclei. These nuclei receive afferents from a wide area of cerebral cortex, especially from the prefrontal, frontal, and parietal regions. The pontine nuclei in turn project to the cerebellum via the **pontocerebellar tracts** carried in the middle cerebellar peduncles. This two-neuron pathway from the cerebrum to the cerebellum via the pontine nuclei is referred to as the **corticopontocerebellar pathway**.

The inferior cerebellar peduncles primarily carry afferents, all originating from sources other than the pons (namely the spinal cord and brainstem), except the juxtarestiform component, which carries reciprocal connections between the cerebellum and vestibular system (i.e., the afferent vestibulocerebellar tracts and the efferent cerebellovestibular tracts).

The majority of cerebellar efferents are conveyed through the superior cerebellar peduncles. The efferent connections from the cerebellum to the cerebral cortex consist of a two-neuron pathway (like the afferent connections from the cerebral cortex to the cerebellum). The cerebellum projects to the **ventral lateral nuclei of the thalamus** via **cerebellothalamic pathways**. The ventral lateral nuclei in turn give rise to fibers that ascend in the thalamic radiations and terminate in the primary motor and premotor cortices. This two-neuron pathway from the cerebellum to the cerebrum via the thalamus is referred to as the **cerebellothalamocortical pathway**. Superior cerebellar peduncle efferents directed to the brainstem project to the red nuclei of the midbrain via **cerebellorubral pathways**. While the superior cerebellar peduncles primarily carry efferents, the exception is the afferent ventral spinocerebellar tracts.

Cerebellar Circuitry

The cerebellar cortex consists of three layers of cells: from outermost to innermost, the molecular, Purkinje, and granular layers. Five cell types are distributed in these three layers: Purkinje cells, granule cells, Golgi cells, stellate cells, and basket cells. The cerebellar cortex has a relatively simple geometric cytoarchitecture that is uniform throughout its entire structure; this contrasts with the varying cytoarchitecture across the cerebral cortex that forms the basis for the Brodmann areas (see Chapter 13, "The Cerebral Cortex"). The homogeneous cytoarchitecture of the cerebellar cortex suggests that all regions perform the same fundamental computational operation (the **universal cerebellar transform**), but what distinguishes different cerebellar regions is their afferent and efferent connections.

The **Purkinje layer** is the middle layer; it is occupied by the cell bodies of **cerebellar Purkinje cells**. These cells are one of the most distinctive cell types in the mammalian brain. The Purkinje cell bodies are large and are arranged in a single row within the Purkinje cell layer. Their dendritic trees, which lie within the molecular layer, are extensive but flattened in one dimension, so they are shaped like a large fan. The dendritic trees are oriented within a plane perpendicular to the long axis of the folium in which they reside and are oriented in parallel to each other.

The **granular layer** is the innermost layer; it is occupied by numerous small, tightly packed granule cells, as well as Golgi cells. **Cerebellar granule cells** are the most abundant class of neurons in the mammalian brain, accounting for most of neurons in the cerebellum and for the fact that the cerebellum contains more than 50% of all neurons in the entire brain.

The **molecular layer** is the outermost layer; it is occupied by the cell bodies of stellate cells and basket cells, as well as the dendrites of Purkinje cells and axons of granule cells. The axons of the granule cells travel vertically from the granule cell layer toward the cortical surface, and within the molecular layer they bifurcate in a distinctive T-shape, sending two horizontal branches known as **parallel fibers** in opposite directions. The parallel fibers run parallel to the folds of the cerebellar cortex, passing at right angles through the Purkinje cell dendritic net, making excitatory synapses with Purkinje cells along the way. Thus, the two-dimensional arbors of the Purkinje cell dendrites are oriented perpendicular to the parallel fibers. Consequently, Purkinje cells receive more synaptic inputs than any other type of cell in the brain.

All inputs to the cerebellum terminate in the granular and molecular layers of the cerebellar cortex. Purkinje cells are the sole output from the cerebellar cortex. The Purkinje cells of the anterior and posterior lobes project from the cerebellar cortex to the deep cerebellar nuclei, which in turn give rise to most of the cerebellar output fibers. The Purkinje cells of the flocculonodular lobe bypass the deep cerebellar nuclei and project directly from the cerebellar cortex to the medial and lateral vestibular nuclei of the brainstem.

Blood Supply

The cerebellum receives its blood supply from three pairs of arteries that are branches of the vertebrobasilar system (Figure 9.5). Moving in a rostral-caudal direction these are the **superior cerebellar artery** (SCA), **anterior inferior cerebellar artery** (AICA), and **posterior inferior cerebellar artery** (PICA). The SCAs arise from each side of the upper basilar artery; they supply the superior surface of the hemispheres and vermis, and the superior cerebellar peduncles. The AICAs arise from each side of the lower basilar artery; they supply the anterior portion of the inferior cerebellum and the middle cerebellar peduncles. Each PICA arises from a vertebral artery; the PICAs supply the posterior inferior portion of the cerebellum and the inferior cerebellar peduncles. Cerebellar ischemic stroke is commonly caused by vascular disease within the vertebrobasilar arteries. Signs and symptoms vary, depending on the location and size of the infarction and whether other structures receiving posterior circulation are also ischemic (see "Cerebellar Infarction," below). Because the SCA, AICA, and PICA also supply brainstem structures, infarcts within their territories may produce a combination of cerebellar and brainstem damage.

Venous drainage of the cerebellum occurs by the **superior** and **inferior cerebellar veins**, which in turn drain into the transverse sinuses and the superior petrosal sinuses.

Functional Divisions of the Cerebellum

The cerebellum is divided into three functional zones based on its afferent input (see Figure 9.3): the vestibulocerebellum, the spinocerebellum, and the cerebrocerebellum. These functional divisions form the basis for understanding the clinical manifestations of cerebellar pathology.

The Vestibulocerebellum

The **vestibulocerebellum** is composed of the flocculonodular lobe (see Figure 9.1). This functional division has reciprocal connections with the vestibular nuclei of the vestibular system.

The vestibular system conveys information about the orientation of the body with respect to gravity, which is key to our awareness of body position, motion, and balance (**equilibrium**). As described in Chapter 6 ("The Peripheral Nervous System"), the vestibular apparatus, located within the inner ear, consists of three semicircular canals connected to two otoliths, the saccule and utricle. This sensory organ transduces and encodes signals about head motion and position relative to gravity, which is critical for posture, balance when stationary (static equilibrium) and moving (dynamic equilibrium), the fine control of visual gaze, spatial orientation, navigation, and the coordination of eye, head, and neck movements.

The otoliths detect static orientation of the head relative to gravity and linear movement. The utricle detects movement in the horizontal plane (e.g., walking forward, walking backward, stepping leftward or rightward). The saccule detects movement in the vertical plane (e.g., moving upward or downward when jumping on a trampoline). The three semicircular canals signal rotational movement of the head; they are oriented along three planes: horizontal (axial), sagittal, and frontal (coronal). Rotation of the head within the horizontal plane consists of movement of the head from side to side (e.g., shaking your head "no"), rotation of the head in the sagittal plane consists of movement of the head up and down (e.g., nodding your head "yes"), and rotation of the head in the frontal plane

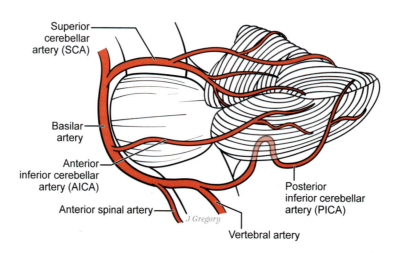

FIGURE 9.5. Blood supply to the cerebellum is provided by three pairs of arteries originating from the vertebrobasilar system: the superior cerebellar artery (SCA), the anterior inferior cerebellar artery (AICA), and the posterior inferior cerebellar artery (PICA).

TABLE 9.1 The Cerebellar Peduncles and Their Pathways

PEDUNCLE/PATHWAY	ORIGIN	DESTINATION
MIDDLE CEREBELLAR PEDUNCLE		
Pontocerebellar	Pontine nuclei	Neocerebellar cortex and dentate nuclei
INFERIOR CEREBELLAR PEDUNCLE		
Restiform Body		
Dorsal spinocerebellar	Spinal cord (lower extremity)	Related somatotopic regions of cerebellar cortex and nuclei
Rostral spinocerebellar	Spinal cord (upper extremity and trunk)	Related somatotopic regions of cerebellar cortex and nuclei
Cuneocerebellar	Cuneate nuclei (upper extremity and trunk)	Related somatotopic regions of cerebellar cortex and nuclei
Olivocerebellar	Inferior olivary nucleus	Cerebellar cortex and deep cerebellar nuclei
Reticulocerebellar	Lateral reticular nucleus	Cerebellar hemispheres and vermis
Trigeminocerebellar	Trigeminal nucleus	Related somatotopic regions of cerebellar cortex and nuclei
Juxtarestiform Body		
Vestibulocerebellar	Vestibular nuclei	Flocculonodular lobe, caudal uvula, and fastigial nucleus
Cerebellovestibular	Fastigial nucleus, vermis, and vestibulocerebellar cortex	Vestibular nuclei
SUPERIOR CEREBELLAR PEDUNCLE		
Cerebellothalamic	Dentate and interposed nuclei	Ventrolateral thalamic nucleus
Cerebellorubral	Dentate and interposed nuclei	Red nucleus
Ventral spinocerebellar	Spinal cord (lower extremity)	Related somatotopic regions of cerebellar cortex and nuclei

Note: Efferent pathways are italicized.

consists of tilting the head from left to right and vice versa (e.g., moving the head to touch the shoulders). Vestibular signals are conveyed to the CNS by the vestibular branch of the eighth cranial nerve. The vestibular afferents terminate in four **vestibular nuclei** (superior, lateral, medial, and descending) located in the medulla and caudal pons. The vestibulocerebellum (flocculonodular lobe) receives its primary input from the vestibular nuclei.

Purkinje cell axons from the flocculonodular lobe send inhibitory projections directly to the medial and lateral vestibular nuclei (i.e., without first synapsing on deep cerebellar nuclei). The medial and lateral vestibular nuclei in turn modulate descending motor signals of the medial and lateral **vestibulospinal tracts**, respectively, in response to vestibular signals about changes in the orientation of the head. The vestibulospinal tracts control the axial (trunk and head) and limb extensor muscles that control posture and keep us upright. They are critical for maintaining balance during standing and walking. The **medial vestibulospinal tract** is a bilateral tract that terminates at the cervical levels of the spinal cord and innervates the muscles of the neck that mediate head position. The **lateral vestibulospinal tract** terminates at all levels of the spinal cord and innervates the ipsilateral proximal limb and axial muscles, providing extensor tone to resist the pull of gravity that is necessary for the control of posture and balance.

The flocculonodular lobe also receives visual system inputs from the superior colliculi and the striate cortex via the pontine nuclei. It functions to coordinate eye movements with head movements. Purkinje cell axons from the vestibulocerebellum send inhibitory projections directly to the medial vestibular nuclei. These nuclei use vestibular information to control involuntary eye movements made in the service of stabilizing the retinal image as we move about the world. Without such eye movements, the retinal image would blur, and we would not be able to see clearly while moving. The vestibular system detects head movements and initiates reflexive eye movements in the direction opposite to head movement in the service of stabilizing gaze and the retinal images; this is known as the **vestibulo-ocular reflex**. In a mirror, look yourself in the eyes and move your head from left to right. When you move your head to the left your eyes move to the right, and when you move your head to the right your eyes move to the left. These eye movements are induced by head rotation and compensate for the head movement in the service of maintaining your visual fixation; they even occur in the absence of visual stimulation (i.e., in the dark or with eyes closed). They are reflexive and therefore involuntary; thus we are unaware of them.

The Spinocerebellum

The **spinocerebellum** is composed of the vermis and intermediate zones of the anterior and posterior lobes, as well as their connections to the fastigial and interposed nuclei, respectively. This functional division has reciprocal connections with the spinal cord. The primary functions of the spinocerebellum are coordination of movement, maintenance of balance, and control of posture.

The spinocerebellum receives its primary input from the spinal cord via the dorsal and ventral **spinocerebellar tracts**, which convey **proprioceptive** (body position and movement) signals from muscle and joint receptors and are critical for movement coordination. The output of the spinocerebellum is directed to the spinal cord. The deep cerebellar nuclei of the spinocerebellum (i.e., the fastigial and interposed nuclei) project to cortical and brainstem regions, giving rise to corticospinal, rubrospinal, vestibulospinal, and reticulospinal descending motor pathways. The spinocerebellum uses somatosensory input to modulate descending motor command signals from cortical and brainstem motor areas to produce adaptive motor coordination.

The vermis and fastigial nuclei coordinate the movements of the central body (e.g., trunk, head, proximal limbs). The vermis receives inputs from spinocerebellar projections conveying information from the trunk. The vermis projects to the fastigial nuclei, which in turn project bilaterally to the lateral vestibular nuclei and brainstem reticular formation nuclei that modulate descending motor signals directed to the axial muscles. The vermis thereby regulates coordination of the trunk musculature, regulating the accuracy of trunk, leg, and head movements, which are critical for the control of posture and locomotion.

The intermediate zones and interposed nuclei coordinate movements of the limbs (i.e., arms, legs, hands, feet). The intermediate zones receive inputs from spinocerebellar projections conveying information from the distal limbs. The intermediate zones project to the interposed nuclei, which in turn project to the red nuclei and motor cortex (via the ventral thalamus), modulating descending motor signals carried by the dorsolateral rubrospinal and corticospinal tracts. The intermediate (paravermal) zones thereby regulate the accuracy of voluntary movements of the arms and hands (e.g., reaching and grasping movements), as well as legs and feet.

Topograghic maps exist within the spinocerebellum. Axial portions of the body (head, neck, trunk, shoulders, and hips) are represented within the vermis. The limbs are represented in the intermediate zones, and each half of the body is represented in the ipsilateral cerebellar cortex. The somatotopic maps are not nearly as precise as those of the cerebral cortex; they are fractured, so that adjacent areas of cerebellar cortex may contain representations of noncontinuous areas of the body.

The Cerebrocerebellum

The **cerebrocerebellum**, also known as the **pontocerebellum** or **neocerebellum**, is the largest functional subdivision of the human cerebellum. It comprises the lateral hemispheres and the dentate nuclei (see Figure 9.4). This functional division has extensive reciprocal connections with the cerebral cortex; afferents from the cerebral cortex arrive to the cerebellum via the pontine nuclei, while cerebellar efferents are directed back to cerebral cortex via the ventral lateral nucleus of the thalamus. This region functions primarily in the service of planning and timing for the coordination of movement of the arms and hands. It plays a role in motor learning of the sequential components of skilled movements which pertain to the upper extremities predominantly (e.g., learning to play piano or hit a baseball) by fine-tuning motor programs to increase movement accuracy with practice. It also plays a role in cognitive function.

The cerebrocerebellum is not topographically organized. The major afferents to this region originate from motor association cortex and terminate in the cerebellar cortex via the pontine nuclei and the middle cerebellar peduncles. These **cortico-ponto-cerebellar** inputs provide the cerebellum with information about volitional movements that are anticipated or are in progress. The major efferents of this region are directed to primary motor and premotor cortex via the dentate nucleus, superior cerebellar peduncles, and ventral lateral nucleus of the thalamus. Through this **cortico-ponto-cerebellar-thalamocortical** circuitry, the cerebrocerebellum modifies activity in the motor cortex and descending corticospinal pathways, and fine-tunes the precision of force, direction, and extent of volitional movements.

The lateral zones receive input from, and control output to, the ipsilateral side of the body. Cerebellar lesions therefore result in ipsilateral motor deficits, unlike lesions of the motor cortex of the cerebrum, which result in contralateral deficits. This ipsilateral motor control is due to the two decussations in the circuit interconnecting the cerebellum and the cerebral cortex (Figure 9.6). The pontine nuclei receive input from ipsilateral cerebral cortex and project to the contralateral cerebellar hemisphere. Each cerebellar hemisphere projects to the contralateral thalamus, which in turn projects to the ipsilateral cortex. For example, the left cerebral cortex projects to pontine nuclei on the left side, which project to the right cerebellar hemisphere, which projects to the left thalamus, which projects to the left cerebral cortex. Thus, a lesion to the right cerebellar

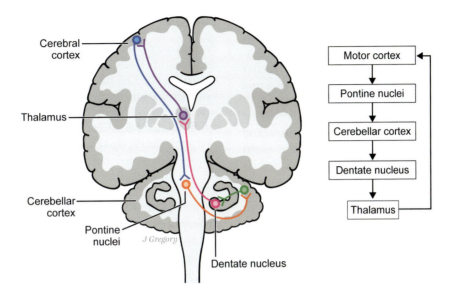

FIGURE 9.6. Ipsilateral motor control by the cerebellum results from two decussations in the cerebrocerebellar circuit interconnecting the cerebellum and the cerebral cortex; from the pontine nuclei to the cerebellar cortex of the contralateral cerebellar hemisphere, and from the dentate nucleus to the contralateral thalamus.

hemisphere results in disordered movement on the right side of the body.

The cerebrocerebellum is also involved in non-motor functions. In humans, the cerebellar hemispheres receive afferents from *all* areas of the cerebral cortex by way of the pontine nuclei. The cerebellar hemispheres project to the dentate nuclei, which in turn send prominent projections to the frontal lobes. This anatomy suggests that the cerebellar hemispheres may also play a role in cognitive functions. Functional imaging studies have shown heightened activity in this region during cognitive tasks, including problem-solving tasks and some language operations (e.g., generation of appropriate verbs from a list of nouns).

Signs and Symptoms of Cerebellar Damage

The main roles of the cerebellum are the maintenance of balance and equilibrium, and coordination of complex motor activity. It also plays a role in motor learning of skilled movements by fine-tuning motor programs and increasing movement accuracy and consistency. The cerebellum does not initiate descending motor commands or decide which movements to execute, but rather calibrates the detailed form of movements by regulating central motor pathways by side loop circuits that modify the activity within the main descending pathways to make movements more adaptive and accurate. Thus, cerebellar lesions do not produce paralysis or paresis (weakness); rather, they cause movements to become erratic in amplitude and direction.

Most movements, even simple ones, involve multiple muscle groups (agonist and antagonist) acting in a coordinated fashion. The cerebellum increases the spatial and temporal accuracy of the individual components of movements by regulating the rate, rhythm, and force of muscle contractions to produce smooth, fluid movements. A disturbance in this coordination results in movements that have an irregular, wavering course. The archetypal sign of cerebellar dysfunction is **ataxia** (Greek *ataxia*, "without order/arrangement"), a movement disorder characterized by lack of coordination.

The cerebellum influences descending motor signals by comparing inputs from the motor cortex that signal motor commands, to sensory feedback inputs signaling the actual movement that is in progress, and correcting discrepancies by sending corrective feedback signals to upper motor neurons. The task of comparing motor intent to actual movement requires extensive signal input; cerebellar afferent (input) fibers are far more numerous than efferent (output) fibers by a ratio of 40:1.

The lateral hemispheres of the cerebellum are primarily involved in coordination of the appendicular (limb) musculature, whereas midline regions and flocculonodular lobe are primarily involved in maintaining balance and coordination of the axial musculature using vestibular, proprioceptive, and visual feedback. Thus, the specific signs and symptoms of cerebellar ataxia are related to the location of cerebellar pathology. Lesions of a cerebellar hemisphere or its connections affect coordination of the ipsilateral distal limb (because there are two decussations in the circuit), while movement of the trunk is preserved. Midline lesions of the vermis and/or flocculonodular lobe affect whole-body posture, locomotion, and balance, while movement of the limbs is preserved. Unilateral midline cerebellar damage causes bilateral motor deficits because the medial motor system of the cerebellum influences the trunk muscles bilaterally. Diffuse cerebellar lesions produce generalized signs and symptoms affecting the body symmetrically.

While ataxia is often due to cerebellar pathology, it can also be due to disruption of proprioceptive and vestibular sensory systems that provide afferent input to the cerebellum. Therefore, ataxias are classified as cerebellar or sensory in origin (i.e., **cerebellar ataxia**, **proprioceptive ataxia**, **vestibular ataxia**). On clinical examination it is also important to distinguish ataxia from other motor abnormalities such as weakness and apraxia. Weakness occurs with upper motor neuron pathology, lower motor neuron pathology, **neuromuscular disease** (i.e., affecting motor neuron to muscle cell communication at the neuromuscular junction), and **myopathy** (muscle disease). **Apraxia** is a loss of the ability to perform learned skilled movements despite normal sensation, muscle strength, coordination, and comprehension; it is a disorder of higher order motor system function at the cortical level, exclusive of primary motor cortex (see Chapter 24, "The Motor System and Motor Disorders").

The **cerebellar motor syndrome** includes a broad array of signs, including postural and gait abnormalities, limb ataxia, tremor, speech deficits, and nystagmus. The neurological examination of cerebellar function aims to reveal such disturbances.

Truncal Ataxia

Cerebellar control of **posture** and **balance** is mainly accomplished by the vermis, its projections to the fastigial nuclei, its projections to the brainstem reticular formation and vestibular system, and the reticulospinal and vestibulospinal pathways that project to spinal cord regions controlling the trunk musculature. Midline cerebellar lesions and vermal atrophy produce **truncal ataxia**, which is characterized by inability to maintain a static upright posture during sitting, standing, and walking. This results in abnormalities in **station** (standing) and **gait** (walking), and imbalance that results in frequent falls. To compensate for balance difficulties, those with truncal ataxia increase the base of their support by using a broad-based stance and gait (i.e., with the feet widely separated). Limb movements are otherwise normal. A common cause of truncal ataxia is **medulloblastoma**, a cancerous tumor that grows in the cerebellum of children (and is the most common malignant brain tumor of childhood). Presenting signs often consist of unsteady station and gait, with severe postural sway with the eyes opened and no change when the eyes are closed.

STATION

Truncal ataxia is characterized by inability to maintain a static posture while standing without support. Balance abnormalities during standing are characterized by postural sway (**titubation**). Persons with truncal ataxia use a wide-based stance, a stereotyped postural strategy to compensate for reduced balance. Those with the most severe forms of truncal ataxia make oscillating body movements while sitting upright, as they are unable to maintain posture without support; this worsens when the person is seated with arms stretched out in front of them.

A classic test for **ataxia of stance** is the ability to stand in **Romberg position** (i.e., heels together) with eyes open and without swaying from side to side. Excessive sway to both sides occurs with conditions that affect the midline cerebellum (such as alcohol intoxication). Persons with truncal ataxia also have a reduced ability to make reflexive postural adjustments in response to unexpected sudden perturbations that challenge postural equilibrium (e.g., being pushed at shoulder level) in order to restore balance.

GAIT

Persons with truncal ataxia also exhibit **gait ataxia**, a type of gait disorder (abnormal pattern of walking). Gait ataxia is characterized by staggering movements, irregular foot trajectories, a widened base of support, and a veering path of movement. Those with gait ataxia give the impression of being drunk; they may hold on to walls or furniture for stability while walking.

Gait is tested by examining the ability to walk in a straight line, perform heel-to-toe walking (**tandem gait**), walking on toes only, walking on heels only, and hopping in place on each foot.

Appendicular Ataxia

Lesions affecting the cerebrocerebellar system (i.e., the cerebellar hemispheres and their afferent and efferent pathways) result in the **neocerebellar syndrome**, which is characterized by a selective ataxia of the upper and lower limbs that is known as **appendicular ataxia** or **limb ataxia**. In humans, cerebellar disorders most commonly consist of a neocerebellar syndrome because the neocerebellum makes up most of the mass of the cerebellum. Unilateral lesions result in unilateral limb ataxia ipsilateral to the cerebellar lesion. Limb ataxia results in functional impairments such as clumsiness with writing, buttoning clothes, or picking up small objects.

Limb ataxia manifests as intention tremor, dysmetria, decomposition of movement, and dysdiadochokinesia (see below). These ataxic symptoms are all attributed to a fundamental defect in control of the rate, range, and force of movements. They are superimposed on volitional movements that are otherwise intact. Within the neurological examination, cerebellar hemispheric function is assessed by tests of limb coordination.

INTENTION TREMOR

Intention tremor, also known as **cerebellar tremor**, is a slow (3–5 Hz), high-amplitude (i.e., easily visible) oscillation of the extremities that occurs with purposeful movements directed toward a target, such as pointing, reaching, and grasping (e.g., pressing a button, reaching to pick up a glass). The tremor oscillations become larger (i.e., the tremor is accentuated) at the endpoint of the movement. This symptom is due to defective feedback control from the cerebellum on cortically initiated movement. The tremor is not present at rest, in contrast to the resting tremor of Parkinson's disease (see Chapter 11, "The Basal Ganglia").

DYSMETRIA

Dysmetria is an ataxic phenomenon in which movements directed to a target (i.e., point-to-point movements) are too large or too small, resulting in undershooting (**hypometria**) or overshooting (**hypermetria, past-pointing**) and missing the target. Dysmetria may affect hand, arm, leg, and eye movements (both saccadic and pursuit).

DECOMPOSITION OF MOVEMENT

Most movements require coordinated activity of multiple muscle groups and joints in order to produce a smooth trajectory of the body part through space. With cerebellar lesions, the timing of different parts of the movement is adversely affected. This results in **decomposition of movement** (**asynergy**) in which complex movements are broken down into components and are performed in jerky stages or in a series of single simple movements that may appear mechanical and puppet-like.

DYSDIADOCHOKINESIA

Dysdiadochokinesia (**dysrhythmokinesis**) is a reduction in the ability to perform **rapidly alternating movements** using antagonistic muscle groups. It is characterized by irregular rhythm (rate) of successive movements, irregular amplitude (range) of successive movements, and slowness of movement. It is due to impaired timing of initiation and stopping of each phase of a repetitive movement.

Joseph Babinski (1857–1932) noted that patients with cerebellar disease have no difficulties pronating or supinating the forearm as a single movement but have great difficulty performing the same movements in rapid succession and do so in a markedly slowed fashion. Babinski coined the term *diadochokinesis* (*diadochokinesia*) from the Greek *diadochos* ("succeeding") and *kinesis* ("movement"), referring to the ability to make antagonistic movements in quick succession.

Tests for dysdiadochokinesia include rapid repetitive pronation and supination of the hand, rapid repetitive flexion and extension of the fingers (i.e., opening and closing of the fist), rapid tapping of the fingers, and imitating screwing in a light bulb. The disorder is more prominent with rapidly alternating movements involving multiple joints, such as imitation of screwing in a light bulb, which involves the shoulder and wrist joints. Tests for dysdiadochokinesia in speech involve repetition of syllables (most commonly, pa-ta-ka) that involve rapid succession of lip, tongue, and soft palate movements.

TESTS FOR LIMB ATAXIA

A classic test for limb ataxia is the **finger-to-nose test** in which the patient rapidly and alternately touches their own nose and the examiner's finger as quickly as possible; the examiner may increase the difficulty of the task by moving their finger between touches. The examiner observes whether the movements are smooth and accurate, or whether there is dysmetria and/or intention tremor. Other tests include rapid repeated tapping of the thumb and forefinger together, clapping hands, and clapping hands with one hand alternating between palmar and dorsal surface making the contact. In examining limb coordination, rhythmicity and speed of the movement are evaluated.

Ataxic Dysarthria

Ataxic dysarthria is a speech disorder that is due to damage to the cerebellum, or its input or output pathways. It is characterized by slow, effortful articulation and staccato prosody (speech rhythm), with the words broken down into their individual syllables, often separated by a noticeable pause. Ataxic dysarthria is also referred to as **scanning speech** or **staccato speech**; it is viewed as a form of decomposition of movement.

Nystagmus

Nystagmus is an involuntary (reflexive), repetitive, to-and-fro oscillation of the eyes. **Physiological nystagmus** occurs normally in a variety of circumstances. For example, **gaze-evoked nystagmus** is induced when attempting to maintain an extreme gaze position (e.g., move your eyes as far to the left as possible). **Vestibular nystagmus** is induced by rotation of the body in space. **Optokinetic nystagmus** is induced by full-field visual motion when the head remains stationary (e.g., looking out the window as a passenger in a car or train). It consists of slow-phase pursuit eye movements that track an object (e.g., individual telephone poles on the side of the road), followed by a fast-phase saccadic return movement triggered when the object moves out of the field. Nystagmus also occurs with a variety of pathological conditions, including but not limited to lesions affecting the vestibular organ of the inner ear, the central connections of the vestibular system to the cerebellum, the vestibulocerebellum itself, or the connections of the vermis to ocular motor nuclei by way of the vestibular nuclei and reticular formation.

In the neurological examination, the examiner assesses whether eye movements pursue a moving object smoothly, whether the nystagmus is greater than normal or sustained when the eye is at maximum deviation, and whether nystagmus occurs at lesser deviations.

Cerebellar Cognitive Affective Syndrome

Although controversial, a growing body of evidence indicates that the cerebellum contributes to the regulation of a range of cognitive and affective functions, and that cerebellar pathology may be associated with non-motor signs and symptoms. Selective motor and cognitive deficits following cerebellar damage are believed to reflect the disruption of distinct cerebro-cerebellar motor and cognitive circuits. Lesions of the cerebellar hemispheres may be associated with a **cerebellar cognitive-affective syndrome**, also known as **Schmahmann's syndrome**.

The cerebellar cognitive-affective syndrome is described as involving deficits in four domains: (1) language, (2) executive function, (3) visual-spatial processing, and (4) affective regulation. The language deficits consist of anomia, agrammatism, dysprosody, and reduced verbal fluency (which occurs in the absence of ataxic dysarthria). Right lateralized posterolateral cerebellar lesions are associated with reduced verbal fluency, reflecting interconnections with left cerebral hemisphere language areas. The executive function deficits observed in patients with focal cerebellar injury include reduced performance on tests of working memory, multitasking, problem-solving, planning, abstract thinking, and mental flexibility. Visuospatial deficits are typically characterized by a disorganized approach to drawing, and thus may reflect executive function deficits in planning and organization. Affective dysregulation is characterized by difficulty modulating affect and flattened affect. Persons with cerebellar damage also have trouble with cognitive timing, such as judging the relative duration of time intervals (i.e., judging whether one time interval is longer or shorter than an immediately previous interval) and the relative velocity of moving stimuli (i.e., judging whether one moving stimulus is moving faster or slower than an immediately previous moving stimulus). The cerebellar cognitive-affective syndrome is thus conceptualized as a "dysmetria of thought," analogous to the dysmetric movements that characterize the cerebellar motor syndrome.

It is difficult to evaluate the controversial hypothesis that the cerebellum plays a critical role in cognitive processes, because motor deficits impact cognitive test performance. Motor deficits such as dysarthric or ataxic speech impact performance on tests requiring timed verbal responses, limb ataxia impacts performance on tests requiring timed motor responses, and nystagmus impacts performance on tests requiring visual scanning. Furthermore, cerebellar lesions due to stroke, trauma, tumors, and developmental anomalies often affect other brain structures, and treatments such as cerebellar resection, chemotherapy, and radiation may themselves produce cognitive and/or affective changes. Thus, the cerebellar cognitive-affective syndrome is not characterized by a distinct profile of cognitive and affective disturbances; the nature and degree of signs and symptoms depend on the location and extent of the damage within the cerebellum, the acuteness of the pathology, and whether other brain structures are affected.

Cerebellar Pathology

Common causes of cerebellar dysfunction are stroke, tumors, trauma, cerebral palsy, multiple sclerosis, neurodegenerative disease, neurodegeneration due to chronic alcoholism, and acute alcohol intoxication. Congenital malformations affecting the cerebellum include Chiari malformations and Dandy-Walker syndrome.

Cerebellar Infarction

The most common signs and symptoms of cerebellar ischemia are a feeling of being off-balance or dizzy and an abnormal gait, followed by dysarthria, limb incoordination, and nystagmus. Cerebellar infarcts are often associated with good recovery and relatively minor long-term neurologic dysfunction; however, acutely, they pose a threat because edema (swelling) within the confines of the posterior cranial fossa can lead to coma or death if not treated quickly.

Massive cerebellar infarcts often cause cerebellar edema and result in increased pressure within the posterior cranial fossa. This can lead to compression of the fourth ventricle, obstructive hydrocephalus, brainstem compression, brainstem infarction due to compression of brainstem arteries, and downward herniation through the foramen magnum or upward transtentorial herniation. Mass effect within the posterior cranial fossa is potentially fatal, and when there is a threat of brainstem compression or herniation, short-term neurosurgical management may be required. This is accomplished by **ventriculostomy (external ventricular drain, extraventricular drain)** to divert fluid from the ventricular system (and simultaneously monitor intracranial pressure), or suboccipital **decompressive craniectomy**, a neurosurgical procedure in which part of the skull is removed to manage a medically intractable rise in intracranial pressure.

Tumors

The cerebellum, brainstem, and fourth ventricle occupy the posterior cranial fossa. Posterior fossa tumors are considered as some of the most critical brain lesions because of the potential for brainstem compression, herniation, and potentially death. The cerebellum is the most common site of presentation of CNS tumors in children; cerebellar tumors are rare in adults. Medulloblastomas and cerebellar astrocytomas account for the majority of tumors affecting the cerebellum.

Symptoms of cerebellar tumors may be caused by focal compression, increased intracranial pressure (i.e., headache, vomiting, and blurred or double vision), and hydrocephalus due to obstruction of the fourth ventricle. Midline lesions of the anterior and posterior lobes cause truncal and gait ataxia. Lesions involving the lateral cerebellar hemispheres produce limb ataxia. Medulloblastomas typically arise within the vermis, and typically present with vomiting, headache, and an ataxic gait. In adults, cerebellar tumors frequently arise in the **cerebellopontine angle cistern**, the triangular space between the anterior surface of the cerebellum and the lateral surface of the pons that is filled with cerebrospinal fluid. Cerebellopontine angle tumors affect the lateral hemispheres and present with limb ataxia.

Brainstem tumors that infiltrate the cerebellar peduncles or compress the cerebellum also cause cerebellar symptomatology.

Acute Alcohol Intoxication and Chronic Alcoholism

ACUTE ALCOHOL INTOXICATION

Acute alcohol intoxication produces short-term cerebellar dysfunction and ataxia. The cellular and molecular mechanisms, however, are unclear. The **Standardized Field Sobriety Test** is a battery of tests used by police officers during a traffic stop to determine if a person suspected of impaired driving is intoxicated. The battery includes the horizontal gaze nystagmus test, the walk-and-turn test, and the one-leg stand test. In the horizontal gaze nystagmus test, the officer asks the person to follow an object, such as a pen or small flashlight, that moves slowly in a horizontal plane, with their eyes. With alcohol intoxication, the eyes cannot follow a moving object smoothly, and there is distinct and sustained nystagmus that is far greater than normal physiologic nystagmus. In the walk-and-turn test, the person is directed to take nine steps by heel-to-toe walking (i.e., tandem gait) along a straight line, turn on one foot, and return in the same manner in the opposite direction. In the one-leg stand test, the person is instructed to stand on one foot and count aloud by ones for 30 seconds.

CHRONIC ALCOHOLISM

Chronic alcoholism causes degeneration of the vermis and is a common cause of acquired ataxia. In early stages, alcoholic cerebellar degeneration occurs predominantly within the anterior vermis. As the disease progresses, lesions spread to the posterior vermis, and at later stages to the lateral hemispheres of the anterior lobe.

The Spinocerebellar Ataxias

The **spinocerebellar ataxias** (SCAs) are a group of genetic (hereditary) disorders characterized by ataxia due to degeneration of the cerebellum and sometimes including the spinal cord. They manifest as a slowly progressive cerebellar syndrome. The SCAs are classified according to the responsible gene mutation; most are inherited in an autosomal dominant manner. There are more than 40 specific spinocerebellar ataxias; they are named with the prefix "SCA" followed by a number reflecting the order of identification (e.g., SCA1).

Congenital Malformations

Congenital malformations of the cerebellum are usually sporadic (non-hereditary) and often occur as part of complex malformation syndromes that affect other parts of the CNS.

CHIARI MALFORMATION

Chiari malformation is a structural defect of the cerebellum. It is usually due to a congenital malformation in which the posterior cranial fossa is abnormally small relative to the size of the cerebellum (i.e., there is a mismatch between the size and contents of the posterior fossa), and the cerebellar tonsils are located below the foramen, protruding into the spinal canal. The severity of Chiari malformations varies widely, and they are traditionally classified according to the degree to which the cerebellar tonsils protrude through the foramen magnum. Specific symptoms generally reflect dysfunction of the cerebellum, brainstem, lower cranial nerves, and spinal cord.

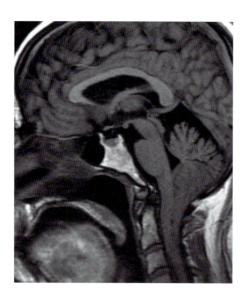

FIGURE 9.7. Chiari I malformation. Sagittal T1-weighted MRI shows cerebellar tonsillar descent with peg-like morphology (the tonsils are pointed rather than rounded).

In **Chiari malformation type I**, the most common variant, the cerebellar tonsils only protrude below the foramen magnum (Figure 9.7). Chiari I malformations are often asymptomatic, and in many people the malformations are discovered only incidentally through a neuroimaging study performed during diagnosis for another condition. In "asymptomatic" cases, adults may present with occipital headache that often worsens with exertion. Most cases are congenital (primary), but acquired Chiari I (secondary) may occur with supratentorial mass (space-occupying) lesion, hydrocephalus, or history of multiple lumbar punctures. **Posterior fossa decompression surgery** is performed in symptomatic patients.

Type II and III Chiari malformations are congenital defects in which the neural tube fails to close normally during embryologic development. In **Chiari malformation type II**, also known as **Arnold-Chiari malformation** and **classic Chiari malformation**, both the cerebellum and the brainstem protrude below the foramen magnum. This variant is typically seen in infants with **spina bifida** and is often associated with hydrocephalus and significant neurological symptoms. **Chiari malformation type III** is characterized by an **encephalocele**, in which a portion of the brain and meninges protrudes through an abnormal opening in the skull; in this case it is the cerebellum that protrudes through the back of the skull. Chiari malformation type III is very rare and is associated with severe neurological deficits.

Chiari malformation type IV is a rare congenital condition characterized by cerebellar hypoplasia (i.e., incomplete development), although the cerebellum is located in its normal position.

DANDY-WALKER SYNDROME

Dandy-Walker syndrome is a rare, congenital malformation of the cerebellum characterized by agenesis (i.e., failure to develop) or hypoplasia of the cerebellar vermis, hydrocephalus due to cystic dilatation of the fourth ventricle, and enlargement of the posterior fossa. Symptoms often occur in early infancy and include slow motor development and progressive enlargement of the skull. In older children, there may be symptoms of increased intracranial pressure.

Summary

The cerebellum lies within the posterior cranial fossa and below the tentorium, along with the brainstem. It consists of three lobes: anterior, posterior, and flocculonodular. The anterior and posterior lobes consist of a midline vermis that is flanked on each side by two cerebellar hemispheres. Three pairs of cerebellar peduncles carry afferents to and efferents from the cerebellum: superior, middle, and inferior. The cerebellum is divided into three functional zones based on anatomical connections: vestibulocerebellum, spinocerebellum, and cerebrocerebellum.

The cerebellum plays a critical role in movement. It regulates central motor pathways by side loop circuits that modify the activity within the main descending pathways, and plays a role in the maintenance of balance, posture, coordination of complex motor activity, and motor learning.

The archetypal sign of cerebellar dysfunction is ataxia in which movements are erratic in amplitude and direction. Midline cerebellar lesions and vermal atrophy produce truncal ataxia, which is characterized by the inability to maintain a static upright posture during sitting, standing, and walking. This results in abnormalities in station and gait. Lesions affecting the cerebellar hemispheres produce limb ataxia, which is characterized by intention tremor, dysmetria, decomposition of movement, and dysdiadochokinesia. Unilateral cerebellar hemisphere lesions produce ipsilateral unilateral limb ataxia, due to the fact that there are two decussations in the circuit interconnecting the cerebellum and cerebral cortex. Other consequences of cerebellar damage include ataxic dysarthria and nystagmus. Lesions of the cerebellar hemispheres are also associated with a cerebellar cognitive-affective syndrome, although this is controversial.

Causes of cerebellar dysfunction include stroke, tumors, multiple sclerosis, neurodegenerative disease, neurodegeneration due to chronic alcoholism, and acute alcohol intoxication. Congenital malformations affecting the cerebellum include Chiari malformations and Dandy-Walker syndrome.

Additional Reading

1. Brodal P. The cerebellum. In: Brodal P, *The central nervous system: structure and function*. 5th ed. Oxford University Press; 2016. https://oxfordmedicine.com/view/10.1093/med/9780190228958.001.0001/med-9780190228958-chapter-24. Accessed July 4, 2022.
2. Blumenfeld H. Cerebellum. In: Blumenfeld H, *Neuroanatomy through clinical cases*. 3rd ed., pp. 699–739. Sinauer Associates; 2021.
3. Buckner RL. The cerebellum and cognitive function: 25 years of insight from anatomy and neuroimaging. *Neuron*. 2013;80(3):807–815. doi:10.1016/j.neuron.2013.10.044
4. D'Angelo E. Physiology of the cerebellum. *Handb Clin Neurol*. 2018;154:85–108. doi:10.1016/B978-0-444-63956-1.00006-0
5. De Smet HJ, Paquier P, Verhoeven J, Mariën P. The cerebellum: its role in language and related cognitive and affective functions. *Brain Lang*. 2013;127(3):334–342. doi:10.1016/j.bandl.2012.11.001
6. Koziol LF, Budding D, Andreasen N, et al. Consensus paper: the cerebellum's role in movement and cognition. *Cerebellum*. 2014;13(1):151–177. doi:10.1007/s12311-013-0511-x
7. Mariën P, Borgatti R. Language and the cerebellum. *Handb Clin Neurol*. 2018;154:181–202. doi:10.1016/B978-0-444-63956-1.00011-4
8. Molinari M, Masciullo M, Bulgheroni S, D'Arrigo S, Riva D. Cognitive aspects: sequencing, behavior, and executive functions. *Handb Clin Neurol*. 2018;154:167–180. doi:10.1016/B978-0-444-63956-1.00010-2
9. Rogers JM, Savage G, Stoodley MA. A systematic review of cognition in Chiari I malformation. *Neuropsychol Rev*. 2018;28(2):176–187. doi:10.1007/s11065-018-9368-6
10. Schmahmann JD. The role of the cerebellum in cognition and emotion: personal reflections since 1982 on the dysmetria of thought hypothesis, and its historical evolution from theory to therapy. *Neuropsychol Rev*. 2010;20:236–260. doi:10.1007/s11065-010-9142-x
11. Schmahmann JD. The cerebellum and cognition. *Neurosci Lett*. 2019;688:62–75. doi:10.1016/j.neulet.2018.07.005
12. Sokolov AA, Miall RC, Ivry RB. The cerebellum: adaptive prediction for movement and cognition. *Trends Cogn Sci*. 2017;21(5):313–332. doi:10.1016/j.tics.2017.02.005
13. Stoodley CJ, Schmahmann JD. Functional topography of the human cerebellum. *Handb Clin Neurol*. 2018;154:59–70. doi:10.1016/B978-0-444-63956-1.00004-7
14. Vanderah TW, Gould D. Cerebellum. In: Vanderah TW, Gould D, *Nolte's the human brain: an introduction to its functional anatomy*. 8th ed., pp. 469–494. Elsevier; 2021.

The Diencephalon

Introduction

The **diencephalon** lies at the central core of the cerebrum, surrounding the third ventricle and surrounded by the cerebral hemispheres (telencephalon). The major structures of the diencephalon are the thalamus, which accounts for about 80% of the diencephalon, and the hypothalamus, which lies inferior to the thalamus (Figure 10.1). No component of the thalamus is visible from the external surface of the brain, and only a small portion of the hypothalamus is visible from the ventral surface. The thalamus and hypothalamus are each composed of many nuclei that perform many different functions. This chapter describes basic thalamic anatomy, thalamic nuclei and their functions, thalamic stroke syndromes, basic hypothalamic anatomy, hypothalamic control of lower autonomic and endocrine centers, hypothalamic control of motivated behaviors required for survival and reproduction, and hypothalamic syndromes.

Other components of the diencephalon are the metathalamus, subthalamus, and epithalamus. The **metathalamus** lies caudoventrally to the main body of the thalamus; it consists of the medial and lateral geniculate bodies, which are components of the auditory and visual pathways, respectively. The medial and lateral geniculate bodies are also referred to as the medial and lateral geniculate nuclei, respectively, and are usually included in descriptions of thalamic nuclei. The **subthalamus** lies immediately ventral to the thalamus and dorsal to the hypothalamus. It includes the **subthalamic nucleus**, which has connections to the basal ganglia and is involved in movement. The **epithalamus** lies superior and posterior to the thalamus; it consists of the pineal gland and habenular nuclei. The **pineal gland** plays a role in regulating the body's **circadian** (24-hour) rhythms through the synthesis and release of **melatonin**. The function of the habenular nuclei is not understood, but there is some evidence that they play a role in emotional and social behaviors.

Thalamus

Anatomy

The **thalamus** ("inner chamber"; plural, thalami) is an ellipsoid-shaped mass of gray matter (Figure 10.2). It is a paired structure, and the two thalami lie on either side of the third

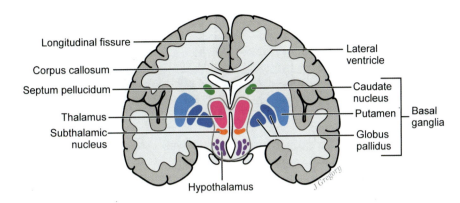

FIGURE 10.1. Coronal section through the cerebrum. The diencephalon lies at the central core of the cerebrum, surrounding the third ventricle. The major structures of the diencephalon are the thalamus and the hypothalamus.

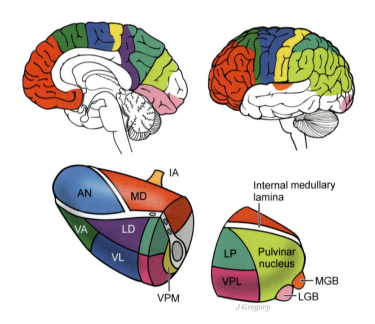

FIGURE 10.2. The two thalami lie on either side of the third ventricle, interconnected medially by the massa intermedia, also known as the interthalamic adhesion (IA). Each thalamus is divided into several distinct nuclei named according to their relative positions, using the six terms of orientation: dorsal (D), ventral (V), anterior (A), posterior (P), medial (M), and lateral (L). The internal medullary lamina forms a Y-shaped partition that traverses the long axis of the thalamus and divides it into three major nuclear regions: the anterior, medial, and lateral nuclear groups. The metathalamus consists of the lateral geniculate bodies (LGB) and the medial geniculate bodies (MGB), components of the visual and auditory pathways, respectively.

ventricle, inferior to the bodies of the lateral ventricles. The thalami are interconnected medially by the **massa intermedia (interthalamic adhesion)**, which runs like a bridge through the third ventricle. This structure is *not* a commissure; each thalamus functions separately within each cerebral hemisphere. Very little is known about the role of this structure. Interestingly, the massa intermedia is absent in approximately 30% of people without apparent consequence; this normal variant is unrelated to age, ethnicity, and race, but it has been observed more frequently in schizophrenia spectrum disorder and bipolar disorder. In contrast to the absence of massa intermedia, double massa intermedia occurs rarely.

The thalami are surrounded by the basal ganglia, with the head of the caudate nucleus lying anterior and superior to each thalamus, and the lenticular nucleus (globus pallidus and putamen) lying lateral to each thalamus. Horizontal sections show that: (1) the two thalami are separated by the third ventricle (i.e., the third ventricle forms their medial boundary); (2) the head of each caudate nucleus lies anterior to each thalamus; and (3) the lateral boundary of each thalamus is formed by the posterior limb of the internal capsule, which separates the thalamus from the more laterally lying lenticular nucleus (Figure 10.3).

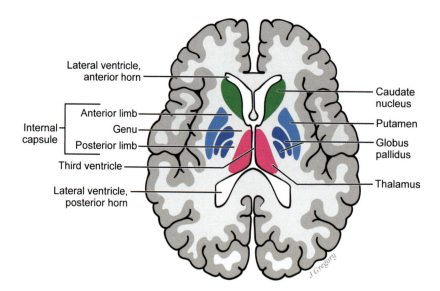

FIGURE 10.3. Horizontal cross section of the cerebral hemispheres. The two thalami are separated by the third ventricle. The internal capsule is a large V-shaped fiber bundle with an anterior limb, genu (bend), and posterior limb. Each thalamus lies medial to the posterior limb of the internal capsule. The lenticular nucleus (globus pallidus and putamen) lies lateral to the thalamus, and the head of the caudate lies anterior to the thalamus.

NUCLEAR GROUPS

Each thalamus is made up of multiple nuclei, many of which are named according to their relative positions using the six terms of orientation (anterior, posterior, dorsal, ventral, medial, and lateral). The long axis of the thalamus is traversed by the **internal medullary lamina**, a band of myelinated axons that forms a Y-shaped partition and divides the thalamus into three major nuclear regions: the anterior, medial, and lateral nuclear groups (see Figure 10.2). The lateral group is subdivided into dorsal and ventral divisions. The internal medullary lamina also encloses a group of nuclei known collectively as the **intralaminar nuclei**; these are the centromedian, parafascicular, paracentral, and central lateral nuclei. The lateral surface of the thalamus is covered by a curved sheet of myelinated fibers called the **external medullary lamina**. Just lateral to the external medullary lamina of each thalamus lies a reticular nucleus (which is not related to the brainstem reticular formation).

THALAMIC CONNECTIONS

There are almost no connections between the various nuclei of the thalamus. The majority of thalamic efferents project to the cerebral cortex. Nearly every nucleus of the thalamus sends axons to the cerebral cortex, and nearly every part of the cortex (except some parts of the temporal lobe) receives afferent fibers from the thalamus and has a corresponding thalamic source nucleus. Almost all information received by the cortex is first relayed through synapses in the thalamic nuclei; thus, the thalamus acts as a relay center, receiving information from a variety of sources and distributing it to the cerebral cortex. For this reason, the thalamus is often referred to as the gateway to the cerebral cortex. Because the thalamus provides so much of the input to the cerebral cortex, the thalamus drives cortical activity.

There is reciprocal innervation between the thalamus and cortex. Every thalamocortical projection is copied faithfully by a reciprocal corticothalamic connection. The corticothalamic and thalamocortical connections are composed of myelinated nerve fibers, known as the **thalamic radiations** because they radiate out from the thalami in a fanning pattern.

Blood Supply

The arterial blood supply to the thalamus is provided by small perforating end-arteries arising from the posterior circulation (i.e., vertebrobasilar system) that result in four thalamic vascular territories: anterior, inferolateral, paramedian, and posterior. There is, however, some anatomical variation among individuals with respect to the parent vessel from which each branch arises, the number and position of arteries, and the nuclei supplied by each vessel.

The **anterior thalamic territory** is supplied by the tuberothalamic artery (anterior thalamic perforating artery, polar artery); this artery arises from the posterior communicating artery, a branch of the posterior cerebral artery (PCA). The **inferolateral thalamic territory** is supplied by several inferolateral arteries (thalamogeniculate arteries) that arise from the PCA. The **paramedian thalamic territory** is supplied by several paramedian arteries (thalamoperforating arteries) that arise from the PCA.

The **posterior thalamic territory** is supplied by the lateral branches of the posterior choroidal arteries, which arise from the PCA.

Thalamic Nuclei and Thalamic Function

The thalamic nuclei are classified into three types based on their cortical connections: specific relay nuclei, association nuclei, and nonspecific nuclei. The **specific relay nuclei** receive well-defined inputs from specific ascending sensory pathways and project to sharply defined, functionally distinct cortical regions that play a role in sensation or movement. The **association nuclei** have reciprocal connections with cortical association areas, which are areas of cortex that are neither sensory nor motor. The **nonspecific thalamic nuclei** form diffuse, reciprocal connections with large regions of cerebral cortex.

The thalamus receives and transmits signals from many neural pathways and plays critical roles in sensory and motor functions, arousal, cognition, and emotion. Lesions of the thalamus therefore result in complex syndromes involving various combinations of deficits. An additional source of neurological deficit following focal thalamic pathology is **diaschisis**, a loss of function within a brain region that is connected to, but distant from, the damaged area due to **deafferentation** (interruption or destruction of the afferent inputs). Given the strong thalamocortical interconnections, focal thalamic pathology may result in focal cortical dysfunction. Physiological evidence of diaschisis effects comes from functional neuroimaging studies by positron emission tomography (PET) and single-photon emission computed tomography (SPECT) showing depressed levels of metabolic activity in the cerebral hemispheres following discrete thalamic infarction. Functional neuroimaging studies have also shown diaschisis effects following **thalamotomy**, the selective neurosurgical ablation of part of the thalamus (the ventrolateral nucleus) to treat a variety of movement disorders such as Parkinson's disease, essential tremor, and dystonia.

SPECIFIC RELAY NUCLEI

The thalamus receives, processes, and relays most sensory information from lower brain structures en route to the cerebral cortex, excepting olfactory (sense of smell) signals, which reach the olfactory cortex without a thalamic relay. The **specific sensory relay nuclei** are the **lateral geniculate nucleus** (LGN) of the visual pathways, **medial geniculate nucleus** (MGN) of the auditory pathways, and **ventral posterior nucleus** (VP) of the somatosensory pathways. The VP nucleus has two major subdivisions: the **ventral posterolateral nucleus** (VPL), which receives sensory information from the body (trunk and limbs) via the dorsal column-medial lemniscal and spinothalamic pathways; and the **ventral posteromedial nucleus** (VPM), which receives sensory information from the head and face via the trigeminal nerve. A parvocellular (small cell) division of the VPM subserves gustation (taste). The specific sensory relay nuclei project directly to their respective primary sensory cortices; the LGN projects to primary visual cortex, the MGN projects to primary auditory cortex, and the VP projects to primary somatosensory cortex. Not surprisingly, focal lesions of sensory relay nuclei produce selective sensory deficits (e.g., a lesion of the LGN produces blindness in part of the visual field).

The **thalamic pain syndrome (Déjerine-Roussy syndrome)** is a rare post-stroke central neuropathic pain syndrome that occurs with infarction in the VPL relay nucleus of the somatosensory pathway. It is a type of **central poststroke pain syndrome**, pain that occurs with a stroke lesion disrupting any portion of the spinothalamic tract.

The thalamus also plays a role in motor control. The **specific motor relay nuclei** receive signals from the basal ganglia and cerebellum, and they project to the motor cortex, acting as a relay in feedback circuits that modify the outgoing signals from the motor cortex. The motor relay nuclei are the **ventral lateral nucleus** (VL), which receives its major input from the cerebellum, and the **ventral anterior nucleus** (VA), which receives its major input from the basal ganglia. Lesions affecting these nuclei play a role in motor disturbances such as dystonia, ataxia, and mild motor weakness that may occur with thalamic stroke and other thalamic pathologies.

ASSOCIATION NUCLEI

The association nuclei have reciprocal connections with cortical association areas; they play a role in the "higher-order" (non-sensory, non-motor) functions of memory, cognition, behavior, and emotion. The contributions of the thalamus to these functions are supported by studies of patients with focal thalamic lesions and memory, cognitive, affective, and/or behavioral disturbances. The precise role of the various thalamic nuclei in these functions is often unclear because naturally occurring lesions in humans rarely involve an individual thalamic nucleus. It is also unknown to what extent the changes in higher-order functions following focal thalamic lesions reflect diaschisis.

The **anterior nuclei** receive afferents from the mammillary bodies of the hypothalamus via the mammillothalamic tract and from the hippocampal formation via the fornix, structures that are critical to memory. Not surprisingly, the anterior nuclei of the thalamus also play a role in memory. Thalamic lesions involving the anterior thalamus produce amnesia, a severe impairment in the ability to acquire and retain new information with otherwise preserved cognitive function (see Chapter 26, "Memory and Amnesia"). Furthermore, amnesia caused by thalamic lesions is associated most often with lesions of the anterior thalamus. Amnesia that is caused by a thalamic lesion is referred to as **thalamic amnesia** or **diencephalic amnesia**. Several studies have reported selective impairment in verbal memory following infarction in the anterior nuclei of the left hemisphere (see Chapter 26, "Memory and Amnesia," for discussion of material-specific amnesia).

Korsakoff's syndrome is a severe form of diencephalic amnesia (see Chapter 26, "Memory and Amnesia"). Focal diencephalic lesions are caused by thiamine (vitamin B_1) depletion, which often occurs with chronic alcoholism. The pathology of Korsakoff's syndrome involves damage to multiple sites and fiber tracts; most strongly implicated are the anterior nuclei of the thalamus, the mammillary bodies of the hypothalamus, and the mammillothalamic tract that provides a unidirectional link from the mammillary bodies to the anterior nuclei; however, it is unknown why these areas are selectively vulnerable to neural damage caused by thiamine deficiency.

The **dorsomedial nucleus** (DM), also known as the **mediodorsal nucleus** (MD), has strong reciprocal connections with the prefrontal cortex, including orbital, medial prefrontal, lateral prefrontal, and anterior cingulate. Frontal behavioral syndromes following thalamic lesions usually involve the DM. It has been proposed that the connections of the DM segregate into distinct functional circuits, such that the zone connected reciprocally with the anterior cingulate cortex is involved in motivation, the zone connected reciprocally with orbitofrontal cortex is involved in behavioral inhibition, and the zone connected reciprocally with dorsolateral prefrontal cortex is involved in executive functions. **Apathy** (loss of initiative and interest in others, poor motivation, flattened affect) is the most common affective feature of unilateral or bilateral vascular thalamic lesions, often occurring with ischemic stroke in the territory supplied by the paramedian arteries which includes the DM. The DM also appears to play a role in memory. Its contribution to the amnesia of Korsakoff's syndrome has been controversial; however, studies of patients with amnesia following thalamic infarcts have shown that isolated lesions of this nucleus can result in amnesia.

The **pulvinar** is the largest of the thalamic association nuclei. It has strong connections to secondary visual areas and parietotemporal association areas, and it is an important component of the visual attention network. Unilateral lesions of the pulvinar result in a contralateral **neglect syndrome** (reduced awareness of stimuli on one side of space that is not attributable to sensory loss), resembling symptoms of posterior parietal cortex lesions (see Chapter 15, "The Parietal Lobes and Associated Disorders").

The thalamus also plays a role in language processing, and thalamic lesions may result in **thalamic aphasia**, a type of subcortical aphasia (see Chapter 26, "Language and the Aphasias"). There is a laterality effect, like that observed with lesions of the cerebral cortex; thalamic aphasia is usually associated with damage (e.g., infarcts) in the left anterior thalamus, as cortical aphasia is usually associated with damage to the left hemisphere. The aphasia may be due to cortical diaschisis resulting from decreased thalamic input. PET and SPECT studies in patients with thalamic aphasia have shown decreased metabolism and decreased blood flow in the cerebral cortex of the left hemisphere.

NONSPECIFIC THALAMIC NUCLEI

The **nonspecific thalamic nuclei** project diffusely throughout the cerebral cortex; they include the **intralaminar**, **midline**, and **reticular nuclei**. This diffuse input controls the level of cerebral activity by acting as a "pacemaker" for the electrical rhythm of the cerebral cortex, thereby establishing the fine-tuning of conscious states, arousal, alertness, and selective (focused) attention. These nuclei also play a role in the pathophysiology of epilepsy and disorders of consciousness.

The nonspecific thalamic nuclei are a critical component of the **ascending reticular activating system** (ARAS) underlying wakefulness and awareness (see Chapter 25, "Disorders of Consciousness"). The idea of an ascending arousal system promoting wakefulness originated from clinical observations of patients during an epidemic of encephalitis from 1915 to 1926. The encephalitis, known as **sleeping sickness** or **encephalitis lethargica**, was characterized by excessive sleep and damage in the midbrain and diencephalon, as described originally by neurologist **Constantin von Economo** (1876–1931). The distinction between specific and nonspecific thalamic nuclei is anatomically based, but electrophysiological characteristics of thalamic nuclei support this dichotomy. High-frequency stimulation of nonspecific thalamic nuclei results in desynchronization of the cortical electroencephalogram (EEG) and concomitantly elicits arousal.

Thalamic Pathology

Thalamic lesions occur with a wide variety of pathologies but are often accompanied by extra-thalamic lesions, obscuring the analysis of thalamic structure-function relationships. The earliest reports of the cognitive and behavioral consequences of thalamic lesions were of patients with thalamic hemorrhage. Hemorrhage, however, is seldom confined to the thalamus; it often includes remote effects on adjacent structures, such as pressure from the hemorrhagic mass and edema, as well as toxic effects of blood products. Cognitive and behavioral deficits have been reported with thalamic tumors, but these lesions also produce pressure effects on adjacent structures.

Studies of patients with ischemic stroke of the thalamus provide the most precise anatomic-clinical correlations and the greatest insights into the functional properties of the human thalamus. Infarcts within the individual vascular territories destroy thalamic nuclei in specific combinations and disrupt their reciprocal cortical-thalamic connections, resulting in **thalamic stroke syndromes**. The thalamic stroke syndromes are not attributable to individual thalamic nuclei because even small, focal ischemic lesions are seldom confined within nuclear boundaries.

Prognosis after thalamic infarction is favorable relative to infarctions of the cerebral cortex or other subcortical structures, due to the low incidence of mortality and good recovery from motor deficits. Long-term cognitive, mood, and personality changes do occur following thalamic infarct; however, the incidence of these sequelae is unknown.

ANTERIOR TERRITORY INFARCTS

The anterior territory includes parts of the anterior, ventral, and intralaminar regions of the thalamus; thus, infarcts in this territory affect a wide array of nuclei. Clinically, such infarcts produce a range of severe cognitive and behavioral deficits. Acutely, there are deficits in arousal and orientation. Chronic clinical features include severe amnesia, executive dysfunction (poor planning and motor sequencing, perseverative thinking, perseverative behavior), and persistent personality changes (apathy, lack of spontaneity). Sensory disturbances are rare, and when present they are minimal and transient. Motor disturbances consist of mild to moderate contralateral weakness or clumsiness.

INFEROLATERAL TERRITORY INFARCTS

The **inferolateral territory** includes the VP and VL nuclei supplied by the principal inferolateral branches, the MGN supplied by the medial inferolateral branches, and the pulvinar nucleus supplied by the inferolateral pulvinar branches. Inferolateral territory infarcts produce contralateral hemisensory loss, hemiparesis, hemiataxia, and central pain syndrome. Small-vessel disease in the penetrating inferolateral arteries can have a variety of presentations due to variable involvement of the vessels. Pure sensory stroke affecting the head and neck occurs with infarction of the VPM, while pure sensory stroke affecting the trunk and extremities occurs with infarction of the VPL. Such sensory deficits may be accompanied by post-stroke thalamic pain. There may also be ataxia if the infarction involves the VL, which conveys cerebellar fibers to the motor cortices.

PARAMEDIAN TERRITORY INFARCTS

The paramedian arteries arise from the first segment of the PCA and supply the intralaminar nuclear group and most of the DM nucleus. Unilateral thalamic infarction in the territory of the paramedian artery produces amnesia and hypersomnolence or coma. The memory deficit reflects injury to the DM nucleus; there is evidence of a left specialization for verbal memory that parallels the hippocampi (see Chapter 26, "Memory and Amnesia"), such that infarcts in the paramedian territory of the left thalamus result in verbal learning and memory deficits. Impaired arousal with fluctuating level of consciousness is a prominent feature in the early phase following paramedian territory infarct. It likely reflects damage within the intralaminar nuclei, part of the nonspecific thalamic nuclei that project diffusely throughout the cerebral cortex and control the level of cerebral activity.

Rarely, both paramedian thalami receive arterial blood flow from a single unilateral vessel arising from the first segment of one PCA. Occlusion of this vessel, usually due to embolism, produces bilateral paramedian thalamic infarctions and results in **thalamic dementia**. It is characterized by severe memory impairment with prominent disorientation in time, and a frontal behavioral syndrome with prominent apathy as well as perseveration and confabulation (see Chapter 17, "The Frontal Lobes and Associated Disorders"). This is a type of **strategic infarct dementia**, a severe cognitive deficit caused by a single small infarct located in a "strategic" brain region that plays a critical role in cognition and behavior.

POSTERIOR TERRITORY INFARCTS

The posterior territory supplies the geniculate nuclei, parts of the intralaminar region, and the pulvinar. Infarcts involving this territory result in a variety of signs and symptoms, depending on whether the lateral posterior choroidal arteries or medial posterior choroidal arteries are affected. These include binocular visual field deficit due to LGN damage (see Chapter 14, "The Occipital Lobes and Visual Processing"). Unilateral lesions of the right pulvinar result in a contralateral neglect syndrome. Occasionally, such lesions produce thalamic aphasia.

Hypothalamus

Anatomy

The **hypothalamus** lies at the base of the brain, just ventral to the thalamus. It is small, weighing approximately 4 grams and making up less than 1% of the total human brain mass. A small portion of the hypothalamus is visible from the ventral surface of the brain, lying posterior to the **optic chiasm** (an X-shaped structure formed by the crossing of the optic nerves as they enter the brain) and within the **interpeduncular fossa** (a depression between the cerebral peduncles of the midbrain); the visible structures are the mammillary bodies, tuber cinereum, and

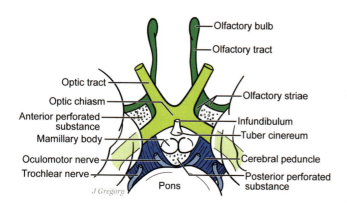

FIGURE 10.4. A small portion of the hypothalamus is visible from the ventral surface of the brain, lying posterior to the optic chiasm within the interpeduncular fossa.

pituitary stalk (Figure 10.4). The **mammillary bodies** are a pair of spherical masses, so named because of their breast-like shape. The **tuber cinereum** is a gray matter eminence situated between the optic chiasm anteriorly and the mammillary bodies posteriorly. The **pituitary stalk** (infundibulum) connects the hypothalamus to the pituitary gland.

The hypothalamus is composed of many distinct nuclei, arranged symmetrically around the floor and lower medial surface of the third ventricle. Their positions within the hypothalamus are described with respect to two axes, rostral-caudal (anterior-posterior) and medial-lateral. Three regions are defined with respect to the anterior-posterior axis: anterior, middle, and posterior. The **anterior region (supraoptic region)** lies above the optic chiasm. The **middle region (tuberal region)** lies above (and includes) the tuber cinereum. The **posterior region (mammillary region)** lies above the mammillary bodies. Three zones are defined with respect to the medial-lateral axis: the **lateral zone**, **medial zone**, and **periventricular zone**, which is the medial-most and lies immediately adjacent to the third ventricle. There are 11 major nuclei in the hypothalamus (Table 10.1).

The hypothalamus receives and gives rise to both neural and humoral (fluid) inputs and outputs. Neural input to the hypothalamus arrives from a wide variety of sources, many of which are considered "limbic" structures; afferent sources include the hippocampus, amygdala, orbitofrontal cortex, midline thalamic nuclei, basal forebrain, septal region, and brainstem afferents via the

TABLE 10.1 Major Nuclei of the Hypothalamus, Their Locations, and Functions

NUCLEUS	REGIONS(S)	ZONE(S)	FUNCTIONS
Preoptic	Anterior	Lateral, medial	Thermoregulation Sexual behavior
Anterior	Anterior	Medial	Thermoregulation Sexual behavior
Suprachiasmatic	Anterior	Medial	Circadian rhythms
Supraoptic	Anterior	Lateral, medial	Fluid balance (via posterior pituitary) Childbirth
Paraventricular	Anterior, middle	Medial, periventricular	Milk ejection (via posterior pituitary) Childbirth Autonomic control Anterior pituitary control
Dorsomedial	Middle	Medial	Emotion (rage)
Ventromedial	Middle	Medial	Appetite Body weight Insulin regulation
Arcuate	Middle	Medial, periventricular	Anterior pituitary control Feeding
Lateral Complex	Middle	Lateral	Appetite Body weight
Posterior	Posterior	Medial	Thermoregulation
Mammillary	Posterior	Medial	Emotion Short-term memory

reticular formation. The four major targets of neural output from the hypothalamus are the anterior thalamic nucleus, midbrain reticular nucleus, amygdaloid complex, and autonomic centers of the brainstem and spinal cord.

Function

The overall function of the hypothalamus is to maintain **homeostasis**, a steady state of the internal environment that is necessary for optimal physiological function. Physiological homeostasis promotes survival of the individual and species.

From body temperature to blood pressure to levels of certain nutrients, many physiological parameters are regulated to maintain levels within a homeostatic range despite changes in environment, diet, or level of activity. Each parameter is controlled by one or more homeostatic mechanisms, typically by a negative feedback loop that counteracts deviation from a physiological set point (the optimal functional level). Signals from the internal and external environments are conveyed to the hypothalamus by specific sensors throughout the body. The hypothalamus monitors these signals for deviation from a set point, which reflects the presence of a physiologic stressor. Deviations from set point values alter the discharge rate of set point neurons within specific hypothalamic nuclei, resulting in altered hypothalamic efferent outflow to counteract the physiological stressor and restore homeostasis. The hypothalamus accomplishes this by directing coordinated autonomic, endocrine, and somatic behavioral responses related to basic survival behaviors known as "the four Fs": feeding, fighting, fleeing, and fornicating.

Autonomic Function

The autonomic nervous system (ANS) is the visceral efferent division of the peripheral nervous system. It maintains stability of the body's internal environment for optimal function. The ANS works through innervation of smooth muscle, cardiac muscle, and the exocrine glands. **Exocrine glands** secrete their chemical products through ducts; these include the salivary glands, sweat glands, lacrimal glands, and mammary glands. (The liver and pancreas act as both exocrine and endocrine glands, secreting their products into the gastrointestinal tract through ducts, and hormones directly into the bloodstream).

The ANS regulates the function of the internal organs and vegetative processes concerned with maintenance of life, such as circulation, respiration, body temperature, and digestion. The four **primary vital signs** of pulse (heart rate), blood pressure, breathing (respiratory) rate, and body temperature reflect autonomic function. The autonomic system largely functions involuntarily in a reflexive manner (without conscious control). Autonomic reflexes are established at the brainstem and spinal cord levels, and they may be carried out without input from neural structures above the brainstem.

The hypothalamus acts as a higher center for the control of lower autonomic centers. **Sir Charles Sherrington** (1857–1952) referred to the hypothalamus as the "head ganglion of the autonomic nervous system." Hypothalamic influences over the autonomic system are mediated by direct innervation of parasympathetic preganglionic neurons in the brainstem and sacral spinal cord, direct connections to sympathetic preganglionic neurons in the thoracolumbar spinal cord, and indirect connections to those centers via brainstem autonomic relay nuclei.

THE AUTONOMIC NERVOUS SYSTEM AND EMOTIONAL EXPRESSION

In the 1872 publication of *The Expression of Emotions in Man and Animals*, **Charles Darwin** (1809–1882) presented the idea that autonomic responses are an intrinsic part of the experience and expression of emotions. The hypothalamus is interconnected to both the limbic structures and autonomic systems, functioning as an interface between the two regions. Although the ANS operates autonomously without conscious awareness, its activity is influenced by emotions. This modulation of the ANS by higher brain centers occurs via the hypothalamus. The hypothalamus is the link through which emotions evoked by external stimuli produce autonomic responses that are physical expressions of emotion, such as the blush of embarrassment, tachycardia of fright, emotional fainting (a strong vasovagal reaction), salivation in response to the sight of a favored food, and the fight-or-flight defense reaction (see "The Fight-or-Flight Defense Reaction," below).

Endocrine Function and the Pituitary Gland

The hypothalamus is the primary regulator of endocrine functions; it controls the endocrine glands via the **pituitary gland** (**hypophysis**), a pea-sized structure at the base of the brain that sits in a depression in the sphenoid bone known as the **sella turcica**, which acts as a protective bony enclosure (Figure 10.5). The hypothalamus is connected to the pituitary by the **pituitary stalk** (**pituitary infundibulum**), which lies just posterior to the optic chiasm. The hypothalamus and pituitary act together as the command center of the endocrine system. The pituitary gland is referred to as the master gland because it regulates the secretion of hormones from the endocrine glands.

Endocrine glands are ductless; they secrete their chemical product, **hormones**, directly into the bloodstream. Hormones circulate and elicit a response at their target tissues. Hormones are like neurotransmitters in that both are chemical messengers; they differ in that they are transmitted through blood and act as long-range

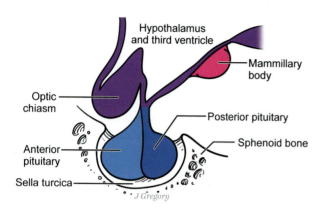

FIGURE 10.5. The pituitary gland lies ventral to the hypothalamus within the sella turcica, a depression within the sphenoid bone. The pituitary gland has two parts, the anterior lobe (adenohypophysis) and posterior lobe (neurohypophysis).

chemical messengers that are generally slow-acting, while neurotransmitters are transmitted across the synaptic cleft and act as short-range fast-acting chemical messengers. In some cases, hormones are synthesized and released into the circulation by neurons, a process known as **neurosecretion**. Neurons showing this phenomenon are **neurosecretory cells**, and their products are **neurohormones**.

The pituitary gland has two lobes, the posterior lobe and the anterior lobe. The release of hormones from these lobes is controlled via the hypothalamus; in turn, the hypothalamus is subject to hormonal feedback control.

POSTERIOR LOBE

The **posterior lobe** of the pituitary (**neurohypophysis, neural lobe**) is composed of neural tissue. It consists of the axon terminals of neurons that originate from the supraoptic and paraventricular nuclei of the hypothalamus and project to the posterior pituitary via the **hypothalamic-hypophysial tract**. Their axons traverse the pituitary stalk and end along blood vessels in the posterior pituitary gland, where they secrete their peptide neurohormones directly into the systemic circulation. The posterior pituitary does not produce any hormones, but it stores and releases two hormones made in the hypothalamus, namely antidiuretic hormone and oxytocin.

Antidiuretic hormone (ADH) is produced predominantly by the **supraoptic nuclei** (which lie dorsal to the optic chiasm) and is released into the blood vessels of the posterior pituitary. This hormone regulates the balance of water and sodium by controlling the amount of water reabsorbed by the tubules of the kidneys. The more water that is reabsorbed and returned to the body, the less that is excreted in the urine. High levels of ADH increase water reabsorption and reduce urinary output, while low levels decrease water reabsorption and increase urinary output.

ADH release is controlled by negative feedback and varies with blood **osmolality**, the blood solute concentration (i.e., the amount of all chemical solute particles that are dissolved in the liquid component of blood). Osmolality of the blood plasma is monitored by **osmoreceptor cells** of the supraoptic nucleus, which in turn control ADH release into the systemic circulation. High blood osmolarity, as occurs with dehydration (reduced fluid intake or excess fluid loss due to heavy sweating, vomiting, or diarrhea) or following a salty meal, stimulates osmoreceptors and signals the release of ADH. With excessive hydration, ADH production decreases and urine output increases. Alcohol consumption inhibits ADH release, resulting in increased urine production that may lead to dehydration and a hangover.

Oxytocin is produced predominantly by the paraventricular nuclei and is released into the blood vessels of the posterior pituitary. This hormone is best known for its role in female reproduction, particularly childbirth and breastfeeding. When fetal development is complete toward the end of pregnancy, the number of oxytocin hormone receptors increases in the smooth muscle cells of the uterus, making the uterus more sensitive to oxytocin's effects. Oxytocin stimulates uterine contractions and dilation of the cervix during childbirth. It is released increasingly throughout the duration of childbirth via a positive feedback mechanism, whereby cervical stretching stimulates further oxytocin synthesis and release, which in turn increases the intensity and effectiveness of uterine contractions and prompts further dilation of the cervix until birth. Following birth, blood levels of oxytocin decrease immediately, and then the hormone plays a role in the **milk ejection reflex**. As the newborn begins suckling, sensory receptors in the nipples transmit signals to the hypothalamus and trigger oxytocin release from the paraventricular nuclei into the bloodstream, causing contraction of the mammary gland milk ducts and ejection of milk into the infant's mouth.

ANTERIOR LOBE

The **anterior lobe** of the pituitary (**adenohypophysis, glandular lobe**) is composed of glandular tissue which manufactures hormones. The secretion of hormones from this lobe is regulated by *other* hormones that are manufactured by the hypothalamus. The arcuate and periventricular nuclei of the hypothalamus project via the **tuberoinfundibular pathway** to the uppermost region of the pituitary stalk and end along a plexus of small blood vessels known as the **hypophyseal portal system**. The hypothalamic hormones are released into the portal system and are transported to the anterior pituitary; only then are the hormones produced by the anterior pituitary released into the general circulation. The hypothalamic hormones are of two types: releasing hormones that stimulate the secretion of hormones from the anterior pituitary, and inhibiting hormones that inhibit secretion. Thus, this pathway involves three steps: (1) the hypothalamus sends signaling molecules (releasing or inhibiting factors) to the anterior pituitary; (2) these in turn stimulate or inhibit the synthesis and release of anterior pituitary hormones into the circulation; and (3) anterior pituitary hormones circulate and act on their target glands to release their hormones into the general circulation.

The **hypothalamic-pituitary-adrenal** (HPA) **axis** functions as a chronic stress response system. This pathway involves: (1) hypothalamic release of corticotropin-releasing factor (CRF), which triggers (2) secretion of adrenocorticotrophic hormone (ACTH) from the anterior pituitary into the general circulation, which triggers (3) secretion of **cortisol** from the adrenal glands. Cortisol affects many different processes, as most cells of the body have cortisol receptors. One of its functions is to reduce inflammation; corticosteroid drugs such as cortisone, hydrocortisone, and prednisone mimic the effects of cortisol and are effective anti-inflammatory agents.

The **hypothalamic-pituitary-thyroid** (HPT) **axis** functions in the regulation of metabolism. This pathway involves: (1) hypothalamic release of thyrotropin-releasing hormone (TRH), which triggers (2) the anterior pituitary to secrete thyroid-stimulating hormone (TSH, thyrotropin) into the general circulation, which stimulates (3) the thyroid gland to secrete **thyroid hormones** (T3 and T4). Thyroid hormones stimulate metabolism in most body tissues; they increase basal metabolic rate, resulting in increased oxygen consumption and rates of adenosine triphosphate (ATP) hydrolysis, resulting in increased body heat production. Any condition that results in thyroid hormone deficiency causes **hypothyroidism**, characterized by lethargy, fatigue, cold intolerance, weakness, hair loss, and reproductive failure. Any condition that results in excess thyroid hormone causes **hyperthyroidism**, characterized by increased metabolism, insomnia, high heart rate, and anxiety; Graves' disease is the most common cause of hyperthyroidism.

The **hypothalamic-pituitary-gonadal** (HPG) **axis** controls sexual development, reproduction, and aging. This pathway involves: (1) hypothalamic release of **gonadotropin-releasing factor** (GnRF), which triggers (2) the anterior pituitary to secrete **gonadotropins**, hormones that regulate the function of the gonads. These are luteinizing hormone (LH) and follicle-stimulating hormone (FSH). In females, LH stimulates the production of estrogens and progesterone by the ovaries; it also triggers ovulation. In males, LH stimulates production of testosterone by the testes. FSH stimulates the production and maturation of gametes (ova in females and sperm in males). Puberty is initiated by hypothalamic production and release of GnRH. Gonadotropins regulate reproductive function throughout life, as well as the onset and cessation of reproductive capacity in women. Symptoms of abnormal gonadotropin levels include hypogonadism, erectile dysfunction, decreased libido, disturbed menstrual cycles (oligomenorrhea or amenorrhea), and infertility.

The **hypothalamic-pituitary-growth** (HPGr) **axis** functions to promote protein synthesis and tissue building. This pathway involves: (1) hypothalamic release of growth hormone-releasing hormone (GHRH) and growth hormone-inhibiting hormone (GHIH, **somatostatin**), which triggers (2) the anterior pituitary to secrete **growth hormone** (GH, **somatotropin**) into the general circulation, which stimulates (3) the growth of essentially all body tissues, especially bone. Secretion of abnormally large amounts of GH results in excessive growth. In children (before bone growth plates have closed), excessive GH results in **gigantism**, characterized by excessive growth in height, muscles, and organs, causing the child to be extremely large for their age. In adults (after bone growth plates have closed), excessive GH results in **acromegaly**, characterized by abnormal enlargement of the bones in the face, hands, and feet. Abnormally low levels of GH in children causes **pituitary dwarfism** (growth hormone deficiency).

The **hypothalamic-pituitary-prolactin** (HPP) **axis** controls the release of **prolactin**. Prolactin stimulates the breasts to produce milk; it is secreted in large amounts during pregnancy and breastfeeding. In nonpregnant women, prolactin secretion is inhibited by prolactin-inhibiting hormone (PIH, dopamine) released from the hypothalamus. During pregnancy, (1) the hypothalamus releases prolactin-releasing hormone (PRH), which triggers (2) the anterior pituitary to secrete **prolactin** into the general circulation, which stimulates (3) milk production in the mammary glands. Excess prolactin can cause the production of breast milk in men and women who are not pregnant or breastfeeding.

Motivated Behaviors

One of the most remarkable features of the central nervous system is its ability to detect changing internal needs and then generate specific somatic behavioral responses to restore homeostasis. The hypothalamus plays a central role in the neural processes underlying motivated behaviors required for survival and reproduction, namely hunger, thirst, aggression, and sex. It is the link through which conditions of the internal environment invoke motivational states that direct motivated behaviors that promote survival of the individual and species. A **motivational state**, or **drive**, is an internal condition of urges or impulses based upon bodily needs. The motivational state (e.g., hunger, thirst) directs an organism's actions toward specific environmental goals (e.g., food, water). **Motivated behaviors** (e.g., feeding, drinking) are behavioral homeostatic mechanisms to satisfy bodily needs, although they are typically experienced as "voluntary." While drives are not observable directly, they can often be inferred from behavior.

Drives increase the likelihood that an organism will engage in specific behaviors required for survival and reproduction. The specific actions taken to satisfy a drive are a means to a clearly defined end; they are purposive, consciously controlled behaviors that are variable (i.e., not fixed or stereotyped) and that allow the organism to take efficient steps to provide for the need. The drives that give intent and intensity to behavior, however, are imposed by unconscious processes in response to physiological cues associated with survival needs.

The hypothalamus plays a critical role in the neural circuitry underlying homeostatic motivated behaviors. Feeding and drinking can be evoked by electrical and chemical stimulation of discrete hypothalamic regions, even in fully replete animals. Similarly, social behaviors such as mating and aggression, which are influenced strongly by hormonal state, can be evoked from hypothalamic regions that contain neurons responsive to peripherally produced sex steroid hormones.

The motivational state of an animal modulates the perceived reward value (**hedonic value**) of motivationally relevant sensory stimuli. The hunger state modulates the appeal of a specific food. The greater the degree of hunger, the lower the perceived palatability threshold (i.e., when you are hungry, you will eat almost anything). The greater the degree of satiety, the higher the perceived palatability threshold (i.e., when you are full, the only thing you will eat is dessert). Similarly, the sexual arousal state influences the appeal of a potential mate.

HUNGER AND FEEDING

In the 1930s and 1940s, experimental studies demonstrated dramatically that hunger and food intake are under the control of the lateral hypothalamus (LH) and ventromedial hypothalamus (VMH). Lesions of the VMH produce **hyperphagia** and severe obesity. Lesions of the LH produce **aphagia** and starvation; animals with LH lesions die unless they are force-fed. Electrical stimulation of these hypothalamic nuclei produces the opposite effects; LH stimulation induces feeding, while VMH stimulation stops feeding in a hungry animal. These findings led to the view that the LH functions as a **hunger center** and the VMH functions as a **satiety center**.

It is now known that metabolic homeostatic mechanisms regulating body weight—by controlling food intake, energy expenditure, and energy storage—are very complex. The hypothalamus plays a central role; it receives afferent messages from adipose tissue, gastrointestinal tract, liver, and pancreatic cells, and it also sends efferent messages to these organs.

THIRST AND DRINKING

Water is essential to the function of every living cell. The average adult human body is 50%–65% water. Water balance is the relation of the total amount of water entering the body through ingestion of liquids and food, to the total output of water lost by way of the kidneys, bowels, lungs (through respiration), and skin (through perspiration). The evolution of mechanisms for monitoring the water content of the body, adjusting fluid output, and transforming physical signals into **thirst**, the motivation to find and consume water, is critical to survival.

The physiological system for maintaining fluid balance is complex. It involves: (1) monitoring both the blood osmolality and blood volume, and (2) triggering a coordinated set of autonomic, neuroendocrine, and behavioral responses that maintain constancy of these quantities when necessary. Increased blood osmolality is the most important homeostatic stimulus for thirst in everyday life.

Chronic deficiency of ADH results in **diabetes insipidus**, characterized by **polyuria** (excessive urination), dehydration, and **polydipsia** (excessive thirst). Because ADH levels are insufficient to trigger water reabsorption by the kidneys, increased fluid consumption fails to effectively decrease blood solute concentration, resulting in polydipsia. **Nephrogenic diabetes insipidus** is due to resistance of the kidney to the effects of ADH. **Central diabetes insipidus** is due to a hypothalamic-pituitary disorder, caused by lesions of the supraoptic nucleus or posterior pituitary and lack of ADH. Nephrogenic diabetes insipidus is treated by adequate water intake, diuretics, and a low-salt diet; central diabetes insipidus is treated by hormone replacement.

The neural circuits responsible for the generation of thirst are not understood well, but the anterior hypothalamus plays a critical role. In animals and humans, lesions of the anterior hypothalamus abolish thirst and produce **adipsia** (absence of drinking behavior), despite severe dehydration and a normal ADH response. Electrical stimulation of the anterior hypothalamus in animals elicits polydipsia, even when animals are overhydrated. Because this area is contiguous with the supraoptic nucleus, lesions may affect both brain areas, resulting in a combination of adipsia and deficient ADH production, and leading to **adipsic diabetes insipidus** and severe **hypernatremia** (elevated serum sodium concentration).

In humans, the subjective feeling of thirst correlates with increases in blood osmolality. Osmoreceptors give rise to the subjective experience of thirst; a 1% increase in blood osmolality can trigger the thirst sensation. However, thirst can be quenched within seconds of drinking, long before the ingested water has time to be absorbed and alter blood osmolality. Thus, the mechanism for thirst termination is not the reverse of the mechanism for thirst generation. Rather, thirst is terminated by sensory signals from the oropharynx (back of the throat) that are transmitted to brain thirst neurons. These neurons decide when to terminate drinking by comparing the physiologic need for water, signaled by osmoreceptors, with the amount of water recently consumed, as conveyed by oropharyngeal signals of fluid intake.

THE FIGHT-OR-FLIGHT DEFENSE REACTION

The **fight-or-flight response** refers to a behavioral adaptation that has evolved in many animals that increases chances of survival in threatening situations. The perception of threat automatically triggers an acute stress response by the sympathetic nervous system that prepares the organism to either fight or flee, thereby protecting itself from possible harm. The concept of a fight-or-flight response was developed by American physiologist **Walter B. Cannon** (1871–1945) to describe an animal's physiological response to both physical threats (e.g., blood loss from trauma) and psychological stressors (e.g., antagonistic encounter between members of the same species).

The fight-or-flight response involves a rapid and widespread sympathetic activation to prepare the body for emergency, initiated by release of adrenaline and noradrenaline from the **adrenal medulla** (the inner part of the adrenal gland) into the circulation. It accelerates heart rate, increases blood pressure, and redistributes blood away from skin and intestine to the brain, heart, and skeletal muscle. It relaxes the smooth muscles of the bronchi, producing bronchodilation and thereby increasing oxygen availability. It causes breakdown of liver glycogen into glucose for immediate use. It relaxes smooth muscles of the intestinal tract, inhibiting peristalsis, and it closes sphincter muscles, inhibiting elimination processes. It dilates pupils, and it causes sweating and **piloerection** (goosebumps, which helps conserve heat, especially in more hirsute animals than humans).

The adrenal medulla is part of the sympathetic branch of the ANS; the **sympathomedullary** pathway is a connection between the sympathetic nervous system and adrenal medulla. Sympathetic preganglionic neurons in the thoracolumbar spinal cord project directly to the adrenal medulla and stimulate it to secrete the hormones adrenaline and noradrenaline into the general circulation. This pathway is activated by descending projections from the hypothalamus. As described above, the hypothalamus is also the link through which limbic brain structures involved in emotion processing and physical/social threat detection trigger the ANS fight-or-flight response.

AGGRESSION AND SHAM RAGE

Aggression is an innate social behavior that is prevalent across animal species. It is essential to the survival of individuals and groups, especially with respect to resource competition, defense, and protection of kin. A variety of lesion and electrical brain stimulation experiments have shown that the hypothalamus is essential for expressing aggressive behaviors.

In 1928, American physiologist **Phillip Bard** (1898–1977) reported the results of a series of experiments revealing that the hypothalamus is critical for the coordination of autonomic and somatic components of emotional behavior. The experiments were performed in cats and involved surgical transections at different levels of the brain. Bard found that with transections disconnecting both cerebral hemispheres (i.e., cerebral cortex, underlying white matter, and basal ganglia) from the diencephalon, the animals spontaneously behaved as if they were enraged. They exhibited typical autonomic responses of anger seen in normal cats, such as increased blood pressure and heart rate, pupil dilation, piloerection (erection of the hairs on the back and tail), and adrenal secretion. They also exhibited typical somatic motor components of anger, such as arching the back, extending the claws, lashing the tail, hissing, spitting, snarling, and biting. This behavior was called **sham rage** because it was undirected (had no obvious target), was elicited by innocuous stimuli (e.g., simply touching the tail), and subsided rapidly upon stimulus removal. This postoperative behavior contrasted to preoperative behavior in which the animals behaved in a calm and friendly manner. Further studies showed that sham rage behaviors were dependent on neural circuitry of the hypothalamus; transections just caudal to the hypothalamus at the midbrain-diencephalic junction did not produce sham rage.

Bard suggested that while the subjective experience of emotion may depend on an intact cerebral cortex, the expression of coordinated emotional behaviors does not. He concluded that sham rage was due to a loss of inhibitory control exerted by higher brain structures on the hypothalamus areas involved in aggression, causing those areas to become hyperactive.

Complementary studies by **Walter Hess** (1881–1973) showed that electrical stimulation of the medial hypothalamus in awake, freely moving cats led to a rage response and attack behavior. Later studies have shown more specifically that within the medial hypothalamus, the ventrolateral part of the ventromedial nucleus is a key structure for driving aggressive behavior. Silencing this area abolishes naturally occurring inter-male attack; activation of this area promotes the attack of suboptimal targets such as females and inanimate objects.

SEXUAL BEHAVIOR

Sexual behavior is regulated by complex neural mechanisms that are not well understood, but estrogen-sensitive and androgen-sensitive neurons in the medial preoptic and ventromedial nuclei play a role. The **medial preoptic**

nucleus is the primary brain region controlling male sexual behavior. It contains a subnucleus known as the **sexually dimorphic nucleus** that is larger and has more neurons in males than females. It contains neurons that are responsive to sex hormones, and neurons that fire during sexual stimulation. Lesions of this nucleus disrupt male sexual behavior; hormonal, chemical, or electrical stimulation of this nucleus can induce or enhance male sexual behavior. The medial preoptic nucleus sends a major projection to the VMH. Lesions in the VMH prevent male (mounting) and female (lordosis) sexual behavior. Injections of estradiol into the VMH elicit copulatory behavior in female rats.

Thermoregulation

Thermal homeostasis is essential for survival in mammals and is one of the most critical functions of the nervous system. **Homeothermic** animals maintain a relatively constant body temperature regardless of the ambient temperature by **thermoregulation**, a homeostatic process that balances heat production and heat loss. Heat production occurs as a biproduct of metabolism and increases with physical activity such as exercise. Heat loss occurs through **radiation** (losing heat to the environment), **conduction** (transferring heat energy to an object, such as by holding a glass of ice water or lying on cold ground), **convection** (losing heat to the environment by air movement, such as wind), **evaporation** (transfer of heat by evaporation of sweat or water on the skin), and **respiration** (releasing warmed air into the surrounding environment).

Humans have a normal and nominal core (internal) temperature of 37 degrees Celsius (98.6 degrees Fahrenheit). Significant deviations from this set point undermine cellular metabolism and lead to organ failure and death. Thus, the overall goal of thermoregulation is to keep the core body temperature as close to the set point as possible. The hypothalamus acts as the body's thermostat. It receives and integrates afferent information from distributed temperature sensors and maintains temperature homeostasis through efferent control of thermoregulatory responses that defend the body temperature against change.

AFFERENT SENSING

Thermoreception (**thermoception**), the sense of temperature, is conveyed by specialized sensory cells known as **thermoreceptors** located in both the peripheral and central nervous systems.

Peripheral thermoreceptors are located within the skin and viscera. There are two types of peripheral thermoreceptors: warmth receptors and cold receptors. The sensory afferents carrying signals from the body enter the central nervous system via the spinal nerves and travel in the spinothalamic tract, while those carrying signals from the head enter via the trigeminal nerve and travel in the trigeminothalamic tract. **Cutaneous thermoreceptors** are located within the skin and report the temperature of the body surface. **Visceral thermoreceptors** are located within the viscera and report the temperature of the body core.

Central thermoreceptors are neurons that sense the core temperature of the brain. Cold- and heat-sensitive central thermoreceptors are located within the **preoptic area** (POA) of the anterior hypothalamus.

CENTRAL CONTROL

The POA receives information about the ambient temperature from the skin and core temperature of the body and brain, and in turn it activates efferent signals that control thermal effector responses to prevent the core temperature from deviations from the set point. The POA acts as a thermostat to maintain body temperature by means of feedback. If body temperature exceeds the set point, the POA initiates responses to dissipate heat. If body temperature falls below the set point, the POA initiates responses to generate and conserve heat.

The POA acts as a **heat loss center**, initiating autonomic responses to promote the dissipation of heat, including cutaneous vasodilation and sweating. Stimulation of neurons within the heat loss center elicits skin vasodilation and sweating. Lesions of the heat loss center produce **hyperthermia** (increase in body temperature) because the neurons that initiate cutaneous vasodilation and sweating to counteract body temperature elevation are no longer functional.

The posterior hypothalamus acts as a **heat gain center** (heat production center), initiating autonomic responses to conserve and generate heat; these include cutaneous vasoconstriction, piloerection, and shivering. Stimulation of neurons within the heat gain center elicits skin vasoconstriction, piloerection, and shivering. Lesions of the heat gain center produce **hypothermia** (decrease in body temperature) because the neurons that initiate cutaneous vasoconstriction, piloerection, and shivering to counteract body temperature reduction are no longer functional.

EFFERENT RESPONSES

Body temperature is regulated by two types of mechanisms, physiological and behavioral.

Physiological thermoregulatory responses are primarily autonomic. The primary responses to cold exposure involve heat conservation by constriction of blood vessels and **thermogenesis** by brown adipose tissue and skeletal muscle shivering. Exposure to warmth triggers a complementary set of autonomic responses, including suppression of thermogenesis, facilitation of heat loss through sweating to allow for water evaporation to cool the skin surface, and dilation of blood vessels at the body surface to allow for dissipation of heat through the skin.

Thermoregulatory behaviors are stereotypical, motivated, somatic motor acts directed toward minimizing or optimizing heat transfer from the body to the environment (i.e., altering the rate of heat loss or absorption). Behaviors directed toward body temperature control include cold- and warm-seeking (e.g., moving to a preferred microenvironment); postural changes (e.g., huddling or cuddling with others in the cold, spreading the limbs in the heat); and alterations in movement (e.g., increased non-shivering movements to generate heat in the cold, decreased movement in heat). In humans, such behaviors include wearing light or heavy clothing, and using heating, fans, or air conditioning. Motivated thermoregulatory behaviors likely involve limbic emotional and dopaminergic reward systems in the brain (see Chapter 12, "Limbic Structures"). The state of being too hot or too cold is a physical and emotional stressor that motivates behaviors aimed at seeking or producing a more comfortable ambient temperature. Achievement of this goal is accompanied by a sense of satisfaction (reward) of being in a more pleasant thermal environment (i.e., a thermoneutral zone that does not threaten homeostasis).

Effector responses are engaged hierarchically; different responses are recruited at different temperature thresholds in order of increasing energy costs. Thermoregulatory behaviors are engaged prior to thermoregulatory autonomic responses, triggered primarily by cutaneous thermal receptors that provide the first indication of a potential threat to core temperature homeostasis (for example, you put a sweater on before you start shivering). Autonomic responses are engaged in stereotyped sequences.

The cold defense is aimed at reducing heat loss. Cold defense thermoregulatory behaviors include huddling, voluntary physical activity (hand rubbing, pacing), sheltering next to a heat source, and wearing warm clothing. Cold defense physiologic autonomic responses include cutaneous vasoconstriction to conserve body core heat, and shivering (involuntary contraction of skeletal muscles), which generates heat. Furry animals use piloerection to trap air close to the skin for heat conservation.

The heat defense is aimed at decreasing heat production and increasing heat loss. Behavioral responses to heat include wearing loose and light clothing, fanning, resting, lying down with limbs spread out, drinking cold liquids, and seeking immersion in water. Heat defense physiological autonomic responses include cutaneous vasodilation to facilitate heat loss by conducting heat from the body core to the body surface, and evaporative cooling through sweating. Some species (e.g., dogs) use panting. The **primary motor center for shivering** is located within the dorsomedial portion of the posterior hypothalamus. Shivering is normally inhibited by signals from the POA heat loss center but is excited by cold signals.

Circadian Rhythms

The hypothalamus plays a role in light-stimulated circadian rhythms such as the sleep-wake cycle, body temperature cycle, and blood hormone levels. Light stimuli from the retina are transmitted via **retinohypothalamic** fibers to the **suprachiasmatic nucleus**, which serves as the pacemaker, or "master clock," of the body. Suprachiasmatic lesions disrupt the sleep-wake cycle and other circadian rhythms (temperature, hormone levels).

Hypothalamic Syndromes

A wide spectrum of pathological processes may affect the structure and/or function of the hypothalamus. Hypothalamic lesions or disease result in **hypothalamic syndromes**, constellations of autonomic, endocrine, and neurobehavioral signs and symptoms. These include disturbances of water balance (central diabetes insipidus), sugar and fat metabolism (hyperphagia, **hypothalamic obesity**), thermoregulation (sweating, vasodilation, vasoconstriction, shivering), sexual development (precocious puberty, hypogonadism), growth (gigantism, acromegaly), and circadian rhythms. Less commonly, hypothalamic lesions may result in violent aggressive behavior known as **hypothalamic rage**.

Summary

The diencephalon lies at the central core of the cerebrum, surrounding the third ventricle and surrounded by the cerebral hemispheres. Its major structures are the thalamus and the hypothalamus, both of which are composed of multiple nuclei that perform many different functions.

The majority of thalamic efferents project to the cerebral cortex, and nearly every region of the cortex receives thalamic input. In addition, there are reciprocal connections from the cortex back to the thalamus. These thalamocortical and corticothalamic connections are made through the thalamic radiations, a massive radiating fiber bundle. The thalamic nuclei are classified into three types: (1) specific relay nuclei that receive well-defined inputs from specific ascending sensory or motor pathways and project to sharply defined, functionally distinct cortical regions; (2) association nuclei that have reciprocal connections with cortical association areas; and (3) nonspecific nuclei that project diffusely throughout the cerebral cortex and receive reciprocal inputs from a wide expanse of cortex.

Focal lesions of sensory relay nuclei produce selective sensory deficits, while focal lesions of motor relay nuclei result in motor disturbances. Thalamic lesions involving the association nuclei produce memory, cognitive, affective, and/or behavioral disturbances. Lesions involving the anterior nuclei often result in thalamic amnesia. Lesions involving the dorsomedial nucleus, which has strong reciprocal connections with the prefrontal cortex, result in frontal behavioral syndromes, especially apathy. Unilateral lesions involving the pulvinar, which has strong connections to secondary visual areas and parietotemporal association areas, result in a contralateral visual neglect. Infarcts in the left anterior thalamus produce thalamic aphasia. The nonspecific thalamic nuclei are a critical component of the ascending reticular arousal system underlying wakefulness and awareness; they play a role in the pathophysiology of epilepsy and disorders of consciousness. Focal thalamic lesions most often are due to ischemic stroke. Infarcts within the individual vascular territories destroy thalamic nuclei in specific combinations, resulting in thalamic stroke syndromes.

The overall function of the hypothalamus is to promote survival of the individual and the species through homeostatic mechanisms and coordinated autonomic, endocrine, and somatic behavioral responses related to basic survival behaviors known as "the four Fs": feeding, fighting, fleeing, and fornicating. The hypothalamus acts as a higher center for the control of lower autonomic centers and is the primary regulator of endocrine functions. It plays a central role in the neural processes underlying motivational states and motivated behaviors such as hunger and feeding, thirst and drinking, fight-or-flight defense reactions, and sexual behavior. Hypothalamic pathologies result in hypothalamic syndromes, constellations of autonomic, endocrine, and neurobehavioral signs and symptoms.

Additional Reading

1. Bordes S, Werner C, Mathkour M, et al. Arterial supply of the thalamus: a comprehensive review. *World Neurosurg.* 2020;137:310–318. doi:10.1016/j.wneu.2020.01.237
2. Brodal P. The central autonomic system: the hypothalamus. In: Brodal P, *The central nervous system: structure and function*. 5th ed. Oxford University Press; 2016. https://oxfordmedicine.com/view/10.1093/med/9780190228958.001.0001/med-9780190228958-chapter-30. Accessed July 4, 2022.
3. Blumenfeld H. Pituitary and hypothalamus. In: Blumenfeld H, *Neuroanatomy through clinical cases*. 3rd ed., pp. 795–821. Sinauer Associates; 2021.
4. Dalrymple-Alford JC, Harland B, Loukavenko EA, Perry B, Mercer S, Collings DA. Anterior thalamic nuclei lesions and recovery of function: relevance to cognitive thalamus. *Neurosci Biobehav Rev.* 2015;54:145–160. doi:10.1016/j.neubiorev.2014.12.007
5. Danet L, Barbeau EJ, Eustache P, Planton M, Raposo N, Sibon I, et al. Thalamic amnesia after infarct: The role of the mammillothalamic tract and mediodorsal nucleus. *Neurology.* 2015 Dec 15;85(24):2107–2115. doi: 10.1212/WNL.0000000000002226. Epub 2015 Nov 13. PubMed PMID: 26567269; PubMed Central PMCID: PMC4691690.
6. De Witte L, Brouns R, Kavadias D, Engelborghs S, De Deyn PP, Mariën P. Cognitive, affective and behavioural disturbances following vascular thalamic lesions: a review. *Cortex.* 2011 Mar;47(3):273–319. doi: 10.1016/j.cortex.2010.09.002. Epub 2010 Sep 26. Review. PubMed PMID: 21111408.
7. Kruk MR, van der Poel AM, de Vos-Frerichs TP. The induction of aggressive behaviour by electrical stimulation in the hypothalamus of male rats. *Behaviour.* 1979;70(3-4):292–322. doi:10.1163/156853979x00106
8. Nelson RJ, Trainor BC. Neural mechanisms of aggression. *Nat Rev Neurosci.* 2007;8(7):536–546. doi:10.1038/nrn2174
9. Oomura Y, Yoshimatsu H, Aou S. Medial preoptic and hypothalamic neuronal activity during sexual behavior of the male monkey. *Brain Res.* 1983;266:340–343. 10.1016/0006-8993(83)90666-2
10. Schmahmann JD. Vascular syndromes of the thalamus. *Stroke.* 2003;34(9):2264–2278. doi:10.1161/01.STR.0000087786.38997.9E
11. Squire LR, Amaral DG, Zola-Morgan S, Kritchevsky M, Press G. Description of brain injury in the amnesic patient N.A. based on magnetic resonance imaging. *Exp Neurol.* 1989;105(1):23–35. doi:10.1016/0014-4886(89)90168-4
12. Vanderah TW, Gould D. The thalamus and internal capsule: getting to and from the cerebral cortex. In: Vanderah TW, Gould D, *Nolte's the human brain: an introduction to its functional anatomy.* 8th ed., pp. 372–394. Elsevier; 2021.

The Basal Ganglia

Introduction

The basal ganglia are a group of subcortical nuclei that primarily play a role in motor function, especially in governing the speed and spontaneity of movement. Basal ganglia pathology gives rise to movement disorders. The basal ganglia also play a role in non-motor functions, although in ways that are less understood. Persons with movement disorders due to basal ganglia pathology may have prominent cognitive and emotional disturbances, even at early stages of the disease. This chapter describes the anatomy of the basal ganglia, clinical syndromes resulting from basal ganglia pathology, and diseases prominently involving the basal ganglia, such as Parkinson's disease and Huntington's disease.

Anatomy

The **basal ganglia** were first described by the seventeenth-century neuroanatomist **Thomas Willis** (1621–1675) as a group of subcortical gray matter masses located deep within the basal regions of cerebral hemispheres, namely the **caudate nucleus**, **putamen**, and **globus pallidus**. The term *basal ganglia*, however, is actually a misnomer because the term *ganglia* refers to collections of nerve cell bodies that lie outside of the central nervous system (see Chapter 1, "Neurons, Glia, and Basic Neuroanatomy"). In addition, while the term *basal ganglia* in the strictest sense refers to deep nuclear structures of the telencephalon, the definition has expanded to include functionally related nuclei, including the **substantia nigra** located in the midbrain and the **subthalamic nucleus** located in the diencephalon.

It was once believed that the basal ganglia had significant direct efferent connections with the spinal cord that were organized in parallel with the pyramidal (corticospinal) tracts. The basal ganglia were therefore referred to as the **extrapyramidal system**, referring to a descending motor system outside of the pyramidal system. The main efferent connections of the basal ganglia, however, are not directed to the spinal cord; rather, they are directed upstream to the motor cortex and other cortical areas via the thalamus. Thus, the term *extrapyramidal system* is falling into disuse. For historical

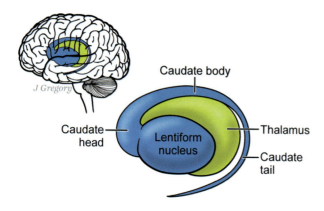

FIGURE 11.1. The basal ganglia. The caudate ("tailed") nucleus is an elongated C-shaped structure with head, body, and tail components. The lentiform (lenticular) nucleus consists of the putamen laterally and the globus pallidus medially.

reasons, however, the term **extrapyramidal syndrome** is used frequently to refer to movement disorders that arise from basal ganglia pathology.

The caudate ("tailed") nucleus is an elongated C-shaped structure that has head, body, and tail components (Figure 11.1). The head of the caudate lies anterior to the thalamus and bulges into the lateral ventricle. In degenerative diseases involving the caudate (e.g., Huntington's chorea), that characteristic bulge is lost. The body of the caudate arches over and around the thalamus and forms the lateral wall of the body of the lateral ventricle. The tail of the caudate extends forward in the roof of the temporal horn and terminates against the amygdala.

The putamen ("shell-shaped") is the largest of the basal ganglia structures, and it is the most lateral of the basal ganglia nuclei (see Figure 10.1 in Chapter 10). The globus pallidus ("pale globe") is so named because it is lighter in color than the caudate and putamen due to containing many more myelinated fibers and fewer cell bodies. The globus pallidus has two functionally distinct components: an external segment located laterally, and an internal segment located medially. The putamen and globus pallidus together form a complex known as the **lentiform nucleus** (**lenticular nucleus**), so named because it is shaped like a lens. In both coronal and horizontal (axial) sections, the lenticular nucleus appears to have three components: from lateral to medial, these are the putamen, globus pallidus external segment, and globus pallidus internal segment (see Figures 10.1 and 10.3 in Chapter 10).

In horizontal cross sections of the cerebral hemispheres, the **internal capsule** (which carries many connections to and from the cortex) appears as a large V-shaped fiber bundle with an **anterior limb**, **genu** (bend), and **posterior limb** (see Figure 10.3 in Chapter 10). The lentiform nucleus is located lateral to the internal capsule, the head of the caudate lies medial to the anterior limb of the internal capsule, and the thalamus lies medial to the posterior limb of the internal capsule.

The caudate and putamen are often referred to together as the **neostriatum** or simply the **striatum**. They are histologically similar and may be thought of as a single large nucleus, with the caudate and putamen components separated by the internal capsule but joined in some places by striations (stripes) consisting of cellular bridges (hence the name *striatum*). The neostriatum is the principal input element to the basal ganglia; it projects to the globus pallidus (also known as the paleostriatum), which serves as the main output from the basal ganglia. The **ventral striatum** comprises the nucleus accumbens and olfactory tubercle; it does not play a role in motor function (see Chapter 12, "Limbic Structures").

The substantia nigra (Latin, "black substance") is a paired midbrain nucleus, so named because it is pigmented black by **melanin**. It has two components: the **pars compacta** where the neurons are packed densely, and the **pars reticulata** where the neurons are packed much less densely. The pars compacta contains the melanin pigmented neurons and is the origin of the dopaminergic **nigrostriatal tract**.

The subthalamic nucleus is the largest nucleus of the subthalamus. Functionally, it is considered a component of the basal ganglia; it receives projections from the globus pallidus, substantia nigra, and other structures, and it sends projections to the globus pallidus and substantia nigra.

Blood Supply

The main source of blood supply to the basal ganglia (and the internal capsule) are the **striate arteries**, a collection of small-caliber penetrating arteries that branch off the anterior and middle cerebral arteries at the anterior circle of Willis. These small arteries create many small holes as they penetrate the brain tissue, forming the **anterior perforated substance**, an area of gray matter within the interpeduncular fossa (the depression between the cerebral peduncles on the inferior surface of the brain, within the circle of Willis).

The **medial striate artery** (recurrent artery of Heubner) branches off the anterior cerebral artery and curves back on itself; it supplies the caudate head (as well as the internal limb of the internal capsule and other structures). The **lenticulostriate** arteries branch off the middle cerebral artery, and they are divided into two groups: the **medial lenticulostriate arteries** supply the more medial structures such as the globus pallidus, whereas the **lateral lenticulostriate arteries** supply the more lateral structures such as the putamen.

These arteries are all **end arteries** (terminal arteries), meaning that they are the only supply of oxygenated blood to a portion of tissue. The lack of collateral blood supply makes these vascular territories susceptible to lacunar infarcts, as well as vulnerable to oxygen deprivation (they have a low threshold for hypoxic damage, as occurs with carbon monoxide poisoning). These small vessels are also particularly susceptible to damage from chronic hypertension and rupture, leading to intracerebral hemorrhage.

Basal Ganglia Circuits

The basal ganglia, like the cerebellum, are motor system components that influence the activity of motor cortex through multisynaptic side-loops of fiber connections with the cerebral cortex. Both structures receive input from the cerebral cortex and send output to the cerebral cortex via the thalamus. The basal ganglia and cerebellar side-loop circuits play a role in movement regulation by way of connections with motor cortex, but they also play a role in cognition and affect by way of connections with non-motor cortex. The basal ganglia and cerebellar side-loop circuits are distinct; they travel through different thalamic nuclei and serve distinct functions. Injury and disease to these regions both result in motor dysfunction, but the nature of the deficits differs.

The basal ganglia receive most of their input from the cerebral cortex itself, and in turn, most of the basal ganglia output is directed back to the cerebral cortex via the thalamus by direct and indirect pathways (Figure 11.2). The direct pathway consists of a **cortico-striato-thalamo-cortical loop**. This pathway has excitatory effects on the cortex and facilitates the initiation and execution of voluntary movement. The cerebral cortex projects to the striatum (caudate and putamen) via excitatory corticostriate fibers; the striatum projects to the internal segment of the globus pallidus via inhibitory fibers; the internal segment of the globus pallidus projects to the thalamus via fibers which exert a tonic (continuous) inhibition; and the thalamus projects to the cerebral cortex by excitatory thalamocortical fibers. This cortico-striato-thalamo-cortical pathway has the net effect of releasing the thalamus from tonic inhibition, which results in cortical excitation and increased motor activity. The indirect pathway has inhibitory effects on the cortex and prevents activation of motor cortical areas that would compete with voluntary movements.

There are multiple versions of this cortico-striato-thalamo-cortical loop that are based on topographic patterns of connectivity. Each loop utilizes a different "parent" cortical area, different regions of the striatum and globus pallidus, and different regions of the thalamus; thus, the loops are segregated. Each loop supports differing domains of behavior, and consequently distinct **syndromes** can occur with selective damage to the cortical-basal ganglionic loops.

The Sensorimotor Basal Ganglia Circuit

The sensorimotor loop has been studied extensively and is best understood because of its role in the pathophysiology of movement disorders. It explains how basal ganglia dysfunction can result in both **hypokinesia**, in which there is a decrease in movement, and **hyperkinesia**, in which there are excessive normal movements, abnormal movements, or a combination of the two.

The sensorimotor circuit begins in somatotopically organized cortical sensorimotor areas (somatosensory, primary motor, premotor, and supplementary motor areas); corticofugal fibers (originating from the cortex) project to the putamen. The putamen in turn projects to the internal segment of the globus pallidus, which then projects to the ventral anterior (VA), ventrolateral (VL), and centromedian nuclei of the thalamus via pallidothalamic fibers carried in the lenticular fasciculus. The VA and VL nuclei project back to the cerebral cortex, especially to the supplementary motor area, premotor area, and prefrontal cortex via thalamocortical fibers.

The Associative Basal Ganglia Circuit

The associative circuit begins in the cortical association areas including dorsolateral prefrontal, lateral orbitofrontal, posterior parietal, and temporal association

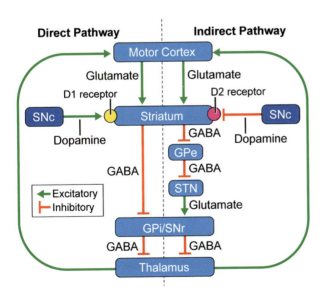

FIGURE 11.2. The basal ganglia receive most of their input directly from the cerebral cortex. In turn, most of the basal ganglia output is directed back to the cerebral cortex via the thalamus by direct and indirect pathways. Substantia nigra pars compacta (SNc), substantia nigra pars reticulata (SNr), globus pallidus external segment (GPe), globus pallidus internal segment (GPi), subthalamic nucleus (STN).

cortices. Corticofugal fibers project with topographic specificity to various regions of the head and tail of the caudate, as well as the putamen.

The projections from the prefrontal cortex go to the head of the caudate; these projections are particularly abundant and account, in part, for the large size of the caudate head. The region of the globus pallidus that receives information from the cortical association areas and caudate nucleus projects back to the prefrontal areas by way of the **dorsomedial nucleus** of the thalamus.

The caudate appears to be involved primarily in cognitive function; few caudate neurons respond to movement or position. This prefrontal circuit plays a role in **executive functions**, top-down cognitive and behavioral control processes for planning and executing subtasks related to goal-directed behavior and adaptation to a range of environmental demands and changes. The executive functions are largely mediated by the frontal lobes (see Chapter 17, "The Frontal Lobes and Associated Disorders").

Clinical Syndromes Resulting from Basal Ganglia Pathology

As noted earlier, our knowledge regarding the function of the basal ganglia is to a large extent based on the motor abnormalities produced by basal ganglia pathology. Basal ganglia pathology does not produce paralysis or paresis, as lesions of the descending corticospinal and corticobulbar motor pathways do; rather, it affects the speed and spontaneity of movement.

There are two classic syndromes of basal ganglia disorders: those dominated by hypokinesia, and those dominated by hyperkinesia. Parkinson's disease is the prototypical basal ganglia disorder characterized by hypokinesia. Huntington's disease is the prototypical basal ganglia disorder characterized by hyperkinesia. The hypokinetic disorders are characterized by a loss of dopamine effects in the basal ganglia, while the hyperkinetic disorders are characterized by excessive dopaminergic activity in the basal ganglia.

Hypokinesia describes a spectrum of movement abnormalities that involve abnormally decreased movement (when there is no paresis). **Akinesia** is an inability to initiate movement. **Bradykinesia** is a slowness of movement.

Hyperkinetic disorders include **dyskinesias**, which are abnormal, purposeless, and involuntary movements. There are a variety of dyskinesias: tremor, chorea, athetosis, and hemiballismus.

Tremor is a rhythmic, involuntary movement that is symmetric about a midpoint within the movement, with both portions of the movement occurring at the same speed. There are several types of tremors (see Chapter 30, "The Neurological Examination"). **Resting tremor** occurs at rest and diminishes during voluntary movement. By contrast, **intention tremor** occurs only when the person performs an intentionally initiated movement, such as occurs with cerebellar pathology.

Chorea (Greek, "to dance") is characterized by successive, involuntary, brisk movements that may resemble fragments of purposeful voluntary movements. Choreiform movements tend to affect the face, tongue, and distal limbs. **Athetosis** (Greek, "without position") is defined as constant, slow, smooth, non-rhythmic involuntary movements that are sinuous and writhing in form. It usually affects the distal limbs (hands or feet) but can affect the trunk, neck, face, or tongue. Dyskinesias that are intermediate forms between chorea and athetosis are referred to as **choreoathetosis**. Both chorea and athetosis are caused by striatal lesions, although it is unknown why a particular lesion induces one state rather than the other.

Hemiballismus (*ballismus, ballism*, derived from the Greek verb meaning "to throw") is a rare unilateral movement disorder characterized by vigorous, irregular, poorly patterned, high-amplitude movements of one arm or leg that look like wild flailing. It is caused by lesions in the contralateral subthalamic nucleus. Hemiballismus is most often seen in older people who have had a stroke in a small perforating branch of the posterior cerebral artery.

Tardive dyskinesia is a delayed side effect of long-term use of antipsychotic medications (especially first-generation antipsychotics such as chlorpromazine) and other dopamine receptor antagonists (see Chapter 3, "Chemical Neurotransmission and Neuropsychopharmacology"). It is characterized by dyskinesias of the jaw, lips, and tongue most prominently. Manifestations include facial grimacing, sticking out the tongue, sucking, or fish-like movements of the mouth. In some cases, there may be chorea or athetosis of the arms and/or legs. This is an extrapyramidal syndrome that may or may not resolve after the causal medicine has been discontinued.

Parkinson's Disease

Parkinson's disease is the most common basal ganglia disorder and the second-most common neurodegenerative disease after Alzheimer's disease. It affects approximately 1% of the population over the age of 55. While Parkinson's disease mainly affects older adults (95% of cases are diagnosed at age 60 or greater), it can occur as young as age 35. Men are 1.5 times more likely to develop Parkinson's disease than women.

Parkinson's disease is caused by selective degeneration of the dopaminergic nigrostriatal tract, which originates in the midbrain's substantia nigra, pars compacta, and projects to the striatum (caudate nucleus and putamen).

The four cardinal motor symptoms of Parkinson's disease are: (1) resting tremor, (2) rigidity, (3) hypokinesia/bradykinesia, and (4) postural instability and gait impairment.

Resting Tremor

The **resting tremor** of Parkinson's disease is a slow, 4–6 Hz, rhythmic, alternating, involuntary movement of regular frequency and amplitude that affects the hands, head, and lips. The tremor diminishes during voluntary movement, and hence it is termed a "resting" tremor. The resting tremor disappears during sleep and general anesthesia; it increases during emotional stress or anxiety. The most common resting tremor in Parkinson's disease affects the fingers and is called a pill-rolling tremor, because it looks like the person is rolling a small object between their thumb and index finger.

Tremor is the symptom that usually leads to a diagnosis of Parkinson's disease, and it is present in about 70% of persons with the disease. At disease onset, motor symptoms are often asymmetric (i.e., affecting one side of the body more than the other); this is associated with asymmetric nigrostriatal degeneration. In cases with asymmetric onset, the typical progression of tremor begins in a single hand, progresses to the ipsilateral leg, and then to the contralateral limbs.

Rigidity

Rigidity (**hypertonia**) is an increase in muscle tone due to the continuous, passive partial contraction of muscles during rest in both flexor and extensor muscles. It manifests as an increased resistance to passive joint movements on neurological examination (i.e., the movement is externally imposed by an examiner and felt as stiffness by the examiner). It is the result of excessive supraspinal drive (upper motor neuron facilitation) acting on alpha motor neurons. Increased muscle tone may result in a form of **dystonia** in which involuntary muscle contractions affecting only some muscles result in an abnormal posture that is relatively fixed. In Parkinson's disease, rigidity predominately affects the flexor muscles of the neck, trunk, and limbs. This results in the typical flexed (stooped) posture of Parkinson's disease. On neurological examination there is **cogwheel rigidity**; the resistance (stiffness) to passive movement felt by the examiner is interrupted by brief relaxations through the movement.

Hypokinesia and Bradykinesia

Hypokinesia refers to paucity and reduced amplitude of movements. Common examples of hypokinesia in Parkinson's disease are reduced postural adjustments when seated and an absence of arm movements that are normally associated with walking. Other manifestations of hypokinesia are **hypomimia** (a mask-like, expressionless face due to reduced facial movements and animation), infrequent eye blinking, and **micrographia** (small handwriting). It also includes reduced amplitude of movements such as **shuffling gait** (small, shuffling steps) and **hypophonia** (soft voice).

Bradykinesia is a slowness of movement. Manifestations of bradykinesia in Parkinson's disease include slow, shuffling gait and slow speech. Bradykinesia also manifests as reduced ability to initiate movement. This symptom is very distressing to those with Parkinson's disease because movement initiation requires a high degree of concentration, and the mental effort can be exhausting.

Hypokinesia and bradykinesia are fundamental deficits. They are not simply the result of rigidity, as those with Parkinson's disease whose rigidity is not pronounced can have hypokinesia or bradykinesia.

Postural Instability and Gait Impairment

Persons with Parkinson's disease have postural instability, and consequently they fall easily when turning or when thrown off balance. This is due to reduced postural reflexes; they fail to make quick postural and righting adjustments to correct imbalance. Postural instability is revealed on neurological examination by the **pull-test**, which examines the patient's tendency to fall; the examiner stands behind the patient and applies a quick, forceful pull on the shoulders. Those with postural instability take several (more than two) steps backward to recover their balance, a phenomenon known as **retropulsion**, or show a tendency to fall if not caught by the examiner.

The classic gait of Parkinson's disease is slow, with short, shuffling steps, poor foot clearance, and decreased arm swing. There may also be **festination**, in which the gait involuntarily quickens, and the person appears to be falling forward continually (**anteropulsion**). Freezing may occur, particularly in narrow or crowded spaces. Turning around is slow and is performed with multiple small steps without the normal twist of the torso (i.e., **en bloc turning**).

Myerson's Sign

Patients with Parkinson's disease often exhibit **Myerson's sign** (**glabellar reflex**) on neurological examination early in the course of the disease. This is a primitive reflex and frontal release sign (see Chapter 30, "The Neurological Examination") in which the patient is unable to suppress blinking to repeated tapping of the glabella (center of the brow above the nose).

Non-Motor Features

Parkinson's disease is also associated with non-motor symptoms, including **dysautonomia** (autonomic dysfunction; see Chapter 6, "The Peripheral Nervous System"), cognitive impairment, neuropsychiatric disorders, and sleep-related disorders.

COGNITIVE IMPAIRMENT

Parkinson's disease may be associated with no cognitive deficits, mild cognitive deficits, or frank dementia that is referred to as **Parkinson's disease dementia**. A common cognitive symptom in those with Parkinson's disease is **bradyphrenia**, a slowness of thought that is apparent as prolonged information-processing time and delayed response time. Greater degrees of cognitive impairment generally involve attention and frontal-executive functions. The cognitive deficits of Parkinson's disease may be uncorrelated with the degree of basal ganglia involvement or the severity of motor symptoms. When tremor predominates, cognitive symptoms and dementia tend to occur less frequently; when bradykinesia and rigidity are the main symptoms, cognitive symptoms and dementia tend to occur more frequently.

Parkinson's disease dementia is a form of **subcortical dementia**, a syndrome that occurs with pathologies that primarily involve subcortical structures (thalamus, basal ganglia, and related brainstem nuclei) with relative sparing of the cerebral cortex. The subcortical dementia syndrome is characterized by deficits in cognitive processing speed, attention (working memory), executive functions, and memory (retrieval deficit with preserved recognition); it is often associated with apathy and depression. Clinical diagnostic criteria for Parkinson's disease dementia require both of the following core features: (1) a diagnosis of Parkinson's disease; and (2) dementia of insidious onset and slow progression in the presence of Parkinson's disease, defined by (a) an impairment of more than one domain of cognition, (b) which represents a decline from premorbid functioning, and (c) that is not ascribable to motor or autonomic dysfunction.

NEUROPSYCHIATRIC DISORDERS

A wide range of neuropsychiatric disorders may occur with Parkinson's disease, including depression, anxiety, apathy, delusions (most commonly paranoid), and hallucinations (usually visual, complex, and well formed). There may be considerable difficulty in distinguishing somatic symptoms of depression from those of Parkinson's disease, such as slowness of movement and thinking, and hypomimia which may give the impression of flat affect.

SLEEP-RELATED DISORDERS

Sleep disturbances are common in Parkinson's disease, including insomnia, excessive daytime sleepiness, and parasomnias. Insomnia and excessive daytime sleepiness both involve a disturbance in the regulation of sleep and wakefulness and the balance between the two. Insomnia is the most common sleep disturbance among those with Parkinson's disease (as in the general population); it is most often due to a disorder of sleep maintenance, but may also occur as a disorder of sleep onset or early morning awakening. While the prevalence of insomnia increases with age, and Parkinson's disease is age-related, the prevalence of insomnia in Parkinson's disease is greater than expected based on aging alone. Excessive daytime sleepiness is an inability to maintain wakefulness and alertness during the day that results in lapses into drowsiness or sleep, including sudden onset of sleep during periods of inactivity or low activity.

Parasomnias are behavioral phenomena that occur during sleep or the transitions between sleep and waking, and include confusional arousals, extremely vivid dreams, nightmares, nocturnal hallucinations, and somnambulism (sleep walking). **Rapid eye movement (REM) sleep behavior disorder**, which is characterized by dream enactment behaviors such as talking, laughing, shouting, gesturing, flailing, punching, kicking, sitting up, running movements, and other complex motor behaviors during sleep, is the most common cause of parasomnias in Parkinson's disease. These enactment behaviors parallel vivid action-filled dream content which often involves aggressive themes like being chased or defense against attack by animals or people. Persons with this disorder may injure themselves and/or their bed partner during sleep, which may lead bed partners to change to sleeping alone. Normally during REM sleep there is skeletal muscle atonia and paralysis (except for the extraocular muscles); in REM sleep behavior disorder, REM sleep and dreaming occur without the skeletal muscle atonia and motor inhibition that normally accompany REM sleep. There is an abnormal disinhibition of the skeletal motor system due to an abnormality within brainstem nuclei controlling muscle tone during REM sleep, leading to dream enactment behaviors. The diagnosis of REM sleep behavior disorder requires evidence of REM sleep without atonia; this is obtained by **polysomnography** (sleep study). Polysomnography records the patient's brain waves, the oxygen level in their blood, heart rate, and breathing, as well as eye and leg movements during the study.

Pathology

In the late 1950s, it was discovered that Parkinson's disease is associated with a loss of pigmentation in the substantia nigra. It was later found that Parkinson's disease is also accompanied by a selective reduction in dopamine (DA) in the striatum (caudate and putamen). It is now known that the loss of pigmentation in the substantia nigra and decreased striatal DA are the consequence of a selective neuronal degeneration of the dopaminergic nigrostriatal pathway.

The nigrostriatal tract originates in the substantia nigra, pars compacta, and projects to all the striatum (see Figure 3.4 in Chapter 3). The striatum in turn projects back to the substantia nigra, pars reticulata, directly via striatonigral projections and indirectly via striatopallidal-pallidonigral projections. The substantia nigra, pars reticulata, in turn projects to the thalamus and then cortex. Dopamine excites neurons of the direct pathway and inhibits neurons of the indirect pathway, with a net excitatory action on the thalamus. Loss of DA results in net inhibition of the thalamus, leading to the paucity of movements in Parkinson's disease.

Brain magnetic resonance imaging (MRI) in idiopathic Parkinson's disease is often nondiagnostic. The structural abnormalities in idiopathic Parkinson's disease (putaminal atrophy, smudging of the substantia nigra) are subtle, nonspecific, and occur in a minority of Parkinson's disease patients. Degeneration of this pathway can, however, be evaluated in vivo using a biomarker for nigrostriatal cell loss. The **DA transporter** (DAT) is a membrane protein located on the presynaptic terminals of dopaminergic projections. It plays a role in regulating extracellular DA concentration and is a marker for DA axon terminal integrity. The protein can be radiolabeled by a radioligand and visualized with single-photon emission computed tomography (SPECT) imaging in a technique known as the **DAT-SPECT scan** (Figure 11.3). This technique plays an important role in Parkinson's disease diagnosis, as other neurodegenerative and non-neurodegenerative causes of parkinsonism do not involve nigrostriatal dopaminergic degeneration.

Lewy bodies are the pathological hallmark of Parkinson's disease and are revealed only by histological examination of brain tissue. Lewy bodies are abnormal **neuronal inclusions** that appear under the microscope as spherical structures with radiating filaments that consist primarily of **alpha-synuclein**. Alpha-synuclein is a normal brain protein, but abnormalities of alpha-synuclein metabolism result in abnormal aggregates (clumps) of the protein. Thus, Parkinson's disease is classified as an alpha-synuclein pathology (**synucleinopathy**). The underlying cause of abnormal alpha-synuclein processing in Parkinson's disease is unknown.

Genetics of Parkinson's Disease

Most cases of Parkinson's disease are **sporadic**, occurring as single, scattered cases without evidence for an environmental or heritable etiology. Approximately 5%–15% of cases of Parkinson's disease, however, are familial, and several genes have been linked to **monogenic** Parkinson's

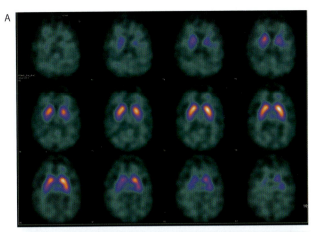

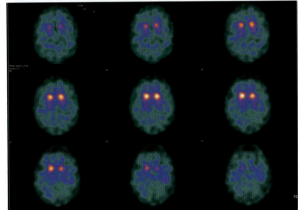

FIGURE 11.3. Dopamine transporter imaging (DAT-scan) in (A) normal brain and (B) brain with Parkinson's disease. The scan visualizes tracer that binds to the dopamine transporter molecule located on dopamine neuron axon terminals. In Parkinson's disease there is degeneration of the nigrostriatal dopaminergic neurons and fewer dopaminergic terminals in the striatum. The resulting lower density of dopamine transporter molecule is apparent as less signal (brightness) in the striatum.

disease (involving a single gene mutation). The genes are all located on **autosomes** (chromosomes other than the sex chromosomes), and are transmitted in either an **autosomal dominant** fashion, in which transmission of one mutated allele of the gene is sufficient to cause the disease, or in an **autosomal recessive fashion**, in which both alleles of the gene must be mutated (i.e., it is inherited from both parents), as evidenced by **pedigree analysis**. The specific chromosomal regions involved are termed *PARK* and are numbered in chronological order of their identification (*PARK1*, *PARK2*, *PARK3*, etc.).

Treatment

PHARMACOTHERAPY

The discovery that Parkinson's disease was due to selective degeneration of the dopaminergic nigrostriatal pathway and a deficiency in DA led to the development of a pharmacological treatment based on enhancing neurotransmitter synthesis by **precursor loading**. **Levodopa (L-DOPA)** is a precursor in the biosynthetic pathway of DA, so providing exogenous levodopa allows the surviving dopaminergic neurons to synthesize more DA. It is believed that the increase in DA restores the normal balance between inhibition and excitation within the striatum. Dopamine replacement by precursor loading alleviates the symptoms of Parkinson's disease, particularly rigidity and akinesia.

Parkinson's disease is treated with a combination of levodopa and carbidopa. When administered orally, levodopa is broken down in the bloodstream by the enzyme DOPA decarboxylase. In order to prevent breakdown of levodopa in the periphery, it is administered in combination with carbidopa. **Carbidopa** competes with levodopa for the active site on DOPA decarboxylase, thereby inhibiting the peripheral metabolism of levodopa. Levodopa crosses the blood-brain barrier, but carbidopa does not. Thus, co-administration of carbidopa allows a greater proportion of levodopa to cross the blood-brain barrier and have central effects. Thus **levodopa-carbidopa** enhances the synthesis of DA by a combination of two different mechanisms: by increasing the amount of precursor centrally and by inhibiting its breakdown in the periphery.

Excess levels of levodopa, however, can result in choreiform and athetoid movements. Sustained levodopa treatment almost invariably leads to **on-off fluctuations** (i.e., fluctuation between "on" time when medication is effective and "off" time when medication is ineffective, and the patient experiences a re-emergence of Parkinson's symptoms). It is believed that these motor fluctuations to levodopa therapy increase with progressive degeneration of the nigrostriatal dopaminergic pathway and a loss of DA storage capacity, so that levodopa has a more pulsatile impact on postsynaptic dopamine receptors (in keeping with its short, 90-minute half-life). Dopamine agonists that mimic DA without increasing DA production or inhibit the enzymatic breakdown of DA may be used in combination with levodopa to reduce the on-off effect of levodopa.

Pharmacological treatment for Parkinson's disease is a **symptomatic therapy**; the goal is to control signs and symptoms for as long as possible while minimizing adverse effects. It is not a **disease-modifying (neuroprotective) therapy**, as it does not stop or diminish the neurodegeneration. While levodopa precursor loading is a remarkably effective treatment that enhances DA synthesis in surviving nigrostriatal neurons, its effectiveness depends on having surviving dopaminergic nigrostriatal cells. As the neurodegeneration progresses, pharmacotherapy becomes less effective in controlling debilitating symptoms.

Adverse side effects of dopamine agonists include **impulse control disorders** such as hypersexuality, gambling, binge eating, compulsive buying/shopping, punding, and hobbyism. **Punding** is a distinctive stereotyped behavior that was first described in amphetamine and cocaine addicts in the 1970s. It is characterized by excessive repetition of non-goal-oriented activities involving manipulation of objects, such as continual handling, sorting, counting, or lining up of objects. **Hobbyism** is an intense time-consuming preoccupation with specific activities (e.g., endlessly working on artistic endeavors or writing, compulsive Internet use).

ABLATIVE SURGERY

To address the limitations of pharmacologic therapy, namely loss of clinical efficacy over time and debilitating side effects of dyskinesias and motor fluctuations, minimally invasive surgery for Parkinson's disease was developed in the late 1980s. The surgical treatment involves creating a therapeutic lesion within dysfunctional brain circuits, targeting the internal segment of the globus pallidus (**pallidotomy**), ventral intermediate nucleus of the thalamus (**thalamotomy**), or subthalamic nucleus (**subthalamotomy**) by a **stereotactic** technique that allows for precise three-dimensional targeting. Ablative surgery, however, has largely been replaced by deep brain stimulation as a treatment for Parkinson's disease.

DEEP BRAIN STIMULATION

Deep brain stimulation is a neurosurgical procedure that involves the implantation of electrodes into specific targets within the brain and the delivery of constant or intermittent electricity from an implanted battery source. Stimulation by high-frequency trains of electrical pulses interfere with signal transmission in pathological neural circuits, creating a functional lesion that can be reversed by turning the stimulation off. Targets of deep brain stimulation for Parkinson's disease are the internal segment of the globus pallidus or subthalamic nucleus. If the Parkinson's disease symptoms are primarily unilateral, a stimulating electrode is implanted in the hemisphere contralateral to the side of the body most prominently affected by motor symptoms. Bilateral motor symptoms are treated with bilateral deep brain stimulation.

Other Causes of Parkinsonism

Parkinsonism is a clinical syndrome of movement disorder characterized by tremor, bradykinesia, and rigidity. Parkinson's disease accounts for approximately 70% of cases of parkinsonism, but other neurodegenerative diseases and non-neurodegenerative neurological insults that do not involve nigrostriatal degeneration may also result in parkinsonism. Parkinsonism can occur secondary to vascular lesions, drugs (e.g., dopaminergic antagonists such as haloperidol or prochlorperazine, reserpine, phenothiazines, butyrophenones, and the MPTP toxin contained in a synthetic heroin-like drug), other toxic insults (e.g., carbon monoxide poisoning, anoxia, mercury poisoning), chronic traumatic encephalopathy (dementia pugilistica), and even psychogenic causes.

Parkinson Plus Syndromes

The **Parkinson plus syndromes**, also known as **disorders of multiple system degeneration**, are a group of neurodegenerative diseases that involve atypical parkinsonism with insidious onset and gradually progressive course. They feature the classical parkinsonian symptoms of rigidity, akinesia/bradykinesia, and postural instability, but they differ from Parkinson's disease in that they (1) tend to lack a resting tremor; (2) have atypical features such as early dementia, early gait instability, gaze palsies, or early orthostatic hypotension; (3) have a more rapid course; and (4) respond poorly to traditional Parkinson's disease therapies. Parkinson plus syndromes associated with cognitive decline and neuropsychiatric symptoms include diffuse Lewy body disease, progressive supranuclear palsy, and corticobasal syndrome. Multisystem atrophy is a Parkinson plus syndrome characterized by a combination of autonomic and motor symptoms; it is not associated with cognitive decline or neuropsychiatric symptoms.

The Parkinson plus syndromes are often difficult to diagnose initially, as they are difficult to differentiate from Parkinson's disease and from each other. Therefore, poor response to levodopa is often used as evidence supporting or ruling out Parkinson's disease as the underlying cause of parkinsonism. Persons with parkinsonism that is based on nigrostriatal degeneration benefit from dopaminergic medication; however, persons with parkinsonism due to postsynaptic etiologies or non-degenerative etiologies do not.

Many of the most common degenerative diseases have pathologic accumulations of normal cellular proteins within vulnerable neuronal populations, and therefore they may be classified based on their molecular pathology. Most degenerative parkinsonian disorders fall into one of two molecular classes: alpha-synucleinopathies or tauopathies.

As previously described, the synucleinopathies are characterized by abnormal aggregates of alpha-synuclein protein in the cytoplasm of selective populations of neurons and glia. The **tauopathies** are characterized by the abnormal accumulation of the protein tau in certain nerve cells; tau is a microtubule-associated protein component of the neurofilaments that are part of the neuronal cytoskeleton. Both the alpha-synucleinopathies and tauopathies may or may not be inherited genetically.

Diffuse Lewy Body Disease

Diffuse Lewy body disease, also referred to as **dementia with Lewy bodies** and **cortical Lewy body disease**, is characterized by cognitive decline and parkinsonism. Onset of motor and cognitive symptoms is within one year of each other, differentiating diffuse Lewy body disease from Parkinson's disease dementia in which motor signs and symptoms precede cognitive signs and symptoms by at least one year. Diffuse Lewy body disease is responsible for 20% of all dementias, second to Alzheimer's disease in prevalence. Like Parkinson's disease, diffuse Lewy body disease is a synucleinopathy. It is characterized anatomically by the presence of Lewy bodies in neurons localized in the substantia nigra and throughout the cerebral cortex, detectable only in postmortem brain histology.

DIAGNOSIS

Diffuse Lewy body disease is diagnosed based on clinical features and biomarkers. Clinical features are categorized as core or supportive; biomarkers are categorized as indicative or supportive.

Dementia is an essential feature. It is defined as a progressive cognitive decline of sufficient magnitude to interfere with normal social or occupational functions, or with usual daily activities. The neuropsychological profile of diffuse Lewy body disease is characterized by visuoperceptual, attentional, and frontal executive impairments most prominently in the early stages. On neuropsychological and neurological examinations, the inability to draw or copy simple diagrams or construct simple figures is prominent. Memory problems are relatively mild and are limited to encoding/acquisition, with relatively preserved delayed recall in the early stages. The rate of progression of cognitive symptoms in diffuse Lewy body disease tends to be more rapid than in Alzheimer's disease.

There are four core clinical features of diffuse Lewy body disease: fluctuating cognition, visual hallucinations, REM sleep behavior disorder, and one or more cardinal features of parkinsonism (bradykinesia, resting tremor, rigidity). Fluctuating cognition is characterized by pronounced variations in attention and alertness. Attentional fluctuations manifest as lapses in attention that are short-lived and are unrelated to situational demands, with reduced awareness of surroundings. At times the person is lucid and eloquent, and at other times appears confused and makes comments that are out of context. Fluctuations in alertness manifest as daytime hypersomnolence and confusion on waking. The visual hallucinations are recurrent and typically are well-formed and detailed; most commonly they involve the perception of people or animals that are especially seen against patterned backgrounds (e.g., in the branches of trees, in patterns of drapes, carpeting, and upholstery). Fluctuating cognition, visual hallucinations, and REM sleep behavior disorder typically occur early and may persist throughout the course of the disease. REM sleep behavior disorder is closely associated with the synucleinopathies of Parkinson's disease, dementia with Lewy bodies, and multiple system atrophy, and may begin years to decades before the onset of motor, cognitive, or autonomic dysfunction, respectively. When the disorder presents on its own, it is often referred to as idiopathic; however, up to 90% of such patients will eventually develop one of the synucleinopathies, so it is better conceptualized as a prodromal feature.

In diffuse Lewy body disease, the onset of spontaneous cardinal features of parkinsonism (bradykinesia, resting tremor, rigidity) either coincide with or occur within one year of the onset of cognitive signs and symptoms. Clinicians and researchers use the one-year rule to differentiate diffuse Lewy body disease and Parkinson's disease dementia; if cognitive symptoms appear at the same time as or at least a year before parkinsonism, the diagnosis is diffuse Lewy body disease; if cognitive problems develop more than a year after the onset of movement abnormalities, the diagnosis is Parkinson's disease dementia.

Supportive clinical features include severe sensitivity to antipsychotic medications; postural instability; repeated falls; syncope or other transient episodes of unresponsiveness; severe dysautonomia (e.g., constipation, orthostatic hypotension, urinary incontinence); hypersomnia; hyposmia (reduced sense of smell); hallucinations in other modalities; systematized delusions; apathy, anxiety, and depression. Those with diffuse Lewy body disease whose hallucinations are treated with typical antipsychotics (neuroleptics) are at risk for developing **neuroleptic malignant syndrome**, which is characterized by catatonia, loss of cognitive function, and/or life-threatening muscle rigidity. Thus, typical antipsychotics are avoided, and atypical antipsychotic medications with the fewest extrapyramidal side effects are used to alleviate hallucinations when required.

Indicative biomarkers consist of the following: (1) reduced SPECT or positron emission tomography (PET) dopamine transporter uptake in the basal ganglia; (2) evidence of sympathetic denervation of cardiac muscle which underlies some symptoms of dysautonomia (i.e., abnormally low cardiac sympathetic nerve endings reflected by myocardial scintigraphy); and (3) polysomnography confirmation of REM sleep without atonia. Supportive biomarkers consist of the following: (1) relative preservation of medial temporal lobe structures on CT or MRI brain imaging; (2) generalized low uptake on SPECT cerebral perfusion or PET cerebral metabolism scan with reduced occipital activity; and (3) prominent posterior slow-wave activity on electroencephalography.

Probable diffuse Lewy body disease is diagnosed if (a) two or more core clinical features of diffuse Lewy body disease are present, with or without the presence of indicative biomarkers; or (b) only one core clinical feature is present, but with one or more indicative biomarker. Possible diffuse Lewy body disease is diagnosed if (a) only one core clinical feature of diffuse Lewy body disease is present, with no indicative biomarker evidence, or (b) one or more indicative biomarkers is present but there are no core clinical features.

TREATMENT

The treatment of diffuse Lewy body disease includes cholinergic agonists (medications that boost acetylcholine) for cognitive impairment, atypical antipsychotic medication to alleviate hallucinations, and levodopa/carbidopa to improve parkinsonism (although this can aggravate hallucinations).

Progressive Supranuclear Palsy

CLINICAL PRESENTATION

Progressive supranuclear palsy (Steele-Richardson-Olszewski syndrome) is a rare neurodegenerative disorder that typically presents as a gradually progressive akinetic-rigid form of parkinsonism predominantly affecting the axial muscles (i.e., of the trunk and neck). The motor abnormalities are typically symmetric (affecting both sides of the body equally). The most common initial manifestations of progressive supranuclear palsy are gait difficulty and falls; tremor is usually absent. As the disease progresses, the parkinsonism worsens and other neurologic manifestations emerge, including oculomotor abnormalities, cognitive decline, dysarthria, and dysphagia (difficulty swallowing).

The gradually progressive rigidity (stiffening and extension) of the axial muscles results in an axial dystonia characterized by a stiff, hyper-erect posture. There may be dystonia of the neck in which the head is drawn back due to contraction of the neck extensor muscles (**retrocollis**). As the disease progresses, the axial rigidity evolves to progressive gait freezing, with sudden and transient motor blocks or start hesitation. This results in postural instability due to a lack of postural reflexes, manifesting as spontaneous loss of balance while standing and unprovoked falls, usually backward. On neurological examination a diagnostic feature of axial rigidity is resistance to passive movement of the neck. Postural instability is revealed on neurological examination by the pull-test. In addition, bradykinesia, reduced eyeblink rate, and marked micrographia are often present.

A hallmark feature of progressive supranuclear palsy is **vertical gaze paresis**, an oculomotor disorder characterized by slow, restricted, or absent voluntary vertical saccadic eye movements (although this may not be present early in the course of the disease). Downward gaze is particularly affected or affected before upward gaze. By contrast, reflexive vertical eye movements may be intact, and full movements can be obtained if the eyes are fixated on a target and the head is moved passively (**doll's eye test**). This profile of impaired and spared vertical eye movements indicates that the lesion is above the cranial nerve nuclei (i.e., suprabulbar, supranuclear) that are responsible for eye movements, affecting the corticobulbar projections to the brainstem cranial nerve nuclei controlling the extraocular muscles. This condition is therefore also referred to as **supranuclear ophthalmoplegia** or **supranuclear vertical gaze palsy**. The deficit in downward gaze disrupts reading and contributes to frequent falls, and in combination with dysphagia leads to the **dirty-tie sign**, as patients cannot look down while eating and see that they are spilling food on their clothes. The face often looks stiff and immobile, with a wide-eyed stare or look of surprise or astonishment, due to a combination of rare blinking, facial dystonia, and gaze abnormality.

Cognitive symptoms typically are present within 2 years of motor symptom onset, and dementia occurs in 50%–80% of cases. The neurocognitive profile of progressive supranuclear palsy is characterized by bradyphrenia and executive dysfunction (e.g., impaired abstract thinking and shifting conceptual set) most prominently. There also is a prominent apathy (loss of interest, emotional unresponsiveness). The dementia evolves in association with the progressive neuronal degeneration. There is atrophy of the frontal lobes due to the pathology affecting the frontal lobes directly; it is not merely due to frontal deafferentation due to subcortical pathology affecting striatothalamocortical circuits. The degree of cognitive dysfunction parallels the degree of frontal atrophy.

Supportive features of progressive supranuclear palsy include **pseudobulbar palsy**, manifest as a combination of dysphagia and dysarthria, with a characteristic hoarse groaning voice due to bilateral degeneration of the corticobulbar tracts (above the brainstem cranial nerve nuclei). There may also be **pseudobulbar affect**, also known as "emotional incontinence," as occurs with other forms of pseudobulbar palsy (see Chapter 24, "The Motor System and Motor Disorders").

NEUROPATHOLOGY

Progressive supranuclear palsy is a **tauopathy**; the defining histopathologic feature is the intracerebral aggregation of the microtubule-associated protein tau. Normally, tau is phosphorylated, and the level of tau phosphorylation is regulated by multiple enzymes (tau kinases and tau phosphatases). In tauopathies, the tau protein is hyperphosphorylated; hyperphosphorylated tau is resistant to proteolysis (enzymatic breakdown of protein), accumulates, and aggregates, forming neurofibrillary tangles. Definitive diagnosis therefore requires direct examination of brain tissue postmortem. The pathology affects the brain bilaterally, and particularly affects the subthalamic nucleus, globus pallidus, striatum, substantia nigra, and red nucleus. There are no effective treatments for progressive supranuclear palsy; the research is focused on developing disease-modifying therapies that reduce brain levels of toxic forms of tau.

RADIOGRAPHIC FEATURES

Neuroimaging by MRI in patients with progressive supranuclear palsy shows atrophy and signal increase in the midbrain, enlargement of the cerebral aqueduct and third ventricle, and degeneration of the red nucleus. With disease progression, frontal and temporal lobe atrophy may become apparent. Atrophy of the midbrain tegmentum is seen on MRI midsaggital cuts as the **hummingbird sign** (**penguin sign**), and on horizontal cuts as the **Mickey Mouse sign** (see Figure 8.5 in Chapter 8). These imaging features are not specific for progressive supranuclear palsy; they are suggestive of the diagnosis, rather than **pathognomonic** (i.e., specifically characteristic or indicative of a particular disease).

PET imaging studies using fluorodeoxyglucose (FDG) as a radiotracer to measure tissue metabolism typically show fronto-subcortical hypometabolism in progressive supranuclear palsy patients. DAT-SPECT imaging studies show reduced tracer uptake in the striatum bilaterally. This is useful for distinguishing progressive supranuclear palsy from mimics such as cerebrovascular disease and normal pressure hydrocephalus, but it cannot distinguish between progressive supranuclear palsy and Parkinson's disease. PET imaging with a radioligand to label and visualize tau accumulation in the living brain, however, does distinguish between progressive supranuclear palsy and Parkinson's disease; it is currently available as a research tool and is being developed for use as a clinical tool.

CLINICAL SUBTYPES

Progressive supranuclear palsy is now known to be a four-repeat (4R-) tauopathy, so named because it has four microtubule-binding domains, and consequently its definition has expanded to comprise a spectrum of motor and behavioral syndromes associated with 4R-tauopathies. Clinical subtypes of progressive supranuclear palsy include classic progressive supranuclear palsy (progressive supranuclear palsy–Richardson syndrome), brainstem variants (progressive supranuclear palsy–predominant parkinsonism, progressive supranuclear palsy–pure akinesia with gait freezing), and cortical variants (progressive supranuclear palsy–corticobasal syndrome, progressive supranuclear palsy–behavioral variant of frontotemporal dementia, progressive supranuclear palsy–progressive nonfluent aphasia).

Corticobasal Syndrome

Corticobasal syndrome is an atypical parkinsonism characterized by rigidity, bradykinesia/akinesia, and dystonia (involuntary muscle contractions that result in abnormal movements and postures) that typically begins unilaterally. The asymmetric akinetic-rigid syndrome is accompanied by variable admixtures of tremor (postural or action), myoclonus (brief involuntary muscle spasms that cause jerky movements), cortical sensory loss, and the cortical motor signs of asymmetric limb apraxia and alien limb syndrome. Cortical sensory deficit consists of an isolated loss of discriminative sensation (stereognosis, graphesthesia, position sense) involving only part of the body (see Chapter 15, "The Parietal Lobes and Associated Disorders"). **Limb apraxia** is characterized by the inability to perform voluntary skilled movements despite normal motor strength and coordination (see Chapter 24, "The Motor System and Motor Disorders"). In **alien limb syndrome**, the person is unaware of or unable to control the movement of a limb (see Chapter 24, "The Motor System and Motor Disorders"). Affected individuals may first become aware of the disorder when they have difficulty performing manual tasks; they describe their actions as stiff and clumsy. The parkinsonism eventually progresses to involve both sides of the body; it is levodopa resistant.

Corticobasal syndrome was first described in the late 1960s. Eventually it was found to be associated with progressive neurodegenerative disease involving the cerebral cortex and basal ganglia and a specific form of tauopathy, and was initially referred to as **corticobasal degeneration** or **corticobasal ganglionic degeneration**. The terms *corticobasal syndrome* and *corticobasal degeneration* have been used interchangeably, but this is not accurate; corticobasal syndrome refers to the clinical syndrome, while corticobasal degeneration refers to the histopathology. The clinical diagnostic criteria for probable corticobasal syndrome require (1) asymmetric presentation of at least two of the following three signs: (a) limb rigidity or akinesia, (b) limb dystonia, (c) limb myoclonus; plus (2) at least two of the following three signs: (a) orobuccal or limb apraxia, (b) cortical sensory deficit, (c) alien limb phenomena (more than simple levitation).

Corticobasal syndrome is associated with a spectrum of pathologies; only about half of patients diagnosed have corticobasal degeneration pathology at autopsy. Other histopathologies include Alzheimer's disease, progressive supranuclear palsy, and frontotemporal degeneration with TDP-43 (FTD-TDP43). Furthermore, corticobasal degeneration pathology may present with other clinical syndromes, such as behavioral variant frontotemporal dementia or temporal variant frontotemporal dementia (primary progressive aphasia).

Initially, corticobasal syndrome was considered to affect primarily the basal ganglia and frontoparietal cortex, accounting for the parkinsonism and apraxia. Structural neuroimaging usually revealed asymmetrical frontoparietal cortical atrophy, and the laterality of atrophy correlated with the laterality of clinical impairment (i.e., the clinical signs were contralateral to the atrophy). Functional neuroimaging studies revealed asymmetrical cortical and subcortical hypoperfusion on SPECT imaging and hypometabolism on PET imaging that was usually more extensive than the degree of atrophy revealed by MRI.

More recent studies have shown that cognitive and behavioral changes are usually present before the onset of motor symptoms. Variable involvement of frontal, parietal, and temporal cortices results in variable cortical signs beyond apraxia and alien limb, including psychomotor slowing, neglect, agnosia, global cognitive impairment, behavioral changes consisting of apathetic behavioral variant frontotemporal dementia (see Chapter 17, "The Frontal Lobes and Associated Disorders"), aphasia consisting of progressive nonfluent aphasia language variant of frontotemporal dementia (see Chapter 27, "Language and the Aphasias"), and frontal release signs (see Chapter 30, "The Neurological Examination").

Multisystem Atrophy

Multisystem atrophy, formerly called Shy-Drager syndrome, is a rare, neurodegenerative neurological disorder characterized by a combination of dysautonomia and motor system dysfunction.

The autonomic nervous system regulates bodily functions such as blood pressure, heart rate, breathing, salivation, perspiration, micturition (urination), digestion, and some sexual functions. **Dysautonomia** is an umbrella term referring to any disorder of autonomic nervous system function; it generally involves a failure of sympathetic or parasympathetic divisions of the autonomic system, but may involve excessive autonomic activity (see Chapter 6, "The Peripheral Nervous System"). A hallmark feature of sympathetic failure is **orthostatic hypotension**, a drop in blood pressure producing fainting or dizziness when one transitions to an upright position, such as from lying to sitting or from sitting to standing. During such postural transitions, gravity causes blood to redistribute from the upper body to the lower body. Normally, sympathetic autonomic reflexes counteract the fall in blood pressure. Orthostatic hypotension is a common feature of multisystem atrophy. Other signs and symptoms of dysautonomia that may occur in multisystem atrophy include urinary and bowel dysfunction (constipation, incontinence); reduced production of sweat, tears, and saliva; impaired body temperature control; and male impotence.

There are two clinical forms of multisystem atrophy based on the dominant motor symptoms: multisystem atrophy with predominant parkinsonism, and multisystem atrophy with cerebellar features. In multisystem atrophy with predominant parkinsonism, there is a predominance of parkinsonian signs and symptoms; this variant is also referred to as **striatonigral degeneration**. In multisystem atrophy with cerebellar features, ataxia is the major clinical feature; this variant is also known as **olivopontocerebellar atrophy**. In multisystem atrophy with cerebellar features, there is a selective loss of myelinated transverse pontocerebellar fibers and neurons in the pontine raphe with preservation of the pontine tegmentum and corticospinal tracts. This variant is associated with the "**hot-cross bun**" sign, a cruciform T2-hyperintense signal on axial MRI of the pons that is due to gliosis of pontocerebellar fibers. This sign is most often observed in those with multisystem atrophy with cerebellar features, but it is not pathognomonic as it occurs in other conditions, including spinocerebellar ataxia, progressive multifocal leukoencephalopathy, and paraneoplastic cerebellar degeneration.

Multisystem atrophy is a sporadic disease; symptom onset typically is between 40 and 60 years of age. It is a synucleinopathy, but unlike Parkinson's disease and diffuse Lewy body disease, intracellular deposits of the abnormal protein are found in both neurons and oligodendroglia.

Huntington's Disease

Huntington's disease is a rare hereditary disorder characterized by choreiform movements in combination with a progressive dementia. The movements often predominantly involve the hands, feet, face, and mouth. The ability to carry out continuous voluntary movements may be impeded by the superimposition of chorea. Choreiform movements disappear during sleep and often are exacerbated by anxiety and emotional distress.

The focus of neurodegenerative atrophy in Huntington's disease occurs initially within the head of the caudate, although other degenerative changes are present throughout the basal ganglia. Normally, the head of the caudate bulges into the lateral ventricle, but with caudate neurodegeneration this characteristic bulge is lost, resulting in "box car" ventricles (Figure 11.4). As the disease progresses, cortical atrophy also develops, particularly in the frontal and temporal lobes. Brain weight may decrease by up to 30% in advanced cases.

The atrophy particularly affects acetylcholine and GABA-containing interneurons in the striatum and cerebral cortex. Symptoms are viewed as the result of excessive dopamine release within the striatum resulting from a loss of inhibitory interneurons that synapse on dopaminergic

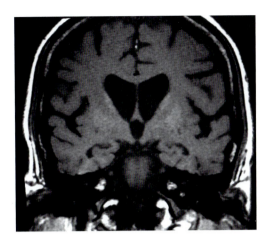

FIGURE 11.4. Coronal brain MRI in a patient with Huntington's disease. Normally, the head of the caudate bulges into the frontal horns of the lateral ventricle. In Huntington's disease the caudate atrophies, creating the "box car" appearance of the frontal horns.

cells. Chorea and athetosis are treated with dopamine antagonists.

Huntington's disease has an autosomal dominant pattern of transmission of a gene located on chromosome 4; the disease is diagnosed by genetic testing. The gene is called **huntingtin** and codes for a protein that is also called huntingtin. The genetic mutation has an abnormally high number of cytosine-adenine-guanine (CAG) trinucleotide repeats, resulting in an abnormal protein that gradually damages brain neurons. The number of CAG repeats determines age of onset (< 28 normal, 28–35 intermediate, 36–40 reduced penetrance, > 40 full penetrance). Huntington's disease typically occurs after the third decade of life (onset in 30s–50s). Juvenile Huntington's disease has an onset before age 20; it is rare and occurs with CAG repeats greater than 55.

Huntington's disease is associated with cognitive and mood changes. The most salient cognitive changes are in psychomotor speed, executive functions (i.e., planning, organization, and cognitive flexibility), and memory. Memory test performance is characterized by poor use of strategies during learning, and poor free recall in the context of relatively preserved cued recall and recognition memory, in keeping with the disruption to striatal-frontal pathways. The most common neuropsychiatric symptoms in Huntington's disease are apathy, loss of motivation and drive, irritability, loss of anger control, and depression.

Cognitive and mood changes may occur years before the onset of motor symptoms. In the early stages of Huntington's disease, especially if cognitive and/or mood symptoms predate the onset of definitive chorea, the patient may be diagnosed as having a psychiatric disorder. Prospective longitudinal studies of those in the prodromal phase of Huntington's disease (i.e., with positive family history and positive genetic testing but without significant motor signs) have found psychomotor slowing and depressed attention and working memory as the most common cognitive deficits. These cognitive changes may be evident for many years prior to the onset of motor signs and are related to striatal volume loss on MRI.

Wilson's Disease

Wilson's disease (**hepatolenticular degeneration**) is a rare, inherited, autosomal recessive metabolic disorder affecting copper metabolism. It is characterized by deposition of copper in organs, particularly in the brain, liver, and cornea. It presents between the first and third decades of life, with a variety of hepatic and neurological symptoms, prominent psychiatric disturbances, and brownish outer corneal deposits of copper known as **Kayser-Fleischer rings**. Toxic accumulations of copper in the liver cause hepatitis and cirrhosis, and toxic accumulations of copper in the brain cause progressive degeneration of the basal ganglia.

Typical neurologic manifestations of Wilson's disease include progressive rigidity, athetosis, and a coarse postural tremor (asterixis). **Asterixis** is also known as **flapping tremor** and **wing-beating tremor**. When the wrist is extended (i.e., bent upward at the wrist), the patient has jerking movements of the outstretched hands that resemble a bird flapping its wings. The tremor is caused by abnormal function of the diencephalic motor centers that regulate the muscles involved in maintaining position. Wilson's disease is also associated with a facial dystonia called **risus sardonicus** that is characterized by a fixed open "grin" and raised eyebrows that appears like a sardonic facial expression to the lay observer.

Treatment for Wilson's disease typically includes **copper chelating agents** which can arrest progression of the disorder.

Wilson's disease is also associated with cognitive decline and neuropsychiatric disorder. If untreated, Wilson's disease results in dementia, while in treated patients there may be abnormalities mainly in the domains of attention, executive functions, and memory (encoding). Neuropsychiatric symptoms usually occur several years after onset and are wide-ranging, including depression, irritability, aggressiveness, paranoia, and hallucinations, and may be reversed after several years of treatment.

Sydenham's Chorea

Sydenham's chorea, also known as rheumatic chorea or chorea minor and historically known as St. Vitus' dance, is a neurological disorder of childhood and adolescence and is the most common acquired movement disorder of this age group. It typically occurs between 5 and 15 years of age and affects girls more often than boys. It results from untreated streptococcal infection by the same bacterium that causes rheumatic fever. Sore throat and fever precede neurological symptom onset, usually by several weeks but up to as long as 6 months. The anti-streptococcal antibodies cross-react with striatal neurons. In other words, the streptococcus bacteria and a striatal neuron molecule appear similar to the immune system; the antibody recognizes both molecule types as antigens and attacks them. Damage to the striatal neurons is responsible for Sydenham's chorea.

Sydenham's chorea is characterized by rapid, irregular (i.e., non-rhythmic), jerky, involuntary movements primarily affecting the face, hands, and feet. The neurologic symptoms have an abrupt onset (sometimes within a few hours). Typical choreic movements include repeated wrist hyperextension, grimacing, lip pouting, fingers moving as if playing a piano, motor impersistence (e.g., inability to sustain tongue protrusion or eye closure), "milk maid grip sign" (fluctuating grip strength as if milking a cow), irregular gait (giving the appearance of skipping or dancing), and reduced fine motor control resulting in dysarthria and deterioration of handwriting. Additionally, there are neuropsychiatric symptoms (emotional lability, impulsive or obsessive-compulsive behaviors, and attention deficit).

Sydenham's chorea typically improves spontaneously and gradually; signs and symptoms typically last 12 to 15 weeks.

Basal Ganglia Calcification

Basal ganglia calcification identified through neuroimaging is most often observed in the elderly, where it occurs idiopathically (without an identified cause) and is considered a normal incidental finding. In these cases, the globus pallidus is most often affected.

In younger people (under age 40), basal ganglia calcification is pathological. There are a wide variety of etiologies, including metabolic disorders that affect calcium homeostasis (e.g., parathyroid disorders), toxic (e.g., carbon monoxide poisoning, lead poisoning, and mineralizing microangiopathy due to radiation therapy and chemotherapy), infectious (e.g., CNS tuberculosis, CNS toxoplasmosis, neurocysticercosis), birth hypoxia, and genetic disorders (e.g., Down syndrome). **Fahr's disease** (**familial idiopathic basal ganglia calcification**) is a rare neurodegenerative disorder that is characterized by bilateral symmetrical calcifications in the basal ganglia and other brain regions (e.g., dentate nucleus of the cerebellum). It may be asymptomatic or may be associated with dementia and extrapyramidal motor symptoms.

Neurodegeneration with Brain Iron Accumulation

Neurodegeneration with brain iron accumulation (NBIA), previously known as **Hallervorden-Spatz disease**, comprises a group of neurodegenerative disorders characterized by prominent extrapyramidal features, dementia, and radiographic evidence of iron deposition in the basal ganglia. It is associated with (but not pathognomonic for) the MRI finding of symmetric bilateral abnormal low signal on T2-weighted MRI (due to abnormal accumulation of iron) in the globus pallidus with a longitudinal stripe of high signal (due to gliosis and spongiosis).

In most cases it is due to due to a rare autosomal recessive genetic disorder, typically presenting with dystonia, dysarthria, rigidity, and choreoathetosis within the first decade of life, progressing, and culminating in early death. There are, however, atypical sporadic adult-onset forms with slower progression and much greater clinical heterogeneity. Approximately one-third of adult-onset cases present as a frontotemporal dementia with a combination of cognitive decline (especially executive dysfunction) and change of personality (disinhibition, emotional blunting, and socially inappropriate behavior). In these cases, FDG-PET functional neuroimaging shows frontotemporal hypometabolism.

Summary

The basal ganglia comprise the caudate nucleus, putamen, globus pallidus, subthalamic nucleus, and substantia nigra. The basal ganglia, like the cerebellum, are motor system components that influence the activity of motor cortex through side-loop circuits. Most input to the basal ganglia originates from the cerebral cortex, and most output is directed to the cerebral cortex via the thalamus. Our knowledge regarding the function of the basal ganglia is based largely on diseases that predominantly affect these structures. There are two classic syndromes of basal ganglia disorders: hypokinesis and hyperkinesis. The prototypical hypokinetic basal ganglia disorder is Parkinson's disease, and the prototypical hyperkinetic basal ganglia disorder is Huntington's disease. While neurodegenerative diseases affecting the basal ganglia give rise to movement disorders, so may other etiologies (e.g., stroke).

The basal ganglia also play a role in non-motor functions. Persons with movement disorders due to basal ganglia pathology may have prominent cognitive and emotional disturbances, even at early stages of the disease. The caudate appears to be involved primarily with cognitive functions; it is part of an associative circuit in which projections from the prefrontal cortex go to the head of the caudate. Degeneration of the caudate, as occurs with Huntington's disease, gives rise to a subcortical dementia characterized by prominent executive dysfunction. Other movement disorders due to basal ganglia pathology also may feature dementia, including Parkinson's disease and the Parkinson plus syndromes.

Additional Reading

1. Aarsland D, Creese B, Politis M, Chaudhuri KR, Ffytche DH, Weintraub D, Ballard C. Cognitive decline in Parkinson disease. *Nat Rev Neurol.* 2017 Apr;13(4):217–231. doi:10.1038/nrneurol.2017.27
2. Brodal P. The Basal Ganglia. In: Brodal P, *The central nervous system: structure and function.* 5th ed. Oxford University Press; 2016. https://oxfordmedicine.com/view/10.1093/med/9780190228958.001.0001/med-9780190228958-chapter-23. Accessed July 4, 2022.
3. Blumenfeld H. Basal Ganglia. In: Blumenfeld H, *Neuroanatomy through clinical cases.* 3rd ed., pp. 741–793. Sinauer Associates; 2021.
4. Blumenfeld Z, Brontë-Stewart H. High frequency deep brain stimulation and neural rhythms in Parkinson's disease. *Neuropsychol Rev.* 2015;25:384–397. doi:10.1007/s11065-015-9308-7
5. Boxer AL, Yu JT, Golbe LI, Litvan I, Lang AE, Höglinger GU. Advances in progressive supranuclear palsy: new diagnostic criteria, biomarkers, and therapeutic approaches. *Lancet Neurol.* 2017 Jul;16(7):552–563. doi:10.1016/S1474-4422(17)30157-6. Epub 2017 Jun 13. PMID: 28653647; PMCID: PMC5802400
6. McKeith IG, Boeve BF, Dickson DW, et al. Diagnosis and management of dementia with Lewy bodies: fourth consensus report of the DLB Consortium. *Neurology.* 2017;89(1):88–100. doi:10.1212/WNL.0000000000004058
7. Nelson AB, Kreitzer AC. Reassessing models of basal ganglia function and dysfunction. *Annu Rev Neurosci.* 2014;37:117–135. doi:10.1146/annurev-neuro-071013-013916
8. Rossi PJ, Gunduz A, Okun MS. The subthalamic nucleus, limbic function, and impulse control. *Neuropsychol Rev.* 2015;25:398–410. doi:10.1007/s11065-015-9306-9
9. Snowden JS. The neuropsychology of Huntington's disease. *Arch Clin Neuropsychol.* 2017;32(7):876–887. doi:10.1093/arclin/acx086
10. Tröster AI. Neuropsychological characteristics of dementia with Lewy bodies and Parkinson's disease with dementia: differentiation, early detection, and implications for "mild cognitive impairment" and biomarkers. *Neuropsychol Rev.* 2008;18(1):103–119. doi:10.1007/s11065-008-9055-0
11. Vanderah TW, Gould D. Basal nuclei. In: Vanderah TW, Gould D, *Nolte's the human brain: an introduction to its functional anatomy.* 8th ed., pp.450–467. Elsevier; 2021.
12. Ward P, Seri S, Cavanna AE. Functional neuroanatomy and behavioural correlates of the basal ganglia: evidence from lesion studies. *Behav Neurol.* 2013;26(4):219–223. doi:10.3233/BEN-2012-120264

12

Limbic Structures

Introduction

The term *limbic system* refers to a group of forebrain structures that lie beneath the neocortical surface, bordering the neocortex and hypothalamus. Based on this positioning, it was believed that these structures formed the anatomic basis linking subjective emotional experience, mediated by the neocortex, to emotional behaviors, mediated by the hypothalamus and its connections. Modern neuroscience, however, has shown that limbic structures do not serve a unified function and that some limbic structures do not play a predominant role in emotional function. This chapter presents the history of the concept of the limbic system and discusses the function of limbic structures in emotional behaviors and motivational drives, namely the cingulate gyrus, amygdala, nucleus accumbens, and septal nuclei. It also discusses the clinical relevance of these structures, including anterior cingulotomy, Klüver-Bucy syndrome, aggression, fear and anxiety, reward and pleasure, limbic encephalitis, and deep brain stimulation for psychiatric disorders.

History

In 1878, **Paul Broca** (1824–1880) referred to the ring of cortex at the margin of each cerebral hemisphere as **limbic cortex** (from the Latin *limbus*, "border"). This ring of cortex, seen in midsaggital view of the brain, begins anteriorly just inferior to the rostrum of the corpus callosum as the subcallosal area of the cingulate gyrus, follows the corpus callosum as the cingulate gyrus, narrows posteriorly as the isthmus of the cingulate gyrus, and merges into the parahippocampal gyrus, a cortical region surrounding the hippocampus (Figure 12.1). Broca referred to the cingulate and parahippocampal gyri collectively as *le grand lobe limbique* ("the great limbic lobe"), although it is not considered to be a true lobe.

Broca was impressed by the connections between the olfactory system and "limbic lobe" and suggested that the function of this brain region was to process olfactory information. Consequently, the limbic lobe was also called the *rhinencephalon* ("smell brain"). Subsequent anatomical and physiological studies, however, showed that limbic

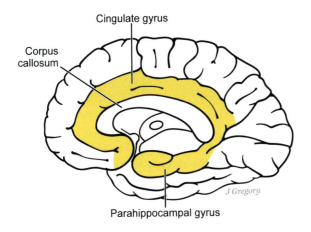

FIGURE 12.1. Limbic cortex consists of the cingulate and parahippocampal gyri.

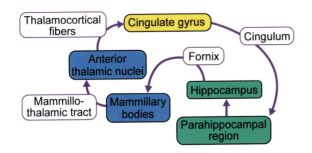

FIGURE 12.2. The Papez circuit.

cortex does not receive its major input form the olfactory system, and the term *rhinencephalon* is now used—more appropriately—for olfactory system neural structures.

In 1937, **James Papez** (1883–1958) proposed that the limbic lobe was the anatomic basis of emotional reactions and expressions. He believed that the hippocampus was interposed between the cingulate gyrus and the hypothalamus, and that the cingulate gyrus and its widespread neocortical connections were the basis for experiencing emotions, whereas the hypothalamus and its connections were responsible for the outward manifestations of emotions. Papez performed a retrograde trans-neuronal tracing study by injecting rabies virus into cat hippocampus and monitoring its progression through the brain over time. He revealed a circuit of connections starting and ending in the hippocampus that came to be known as the **Papez circuit** (Figure 12.2). Papez believed that this circuit was the central anatomical substrate of emotional experience.

This circuit provided the basis for the concept of the limbic system, defined as the limbic lobe and anatomically related structures that underlie the neurological regulation of emotion. Over time, more structures have been associated with the limbic system, and there is no universal agreement on the specific components (although most designations include the cingulate and parahippocampal gyri, hippocampal formation, amygdala, nucleus accumbens, and septal nuclei). In addition, the hippocampus, parahippocampal gyrus, and Papez circuit itself are now known to play a major role in memory formation, rather than emotion. Because the structures that have been subsumed under the heading of *limbic system* do not constitute a system with a unified function, the term is falling into disuse, and the term *limbic structures* is preferred. The neurocircuitry of emotion and memory is therefore usually studied separately. This chapter focuses on the limbic structures playing a prominent role in emotion, namely the cingulate gyrus, amygdala, nucleus accumbens, and septal nuclei (see Chapter 26, "Memory and Amnesia," for discussion of the limbic structures playing a prominent role in memory).

The Cingulate Cortex

The **cingulate cortex** is located medially within each cerebral hemisphere and is best viewed in the midsaggital plane of section. It consists of the cingulate gyrus and cortical tissue buried within the cingulate sulcus. As described above, the **cingulate gyrus** is a C-shaped gyrus that immediately borders the corpus callosum; the **cingulate sulcus** immediately overlies the cingulate gyrus. The **cingulum** (cingulum bundle) is a C-shaped collection of white matter tracts that run within the cingulate and parahippocampal gyri and carry the afferent and efferent connections to those cortical areas. It was once believed that limbic cortex played a role in emotion, but now it is known that different regions have different functional specializations.

The cingulate cortex has extensive anatomical connections with other brain structures, and its functions are varied and complex. Sources of afferent input include association areas of the frontal, parietal, and temporal lobes (see Chapter 13, "The Cerebral Cortex," for a description of association cortex), as well as subcortical nuclear structures such as the septal nuclei and thalamic nuclei (mediodorsal, anterior). Efferent targets include association areas of frontal, parietal, and temporal lobes, as well as subcortical nuclear structures such as the amygdala and septal nuclei. In general, the cingulate cortex is thought of as a connecting hub of signals pertaining to emotion, sensation, and action due to its anatomical connections to brain areas closely associated with these functions.

The cingulate cortex is divided into four functionally distinct regions: anterior cingulate cortex, midcingulate cortex, posterior cingulate cortex, and retrosplenial cortex (i.e., posterior to the **splenium**, the posterior-most portion of the callosum). Further subdivisions are also defined based on anatomic features. The **anterior cingulate cortex** is mainly involved with emotion processing (including regulation of overall affect, assigning emotions to internal and external stimuli, and pain perception) and regulating the endocrine and autonomic responses to emotions. The **midcingulate cortex** is involved mainly in cognitive processing. The **cingulate motor area**, a subdivision of this midcingulate region, is a higher-order motor area located within the cingulate sulcus. It receives signals about emotional state from limbic structures and translates them into motor commands that are executed by the primary and supplementary motor cortices. The **posterior cingulate cortex** plays a role in visuospatial orientation. The **retrosplenial cortex** has reciprocal connections with the hippocampal formation, parahippocampal region, and anterior and lateral dorsal thalamic nuclei; it plays a role in memory.

Blood Supply

The cingulate cortex receives its blood supply from the **pericallosal artery**, a branch of the **anterior cerebral artery** that directly extends from that vessel and runs within the callosal sulcus, which separates the cingulate gyrus and the corpus callosum (see Figure 5.3 in Chapter 5). The pericallosal artery gives rise to many small cortical branches that supply the anterior cingulate gyrus; the **callomarginal artery** runs within the cingulate sulcus and supplies the middle portion of cingulate gyrus (below the paracentral lobule); and the **precuneal artery** supplies the posterior cingulate gyrus.

Cingulotomy

Anterior cingulotomy is a form of **psychosurgery** (brain surgery performed to treat psychiatric disorders). Cingulotomy was introduced in 1948 as an alternative to prefrontal lobotomy, based on Papez's hypothesis that the cingulum was a major component of an anatomic circuit believed to play a significant role in emotion (i.e., the Papez circuit). Today, cingulotomy is used as treatment for chronic intractable pain, especially oncological pain. The procedure involves bilateral stereotactic ablation of the anterior cingulate gyri. It is believed that therapeutic efficacy results specifically from severing cingulum bundle fibers, and that this affects the perception of pain and its emotional consequences. Other applications of bilateral cingulotomy include treatment of major depression and obsessive-compulsive disorder.

The Amygdala: Fear, Anxiety, and Aggression

The amygdala is a critical component of the neural circuitry underlying the processing and expression of fear, anxiety, and aggression. Damage to the amygdala can lead to dramatic deficits in emotional reactivity. Not surprisingly, both structural and functional changes to this region have been implicated in a variety of psychiatric disorders.

Anatomy

The amygdala (Latin, "almond") is an almond-shaped structure located in the anterior medial temporal lobe, just anterior to the hippocampus. It also lies anterior to the anterior tip of the inferior (temporal) horn of the lateral ventricle (Figure 12.3).

The amygdala is a complex of multiple nuclei that are divided into three regions: **basolateral** (consisting of the lateral, basal, and accessory basal nuclei), **corticomedial** (consisting of the cortical and medial nuclei), and **central** (consisting of the central nucleus). The basolateral complex is considered the "amygdala proper"; it receives strong sensory inputs directly from visual, auditory, and somatosensory cortices. This region is especially well developed in humans. The corticomedial region, which is poorly developed in humans, receives olfactory input. There are extensive intrinsic connections within the amygdaloid complex, but in general the basolateral and corticomedial groups pass information to the central nucleus, the main output channel. In addition, the right and left **amygdalae** are interconnected by the **anterior commissure**. Efferents leave the amygdala by two major pathways, the **stria terminalis** which projects to the septal area and hypothalamus, and the **ventral amygdalofugal pathway** which projects to the septal area, hypothalamus, and medial dorsal thalamic nuclei.

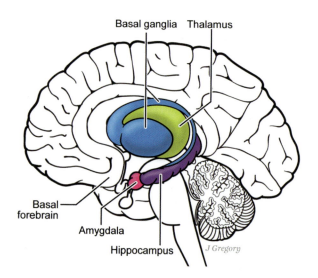

FIGURE 12.3. The amygdala is an almond-shaped structure located in the anterior medial temporal lobe, just anterior to the hippocampus and tail of the caudate.

Klüver-Bucy Syndrome

One of the earliest and most important observations in the study of the neurology of emotion came in 1939, when **Heinrich Klüver** (1897–1979) and **Paul Bucy** (1904–1992) reported that bilateral ablation of the anterior temporal lobe (i.e., bilateral temporal lobectomy) changed typically aggressive rhesus monkeys into docile animals with decreased emotionality. The animals became tame and exhibited a loss of aggressive, fear, and defensive reactions when confronted with stimuli that normally evoke fear, such as snakes and other threats from humans and other animals; they did not exhibit the motor and vocal reactions generally associated with anger and fear. They also exhibited hypersexuality (greatly increased autoerotic activity and indiscriminate partnerships), hyperorality (tendency to examine all objects by mouth), visual agnosia ("psychic blindness"), and **hypermetamorphosis** (a strong tendency to attend to and react to visual stimuli). This constellation of neurological signs is known as **Klüver-Bucy syndrome** and is seen in pathological states that destroy the anterior and medial temporal lobes bilaterally.

The full Klüver-Bucy syndrome requires bilateral removal of the amygdala and inferior temporal neocortex. Later studies demonstrated that removal of the temporal neocortex alone did not produce tameness and loss of fear (i.e., change in reactivity to reinforcing and aversive stimuli), but bilateral **amygdalectomy** did produce these changes. Lesions in the temporal cortex accounted for the visual agnosia symptoms of the full syndrome (see Chapter 14, "The Occipital Lobes and Visual Processing"). Many of the symptoms of Klüver-Bucy syndrome are believed to be due to a general defect in evaluating sensory stimuli in terms of their emotional or motivational significance (i.e., reward and punishment), resulting in socially inappropriate behavior.

Klüver-Bucy syndrome also occurs in humans, although the full syndrome is very rare. This is likely because the bilateral temporal lobe lesions in humans are usually less extensive than the total temporal lobe ablations that were performed in experimental animal studies. Human Klüver-Bucy syndrome was first reported following a bilateral temporal lobectomy for the treatment of seizures; it occurs with a wide variety of pathologies involving bilateral mesial temporal lobe pathology, but most commonly herpes simplex viral encephalitis (see "Limbic Encephalitis," below). The emotional changes reported in human Klüver-Bucy syndrome are characterized by blunted or flattened affect, apathy, and even a "pet-like" compliance. In some cases, there are alterations in sexual behavior, such as inappropriate copulation and masturbation, or more commonly, inappropriate sexual overtures and comments. In humans, the syndrome is usually accompanied by amnesia due to involvement of the hippocampi. It may also include seizures due to mesial temporal involvement (see Chapter 19, "Epilepsy"), and aphasia due to involvement of the lateral aspects of the temporal lobe in the language-dominant hemisphere (see Chapter 27, "Language and the Aphasias").

Aggression

Following the lesion studies that produced Klüver-Bucy syndrome, further studies of amygdala function were performed using electrical stimulation. These studies found that electrical stimulation of the basolateral amygdala in cats elicits rage and aggressive reactions characterized by pupillary dilation, piloerection, growling, and hissing.

Aggressive reactions or rage attacks may be seen in patients with temporal lobe tumors, and they are related primarily to involvement of the anteromedial temporal lobe. Aggressive and violent behaviors may also occur with temporal lobe epilepsy, either during the seizure (referred to as ictal rage and aggression) or immediately following a seizure (referred to as post-ictal rage and aggression). Seizure-related aggressive acts are unprovoked (spontaneous), stereotyped, nondirected, inappropriate to the situation, and associated with confusion. Amygdalectomy, a psychosurgical procedure involving stereotactic lesions of the amygdala, has been used in the past to treat patients with severe aggressive and violent behavioral disturbances, reportedly with some success.

Unfortunately, human aggression and violence are common and represent a significant public health concern. The underpinnings of human aggression are of course multifactorial and include socioeconomic, cultural, political, and psychological factors. However, research indicates that some forms of pathological aggression have a neurobiological basis.

When a threat is dangerous and imminent, impulsive aggression may be a normal defensive reaction. **Impulsive aggression** (also known as hostile, reactive, or affective aggression) is an abrupt, unpremeditated response to a perceived threat or provocation; by contrast, instrumental aggression (also known as predatory or proactive aggression) is premeditated, goal-oriented, and emotionless. When an impulsive aggressive response is exaggerated in relation to the emotional provocation that occurs, however, it is pathological. It is hypothesized that in some individuals, repetitive acts of impulsive aggression are neurobiologically based and involve both hyper-responsivity of the amygdala and a failure of prefrontal regulatory influences. In other words, the amygdala performs affective evaluation of stimuli and has a lower threshold for triggering anger; this, in combination with poor top-down modulation by the prefrontal cortex, leads to pathological impulsive aggression.

Fear and Anxiety

In a world where there is danger, one of the crucial tasks of the nervous system is to identify what is beneficial to survival and what is detrimental. Fear is the emotional response to real or perceived imminent threat. Anxiety is anticipation of future threat. Fear and anxiety are defensive, emotional responses that may be essential to an organism's survival. They allow for integration of sensory inputs that identify potential dangers and threats, and trigger physical reactions (e.g., autonomic responses such as increased heart rate and sweating, somatic responses such as freezing or fleeing).

Using conditioned fear as a behavioral model, consensus neuroscientific evidence on the neuronal circuitry underlying fear and anxiety has concluded that the amygdala is critical for both the learning and expression of conditioned fear responses. Fear conditioning is a form of classical (Pavlovian) conditioning in which animals (including humans) *learn* to fear specific stimuli within the environment that predict aversive events. The conditioned fear paradigm involves the repeated pairing of a non-threatening stimulus such as a soft tone (conditioned stimulus) with a noxious stimulus such as mild foot shock (unconditioned stimulus). After several pairings, the presentation of the tone alone is sufficient to elicit fear, which manifests behaviorally as a conditioned fear response such as freezing (a cautious period of watchful immobility). Conditioned fear provides a critical survival function that activates a range of defensive behaviors, protecting the animal against potentially dangerous environmental threats.

The most famous example of learned fear in humans is "**The Little Albert Experiment**," published in 1920 by John Watson and Rosalie Rayner. As behaviorists, Watson and Rayner believed that many behaviors were learned rather than innate, including most fear responses. Their subject was 11-month-old "Albert B." Watson and Rayner tested Albert's reactions to a wide variety of stimuli, including various animals such as a white laboratory rat, a dog, and a rabbit. He showed no signs of fear in response to the animals' presence, establishing that fear of the animals was not innate. Watson and Rayner then tested whether it was possible for the infant to learn fear by classical Pavlovian conditioning. When they presented him with a white rat, Albert showed no fear and reached out to touch the rat. Just as he touched the rat, the experimenters struck a steel bar with a hammer behind his head (so that he did not see it), creating a very loud noise, a fear-inducing stimulus akin to thunder. Albert exhibited an innate fear response to the loud sound, jumping violently away, burying his face in a mattress, and whimpering. After a short period of time, he reached out to the rat again, and the procedure was repeated one more time.

One week later, the experimenters presented the white rat to Albert again, but rather than reaching for the rat, he avoided it and scrambled away crying. This study demonstrated that while some fear responses are innate, others are learned.

It is now understood that the amygdala plays a central role in fear by mediating sensory-affective associations between the unconditioned and conditioned stimuli. Destruction of the amygdala prevents the establishment of conditioned fear responses in a wide variety of conditioned fear paradigms.

On the input side, the lateral nucleus of the amygdala receives afferents from sensory cortical areas and sensory thalamic nuclei carrying information about the external environment. The signals are further processed by the basolateral nucleus, which in turn projects to the central nucleus. On the output side, signals from the central nucleus are then directed to pathways regulating facets of emotional expression. Projections to the dorsal motor nucleus of the vagus nerve mediate parasympathetic responses. Projections to the lateral hypothalamus mediate sympathetic responses. Projections to the central gray region mediate conditioned freezing and conditioned stress-induced analgesia (reduced pain perception). Projections to the pons mediate fear-potentiated startle. Lesions of these efferent target sites disrupt the generation of individual conditioned emotional responses, leaving others intact. However, ablation of the central nucleus of the amygdala produces global deficits in conditioned fear responses. The central nucleus of the amygdala therefore serves to coordinate divergent emotional response output during fear conditioning. In support of this concept, functional neuroimaging studies performed in healthy volunteers have shown that amygdala activity increases during fear conditioning. Since the amygdala mediates sensory-affective associations, the symptoms of amygdala lesions are interpreted as a general defect in evaluating sensory stimuli in terms of their emotional or motivational significance.

The amygdala also appears to play a central role in the neurocircuitry of anxiety. Electrical stimulation of the amygdala in humans generates feelings of anxiety. Patients with anxiety disorders demonstrate hyperexcitability of the amygdala in response to evocative stimuli.

PATIENT S.M.

Focal bilateral amygdala lesions are extremely rare in humans, but they are associated with a marked absence of subjective fear and fear responses. The most intensively studied case is **patient S.M.** S.M. is a woman with exclusive and complete bilateral amygdala damage since late childhood due to **Urbach-Wiethe disease**, a very rare genetic disorder with autosomal recessive inheritance (its expression requires two copies of the genetic mutation, one inherited from each parent), which produces bilateral symmetric calcification of the anterior medial temporal lobes, especially the amygdalae, in 50%–75% of cases. Studies have shown that S.M. has little to no capacity to experience fear; all other emotional faculties are intact, including happiness and sadness. She reports no subjective experience of fear during exposure to a variety of fear-provoking stimuli, including life-threatening traumatic events, handling of snakes and tarantulas, a walk through a haunted house attraction, fear-inducing horror film clips; rather, she expresses interest, curiosity, and excitement. She also has a severe inability to recognize fearful facial expressions. Studies of other patients with similar lesions have reported similar findings. Subsequent studies have shown that S.M. and other patients with focal bilateral amygdala lesions do, however, react with panic to simulation of suffocation (inhalation of 35% carbon dioxide) and make an important distinction between fear triggered by external threats from the environment versus fear triggered internally by physiological events.

ICTAL FEAR AND ANXIETY

The most common affective symptoms of temporal lobe seizures are fear and anxiety. As with other focal seizure symptoms, the onset of **ictal fear** is paroxysmal and usually lasts 30–120 seconds; in contrast, panic attacks often build up gradually and last over 5–10 minutes. Differentiation between spontaneous ictal versus reactive ictal fear can be difficult. **Spontaneous ictal fear** is a sudden, unprovoked emotion; there is no frightening thought or stimulus leading to the fear response. **Reactive fear** is a fear response that occurs when a person with a seizure disorder realizes that a seizure is impending. Ictal fear and anxiety are associated with anteromedial temporal (and less often cingulate) seizure foci.

The Nucleus Accumbens: Reward and Pleasure

The nucleus accumbens is a key component of the neural circuitry underlying reward and pleasure. It plays a central role in a positive emotional response pathway, which is counterbalanced by a negative emotional response pathway mediated by the amygdala. This brain region plays a key role in addiction.

Anatomy

The nucleus accumbens is a small nucleus located in the **basal forebrain region**, an area of the forebrain located anterior and ventral to the striatum that is composed of a mixture of gray and white matter (Figure 12.4). The nucleus accumbens is also referred to as the **ventral striatum**.

The nucleus accumbens is divided into two anatomically and functionally distinct areas: a central core surrounded by an outer shell. It receives dopaminergic fibers from the midbrain **ventral tegmental area**. Other inputs come from the prefrontal cortex, basolateral part of the amygdala, midline and intralaminar nuclei of the thalamus, and ventral hippocampus. The nucleus accumbens sends efferents to the ventral tegmental area, substantia nigra, and pontine reticular formation.

Reward

BRAIN STIMULATION REWARD

In 1954, **James Olds** (1922–1976) discovered that rats would self-administer pulses of electrical stimulation via chronically implanted stimulating electrodes in their brains. This discovery occurred fortuitously when Olds was screening for aversive effects of stimulation described by **Neal Miller** (1909–2002) in an experiment studying the

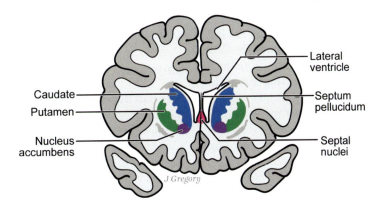

FIGURE 12.4. Coronal section through the rostral forebrain. The nucleus accumbens and septal nuclei are part of the basal forebrain nuclei.

effects of reticular formation stimulation on learning and memory. One of the electrodes missed its target and ended up in the basal forebrain. James Olds and **Peter Milner** (1919–2018) applied a brief series of stimulation pulses to the brain whenever the animal entered one corner of the test chamber. If the stimulation was aversive, then the animal was expected to learn to avoid that corner. With each burst of stimulation, however, the animal did leave the corner but then returned to the corner shortly afterward, suggesting that the stimulation was both aversive and rewarding. As Olds put it, the animal was "coming back for more."

Olds and Milner then implanted electrodes deliberately into the basal forebrain region and placed the animals in an experimental chamber with a lever that initiated a brief series of electrical pulses when pressed. The animals readily pressed the lever, sometimes thousands of times per hour. This self-administration of electrical stimulation to a specific brain region is called **intracranial self-stimulation**. It is believed that the electrical stimulation is self-administered because it is rewarding (reinforcing).

Most research on intracranial self-stimulation has been performed in rats, but the findings generalize to a wide variety of other species, including cats, dogs, and monkeys. The intracranial self-stimulation findings also extend to humans. Robert Heath (1915–1999) was the first to experiment with human brain stimulation, beginning in the 1950s through the 1970s, at Tulane University. Although these studies were not well-controlled and crossed acceptable ethical boundaries even for that time, they remain informative. Patients with severe neuropsychiatric disorders were implanted with chronic surface and deep electrodes for both recording and stimulation, as well as intracerebral cannulas for chemical stimulation. Heath discovered that humans would self-administer brain stimulation to some sites, and that they described the effects as pleasurable.

A variety of brain regions support intracranial self-stimulation, including the ventral tegmental area in the midbrain tegmentum, the lateral hypothalamus, and the nucleus accumbens in the basal forebrain. The region supporting the strongest intracranial self-stimulation is the **medial forebrain bundle** as it passes through the lateral hypothalamus. This bundle consists of multiple fiber types, including long ascending dopaminergic, noradrenergic, and serotonergic fibers interconnecting forebrain and midbrain structures, and short axons connecting adjacent regions. Activity within the subset of the dopaminergic fibers that travel from the ventral tegmental area to the nucleus accumbens is responsible for the reinforcing effects of medial forebrain bundle stimulation.

The distribution of reinforcing electrode sites was found to correspond to the distribution of neurons containing catecholamine (CA), suggesting that one or both CAs (i.e., dopamine or norepinephrine) are involved in reinforcement. Pharmacological studies indicated that dopamine (DA) was involved in reinforcement. Dopamine receptor antagonists (e.g., antipsychotic medications) blocked both **brain stimulation reward** and the reinforcing properties of natural rewards, as indicated by experimental studies finding that DA receptor antagonists suppress operant responding for brain stimulation reward, food, and water. This suppression of behavior was not attributable to motor effects; the operant response gradually declined because it was no longer rewarding. Furthermore, DA receptor antagonist pretreatment prevents the establishment of a conditioned place preference using food or electrical stimulation of the brain as the reward. Thus, DA is necessary for the establishment of a conditioned place preference. Because the animals were tested in a drug-free condition, the effects could not be attributable to motor side effects.

There are two main dopaminergic nuclei: the substantia nigra, which gives rise to the nigrostriatal pathway; and the ventral tegmental area, which gives rise to the mesolimbic and mesocortical pathways. The **mesolimbic pathway** begins in the ventral tegmental area and projects through the medial forebrain bundle to the amygdala, lateral septum, bed nucleus of the stria terminalis, hippocampus, and nucleus accumbens.

The mesolimbic dopaminergic projection from the ventral tegmental area to the nucleus accumbens is a critical component of brain reward circuitry (see Figure 3.4 in Chapter 3). Microinjections of DA antagonists into the nucleus accumbens reduce the reinforcing effects of medial forebrain bundle stimulation. Selective lesions of this pathway, achieved by microinjection of a neurotoxin selective for DA neurons into the ventral tegmental area, medial forebrain bundle, or nucleus accumbens, reduce the reinforcing effects of medial forebrain bundle stimulation. Dopamine agonists, such as amphetamine and cocaine, potentiate brain stimulation reward. Studies using **microdialysis** (a technique used to assess the chemistry of the extracellular space of living tissues) have shown that self-stimulation in the ventral tegmental area or medial forebrain bundle produces DA release within the nucleus accumbens. The amount of DA released is directly related to the effectiveness of the stimulation as a reinforcer.

DRUG REWARD

All drugs that are known to be addictive in human beings, including opiates, cocaine, amphetamine, tetrahydrocannabinol (THC), and even nicotine and caffeine, are self-administered in a variety of species, including rats and monkeys. Mechanisms of drug reward therefore can be studied in animals using drug self-administration behavioral paradigms in which an operant response (e.g., a bar press) results in a fixed dose of drug. The reinforcing effects of these drugs are mediated by DA; pretreatment with DA antagonist reduce the drug's reinforcing effect and produces higher rates of operant responding to achieve the optimal dose.

Dopamine agonists such as amphetamine and cocaine derive their reinforcing effects by acting at DA receptors in the nucleus accumbens. Microinjection of DA antagonists into the nucleus accumbens blocks systemic amphetamine and cocaine self-administration. Selective destruction of the dopaminergic mesolimbic projection from the ventral tegmental area to nucleus accumbens (by microinjection of a neurotoxin selective for DA neurons) also blocks drug reward.

NATURAL REWARD

Electrical stimulation of the medial forebrain bundle and drugs of abuse are believed to be rewarding because they activate the same brain regions activated by natural reinforcers such as food, water, and sexual contact. An extensive body of literature indicates that any treatment leading to stimulation of DA receptors in the nucleus accumbens is reinforcing.

The dopaminergic mesolimbic pathway from the ventral tegmental area to nucleus accumbens is responsible for the reinforcing effects of natural appetitive stimuli. Microinjection of DA antagonists into the nucleus accumbens blocks the reinforcing effects of natural appetitive stimuli. Microdialysis studies have shown that DA is released into the nucleus accumbens with natural reinforcers, such as when a thirsty animal drinks, a sodium-depleted animal ingests salt, or a hungry animal eats.

Studies recording electrical activity of DA neurons within the ventral tegmental area have found that these neurons are activated not only by primary reinforcers such as food, but also by conditioned reinforcers established by classical conditioning. For example, a stimulus such as saccharin is naturally rewarding and causes release of DA in the nucleus accumbens, but if it has been conditioned to be aversive by pairing it with lithium chloride (a nondeadly poison), a saccharin stimulus presented alone will result in a decrease of DA release in the nucleus accumbens. Thus, the same stimulus can have very different effects on the activity of mesolimbic DA neurons, depending on the animal's previous experience with the stimulus, suggesting that mesolimbic DA operates on the general property of reward, rather than on the rewarding effects of specific stimuli.

Dopamine antagonists within the nucleus accumbens (or destruction of DA terminals within the nucleus accumbens) abolish the reinforcing effects of medial forebrain bundle stimulation and drugs of abuse, and they also extinguish the expression of behaviors reinforced by natural stimuli. For example, microinjections of DA antagonist into the nucleus accumbens cause rats to stop lever-pressing for a preferred food without interfering with their hunger, as they will eat a less preferred food that is freely available.

Some evidence indicates that the mesolimbic dopamine pathway in the reward system plays a role in human romantic love. Functional magnetic resonance imaging (MRI) studies have found increased activity within the ventral tegmental area when people who are in the early stage of intense romantic love look at photographs of their beloved, versus photographs of friends of similar age, sex, and relationship duration. Romantic love is also distinguished from sex drive, as functional imaging studies of human sexual arousal show different patterns of regional activation.

The Septal Region

Anatomy

The term *septum* is an anatomical term that refers to a thin wall-like partition (e.g., nasal septum). In neuroanatomy, the **septum pellucidum** (Latin, "translucent wall") is a thin, almost transparent, two-layered membrane that runs down the midline from the corpus callosum to the fornix and separates the anterior horns of the lateral ventricles (see Figure 12.4). The **septum verum** ("true septum") is located immediately beneath the base of the septum pellucidum, forming part of the medial wall of the lateral ventricle anterior horns. It is a gray matter region that contains the **septal nuclei**, which are part of the basal forebrain nuclei.

The septal nuclei have significant connections with the amygdala via the diagonal band of Broca and stria terminalis, the hypothalamus and ventral tegmental area via the medial forebrain bundle, and the hippocampus via the fornix. The **septohippocampal pathway** is the subset of fornix fibers that project from the septal nuclei to the hippocampus.

Function

The functions of the septal region are poorly understood, and most of what is known comes from animal studies. Lesions of the septal nuclei in animals alter sexual and foraging behavior and reduce aggressive behaviors. Electrical stimulation produces aggression.

Studies of the neurological control of human sexual behavior are scarce, but some data indicate that the septal region plays a role. Robert Heath reported that (1) recordings from the septal region during sexual arousal (whether by intercourse or masturbation) showed a specific pattern of electroencephalograph (EEG) activity during orgasm; (2) chemical stimulation by acetylcholine delivered directly to the septal region elicited euphoria and sexual orgasm, and the same EEG pattern change recorded from the septal region; and (3) electrical stimulation of the septal region elicited orgasms and a compulsion to masturbate.

Limbic Encephalitis

Encephalitis is an acute inflammation (swelling) of the brain. It is generally caused by either infectious agents or antibodies that attack specific neuronal proteins (see Chapter 22, "Brain Infections").

Limbic encephalitis is a relatively localized inflammation that most prominently involves limbic structures, particularly the mesial temporal lobe regions. The cardinal sign of limbic encephalitis is a severe impairment of short-term memory due to involvement of the hippocampi. The anterograde amnesia is often accompanied by psychiatric symptoms and seizures. Temporal lobe involvement is evidenced by neuroimaging and/or focal EEG abnormalities. The underlying cause is either infectious or autoimmune.

Herpes Simplex Virus Limbic Encephalitis

Herpes simplex virus encephalitis (HSVE) is caused by the herpes simplex virus type 1 (see Chapter 22, "Brain Infections"). This virus has a predilection for the inferior-medial temporal lobes and orbitofrontal regions of the brain. The infection begins unilaterally and then spreads to the contralateral hemisphere. Consequently, neuroimaging shows either unilateral or asymmetric bilateral involvement of the inferior-medial temporal lobes and orbitofrontal regions (Figure 12.5).

Herpes simplex virus encephalitis is associated with high rates of mortality and morbidity. Bilateral lesions are associated with worse outcomes. Survivors often have severe memory disorder, extreme alteration of personality, and marked behavioral disorder. Several cases of Klüver-Bucy syndrome due to bilateral damage of the amygdalae have been reported.

Autoimmune Limbic Encephalitis

In **autoimmune encephalitis**, the body's immune system attacks healthy neurons by manufacturing antibodies to specific proteins found on the surface of nerve cells (e.g., specific neurotransmitter receptors or ion channels) or within nerve cells, leading to brain inflammation. The presence of antineuronal antibodies in blood or cerebrospinal fluid supports an autoimmune pathogenesis. Antibody-mediated **encephalitides** (plural of *encephalitis*) are classified according to (1) the location of neuronal antigens (i.e., cell surface or intracellular) and neural regions affected; (2) whether they are paraneoplastic (i.e., associated with systemic cancer) or non-paraneoplastic; and (3) the specific antigen that is targeted.

In **autoimmune limbic encephalitis**, there is a predilection for antigens within limbic structures, resulting in a localized antibody-mediated attack and localized inflammation. This leads to a subacute onset of anterograde amnesia, confusion, psychiatric and behavioral changes, and seizures. Brain MRI shows lesions within the mesial temporal lobes. Other autoimmune encephalitides target different brain regions. **Rhombencephalitis** refers to inflammatory diseases affecting the hindbrain (brainstem and cerebellum).

There are two broad categories of autoimmune encephalitis: paraneoplastic and non-paraneoplastic. **Paraneoplastic encephalitis** is associated with systemic cancer; in **non-paraneoplastic encephalitis** there is no identified cancer. Both are due to an immune system response to the neoplasm, in which the body produces antibodies to an antigen that is shared by tumor and neuronal cells. In more than half of patients, the neurologic syndrome develops before the cancer diagnosis is known. Treatment of paraneoplastic autoimmune encephalitis involves neoplasm removal and immunotherapy.

The diagnosis of **paraneoplastic limbic encephalitis** requires: (1) subacute onset of memory loss, seizures, and psychiatric symptoms; (2) neuroimaging or EEG evidence of limbic structure involvement; and (3) cancer diagnosis within 5 years of onset of the limbic encephalitis. The most common associated neoplasm types are small cell lung carcinoma, testicular germ cell tumors, thymomas (tumor of the thymus gland), and Hodgkin lymphoma. Autoimmune limbic encephalitis is associated with several specific antibodies directed at specific proteins, including voltage-gated potassium channel complex antibodies, AMPA receptor antibodies, and GABA receptor antibodies.

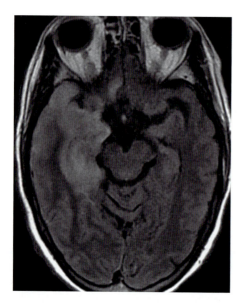

FIGURE 12.5. Herpes simplex viral encephalitis with asymmetric involvement of the right temporal lobe and right gyrus rectus.

Deep Brain Stimulation for Psychiatric Disorders

Deep brain stimulation (DBS) is an intervention in which specific brain structures are electrically stimulated by neurosurgically implanted electrodes. The mechanism of DBS is not well understood (e.g., whether it has inhibitory effects, excitatory effects, or disrupts signal transmission), but it is believed that its clinical efficacy derives from modulation of dysfunctional neuronal circuits. It is primarily used to treat movement disorders such as Parkinson's disease, essential tremor, and dystonia, and more recently has been approved for use in epilepsy and obsessive-compulsive disorder (OCD). As indicated above, experimental neuromodulation via brain electrodes in patients with psychiatric disorders was first attempted during the 1950s by Heath. Its potential as a therapeutic intervention for various treatment-refractory psychiatric conditions has prompted renewed interest in this area of investigation.

Addiction

Substance abuse and addiction pose serious public health problems, contributing to domestic violence, child abuse, assaults, motor vehicle accidents, sexually transmitted diseases, crime, homicide, accidental death from overdose, and suicide. Addiction risk is related to complex interactions among multiple biological and environmental factors. Genetic factors account for approximately 50% of the risk; the specific genes implicated can influence the biological response to substances of abuse or their metabolism. Current therapies consisting of psychosocial and/or pharmacological interventions have high relapse rates (up to 50%–70%). Therefore, additional treatments are needed.

Neurosurgical ablation procedures for addiction were attempted during the 1960s and 1970s, including cingulotomy, hypothalamotomy, and resection of the substantia innominata, but had limited efficacy. For the past 15 years, DBS has been investigated as a potential treatment for addiction. Most of this research focuses on modulation of the mesolimbic dopaminergic pathway by DBS of the nucleus accumbens, theoretically to reverse neuroplastic changes that occur in the brain circuit in response to addiction.

Numerous preclinical studies have shown that DBS within the nucleus accumbens decreases drug self-administration for a variety of drugs of abuse. The self-administration behavioral paradigm is an animal model of human addiction and the gold standard for screening potential treatments for drug addiction. Animals are placed in a chamber with access to the drug, such as by drinking alcohol from a bottle or pressing a lever to administer intravenous injection of a drug (e.g., cocaine, heroin). Animals self-administer until they reach a steady state of drug intake; this self-administration phase is analogous to the **binge/intoxication stage** of addiction in humans. Behavior may also be studied during an abstinence phase in which drug is no longer available, analogous to the **withdrawal stage** in humans, and a reinstatement phase in which drug becomes available again, analogous to the **relapse stage** in humans. Several animal studies have shown that DBS for 30–60 minutes per day during the binge/intoxication, withdrawal, or relapse stage reduces drug self-administration behavior for a wide variety of substances.

Initial studies investigating DBS for addiction in humans were prompted by an accidental observation that in some patients, DBS within the nucleus accumbens for OCD or depression induced relief from addiction to alcohol or smoking without reducing OCD symptoms. Following these reports, several small case series and reports have shown that nucleus accumbens DBS decreases drug use. Larger-scale randomized controlled trials are necessary before this approach advances from an experimental to an evidence-based treatment.

Major Depressive Disorder

There is some evidence that DBS has antidepressant effects, and it is currently used as an experimental treatment for treatment-resistant major depression. As the pathophysiology of depression is poorly understood, a variety of brain structures have been targeted. This is based on theoretical models of neural circuit dysfunction underlying anhedonia and dysphoria in mood disorders, neuroimaging studies in humans, observations of improved mood as a positive side effect in patients undergoing DBS for other diagnoses, and DBS effects in animal models of depression. Targets examined include the subcallosal cingulate gyrus, nucleus accumbens, and medial forebrain bundle, but further research is necessary. The medial forebrain bundle is a promising target, particularly for anhedonia.

Summary

The concept of the *limbic system* as an anatomical entity that consists of a ring of cortex at the margin of each cerebral hemisphere (limbic cortex) and its connections to deep brain nuclei and that serves as the neural basis of emotion regulation has not been supported by modern neuroscience; thus the term *limbic structures* is now preferred. Limbic structures that do play a clear role in emotion-related behaviors and motivational drives include the cingulate cortex, amygdala, nucleus accumbens, and septal nuclei.

The cingulate cortex consists of the cingulate gyrus and cortical tissue buried within the cingulate sulcus. It has extensive anatomical connections with other brain structures, and its functions are varied and complex. The cingulate cortex is divided into four functionally distinct regions; the anterior cingulate cortex is mainly involved with emotion processing. Anterior cingulotomy is a form of psychosurgery that involves bilateral stereotactic ablation of the anterior cingulate gyri; it is used as a treatment for chronic intractable pain (especially oncological pain), major depression, and OCD.

The amygdala is a critical component of the neural circuitry underlying the processing and expression of fear, anxiety, and aggression. One of the most important observations in the study of the neurology of emotion is Klüver-Bucy syndrome in which bilateral temporal lobectomy changed normally aggressive rhesus monkeys into tame animals with a loss of aggressive, fear, and defensive reactions. Later studies demonstrated that more selective lesions of the amygdala (bilateral amygdalectomy) were responsible for these changes. Klüver-Bucy syndrome also occurs in humans with bilateral mesial temporal lobe pathology. Electrical stimulation of the basolateral amygdala elicits aggressive reactions in animals; rage attacks may also occur in humans with tumors or seizures involving the anteromedial temporal lobe. Patient S.M. is a woman with exclusive and complete bilateral amygdala damage since late childhood; she has almost no capacity to experience fear, yet all other emotional faculties are intact. The most common affective symptoms of temporal lobe epilepsy are ictal fear and anxiety.

The nucleus accumbens, located in the basal forebrain, is a key component of the neural circuitry underlying reward and pleasure and is clinically significant because of its key role in addiction. The mesolimbic dopaminergic projection from the ventral tegmental area to nucleus accumbens is a critical component of brain reward circuitry. All drugs that are addictive in human beings are self-administered in a variety of mammals; the reinforcing effects of these drugs are mediated by DA release into the nucleus accumbens. Selective destruction of the dopaminergic mesolimbic projection from the ventral tegmental area to nucleus accumbens blocks drug reward.

The septal nuclei are located within the basal forebrain. The functions of the septal nuclei are poorly understood, but in animal studies, lesions of this region alter sexual and foraging behavior and reduce aggressive behaviors, and electrical stimulation elicits aggression.

Limbic encephalitis is a relatively localized inflammation that most prominently involves the mesial temporal lobe. The cardinal sign of limbic encephalitis is a severe impairment of short-term memory due to involvement of the hippocampi, but it is often accompanied by psychiatric symptoms and seizures.

Deep brain stimulation of limbic structures is being investigated as a potential treatment for addiction and treatment-resistant major depression (particularly for anhedonia).

Additional Reading

1. Adolphs R, Tranel D, Damasio H, Damasio A. Impaired recognition of emotion in facial expressions following bilateral damage to the human amygdala. *Nature.* 1994;372(6507):669–672. doi:10.1038/372669a0
2. Amaral D, Adolphs R. *Living without an amygdala.* Guilford Press; 2016.
3. Baumeister AA. Serendipity and the cerebral localization of pleasure. *J Hist Neurosci.* 2006 Jun;15(2):92–98. doi:10.1080/09647040500274879. PMID: 16608738.
4. Brodal P. The amygdala and other neuronal groups with relation to emotions. In: Brodal P, *The central nervous system: structure and function.* 5th ed. Oxford University Press; 2016. https://oxfordmedicine.com/view/10.1093/med/9780190228958.001.0001/med-9780190228958-chapter-31. Accessed July 4, 2022.
5. Calder AJ, Lawrence AD, Young AW. Neuropsychology of fear and loathing. *Nat Rev Neurosci.* 2001;2(5):352–363. doi:10.1038/35072584
6. Choi KS, Riva-Posse P, Gross RE, Mayberg HS. Mapping the "depression switch" during intraoperative testing of subcallosal cingulate deep brain stimulation. *JAMA Neurol.* 2015 Nov;72(11):1252–12560. doi:10.1001/jamaneurol.2015.2564. PMID: 26408865; PMCID: PMC4834289
7. Feinstein JS, Buzza C, Hurlemann R, et al. Fear and panic in humans with bilateral amygdala damage. *Nat Neurosci.* 2013;16(3):270–272. doi:10.1038/nn.3323
8. Heath RG, Cox AW, Lustick LS. Brain activity during emotional states. *Am J Psychiatry.* 1974 Aug;131(8):858–862. doi:10.1176/ajp.131.8.858. PMID: 4209918.
9. Heath RG, Monroe RR, Mickle WA. Stimulation of the amygdaloid nucleus in a schizophrenic patient. *Am J Psychiatry.* 1955;111(11):862–863. doi:10.1176/ajp.111.11.862
10. Klüver H, Bucy PC. Preliminary analysis of functions of the temporal lobes in monkeys [1939]. *J Neuropsychiatry Clin Neurosci.* 1997;9(4):606–620. doi:10.1176/jnp.9.4.606
11. Koob GF, Volkow ND. Neurobiology of addiction: a neurocircuitry analysis. *Lancet Psychiatry.* 2016;3(8):760–773. doi:10.1016/S2215-0366(16)00104-8
12. Kozlowska K, Walker P, McLean L, Carrive P. Fear and the defense cascade: clinical implications and management. *Harv Rev Psychiatry.* 2015;23(4):263–287. doi:10.1097/HRP.0000000000000065
13. LeDoux J. The amygdala. *Curr Biol.* 2007;17(20):R868–R874. doi:10.1016/j.cub.2007.08.005
14. Papez JW. A proposed mechanism of emotion [1937]. *J Neuropsychiatry Clin Neurosci.* 1995;7(1):103–112. doi:10.1176/jnp.7.1.103
15. Parvizi J, Rangarajan V, Shirer WR, Desai N, Greicius MD. The will to persevere induced by electrical stimulation of the human cingulate gyrus. *Neuron.* 2013 Dec 18;80(6):1359–1367.

doi:10.1016/j.neuron.2013.10.057. Epub 2013 Dec 5. PMID: 24316296; PMCID: PMC3877748.
16. Rolls ET. Limbic systems for emotion and for memory, but no single limbic system. *Cortex*. 2015;62:119–157. doi:10.1016/j.cortex.2013.12.005
17. Terzian H, Ore GD. Syndrome of Klüver and Bucy; reproduced in man by bilateral removal of the temporal lobes. *Neurology*. 1955;5(6):373–380. doi:10.1212/wnl.5.6.373
18. Tranel D, Hyman BT. Neuropsychological correlates of bilateral amygdala damage. *Arch Neurol*. 1990;47(3):349–355. doi:10.1001/archneur.1990.00530030131029
19. Vanderah TW, Gould D. Drives and emotions: the hypothalamus and limbic system. In: Vanderah TW, Gould D, *Nolte's the human brain: an introduction to its functional anatomy*. 8th ed., pp. 549–572. Elsevier; 2021.
20. Volkow ND, Boyle M. Neuroscience of addiction: relevance to prevention and treatment. *Am J Psychiatry*. 2018;175(8):729–740. doi:10.1176/appi.ajp.2018.17101174
21. Watson JB, Rayner R. Conditioned emotional reactions [1920]. *Am Psychol*. 2000;55(3):313–317. doi:10.1037//0003-066x.55.3.313

The Cerebral Cortex

Introduction

The cerebral cortex is the outermost layer of gray matter that covers the two cerebral hemispheres. It is especially well developed in humans and is the part of the brain that is most closely associated with cognition and voluntary behavior. A basic knowledge of cortical anatomy is critical to understanding disorders of higher brain functions. This chapter discusses the anatomy of the cerebral cortex and its connections, history of the concept of cortical localization, functional maps, the concept of hierarchical cortical function, and Brodmann's cytoarchitectural map. In addition, it describes conditions of clinical relevance, such as cortical disconnection syndromes, corpus callosotomy, and agenesis of the corpus callosum.

Comparative Neuroanatomy

Comparative neuroanatomy is the study of similarities and differences of nervous system structure across species. In non-mammalian vertebrates, the cerebral surface is composed of a simple layered structure called the **pallium**. The pallium is concerned primarily with olfaction, one of the most important senses for non-mammalian vertebrates. The main receptive area for senses other than olfaction is the thalamus; the basal ganglia are the main motor control area.

Comparing the brains of mammals vs. non-mammalian vertebrates, the hindbrain and midbrain are very similar, but the forebrain (cerebrum) differs dramatically in that the mammalian forebrain is enlarged and has evolved a 6-layered neocortex.

Comparing the brains of different mammals, as we move phylogenetically closer to humans the number of cells and surface area of the cerebral cortex increases and the cortex becomes more convoluted. Thus, in lower mammals the cortical surface is smooth (**lissencephaly**), while in higher mammals the cortical surface is convoluted (**gyrencephaly**). A large cerebral cortex is characteristic of the primate brain. This enlargement is greatest in the highly convoluted cerebral cortex of humans (especially within the frontal lobes), where two-thirds of its surface is buried within sulci and fissures. Thus, the cerebral cortex is the brain structure that most strongly distinguishes mammals from other vertebrates, primates from other mammals, and humans from other primates.

In human evolution, expansion of the cerebral cortex is also the feature that most strongly distinguishes modern humans from our hominin ancestors. Over the course of approximately 4 million years of evolution, from *Australopithecus afarensis* (one of the earliest known hominin species related to precursors of the *Homo* genus) to *Homo sapiens* (modern humans), brain volume increased from approximately 450 cm^3 to about 1,300 cm^3. Most of this increase in brain size is due to cerebral cortical expansion, as anthropologists have inferred from changes in skull shape over time.

This dramatic development of the cerebral cortex is generally thought to be the key to the increased cognitive capacities of humans relative to other animals and our hominin ancestors, underlying our vastly enhanced abilities to store, process, and convey information and the development of language, complex social structures, culture, and technology. The cerebral cortex has therefore been referred to as the "organ of civilization."

Cortical Anatomy

Human cerebral cortex is only about 3 mm thick but has a total surface area of approximately 2.6 square feet (2,415 cm^2). It is estimated to contain 14–16 billion neuron cell bodies and constitutes more than half of all the central nervous system (CNS) gray matter. The larger and deeper infoldings are generally called **fissures**, whereas the majority of infoldings are less deep and are called **sulci**. The folds of neural tissue between these grooves are called **gyri**. The cortex is essentially a sheet of tissue that is crumpled up.

In normal aging beyond age 60, there is mild, diffuse cortical atrophy as sulci become wider and more prominent and gyri become thinner. In cortical neurodegenerative diseases, cortical atrophy is especially prominent.

Major Landmarks and Lobes

The most prominent infoldings on the surface of the brain are relatively invariant from one person to another, and therefore are used as landmarks to divide the cerebral cortex into various regions. There are four principal sulci: the longitudinal fissure, the central sulcus, the lateral sulcus, and the parieto-occipital sulcus (Figure 13.1). The **longitudinal fissure**, also known as the **sagittal fissure** or the **interhemispheric fissure**, divides the cerebral cortex into two symmetrical **hemispheres** (though they are not literally hemispheres). The **central sulcus**, also called the **central fissure** or the **Rolandic fissure**, runs in a superior-inferior direction on the lateral side of each hemisphere, dividing the cerebrum into an anterior half and posterior half. It does not run deeply over the medial surface. The **lateral sulcus**, also called the **lateral fissure** or the **Sylvian fissure**, runs in an anterior-posterior direction on the lateral surface for about half the length of each hemisphere. The **parieto-occipital sulcus** runs on the mesial surface of the brain.

Each hemisphere is divided into four lobes: **frontal, parietal, temporal,** and **occipital**. The lobes are named after the overlying bones of the cranium. The cerebral cortex can roughly be divided into two major functional parts: the "posterior lobes" (i.e., parietal, occipital, and temporal lobes) and the frontal lobes. This division is based on the concept that the posterior lobes are concerned mainly with sensing and perceiving, while the frontal lobes are concerned mainly with action and are the origin of all voluntary (i.e., consciously controlled) behavior.

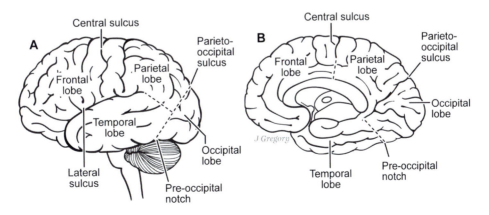

FIGURE 13.1. The boundaries of the four lobes. (A) Lateral surface. (B) Medial surface.

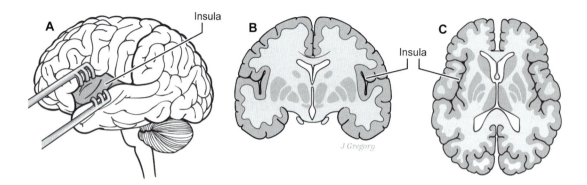

FIGURE 13.2. The insular cortex is not visible from the external surface of the brain. (A) Lateral view. (B) Coronal view. (C) Axial view.

The extreme ends of each hemisphere are called **poles**: the frontal poles, the temporal poles, and the occipital poles (the parietal lobes do not have poles). The boundaries between the lobes are defined by the four major sulci and the **preoccipital notch**, an indentation located about 5 cm anterior to the occipital pole.

Two cortical areas are not visible from the external surface of the brain: the insular cortex and the cingulate cortex. The **insular cortex (island of Reil)** is covered by parts of the frontal, parietal, and temporal lobe around the lateral sulcus that are known as the **operculum** (Figure 13.2). The insular cortex is surrounded by the **circular sulcus (limiting sulcus)**, and divided by the **central sulcus** into an **anterior lobule** and **posterior lobule**. The **cingulate cortex** is seen from the medial surface of the brain. The insula and cingulate have been referred to as lobes (i.e., the insular lobe, the cingulate lobe), but this is not accurate.

The following sections examine cortical anatomy in greater detail from the lateral, medial, and inferior surfaces.

Lateral Surface

The lateral surface is convex in shape and therefore is also referred to as the **convexity**. The central and lateral sulci are the most prominent landmarks.

On the lateral surface, the boundaries of the four lobes are defined as follows (see Figure 13.1). The frontal lobe lies anterior to the central sulcus and superior to the lateral sulcus. The occipital lobe is demarcated from the parietal and temporal lobes by the **lateral parietotemporal line**, an imaginary line from the endpoint of the parieto-occipital sulcus to the preoccipital notch. The temporal lobe lies inferior to the lateral sulcus and anterior to the occipital lobe. The parietal lobe lies posterior to the central sulcus, anterior to the parieto-occipital sulcus, and superior to the lateral sulcus. The **occipitotemporal line**, an imaginary line that runs from the posterior end of the lateral sulcus to meet the lateral parietotemporal line at a right angle, separates the temporal and parietal lobes. The sulci and gyri on the lateral surface of the cerebral cortex are shown in Figures 13.3 and 13.4.

On the lateral surface of the frontal lobe, the **precentral sulcus** runs anterior and parallel to the central sulcus. The central and precentral sulci define the **precentral gyrus**. Anterior to this gyrus, two sulci run in an approximately anterior-posterior direction: the **superior frontal sulcus** and the **inferior frontal sulcus**. These two sulci demarcate the boundaries of the **superior frontal gyrus**, the **middle frontal gyrus**, and the **inferior frontal gyrus**. The inferior frontal gyrus itself is subdivided into three parts. From anterior to posterior these are: **pars orbitalis**, **pars triangularis**, and **pars opercularis**. In summary, the lateral surface of the frontal lobe is composed of four gyri: the precentral gyrus, the superior frontal gyrus, the middle frontal gyrus, and the inferior frontal gyrus.

The temporal lobe's lateral surface has two sulci that run in an approximately anterior-posterior direction: the **superior temporal sulcus** and the **inferior temporal sulcus**. These sulci demarcate the **superior temporal gyrus**, the **middle temporal gyrus**, and the **inferior temporal gyrus**. This naming system is similar to that of the frontal lobe.

The parietal lobe's lateral surface has a sulcus that runs posterior and parallel to the central sulcus and is known as the **postcentral sulcus**. The central and postcentral sulci define the **postcentral gyrus**. Posterior to the postcentral gyrus, the **intraparietal sulcus** runs in an approximately anterior-posterior direction. It divides the parietal lobe into the **superior parietal lobule** and the **inferior parietal lobule**. The inferior parietal lobule is composed of the **supramarginal gyrus** and the **angular gyrus**. Thus, the lateral surface of the parietal lobe is composed of the postcentral gyrus, the superior parietal lobule, and the inferior parietal lobule.

On the lateral surface of the occipital lobe, the **lateral occipital sulcus** runs in an approximately anterior-posterior direction. The anatomy of the lateral surface of the occipital lobe is somewhat variable from individual to

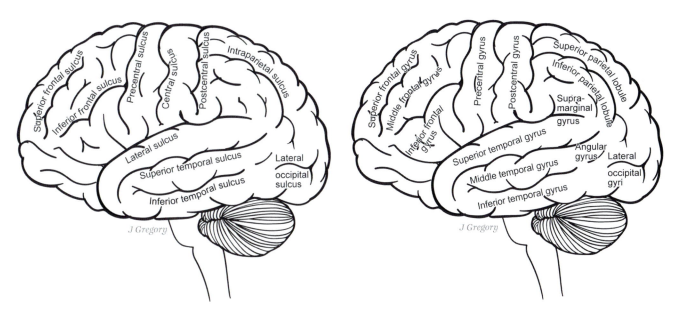

FIGURE 13.3. The lateral surface of the cerebral cortex; sulci.

FIGURE 13.4. The lateral surface of the cerebral cortex; gyri.

individual in that there may be multiple lateral occipital sulci. In either case, the surrounding gyri are known as the **lateral occipital gyri**.

Medial Surface

On the medial surface of the brain, the boundaries of the four lobes are defined as follows (see Figure 13.1). The frontal lobe lies anterior to the central sulcus and an imaginary line drawn to the corpus callosum. The parietal lobe is located posterior to this and anterior to the parieto-occipital sulcus. The occipital lobe is demarcated from the more anterior parietal and temporal lobes by the **basal parietotemporal line**, an imaginary line connecting the mesial endpoint of the parieto-occipital sulcus to the preoccipital notch line.

The sulci and gyri on the medial surface of the cerebral cortex are shown in Figures 13.5 and 13.6. In a midsaggital section, the most striking feature of the medial surface of each hemisphere is the **corpus callosum** ("tough body"), a thick band of white matter that is approximately 10 cm in length. It lies immediately superior to the lateral ventricles, forming a roof over them. The **septum pellucidum** is a thin, triangular membranous tissue located in the midsagittal plane that stretches vertically between the corpus callosum and the fornices; it separates the right and left lateral ventricles at the anterior horns (i.e., it forms the medial walls of the lateral ventricles at the anterior horns).

The corpus callosum is bounded superiorly by the **callosal sulcus**. The callosal sulcus, together with the **cingulate sulcus**, forms the boundaries of the **cingulate gyrus**. The cingulate gyrus and the cortex buried within the cingulate sulcus together are referred to as **cingulate cortex** or **limbic cortex**.

The **cingulate sulcus** lies above the cingulate gyrus; it begins in the subcallosal area and has two branches, the **paracentral sulcus** and the **marginal sulcus** (also known as the **marginal ramus of the cingulate sulcus**). The area between the paracentral sulcus and the marginal sulcus is the **paracentral lobule**; this region is the medial surface of the pre- and postcentral gyri. The area between the marginal sulcus and the parieto-occipital sulcus is the **precuneus**; this region is the medial extension of the superior parietal lobule.

Running within the middle of the occipital lobe, the **calcarine fissure** divides the occipital lobe into the **cuneus** lying above the fissure, and the **lingual gyrus** (tongue-shaped gyrus) lying below the fissure. Within the occipital lobe and immediately inferior to the calcarine sulcus is the **collateral sulcus**. The calcarine and collateral sulci demarcate the **lingual gyrus.**

Within the temporal lobe, the **collateral sulcus** courses anteriorly, and often merges with the **rhinal sulcus**. These sulci demarcate the **parahippocampal gyrus** of the temporal lobe. The parahippocampal gyrus is continuous with the cingulate gyrus via the **isthmus** (a narrow connection); the cingulate and parahippocampal gyri have been referred to collectively as the **limbic lobe** because they form a border between the neocortex and the subcortical structures; however, these structures do not constitute a true lobe. The parahippocampal gyrus of the temporal lobe also is continuous with the **lingual gyrus** within the occipital lobe; together these are often referred to as the **medial occipitotemporal gyrus**. The **occipitotemporal sulcus** lies lateral to collateral sulcus; these two sulci form the boundaries of the **lateral occipitotemporal gyrus.**

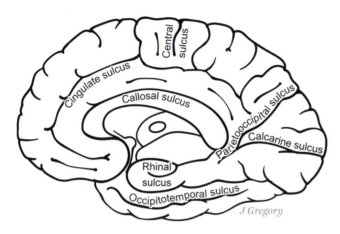

FIGURE 13.5. The medial surface of the cerebral cortex; sulci.

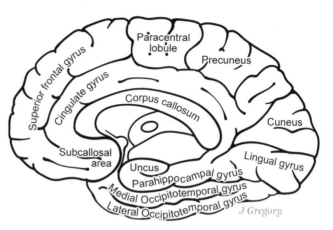

FIGURE 13.6. The medial surface of the cerebral cortex; gyri.

On the medial surface of the frontal lobe, we see the medial aspect of the superior frontal gyrus (sometimes referred to as the *medial frontal gyrus*, although it is not a separate gyrus) and the **subcallosal area**.

Inferior Surface

The inferior surface of the brain is composed of the orbital surface of the frontal lobe and the inferior surfaces of the temporal and occipital lobes (Figure 13.7).

The orbital surfaces of the frontal lobes are marked by the **olfactory bulbs** and the **olfactory tracts**. An **olfactory sulcus** lies beneath each olfactory bulb and tract. Mesial to the olfactory sulcus lies the **gyrus rectus** (straight gyrus), and lateral to the olfactory sulcus lies the **orbital gyri**. The inferior surface of the temporal lobe is composed of the parahippocampal gyrus medially and the occipitotemporal gyrus/gyri laterally. The anterior end of the parahippocampal gyrus bends to form a hook-like structure known as the **uncus**.

Cortical Connections

The fibers comprising the brain's white matter are classified into three types based on their connections: projection fibers, commissural fibers, and association fibers. Commissural and association fibers play a particularly important role in cortical connectivity and cortical functions.

Projection Fibers

Projection fibers consist of long connections between different structures of the CNS. **Cortical projection fibers** connect the cerebral cortex to subcortical structures. Tracts are often named after the two CNS regions that they connect; the prefix of the tract name indicates the site of origin, and the suffix reflects the site of termination (e.g., the corticospinal tract).

Projection fibers that originate from subcortical structures and terminate in the cerebral cortex are **corticopetal** (**corticoafferent**). The majority of corticopetal fibers arise from the thalamus and are carried by **thalamocortical** pathways. Projection fibers that travel from the cerebral cortex to subcortical structures are **corticofugal** (**corticoefferent**). These fibers are the axons of pyramidal cells (see "Cortical Cells," below), and they originate from virtually all parts of the cerebral cortex. Corticofugal fibers

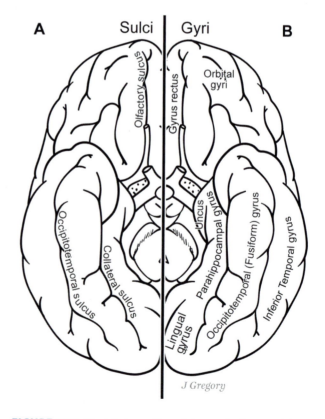

FIGURE 13.7. The inferior surface of the cerebral cortex; (A) sulci and (B) gyri.

include **corticostriate** (from the cortex to the striatum of the basal ganglia), **corticothalamic** (from the cortex to the thalamus), **corticohypothalamic** (from the cortex to the hypothalamus), **corticopontine** (from the cortex to the pons), **corticobulbar/corticonuclear** (from the cortex to brainstem cranial nerve motor nuclei), **corticoreticular** (from the cortex to the reticular formation), and **corticospinal** (from the cortex to the spinal cord) pathways.

Afferent and efferent nerve fibers passing between the cerebral cortex and subcortical structures form the **corona radiata**, a radiating crown of projection fibers to and from every part of the cerebral cortex (Figure 13.8). A subset of the corona radiata fibers consist of the **thalamic radiations**. Virtually every nucleus of the thalamus sends axons to the cerebral cortex, and virtually every part of the cortex receives afferent fibers from the thalamus. Thus, almost all of the information that the cortex receives is first relayed through the thalamus. For this reason, the thalamus is often called the "gateway to the cerebral cortex." Since the thalamus provides so much of the input to the cerebral cortex, the thalamus actually drives cortical activity. Without the thalamus, the cortex would have little to do.

Moving caudally, the corona radiata projections aggregate into a compact bundle known as the **internal capsule** (see Figure 13.8). The internal capsule carries thalamocortical, corticothalamic, corticopontine, corticobulbar, and corticospinal projections. Horizontal sections through the internal capsule show that it is V-shaped, with an **anterior limb**, a **genu** (knee), and a **posterior limb** (see Figure 10.3 in Chapter 10). The internal capsule is flanked by the putamen and globus pallidus laterally. Medially, the anterior limb of the internal capsule is flanked by the caudate nucleus, and the posterior limb is flanked by the thalamus.

As the internal capsule fibers pass inferiorly beyond the thalamus, they form the **cerebral peduncles** (Latin, "foot"), a pair of cylindrical bodies that occupy the ventral portion of the midbrain (see Figure 13.8; see also Figure 8.2 in Chapter 8). Other names for the cerebral peduncles are *crus cerebri* (leg of the cerebrum) and *basis pedunculi*. The cerebral peduncles carry corticopontine, corticobulbar, and corticospinal fibers (which decussate at the transition from the medulla to the spinal cord).

Commissural Fibers

A **commissure** is a point of junction between two parts. Brain commissures are collections of axons within the CNS that interconnect symmetrical structures between the two hemispheres of the brain. Two commissures interconnect the two halves of the cerebral cortex: the corpus callosum and the anterior commissure (Figure 13.9). These carry interhemispheric corticocortical fibers. Three other brain commissures that interconnect subcortical structures are the posterior commissure, the hippocampal commissure (commissure of the fornix), and the habenular commissure. Neural fibers within the commissures, of course, carry information only in one direction, but commissures contain fibers that convey information in each direction (i.e., from right to left and from left to right).

Most cortical commissural fibers are contained within the **corpus callosum**. The vast majority of these callosal fibers connect **homologous** regions (i.e., in corresponding locations) of the two hemispheres, but a small number

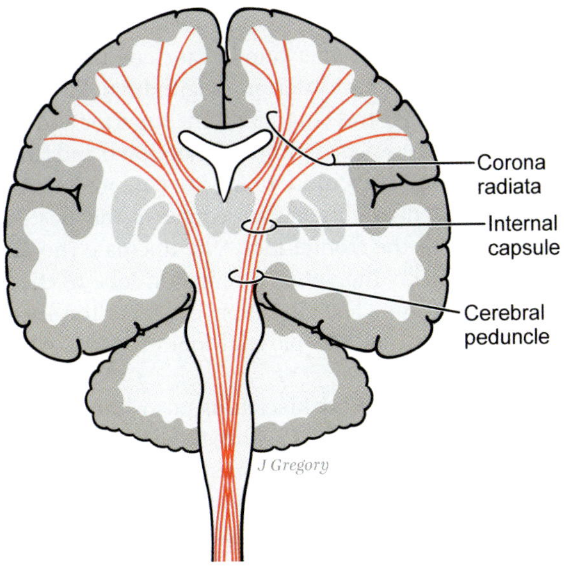

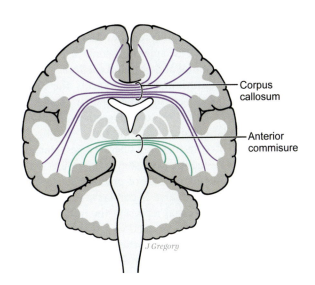

FIGURE 13.8. Corona radiata projections aggregate into the internal capsule. As the internal capsule fibers pass inferiorly beyond the thalamus, they form the midbrain cerebral peduncles.

FIGURE 13.9. Two commissures interconnect the two halves of the cerebral cortex: the corpus callosum and the anterior commissure.

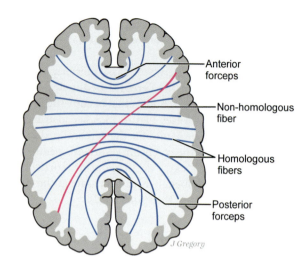

FIGURE 13.10. The vast majority of the corpus callosum fibers connect homologous regions of the two hemispheres, but a small number connect non-homologous cortical regions.

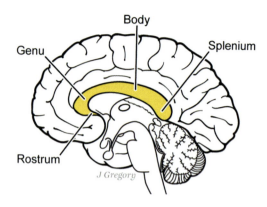

FIGURE 13.11. The corpus callosum has four components: the rostrum, the genu, the body, and the splenium.

connect **non-homologous** cortical regions (Figure 13.10). Fibers interconnecting the medial aspects of the frontal and occipital lobes take a curved pathway known as the **anterior forceps (forceps minor)** and **posterior forceps (forceps major)**.

The corpus callosum is especially prominent in the human brain, reflecting advances in development of the neocortex relative to our early evolutionary relatives. It is composed of approximately 200–300 million fibers and accounts for approximately 2%–3% of all cortical fibers. Approximately half of these fibers are small and unmyelinated.

The corpus callosum has four components: the rostrum, the genu, the body, and the splenium (Figure 13.11). The **rostrum** (nose) and **genu** (knee) contain fibers interconnecting the prefrontal cortices, the **body** contains fibers from posterior frontal (premotor and motor areas) and parietal cortex, and the **splenium** contains fibers that interconnect the temporal and occipital cortices.

The **anterior commissure** interconnects the middle and inferior temporal gyri, and the olfactory bulbs (see Figure 13.9). It is best seen on coronal sections, just below the interventricular foramen.

The density of commissural fibers varies considerably among cortical regions, and some regions are almost or totally devoid of them. Areas of primary sensory and motor cortex (see "Primary, Secondary, and Tertiary Cortical Zones," below) that represent the midline of the body have dense callosal connections. This observation has given rise to the **zipper hypothesis of callosal function**, which states that the corpus callosum knits together the representations of the midpoints of the body and space that are divided by the longitudinal fissure. Areas of primary sensory and motor cortex that represent distal portions of the limbs (i.e., the hands and feet) lack callosal connections; this lack of connectivity is believed to be essential to the ability of the hands and feet to function independently of each other. Most primary visual cortex is devoid of interhemispheric connections, except for the portion representing the boundary between the right and left visual fields.

Association Fibers

Association fibers are intrahemispheric corticocortical fibers; they connect cortical regions within the same hemisphere (Figure 13.12). Association fibers are divided into short and long groups. **Short association fibers** interconnect adjacent gyri. They lie immediately beneath the cortex and are U-shaped; thus they are also called **U-fibers** or **arcuate** (arc-shaped) **fibers**.

Long association fibers generally interconnect cortical regions in different lobes within the same hemisphere in bundles of myelinated fiber bundles known as **fasciculi**. These long association fiber bundles include the superior longitudinal fasciculus, inferior longitudinal fasciculus, uncinate fasciculus, and cingulum. The superior and inferior longitudinal fasciculi connect the occipital and frontal lobes. The arcuate fasciculus, which is a component of the superior longitudinal fasciculus, connects the posterior temporal area to the frontal lobe. The uncinate (hooked) fasciculus connects the anterior temporal lobe to the inferior frontal gyrus. The cingulum connects the medial regions of the frontal and parietal lobes to the parahippocampal gyrus.

The **vertical occipital fasciculus**, also known as the **perpendicular fasciculus**, is a fiber bundle that connects dorsolateral and ventrolateral cortical regions *within* the occipital lobe. This fasciculus was originally described by **Karl Wernicke** (1848–1905) in 1881, but because it contradicted a general principle of cortical white matter organization put forth by **Theodor Meynert** (1833–1892) that long association fibers were oriented in an anterior-posterior direction, the discovery of this fasciculus was

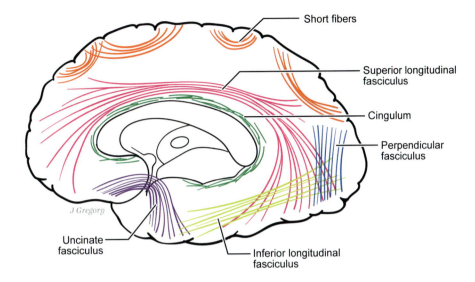

FIGURE 13.12. Association fibers are intrahemispheric corticocortical fibers; they connect cortical regions within the same hemisphere. Short association fibers interconnect adjacent gyri and lie immediately beneath the cortex. Long association fibers generally interconnect cortical regions in different lobes within the same hemisphere; they are collected into fasciculi.

controversial and eventually came to be disregarded, only to be rediscovered a century later.

Cortical Localization

History

The concept of cortical localization began with **Franz Josef Gall** (1758–1928), best known for developing and promoting **phrenology** in the late eighteenth and early nineteenth centuries, which simplistically postulated the existence of distinct brain regions with specific functions that could be identified and evaluated by palpating and measuring the skull. While we now know this to be pseudoscience, Gall and his collaborator **Johann Spurzheim** (1776–1832) made several important and positive contributions to neuroanatomy through dissection methods, including distinguishing gray and white matter and proving the decussation (crossing) of descending motor pathways in the medullary pyramids (two fiber bundles on the ventral surface of the medulla that appear as triangular-shaped swellings, especially in cross section). Gall's phrenology also was the precursor to the concept of **cerebral localization**, which survives to this day as an important underpinning of neurology and neuropsychology.

Gall theorized that the cerebral cortex is a collection of separate organs, each of which controls a separate innate faculty, and whose individual development is reflected in the bumps and depressions of the overlying skull (i.e., that the development of specific cortical regions correlates with specific talents or behaviors, and that skull features indicate underlying brain development). Gall attempted to associate functions with structures primarily by examining the crania of individuals from the extremes of society (great writers, statesmen, people with special talents, criminals, people with psychiatric disorders). To aid with his correlations, he collected skulls from individuals with known mental characteristics (over 300) and skull casts from living persons (over 120). Gall localized 27 faculties to different parts of the human cerebral cortex, including wisdom, passion, sense of satire, destructiveness, acquisitiveness, and ideality.

During the early nineteenth century, criticisms of Gall's methods led to a backlash against localism in mainstream contemporary scientific thought. **Jean-Pierre Flourens** (1794–1867) carried out lesion experiments that were interpreted as disproving Gall's theory of localization and supporting the competing theory, **cortical equipotentiality** (**holism**), which stated that the cortex functions as a whole and that all parts of the cortex serve all functions. This conclusion was based on his failure to find differences following lesions in different parts of the cortex, and the observation that lost functions recovered over time. He argued that loss of function correlated with the extent of cortical ablation, but not location. Flourens's observations, however, were based on studies performed on birds and frogs (vertebrates with less cortical dependency than higher mammals) and behaviors such as eating and wing-flapping. Gall's ideas of cortical localization of function now seem seminal, though his reliance on skull measurements was flawed. In contrast, Flourens was a laboratory scientist who used valid methods, but due to his poor choices of animal models for the human brain, he ultimately ended up with empirical support for an incorrect theory. Nevertheless, both of these early neuroscientists made important contributions to the study of cortical function.

Because the theory of localization of cortical function fell into disfavor in the early nineteenth century, early observations supporting the theory were largely ignored. For example, **Jean-Baptiste Bouillard's** (1796–1881) 1825 case series report associated frontal lobe lesions with loss of speech, and **Marc Dax**'s (1770–1837) 1836 case series report associated left-hemisphere damage with right hemiplegia and aphasia. In 1861, however, **Paul Broca** (1824–1880), a French surgeon and anatomist, presented findings of two patients who lost the ability to speak after injury to the posterior inferior frontal gyrus at two scientific society meetings in Paris.

The first patient was a 51-year-old man named **Louis Victor Leborgne** (1809–1861), who had a progressive loss of speech for over 20 years with preserved language comprehension, as well as other neurological problems (epilepsy, right hemiplegia). Leborgne could only produce a single syllable, "tan," with every attempt to speak, and he therefore became known as "**Tan**" in the hospital where he was institutionalized for many years. Leborgne came under Broca's care because he developed gangrene. Tan died several days later, and Broca performed an autopsy which revealed a lesion in the posterior third (inferior) frontal gyrus. Broca termed the loss of articulated speech as **aphemia**, which is now known as **Broca's aphasia**.

Several months later, Broca encountered a second patient with loss of speech. **Lazare Lelong** was an 84-year-old man who had a stroke one year earlier. On autopsy he was found to have a lesion in the posterior third (inferior) frontal gyrus, confirming the localization of speech to this area. Broca continued to collect cases of speech disturbance. In 1863, he published an additional series of eight cases establishing that the causative lesions not only invariably encompassed the "third convolution of the frontal lobe" (i.e., the inferior frontal gyrus), but they also occurred exclusively within the left hemisphere.

Broca's methods and findings were groundbreaking and represent a turning point in the history of our understanding of brain function. His landmark work not only established the localization of language, but also introduced the method of **lesion-deficit analysis** for the systematic study of the functional organization of the human brain. This method investigates the relationship between deficits exhibited in clinical cases prior to death, and postmortem lesion localization. Clinical-anatomical postmortem correlation studies formed the foundation of modern clinical neuropsychology and behavioral neurology, and the basis of investigation into cerebral localization for decades. Paul Broca is therefore considered to be "the father of cerebral localization." It is only relatively recently, with the advent of computerized tomography (CT) and magnetic resonance imaging (MRI), that lesion localization has become possible before death (and autopsy) and quantitative. Broca's concepts and methods were soon applied to other higher functions and experimental work on animals.

In the 1870s, **Edward Hitzig** (1838–1907) and **Gustav Fritsch** (1838–1927) performed experiments involving electrical stimulation of the cortex in dogs. They found that motor functions are also localized, and that motor cortex is organized by a topographic representation of the body (**somatotopy**). This confirmed **John Hughlings Jackson**'s (1835–1911) hypothesis that the human brain contains a map of the body's musculature, which was based purely on his observations that in some motor epilepsies the involuntary muscular contractions systematically move across the body, and that this must reflect spread of the seizure discharge within a cortical somatic motor map (see Chapter 19, "Epilepsy").

Little was known about the function of the human cerebral cortex until World War I and World War II, when many soldiers sustained bullet wounds to discrete regions of the brain. The relationship between lesion location and functional changes was studied systematically, most notably by **Kurt Goldstein** (1878–1965) in Germany during World War I and **Alexander Romanovich Luria** (1902–1977) in Russia during World War II. These and other clinical pathology studies in humans with focal cortical lesions of various etiologies, as well as experimental studies of behavioral effects of focal cortical lesions in animals, further established that different parts of the cerebral cortex are functionally specialized. This led to the concept of pathognomonic cortical signs.

A **sign** is any objective evidence of a disease; it can be detected by those other than the individual affected by the disease (e.g., hemiplegia). This stands in contrast to **symptoms**, which are subjective phenomena experienced by the affected individual but not observable by others (e.g., headache). Signs and symptoms give hints as to the underlying **etiology** (cause) and diagnosis. Signs and symptoms that are *so* characteristic for a particular disease that their presence can be used to make a definitive diagnosis are **pathognomonic**. A **pathognomonic cortical sign** is one that indicates the lesion location. For example, the clinical syndrome of Broca's aphasia is pathognomonic for a lesion in the posterior inferior frontal gyrus in the language-dominant hemisphere (usually the left hemisphere).

Functional Maps of the Cerebral Cortex

Functional maps of the cerebral cortex are created by studying the behavioral, cognitive, and phenomenological effects of focal cortical lesions, focal electrical stimulation, electrical recordings from cortical regions, and, most recently, functional neuroimaging.

The best known of the many functional maps is that of the **motor and somatosensory cortex** by Canadian

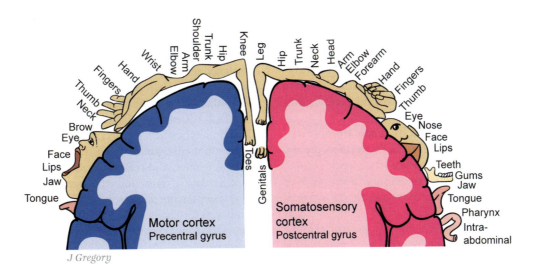

FIGURE 13.13. Somatotopic organization of the primary somatosensory cortex (shown in pink) and primary motor cortex (shown in blue).

neurosurgeon **Wilder Penfield** (1891–1976), made in the early 1950s. During the course of brain surgery, he applied small points of electrical stimulation to the cortical surface in conscious people who were undergoing neurosurgery with local anesthesia. Penfield distinguished motor and sensory cortex based on elicited responses and established that there is a point-to-point correspondence between body parts and cortical areas. Penfield is sometimes mistakenly considered to be the discoverer of the map in motor cortex, when in fact **David Ferrier** (1843–1928) first mapped motor cortex in monkeys approximately 70 years earlier. Penfield was, however, the first to apply the concept of the **homunculus** (*little man within the man*) to the brain; this concept was based on the sixteenth-century notion that sperm cells each contained a little person. Penfield also was the first to draw the homunculus stretched over the cortical surface to illustrate the mappings of the body onto somatosensory and motor cortices (Figure 13.13).

Primary, Secondary, and Tertiary Cortical Zones

Projection maps show how neurons spanning different brain structures are connected, and they are constructed by tracing axonal connections. An example of a projection map is one that shows the connections between the thalamus and cerebral cortex (see Figure 10.2 in Chapter 10). The projections from the thalamus to the cortex can be broadly classified as specific (i.e., projecting to a specific and sharply defined area) or nonspecific (i.e., projecting diffusely to a large area). Almost every area of cortex, except for some parts of the temporal lobe, receives input from a specific thalamic source nucleus. The specific projections include those from **sensory relay nuclei of the thalamus** that receive projections from specific sensory pathways (e.g., visual, auditory, somatosensory) to a functionally distinct area of cerebral cortex.

Areas of cortex that receive direct input from the major sensory pathways are known as **primary sensory areas**. They are localized cortical regions first receiving sensory information from subcortical regions and are therefore the first stop for cortical sensory processing. **Primary visual cortex** (V1) is located in the occipital lobe, mostly within the walls of the calcarine sulcus. **Primary auditory cortex** (A1) is located in the temporal lobe on **Heschl's gyrus** (also known as the transverse temporal gyrus), which is entirely hidden within the Sylvian fissure. **Primary somatosensory cortex** (S1) is located on the postcentral gyrus in the parietal lobe. **Primary gustatory cortex** (G1) is localized to the opercular region of the postcentral gyrus. **Primary olfactory cortex** (O1) is located on the ventral surface of the temporal lobe within the uncus. Not surprisingly, primary sensory areas possess high **modal specificity**—they respond only to stimuli within one sensory modality.

Primary sensory areas also are **topographically** organized, containing maps of the sensory receptor surface they represent (except for olfactory cortex). In the somatic sensory system, this organization is termed **somatotopy**; in the visual system, it is called **retinotopy** (see Chapter 16, "The Temporal Lobes and Associated Disorders," for a description of cochleotopy). In the somatosensory and visual systems, the topographic maps allow us to localize stimuli on the skin surface or in visual space, respectively. Receptor cells, however, are not uniformly distributed throughout the sensory surface, and the amount of neural tissue devoted to representing a particular portion of the receptor surface is correlated with the receptor density of that region. Regions of the sensory surface in which receptors are densely packed gather a lot of information and therefore provide for high spatial resolution. They also convey a lot of information to the CNS via many afferent fibers. Consequently, the maps of the sensory surfaces are distorted such that regions of the sensory surface with high resolution have large representations (e.g., the hands and face regions within the somatosensory homunculi).

Primary somatosensory cortex has a **somatotopic** representation. Primary visual cortex has a **retinotopic** organization. Primary auditory cortex has a **cochleotopic** organization. Lesions of primary sensory areas produce profound sensory deficits (e.g., hemianesthesia, cortical blindness).

Areas of cerebral cortex outside of primary sensory and motor areas are generically referred to as **association cortex**. Of these, there are two main types: unimodal and multimodal association areas.

Unimodal association areas, also known as **secondary zones**, receive their input from a single primary cortical sensory area. These secondary areas are located adjacent to their primary projection area. The input from primary to secondary zones arrives via association fibers and provides a substrate for sequential (serial) processing of information. Each sensory system has its own unimodal association area(s). **Visual association cortex (secondary visual cortex)** surrounds primary visual cortex (V1), occupying the entirety of the occipital lobes exclusive of primary visual cortex, as well as posterior aspects of the parietal lobe and temporal lobe (areas V2, V3, V4, and V5). **Auditory association cortex (secondary auditory cortex)** forms much of the superior temporal gyrus (A2). **Somatosensory association cortex (secondary somatosensory cortex)** is located posterior to the postcentral gyrus in the superior parietal lobule (S2). Compared to primary areas, secondary areas perform a higher level of processing than primary areas, in that the outputs of neurons in primary areas serve as their inputs. Accordingly, they process more complex aspects of the sensory modality and have less topographically specific maps of the sensory surface that they represent. It is thought that modality-specific information becomes integrated into more general functional units within these secondary sensory areas. It is useful to think of primary zones as more concerned with sensations, while secondary zones are more concerned with perceptions and knowledge, or **gnosis**. Lesions affecting primary zones result in *sensory* deficits, while lesions affecting secondary zones result in *perceptual* disorders, or **agnosias** (see Chapter 14, "The Occipital Lobes and Visual Processing").

Multimodal (heteromodal) association areas, also known as **tertiary zones**, surround secondary areas and function to integrate diverse sources of information; these zones are not modality-specific. There are three major tertiary zones: parietal-temporal-occipital association cortex, prefrontal association cortex, and limbic association cortex. The **parietal-temporal-occipital association cortex**, located at the junction of the parietal, temporal, and occipital lobes, integrates information from multiple sensory modalities (somatosensation, audition, vision) and is concerned with higher perceptual functions and language. **Prefrontal association cortex** occupies the rostral part of the frontal lobe, anterior to secondary and primary motor cortices. It is clearly implicated in the planning of behavior (as well as other functions). **Limbic association cortex** is located on medial and inferior surfaces of the brain in the parietal, frontal, and temporal lobes: orbitofrontal cortex, the cingulate gyrus, and the parahippocampal gyrus. These components play a role in motivation, emotion, and memory (see Chapter 12, "Limbic Structures"). Lesions in tertiary zones result in frank cognitive and behavioral disorders.

In the sensory systems, information flows from primary sensory cortex to secondary sensory areas to tertiary association cortex. In the motor system, information flows in the opposite direction, from tertiary association cortex to secondary motor areas to primary motor cortex. Primary motor cortex, the site from which voluntary movement is initiated and the origin of the corticospinal tract, occupies the precentral gyrus of the frontal lobe. Secondary motor areas lie anterior to the precentral gyrus and receive much of their input from prefrontal association cortex.

The concept of hierarchical cortical function was first put forth by Alexander Luria, who is also considered the originator of neuropsychological assessment. He conceived of the cerebral cortex as a hierarchy of three basic functional levels, with primary sensory areas at the bottom, sensory association areas at the middle, and higher-order association areas at the top of the hierarchy. This scheme assumes both that the brain processes information serially and that serial processing is hierarchical, so that each level of processing adds complexity to the processing of earlier levels. These assumptions are not fully consistent with more recent anatomical and physiological findings. Information processing within the cortex is not strictly hierarchical; all cortical areas have reciprocal connections and therefore do not simply operate in a feed-forward fashion. Each cortical zone has connections with many cortical areas, and cortical processing is also often relayed to subcortical structures. The contemporary view of cortical function is that both parallel and serial processing occur within distributed hierarchical systems. Luria's concept of hierarchical cortical processing nevertheless serves as a useful clinical heuristic for understanding the effects of cortical lesions.

Over the course of mammalian evolution, primary areas occupied proportionately less of the cortical surface, and association cortex occupied proportionately more. In primates, the association areas constitute the majority of the cortex, and this is especially so in humans. Association areas are also the last to mature in ontogenetic development.

Cytoarchitecture of the Cerebral Cortex

The cerebral cortex varies across the cortical sheet in cell type, layering pattern, and columnar organization. These variations are displayed in what are known as cytoarchitectural maps.

Cortical Cells

The cells of the cerebral cortex may be divided generally into pyramidal cells and nonpyramidal cells.

Pyramidal cells are the most numerous cortical cell type, constituting about two-thirds of all cortical neurons. They have a pyramid-shaped cell body, a single apical dendrite that extends from the apex of the cell body toward the cortical surface (and oriented perpendicular to it), and several basal dendrites that spread out from the base of the cell body. Their cell bodies, measured from the base of the cell to the origin of the apical dendrite, range in size over a factor of 10, from less than 10 μm in diameter, to more than 100 μm in the case of the **giant pyramidal cells** (**Betz cells**) of the primary motor cortex. Pyramidal cells are the major output neurons of the cortex; their axons form the projection, commissural, and association pathways of the cortex. Pyramidal cell axons also give rise to intrinsic axon collaterals which course through the cortical gray matter and form local connections within a cortical region.

Nonpyramidal cells usually have short axons that remain within the cortex, and thus they serve as local interneurons. These include stellate (star-shaped), horizontal, basket, fusiform, chandelier, bouquet, and Martinotti cells.

Cortical Layers

Microscopic examination of the cerebral cortex reveals that different neuron types tend to be organized in layers. This is apparent with any of several neural tissue staining methods. The **Golgi stain** reveals the arborization shape of cortical neurons by completely staining a small percentage of them. The **Nissl method** stains the cell bodies of all neurons, showing their shapes and packing densities. The **Weigert method** stains myelin, revealing horizontally oriented bands and vertically oriented cortical afferents and efferents. All these methods in combination help elucidate the layered structure of the cortex.

In general, there are six laminae parallel to the surface of the neocortex. They are numbered 1–6, from external surface to internal. Traditionally, Roman numerals are used to label the cortical layers, cranial nerves, and subdivisions of other structures, but Arabic numerals are also used. Each cortical layer has its own afferent and efferent connections, and so different layers carry out distinct information-processing functions.

Layer 4 (the **internal granular layer**) contains densely packed small stellate cell bodies. These stellate cells are a type of **granule cell**, a term that applies to a variety of neuron types with small cell bodies (approximately 10 μm) that have a granular appearance with cell body staining. In primary sensory cortex, layer 4 is thick; it receives abundant and precise, topographically organized inputs from thalamic sensory relay nuclei. Primary motor cortex lacks a distinct densely packed granule cell layer 4; it is therefore referred to as **agranular cortex**, in contrast to **granular cortex**.

Layer 5 (the **internal pyramidal/ganglionic layer**) contains the largest pyramidal cells. These pyramidal cells project to regions outside the cortex, primarily the striatum, brainstem, and spinal cord. Layer 5 is particularly well developed in primary motor cortex, and its projections form the **corticostriate**, **corticobulbar**, and **corticospinal tracts**. The giant pyramidal cells of Betz, the largest neurons in the CNS, are located in layer 5 of primary motor cortex and project through the corticospinal tract to synapse directly on lower motor neurons.

Layer 4 is used as a reference point to define the **supragranular layers** 1–3 (i.e., above the internal granular layer) and the **infragranular layers** 5 and 6 (i.e., below the internal granular layer). The supragranular layers mainly receive inputs to the cortex, while the infragranular layers mainly give rise to the outputs of the cortex. Layer 1 (the **molecular/plexiform layer**) is rich in fibers (axons and apical dendrites of pyramidal cells) and contains few cell bodies. Layer 2 (the **external granular layer**) consists of densely packed small cell bodies that act in local circuits. Layer 3 (the **external pyramidal layer**) contains medium-sized pyramidal cells that send axons primarily to other cortical regions, forming association and commissural fiber connections. Supragranular layers 2 and 3, receiving association fibers, predominate in secondary and tertiary areas. Layer 4 (the internal granular layer) receives thalamocortical connections, and layer 5 (the internal pyramidal/ganglionic layer) contains pyramidal cells that project to regions outside the cortex. Layer 6 (**the multiform/polymorphic layer**) contains many small pyramidal cells; it is the major source of corticothalamic fibers.

An additional feature of the layering pattern in neocortex is two myeloarchitectural structures, the **bands of Baillarger**. These two bands of myelinated fibers course horizontally in layers 4 and 5 and are known as the **outer (external) band of Baillarger** and the **inner (internal) band of Baillarger**, respectively. In the primary visual cortex (V1), the external band of Baillarger is enlarged and is called the line (stria) of Gennari.

Types of Cerebral Cortex Based on Microstructure of Cortical Layers

There are two main categories of cerebral cortex based on layering pattern: neocortex and allocortex.

Neocortex is of recent phylogenetic development; it appeared late in vertebrate evolution and is characteristic of mammals. In humans, about 90% of the cerebral cortex is neocortex. The defining feature of neocortex is that during early development (i.e., in the third trimester of gestation), it develops as a six-layered structure with a standard cellular arrangement that is repeated across the entire sheet. Because of its uniformity in early development, it is also referred to as **isocortex** (iso = uniform, unchanging) and **homogenetic** cortex.

During development, some neocortical areas retain this layering pattern, but some lose their six-layered structure. Areas of neocortex that retain the fetal six-layered pattern into adulthood are classified as **homotypical** cortex; these include the association areas. Areas of neocortex in which the layering pattern is altered are classified as **heterotypical** cortex. These include primary visual cortex in which both layers 3 and 4 divide into two or three sublayers, and primary motor cortex in which layer 4 is almost absent.

Allocortex, also known as **heterogenetic** cortex, is defined as cortex that does not go through a six-layered stage. Allocortex is phylogenetically older than neocortex; it is characteristic of the non-mammalian vertebrate brain and accounts for about 10% of the total cortex in humans. Two subtypes of allocortex have been defined, paleocortex and archicortex, although textbooks vary in their definitions of these two types. In humans, **paleocortex** (Greek, "old cortex") is found on the inferior and medial aspects of the temporal lobe and makes up the primary olfactory cortex located on the uncus of the parahippocampal gyrus, and **archicortex** (Greek, "ancient cortex") is found in the hippocampus. It is worth noting here that the hippocampus, because it has a layered structure, is not really a subcortical structure, but rather a cortical (layered) structure that lies beneath the surface of the cerebral cortex because of infolding (just as the cingulate gyrus and insula).

Mesocortex is intermediate between isocortex and allocortex, consisting of 4–5 layers. It is found in much of the cingulate gyrus, entorhinal cortex, parahippocampal gyrus, and orbital cortex.

Cytoarchitectural Maps

Histological examination reveals that as we move across the sheet of the cerebral cortex, the tissue varies in absolute thickness, layer thickness, cell-type composition of each layer, cell density in each layer, and nerve fiber lamination. These differences form the basis of **cytoarchitectural maps**. In many cases, a cytoarchitectonically defined area is also unique in its afferent and efferent connections, as well as the physiological response properties of its cells and its function.

The best known of these cytoarchitectural maps is one published by the German histologist **Korbinian Brodmann** (1868–1918) in 1909, in which 52 cytoarchitecturally distinct regions were distinguished. In Brodmann's map, the 52 cytoarchitecturally distinct regions are known as **Brodmann areas**, and they are identified simply by the numbers 1–52. The numbers themselves have no meaning; they simply reflect the order in which he studied the areas. Brodmann mapped the cortex on the gyral surfaces and the insular cortex. He did not map cortex that lies within the walls of the sulci. The areas Brodmann defined were based solely on their neuronal organization, but we have since found that they are related to specific information-processing roles and specific cortical functions. For example, area 17 corresponds to primary visual cortex (V1), located in the occipital lobe mostly within the walls of the calcarine sulcus. Area 41 corresponds to primary auditory cortex (A1), located in the temporal lobe on Heschl's gyrus. Brodmann areas 1, 2, and 3 correspond to primary somatosensory cortex (S1) located on the postcentral gyrus in parietal lobe. Brodmann area 4 (BA4) corresponds to primary motor cortex (M1), located in the precentral gyrus. Recall that primary motor cortex is agranular; for Brodmann this was an important anatomic landmark, and historically it helped spur the development of theories of localization of function.

Brodmann's map has been refined with the development of techniques such as single-cell recording, and in some cases two or more distinct cytoarchitectural areas have been identified within a single Brodmann area. Thus, the map is continually updated and now consists of a mixture of numbers, letters, and names (e.g., BA3 is now subdivided into areas 3a and 3b).

Columnar Organization

In addition to its layered organization, the cerebral cortex has a vertical organization perpendicular to the surface. This is reflected in the vertically oriented apical dendrites of pyramidal cells and the axons of some intracortical cells. This anatomic vertical organization results in **columns** or **modules** of cells that have related functions. This columnar organization was revealed by electrophysiological studies of sensory cortices that examined the response properties of cells at different layers within the column to sensory stimuli. All of the cells along a vertical path have the same **receptive field** (i.e., they all monitor the same loci on the sensory receptor surface) and **response properties** (i.e., they all respond to the same sensory features). The columnar organization of the cerebral cortex was first discovered by **Vernon Mountcastle** (1918–2015) in somatosensory cortex, who proposed that the cortical column is the basic functional unit throughout the entire cerebral cortex. **David Hubel** (1926–2013) and **Torsten Wiesel** (1924–) subsequently discovered that visual cortex also has a columnar organization and described in detail the functional architecture of the columns.

Cortical Disconnections

Cortical disconnection is the severing of white matter connections between two areas of the cerebral cortex without damaging the cortical areas themselves; this occurs with discrete focal brain lesions, particularly ischemic stroke. **Intrahemispheric disconnections** occur with lesions in the long association fiber bundles. In cross section, these fasciculi are apparent as discrete collections of axons; viewed this way, it is easy to appreciate how relatively small lesions can produce a disconnection syndrome (Figure 13.14). **Interhemispheric disconnections** occur with lesions in a cortical commissure; these lesions disrupt information transfer between the two hemispheres (e.g., **callosal disconnection**). The **cortical disconnection syndromes** are a group of clinical syndromes resulting from interruption of cortical association fibers or cortical commissural fibers.

The four classical disconnection syndromes are: (1) **conduction aphasia**, a selective deficit in repetition of words and sentences in the context of preserved expressive and receptive language (see Chapter 27, "Language and the Aphasias"); (2) **associative visual agnosia**, a selective deficit in the ability to recognize objects visually (see Chapter 14, "The Occipital Lobes and Visual Processing"); (3) **apraxia**, a selective loss of the ability to perform skilled movements (see Chapter 24, "The Motor System and Motor Disorders"); and (4) and **pure alexia**, a selective loss of the ability to read that is not associated with a loss of the ability to write or other aphasic symptoms (see Chapter 14, "The Occipital Lobes and Visual Processing," and Chapter 28, "Alexia, Agraphia, and Acalculia").

American neurologist **Norman Geschwind** (1926–1984) introduced the term **disconnection syndrome** in his seminal 1965 work titled "Disconnexion Syndromes in Animals and Man." Geschwind also broadened the concept of disconnection to refer to the effects of lesions that interrupt information transfer and communication between parts of the brain, whether through broken white matter connections or lesions of association cortex. For example, he described several disconnection syndromes due to disconnection between specific sensory areas and language areas that lead to modality-specific language deficits: tactile anomia, pure word deafness, and pure alexia. **Tactile anomia** (previously referred to as *tactile aphasia*) is a selective inability to name held objects, in the context of normal somatosensation and preserved ability to name objects encountered through other sensory modalities (e.g., vision, audition). **Pure word deafness** is a selective inability to understand spoken words, in the context of normal hearing and otherwise normal language function. As described above, pure alexia is a selective inability to read (i.e., comprehend written language), in the context of normal vision and otherwise normal language function.

Corpus Callosotomy

In 1940 a scientific journal article describing experiments on the spread of epileptic discharge from one hemisphere to the other in the brains of monkeys concluded that the spread occurred largely or entirely by way of the corpus callosum. Subsequent studies in monkeys showed that surgical severing of the corpus callosum resulted in no behavioral changes. These findings paved the way for a surgical treatment for patients with uncontrolled epilepsy: **corpus callosotomy**, also known as **split-brain surgery** (Figure 13.15). Callosotomy is also referred to as **commissurotomy**, although this term refers to surgical section of a commissure (e.g., anterior commissurotomy) and is not specific to the corpus callosum.

The first callosotomies were performed in the early 1940s to control intractable seizures, by neurosurgeon **William Van Wagenen** (1897–1961) in Rochester, New York. Just as the early animal studies found that such surgeries produced minimal effects, the first humans to undergo the procedure showed no apparent effect on everyday behavior, and early postsurgical studies performed by **Andrew Akelaitis** (1904–1955) also showed no perceptual or motor deficits. However, given the variable success in relieving seizures, Van Wagenen stopped performing these procedures.

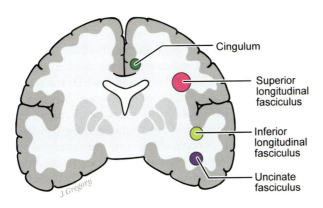

FIGURE 13.14. Intrahemispheric disconnections occur with lesions in the long association fiber bundles. In cross section, these fasciculi are apparent as discrete collections of axons; viewed this way, it is easy to appreciate how relatively small lesions can produce a disconnection syndrome.

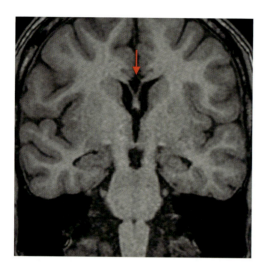

FIGURE 13.15. Postoperative coronal MRI demonstrating a corpus callosotomy.

The animal research on callosotomy continued, and in the 1950s **Ronald Myers** and **Roger Sperry** (1913–1994) made some astonishing discoveries that marked a turning point in our understanding of corpus callosum function. They showed that visual information presented to one hemisphere in callosotomized cats was not available to the other hemisphere. In order to restrict visual information to a single hemisphere, the decussating fibers of the optic chiasm were also severed. Normally, the right visual fields of both eyes project to the left hemisphere, and the left visual fields of both eyes project to the right hemisphere (see Chapter 14, "The Occipital Lobes and Visual Processing"). Transection of the decussating fibers of the optic chiasm results in the left visual field projecting to the right hemisphere by way of the right eye only, and the right visual field projecting to the left hemisphere by way of the left eye only.

The cats were trained on various visual discrimination tasks (e.g., press a lever when a circle appears on the screen, don't press the lever when a square appears on the screen). When a normal cat was trained on the task with one eye occluded by a patch, it could perform the task during the test phase using either eye because the information during training is transmitted to both hemispheres. When a callosotomized cat was trained with one eye occluded, this limited the information to the hemisphere on the side of the non-occluded eye. During the test phase, if the patch was switched to the other eye, the cat was completely unable to perform the task. Furthermore, if trained to do the same task all over again, the training took just as long, as if the cat had never learned the task. Myers and Sperry concluded that severing the corpus callosum prevented interhemispheric transfer of information. When training the callosotomized cats on the discrimination task with one eye occluded, they had trained only one-half of the brain.

Based on the animal data showing no major adverse effects from the surgery, two neurosurgeons, **Philip Vogel** and **Joseph Bogen** (1926–2005), started performing callosotomy as a treatment for intractable epilepsy in 1962 in California. **Roger Sperry** (1913–1994) at the California Institute of Technology (CalTech) studied some of these human split-brain subjects and made some remarkable discoveries, for which he won the Nobel Prize in Physiology or Medicine in 1981. In the 1960s, CalTech Ph.D. student **Michael Gazzaniga** (1939–) joined Sperry's lab. Sperry and Gazzaniga studied some of these split-brain patients, examining how the two hemispheres function independently. To understand their findings, it is necessary to appreciate some basic facts of cerebral localization, which were already known from studies of patients with focal cortical lesions: (1) in the vast majority of people the neural structures critical to language are localized in the left hemisphere (see Chapter 27, "Language and the Aphasias"); (2) the right half of visual space (i.e., the right visual field) is represented in the visual cortex of the left hemisphere, and the left half of visual space (i.e., the left visual field) is represented in the visual cortex of the right hemisphere (see Chapter 14, "The Occipital Lobes and Visual Processing"); (3) the motor cortex of the right hemisphere controls movement of the left side of the body, and the motor cortex of the left hemisphere controls movement of the right side of the body; and (4) tactile information also has a contralateral representation in somatosensory cortex, so that touch information from the right side of the body is projected to somatosensory cortex of the left hemisphere, and touch information from the left side of the body is projected to somatosensory cortex of the right hemisphere.

Sperry and Gazzaniga demonstrated a variety of disconnection phenomena in callosotomy patients. When an image of an object was briefly flashed to the left visual field (right hemisphere), the patients could not name the object; they were, however, able to perform the task when the stimulus was presented to their right visual field (processed in the left hemisphere). An inability to name objects seen is known as **visual anomia**; it is due to a visual-verbal disconnection. When printed words were flashed to their left visual field (right hemisphere), the callosotomy patients could not read what they had been shown; they were, however, able to perform the task when the stimulus was presented to their right visual field (left hemisphere). A selective inability to read material in one-half of the visual field is known as **hemi-alexia**; it also is due to a visual-verbal disconnection. When asked to perform hand movements to verbal commands, they could not do so with their left hand (right hemisphere), but could do so with their right hand (left hemisphere). This is known as **left-sided apraxia** to verbal command; it is due to a verbal-motor disconnection. When asked to write (i.e., to verbal command), the callosotomy patients could not do so with their left hand (right hemisphere), but could do so with their right hand (left hemisphere). This is known as **left-handed agraphia**; it also is due to a verbal-motor disconnection. When blindfolded and presented with objects to manipulate only with their right hand (left hemisphere), the callosotomy patients were well able to identify the objects by name, but when performing the task with only their left hand (right hemisphere), they were not able to name or describe what they had touched. This is known as **unilateral tactile anomia**; it is due to a tactile-verbal disconnection. The patients did, however, accurately perceive the object, as demonstrated by their ability to perform a tactile multiple-choice matching task. These findings show that the corpus callosum plays a critical role in the communication between the two hemispheres, and that severing this connection isolates the right hemisphere from the language abilities of the left hemisphere.

Hemispheric disconnection syndromes can also occur with naturally occurring focal lesions (e.g., ischemic stroke) that result in a partial disconnection of callosal fibers.

Agenesis of the Corpus Callosum

Agenesis of the corpus callosum is a neurodevelopmental abnormality in which the corpus callosum completely or partially fails to develop (Figure 13.16). It is a congenital malformation that may be due to genetic or environmental factors, in which there is a disruption of brain cell migration during fetal development. Agenesis of the corpus callosum is often seen in association with other neurodevelopmental abnormalities, but it also can occur without other neurodevelopmental abnormalities. It may be completely asymptomatic, with no behavioral or cognitive manifestations, only to be discovered incidentally on MRI or postmortem examination. Studies using sophisticated procedures to examine for disconnection effects demonstrate that individuals with agenesis of the corpus callosum may exhibit deficits performing bimanual tasks (integration between hands) but fail to show other callosal disconnection effects. This is thought to be attributable to the development of alternative compensatory pathways such as an enlarged anterior commissure or Probst bundles.

Probst bundles (**longitudinal callosal fascicles**) are white matter fibers that would normally cross the corpus callosum, but instead run parallel the interhemispheric fissure, leading to widely spaced lateral ventricles (Figure 13.17). This produces what is known as the **"racing car" configuration** that is apparent on axial MRI and CT images, because it is reminiscent of a Formula One racing car seen from above, with the tires represented by the widely spaced frontal and occipital horns and the dilated atria (Figure 13.18).

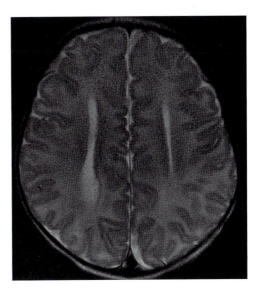

FIGURE 13.17. Agenesis of the corpus callosum. Axial image demonstrating Probst bundles (longitudinal callosal fascicles), white matter fibers that would normally cross the corpus callosum but instead run parallel to the interhemispheric fissure, leading to widely spaced lateral ventricles.

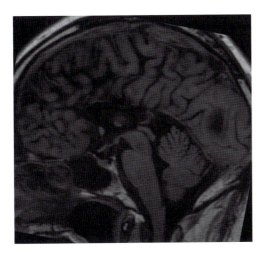

FIGURE 13.16. Sagittal image demonstrating agenesis of corpus callosum.

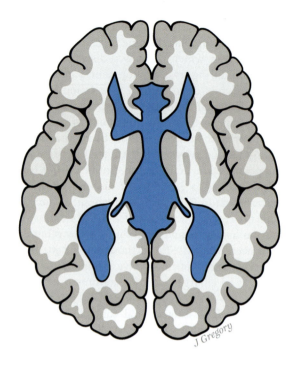

FIGURE 13.18. Agenesis of the corpus callosum; the "racing car" sign seen on axial MRI and CT images.

Summary

The cerebral cortex is the outermost layer of gray matter that covers the two cerebral hemispheres. In humans, it is only about 3 mm thick, but it is estimated to contain 14–16 billion neuron cell bodies and to constitute more than half of all the CNS gray matter. The cerebral cortex communicates by projection fibers to other CNS structures, by commissural fibers to cerebral cortical regions of the opposite hemisphere, and by association fibers to cerebral cortical regions within the same hemisphere. The cytoarchitecture of the cerebral cortex varies across regions, and in many cases these areas are unique in their afferent and efferent connections, and their function. This forms the basis for cortical localization of function.

The cerebral cortex plays a role in sensory and motor functions as a component of sensory and motor circuits (primary and secondary areas). Lesions in these areas produce sensory or motor deficits, respectively. In humans, large regions of cerebral cortex are dedicated to integrating and processing diverse information from within and without the cortex and play an essential role in cognition and behavior. Lesions in these association (tertiary) areas result in cognitive and/or behavioral disorders. Additional knowledge has come from clinical and experimental studies of the effects of cortical disconnection in which the white matter connections (interhemispheric or intrahemispheric) between two areas of the cerebral cortex have been severed without damaging the cortical areas themselves, forming the basis of the cortical disconnection syndromes. Much has been learned about the localization of function within the cerebral cortex based on clinical and experimental studies of the effects of focal cortical lesions. It is worth noting that our knowledge of cortical function is constantly being supplemented, refined, and updated in a rapidly advancing field by experimental studies that use techniques not touched upon in this chapter, including newer methods of focal electrical stimulation and recording, single- and multicellular recording, optogenetics, and functional neuroimaging.

Additional Reading

1. Absher JR, Benson DF. Disconnection syndromes: an overview of Geschwind's contributions. *Neurology*. 1993;43(5):862–867. doi:10.1212/wnl.43.5.862
2. Benjamin S, MacGillivray L, Schildkrout B, Cohen-Oram A, Lauterbach MD, Levin LL. Six landmark case reports essential for neuropsychiatric literacy. *J Neuropsychiatry Clin Neurosci*. 2018;30(4):279–290. doi:10.1176/appi.neuropsych.18020027
3. Brodal P. The cerebral cortex: intrinsic organization and connections. In: Brodal P, *The central nervous system: structure and function*. 5th ed. Oxford University Press; 2016. https://oxfordmedicine.com/view/10.1093/med/9780190228958.001.0001/med-9780190228958-chapter-33. Accessed July 4, 2022.
4. Brown WS, Paul LK. The neuropsychological syndrome of agenesis of the corpus callosum. *J Int Neuropsychol Soc*. 2019;25(3):324–330. doi:10.1017/S135561771800111X
5. Dronkers NF, Plaisant O, Iba-Zizen MT, Cabanis EA. Paul Broca's historic cases: high resolution MR imaging of the brains of Leborgne and Lelong. *Brain*. 2007 May;130(Pt 5):1432–1441. doi: 10.1093/brain/awm042.
6. Gazzaniga M. *The bisected brain*. Appleton Press; 1970.
7. Gazzaniga MS, Bogen JE, Sperry RW. Some functional effects of sectioning the cerebral commissures in man. *Proc Natl Acad Sci USA*. 1962;48(10):1765–1769. doi:10.1073/pnas.48.10.1765
8. Gazzaniga MS, Bogen JE, Sperry RW. Observations on visual perception after disconnexion of the cerebral hemispheres in man. *Brain*. 1965;88(2):221–236. doi:10.1093/brain/88.2.221
9. Gazzaniga MS, Bogen JE, Sperry RW. Dyspraxia following division of the cerebral commissures. *Arch Neurol*. 1967;16(6):606–612. doi:10.1001/archneur.1967.00470240044005
10. Goldberg E. Gradiental approach to neocortical functional organization. *J Clin Exp Neuropsychol*. 1989;11(4):489–517. doi:10.1080/01688638908400909
11. Geschwind N, Kaplan E. A human cerebral deconnection syndrome: a preliminary report. 1962. *Neurology*. 1998;50(5). doi:10.1212/wnl.50.5.1201-a
12. Geschwind N. Disconnexion syndromes in animals and man: Part I [1965]. *Neuropsychol Rev*. 2010;20(2):128–157. doi:10.1007/s11065-010-9131-0
13. Geschwind N. Disconnexion syndromes in animals and man. II. *Brain*. 1965;88(3):585–644. doi:10.1093/brain/88.3.585

14. Ledoux JE, Risse GL, Springer SP, Wilson DH, Gazzaniga MS. Cognition and commissurotomy. *Brain*. 1977;100 Pt 1:87–104. doi:10.1093/brain/100.1.87
15. Luria AR. *Higher cortical functions in man*. Basic Books; 1966.
16. Mancuso L, Uddin LQ, Nani A, Costa T, Cauda F. Brain functional connectivity in individuals with callosotomy and agenesis of the corpus callosum: a systematic review. *Neurosci Biobehav Rev*. 2019;105:231–248. doi:10.1016/j.neubiorev.2019.07.004
17. Mesulam MM. Fifty years of disconnexion syndromes and the Geschwind legacy. *Brain*. 2015;138(Pt 9):2791–2799. doi:10.1093/brain/awv198
18. Sperry RW. Cerebral organization and behavior: the split brain behaves in many respects like two separate brains, providing new research possibilities. *Science*. 1961;133(3466):1749–1757. doi:10.1126/science.133.3466.1749
19. Sperry RW. Hemisphere deconnection and unity in conscious awareness. *Am Psychol*. 1968;23(10):723–733. doi:10.1037/h0026839
20. Seymour SE, Reuter-Lorenz PA, Gazzaniga MS. The disconnection syndrome: basic findings reaffirmed. *Brain*. 1994;117 (Pt 1):105–115. doi:10.1093/brain/117.1.105
21. Stone JL. Paul Broca and the first craniotomy based on cerebral localization. *J Neurosurg*. 1991;75(1):154–159. doi:10.3171/jns.1991.75.1.0154
22. Thiebaut de Schotten M, Dell'Acqua F, Ratiu P, et al. From Phineas Gage and Monsieur Leborgne to H.M.: revisiting disconnection syndromes. *Cereb Cortex*. 2015;25(12):4812–4827. doi:10.1093/cercor/bhv173
23. Vanderah TW, Gould D. Cerebral cortex. In: Vanderah TW, Gould D, *Nolte's the human brain: an introduction to its functional anatomy*. 8th ed., pp. 513–545. Elsevier; 2021.

14

The Occipital Lobes and Visual Processing

Introduction

The occipital lobes process only visual information, but visual cortex (i.e., cortical areas that process visual information) extends beyond the occipital lobes and into the inferior temporal and posterior parietal lobes. Cerebral lesions that affect vision do so in one of two general ways: (1) by affecting all aspects of vision within a localized region of the visual field, or (2) by affecting a specific modality of visual processing without regard to location within the visual field. Lesions of the visual pathways and/or primary visual cortex result in local areas of blindness within specific regions of the visual field. Lesions to cortical visual areas beyond primary visual cortex result in disorders of higher-order visual processing, often without regard to location within the visual field. This chapter discusses basic anatomy of the occipital lobes and visual pathways, visual field defects, disorders of higher-order visual processing, and visual disconnection syndromes.

Basic Anatomy of the Occipital Lobes

The occipital lobes are the posterior-most lobes of the brain. They lie above the tentorium, which separates them from the underlying cerebellum.

On the lateral surface of the brain, each occipital lobe is demarcated from the more anterior parietal and temporal lobes by two landmarks: the parieto-occipital sulcus and the pre-occipital notch, and an imaginary line connecting the lateral endpoint of the parieto-occipital sulcus to the pre-occipital notch (**the parieto-temporal lateral line**). The anatomy of the lateral surface of the occipital lobe is especially variable from person to person, with one to three lateral occipital sulci running through it in an anterior-posterior direction, resulting in two to four lateral occipital gyri (see Figure 13.1 in Chapter 13).

On the medial surface of the brain, the occipital lobe is demarcated from the more anterior parietal and temporal lobes by the same two landmarks: the parieto-occipital sulcus and the pre-occipital notch, and an imaginary line connecting the mesial endpoint of the parieto-occipital sulcus to the pre-occipital notch (**the parieto-temporal basal line**; see Figure 13.1 in Chapter 13). The **calcarine** (spur-shaped) **fissure** divides

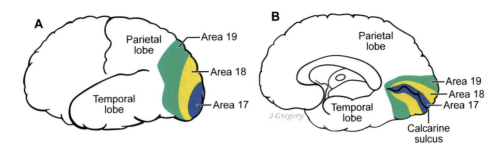

FIGURE 14.1. Occipital lobe Brodmann areas 17, 18, and 19. Area 17, primary visual cortex, surrounds the calcarine fissure. (A) Lateral view. (B) Medial view.

the occipital lobe into a dorsal **cuneus** (wedge) and a ventral **lingual gyrus** (*lingua*, "tongue"; see Figure 13.6 in Chapter 13). The lingual gyrus (along with the cingulate gyrus) is continuous with the parahippocampal gyrus of the temporal lobe. Lateral to the lingual gyrus is the **fusiform** (spindle-shaped) **gyrus**, which is also known as the **occipitotemporal gyrus** because it runs from the occipital lobe through the temporal lobe (see Figure 13.7 in Chapter 13). The occipitotemporal gyrus is bounded laterally by the occipitotemporal sulcus and medially by the collateral sulcus. The occipitotemporal gyrus is sometimes differentiated into two gyri, in which case they are named the medial and lateral occipitotemporal gyri.

The occipital lobe consists of several cytoarchitecturally distinct regions: Brodmann areas 17, 18, and 19 (Figure 14.1). Area 17 is located primarily within the medial walls of the hemispheres, buried within and surrounding the calcarine fissure, and extending over the occipital pole. Area 18 immediately surrounds area 17, and area 19 bounds area 18 and borders the temporal and parietal lobes.

Functionally, area 17 is **primary visual cortex**; it is therefore also referred to as V1. Areas 18 (V2) and 19 (V3, V4, V5) are secondary visual areas. Additional secondary visual areas extend beyond the boundary of the occipital lobe, into the posterior parietal lobes and inferior temporal lobes.

In general, the neocortex is a six-layered structure, with layer 4 being the principal target of thalamocortical afferents. Area 17 receives its main input from the lateral geniculate nucleus (LGN). In area 17, layer 4 is differentiated into four sublayers: 4A, 4B, 4Cα, and 4Cβ. A band of myelinated axons running horizontally in sublayer 4B forms a thin white stripe (striation), which can be seen with the naked eye when viewing tissue sections stained for cell bodies (e.g., cresyl violet). This band of fibers is known as the **stria of Gennari** and is an enlarged external band of Baillarger (see Chapter 13, "The Cerebral Cortex"). The transition from area 17 to area 18 is demarcated by the sudden disappearance of the layer 4 sublayers and this striation. The stria (stripe, band, line) of Gennari is unique to area 17; thus primary visual cortex is also known as **striate cortex**.

Area 18 is known as **parastriate cortex** (*para*, "beside/alongside"), while area 19 is known as **peristriate cortex** (*peri*, "around, about"). The term **extrastriate cortex** is a general term used to refer to secondary visual areas within and beyond the occipital lobe (i.e., visual cortex exclusive of striate cortex).

The Retina

Light is dually defined as photons and as electromagnetic radiation having wavelengths in the range of 380–760 nanometer range, the range that humans respond to (other species respond to different frequency ranges). Light energy from the environment is transduced into neural energy by photoreceptor cells in the **retina**, the light-sensitive sheet of neural tissue that lines the posterior inner surface of the eye. The cornea and lens together focus light to form images on the retina, a multilayered structure containing five types of neural cells arranged into five discrete layers: photoreceptors, horizontal cells, bipolar cells, amacrine cells, and retinal ganglion cells. The **photoreceptors** transduce (convert) light energy into graded electrical potentials; they carry out the first stage of retinal signal processing. The horizontal, bipolar, and amacrine cells carry out intermediate-stage signal processing. The retinal ganglion cells are the last stage of retinal signal processing; their axons make up the **optic nerve** and carry information to the brain. The optic nerve exits the retina at a location called the **optic disc**.

There are two main classes of photoreceptors: **rods**, which mediate night/low illumination (**scotopic**) vision; and **cones**, which mediate daytime (**photopic**), color, and fine detail vision. The density of rods and cones varies across the retina, with the cones predominant in a central region that is used to perform tasks that require fine spatial vision (e.g., reading) and color discrimination. Rods are far more numerous than cones and are situated mostly in the retinal periphery. They are specialized for catching photons and allow us to see in very dim illumination.

Located centrally in the retina lies the **macula**, a small, pigmented area with a high density of photoreceptors that is very important functionally because it provides detailed vision. Roughly in the center of the macula is the **fovea**, a small depression that contains the highest density of photoreceptors in the retina, and not surprisingly provides the highest level of visual acuity.

Assuming a Cartesian coordinate system with the origin at the fovea and the horizontal and vertical midlines as axes, the region above the horizontal midline is the **superior hemiretina**, the region inferior to the horizontal midline is the **inferior hemiretina**, the region medial to the vertical midline is the **nasal hemiretina** (toward the nose), and the region lateral to the vertical midline is the **temporal hemiretina** (toward the temporal bone). The retina can thus be divided into four quadrants: superior temporal, superior nasal, inferior nasal, and inferior temporal.

The Receptive Field

An important concept in sensory physiology is that of the **receptive field**. For any neuron in a sensory pathway, the receptive field is the region of the sensory surface that influences the activity of that neuron. Each retinal ganglion cell receives input, indirectly through the intervening cell layers, from a pool of photoreceptors (Figure 14.2). The region of the retina where those photoreceptors are located constitutes the receptive field of the retinal ganglion cell. This concept of receptive field applies to visually responsive neurons anywhere in the visual pathway.

Each optic nerve contains over one million fibers. This sounds like a lot until one considers the neural convergence that characterizes retinal processing: there are at least 12 times as many photoreceptors as ganglion cells in each eye. The degree of convergence from photoreceptors to retinal ganglion cells varies across the retina, with the photoreceptor to ganglion cell ratio in a one-to-one ratio in the fovea and increasing with distance from the fovea. Thus, retinal ganglion cells serving the central retina have much smaller receptive fields that those serving the peripheral retina.

The fovea conveys the most detailed information about the visual scene and is the region of the retina with the highest acuity/spatial resolution because the photoreceptor receptor density is greatest, and convergence is lowest there. **Foveal vision** provides our sharpest vision that is necessary for activities requiring visual detail. We make head and eye movements in order to foveate objects that we want to examine visually.

Pattern Processing within the Retina

A surprising fact about the retina is that the photoreceptor layer is the posterior-most layer; thus light must travel through the four non-photoreceptor layers before it strikes the photoreceptors. Indeed, the image impinging on the photoreceptor layer is of suboptimal optical quality, and among the important roles of retinal and higher visual processing is the improvement of that image and the construction of a visual world that appears sharp and of high resolution.

It is important to recognize that the retina is a complex nervous system structure that carries out a great deal of visual-information processing and is not merely a means of conveying optical information. Photoreceptors code and convey information about light levels and wavelengths that comprise the image falling onto the retina into a neural code that is analogous to pixel coding in a digital camera. The horizontal and amacrine interneurons are responsible for lateral inhibitory interactions (lateral inhibition), which serve to sharpen and enhance contrast in the neural image. Consequently, retinal ganglion cells do not simply signal the amount of light falling within their receptive field, as photoreceptors do. Their receptive fields are shaped with a concentric configuration in a pattern known as **center-surround antagonism**, in which the center and surround portions of the receptive field have opposite responses to light (Figure 14.3). In some retinal ganglion cells, light falling in the center part of the receptive field increases the cell-firing rate, while light falling in the ring-shaped surround diminishes the cell-firing rate (center-on, surround-off). Other retinal ganglion cells have the opposite profile, where light falling in the center part of the receptive field decreases the cell-firing rate, while light falling in the ring-shaped surround increases the cell-firing rate (center-off, surround-on). Consequently, retinal ganglion cells respond best to contrast; similar levels of light falling on the center and surround regions of the receptive field cancel each other out and result in no neural response. Thus, the nature of the signal processing changes at each level of processing, from the photoreceptors reporting on light level (rate of photo absorption) to the ganglion cells reporting on pattern contrasts.

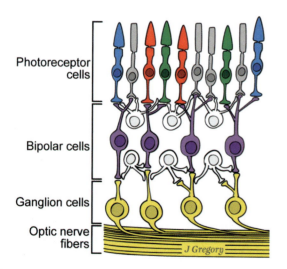

FIGURE 14.2 Each retinal ganglion cell receives input, indirectly through the intervening cell layers, from a pool of photoreceptors. Retinal ganglion cell axons make up the optic nerves.

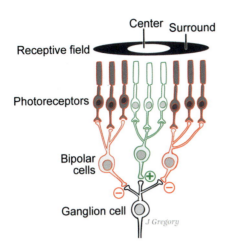

FIGURE 14.3. The retinal ganglion cell receptive fields are concentric, with center-surround antagonism, and therefore respond best to contrast between their receptive field center and surround. When a spot of light falls on the entire receptive field, inhibitory input from the surround cancels the excitatory input from the center, and the ganglion cell has no response. When a smaller spot of light falls only on the center of the receptive field, excitatory input from the center is not canceled by inhibitory input from the surround, and the ganglion cell has a strong response.

Early in the sensory pathway, neurons have simple response properties (e.g., light level and wavelength), while later in the pathway, neurons have more complex response properties (e.g., contrast). This principle applies to higher levels of processing within the visual pathways as well, where different properties of a visual pattern may be processed separately by different pathways and cortical regions; for example, the visual system has separate pathways for processing form, color, and movement. These principles also apply more broadly to sensory systems in general.

Color Coding

Three different cone types are responsible for color vision. They differ in their sensitivity to different wavelengths of light due to differences in the light-absorbing molecule (cone opsin) they contain. Each cone opsin absorbs light best in a different part of the visual spectrum, making cones selectively sensitive to short-, medium-, or long-wavelength light (which alone appear blue, green, or red, respectively). Genetic defects that result in anomalies in one or more of the three types of cones result in **color vision deficiencies**.

As with pattern processing, substantial processing of color perception takes place in the retina. The three-opsin code of photoreceptors gets translated into an opponent-color system at the level of the retinal ganglion cells, where the ganglion cells respond specifically to pairs of primary colors, either red-green or blue-yellow, in a center-surround fashion (e.g., excited by red and inhibited by green in the center, and excited by green and inhibited by red in the surround). Blue-yellow ganglion cells respond to yellow light because that portion of their receptive field receives input from both red and green cones.

The Visual Field

The **visual field** is the field of view of the external world seen by a single eye, without eye or head movements. It is measured using **visual perimetry**, which assesses basic visibility of small spots of light at a sample of positions in space. Clinical perimetric visual fields are expressed in visual angular units, rather than in retinal coordinates. The monocular visual field is the field of view of an individual eye. The binocular visual field is the field of view when both eyes are open. In humans, since both of our eyes are forward facing (as opposed to sideways facing, like in rabbits), the monocular visual fields of each eye overlap, and most points in the visual space are seen by both eyes. It is important to recognize that the terms *visual field* and *receptive field* are not synonymous. The receptive field is a property of an individual sensory neuron. It is the particular region of the sensory surface (e.g., the retina) and its associated region of the sensory space (e.g., that portion of visual field) in which a stimulus will trigger the firing of that neuron. The receptive field is, in a sense, the visual field of an individual neuron.

The monocular visual field can be divided into superior and inferior hemifields, temporal and nasal hemifields, and four quadrants: superior temporal, superior nasal, inferior nasal, and inferior temporal (Figure 14.4). The hemifields and quadrants of the visual field are defined with respect

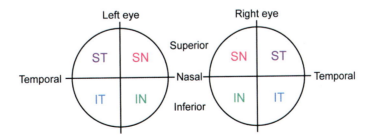

FIGURE 14.4. The visual field of each eye can be divided into superior and inferior hemifields, and temporal and nasal hemifields, resulting in four quadrants: superior temporal (ST), superior nasal (SN), inferior nasal (IN), and inferior temporal (IT). The hemifields and quadrants of the visual field are defined with respect to the point of visual fixation; the vertical and horizontal meridians intersect at the fixation point.

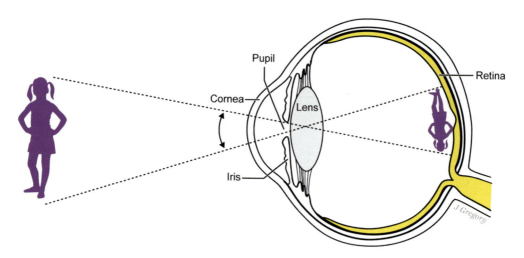

FIGURE 14.5. The optical system of the eye inverts the visual image upon the retina, so that the image that falls on the retina is upside down, and right and left are reversed.

to the point of visual fixation, the point in visual space on which one's visual gaze is centered and with which the fovea is aligned, and the vertical and horizontal meridians that intersect at the fixation point.

The optics of the eye are such that images of objects in space are rotated by 180 degrees when projected on the retina, so that retinal imagery is upside down, and right/left reversed with respect to the external world (Figure 14.5). Consequently, the relationship between the visual field and the retina is such that: (1) light from the superior half of the visual field is projected onto the inferior half of each retina; (2) light from the inferior half of the visual field is projected onto the superior half of the retina; (3) light from the temporal hemifields is projected onto the nasal hemiretinas; and (4) light from the nasal hemifields is projected onto the temporal hemiretinas. In terms of quadrants: (1) objects in the superior temporal field are imaged on the inferior nasal retina; (2) objects in the inferior temporal field are imaged on the superior nasal retina; (3) objects in the superior nasal field are imaged on the inferior temporal retina; and (4) objects in the inferior nasal field are imaged on the superior temporal retina.

The Visual Pathways

Understanding the anatomy of the visual pathways allows us to interpret the patterns of field defects and localize lesions along the pathways from the eye to the visual cortex.

From the Retina to the Lateral Geniculate Nucleus

The axons of the retinal ganglion cells make up the **optic nerves** and carry the retinal output into the brain (Figure 14.6).

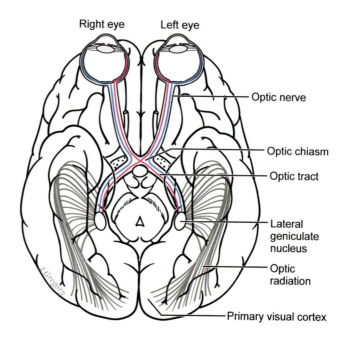

FIGURE 14.6. Bottom-up view of the visual pathways. The optic nerves merge at the optic chiasm, where fibers from the temporal hemiretinas remain ipsilateral and fibers from the nasal hemiretinas decussate to join the fibers from the temporal hemiretina of the other eye as they form the optic tract. Each optic tract projects to a lateral geniculate nucleus, which in turn projects to the primary visual cortex via the optic radiations.

Each optic nerve contains over one million fibers; as noted earlier, there are at least 12 times as many photoreceptors as ganglion cells in each eye. The optic nerves of the two eyes merge at the **optic chiasm**, an X-shaped structure where fibers from the nasal hemiretinas decussate and join the fibers from the temporal hemiretina of the other eye, forming two **optic tracts**. The optic tracts wrap around the cerebral peduncles of the midbrain and then enter the brain.

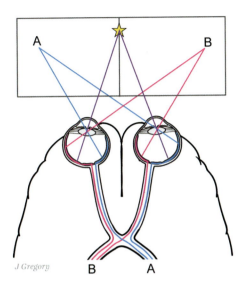

FIGURE 14.7. Top-down view. (A) Images of objects in the left visual field are projected onto the right hemiretinas (nasal hemiretina of the left eye and temporal hemiretina of the right eye) and transmitted to the right hemisphere. (B) Images of objects in the right visual field are projected onto the left hemiretinas (nasal hemiretina of the right eye and temporal hemiretina of the left eye) and transmitted to the left hemisphere.

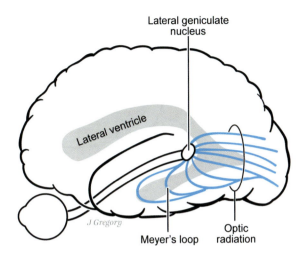

FIGURE 14.8. The optic radiations project from the lateral geniculate nucleus to the primary visual cortex in each occipital lobe. Fibers representing the superior visual fields are carried in the inferior portion of each radiation (Meyer's loop) and fibers representing the inferior visual fields are carried in the superior portion of each radiation.

Information from the temporal hemiretina is transmitted to the ipsilateral hemisphere, while information from the nasal hemiretina is transmitted to the contralateral hemisphere. Consequently, images of objects in the right visual field, which are projected onto the left hemiretinas (i.e., the left eye's temporal hemiretina and the right eye's nasal hemiretina), are transmitted to the left hemisphere (Figure 14.7). Images of objects in the left visual field, which are projected onto the right hemiretinas (i.e., the right eye's temporal hemiretina and the left eye's nasal hemiretina), are transmitted to the right hemisphere.

The bulk (90%) of optic tract fibers synapse in the **lateral geniculate nucleus** (LGN), the principal subcortical structure that carries visual information to the cerebral cortex. The remaining 10% form the extrageniculate pathways (see "The Extrageniculate Pathways," below).

From the Lateral Geniculate Nucleus to the Primary Visual Cortex

From the LGN, **geniculocalcarine fibers** form the **optic radiations** and project to primary visual cortex (V1, BA17). Fibers from the inferior half of the retinas (representing the superior visual fields) are carried in the inferior portion of the geniculocalcarine tract, which swings rostrally in a broad arc over the temporal horn of the lateral ventricle in **Meyer's loop** before turning caudally to reach the occipital pole (Figure 14.8). These fibers terminate in the inferior bank of the calcarine fissure. Fibers from the superior half of the retinas (representing the inferior visual fields) are carried in the superior portion of the geniculocalcarine tract and terminate in the superior bank of the calcarine fissure.

The Retinotopic Map

The spatial relationships among retinal ganglion cells are maintained throughout the visual pathway, resulting in topographic representations of the retina (**retinotopic map**) within the LGN and V1. Retinal ganglion cells project to the LGN in an orderly manner; the entire visual hemifield is mapped onto each of the six layers of the LGN, and the maps of each layer are in register. Within V1, the fovea is represented most posteriorly, and more peripheral regions of the retina are represented progressively more anteriorly (Figure 14.9). The map, however, is distorted. The amount of cortical area devoted to each unit of the sensory surface is not uniform. This distortion reflects the density of sensory input across the sensory surface, analogous to the somatosensory homunculus. Thus, the representation of the macula is disproportionately large, occupying much of the occipital pole.

Secondary Visual Cortex

Beyond V1, there are numerous secondary visual areas that extend throughout the occipital lobe and beyond. In monkeys, over 40 secondary visual areas have been identified.

There are two broad streams of projections from primary visual cortex to extrastriate cortex, identified by **Leslie Ungerleider** (1946–2020) and **Mortimer Mishkin** (1926–2021) (Figure 14.10). The ventral stream is directed

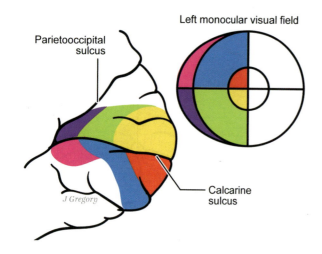

FIGURE 14.9. The retinotopic map within V1. The fovea is represented most posteriorly, and more peripheral regions of the retina are represented progressively more anteriorly. The map is distorted according to the density of sensory input across the retina, so that the representation of the macula is disproportionately large, while representation of the peripheral retina is disproportionately small.

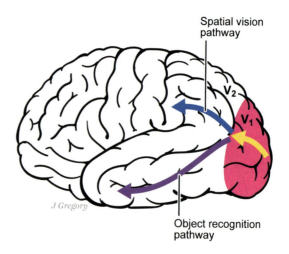

FIGURE 14.10. The visual areas beyond the striate cortex are broadly organized into two pathways: a ventral pathway specialized for object recognition, and a dorsal pathway specialized for spatial vision.

ventrally into the inferior temporal lobe and is informally termed the "what" (object recognition) pathway. The dorsal stream is directed dorsally into the posterior parietal lobe and is known as the "where" (object localization and movement) pathway.

The Extrageniculate Pathways

Approximately 10% of the fibers carried in the optic tracts project to structures outside of the LGN. These projections are collectively referred to as the **extrageniculate pathways**. Their destinations are the superior colliculi, the pretectal nuclei, and the suprachiasmatic nucleus (a midline hypothalamic nucleus that lies above the optic chiasm). Optic nerve projections to the superior colliculi play a role in reflex saccadic eye movements and orienting head and body movements to visual stimuli by way of the tectobulbar and tectospinal pathways. Optic nerve projections to the pretectal nuclei form the afferent link for the pupillary light reflex. Optic nerve projections to the suprachiasmatic nucleus of the hypothalamus (via the retinohypothalamic tract) inform the master circadian rhythm generator about the environmental light-dark cycle (day length) and coordinate the "body clock" to the "peripheral clock" for regulating processes that follow a circadian rhythm, such as the sleep-wake cycle and the secretion of various hormones. For example, in diurnal mammals, melatonin is released from the pineal gland in a circadian rhythm; as its release increases through the evening, so does the propensity for sleep. The rhythmic release of melatonin is regulated by the central circadian rhythm generator—the suprachiasmatic nucleus.

Visual Field Defects and Cerebral Blindness

A variety of patterns of visual field defects are produced by lesions at various levels in the visual pathways, up to and including primary visual cortex. These patterns reveal the way that the visual world is projected onto the primary visual cortex (Figure 14.11).

Unilateral optic nerve section produces **monocular blindness**, blindness in one eye. Selective destruction of decussating fibers within the optic chiasm produces bitemporal hemianopia. **Hemianopia**, also known as **hemianopsia**, is blindness in half the visual field (*hemi* meaning half, *anopia/anopsia* meaning without vision). In **bitemporal hemianopia** the temporal fields of both eyes are blind. These field defects are **heteronymous**, meaning that they stand in opposite relations; there is a loss of vision in opposite halves of the visual field. Since the temporal field of the right eye conveys information from the right half of space, and the temporal field of the left eye conveys information from the left half of space, there is no overlap in the areas of blindness in the two eyes' visual fields. Since the intact hemifields do not overlap, the functional consequence of bitemporal hemianopia is the elimination of stereoscopic depth cues and the consequent reduction of depth perception. This kind of lesion is most often produced by pituitary tumors that compress the chiasm, and surgical removal of pituitary tumors.

Complete destruction of one optic tract, one LGN, one optic radiation, or primary visual cortex (V1) of one hemisphere produces a loss of vision in the entire contralateral visual hemifield of both eyes. This pattern of field defect is called **homonymous hemianopia**. The term *homonymous*

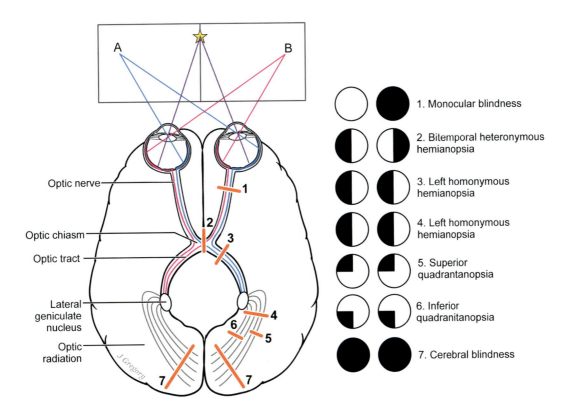

FIGURE 14.11. Visual field defects, top-down view. Unilateral optic nerve section (1) produces monocular blindness. Selective destruction of decussating fibers within the optic chiasm (2) produces bitemporal hemianopsia. Complete destruction of one optic tract (3), one LGN, one optic radiation (4), or striate cortex of one hemisphere produces a contralateral homonymous hemianopia. Unilateral lesions in the lower optic radiation (Meyer's loop) or the inferior bank of the calcarine cortex result in a contralateral superior quadrantanopia (5). Unilateral lesions in the upper optic radiation result in a contralateral inferior quadrantanopia (6). Bilateral lesions of the primary visual cortex or the optic radiations result in cerebral blindness (7).

refers to the fact that both eyes are blind to the same region of the visual field. All homonymous hemianopias are retrochiasmal (they occur in the visual pathway beyond the chiasm). Homonymous hemianopia can be due to either subcortical or cortical lesions.

A partial lesion in the optic radiation or V1 can produce a quadrantic field defect known as **homonymous quadrantanopia**, a loss of vision in a quadrant of the visual field. Since information from the superior half of the visual field is represented in the inferior portions of the visual pathways, unilateral lesions in the lower optic radiation (Meyer's loop) or the inferior bank of the calcarine cortex result in a contralateral superior quadrantanopia, also known as the "pie in the sky" field defect. Such defects occur with unilateral lesions of the temporal lobe, and with some temporal lobectomies and middle cerebral artery (MCA) inferior division infarcts. Unilateral lesions in the upper optic radiation (which passes through the parietal lobe) result in a contralateral inferior quadrantanopia, also known as the "pie on the floor" visual field defect. A common cause of inferior quadrantanopia is infarction in the territory of the MCA superior division. It is more difficult to document inferior quadrantanopia than superior quadrantanopia, however, because it is often associated with hemi-neglect when due to right hemisphere lesions, and with aphasia when due to left hemisphere lesions.

Bilateral lesions of the primary visual cortex or the optic radiations result in **cerebral blindness** (blindness due to a lesion of the visual cortex or underlying white matter). The term *cortical blindness* is also used, but it is inaccurate because often the lesions are in the subcortical white matter. Cerebral blindness usually occurs suddenly from occlusion of the posterior cerebral artery (PCA). Either one or both PCAs may be affected, since both PCAs originate from a single basilar artery. Vascular occlusions within the PCA may cause homonymous hemianopia without affecting central vision subserved by the macula because of collateral circulation from the MCA. This produces macular sparing, where a central zone of vision is preserved across the midline. Homonymous hemianopia and complete cerebral blindness due to PCA stroke are often associated with anterograde amnesia because the PCAs also supply the posterior two-thirds of the hippocampi. Primary visual cortex lesions are also caused by tumors, hemorrhage, infection, and trauma to the occipital poles. Cerebral blindness also may be temporary due to transient ischemia, migraine, seizures, or vasospasm during vertebrobasilar arteriography.

Small lesions in the occipital lobe can produce smaller **scotomas** (localized visual field defects), although most scotomas are caused by eye disorders, including macular degeneration, glaucoma, diabetic retinopathy, optic neuropathy, and optic atrophy.

The optic disc contains no photoreceptors and therefore is insensitive to light, producing a natural scotoma known as the **blind spot**. We do not normally notice its presence even when viewing with one eye, and it is also compensated for by functional retina in the other eye during most binocular viewing. Though scotomas are often portrayed as black spots, they are not experienced as such. For the most part, they are not noticed at all unless they interfere with the performance of a visual task that requires the scotomatous visual field area.

Disorders of Higher-Order Visual Processing: Visual Agnosia

The Concept of Visual Agnosia

Agnosia is a relatively rare disorder characterized by a failure of recognition that is modality-specific and not due to sensory defects, general mental deterioration, or disturbances of level of consciousness, attention, or language. Agnosias occur for the visual, auditory, and tactile modalities.

Visual agnosia is a specific impairment in the ability to recognize visually presented objects, despite normal basic visual functions such as acuity, brightness discrimination, depth perception, and color vision. Those with visual agnosia fail to recognize material presented visually but are successful when allowed to handle objects or to hear their characteristic sound.

The first demonstration of visual agnosia was by **Hermann Munk** (1839–1912), who reported in 1881 that dogs with partial bilateral occipital lobe excisions avoided obstacles placed in their path but failed to recognize/react appropriately to objects that previously frightened or attracted them. Munk called this phenomenon "mindblindness" and explained it as a failure to relate current perceptions to past experiences and an inability to grasp the meaning of visually perceived stimuli. Other terms that have historically been used to describe this phenomenon are *asymbolia* and *imperception*. The term *agnosia* (Greek, "without knowledge") was introduced in 1891 by Sigmund Freud (1856–1939) to describe a disruption in the relationship between things themselves and a person's concept of these things. He contrasted this with aphasia, in which there is a disruption in the relationship between objects and the words to signify them.

Heinrich Lissauer (1861–1891) was the first to distinguish two forms of visual object recognition disorder: apperceptive and associative. This distinction was based on early theories that object perception occurs in two stages. The first stage, apperception, consists of building up a percept from elementary sensory impressions—a piecing together of visual attributes into a whole. The second stage, association, consists of imparting meaning to the content of the perception by linking it to stored semantic memory (i.e., general world knowledge). A defect in the first mechanism results in an apperceptive disorder, while a defect in the second mechanism results in an associative disorder.

While it has become clear that visual object recognition occurs by a process that is far more complex than this two-stage model, the apperceptive–associative distinction remains useful clinically, as **apperceptive agnosia** is defined a failure of recognition due to impaired perception, while **associative agnosia** is defined a failure of recognition due to impaired association of the intact perception with semantic knowledge. The definition of agnosia put forth by **Hans-Lukas Teuber** (1916–1977), one of the founders of neuropsychology, is an object recognition disorder characterized by "a normal percept that has somehow been stripped of its meaning." Apperceptive visual agnosia therefore is not a true agnosia, as it does not fit Teuber's classic definition of agnosia; this terminology nevertheless remains.

Apperceptive Visual Agnosia

Apperceptive visual agnosia is defined as a failure of recognition due to impaired perception. Elementary visual functions that we typically associate with sensation (as contrasted with perception), such as visual acuity, brightness, depth, movement, and color vision, are normal or near normal, as are visual fields, visual scanning, and the ability to maintain visual fixation. Despite these capabilities, persons with apperceptive visual agnosia cannot identify objects. Further examination reveals that they also cannot point to objects named by the examiner, draw objects, describe the formal features of objects or patterns, match to sample, or discriminate even simple shapes.

The fundamental deficit in apperceptive visual agnosia is an inability to process features to the point of developing a percept of the overall structure of an object. There is a breakdown at the stage where sensory features of the stimulus are processed to achieve a structural representation. Thus, the core deficit is specifically in shape (form) identification and discrimination. Those with apperceptive visual agnosia cannot make use of shape information, and they guess at object identity using color and texture cues. They can navigate their environment, report that they can see, but appear to be blind because they cannot recognize, describe, or match objects perceived. The visual system is unable to construct a sensory representation that is adequate for identification.

The lesions associated with apperceptive visual agnosia are typically diffuse, affecting the posterior cerebral hemispheres (occipital, parietal, posterior temporal) bilaterally. Such injuries are most commonly the result of anoxia (e.g., due to carbon monoxide poisoning, cardiac arrest) or bilateral posterior hemispheric strokes. Apperceptive visual agnosia may occur as a phase of recovery from cerebral blindness after these types of injury. Apperceptive visual agnosia is also seen in the context of neurodegenerative posterior cortical atrophy (see Chapter 15, "The Parietal Lobes and Associated Disorders").

Disorders of Higher-Order Visual Processing: Lesions of the Ventral Stream

The ventral stream of information processing is specialized for form and color analysis. Lesions in the inferior occipitotemporal visual association cortex, usually due to PCA infarcts, result in visual object agnosia, prosopagnosia, and achromatopsia.

Associative Visual Agnosia/Visual Object Agnosia

Visual object agnosia (**associative visual agnosia**) is the failure to recognize pictures or objects presented via the visual modality, despite adequate demonstrations that the picture or object has been perceived with sufficient resolution or detail to allow for identification, and with preserved recognition through other modalities.

Intact basic visual perception is evidenced by the person's preserved abilities. They can give detailed descriptions about shape, size, contour, position, and number. They can match to sample items that they fail to identify visually. They can copy complex figures that they cannot identify (although they usually use a slow, laborious, line-by-line strategy). Yet they cannot identify objects presented visually. They cannot describe what the object does or how it works, either verbally or by gesture. They also cannot point to objects named by the examiner. They cannot sort objects and pictures into functional categories. They cannot match real objects with object representations (e.g., photographs, line drawings), and cannot match morphologically different representations of the same object (e.g., matching a line drawing of a wristwatch with a photograph of a wristwatch). This disturbance fits the narrow definition of agnosia defined by Teuber as "a perception stripped of its meaning."

The gradient of difficulty in visual object recognition is especially steep in those with associative visual agnosia. Common objects are recognized more easily than uncommon objects. Viewing an item in use or otherwise in context aids recognition, whereas viewing an item in an unusual use or out of context hinders identification. Partially covering or obscuring an item hinders identification. Viewing an item from an unusual perspective hinders identification. Real objects are better recognized than photographs, and photographs are better recognized than line drawings. With recovery from visual object agnosia, a selective difficulty identifying line drawings may be the only residual disturbance. Object identification errors usually consist of objects that are similar in shape to the target (e.g., pen for asparagus spear), indicating preserved ability to make use of shape information.

It is not uncommon for persons with visual object agnosia to function acceptably in everyday life but fail when tested formally in the controlled conditions of a laboratory or clinical examination. Of course, context cues and redundancy from other senses abound in the familiar environments of everyday life and may mask the difficulty encountered in the impoverished clinical testing situation.

When a patient fails to name to visual confrontation, the underlying deficit may be either visual agnosia or anomia. In contrast to persons with visual object agnosia, those with anomia indicate visual recognition by means other than naming, such as describing and gesturing the use of the object. They perform well on nonverbal tests of visual object recognition, such as matching functionally related objects. Object identification does not improve when the material is presented through another sensory modality.

The recognition defect in visual object agnosia appears to result from a breakdown in the ability of the neural structures mediating visual perception to activate neural structures responsible for the patient's database of stored knowledge (semantic memory) about objects (i.e., their functional, contextual, and categorical properties) that permit their identification. Semantic memory itself, however, is preserved.

Postmortem and neuroimaging studies have localized the lesions causing visual object agnosia to the fiber pathways lying beneath the inferior temporal-occipital junction, particularly affecting the inferior longitudinal fasciculus connecting the occipital association cortex of the fusiform gyrus to temporal lobe structures. The lesions are usually due to bilateral infarcts in the territories of the PCAs, but visual object agnosia may also occur with unilateral damage. The lesions producing visual object agnosia are more localized than those producing visual apperceptive agnosia.

Persons with visual object agnosia often also have prosopagnosia and achromatopsia.

Prosopagnosia

Prosopagnosia is the inability to recognize familiar faces. Persons with this disorder can often recognize others by nonfacial visual cues or nonvisual cues such as voice. The inability is not attributable to other deficits in perception, language, or memory, or a general confusion (i.e., delirium, dementia). The term *prosopagnosia* (Greek, *prosopon*, "face") was coined by the German neurologist **Joachim Bodamer** (1910–1985), who published the first case series in 1947.

Unlike persons with visual object agnosia, those with prosopagnosia recognize faces as faces. They can identify parts of the face. They can identify emotional expressions (unlike those with bilateral amygdala lesions, who can recognize and learn new faces but cannot recognize facial expressions of emotions). Persons with prosopagnosia are able to discriminate and match faces normally. Their defect is in identifying whose face they are viewing. It is not the identity of the person that is lost, but the connection between a particular face and a particular identity. In some cases, the impairment of facial recognition is so severe that persons with prosopagnosia are unable to recognize their own face in the mirror. They often describe faces as all looking similar, unattractive, or having lost their individuality. They learn to identify others using extra-facial cues, including clothing, length of hair, height, stature, voice, and gait.

This disorder is distinguished from **Capgras syndrome**, a delusional disturbance in which the person believes that familiar persons have been replaced by imposters.

Prosopagnosia often coexists with visual object agnosia. It is not, however, attributable to a general failure of visual recognition. It is a distinct clinical entity that is separate from visual object agnosia, as evidenced by the fact that there is a **double dissociation** between prosopagnosia and visual object agnosia. There are many descriptions of prosopagnosic patients who can recognize objects, and visual object agnosia is not always associated with prosopagnosia.

The fact that prosopagnosia can occur as an isolated, pure deficit is consistent with a dedicated cortical area for face recognition, and the **fusiform face area** (in the fusiform gyrus), first described by Justine Sergent (1950–1994) in 1992, seems to be exactly that. Positron emission tomography (PET) and functional magnetic resonance imaging (fMRI) studies have confirmed its existence. Additionally, facial hallucinations (seeing faces where there are none) are associated primarily with increased neurophysiological activity in the fusiform face area. That there exists a dedicated cortical area for face recognition is further supported by the rare clinical phenomenon of **prosopometamorphopsia**, a disorder characterized by distorted visual perception that occurs selectively for faces. The distortions may be described as cartoonish, contorted with displaced features, ugly with prominent eyes and teeth, or morphing into animal faces (e.g., dragon faces or fish heads). The distorted perceptions may involve the whole face or only one side of the face. Prosopometamorphopsia has been reported with stroke, epilepsy, tumor, traumatic brain injury, migraine, and hallucinogens. It has even been induced experimentally by electrical stimulation.

Most patients with persistent prosopagnosia have bilateral lesions affecting the visual association cortices at the occipitotemporal junctions (lingual and fusiform gyri), which are most often caused by embolic infarction in the PCA territory. Associated deficits include achromatopsia and visual object agnosia; when prosopagnosia occurs in isolation, the lesions are less extensive and are confined to the fusiform gyrus.

Acquired Achromatopsia

Acquired achromatopsia is a rare disorder of color vision involving the loss of color perception in part or all of the visual field due to disease or injury of the CNS. It may affect the full visual field, a hemifield (**hemiachromatopsia**), or a quadrant (**quadrant/quadrantic achromatopsia**). Acquired achromatopsia is distinct from congenital (hereditary) achromatopsia in which there is an absence of color vision due to a lack or dysfunction of cone photoreceptors in the retina.

Persons with acquired achromatopsia generally describe their visual world as "in black and white," "all grey," "washed out," or "dirty." They cannot accurately name colors seen or point to named colors. Visual acuity and the ability to distinguish subtle differences in form and depth are well preserved. Visual form perception, depth perception, and motion perception are normal.

Achromatopsic patients perform poorly on strictly visual tasks of color perception, such as pseudoisochromatic plates (Ishihara plates, Munsell Farnsworth 100-Hue Test), hue discrimination, color matching, color sorting, and matching colors to uncolored line drawings of objects. They do, however, perform well on verbal-verbal color-association tasks, such as naming the colors of common objects that are out of sight ("What color is blood?", "Name three blue things"), indicating that their semantic knowledge about colors remains unaffected.

Cerebral achromatopsia results from lesions in the occipitotemporal visual association cortex of the lingual and fusiform gyri. The etiology is usually an embolic stroke in the PCA territory. Functional imaging studies in humans performing tasks requiring inspection of or searching for colored stimuli show activation in the lingual and fusiform gyri. One area in each hemisphere controls color processing for the entire contralateral hemifield. Full-field achromatopsia results from bilateral occipitotemporal lesions; it is often associated with visual object agnosia and prosopagnosia. The purest form of achromatopsia occurs in the left hemifield, with lesions in the middle third of the lingual gyrus (affecting area V4) or the subjacent white matter behind the posterior tip of the lateral ventricle of the right hemisphere, with sparing of the striate cortex and optic radiations. Right-sided hemiachromatopsia (due to left hemisphere lesions) is often associated with pure alexia (see "Pure Alexia," below). Achromatopsia must be distinguished from color anomia, a selective deficit in the ability to identify colors by name (see "Color Anomia," below).

Disorders of Higher-Order Visual Processing: Lesions of the Dorsal Stream

The dorsal stream is specialized for processing information about the location and depth (three-dimensional organization) of objects, object motion, and the visual control of action. Lesions of the posterior parietal lobes result in disorders of higher-order visual processing. These include the visuospatial processing disorders astereopsis and akinetopsia. They also include visuomotor disorders that selectively affect movements under visual guidance, namely optic ataxia and oculomotor apraxia. Visuomotor disorders typically occur with biparietal lesions affecting the posterior parietal lobes, which are often due to MCA-PCA watershed territory infarcts.

Astereopsis

Depth perception results from several sources of information, including stereopsis and monocular depth cues. Stereopsis allows for the perception of depth based on differences in location of the retinal image elements in the two eyes (**binocular disparity**). Monocular depth cues allow for the ability to appreciate the relative location of objects using one eye; many of these cues are used in two-dimensional representational drawings and paintings to portray depth and distance. These include occlusion, relative size, relative height, linear perspective, texture gradient, atmospheric perspective, and shadows.

Astereopsis (**stereoblindness**) is the absence of stereoscopic depth perception in persons with otherwise normal vision in both eyes. Few cases of this disorder acquired after lesions of the visual cortex have been reported. These defects may be revealed through testing random dot stereograms, which provide stereoscopic cues to depth embedded in images lacking global object form information.

Akinetopsia

Akinetopsia (**movement agnosia**, **motion blindness**) is a very rare disorder involving the selective loss of visual movement perception; the ability to perceive movement of auditory and tactile stimuli is preserved. It is characterized by an inability to distinguish between moving and stationary objects. Other visual perceptual abilities are intact. For those with motion blindness, smooth movements of objects appear as discontinuous freeze-frame images, as if viewed through a strobe light; moving objects are seen as a series of stationary images. Akinetopsia results in visuomotor deficits for tasks that depend on visual motion perception, such as reaching for and catching objects.

Akinetopsia was first reported by **Josef Zihl** in 1983. The 43-year-old woman, referred to in the literature as L.M., had a thrombosis of the superior sagittal sinus, which resulted in bilateral, symmetrical lesions in extrastriate visual cortex. She had mild aphasia and acalculia, but her most severe impairment was in motion perception. When she poured liquid into a cup, the "fluid appeared to be frozen, like a glacier." Because she was unable to perceive the movement of the fluid within the cup as it gradually filled, she could not stop pouring at the right time and would pour to the point of overflowing the cup. Crossing the street was very difficult. "When I am looking at the car, first it seems far away. But then when I want to cross the road, suddenly the car is very near." In addition, L.M. complained of trouble following conversations because of difficulty perceiving lip movements and changing facial expressions. Very few cases of akinetopsia have since been reported; thus the disorder is extremely rare.

The selective disturbance of visual motion is due to the anatomical separation of visual motion processing from other visual functions. This occurs after damage in the parieto-occipito-temporal junction involving area V5 (single-cell recording studies in macaque monkeys have shown that all neurons in this area are motion sensitive and most are directionally selective).

There have been cases of patients with tumors in the right posterior parietal lobe (presumably in V5) that caused seizures and ictal illusions of visual movement. One woman described the road as "waving" during her seizures. The illusion of visual movement due to seizure activity is termed **epileptic kinetopsia**.

Optic Ataxia

Optic ataxia (**visuomotor ataxia**) is the inability to direct movements of the limbs to points in space using visual guidance (i.e., pointing, reaching, and grasping). Persons with optic ataxia make large pointing and reaching errors. Movements under proprioceptive guidance without visual cues (e.g., pointing to parts of one's own body, buttoning buttons on one's own clothing) are performed accurately, but visually guided movements (e.g., pointing to a body part on someone else, buttoning buttons on a child's clothing) are not. There is no general ataxia due to cerebellar or proprioceptive deficits; optic ataxia is present only under visual guidance.

Isolated optic ataxia is a common symptom of superior posterior parietal lesions. It can occur after unilateral lesions and is then limited to the contralesional hemifield with a directional bias, such that reaching and pointing deviate to the side of the lesion. Optic ataxia is conceptualized as a disruption in the transformation of retinotopic locations into body-referenced coordinates that serve as a framework for motor guidance.

Oculomotor Apraxia

Oculomotor apraxia (**gaze apraxia**) is a visuomotor disorder in which pursuit and visually guided saccades are impaired, but saccades that are not dependent on vision (i.e., saccades to command, saccades to remembered targets, saccades to sounds, and spontaneous saccades) are preserved. Gaze wanders aimlessly, and targets are found by chance. Even when persons with oculomotor apraxia are told where to look to see an object, they have difficulty directing foveal vision to that spot. Oculomotor reflexes are normal, such as the **menace reflex** (reflex saccades to suddenly appearing eccentric visual targets), the **oculocephalic reflex**, and **optokinetic nystagmus** (see Chapter 30, "The Neurological Examination"). Oculomotor apraxia is not a true apraxia. Apraxia is defined as an inability to perform learned or skilled motor actions to command; saccade initiation is neither a learned nor a skilled action.

Isolated oculomotor apraxia is rare. It results from bilateral posterior parietal lobe lesions affecting the **parietal eye fields** that surround the posterior, medial segment of the intraparietal sulcus. The parietal eye fields generate both pursuit and visually guided saccades; the **frontal eye fields** trigger voluntary nonvisually guided eye movements.

Visual Disconnection Syndromes

Disconnection between visual cortex and language cortex underlies the visual disconnection syndromes of pure alexia and color anomia. Visual disconnection syndromes also occur with lesions in the splenium, which prevent information from the left visual field/right hemisphere from accessing left temporal lobe language structures (see Chapter 13, "The Cerebral Cortex"). Such lesions produce an inability to name objects seen (**visual anomia**, optic aphasia) within the left visual field, as well as inability to read words presented in the left visual field (**hemi-alexia**). These conditions are all due to a visual-verbal disconnection.

Pure Alexia

Alexia is an acquired disorder of reading; **agraphia** is an acquired disorder of writing. Alexia is usually a component of receptive aphasia (aphasic alexia), and usually is associated with agraphia.

Alexia does, however, sometimes occur in pure form, without agraphia and without other aphasic symptoms. **Alexia without agraphia** is also known as **pure alexia**, pure word blindness, agnosic alexia, and word form alexia. **Joseph Jules Déjerine** (1849–1917) first described the case of a patient with a pure reading disorder that he called *verbal blindness*. Unlike the aphasic patient who is alexic, the patient with pure alexia can write spontaneously and can write to dictation (although they cannot read what they wrote). Also, unlike the aphasic patient who is alexic, the patient with pure alexia cannot copy printed material. In mild forms, the patient with pure alexia may have difficulty only with long words.

Pure alexia is a disconnection syndrome. It occurs with lesions in the left ventromedial occipital region that include the splenium of the corpus callosum. Such a lesion disconnects the transfer of visual information from the right hemisphere (left visual field) to the left hemisphere by commissural fibers and disconnects the transfer of visual information from the left hemisphere (right visual field) to left hemisphere language regions by intrahemispheric fibers (Figure 14.12). Pure alexia typically occurs with infarction in the PCA territory of the language-dominant hemisphere, leaving regions concerned with the nonvisual aspects of language intact. It also occurs with surgical removal of the occipital lobe in the language-dominant hemisphere. Such lesions also involve the primary visual cortex; thus right homonymous hemianopsia accompanies the pure alexia. Lesions that involve the splenium but not the intrahemispheric fibers, as occurs in callosotomy, produce alexia that is confined to the left visual field, as the right occipital lobe is disconnected from the left hemisphere. The very rare condition of reversible alexia without agraphia occurs with vasospasm in the region of the PCA due to migraine or vertebral arteriography.

Color Anomia

Color anomia is a selective defect in naming colors to visual confrontation. The deficit is not due to a general defect in naming; persons with color anomia cannot name colors seen, but they have no difficulty naming other visual stimuli. Unlike those with achromatopsia, persons with color anomia do not complain about loss of color vision, and they perform normally on visual-visual tasks of color discrimination and

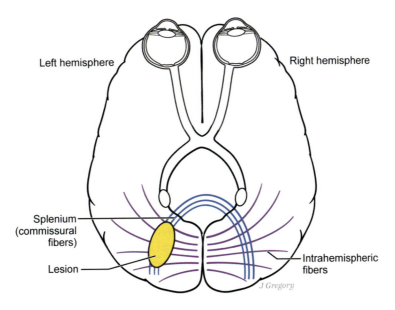

FIGURE 14.12. Top-down view. Pure alexia (alexia without agraphia) occurs with lesions in the language-dominant hemisphere that disconnect visual cortex from more anterior temporal structures involved in language comprehension.

color matching. Persons with color anomia also perform normally on verbal-verbal tasks of color identification (e.g., answering questions such as "What color is grass?"), indicating that their semantic knowledge about colors is preserved. They simply cannot name colors seen.

Color anomia occurs with lesions of the left mesial occipitotemporal region, inferior to the splenium. The underlying mechanism is a disconnection between the visual cortex and language zones within the left hemisphere. Color anomia is usually associated with right homonymous hemianopsia and pure alexia. Almost 100% of cases of color anomia also have pure alexia, while about 50% of pure alexia cases have color anomia.

Visual Hallucinations and Illusions

Hallucinations and illusions both are disorders of perception. A **hallucination** is a nonveridical perceptual experience that arises internally without a corresponding external stimulus; in other words, it is the perception of something that is not real. An **illusion** is a nonveridical perceptual experience of an external stimulus; in other words, it is a perceptual distortion or misinterpretation of something that is real.

Pathologic visual illusions (i.e., not optical illusions) are characterized by alterations in the visual image in size, form/shape, distance, movement, or color. For size illusions, objects appear to be smaller (**micropsia**) or larger (**macropsia**) than they actually are. Form/shape illusions involve the altered/distorted form of objects (**metamorphopsia/dysmorphopsia**) or faces (**prosopometamorphopsia**). For distance illusions, objects appear to be nearer (**pelopsia**) or farther (**teleopsia**) than they are. The illusions of movement involve the perception of stationary objects as moving (**kinetopsia**) or the environment moving rhythmically to and fro (**oscillopsia**). Some ophthalmic conditions and medications can generate altered perception of color across the entire field of vision (**dyschromatopsia**).

Polyopia (multiple simultaneous perceptions of a single object) and **palinopsia** (the persistence of a visual percept after the stimulus has disappeared, also known as "trailing" or "echoing" afterimages) do not easily fit into the categories of illusion or hallucination.

There are two classes of visual hallucinations: simple (unformed) and complex (formed). Simple visual hallucinations include the perception of bright points, flashes, or sparks of light (**photopsia**) and geometric patterns such as grids or tiles (**tessellopsia**), zigzags (**teichopsia/fortification**), and branching structures (**dendropsia**). **Complex visual hallucinations** are defined as formed images of objects or persons and include seeing oneself (**autoscopy**), faces (facial hallucinations), small people (**Lilliputian hallucinations**), animals (**zoonopsia**), and fantasy creatures and scenes. Faces may be realistic, or may have a cartoonish appearance or distorted features.

Visual hallucinations and visual illusions are associated with many pathologies, including migraine, psychiatric disorders, toxic-metabolic disorders, ophthalmological disorders, seizures, brain tumors, stroke, and neurodegenerative disease, and can occur with conditions affecting any part of the visual pathway, from the eye to extrastriate cortex. Simple visual hallucinations may occur with focal pathologies anywhere between the eye and visual cortex within the occipital lobes; thus these hallucinations have little localizing value (i.e., the signs and symptoms give no hint as to the location of the causative pathology along the visual pathway). Lateralization is usually ipsilateral with prechiasmatic lesions, and contralateral with postchiasmatic lesions. Complex visual hallucinations occur with focal pathologies affecting visual association areas in the occipitotemporal or occipitoparietal regions and are localized to the contralateral visual field. Electrical stimulation studies have confirmed that complex visual experiences can be evoked by stimulation in temporal, temporo-occipital, and parietooccipital regions, but not from primary visual cortex.

Dementia with Lewy bodies (see Chapter 11, "The Basal Ganglia") is frequently associated with well-formed complex visual hallucinations, most commonly involving perception of people or animals. Visual hallucinations within the context of dementia with Lewy bodies are associated with reduced blood flow or hypometabolism in the primary visual cortex on functional imaging studies.

Visual Release Hallucinations (Charles Bonnet Syndrome)

Visual release hallucinations are visual hallucinations that occur in the context of severe vision loss; this condition is also known as **Charles Bonnet syndrome**. The hallucinations are exclusively visual, may be formed or unformed, occur when the person is in a state of clear consciousness, are understood to be hallucinations by the person experiencing them, are not distressing, and may disappear when the eyes are closed or moved.

Release hallucinations are defined as hallucinations that occur in the absence of normal sensory input. They are believed to result from de-afferentation of cortical sensory areas and a disinhibition or "release" of processes that are normally held in check by sensory input; in other words, they are "released" by removal of normal sensory input. Visual release hallucinations have been interpreted as the visual analogue of phantom limb phenomenon, where lack of somatosensory input (e.g., due to amputation of a body part or section of a sensory nerve) results in a somatosensory experience generated by the brain, based on the brain's inherent capacity to generate or construct meaningful experience without direct correspondence to sensory stimuli (as occurs in dreams). Visual release hallucinations are therefore also referred to as visual phantoms.

Summary

The occipital lobes are dedicated to visual information processing; however, visual cortex extends beyond the occipital lobes into the inferior temporal and posterior parietal lobes. Brain lesions within the visual pathways from the eyes up to primary visual cortex result in a sensory loss affecting all aspects of vision within the visual field or a restricted region of the visual field (i.e., cerebral blindness and the various patterns of visual field defect). Brain lesions within secondary visual cortical areas result in disorders of higher-order visual processing without regard to location within the visual field. Lesions in the inferior occipitotemporal visual association cortex affecting the ventral stream of visual information processing specialized for form and color analysis result in visual object agnosia, prosopagnosia, and achromatopsia. Lesions of the posterior parietal visual association cortex affecting the dorsal stream of visual information processing specialized for motion and depth result in the visuospatial processing disorders astereopsis and akinetopsia, and visuomotor disorders that selectively affect movements under visual guidance, namely optic ataxia and oculomotor apraxia. Disconnection between visual cortex and language cortex results in the disconnection syndromes of pure alexia and color anomia.

Additional Reading

1. Beauvois MF. Optic aphasia: a process of interaction between vision and language. *Philos Trans R Soc Lond B Biol Sci.* 1982;298(1089):35–47. doi:10.1098/rstb.1982.0070
2. Bruce V, Young A. Understanding face recognition. *Br J Psychol.* 1986;77(Pt 3):305–327. doi:10.1111/j.2044-8295.1986.tb02199.x
3. Calder AJ, Young AW. Understanding the recognition of facial identity and facial expression. *Nature Rev Neurosci.* 2005;6:641–651.
4. Chatterjee A, Coslett HB. Disorders of visuospatial processing. *Continuum (Minneap Minn).* 2010;16(4 Behavioral Neurology):99–110. doi:10.1212/01.CON.0000368263.61286.55
5. Coslett HB, Saffran EM. Preserved object recognition and reading comprehension in optic aphasia. *Brain.* 1989;112 (Pt 4):1091–1110. doi:10.1093/brain/112.4.1091
6. Damasio AR, Damasio H, Van Hoesen GW. Prosopagnosia: anatomic basis and behavioral mechanisms. *Neurology.* 1982;32(4):331–341. doi:10.1212/wnl.32.4.331
7. Damasio AR, Damasio H. The anatomic basis of pure alexia. *Neurology.* 1983;33(12):1573–1583. doi:10.1212/wnl.33.12.1573
8. De Renzi E. Disorders of visual recognition. *Semin Neurol.* 2000;20(4):479–485. doi:10.1055/s-2000-13181
9. Farah M. *Visual agnosia.* 2nd ed. MIT Press; 2004.
10. Ffytche DH, Blom JD, Catani M. Disorders of visual perception. *J Neurol Neurosurg Psychiatry.* 2010;81(11):1280–1287. doi:10.1136/jnnp.2008.171348
11. Ganel T, Goodale MA. Still holding after all these years: an action-perception dissociation in patient DF. *Neuropsychologia.* 2019;128:249–254. doi:10.1016/j.neuropsychologia.2017.09.016

12. Geschwind N, Fusillo M. Color-naming defects in association with alexia. *Arch Neurol.* 1966;15(2):137–146. doi:10.1001/archneur.1966.00470140027004
13. Goodale MA, Milner AD. Separate visual pathways for perception and action. *Trends Neurosci.* 1992;15(1):20–25. doi:10.1016/0166-2236(92)90344-8
14. Goodale MA. Action without perception in human vision. *Cogn Neuropsychol.* 2008;25(7-8):891–919. doi:10.1080/02643290801961984
15. Heywood CA, Kentridge RW. Achromatopsia, color vision, and cortex. *Neurol Clin.* 2003;21(2):483–500. doi:10.1016/s0733-8619(02)00102-0
16. Lhermitte F, Beauvois MF. A visual-speech disconnexion syndrome: report of a case with optic aphasia, agnosic alexia and colour agnosia. *Brain.* 1973;96(4):695–714. doi:10.1093/brain/96.4.695
17. Meadows JC. Disturbed perception of colours associated with localized cerebral lesions. *Brain.* 1974;97(4):615–632. doi:10.1093/brain/97.1.615
18. Milner AD, Cavina-Pratesi C. Perceptual deficits of object identification: apperceptive agnosia. *Handb Clin Neurol.* 2018;151:269–286. doi:10.1016/B978-0-444-63622-5.00013-9
19. Mishkin M, Ungerleider LG, Macko KA. Object vision and spatial vision: Two cortical pathways. *Trends Neurosci.* 1983;6(10):414–417. doi:10.1016/0166-2236(83)90190-X
20. Newcombe F, Ratcliff G, Damasio H. Dissociable visual and spatial impairments following right posterior cerebral lesions: clinical, neuropsychological and anatomical evidence. *Neuropsychologia.* 1987;25(1B):149–161. doi:10.1016/0028-3932(87)90127-8
21. Ungerleider LG, Haxby JV. "What" and "where" in the human brain. *Curr Opin Neurobiol.* 1994;4(2):157–165. doi:10.1016/0959-4388(94)90066-3
22. Wilson BA, Clare L, Young AW, Hodges JR. Knowing where and knowing what: a double dissociation. *Cortex.* 1997;33(3):529–541. doi:10.1016/s0010-9452(08)70234-x
23. Zihl J, von Cramon D, Mai N. Selective disturbance of movement vision after bilateral brain damage. *Brain.* 1983;106 (Pt 2):313–340. doi:10.1093/brain/106.2.313
24. Zihl J, Heywood CA. The contribution of LM to the neuroscience of movement vision. *Front Integr Neurosci.* 2015;9:6. Published 2015 Feb 17. doi:10.3389/fnint.2015.00006

The Parietal Lobes and Associated Disorders

Introduction

The parietal lobes consist of two basic functional zones: an anterior zone of somatosensory cortex, and a posterior zone of association cortex that integrates somatosensory information with visual and auditory information. Lesions to the anterior zone or its connections (including the input pathways) result in a variety of somatosensory disorders. Lesions to the posterior zone result in a wide array of deficits that can be classified as disorders of spatial cognition and disorders of spatial attention. This chapter describes the anatomy of the parietal lobes and disorders associated with parietal lobe lesions, including cortical somatosensory processing disorders, body schema disorders, unilateral spatial neglect, simultanagnosia, Bálint's syndrome, left angular gyrus syndrome, anosognosia, and posterior cortical atrophy.

Basic Anatomy of the Parietal Lobes

The parietal lobe has two main surfaces, one lateral and one medial (see Figure 13.1 in Chapter 13). On the lateral surface, the anterior border of the lobe is formed by the central sulcus, the posterior border of the lobe is formed by an imaginary line drawn from the parieto-occipital sulcus to the pre-occipital notch, and the inferior border of the lobe is formed by an imaginary line drawn from the lateral sulcus (Sylvian fissure) to the posterior line of demarcation at a 90-degree angle. On the medial surface of the lobe, an imaginary line drawn from the central sulcus to the corpus callosum forms the anterior border, and the parieto-occipital sulcus forms the posterior border.

On the lateral surface of the parietal lobe (see Figure 13.3 in Chapter 13) there are two prominent landmark sulci: the postcentral sulcus and the intraparietal sulcus. The **postcentral sulcus** runs parallel and posterior to the central sulcus. These two sulci define the **postcentral gyrus**. Functionally, this region is primary somatosensory cortex. The **parietal operculum**, the superior bank of cortex buried within the lateral sulcus, is the location of primary gustatory cortex.

The **intraparietal sulcus** runs in an anterior-posterior direction and divides the lateral surface of the posterior parietal lobe into a **superior parietal lobule** (BA5 and

BA7) and an **inferior parietal lobule**. The inferior parietal lobule is composed of the crescent-shaped **supramarginal gyrus** (BA40) running around the end of the lateral sulcus, and the **angular gyrus** (BA39) running around the end of the superior temporal sulcus. Within the language-dominant hemisphere, the supramarginal and angular gyri are components of **Wernicke's area**, a region of the brain specialized for language comprehension (see Chapter 27, "Language and the Aphasias"). The cortex within the intraparietal sulcus does not have a Brodmann designation because Brodmann did not map this region.

On the mesial surface (see Figure 13.5 in Chapter 13), the posterior half of the **paracentral lobule** lies within the parietal lobe. This region is the medial extension of the postcentral gyrus, and it contains the somatosensory representation of the lower body. The **precuneus** lies between the **marginal** and **parieto-occipital sulci**, posterior to the paracentral lobule and anterior to the cuneus of the occipital lobe.

Somatosensation

The ascending tracts of the spinal cord convey somatosensory information from skin, muscles, tendons, joints, and viscera to the CNS, some of which does not reach conscious awareness. The somatosensory system is not a unitary system. Rather, it encompasses four major separate somatic senses (sensory modalities) that utilize separate anatomical pathways: tactile sensation (discriminative touch and nondiscriminative touch), proprioception, thermosensation, and nociception (see Chapter 7, "The Spinal Cord"). Somatosensory signals that are transmitted along pathways to the **ventral posterior nucleus** (VPN) of the thalamus and the cerebral cortex give rise to conscious awareness and are referred to as **cortical sensations**. Somatosensory signals that are transmitted along pathways to the cerebellum and brainstem structures without reaching the cerebral cortex do not give rise to conscious awareness and are referred to as **subcortical sensations**.

Discriminative touch (**fine touch**) is a **cutaneous** (skin) sensation that enables us to sense the presence, texture, and shape of objects on the skin surface, as well as their location on and movement across the skin surface. **Nondiscriminative touch** (**crude touch**) enables one to sense the presence of an object but not to localize it. **Proprioception** is the sense of body position and movement (i.e., knowing where one's body parts are in space). **Thermosensation** (**thermesthesia**) refers to temperature sensation (i.e., warmth and cold). **Nociception** is the pain sensation; it is triggered by physical tissue damage due to mechanical stimuli, extreme heat and cold, and chemicals released by damaged tissue.

There are four classes of somatosensory receptors, reflecting the four somatosensory modalities. **Cutaneous (skin) mechanoreceptors** (i.e., Pacinian corpuscles, Merkel disks, Meissner corpuscles, Ruffini endings) signal mechanical displacement of the skin, giving rise to the sensation of touch. **Proprioceptors** signal changes in muscle length (signaled by muscle spindles), muscle tension (signaled by Golgi tendon organs), and joint movements (signaled by joint capsules), giving rise to the sense of body position and movement. **Thermoreceptors** signal temperature of the skin, giving rise to the sensations of hot and cold. **Nociceptors** signal tissue damage, giving rise to pain sensation. In the somatosensory system, the sensory receptors consist of either specialized nerve endings of the primary afferent neurons, as is the case for cutaneous and proprioceptive mechanoreceptors, or free nerve endings, as is the case for thermoreceptors and nociceptors.

The Somatosensory Pathways

Somatosensory information from the body (and posterior head) enters the spinal cord through the dorsal roots of the spinal nerves (see Figure 6.3 in Chapter 6). Somatosensory information from the face and anterior head enters the brain through the fifth (trigeminal) cranial nerve (CN5).

Somatosensory information reaches the primary somatosensory cortex by two main pathways: the dorsal column–medial lemniscal pathway and the spinothalamic pathway.

THE DORSAL COLUMN–MEDIAL LEMNISCAL PATHWAY

The **dorsal column–medial lemniscal system**, also known as the **posterior column–medial lemniscal system**, is the principal ascending pathway for both discriminative touch and proprioception. The name of the pathway derives from the fact that the pathway ascends in the dorsal columns of the spinal cord and the medial lemniscus of the brainstem.

Discriminative touch signals arise from skin mechanoreceptors, and proprioceptive signals arise from muscle, tendon, and joint mechanoreceptors. The primary afferents are unipolar neurons (i.e., neurons that have a single process that extends from the cell body and bifurcates into a dendrite and an axon); they have large caliber peripheral nerve fibers (for rapid signal transmission), cell bodies in the dorsal root ganglia, and central axons that enter the spinal cord via the dorsal roots and ascend in the dorsal (posterior) columns of the spinal cord ipsilaterally. Each dorsal column is composed of a **gracile fasciculus**, located medially, carrying afferents from the lower body (i.e., vertebral levels T7 and below), and a **cuneate fasciculus**, located laterally, carrying afferents from the upper body (i.e., vertebral levels T6 and above; see Figure 7.3 in Chapter 7).

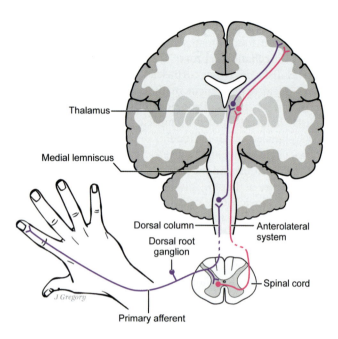

FIGURE 15.1. The somatosensory pathways. Somatosensory information reaches the primary somatosensory cortex by two main pathways: the dorsal column–medial lemniscal pathway conveys discriminative touch and proprioception, and the spinothalamic pathway (part of the anterolateral system) conveys pain, temperature, and non-discriminative touch.

These first-order neurons synapse in the **cuneate** and **gracile nuclei** (also known as the **dorsal column nuclei**) within the medulla ipsilaterally (Figure 15.1). These axons are the longest in the body, with some traveling from the toes up to the medulla.

Second-order neurons decussate immediately within the medulla as **internal arcuate fibers**, so named because they take a curving (arcing) course as they cross the midline of the medulla. These axons then ascend through the brainstem in a large fiber bundle known as the **medial lemniscus**, and synapse in the **ventral posterolateral nucleus** (VPL) of the thalamus, a component of the ventral posterior nuclear complex. From the VPL, third-order neurons project through the internal capsule by a subset of the thalamocortical projections known as the **somatosensory radiations** to **primary somatosensory cortex** (S1) located on the postcentral gyrus of the parietal lobe.

Primary afferents from the skin, muscles, and joints of face and anterior head that mediate fine touch and proprioception travel in the trigeminal nerve (CN5) and terminate in the **main sensory nucleus of the trigeminal** located at mid-pons level. Second-order neurons join the medial lemniscus as it ascends to the thalamus, and synapse in the **ventral posteromedial nucleus** (VPM), a component of the ventral posterior nuclear complex. From the VPM, third-order neurons project through the internal capsule by thalamocortical projections to S1 (i.e., the somatosensory radiations).

THE SPINOTHALAMIC PATHWAY

The **spinothalamic system** conveys pain and temperature sensation to the CNS, as well as **nondiscriminative touch** (crude touch, pressure sensation) that enables us to sense touch without the ability to localize it. The spinothalamic system is part of the **anterolateral (ventrolateral) system**, which includes the **spinoreticular** and **spinotectal tracts**.

The spinothalamic system uses small-caliber peripheral nerve fibers. Primary afferents from the body enter the spinal cord through the dorsal roots and synapse within the dorsal horn gray matter. Second-order neurons decussate within the spinal cord and give rise to the **lateral spinothalamic tract** and the **anterior spinothalamic tract** (also known as the **ventral spinothalamic tract**). The lateral spinothalamic tract travels in the lateral columns of the spinal cord and conveys information about pain and temperature. The anterior spinothalamic tract travels in the anterior (ventral) columns of the spinal cord and conveys information about nondiscriminative touch.

The two spinothalamic pathways ascend in the cord contralateral to their origin and merge into a single spinothalamic tract in the medulla. As these fibers ascend through the brainstem, many give off collaterals that go to the reticular formation. These second-order neurons terminate in the VPL and intralaminar nuclei of the thalami. Third-order thalamocortical fibers ascend to S1.

Primary afferents from the skin of the face and anterior head that mediate pain and temperature are carried by the trigeminal nerve and terminate in the **spinal trigeminal nucleus**. Second-order neurons leave the spinal trigeminal nucleus, join the spinothalamic tract, and synapse in the VPM of the thalamus. Third-order thalamocortical fibers ascend to S1.

Somatosensory Cortex

Primary somatosensory cortex (S1) is located in the parietal lobe, on the postcentral gyrus, and in the depths of the postcentral sulcus. It receives its input from the lower somatosensory pathways (i.e., the dorsal column–medial lemniscal pathway and spinothalamic pathways) via the VPL conveying information about the body, and via the VPM conveying information about the head. In addition to receiving thalamic afferents from the VPM and VPL, S1 receives commissural fibers through the corpus callosum from the contralateral primary somatosensory cortex (although cortical regions that receive inputs from the distal limbs are not connected through the corpus callosum), and short association fibers from adjacent cortical regions (i.e., primary motor cortex in the precentral gyrus, secondary somatosensory cortex, and posterior parietal cortex areas 5 and 7).

The entire body surface is represented topographically in the cortex in a **somatotopic map** known as the **somatosensory homunculus** (see Figure 13.13 in Chapter 13). Because there is a single decussation in both the dorsal column and spinothalamic systems, the cortical representation of the body surface is contralateral. Within S1, the pharynx, tongue, jaw, teeth, and gums are represented in the most ventral portion, followed in ascending order by the face, hand, arm, trunk, and thigh. The medial surface of the postcentral gyrus contains the representations for calf, foot, anus/rectum, and genitals. In homunculus drawings, the somatosensory homunculus is easily distinguished from the motor homunculus because it contains a representation of the genitals, while the motor homunculus does not. The somatosensory map is distorted, as body parts are represented in terms of how richly they are innervated. Thus, body regions with a high density of somatosensory receptors have more extensive cortical representation, reflecting the importance of these parts in sensory function. Face, lips, and hands have particularly extensive representations. The **glabrous** (hairless) skin over the fingers and palms is the most sensitive zone in our somatosensory system; it is the "fovea" of tactile perception and is required for fine manual dexterity.

Primary somatosensory cortex consists of Brodmann areas 1, 2, and 3. Area 1 forms the crown of the postcentral gyrus. Area 2 forms the posterior wall of the postcentral gyrus. Area 3 is subdivided into area 3a which lies in the valley of the central sulcus, and area 3b which lies on the posterior wall of the central sulcus. These cytoarchitectonic areas differ with respect to the kind of information that they process, and within each cytoarchitectonic division the whole body is represented. Thus, there are really four maps of the body in S1. Areas 1 and 3b receive cutaneous tactile input, areas 2 and 3a receive proprioceptive input.

Electrical stimulation of primary somatosensory cortex gives rise to simple but localized sensations on the opposite side of the body, such as numbness, tingling, itching, tickling, warmth, and the sensation of movement. Vibration sense, thermosensation, and nociception are mediated mainly by subcortical structures.

Secondary somatosensory cortex (S2) is located inferiorly, within the parietal operculum (pars opercularis of the parietal lobe) that forms the upper bank of the lateral sulcus. Little is known about the exact extent, structural organization, or function of S2, but this region is divided into several sub-areas that have somatotopic organization.

Both S1 and S2 send inputs to multimodal association cortex located within the posterior parietal region, where somatosensory information converges with inputs from the visual system and other systems involved in attention.

Disorders of Cortical Somatosensory Processing

With **radiculopathies** (i.e., compression of the spinal nerve root), the distribution of sensory loss is **segmental** and follows the corresponding dermatome (see Chapter 6, "The Peripheral Nervous System"). With central lesions involving the spinal cord or brain, there is usually a dissociation in the sensory loss, such that touch sensation and proprioception are affected while thermosensation and nociception are intact, or vice versa. This is due to the selective involvement of either the dorsal column–medial lemniscal or spinothalamic systems.

Lesions of Primary Somatosensory Cortex

Lesions of S1, or its deafferentation by subcortical lesions, result in deficits in discriminative touch and proprioception and impair the discrimination of texture, size, and shape of objects. Complete unilateral lesions of the postcentral gyrus result in a complete loss of discriminative touch (**hemianesthesia**) and proprioception on the contralesional side of the body; partial lesions produce a loss of discriminative touch and proprioception in a restricted area of the body. Less severe S1 lesions result in diminished discriminative touch (**hypoesthesia**) and proprioception. A crude awareness of pain, temperature, and pressure sensation is preserved, as these sensations are perceived at the thalamic level. The ability to determine the source, severity, and quality of such sensations, however, is lost.

Deficits in discriminative touch manifest as elevated **detection threshold**, inability to localize points of touch on the skin, elevated **two-point discrimination** threshold (i.e., the smallest distance between two points that are perceptually distinguished as two), impaired shape perception (**amorphognosia**), and impaired texture perception (**ahylognosia**). Proprioceptive deficits manifest as reduced ability to detect **passive movements** (i.e., movements produced by an external force during muscular inactivity), and clumsy movements due to reduced feedback of position information, a condition known as **afferent paresis** and **sensory ataxia**.

Shape and texture perception allow for **stereognosis**, the ability to identify three-dimensional objects by tactile exploration (also known as haptic perception). Lesions of S1 result in **astereognosis**, an inability to recognize objects by touch. This is also true of lesions that result in S1 deafferentation, such as those involving the peripheral nerves, spinal nerves, dorsal columns of the spinal cord, medial lemniscus of the brainstem, the VPL of the thalamus, and the somatosensory radiations.

Astereognosis encompasses tactile recognition disorders due to primary deficits in basic tactile perception that allow us to determine physical properties such as shape, texture, and weight. Persons with astereognosis are unable to recognize objects by touch because they cannot perceive shape, texture, or weight. In these cases, the astereognosis is analogous to apperceptive visual agnosia; thus it is not a true agnosia.

Disorders of Higher-Order Cortical Somatosensory Processing

Lesions in secondary somatosensory cortex or its connections to association cortex result in deficits in higher-order interpretive tactile perception that are carried out cortically. These processes of course require intact elementary somatosensory submodalities. Higher-level somatosensory function testing includes tactile object recognition, graphesthesia, and tactile naming, assessing for the conditions of tactile agnosia, agraphesthesia, and tactile anomia, respectively. It also includes double simultaneous stimulation testing for neglect in the somatosensory domain (see "Unilateral Spatial Neglect," below).

TACTILE AGNOSIA

Tactile agnosia is a selective impairment of tactile object recognition that occurs in the absence of any impairment of basic somesthetic function or other gross cognitive disorder that would undermine task performance (e.g., hemispatial neglect, aphasia, dementia). Those with tactile agnosia are unable to tactually recognize objects *despite* intact ability to perceive shape and texture. Tactile exploration strategies are also normal. Ability to form a "tactile image" is intact (i.e., the person is able to form a mental image of an object perceived by the sense of touch). Persons with this disorder can decipher the salient somesthetic characteristics of the objects they fail to recognize, as evidenced by their ability to describe the basic sensory features and draw the objects that they cannot identify. When asked to identify objects by palpation, usually their errors are spatially similar to the target object.

Tactile agnosia is interpreted as an inability to associate accurate tactile perceptions of objects with stored knowledge. It occurs when the somatosensory association cortex is disconnected from the semantic memory store located in the inferior temporal lobe by a subcortical lesion of the angular gyrus, affecting pathways of the arcuate fasciculus and/or inferior longitudinal fasciculus. Tactile agnosia is often a unilateral disorder, but there are some cases of bilateral tactile agnosia. It is not a source of great disability.

AGRAPHESTHESIA

Graphesthesia is the ability to identify characters (numbers or letters) "written" on the skin (palm or fingertips) by an examiner using a dull, pointed object (*graph*, "writing"; *esthesia*, "sensing"). This of course is not a task that we do in everyday life, but rather reflects the ability to recognize form patterns by touch and is a measure of cortical somatosensory function. **Agraphesthesia** is usually associated with lesions in the region of the intraparietal sulcus and is believed to be due to a disconnection between the somatosensory association cortex and language-related areas.

TACTILE ANOMIA

Tactile anomia (**tactile aphasia**) is the selective inability to name objects by touch. Tactile object recognition is intact, as evidenced by the fact that those with tactile anomia can demonstrate their recognition nonverbally, such as showing how the object is used (pantomime) and matching to sample (i.e., from a multiple-choice array of objects or pictures of objects).

Tactile anomia is due to lesions that disconnect somatosensory cortices from language areas. It is always unilateral and affects the hand ipsilateral to the language-dominant hemisphere. Tactile anomia occurs with lesions of the caudal body of the corpus callosum and with callosotomy. In left-hemisphere language-dominant persons, when palpating an object with their left hand, the right-hemisphere somatosensory areas are unable to transfer information to the left-hemisphere language areas; thus, the person is unable to name objects touched using their left hand.

Somatosensory Hallucinations and Illusions

Positive Somesthetic Symptoms

In addition to the negative somatosensory symptoms described above, lesions of the somatosensory system can result in positive somesthetic symptoms consisting of illusions (distorted sensations) or hallucinations (perceptual experiences that occur in the absence of external sensory stimuli). Hallucinations are generally classified as either simple (unformed) or complex (formed). Somesthetic illusions and hallucinations also occur in psychotic disorders and use of hallucinogenic drugs.

Somesthetic illusions and simple hallucinations include paresthesia, hyperesthesia, dysesthesia, hyperalgesia, and allodynia. **Paresthesia** is an abnormal (but not unpleasant or painful) sensation that may be spontaneous or evoked by touch, such as tingling, itching, pinching, rubbing, and pins-and-needles sensations. Formication is a feeling of insects crawling under the skin; it commonly occurs in alcohol withdrawal syndrome (delirium tremens). Kinesthetic illusions or hallucinations involve a distorted or false perception of body position and/or movement. **Hyperesthesia** is an exaggerated nonpainful sensation. **Dysesthesia** is an altered sensation or unpleasant distortions of actual sensory stimuli. **Hyperalgesia** is an exaggerated painful sensation. **Allodynia** is painful sensation to nonpainful stimuli. Complex tactile hallucinations, also called spontaneous stereognostic sensations, such as the experience of being touched or of having an object placed in one's hand, result from lesions of S1 and the superior parietal lobule (BA5 and BA7).

Somesthetic hallucinations may also involve internal and visceral sensations (i.e., corporeal sensations) such as bone, muscle, ligament, and joint sensations, as well as pain, nausea, hunger, thirst, and sexual pleasure. **Visceral hallucinations** often accompany complex partial seizures.

The **thalamic pain syndrome (Dejerine-Roussy syndrome)** is characterized by unilateral, agonizing pain ("stabbing, crushing, burning") brought about by the most trivial of cutaneous stimuli. It results from a lesion to the contralateral VPL nucleus of the thalamus. This syndrome is often caused by vascular occlusion of the **thalamogeniculate** branches from the proximal portion of the posterior cerebral artery (PCA). In the acute phase of injury, there is a loss of sensation (numbness) on the affected side, but over time dysesthesia and allodynia emerge.

Parietal Lobe Epilepsy

Parietal lobe epilepsy (PLE) is a rare disorder in which focal seizures originate from the parietal lobe. The most common etiology of PLE is brain tumor. Of brain tumors, those located within the parietal lobe are particularly epileptogenic.

The most common symptom of PLE is subjective somatosensory illusions and hallucinations. PLE is associated with very few objective ictal signs. The most common manifestation of PLE is paresthesia without an evoking sensory stimulus. The paresthesias usually consist of feelings of numbness, "pins and needles," and tingling sensations, but they may also consist of prickling, tickling, crawling, itching, or electric shock sensations, and less commonly pain and thermal (burning or cold) sensations. Seizures originating from the mesial primary sensory area within the paracentral lobule may give rise to genital sensations. The paresthesias are lateralized and contralateral to the hemisphere of seizure onset. The face and hands are more commonly affected due to their disproportionate representation in the sensory homunculus. The seizure activity may spread along the somatosensory cortex, and the ictal sensory experience moves along the skin sequentially according to the sensory homunculus in what is termed a **Jacksonian march**.

The second most common symptom of PLE is somatic illusions and distorted body image. Somatic illusions reported in PLE include feelings of movement (e.g., floating) in a stationary limb, altered or distorted posture (e.g., twisting movement), and altered limb position. Body image illusions reported in PLE include feelings that a body part is enlarged (**macrosomatognosia**), shrunken (**microsomatognosia**), elongated (**hyperschematica**), shortened (**hyposchematica**), absent (**asomatognosia**), or alien (i.e., does not belong to the person).

Phantom Limb

After an amputation, approximately 60%–80% of people experience **phantom limb**, a hallucinatory perception of the missing limb. These phantoms are experienced as belonging to one's body. They are even experienced as moving in coordination with the body (when the person's eyes are open).

Initially the phantom may be experienced exactly as the true limb was experienced before the amputation. Over time, however, the experience may become less natural. Particularly in cases of upper limb phantoms, the representation of the proximal portion of the limb may weaken, eventually vanishing and leading to a strange sensation of a hand belonging to one's body but disconnected from it, or giving rise to "telescoping" in which the phantom arm is perceived as shorter than the other arm or as if a phantom hand resides within the amputation stump. The restriction of the phantom to the most distal portion of the body part reflects the over-representation of the hands and fingers in the cortical maps.

The phantom limb phenomenon occurs not only with amputations, but also with deafferentation due to lesions at various levels of the neuraxis (i.e., peripheral nerve, spinal cord, and subcortical pathway lesions). The source of phantom sensations is unknown, but in several patients a cortical lesion of the parietal lobe abolished a contralateral phantom limb.

Disorders of Spatial Cognition

The posterior parietal cortex plays a major role in processing spatial information by forming mental representations of personal space, peripersonal space, and extrapersonal space. **Personal space** is the space of the body surface. **Peripersonal space** is the space within arm's reach, within which we can directly act on objects. **Extrapersonal space** is space that is beyond one's reach; we can orient our eyes toward and point to or throw things at objects in our extrapersonal space. These mental maps form the basis of a broad range of abilities that are collectively referred to as **spatial cognition**.

Posterior parietal cortex plays an important role in **visuomotor** function (movements under visual guidance) by integrating somatosensory and visual maps. It plays a role in processing **visuospatial** information about spatial properties of visual stimuli and locations of objects in visual space. It also plays a role in spatial cognitive processes that are not bound to the visual domain, including knowledge of the spatial layouts of the external environment (**topographical knowledge**) and the body (**body schema**). Posterior parietal lobe lesions give rise to disorders of visuomotor function, visuospatial processing, topographical orientation, and body schema.

Visuomotor Disorders

Lesions of the superior posterior parietal lobe lead to **visuomotor disorders**. These are characterized by a selective impairment of visually guided movements; movements not performed under visual guidance are normal. The visuomotor disorders are optic ataxia and oculomotor apraxia (see Chapter 14, "The Occipital Lobes and Visual Processing"). **Optic ataxia** is a deficit in reaching, pointing, or other movements directed toward visual targets (e.g., throwing a ball to someone). **Oculomotor apraxia** is an inability to direct gaze to and follow moving visual targets.

Visuospatial Disorders

One function of the parietal association cortex is to construct a map that represents visual space and provides information about where things are in visual space and the spatial relations among objects within that space. **Visuospatial disorders** (sometimes referred to as **visuospatial agnosias**) are based on a disturbance in the brain's ability to compute spatial relations, and manifest as impaired ability to localize objects in space, judge depth, and judge angular orientation. Such disorders can occur with either right or left parietal lobe lesions.

Topographic Disorientation

Topographic disorientation is the inability to find one's way in familiar environments and learn new routes. It is a relatively specific deficit of orientation to the environment; it is not secondary to global amnesia, global dementia, or disorders of visual perception. It encompasses a loss of memory for the spatial layout of familiar environments, as well as inability to locate familiar items such as countries or cities on a map or to learn new spaces. Those with the most severe form of this disorder may lose their bearings even within their own homes. The basic defect underlying topographic disorientation is an inability to retrieve an abstract map of the environment or route (which specifies the spatial relationships defining the position of a place with respect to other places), and to transform those spatial relationships into a route for navigation. Those with the disorder tend to rely on verbal tags (e.g., room numbers, street names), salient landmarks, and verbal cues defining the segments making up a route.

Topographic disorientation and loss of topographic memory are most often associated with right or bilateral parietal lesions. The most frequent etiology is an infarct in the territory of the right PCA. Topographic disorientation is frequently seen in Alzheimer's disease.

Body Schema Disorders

The concept of **body schema** was introduced in 1912 by **Henry Head** (1861–1940) and **Gordon Holmes** (1876–1965). They hypothesized that the brain houses a spatially organized model of the body, based on their observation that some patients with brain disease make gross errors in pointing to parts of their body on verbal command. There are three forms of body schema disturbance: autotopagnosia, finger agnosia, and right-left disorientation. All of these are associated with lesions to the inferior parietal lobe of the language-dominant hemisphere.

AUTOTOPAGNOSIA

Autotopagnosia (**bodily agnosia**) is a selective loss of knowledge about the spatial layout of the human body. It manifests as an inability to localize body parts on the self and others, as well as on models and pictures of the human body. Persons with autotopagnosia can recognize body parts, name the body parts when someone else points at them, describe the function and appearance of body parts, and match and discriminate pictures of body parts. However, they are unable to describe the location of the body parts. Performance on "what" tasks is intact, while performance on "where" tasks is not. The disorder is selective for the human body schema; it does not affect the ability to point on command to parts of objects or body parts of animals.

Assessment for autotopagnosia involves testing a variety of conditions, as it is important to rule out language, motor, and visuospatial deficits as the cause of poor task performance. Verbal command tasks involve asking the examinee to point to the body part named by the examiner on his own body, the examiner's body, a picture, or a model. If the examinee cannot localize body parts on verbal command, then body part localization must be assessed with tasks that do not require language comprehension. These include the examiner pointing to a part on their own body and asking the examinee to point to the corresponding part on their own body or a model; pointing to a body part on a model and asking the examinee to point to the corresponding part on their own body or the examiner's body; and presenting a picture of an isolated body part and asking the examinee to point to the same body part on their own body, the examiner's body, or a model.

Persons with autotopagnosia make three types of errors. The majority are **contiguity errors**, in which they point to an incorrect body part in the vicinity of the designated one. Less frequent are **semantic errors** (also known as **conceptual errors**) in which the person confuses categorically related body parts (e.g., elbow and knee). Errors that cannot be classified as either contiguity or semantic errors, such as aimless searching for the body part in a fumbling, groping manner, are classified as **random errors**. Body parts that do not have definite boundaries (e.g., cheeks) are more difficult to localize than well-defined body parts (e.g., ear, nose).

Autotopagnosia is not simply a manifestation of a visuomotor defect. There is a dissociation between localization in external space (e.g., object localization) and the ability to localize parts of one's own body. Persons with impaired reaching for objects in space (i.e., optic ataxia) typically are able to point accurately and without hesitation to their body parts. Persons with impaired ability to point accurately to body parts have normal ability to accurately point to and reach for objects in space. Autotopagnosia is also not attributable to a more basic sensorimotor disorder; movements under proprioceptive guidance (e.g., buttoning one's shirt buttons) are performed normally.

In cases of pure autotopagnosia, the lesion always involves the posterior parietal lobe of the language-dominant hemisphere. This region is believed to house a representation of body parts and their spatial and functional interrelations. Autotopagnosia is a bilateral condition; localization on both the right and left sides of the body is similarly affected.

FINGER AGNOSIA

Finger agnosia is a specific deficit in localizing the fingers. The examiner touches single fingers on the examinee, and the examinee indicates the fingers touched by name, number (1–5), or pointing to the corresponding finger on a model of a hand. Testing is performed with the hand in view as well as the hand hidden from view. It is most difficult to differentiate the middle three fingers. If the disorder is subtle, it may be revealed by stimulating two fingers (e.g., by light touch) and asking the patient to identify both in order of stimulation.

Finger agnosia is associated with lesions in the region of the angular and supramarginal gyri (i.e., inferior parietal lobule) within the language-dominant hemisphere. Bilateral finger agnosia, when not explained by a basic somatosensory defect, aphasia, or general mental impairment, is indicative of a focal lesion in the left posterior parietal region. Electrical stimulation in the inferior parietal lobule of the language-dominant hemisphere has been found to produce finger agnosia in an epilepsy surgery candidate who underwent preoperative functional mapping in the region of the planned resection (see Chapter 19, "Epilepsy").

RIGHT-LEFT DISORIENTATION

Right-left disorientation (**right-left confusion**) is characterized by difficulty differentiating between the right and left halves of space from the frame of reference of the patient, and the right and left halves of the body on oneself and others. It implies difficulty appreciating spatial concepts in the body's lateral orientation. Other spatial concepts such as up-down and front-back are preserved.

Right-left disorientation is assessed by tasks of increasing difficulty: (1) ability to point to single body parts on oneself (e.g., "Show me your right hand"); (2) double uncrossed commands (e.g., "With your left hand, touch your left ear"); (3) double crossed commands (e.g., "Touch your left ear with your right hand"); (4) single body parts on the examiner sitting across from the patient (e.g., "Point to my left cheek"); and (5) combined orientation (e.g., "With your right hand, point to my left knee"). In mild cases of right-left disorientation, only crossed commands may be impaired. These evaluations are conducted only after verifying that there is no autotopagnosia. In cases where there is an autotopagnosia, right-left disorientation can be ascertained only with tasks that do not also require body part localization, such as by asking the patient to point to objects in extrapersonal space.

Right-left disorientation is usually associated with parieto-occipital injuries involving the supramarginal and angular gyri in the language-dominant hemisphere. It is also seen with diffuse cerebral dysfunction. Right-left disorientation is rare with right hemisphere lesions. Right-left disorientation has been elicited by electrical stimulation of the left posterior inferior parietal lobule in an epilepsy surgery candidate who underwent preoperative functional mapping in the region of the planned resection.

Disorders of Spatial Attention

Lesions of parietal cortex are associated with two neurobehavioral syndromes that are conceptualized as disorders of spatial attention: unilateral spatial neglect and simultanagnosia.

Unilateral Spatial Neglect

Unilateral spatial neglect (**hemispatial neglect, hemi-inattention**) was first described in 1874 by **John Hughlings Jackson** (1835–1911), who reported on a patient who neglected the left half of the page when reading (what we now refer to as a form of **spatial alexia**). In the 1940s, **Andrew Paterson** and **Oliver Zangwill** (1913–1987) defined unilateral spatial neglect as an impairment of the ability to report, respond to, and orient toward (notice or explore) stimuli from one half of space, despite preserved primary motor and sensory functions. It may affect visual, tactual, and/or auditory modalities. When sensory deficits such as hemianopia or hemianesthesia are present, they do not explain the failure of detection.

Manifestations of neglect include shaving one half of the face, wearing a sweater with a sleeve only on one arm, leaving food on the neglected half of a plate, using cutlery from only one side of the plate, colliding with objects in the neglected side of space, reading words from only one half of a page, omitting details from one side of graphic stimuli that are copied or drawn from imagination, and failure to respond to voices and other sounds coming from the neglected side of space.

Hemispatial neglect may appear similar to hemianopia; however, persons with hemianopia are aware of the existence of stimuli in their blind field and make orienting head and eye movements, while those with neglect appear to lack awareness of the continuity of visually presented objects and do not make orienting head and eye movements. This suggests that the core defect in neglect is in spatial attention. When persons with hemispatial neglect are cued to direct their attention to the neglected side, this increases the likelihood that they will detect stimuli in the neglected half of space. This method is commonly used in rehabilitation, such as marking the neglected side of a page with a bright red line to make it more salient and improve reading. Persons with neglect also have anosognosia (i.e., lack of awareness of the disorder) and therefore do not compensate for their impairment, also unlike those with hemianopia. Thus, neglect is more disabling. Even a mild neglect is dangerous because those with this condition are unaware of what they are missing.

The dramatic features of neglect tend to ameliorate within the first several weeks after stroke, but milder features may persist and are a key factor in difficulties regaining independence. In its mildest form, neglect may manifest merely as a response bias in favor of the space on the same side as the lesion (ipsilesional), as evidenced by performance on horizontal line bisection tasks, cancellation tasks, and drawing and copying tasks, or extinction on double simultaneous stimulation. Rehabilitation techniques involve training the patient to actively and consciously visually scan and attend to the neglected half of space.

VARIANTS OF NEGLECT

Neglect is a heterogeneous disorder with many variants, and there are a number of classification systems. In most cases, the left half of space (i.e., the contralesional space) is neglected; the following descriptions will describe left-neglect, although the same concepts apply to the far less common phenomenon of right-neglect.

One dimension along which forms of neglect have been differentiated is the frame of reference (the origin from which space is measured). In a **viewer-centered** (**egocentric**) frame of reference, patients neglect objects to the left of their midline. In an **object-centered** (**allocentric**) frame of reference, patients neglect the left side of an object, regardless of the object's location in space. Object-centered neglect appears to disregard the holistic nature of objects. There is evidence that these two forms of neglect are dissociable, so that patients may have viewer-centered neglect without object-centered neglect, or object-centered neglect without viewer-centered neglect. Some exhibit both forms of neglect.

Forms of neglect have also been differentiated with personal, peripersonal, and extrapersonal space; these various forms are dissociable. Neglect for left personal space can occur without neglect for left peripersonal space, and neglect for left peripersonal space can occur without neglect for left personal space. Similar double dissociations have been observed between neglect of left peripersonal space (**near neglect**) and extrapersonal space (**far neglect**).

Hemispatial neglect may also affect mental imagery; this is known as **representational neglect**. This form of neglect was first demonstrated by **Edoardo Bisiach** and **Claudio Luzzatti** in 1978, when they asked Milanese subjects with left hemi-neglect to describe from memory the Piazza del Duomo, a well-known plaza and cathedral in Milan. Subjects failed to mention buildings on the left half of imagined space, whether they imagined that they were standing on the steps of the cathedral looking out onto the plaza or standing on the opposite side of the plaza looking at the cathedral. Representational neglect may dissociate from peripersonal neglect. Some patients who exhibit neglect on copying tasks may draw complete representations when asked to draw from memory.

As neglect recovers, it typically evolves into **extinction**, a mild form of neglect in which stimuli within the affected hemi-space are perceived when presented unilaterally but not when presented bilaterally (i.e., when a contralateral stimulus is present). In other words, patients with extinction exhibit neglect of contralesional stimuli only in the presence of ipsilesional stimuli. Extinction is tested with **double simultaneous stimulation**. For example, the examiner touches homologous parts of the body on one side, the other side, and both sides simultaneously, and the patient, with eyes closed, identifies which side of the body was touched or whether both sides were touched. The patient with mild neglect detects unilateral stimuli whether presented to the left or right side, but with double simultaneous stimuli the patient extinguishes the stimulus within the neglected hemi-space (i.e., only reports detecting stimuli in the non-neglected hemi-space). Extinction can be tested within visual, tactual, and auditory modalities. There is controversy as to whether extinction is modality-specific (i.e., occurs selectively in the visual, auditory, or somatosensory spheres) or supramodal (i.e., affecting all sensory spheres).

NEUROANATOMICAL BASIS OF NEGLECT

Neglect is most common following infarction of the nondominant right inferior posterior parietal lobule (BA39 and BA40) due to ischemic stroke in the territory of the middle cerebral artery (MCA). The most severe forms occur in the acute stages of MCA infarct involving the posterior two-thirds of the nondominant inferior parietal lobe. The head and eyes are deviated to the right, and patients may fail to bring their eyes to the left of their midline on command, despite preserved ability to make leftward eye movements, as indicated by intact **oculocephalic reflex** (see Chapter 30, "The Neurological Examination").

Neglect also occurs with lesions outside of the right inferior parietal lobe, although it is usually milder and more transient. Other lesion locations include left parietal lobe, right dorsolateral prefrontal area, cingulate gyrus, caudate nucleus, posterior internal capsule, thalamus, superior colliculus, and hypothalamus, suggesting that the posterior parietal lobe and these other structures are components of a neural network that mediates spatially directed attention.

There have been many theories of neglect, but an attentional deficit has received the greatest support. **Kenneth Heilman** (1938–) is a key proponent of this view. He argues that the right hemisphere has a greater dominance for attentional mechanisms when scanning the environment. Heilman has hypothesized that attentional neurons in the right parietal region monitor both halves of space, whereas those in the left hemisphere monitor only the contralateral half of space. Consequently, with left parietal lesions the intact right parietal lobe can attend to both left and right halves of space, but with right parietal lesions the intact left parietal lobe can only attend to the right half of space, producing a left neglect.

Simultanagnosia

Simultanagnosia (**simultagnosia**) is a rare neurological disorder characterized by the inability to visually detect and perceive more than one object at a time and appreciate the visual field as a whole. One of the earliest descriptions of the disorder was published by **Rezső Bálint** (1874–1929) in 1909; the term *simultanagnosia* was coined by **Ilja Wolpert** in 1924.

In simultanagnosia, perceptual awareness is limited to a single object at a time; there is no awareness of the presence of other stimuli. Persons with simultanagnosia visually perceive only one object at a time. They report that objects spontaneously disappear from view as they become aware of another object in the scene; they are described as having piecemeal vision and being unable to "see the forest for the trees." Persons with simultanagnosia describe complex scenes in a slow and fragmentary way, identifying single objects in isolation. They are unable to comprehend the interactions between objects in the scene. They also cannot grasp the overall meaning of the scene or the theme depicted, which becomes apparent when they are asked to interpret the overall meaning of the picture.

As they can visually perceive only one object at a time, persons with simultanagnosia cannot count objects by visual inspection, read, or interpret pictures. The deficit in simultanagnosia, however, is not due to a primary sensory deficit or a general intellectual impairment. Those with simultanagnosia walk cautiously and haltingly with short, slow steps and hands held out in front of them as if groping their way through the dark, and they bump into obstacles. Thus, they appear to be blind as they "do not see," although they can describe minute visual details requiring normal visual acuity and can recognize isolated small visual stimuli with their foveal vision.

Simultanagnosia may be misdiagnosed as cerebral blindness with macular sparing (i.e., visual field loss with preserved vision in the center of the visual field). Persons with cerebral blindness with macular sparing, however, are able to reach accurately, count arrays of objects, and interpret pictures within their narrow visual field. Persons with simultanagnosia may also be misdiagnosed as malingering (feigning deficit for secondary gain/external reward), since they have normal visual acuity, normal single-finger confrontation visual fields, and an otherwise normal neurologic exam.

Bálint interpreted simultanagnosia as a defect in visual attention, such that the spatial window of visual attention, the **attentional field**, is restricted to one item; therefore, the patient cannot perceive more than one object at a time. Bálint observed that his patient could only perceive one object at a time when presented with a complex scene or array, and the size of an object did not affect his patient's ability to perceive it. The patient perceived only one item at a time, whether it was large or small. Since the size of the visual space that the patient responded to varied with the size of the item perceived, the defect was not attributable to a narrowing of the sensory field. It was the "attentional window" that was limited to one object. Since persons with simultanagnosia have a reduced response to stimuli on either side of visual fixation, the disorder has also been referred to as **bilateral visual inattention**.

Michael Posner and Martha Farah have hypothesized that simultanagnosia is rooted in a deficit of disengaging visual attention. Persons with simultanagnosia are described as having **sticky visual attention**. When confronted with several objects, their attention is "stuck" or "locked onto" one object, and they have difficulty disengaging their attention from that object and shifting it toward another object. As a result of this sticky attention, they can perceive only one object at a time.

Isolated (i.e., pure) simultanagnosia results from bilateral lesions at the parieto-occipital junction in the dorsal stream, with sparing of the parietal visuomotor control area. Such lesions usually occur in the context of hypoxic-ischemic encephalopathy following systemic hypotension or respiratory compromise (hypotensive stroke, hypoxia, bilateral PCA occlusions). Simultanagnosia also occurs as a component of Bálint's syndrome.

Bálint's Syndrome

Bálint's syndrome, also known as **Bálint-Holmes syndrome**, is a rare disorder first described in 1909 by Rezső Bálint and Gordon Holmes. It consists of a triad of visuospatial and visuomotor symptoms: simultanagnosia, optic ataxia, and oculomotor apraxia. Optic ataxia is an impairment of visually guided movements, and oculomotor apraxia is an impairment of voluntary eye movements to targets in visual space. Simultanagnosia, optic ataxia, and oculomotor apraxia can occur independently and in pure form. There is a double dissociation between simultanagnosia and oculomotor apraxia, and between simultanagnosia and optic ataxia; thus, in Bálint's syndrome, the simultanagnosia is not secondary to oculomotor apraxia, and the optic ataxia and oculomotor apraxia are not secondary to the inattention of simultanagnosia.

Bálint's syndrome results from bilateral superior parietal lobule lesions affecting the dorsal stream of visual information processing. The most common cause is bilateral borderzone (watershed) infarction in the occipito-parietal region resulting from hypotensive stroke or hypoxia. It is occasionally caused by bilateral PCA occlusions or parafalcine meningiomas in the parietal lobes.

The first reported case of Bálint's syndrome by Bálint and Holmes in 1909 is notable for being the first description of the effects of lesions to the parietal lobes.

Left Angular Gyrus Syndrome and Gerstmann's Syndrome

Left angular gyrus syndrome, also known as **left parietal syndrome**, is a constellation of signs and symptoms that occur with focal lesions of the angular gyrus in the language-dominant hemisphere, including finger agnosia, right-left disorientation, **agraphia** (the acquired loss of writing skills), **acalculia** (the acquired loss of arithmetic skills), **anomia** (word-retrieval failures), and **pure alexia** (the acquired loss of reading skills in the absence of other language deficits).

Gerstmann's syndrome, named after **Josef Gerstmann** (1887–1969), is defined as a tetrad of symptoms: finger agnosia, right-left disorientation, agraphia, and acalculia. The acalculia seen in Gerstmann's syndrome is primary acalculia, characterized by inability to perform calculations mentally or with pen and paper; it is not secondary to poor concentration, language disorders (agraphia, alexia), hemispatial neglect, or executive deficits (e.g., difficulty maintaining order or planning a sequence).

There has been much controversy about the specificity and existence of Gerstmann's syndrome, as the four symptoms do not always co-occur, and other symptoms are often present. When all four components are present, however, it is pathognomonic for (i.e., predicts with a high degree of accuracy) a lesion in the angular gyrus of the language-dominant hemisphere.

Anosognosia

Anosognosia is a lack of awareness or denial of deficit. The term was coined by French neurologist **Joseph Babinski** (1857–1932) in 1914 to describe an unawareness of hemiplegia. However, anosognosia can occur with neurological deficits other than hemiplegia, including cerebral visual losses (see "Visual Anosognosia," below), cognitive deficits, and behavioral deficits. The term can be applied in any circumstance in which there is a significant discrepancy between self-report of level of disability and objective evidence of level of functioning indicating greater disability. In a milder form of the disorder, **anosodiaphoria**, persons may recognize their deficit but be indifferent to it.

Anosognosia is strongly associated with damage to the right parietal and insular cortex. It is rarely observed in connection with left hemisphere damage. Because it often occurs with right parietal damage, anosognosia is often associated with left neglect.

Anosognosia for Hemiplegia (Babinski's Syndrome)

In **anosognosia for hemiplegia**, also known as **Babinski's syndrome**, persons deny hemiplegia or rationalize (make excuses) about the failure to use their paralyzed limbs. They may also have a variety of attitudes toward the paralyzed extremity and its relation to the self. They may have the delusion that the limb does not belong to them and experience the limb as being outside the realm of their own body image. They may develop a dislike or morbid revulsion to the limb, trying to throw it away or conceal it from view. They may even attribute limb ownership to another person.

Anosognosia for hemiplegia is almost always associated with acute, right parietal infarcts. It is often accompanied by left hemianopia and hemianesthesia. It typically resolves.

Visual Anosognosia (Anton's Syndrome)

Visual anosognosia is the unawareness of visual loss in persons with cerebral blindness. It may occur with complete or partial blindness. It is also known as **Anton's syndrome** (named after the neurologist Gabriel Anton). The unawareness ranges from indifference, to failure to recognize the defect, to frank denial of the defect. Those with Anton's syndrome walk into walls, bump into furniture, and attempt to walk outside without taking the precautions that a blind person normally would take. Anton's syndrome is strongly associated with **confabulation**, a memory error in which memory gaps are unconsciously filled with fabricated, misinterpreted, or distorted information. Although persons with Anton's syndrome cannot identify objects on visual confrontation and bump into objects when walking, they often make excuses for their failure to see, such as claiming that the lighting is insufficient, that they need better glasses, or that they see better at home. They confabulate about what they are "seeing," a condition called **confabulatory pseudo-recognition**. This differs from visual release hallucinations of Charles Bonnet syndrome, in which persons experience visual hallucinations in the context of visual impairment but are fully aware that they are blind and that the hallucinations are not real (see Chapter 14, "The Occipital Lobes and Visual Processing").

In Anton's syndrome, the bilateral lesion extends beyond V1 to include visual association areas in parietal and temporal areas. The lesion often extends into inferomedial temporal regions, and consequently Anton's syndrome often occurs with memory impairment.

Posterior Cortical Atrophy (Benson's Syndrome)

Posterior cortical atrophy is focal cortical atrophy syndrome due to neurodegenerative disease focused within the posterior regions of the cerebral cortex (i.e., the occipital and parietal lobes, Figure 15.2). It is characterized by a relatively selective and progressive disruption of higher-order visual processing, with insidious onset and

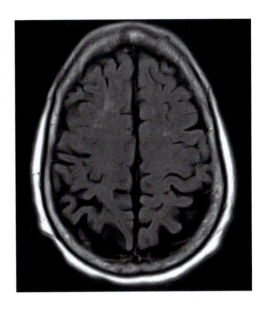

FIGURE 15.2. Posterior cortical atrophy is focal cortical atrophy due to neurodegenerative disease focused in the occipital and parietal lobes.

gradual progression; thus it is also known as **visuospatial dementia**. Posterior cortical atrophy was first described in 1988 by **D. Frank Benson** (1928–1996); thus it is also referred to as **Benson's syndrome**.

The main symptoms of posterior cortical atrophy reflect disturbances of visuoperceptual, visuospatial, and visuomotor function. Initial complaints mostly relate to vision, particularly with respect to reading, writing, walking, and driving. General visual difficulties due to posterior cortical atrophy may include difficulty identifying static objects within the visual field (inability to find objects that are "in plain sight"). Reading difficulties are often due to more generic visual problems, such as inability to track lines of text. Writing difficulties often involve inability to place numbers in columns, trouble filling out forms and writing on lines, and deterioration in handwriting. Walking difficulties involve trouble with uneven ground, stairs, sidewalk borders, and escalators. Driving difficulties often involve reduced ability to judge distances (which leads to problems in driving and parking) and reduced ability to navigate to familiar places due to trouble recognizing landmarks.

Standard optometric and ophthalmological examinations are normal, often leading to a misdiagnosis of malingering. Depression and anxiety are common early features and may lead to misdiagnosis of a primary psychiatric disorder.

Higher-order visual function testing reveals parieto-occipital dysfunction. Manifestations include deficits in visual object recognition. When presented with visual stimuli, persons with posterior cortical atrophy exhibit signs of visual difficulty, such as making positional adjustments of their head, glasses, and the object under scrutiny in an effort to decipher what they are seeing. Basic visuoperceptual processes are impaired, such as shape and pattern perception (as revealed by tests of visual form discrimination), integrating features into objects, and figure-ground discrimination. There may be visuospatial information-processing deficits, such as loss of depth perception and impaired discrimination of length and orientation. The impairment therefore conforms to an apperceptive visual agnosia (see Chapter 14, "The Occipital Lobes and Visual Processing").

Persons with posterior cortical atrophy exhibit marked inability to draw or construct simple configurations. Their approach to copying designs is piecemeal, indicating that they do not perceive the Gestalt organization of the design, and they may attempt to use verbal mediation to guide their response. Visuoperceptual disturbance underlies impairments in constructional ability.

Posterior cortical atrophy may be associated with prosopagnosia, environmental agnosia (failure to recognize familiar places), cerebral achromatopsia or hemiachromatopsia (the selective loss of color vision), simultanagnosia, optic ataxia (mis-reaching for objects), oculomotor apraxia (impaired visually guided eye movements), reverse-size effects (e.g., better ability to read small vs. large print), and other manifestations of reduced "effective visual field." This is due to visual attention and eye movement deficits, visual hemi-inattention, features of Gerstmann's syndrome (i.e., acalculia, agraphia, finger agnosia, right-left disorientation), and "dressing apraxia" (the inability to perform the task of dressing due to inability to mentally formulate the act of placing clothes on the body).

Neuropsychological assessment beyond the visuoperceptual domain requires selection of tasks which minimize visual demands, as visual deficits will have confounding effects on many general neuropsychological tests that use visual stimuli (e.g., tests of psychomotor speed requiring visual search, tests of visual memory, visual confrontation naming). Anterograde memory and language are typically preserved early in the course of posterior cortical atrophy. Insight is also preserved early in the course. The clinical diagnostic criteria for posterior cortical atrophy are presented in Box 15.1.

Age of onset is typically between 50 and 65. There is a gradual progression to a global dementia state. Management of posterior cortical atrophy incorporates visual aids through visual rehabilitation services for blind people and antidepressant medication.

Posterior cortical atrophy is associated with a variety of underlying pathologies, but most cases (80%) that come to autopsy have been found to be due to Alzheimer's pathology (extraneuritic plaques and neurofibrillary tangles) concentrated in primary visual and visual association cortices in the occipital lobes, posterior parietal lobes, inferior temporal-occipital junction regions, and posterior cingulate cortex. It is therefore also known as **biparietal Alzheimer's disease** and **visual variant Alzheimer's**

BOX 15.1 Clinical Diagnostic Criteria for Posterior Cortical Atrophy

Core diagnostic features (all five are required):

- Insidious onset and gradual progression
- Presentation of visual complaints with intact primary visual functions
- Evidence of predominant disorder of complex visual processes on examination (visual agnosia, elements of Bálint's syndrome, dressing apraxia, environmental disorientation)
- Proportionally less deficit in memory and verbal fluency
- Relatively preserved insight with or without depression

Supporting diagnostic features include:

- Presenile onset (before age 65)
- Pure alexia (i.e., no additional disturbances in language)
- Elements of Gerstmann's syndrome (agraphia and acalculia)
- Ideomotor apraxia
- Normal physical exam
- Abnormal investigations by either (a) neuropsychological examination that reveals predominant visuoperceptual deficits, or (b) neuroimaging evidence of predominant occipitoparietal abnormality, especially on functional imaging (i.e., SPECT hypoperfusion, PET hypometabolism)

Crutch SJ, Schott JM, Rabinovici GD, et al. Consensus classification of posterior cortical atrophy. *Alzheimers Dement*. 2017 Aug;13(8):870–884. doi:10.1016/j.jalz.2017.01.014.

disease. Other pathologies associated with posterior cortical atrophy are corticobasal degeneration, dementia with Lewy bodies, Creutzfeldt-Jacob disease, and subcortical gliosis.

Posterior cortical atrophy is associated with: (1) atrophy in occipital and parietal cortices on structural neuroimaging; (2) occipitoparietal hypoperfusion on single-photon emission computerized tomography (SPECT) functional imaging; (3) occipitoparietal hypometabolism on positron emission tomography (PET) functional imaging; (4) cerebrospinal fluid biomarker studies similar to typical Alzheimer's disease (see Chapter 16, "The Temporal Lobes and Associated Disorders"); and (5) positive brain beta-amyloid (the main constituent of plaques) imaging, with deposition predominantly in occipital and parietal lobes.

Summary

The parietal lobes consist of two basic functional zones: an anterior zone of primary and secondary somatosensory cortex, and a posterior zone of association cortex.

Lesions of the primary somatosensory cortex or its inputs result in deficits in discriminative touch and proprioception. Discriminative touch deficits manifest on neurological examination as elevated detection threshold, elevated two-point discrimination threshold, impaired shape perception (amorphognosia), impaired texture perception (ahylognosia), and an inability to recognize objects by touch (astereognosis). Proprioceptive deficits manifest as reduced ability to detect passive movements, and clumsy movements due to poor control from reduced feedback of position information. Lesions in secondary somatosensory cortex and its connections to association cortex result in cortical somatosensory processing disorders: tactile agnosia, agraphesthesia, and tactile anomia.

The posterior parietal association cortex plays a major role in processing spatial information by forming mental representations of space that underlie spatial cognition. Lesions to this region result in a wide range of cognitive and behavioral deficits, including visuomotor disorders (i.e., optic ataxia and oculomotor apraxia), visuospatial disorders (impaired ability to localize objects in space, judge depth, and judge angular orientation), topographic disorientation and loss of topographic memory, body schema disorders (autotopagnosia, finger agnosia, and right-left disorientation), and disorders of spatial attention (unilateral spatial neglect, simultanagnosia).

Bálint's syndrome is the triad of simultanagnosia, optic ataxia, and oculomotor apraxia; it results from bilateral superior parietal lobule lesions. Left angular gyrus syndrome consists of finger agnosia, right-left disorientation, agraphia, acalculia, anomia, and pure alexia; it occurs with focal lesions of the angular gyrus in the language-dominant hemisphere. Anosognosia is a lack of awareness or denial of neurologic deficit; it occurs with damage to the right parietal cortex. Posterior cortical atrophy is focal cortical atrophy syndrome due to neurodegenerative disease focused within the parietal and occipital lobes; it is characterized by deficits in higher-order visual processing, visuomotor functions, visuospatial functions, and spatial cognition.

Additional Reading

1. Benton AL. Gerstmann's syndrome. Arch Neurol. 1992;49(5):445–447. doi:10.1001/archneur.1992.00530290027007
2. Bisiach E, Luzzatti C. Unilateral neglect of representational space. Cortex. 1978;14(1):129–133. doi:10.1016/s0010-9452(78)80016-1
3. Caminiti R, Chafee MV, Battaglia-Mayer A, Averbeck BB, Crowe DA, Georgopoulos AP. Understanding the parietal lobe syndrome from a neurophysiological and evolutionary perspective. Eur J Neurosci. 2010;31(12):2320–2340. doi:10.1111/j.1460-9568.2010.07291.x
4. Chechlacz M. Bilateral parietal dysfunctions and disconnections in simultanagnosia and Bálint syndrome. Handb Clin Neurol. 2018;151:249–267. doi:10.1016/B978-0-444-63622-5.00012-7
5. Coslett HB, Saffran E. Simultanagnosia: to see but not two see. Brain. 1991;114 (Pt 4):1523–1545. doi:10.1093/brain/114.4.1523
6. Coslett HB, Lie G. Simultanagnosia: when a rose is not red. J Cogn Neurosci. 2008;20(1):36–48. doi:10.1162/jocn.2008.20002
7. Crutch SJ, Schott JM, Rabinovici GD, et al. Consensus classification of posterior cortical atrophy. Alzheimers Dement. 2017 Aug;13(8):870–884. doi:10.1016/j.jalz.2017.01.014. Epub 2017 Mar 2. PMID: 28259709; PMCID: PMC5788455.
8. Gainotti G, Trojano L. Constructional apraxia. Handb Clin Neurol. 2018;151:331–348. doi:10.1016/B978-0-444-63622-5.00016-4
9. Goldenberg G. Apraxia and the parietal lobes. Neuropsychologia. 2009;47(6):1449–1459. doi:10.1016/j.neuropsychologia.2008.07.014
10. Heilman KM, Valenstein E. Mechanisms underlying hemispatial neglect. Ann Neurol. 1979;5(2):166–170. doi:10.1002/ana.410050210
11. Hilti LM, Brugger P. Body integrity identity disorder: deranged body processing, right fronto-parietal dysfunction, and phenomenological experience of body incongruity. Neuropsychol Rev. 2011;21(4):320–333. doi:10.1007/s11065-011-9184-8
12. Karnath HO, Perenin MT. Cortical control of visually guided reaching: evidence from patients with optic ataxia. Cereb Cortex. 2005;15(10):1561–1569. doi:10.1093/cercor/bhi034
13. Ogden JA. Autotopagnosia: occurrence in a patient without nominal aphasia and with an intact ability to point to parts of animals and objects. Brain. 1985 Dec;108 (Pt 4):1009–1022. doi:10.1093/brain/108.4.1009. PMID: 4075073.

14. Paterson A, Zangwill OL. Disorders of visual space perception associated with lesions of the right cerebral hemisphere. *Brain*. 1944;67:331–358. doi.org/10.1093/brain/67.4.331
15. Perrine K, Uysal S, Dogali M, Luciano DJ, Devinsky O. Functional mapping of memory and other nonlinguistic cognitive abilities in adults. *Adv Neurol*. 1993;63:165–177. PMID: 8279301.
16. Rode G, Pagliari C, Huchon L, Rossetti Y, Pisella L. Semiology of neglect: an update. *Ann Phys Rehabil Med*. 2017;60(3):177–185. doi:10.1016/j.rehab.2016.03.003
17. Rusconi E. Gerstmann syndrome: historic and current perspectives. *Handb Clin Neurol*. 2018;151:395–411. doi:10.1016/B978-0-444-63622-5.00020-6
18. Salanova V. Parietal lobe epilepsy. *Handb Clin Neurol*. 2018;151:413–425. doi:10.1016/B978-0-444-63622-5.00021-8

16

The Temporal Lobes and Associated Disorders

Introduction

The temporal lobes encompass several anatomical regions with distinct functional roles; consequently, temporal lobe damage can result in a wide array of signs and symptoms. They play a central role involving the special senses of audition, olfaction, and vision, as well as the complex processes of language, memory, and emotion. The principal symptoms of temporal lobe damage are disorders of auditory perception, olfactory perception, visual perception and object recognition, episodic memory, semantic memory, language comprehension, and affective behavior.

The posterior temporal region within the middle and superior temporal gyri is the site of auditory cortex. The inferior surfaces contain olfactory cortex. The ventral temporal lobe plays a role in visual object recognition and color perception. In the language-dominant hemisphere, the posterior temporal region includes Wernicke's area, which is critical for language comprehension (see Chapter 27, "Language and the Aphasias"). The temporal polar cortical areas are critical for semantic memory. Of the mesial temporal structures, the hippocampus plays a role in episodic memory, and the amygdala plays a role in affective behavior. This chapter focuses on describing temporal lobe anatomy and its role in audition, olfaction, and Alzheimer's disease. The role of the temporal lobes in visuoperceptual processing, memory, language, and emotion is discussed in separate chapters due to the scope and importance of these processes (see Chapter 12, "Limbic Structures"; Chapter 14, "The Occipital Lobes and Visual Processing"; Chapter 26, "Memory and Amnesia"; Chapter 27, "Language and the Aphasias").

Basic Anatomy of the Temporal Lobes

The temporal lobes are named after the overlying temporal bones, so named because this is where hair begins to gray in men, indicating aging and the passage of time.

Within each hemisphere, the temporal lobe lies below the Sylvian (lateral) fissure (see Figure 13.1 in Chapter 13). The posterior boundary of the temporal lobe is an imaginary line drawn from the parieto-occipital sulcus to the pre-occipital notch (i.e., **the parieto-temporal lateral line**). The superior boundary consists of the lateral sulcus and an imaginary line that runs from the upper end of the lateral sulcus to join the posterior boundary line at a right angle.

The lateral surface of the temporal lobe is composed of three gyri: the **superior temporal gyrus** (BA22), **middle temporal gyrus** (BA21), and **inferior temporal gyrus** (BA20 and BA37; see Figure 13.4 in Chapter 13). These three gyri are separated by two sulci that run in an anterior-posterior direction: the **superior temporal sulcus** and **inferior temporal sulcus** (see Figure 13.3 in Chapter 13). The **temporal pole** is the anterior end of the temporal lobe (BA38).

Two (but sometimes only one) **transverse temporal gyri of Heschl** (BA41 and BA42) are located on the temporal operculum, the portion of the temporal lobe lying within the Sylvian fissure (lateral sulcus). This area is also known as the **temporal planum (planum temporale)**. This region is notable for significant asymmetry. The temporal planum is larger in the left hemisphere in 70% of *all* people (i.e., without taking into consideration handedness or hemispheric language dominance), and in 100% of people with left-hemisphere language dominance, as determined by functional brain imaging or Wada testing (a procedure performed before epilepsy surgery that is used to determine lateralization of language by injecting a short-acting barbiturate into one carotid artery to put the hemisphere "to sleep" and testing the functions of the "awake" hemisphere; see Chapter 19, "Epilepsy"). The left hemisphere temporal planum can be up to five times larger than the right side, making it the most asymmetrical structure of the human brain.

On the inferior surface of the temporal lobe, from lateral to medial, lie the **inferior temporal gyrus**, the anterior portion of the **lateral occipitotemporal gyrus** (also known as the **fusiform gyrus**), and the **parahippocampal gyrus** (see Figure 13.7 in Chapter 13). The parahippocampal gyrus of the temporal lobe is continuous with the **lingual gyrus** within the occipital lobe; together these are often referred to as the **medial occipitotemporal gyrus**. Posteriorly, the parahippocampal gyrus is also continuous with the cingulate gyrus at the **isthmus** (a narrow structure connecting two larger parts). The anterior end of the parahippocampal gyrus is called the **uncus** (hook) or **uncinate gyrus**; it is also known as the **pyriform** (pear-shaped) **cortex**. The **occipitotemporal sulcus** separates the inferior temporal gyrus from the lateral occipitotemporal gyrus. The **collateral sulcus** separates the lateral occipitotemporal gyrus from the medial occipitotemporal gyrus.

The mesial temporal lobe structures include the **amygdaloid complex** anteriorly and the **hippocampus** posteriorly (see Figure 12.3 in Chapter 12). These structures occupy the floor of the temporal horn of the lateral ventricle.

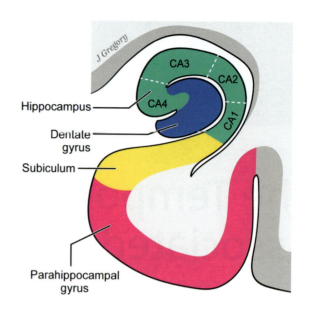

FIGURE 16.1. Coronal section through the medial temporal lobe showing the hippocampus, dentate gyrus, subiculum, and parahippocampal gyrus. The hippocampus is divided into 4 fields, CA1–CA4. In cross section the hippocampus and dentate gyrus look like two interlocking Cs.

The Hippocampal Formation

ANATOMY

The **hippocampal formation** is located within the mesial temporal lobe. It is about 5 cm in length from its anterior end at the amygdala to its posterior end near the splenium of the corpus callosum. The hippocampal formation consists of three components: the hippocampus, dentate gyrus, and subiculum (Figure 16.1). The **hippocampus** (Greek, "seahorse") is so named because when dissected out of the brain with the fornix, it resembles a seahorse. It is also referred to as the **hippocampus proper**, to distinguish it from the hippocampal formation. In cross section, the hippocampus resembles a ram's horn, thus it is also referred to as **Ammon's horn** (*Cornu Ammonis* in Latin), Ammon being an early Egyptian deity with a ram's head. The hippocampus and dentate gyrus look like two interlocking Cs; together they comprise the **hippocampal-dentate complex**. The **hippocampal complex** comprises the hippocampal formation and the adjacent parahippocampal gyrus.

The hippocampus is divided into four fields, CA1–CA4, with CA standing for Cornu Ammonis (see Figure 16.1). CA1, also known as **Sommer's sector**, is the largest in humans. The hippocampus, and the CA1 field in particular, is highly vulnerable to anoxic and ischemic damage. In controlled laboratory experiments examining location and extent of neuropathology following variable intervals of cerebral ischemia or hypoxia, the CA1 region is one of the first brain regions to show pathological changes.

HIPPOCAMPAL CONNECTIONS

There are three afferent pathways into the hippocampal formation: the perforant path, alveus, and fornix. The majority of inputs to the hippocampal formation come from a part of the parahippocampal gyrus known as the **entorhinal cortex** (so named because it is located within the rhinal sulcus), which in turn receives its input from widespread areas of the neocortex, especially the cingulate gyrus and prefrontal cortex. The **perforant path** carries entorhinal inputs into the dentate gyrus, while the **alveus** (**alvear pathway**) carries entorhinal inputs into the hippocampus proper. Other inputs from the septal area, basal forebrain cholinergic nuclei, thalamic and hypothalamic nuclei, and contralateral hippocampus enter the hippocampal formation via the **fornix**, an arch-like white matter tract.

All hippocampal formation efferents are carried in the fornices (Figure 16.2). Each human fornix contains more than one million myelinated axons. The fibers curve upward forming the posterior columns, then course anteriorly and converge forming the body of the fornix, and then diverge and curve inferiorly forming the anterior columns. The efferents terminate in the septal region, nucleus accumbens, mammillary bodies of the hypothalamus, and anterior thalamic nuclei.

The hippocampal formations of the right and left hemispheres are also connected to each other via the **hippocampal commissure** (also known as the **psalterium**, or **commissure of the fornix**), which runs between the two **crura**.

HIPPOCAMPAL FUNCTION

The hippocampal formation plays a major role in learning and memory (see Chapter 26, "Memory and Amnesia"). Bilateral lesions of the hippocampi produce anterograde amnesia, a selective impairment in the ability to learn new information and form new memories. Isolated damage to the hippocampal formation does not abolish recall of older memories because the memory traces are stored in the cerebral cortex after a certain amount of time. Interruption of structures that form circuits with the hippocampal formation (such as the fornices) also produces anterograde amnesia. Early-stage Alzheimer's disease is typically characterized by anterograde amnesia as the earliest and most prominent symptom because the earliest neuropathological changes occur in the hippocampus and entorhinal cortex (see "Alzheimer's Disease," below).

Blood Supply to the Temporal Lobes

The convexity of each temporal lobe receives its blood supply from the inferior branch of the middle cerebral artery. The medial and inferior aspects of the temporal lobe, including the hippocampus, receive their blood supply from the temporal branch of the posterior cerebral artery. As such, ischemic stroke in the middle or posterior cerebral arteries can lead to specific deficits in temporal lobe functions.

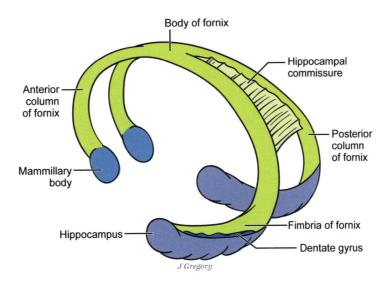

FIGURE 16.2. The fornix is an arch-like white matter tract that carries all efferents from and some afferents to the hippocampal formation. The fibers leave each hippocampus via the fimbria and crura, which converge to form the body of the fornix, and then diverge and curve inferiorly to form the columns of the fornix. The hippocampal formations of the right and left hemispheres are interconnected via the hippocampal commissure.

Audition

Auditory cortex is defined as the region of cerebral cortex devoted exclusively to processing of auditory input. Primary and secondary auditory areas occupy the superior temporal gyrus. Lesions to these areas result in auditory processing disorders.

The Auditory Stimulus

What is sound? We define a sound as what we hear. That definition is based on perception, but it also has a physical existence outside us as a **sound wave**. Sound is composed of variations in pressure within an elastic medium (e.g., air). Sound travels in waves of condensations and rarefactions of pressure that are produced by objects that may collide or vibrate and initiate the propagation of a pressure wave in the surrounding medium. One condensation and one rarefaction make up a single cycle of this traveling wave. Sound waves travel in all directions from a source, not unlike ripples in a body of still water produced by an object thrown into it (except that sound spreads in three dimensions). Of course, media other than air also carry sound; this is why you can hear sounds when you are under water.

Sound waves travel away from the object of origin at the speed of sound, which is approximately 700 miles per hour in air. (Light travels almost a million times that speed, at 186,000 miles per second.) "Pure" sound waves that are used in assessing hearing are sinusoidal and quantified in terms of frequency and amplitude. **Frequency** is measured in cycles per second (cps) or hertz (Hz). The **amplitude** of a sound wave is the degree to which condensations and rarefactions differ from each other; the logarithmic measure of sound intensity is the **decibel (dB) scale**. The human auditory system responds to frequencies within the 30–20,000 Hz range; thus these frequencies qualify as sound waves for humans. Different species, however, are sensitive to different ranges of the sound spectrum.

Sound has three perceptual dimensions: pitch, loudness, and timbre. The frequency of a sound wave is a physical property and is related to the subjective psychological variable called **pitch**. The amplitude of the waveform is related to perceived **loudness**. The loudest sound that humans can hear without discomfort is over 1 trillion times more intense than the softest audible sound. The human dynamic range is 120 dB. Natural sounds have complex waveforms that can be decomposed into a mixture of simple sinusoidal waveforms that vary in frequency, amplitude, and phase (registration). The composition of a waveform (i.e., the particular mixture of frequencies) includes the relative intensities and phase relationships, and it relates to the perceptual variable called **timbre** or sound quality.

The Ear

The ear is divided into outer, middle, and inner components (Figure 16.3). The **outer (external) ear** consists of the **pinna (auricle)** and ear canal (**auditory meatus**). In some species, the pinna possesses muscles that allow it to move. The ear canal terminates at the **tympanic membrane** (eardrum), which is the boundary between the external and middle ear.

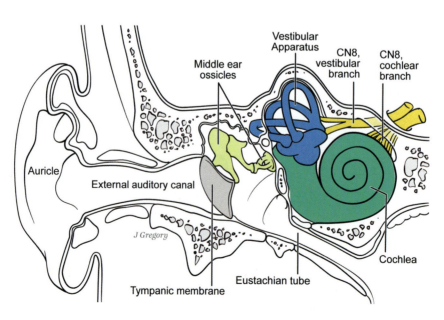

FIGURE 16.3. Both the cochlea and vestibular apparatus are innervated by the vestibulocochlear nerve (CN8). This cranial nerve has a branch that innervates the cochlea (cochlear branch), and a branch that innervates the vestibular apparatus (vestibular branch).

The **middle ear** contains air and connects to the nasopharynx by the eustachian tube, which provides aeration and is responsible for equalizing pressure on both sides of the eardrum. A chain of three bones, or **ossicles**, connects the tympanic membrane to the **oval window** of the cochlea. In sequential order, these bones are the **malleus** (hammer), **incus** (anvil), and **stapes** (stirrup).

The **inner ear**, also known as the **labyrinth**, is located within the **petrous** part of the temporal bone. It contains the **cochlea (auditory labyrinth)**, which is specialized for audition, and the **vestibular apparatus (vestibular labyrinth)**, which is specialized for sensing head position and head movements and plays an important role in balance and cerebellar function. Both the cochlea and vestibular apparatus are innervated by the **vestibulocochlear nerve** (CN8); this cranial nerve has a branch that innervates the cochlea (**cochlear branch**), and a branch that innervates the vestibular apparatus (**vestibular branch**).

The cochlea is a spiral-shaped cavity; its name derives from the Greek word *kokhlos*, meaning "land snail." The cochlea can be visualized as a 1-inch tunnel that spirals for 2¾ turns. Straightened out, the basal end of the cochlear spiral is closest to the stapes, and the apical end is farthest from the stapes. In cross section, the cochlea is roughly circular. Two membranes divide the cochlea into three fluid-filled compartments that each run the full length of the cochlea; the scala vestibuli, the scala media, and the scala tympani. The scala media, also known as the **cochlear duct**, contains the **organ of Corti**. The essential components of the organ of Corti are the **basilar membrane**, **auditory receptor hair cells**, and **tectorial membrane**. The auditory receptor hair cells lie between the basilar and tectorial membranes (Figure 16.4). At their apical end, they have hairs called **cilia** which are rigid and touch the tectorial membrane.

Much of the auditory system's frequency analysis is performed mechanically by the cochlea. Different frequencies

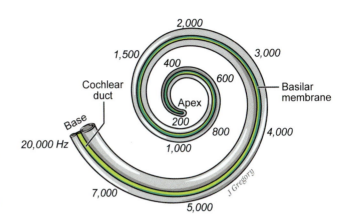

FIGURE 16.5. Sound frequency is coded spatially along the basilar membrane. Low frequencies produce maximal displacement of the basilar membrane near the apex, and high frequencies produce maximal displacement near the base.

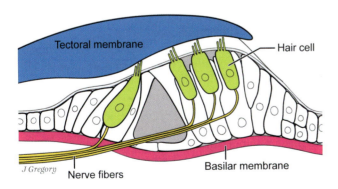

FIGURE 16.4. Within the cochlea, the essential components of the organ of Corti are the basilar membrane, auditory receptor hair cells, and tectorial membrane. Nerve impulses are initiated by displacement of auditory receptor hair cell cilia by movement of the basilar membrane relative to the tectorial membrane.

produce maximal displacement at different portions of the basilar membrane. The basal end of the basilar membrane is displaced by higher frequencies, while the apical end is displaced by lower frequencies. This **frequency tuning** is a consequence of variations in the physical characteristics along the basilar membrane. Frequency is therefore coded spatially along the basilar membrane (Figure 16.5). Since the majority of the auditory nerve cells innervate a restricted portion of the basilar membrane, it is not surprising that they show **frequency selectivity** (i.e., they fire only in response to a restricted range of frequencies).

Sound wave condensations cause inward displacement of the stapes against the oval window and sound wave rarefactions cause the stapes to move outward. These movements result in pressure waves that travel through the fluid of the scala vestibuli and scala tympani, causing displacement of a region of the basilar membrane. As described above, because the physical characteristics of the basilar membrane vary along its length, the displacement is maximal at the region that has a natural resonant frequency equal to the respective sound frequency. Movement of a region of the basilar membrane relative to the tectorial membrane causes bending of the cilia of the hair cells and the generation of receptor potentials. Thus, the hair cells transduce mechanical disturbances of the inner ear fluids, caused by sound waves impinging on the eardrum, into electrical signals (receptor potentials). Auditory hair cells synapse on dendrites of the primary afferents, which make up the cochlear branch of the vestibulocochlear nerve. The primary afferents are bipolar neurons; their cell bodies reside in the **cochlear nerve ganglion** (also known as the **spiral ganglion**).

In summary, (1) sound waves are transmitted through outer ear and middle ear ossicles; (2) the footplate of the stapes transmits the vibrations through the **oval window** into the fluid of the cochlea; (3) the pressure waves travel through the perilymph and fluid vibrations cause the basilar membrane to move relative to the tectorial membrane; and (4) this movement causes receptor hair cell cilia to bend and initiates nerve impulses.

The Auditory Pathways

The auditory pathways are complex, with multiple subcortical nuclei, several decussations, and an ipsilateral representation (Figure 16.6). The pathway consists of the organ of Corti within the cochlea, the cochlear branch of the eighth cranial nerve (CN8), the **dorsal and ventral cochlear nuclei** within the medulla, the **superior olivary nucleus** within the pons, the lateral lemniscus within the pons and midbrain, the **inferior colliculus** within the midbrain tectum, the brachium of the inferior colliculus (inferior brachium), the **medial geniculate nucleus** within the diencephalon, the auditory radiations, the primary auditory cortex, and secondary auditory cortical areas.

The fifth and final synapse of the primary auditory pathway occurs in the primary auditory cortex. **Primary auditory cortex** (A1, BA41) lies hidden within the lateral fissure on the temporal operculum of the superior temporal gyrus, within Heschl's gyri (described above). It receives direct projections from the ascending auditory pathway via the **auditory radiations**. Information from one ear is represented in both hemispheres, with the contralateral hemisphere receiving the majority (80%) of inputs and the ipsilateral hemisphere receiving the minority (20%). Thus, the primary auditory cortices have a bilateral representation.

Just as the body is mapped onto the postcentral gyrus (somatotopy) and the retina is mapped onto striate cortex (retinotopy), the cochlear receptor surface is mapped onto primary auditory cortex (**cochleotopy**). Consequently, a topographic representation of the cochlear receptor surface within primary auditory cortex (**cochleotopic map**) results in a **tonotopic map**. High frequencies are represented medially within the primary auditory cortex of the temporal operculum, while low frequencies are represented laterally.

Secondary auditory cortex (auditory association cortex, A2, BA42) receives projections from the primary auditory cortex. It lies immediately rostral to primary auditory cortex. Secondary auditory cortex plays an important role in the analysis of complex sounds. It also shows lateral specialization. Typically, the left temporal region is specialized for analyzing speech sounds, the phonetic and linguistic features of incoming auditory information. It is a part of Wernicke's area (which comprises part of BA22, as well as BA39/angular gyrus of the parietal lobe). The right temporal region is specialized for identifying nonverbal environmental sounds, speech prosody (melodic nuances that convey emotional meaning), and most aspects of music.

Auditory Deficits

There are two general types of deafness: **conduction deafness**, which is caused by impairment of the mechanisms for transmitting sounds to the cochlea (e.g., otosclerosis); and **nerve deafness** (**sensorineural hearing loss**), which is caused by impairment in the cochlea, auditory nerve, or CNS auditory circuits. The neurological exam can differentiate conduction versus nerve deafness (see Chapter 30, "The Neurological Examination").

Age-related hearing loss (**presbycusis**) is very common in older adults, especially at high frequencies. It is usually due to damage to hair cells of the inner ear, and thus it is a form of sensorineural hearing loss. **Hearing aids** are primarily used to improve hearing in those with this type of hearing loss, particularly when the losses affect frequencies that are important in speech intelligibility (2–4 kHz). Hearing aids amplify sound vibrations entering the ear. The more severe the hearing loss, the greater the need for amplification. Some hearing aid technology may be tuned to compensate for specific frequency loss profiles.

In persons with moderate to profound sensorineural hearing loss due to receptor hair cell damage, **cochlear implants** may be used to send sound signals directly to

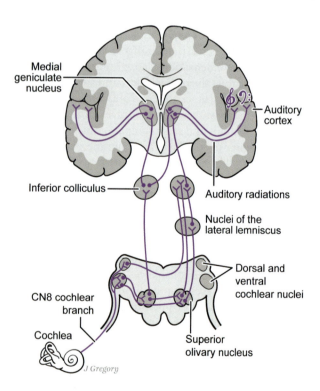

FIGURE 16.6. The auditory pathways are complex, with multiple subcortical nuclei, several decussations, and both contralateral and ipsilateral representations. Within the primary auditory cortex of the temporal operculum, high frequencies are represented medially while low frequencies are represented laterally.

the auditory nerve. This **neuroprosthetic device** has two components: a receiver-stimulator which is surgically implanted in the cochlea, and a speech processor that is worn behind the ear like a hearing aid.

AUDITORY EVOKED POTENTIALS

Evoked potentials are electrical changes in the brain that are evoked by a stimulus, such as a sound or flash of light, and recorded synchronously through macroelectrodes placed on the scalp. Because the recorded electrical signals are of very low amplitude, they must be averaged over many stimuli (more than 500) in order to cancel out random background electrical activity and discern the response. Evoked potentials are used clinically to evaluate conduction within specific sensory pathways (i.e., somatosensory, visual, auditory).

The **auditory evoked potential** reflects electrical activity of the nervous system in response to a brief acoustic stimulus, such as a click or tone pip (a short duration pure tone). It can be decomposed into several parts; short latency components reflect neural activity within the lower levels of the auditory pathways (e.g., auditory nerve), while long latency components reflect neural activity within higher levels (e.g., auditory cortex). The **brainstem auditory evoked potential** comprises the short-latency potentials that occur within the first 10 msec following a transient acoustic stimulus. It consists of five peaks; the amplitude and latency of each potential are measured. The brainstem auditory evoked potential reflects the functional integrity of the auditory pathways, from the nerve to the thalamocortical auditory radiations; thus the term "brainstem auditory potential" is a misnomer, since some components of the electrical response do not originate in the brainstem.

Cortical Auditory Disorders

The cortical auditory disorders are cortical deafness, auditory verbal agnosia, and auditory sound agnosia.

CORTICAL DEAFNESS

Cortical deafness is deafness caused by cortical lesions. Due to the bilateral representation of each ear, total cortical deafness requires bilateral lesions of the auditory radiations or primary auditory cortices. This is most often caused by bilateral embolic stroke affecting Heschl's gyri; however, this is very rare because the arteries supplying these gyri do not share a parent vessel. Thus, few cases of this disorder have been studied. Unilateral lesions of primary auditory cortex produce impaired sound localization and increased auditory thresholds bilaterally, with no marked effect on auditory acuity (e.g., single-double click discrimination).

AUDITORY VERBAL AGNOSIA

Auditory verbal agnosia is a selective inability to understand spoken words in the context of normal hearing and otherwise normal language function. It was originally described by **Hugo Liepmann** (1863–1925), who termed it **pure word deafness**, a term that remains in use despite the fact that there is no hearing loss. Persons with auditory verbal agnosia complain that speech sounds muffled, like a foreign language, or like meaningless noise. Hearing loss is ruled out as the underlying cause of the disorder by normal audiological exam, normal perception of nonverbal sounds, and normal perception of the paralinguistic aspects of speech (e.g., they can recognize who is speaking). Brainstem auditory evoked potentials are normal, indicating intact processing up to the level of the auditory radiations.

Auditory verbal agnosia occurs with cortical and subcortical lesions of the middle and posterior parts of the superior temporal gyrus in the language-dominant (usually left) hemisphere, with sparing of Heschl's gyrus, usually due to a cerebrovascular etiology. The lesion results in a bilateral disconnection of Wernicke's language comprehension area from all acoustic input (i.e., both auditory cortices are disconnected from an intact Wernicke's area). Consequently, there is an inability to associate linguistic sounds with their meanings. Writing to dictation and repetition, which depend on auditory input, are also impaired. All aspects of language that do not depend on audition, such as reading, writing, and speaking, are preserved. Auditory verbal agnosia is usually seen in those recovering from Wernicke's aphasia (see Chapter 27, "Language and the Aphasias"); the disturbances in spoken language expression, reading, and writing resolve, but the disturbance in spoken language comprehension remains. The language disturbance is "pure" in that it affects only comprehension of spoken language.

AUDITORY SOUND AGNOSIA

Auditory sound agnosia (**acoustic agnosia**) is a selective deficit in recognizing and identifying nonverbal sounds. This affects the ability to recognize and identify environmental sounds (e.g., birds singing, keys jangling, water running, dogs barking). Auditory sound agnosia occurs with lesions of the secondary auditory cortex in the superior temporal gyrus of the non-language-dominant (usually right) hemisphere.

Other aspects of acoustic agnosia are receptive amusia and paralinguistic agnosia. **Receptive amusia** is a loss of the ability to appreciate music due to impaired music perception (timbre, tone, pitch, melody) and impaired recognition of familiar melodies. **Paralinguistic agnosia** is a non-linguistic auditory-verbal recognition deficit; it encompasses phonagnosia and receptive aprosodia.

Phonagnosia is an inability to recognize familiar voices; this can be considered as the auditory analog of prosopagnosia. **Receptive aprosodia** (**auditory affective agnosia**) is impaired comprehension of prosody. Variations in prosody (volume, timbre, pitch, and rhythm) often convey information about speakers' emotional states. Persons with receptive aprosodia are unable to judge the emotion conveyed by tone of voice. When presented with neutral sentences spoken in an emotional manner, persons with auditory sound agnosia are unable to recognize the affect conveyed by the tone of voice, although they are able to comprehend the linguistic content of the message. Thus, comprehension of affective tone and linguistic speech content are dissociable; they can be selectively impaired. Receptive amusia, phonagnosia, and receptive aprosodia often co-occur, but they can occur as isolated deficits.

Auditory Illusions and Hallucinations

Auditory illusions are false perceptions of a sound in which the listener hears distortions that are not present in a sound stimulus. Auditory hallucinations (**paracusia**) are perceptions of sounds without identifiable external stimuli. In addition to classification based on the affected sensory modality, hallucinations are classified on the basis of the perceptual content as simple (unformed) hallucinations or complex (formed) hallucinations. Simple auditory hallucinations take the form of distinct or indistinct noises, such as buzzing, clicking, ticking, humming, murmurs, knocking, whispering, blowing, whistles, ringing, sirens, footsteps, clapping, running water, or the sound of motors. Complex auditory hallucinations take the form of well-articulated speech, singing, or music. Electrical stimulation of Heschl's gyrus produces auditory hallucinations.

Auditory illusions and hallucinations have a variety of etiologies, including pathologies affecting the auditory pathways. **Tinnitus** is an extremely common simple auditory hallucination that often occurs even with mild age-related hearing loss; it is most often described as ringing in the ears, but can also include buzzing, roaring, clicking, hissing, or humming. Interestingly, the underlying mechanism(s) are unknown. **Acoustic neuroma**, a benign tumor that involves overproduction of Schwann cells of the eighth cranial nerve (CN8), may produce hearing loss and/or tinnitus. Tumors in cortical auditory areas often produce auditory illusions, such as sounds seeming louder, softer, more distant, or strange. Irritative lesions (i.e., lesions that stimulate neural activity) such as seizure foci and tumors affecting primary and secondary auditory areas often produce auditory hallucinations; these are usually localized as originating in the contralesional auditory space.

Auditory hallucinations also occur with lesions to brainstem structures of the central auditory pathways, and most commonly with acute lesions of the pontine tegmentum, a condition known as **pontine auditory hallucinosis**. These hallucinations occur in association with central auditory processing disorders (e.g., central hearing loss, impaired sound localization); they are believed to be release hallucinations (hallucinations that occur in the absence of normal sensory input).

Auditory verbal hallucinations are usually associated with schizophrenia. Functional neuroimaging studies performed in those actively experiencing such hallucinations at the time of scanning have shown spontaneous activation in the superior posterior left temporal lobe, similar to the activation pattern observed in healthy subjects when listening to speech.

Olfaction

Olfaction, the sense of smell, is one of the **chemical senses** (i.e., a sensory system dedicated to detecting chemical signals). **Gustation**, the sense of taste, is the other main chemical sense. A third chemical sense is the chemoreceptive component of the trigeminal nerve that detects intranasal irritants such as ammonia.

The human olfactory system can discriminate thousands of different odors that can all be classified into six major groups: floral, fruit, spicy, resin, burnt, and putrid. Odorant molecules in the air enter the nasal passages, dissolve in the watery **olfactory mucosa** (**olfactory epithelium**), and bind with **chemoreceptor** protein molecules embedded in the membranes of olfactory receptor neurons. These neurons transduce the chemical signals into neural signals. Their axons enter the cranial cavity through perforations in the cribriform plate of the ethmoid bone as multiple small nerve bundles (**olfactory filaments/fila**) that are collectively referred to as the **olfactory nerve** (CN1; Figure 16.7).

Olfactory nerve axons synapse on the **olfactory bulbs**, which lie on the inferior surface of the frontal lobes (see Figure 6.5 in Chapter 6). From each bulb, second-order neurons give rise to the **olfactory tract**, which runs on the inferior surface of the frontal lobes within the **olfactory sulcus**, also known as the **olfactory groove**. Each olfactory tract divides into **lateral** and **medial olfactory striae**, and each stria projects to multiple brain regions. The main olfactory bulb projections travel via the lateral olfactory stria to **primary olfactory cortex**, located in the pyriform cortex (uncus) and adjoining the entorhinal area of the anterior parahippocampal gyrus. Other regions also receive direct inputs from the olfactory bulb, including the amygdala. The parts of the brain directly involved in olfactory processing are sometimes referred to as the **rhinencephalon** ("nose brain"). The exact roles of the various anatomical regions of the olfactory system are not fully

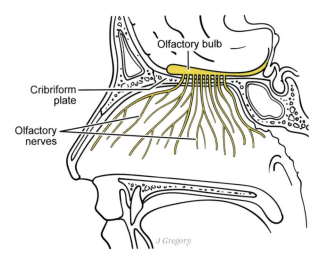

FIGURE 16.7. Olfactory afferents enter the cranial cavity through perforations in the cribriform plate of the ethmoid bone as multiple olfactory filaments that are collectively referred to as the olfactory nerve. Olfactory nerve axons synapse on the olfactory bulbs, which lie on the inferior surface of the frontal lobes.

elucidated, but these diverse projections allow olfactory cues to influence cognitive, visceral, emotional, and homeostatic behaviors.

The olfactory pathways differ from other sensory pathways in several important ways. First, they do not involve a thalamic relay before entering the cerebral cortex: olfactory cortex receives direct input from the olfactory bulb. Second, they project ipsilaterally rather than contralaterally from the sense organ to the brain; signals originating in the left nostril project to the left hemisphere, while signals originating in the right nostril project to the right hemisphere. Finally, olfactory cortex does not contain a spatial map of the receptor surface, as such a map would not provide useful information about stimulus location (i.e., the location of the olfactory signal source is not coded on the olfactory epithelium).

Anosmia

Damage to the olfactory pathways and olfactory cortex produces **anosmia**, a total or partial loss of the sense of smell. Since olfactory sensation also contributes to our experience of taste, persons with anosmia often complain of a loss of pleasure from eating. Anosmia may be tested with **scratch-and-sniff tests**. Anosmia occurs with lesions of the anterior inferior aspect of the brain (although the most common cause is sinus infection). Lesions in this region commonly occur with traumatic brain injury. Temporal lobectomy, in which the anterior portion of one temporal lobe is surgically removed as an intervention for medically refractory seizures of temporal lobe origin, results in partial anosmia with elevated thresholds for detection and identification of odors, and reduced discrimination of odors.

Olfactory Hallucinations

Olfactory hallucinations, also referred to as osmic hallucinations or **phantosmia** (for phantom smell), is the experience of sensing an odor without a specific initiating stimulus. Typically, these hallucinations are experienced as unpleasant, and described as a rotten, burnt, or chemical-like smell. They are associated with several conditions, including traumatic brain injury, brain tumors, and seizures, as well as in circumstances with no known neurological disorder. Focal seizures originating from the temporal lobe often begin with an olfactory hallucination as an aura (the initial symptom and warning sign of an impending seizure), due to onset of abnormal electrical activity within the uncus of the temporal lobe before spreading; these auras are therefore known as **uncinate seizures** (see Chapter 19, "Epilepsy").

Alzheimer's Disease

Alzheimer's disease was first characterized in 1907 by **Aloysius ("Alöis") Alzheimer** (1864–1915), who described the symptoms of a 51-year-old woman, Auguste Deter, who was under his care at the state asylum in Frankfurt, Germany. She had a progressive dementia characterized by profound memory loss. Following her death, he performed an autopsy and a histological examination of her brain tissue using a silver staining technique. Throughout the cerebral cortex he observed small spheric extracellular structures within the cerebral cortex and intracellular flame-shaped fiber-like bundles. These later became known as extraneuritic plaques and neurofibrillary tangles, respectively, and are now known as the hallmark pathologies of the disease. The disease was named after Alzheimer by his mentor **Emil Kraepelin** (1856–1926) from the Munich Medical School, who first used the term "Alzheimer's disease" in his *Handbook of Psychiatry*.

Clinical Presentation

Neurodegenerative disease is typically characterized by insidious onset and gradual progression of signs and symptoms. The initial stage of Alzheimer's disease is typically characterized by **amnesia**, a relatively selective decline in memory for recent information. The onset is insidious, beginning with forgetting minor day-to-day events. Over time, there is a gradually progressive decline in memory for recent information, leading to forgetting recent conversations and repeating questions and comments within a short period of time.

Other cognitive deficits eventually emerge. These usually begin with **anomia** (significant word-finding difficulties). With disease progression, there are declines in ability to navigate in familiar places (topographic disorientation), ability to perform calculations (**acalculia**, see Chapter 28, "Alexia, Agraphia, and Acalculia"), abstract thinking, judgment, and comprehension of complex material, as well as anosognosia (unawareness of deficit). Late in the course, thinking becomes perseverative and tangential, and there are declines in ability to perform skilled movements such as dressing and using tools (**apraxia**; see Chapter 24, "The Motor System and Motor Disorders").

In general, personality is unaffected in the early phases of Alzheimer's disease, but in the middle to late stages, neuropsychiatric symptoms such as **delusions** (firmly held beliefs in things that are not real) may emerge. In Alzheimer's disease and other dementias, delusions of persecution predominate. Common persecutory delusions include beliefs that others are stealing from the person, that others intend to physically harm or mistreat the person or their family, that neighbors or landlords are trying to evict the person, and that the person's spouse has been unfaithful. **Delusions of misidentification**, of which there are several types, are also common in Alzheimer's disease and other neurodegenerative dementias. In **Capgras syndrome**, the person believes that his or her spouse is an imposter. **Reduplicative paramnesia** is the belief that a place or location has been duplicated or relocated; most commonly, persons with this delusion insist their residence is not their real home but an exact replica. In **phantom boarder syndrome**, the person believes that their home is inhabited by unwelcome guests. In Alzheimer's disease and other dementias, the content of the delusions is nonfantastic (i.e., does not involve supernatural, science fiction, or religious themes), unlike the fantastic delusions that often occur in the context of schizophrenia.

Sleep disorders are common in Alzheimer's disease and include excessive awakenings, early morning awakening, excessive daytime sleepiness, napping for more than one hour during the day, and in extreme cases, a complete day/night sleep pattern reversal (e.g., the person may dress to go out in the middle of the night). Sleep disturbances tend to correlate with the severity of cognitive decline. Abnormalities in other circadian rhythms such as body temperature and hormone concentrations are also observed in those with Alzheimer's disease. The **sundowning** phenomenon, an exacerbation of behavioral symptoms of dementia in the late afternoon and evening, may be related to circadian rhythm abnormalities.

In the terminal phase of Alzheimer's disease, there is essentially a complete loss of higher cognitive control. A state of **decortication** eventually develops, with limb and eventually whole-body contractures until the person eventually assumes a fetal position. At this stage, death usually occurs secondary to pneumonia or aspiration.

Epidemiology

Alzheimer's disease is the most prevalent neurodegenerative disease and accounts for the majority (approximately 70%) of dementia cases. Five percent of the population over the age of 65 has clinical Alzheimer's disease (dementia of the Alzheimer type). After age 65, the risk doubles every 4 5 years, and 25%–50% of people over age 85 meet criteria for a diagnosis of Alzheimer's disease. Illness length typically ranges from 5 to 15 years, with a mean duration of 7 years.

Pathology

The major pathological findings in Alzheimer's disease are cortical atrophy, extraneuritic plaques that form outside cells, and neurofibrillary tangles that form inside cells.

Cortical atrophy is evidenced by widened sulci and thinned gyri that are most evident in the superior and middle temporal gyri (Figure 16.8). There is diffuse ventricular dilation, sometimes particularly prominent in the temporal horns. Structural neuroimaging studies reveal a shrunken, walnut-like appearance on high cuts over the cortex, ventricular enlargement, large basal cisterns, and large Sylvian fissures on cuts through the ventricular system. MRI may reveal marked hippocampal atrophy and enlarged temporal horns of the lateral ventricles. The primary motor and sensory cortices are unaffected. The early stage of Alzheimer's disease typically involves a relatively

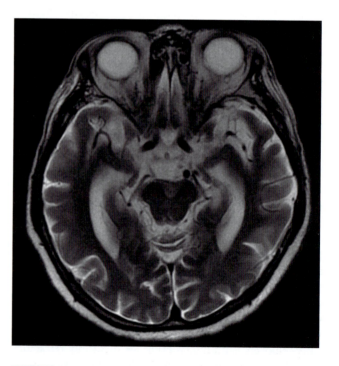

FIGURE 16.8. Axial T2 image of the brain demonstrates bilateral temporal lobe volume loss with enlargement of the temporal horns in Alzheimer's disease.

selective neurodegeneration within the hippocampus, especially in the CA1 and CA2 fields, and the entorhinal cortex of the temporal lobes, accounting for the amnesia. Over time the cell loss becomes more diffuse, spreading to the lateral temporal, parietal, and frontal association cortices, but predominantly affecting the large pyramidal neurons. The more extensive the neocortical atrophy, the more diffuse and severe is the cognitive dysfunction. In advanced stages, there is extreme atrophy of the hippocampi, diffuse cortical atrophy with thinned gyri and widened sulci, and enlarged lateral and third ventricles. Brain weight may be reduced by 20% or more.

Extraneuritic plaques and neurofibrillary tangles are the histopathologic hallmark of Alzheimer's disease. They are identified by postmortem histological tissue examination, providing a tissue diagnosis of the specific disease process as well as clues about disease pathogenesis. **Extraneuritic plaques** are extracellular products of neuronal degeneration, with a characteristic core of **β-amyloid** protein that is mixed with branches of dying nerve cells. It is normal for the body to produce amyloid protein fragments, but normally they are broken down and eliminated. **Neurofibrillary tangles** consist of abnormal tangled bundles of fine fibers within cell bodies that consist primarily of abnormal clumps of **tau** protein. Tau is a normal component of microtubules, but in Alzheimer's disease the tau protein is hyperphosphorylated and aggregates into paired helical filaments that make up the neurofibrillary tangles, causing the microtubule structure to collapse. Thus, Alzheimer's disease is a **tauopathy**.

Longitudinal studies examining cognitive function during life and brain pathology at death have shown that (1) the earliest neuropathological changes occur in the hippocampus and entorhinal cortex, (2) later stages are characterized by extensive neocortical atrophy, and (3) there is a **preclinical stage** of disease when the pathophysiological process has begun but clinical symptoms are not yet manifest. Much research is therefore directed toward understanding the mechanisms by which extraneuritic plaques and neurofibrillary tangles form, and interventions to prevent plaque and tangle formation.

Etiology

The etiology of Alzheimer's disease is unknown. Age is the single greatest risk factor. There is some genetic predisposition to develop Alzheimer's disease. Mutations on three separate genes have been identified in families with an autosomal dominant inheritance pattern of early-onset Alzheimer's disease (i.e., generally before age 60): the amyloid precursor protein (APP) gene on chromosome 21, the presenilin 1 gene on chromosome 14, and the presenilin 2 gene on chromosome 1. These forms of familial Alzheimer's disease, however, are rare and account for only 1%–2% of all cases of the disease.

Most cases of Alzheimer's disease are **sporadic** (i.e., occurring irregularly, without an inherited pattern). There are, however, genetic risk factors for susceptibility to develop the disease. The most potent susceptibility gene for sporadic, late-onset Alzheimer's disease is the ε4 allele of the gene for apolipoprotein E (ApoE), a low-density lipoprotein cholesterol carrier. The ApoE gene is located on chromosome 19; there are three major isoforms of the gene: ε2, ε3, and ε4 (corresponding to the proteins ApoE2, ApoE3, and ApoE4). Unlike the genes associated with early-onset familial Alzheimer's disease, the ApoE ε4 allele is not deterministic; rather, it increases susceptibility to developing Alzheimer's disease. If one copy of the ε4 allele is present, the risk of developing sporadic Alzheimer's disease is increased threefold. If two copies of the ε4 allele are present, the risk of developing sporadic Alzheimer's disease is increased eightfold.

THE AMYLOID HYPOTHESIS

The **amyloid hypothesis** of Alzheimer's disease posits that accumulation of β-amyloid in the brain triggers a neurodegenerative process that leads to memory loss and dementia. According to this hypothesis, abnormal processing of the amyloid precursor protein leads to the abnormal accumulation of β-amyloid deposits (**amyloidosis**) in the brain, which drives the formation of abnormal tau aggregates, which leads to tangle-mediated neuronal injury and neurodegeneration, which then produces cognitive and functional impairment. This hypothesis has been very influential in guiding research aimed at developing disease-modifying treatment strategies for Alzheimer's disease; if amyloid accumulation could be prevented by promoting amyloid clearance, the cascade of events leading to neurodegeneration and cognitive decline also could be prevented.

The amyloid hypothesis, however, has been called into question. There is no clear relationship between amyloid deposition and neuronal degeneration. While plaques and tangles are found throughout the association areas of the cerebral cortex, only neurofibrillary tangles and cell loss correlate with dementia severity; amyloid plaque burden does not. Amyloid plaque pathology accumulates diffusely across neocortex, while neurofibrillary tangle pathology begins in the medial temporal lobes and progresses to adjacent association cortices and beyond in a pattern consistent with the profile of neurocognitive decline. Neurodegeneration can occur before amyloidosis in individuals with prodromal Alzheimer's disease (i.e., the stage when early signs and symptoms of a disease appear but are not yet clinically specific or severe). Thus far, interventions based on the amyloid hypothesis of Alzheimer's disease have been disappointing. Some recent research is focusing on tau-altering pharmacologic interventions.

Diagnosis

Comprehensive diagnostic evaluation of Alzheimer's disease consists of a detailed clinical history, neurologic examination, neuropsychological evaluation, MRI structural brain imaging study, blood chemistry studies, functional brain imaging, electroencephalography (EEG), and biomarker studies. Other potential causes of dementia must be ruled out.

In the early stage of the disease, the clinical history is typically characterized by memory loss as the earliest and most prominent symptom with insidious onset and gradual progression. Neurological exam may show mild impairment on mental status screening, but typically shows no motor or sensory deficits. The neurologist will evaluate for the possibility of medication side effects as a potential cause in patients who present with a clinical history of cognitive signs and symptoms. Neuropsychological examination reveals objective evidence of memory deficit and may also show other milder deficits in word retrieval. Longitudinal examinations performed at one-year intervals reveal a gradually progressive decline. Neuropsychological examination plays an important role in differential diagnosis, distinguishing normal versus abnormal cognitive function, mild cognitive impairment versus dementia, and neurocognitive profiles consistent with Alzheimer's disease versus other etiologies.

The MRI study rules out structural lesions that could account for the clinical presentation. In the early stages of the disease, the structural imaging is often unremarkable for the patient's age. In more advanced stages, the MRI shows enlargement of the lateral and third ventricles with cortical atrophy, as evidenced by thinned gyri and widened sulci. Coronal images show disproportionate hippocampal atrophy and corresponding disproportionate enlargement of the temporal horns of the lateral ventricles.

Blood tests are performed to rule out other potential causes of cognitive decline, such as nutritional deficiencies (e.g., vitamin B_{12}), thyroid function disorder, electrolyte imbalances, anemia, infection, HIV, syphilis, and other diseases and conditions that may be associated with cognitive dysfunction.

Functional imaging studies by positron emission tomography (PET) show biparietal hypometabolism in Alzheimer's disease, and cerebral blood flow single-photon emission computed tomography (SPECT) shows biparietal hypoperfusion. The EEG is often normal in early stages but may show mild diffuse slowing as the disease progresses; EEG is most useful for ruling out metabolic encephalopathy, a diffuse brain dysfunction caused by various systemic derangements.

Biomarker studies that measure cerebrospinal fluid levels of Aβ (the main constituent of plaques) and tau protein (a constituent of neurofibrillary tangles) that are indicative of Alzheimer's disease pathology may also aid in the diagnosis. The specific combination of both low CSF Aβ-42 and elevated CSF phosphorylated tau (P-tau) and a low CSF amyloid/tau ratio are considered to be the biological signature of Alzheimer's disease. Low Aβ levels reflect amyloid deposition, and high tau levels indicate neuronal damage. PET amyloid imaging using agents that bind to Aβ, such as Pittsburgh compound-B (PiB), reveals deposition of amyloid in the brain. Tau-binding agents that can be used with PET imaging have also been developed.

Treatment

There is as of yet no effective disease-modifying treatment for Alzheimer's disease, but there are therapies that can ameliorate some of the cognitive decline.

Acetylcholinesterase inhibitors (AChEIs), also known as **anticholinesterases**, are used to treat the cognitive deficits, especially amnesia, associated with Alzheimer's disease. These include **donepezil**, **galantamine**, and **rivastigmine**. It has long been known that acetylcholine (ACh) plays a critical role in learning and memory; drugs that act as cholinergic antagonists at muscarinic receptors (e.g., **scopolamine**) are well known for their ability to produce amnesia and prevent new learning. AChEIs inhibit the enzymatic breakdown of ACh by binding with acetylcholinesterase (AChE) at the site where ACh normally binds and preventing the hydrolysis of ACh, thereby increasing the level of ACh in the synapse and duration of action. AChE also is present in the axon terminal and regulates presynaptic ACh synthesis, maintaining a steady state between ongoing synthesis and degradation. Thus, AChEIs also increase the availability of presynaptic ACh by curtailing intracellular degradation and increasing synthesis.

Memantine, a partial N-methyl-D-aspartate (NMDA) glutamate receptor antagonist, is often introduced at the moderate stage of dementia in conjunction with AChEIs. The rationale is to limit the excitotoxicity cascade, a multi-step process triggered by the excitatory neurotransmitter glutamate and glutamate receptor activation that results in the degeneration of dendrites and cell death in neurodegenerative disease.

Neuropsychiatric symptoms that may arise from or be concomitant with Alzheimer's disease, such as agitation, aggressive behaviors, and psychosis, are treated with antipsychotic medications. Depression and anxiety are treated with antidepressant and anxiolytic medications.

Summary

The temporal lobes play a role in a wide variety of functions, including audition, olfaction, visual perception, language, memory, and emotion. Consequently, temporal lobe pathology may give rise to disorders of auditory perception, olfactory perception, visual perception and object recognition, episodic memory, semantic memory, language comprehension, and affective behavior. The role of the temporal lobes in visuoperceptual processing, memory, language, and emotion is discussed in separate chapters in this volume due to the scope and importance of these processes.

Auditory cortex is located within the posterior aspects of the middle and superior temporal gyri. The cortical auditory disorders are cortical deafness, auditory verbal agnosia, and auditory sound agnosia. Cortical deafness is very rare; it requires bilateral lesions of the auditory radiations or primary auditory cortices since information from each ear is represented in both hemispheres (with an 80% contralateral and 20% ipsilateral representation). Auditory verbal agnosia is a selective inability to understand spoken words in the context of normal hearing and otherwise normal language function. The causative lesion within the language-dominant (usually left) hemisphere results in a disconnection between both auditory cortices and an intact Wernicke's area. Auditory sound agnosia is a selective inability to recognize and identify nonverbal sounds (e.g., environmental sounds, music perception, paralinguistic aspects of speech). It occurs with lesions of the secondary auditory cortex in the superior temporal gyrus of the non-language-dominant (usually right) hemisphere. Auditory illusions and hallucinations have a variety of etiologies, including pathologies affecting the auditory pathways.

Primary olfactory cortex is located within the uncus and adjoining the entorhinal area of the anterior parahippocampal gyrus. Damage to the olfactory pathways and olfactory cortex produces anosmia; lesions in this region commonly occur with traumatic brain injury and temporal lobectomy. Focal seizures originating from the temporal lobe often begin with an olfactory aura due to onset within the uncus.

Alzheimer's disease is a cortical neurodegenerative disease typically characterized by prominent amnesia with insidious onset and gradual progression. The major pathological findings in Alzheimer's disease are cortical atrophy, extraneuritic plaques, and neurofibrillary tangles. The early stage of Alzheimer's disease typically involves a relatively selective neurodegeneration within the hippocampus and the entorhinal cortex of the temporal lobes, accounting for the amnesia. MRI studies may show enlargement of the lateral and third ventricles with disproportionate hippocampal atrophy and enlargement of the temporal horns of the lateral ventricles. Over time the pathology spreads to the lateral temporal, parietal, and frontal association cortices, and additional cognitive deficits eventually emerge (e.g., anomia, topographic disorientation). While we do know that Alzheimer's disease is a tauopathy, the etiology remains unknown and there is no effective disease-modifying treatment.

Additional Reading

1. Brodal P. The hippocampal formation: learning and memory. In: Brodal P, *The central nervous system: structure and function*. 5th ed. Oxford University Press; 2016. https://oxfordmedicine.com/view/10.1093/med/9780190228958.001.0001/med-9780190228958-chapter-32. Accessed July 4, 2022.
2. Khashper A, Chankowsky J, Del Carpio-O'Donovan R. Magnetic resonance imaging of the temporal lobe: normal anatomy and diseases. *Can Assoc Radiol J*. 2014;65(2):148–157. doi:10.1016/j.carj.2013.05.001
3. McKhann GM, Knopman DS, Chertkow H, et al. The diagnosis of dementia due to Alzheimer's disease: recommendations from the National Institute on Aging-Alzheimer's Association workgroups on diagnostic guidelines for Alzheimer's disease. *Alzheimers Dement*. 2011;7(3):263–269. doi:10.1016/j.jalz.2011.03.005
4. Ochoa-Escudero M, Herrera DA, Vargas SA, Dublin AB. Congenital and acquired conditions of the mesial temporal lobe: a pictorial essay. *Can Assoc Radiol J*. 2015;66(3):238–251. doi:10.1016/j.carj.2014.12.006
5. Sihvonen AJ, Särkämö T, Rodríguez-Fornells A, Ripollés P, Münte TF, Soinila S. Neural architectures of music: insights from acquired amusia. *Neurosci Biobehav Rev*. 2019;107:104–114. doi:10.1016/j.neubiorev.2019.08.023
6. Slevc LR, Shell AR. Auditory agnosia. *Handb Clin Neurol*. 2015;129:573–587. doi:10.1016/B978-0-444-62630-1.00032-9
7. Wixted JT, Squire LR. The medial temporal lobe and the attributes of memory. *Trends Cogn Sci*. 2011;15(5):210–217. doi:10.1016/j.tics.2011.03.005

The Frontal Lobes and Associated Disorders

Introduction

The human frontal lobes are very large, making up approximately one-third of the cerebral hemispheres. They can be divided into two major functional subdivisions: motor cortex and prefrontal cortex. Motor cortex is defined as cortical areas from which motor responses are elicited by electrical stimulation. Frontal lobe cortical areas from which motor responses are *not* elicited by electrical stimulation (i.e., non-motor areas of the frontal lobe) are referred to as prefrontal cortex because they are located anterior to the motor areas. This region is associated with personality and higher levels of cognitive and behavioral control that are collectively referred to as executive functions. This chapter describes basic frontal lobe structural and functional anatomy, the landmark case of Phineas Gage, prefrontal injury and disease syndromes, behavioral variant frontotemporal dementia, and prefrontal lobotomy.

Basic Anatomy of the Frontal Lobes

The frontal lobes lie anterior to the central sulcus and superior to the lateral fissure (see Figure 13.1 in Chapter 13). Each frontal lobe has three surfaces: dorsolateral, orbital, and mesial.

The dorsolateral surface of the frontal lobe has four principal gyri and three principal sulci (see Figures 13.3 and 13.4 in Chapter 13). The four principal gyri are the **precentral gyrus**, **superior frontal gyrus**, **middle frontal gyrus**, and **inferior frontal gyrus**. The inferior frontal gyrus is composed of three sub-regions: the **pars opercularis**, **pars triangularis**, and **pars orbitalis**. The three principal sulci are the **precentral sulcus**, **superior frontal sulcus**, and **inferior frontal sulcus**. The precentral sulcus lies anterior and parallel to the central sulcus; these two sulci form the boundaries of the precentral gyrus. The superior and inferior frontal sulci run in an anterior-posterior orientation; they demarcate the superior, middle, and inferior frontal gyri.

The inferior surfaces of the frontal lobes rest on the orbital plate of the frontal bone; this region is therefore also referred to as the orbital surface of the frontal lobes. The orbital surface of each frontal lobe is composed of four orbital gyri and the gyrus rectus

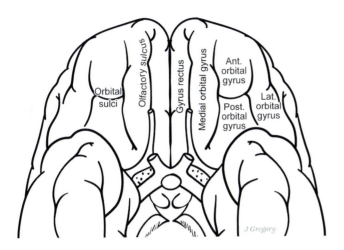

FIGURE 17.1. The orbital surface of each frontal lobe is composed of the gyrus rectus and four orbital gyri (medial, anterior, lateral, and posterior) that are demarcated by the H-shaped orbital sulcus. The gyrus rectus is demarcated from the medial orbital gyrus by the olfactory sulcus.

(Figure 17.1). The four orbital gyri are the **medial orbital gyrus, anterior orbital gyrus, lateral orbital gyrus,** and **posterior orbital gyrus**. These gyri are demarcated by the H-shaped **orbital sulcus**. The **gyrus rectus** (straight gyrus) lies medial to the orbital gyri, separated from the medial orbital gyrus by the **olfactory sulcus**, which lies beneath the olfactory bulb and tract. The gyrus rectus is continuous with the superior frontal gyrus on the medial surface of the brain.

The medial surface of the frontal lobe is composed of the medial portion of the precentral gyrus (which also is the precentral portion of the paracentral lobule), the medial aspect of the superior frontal gyrus, and the anterior portion of the cingulate gyrus (see Figures 13.5 and 13.6 in Chapter 13). The medial aspect of the superior frontal gyrus is sometimes referred to as the "medial frontal gyrus," but it is not really a separate gyrus. The cingulate sulcus lies superior to the cingulate gyrus, separating it from the medial aspect of the superior frontal gyrus. The callosal sulcus lies superior to the corpus callosum, separating it from the cingulate gyrus.

Motor Cortex

History

The concept of a **motor cortex**, a specialized part of the neocortex devoted to causing movement, was originated by British neurologist **John Hughlings Jackson** (1835–1911). Jackson observed that many of his patients had seizures characterized by involuntary movements without altered awareness (i.e., motor seizures). He reasoned that these movements must be caused by abnormal electrical discharge from a brain region specialized for controlling movement. He also observed that the abnormal movements often would spread from an initial body part to others in a systematic fashion (i.e., **Jacksonian motor seizures**), leading him to further deduce that the body's musculature must be represented systematically (i.e., mapped) within the brain region specialized for controlling movement.

This theory gained support in the 1870s with electrical stimulation and ablation studies performed in animals. **Gustav Fritsch** (1838–1927) and **Eduard Hitzig** (1838–1907) were the first to conduct such studies, and they provided the first direct evidence that distinct brain regions located just anterior to the central sulcus control movements on the contralateral side of the body. **David Ferrier** (1843–1928) conducted electrical stimulation studies in monkeys and found that the motor cortex contained a rough map of the body, with the feet represented dorsally and the face represented ventrally within the motor cortex. In the early twentieth century, **Charles Sherrington** (1857–1952) extended these findings in his studies of great apes, conducted primarily in chimpanzees, but also in gorillas and orangutans.

In the 1930s, neurosurgeon **Wilder Penfield** (1891–1976), who had studied with Charles Sherrington, used cortical stimulation in humans undergoing brain surgery to identify functional areas that should be spared when excising abnormal tissue to avoid detrimental neurological sequelae. He showed that the human brain also contains motor cortex in the precentral region, and that motor cortex contains a **motor map** in which the parts of the body used in tasks requiring precision and fine control (e.g., lips and hands) have a disproportionately large representation. This discovery supported John Hughlings Jackson's hypothesis that the human brain has a region specialized for controlling movement in which the body's musculature is represented systematically, and it explained the mechanism of Jacksonian motor seizures in which the epileptic movement progresses from one part of the body to another in a characteristic sequence corresponding to the body representation in the cortex. It also explained why damage to different areas of the frontal lobe results in selective weakness of the contralateral arm or leg.

Based on neuroanatomical studies, clinical observations, and cortical ablation studies performed in monkeys, motor cortex is composed of two distinct functional zones: **primary motor cortex** located within the precentral gyrus, and **secondary motor cortex** (**motor association cortex**) located in the frontal lobes anterior to primary motor cortex (Figure 17.2). These regions are distinct in their cytoarchitecture as well as their function. Primary motor cortex exclusively possesses **giant pyramidal cells of Betz** (Betz cells) and is **agranular** (i.e., it lacks a layer IV). Secondary motor cortex is **dysgranular** (it contains a faint layer IV with fewer cells than granular cortex)

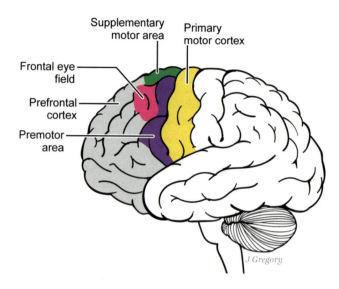

FIGURE 17.2. Secondary motor cortex is located anterior to primary motor cortex and consists of multiple areas including the premotor area, supplementary motor area, and frontal eye field.

and is a transition zone between the immediately posterior agranular primary motor cortex and the immediately anterior prefrontal cortex, which is granular (i.e., it has a fully formed granular layer IV).

Primary Motor Cortex

Primary motor cortex, also known as M1 and consisting of BA4, is located within the **precentral gyrus**. It plays a critical role in the execution of voluntary movements. Inputs to the primary motor cortex come from secondary motor areas, somatosensory regions, cerebellum (via the thalamus), and basal ganglia (via the thalamus). Primary motor cortex relays motor commands to motor neurons (motoneurons) which directly contact muscle cells, encoding features of movement such as force, direction, extent, and speed.

Primary motor cortex contains a systematic representation of the body's contralateral musculature known as the motor map or **motor homunculus** (see Figure 13.13 in Chapter 13). As indicated by the map, the area of cortex that controls a particular body part is proportional to the skill or control involved in moving that part, and is unrelated to the mass of muscle involved. Thus, those body parts that are capable of the finest motor abilities or the greatest degree of control have the largest representation. The hands and lips, not surprisingly, have an extensive representation; more than half of the entire primary motor cortex is concerned with controlling the hands and the muscles of speech. The parts of M1 that represent the trunk and proximal limbs are linked across the two hemispheres by corpus callosum fibers, while M1 regions that represent the hands and feet lack callosal connections. This allows for movements of the hands and feet to be independently controlled by the contralateral motor cortex only.

The musculature of the body is represented contralaterally (i.e., primary motor cortex of the left hemisphere controls the right side of the body, and vice versa). Much of the musculature of the head is represented bilaterally, except for the tongue and lower face. The lower body, from knees to toes, is represented on the medial surface of the precentral gyrus, while the upper body and head are represented on the lateral surface of the precentral gyrus.

Cortical motor areas representing the body give rise to the **corticospinal tracts** that project directly to the spinal cord, while cortical motor areas representing the face and head give rise to the **corticobulbar tracts** that project to the brainstem cranial nerve motor nuclei. These upper motor neuron pathways make direct synaptic contact with lower motor neurons that directly innervate muscle and cause muscle contraction, or interneurons that in turn directly contact lower motoneurons. These pathways descend in the corona radiata, the posterior limb of the internal capsule, and the cerebral peduncles of the midbrain (see Figure 13.8 in Chapter 13). Corticospinal fibers, which control the muscles of the body, continue their descent through the brainstem, forming the medullary pyramids (see Figure 8.2 in Chapter 8). At the caudal-most level of the medulla, most cross the midline in the pyramidal decussations and continue their descent within the lateral columns of the spinal cord as the lateral corticospinal tract, until entering the ventral horn gray matter of the spinal cord and synapsing. A small contingent of corticospinal fibers do not decussate and descend in the ventral columns of the spinal cord as the ventral (anterior) corticospinal tract. In accordance with the somatotopic map within M1, fibers from the medial aspect of the precentral gyrus representing the legs end in the lumbosacral levels of the spinal cord, whereas fibers from more lateral aspects of the precentral gyrus representing the upper body end in the cervical and thoracic levels of the spinal cord. Corticobulbar fibers, which control the muscles of the head (face, tongue, pharynx, and larynx), diverge from the corticospinal fibers in the brainstem, and project to cranial nerve motor nuclei.

Electrophysiological recording studies have shown that primary motor cortex neurons fire with short latency (5–100 msec) before the onset of a movement. Low levels of electrical stimulation applied to points on the primary motor cortex elicit simple movements of individual skeletal muscles.

A unilateral lesion in primary motor cortex (or descending fibers) results in contralateral **hemiplegia** (paralysis of one side of the body) or **hemiparesis** (weakness on one side of the body), due to damage to the corticospinal tract upper motor neurons. Involuntary reflex movements remain intact because the lower motor neurons that directly contact muscle are unaffected. Unilateral damage to the cortical motor area representing the head (or descending fibers) results in paralysis or paresis of the contralateral lower quadrant of the face only; the upper face and other head muscles (tongue, pharynx, and larynx) are not paralyzed because they are under the control of the motor cortex of both hemispheres.

Secondary Motor Cortex

Secondary motor cortex lies immediately anterior to primary motor cortex; it is therefore also referred to as **premotor cortex**. It projects directly to M1 and essentially instructs M1 in what to do. Thus, secondary motor cortex lies above primary cortex in the hierarchy of cortical processing (see Chapter 13, "The Cerebral Cortex"). Secondary motor cortex also contributes fibers to the corticospinal tract, but its projections to and influence on M1 are most important. Secondary motor cortex receives strong inputs from prefrontal and posterior parietal association cortices.

Like primary motor cortex, secondary motor cortex has a high degree of topographic organization. Electrophysiological recording studies have shown that secondary motor cortex usually becomes active in advance of M1 during voluntary movements. Compared to primary motor cortex, electrical stimulation of secondary motor cortex elicits more complex movements involving groups of muscles, and higher currents are required to elicit those movements (primary motor cortex has the lowest stimulus thresholds of any cortical region for eliciting movement).

While primary motor cortex plays a critical role in the execution of movements, secondary motor cortex plays a role in the prior stages of programming patterns of muscle activity. Any goal-directed movement (e.g., reaching for an object) requires the precise sequencing of multiple muscle activities.

Several distinct secondary motor areas have been identified in primates, including the premotor area, supplementary motor area, and frontal eye fields.

THE PREMOTOR AREA

The **premotor area** (PMA) lies just anterior to the primary motor cortex on the lateral surface of the cortex (i.e., the lateral portion of BA6). It receives its principal input from posterior parietal cortex (BA5 and BA7), which is secondary somatosensory cortex and plays a role in localizing objects in space. It also receives input from the basal ganglia by way of the ventral anterior nucleus of the thalamus. The major projections of the PMA are to primary motor cortex and the contralateral premotor area by way of the corpus callosum.

Lesions to the PMA result in deficits in fine motor control contralaterally, without paralysis.

THE SUPPLEMENTARY MOTOR AREA

The **supplementary motor area** (SMA) is located within the medial wall of the frontal lobe, anterior to the primary motor area (i.e., the medial portion of BA6). It contains a crude neuronal representation of the body's musculature and sends many fibers to primary motor cortex. The SMA is important in programming movement sequences (i.e., the temporal organization of movement) and in coordinating bilateral movements.

Cerebral blood flow studies performed in humans performing motor tasks of increasing complexity have shown that the SMA is involved in movement programming, rather than movement execution. During simple tasks, such as repetitive extension and flexion of a digit, blood flow increases dramatically within the contralateral hand areas of primary motor and somatosensory cortices. During a complex sequence of movements involving all the digits, the increase in cerebral blood flow extends to the SMA. When subjects mentally rehearse the sequence without movement, blood flow increases only in the SMA. Thus, increased activity in the SMA does not appear to be related to the movement itself, since it is sufficient that the person simply imagines performing a complex movement to activate the SMA.

There are few clinical case reports of persistent effects of SMA lesions on motor behavior. In the acute phase, there is a global reduction in behavior (**akinesia**) and slow clumsy movements that is particularly pronounced contralateral to the lesion. These manifestations are transient and resolve within several weeks. There is, however, a lasting deficit in the ability to program alternating movements of the hands and bimanual coordination.

These deficits are revealed by sequential motor tasks such as the reciprocal coordination task, the Luria manual sequencing task, and written alternating sequencing tasks. In the **reciprocal coordination task**, also known as the **Ozeretski alternating motor sequence task**, the examinee holds one hand palm down and the other in a fist on a table, and simultaneously and rapidly, repeatedly changes the hand positions in a smooth and steady manner (Figure 17.3). In the **Luria manual sequencing task**, also known as **fist-edge-palm**, the examinee is asked to place their hand on a table surface in three different positions sequentially, and to rapidly repeat the sequence (Figure 17.4): a fist resting horizontally (fist), a palm resting vertically (edge), and a palm resting horizontally (palm). **Written alternating sequencing tasks** include ramparts, alternating cursive "m"s and "n"s across a page, or alternating "+"s and "0"s (Figure 17.5).

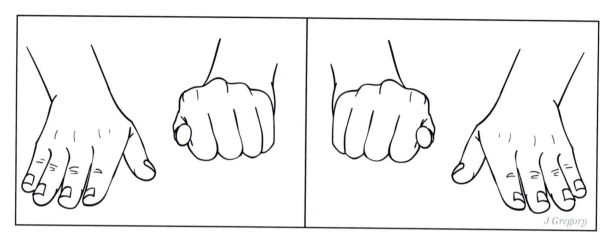

FIGURE 17.3. The reciprocal coordination test. This bimanual coordination task requires the patient to alternate smoothly between two hand positions up to 15 times. The examiner observes for transition hesitations, position errors, or breakdown in the sequence of movements.

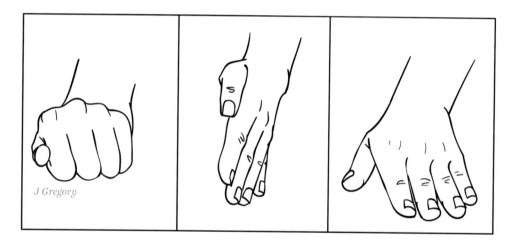

FIGURE 17.4. The Luria manual sequencing task (fist-edge-palm). The examinee is asked to place their hand on a table surface in three different positions sequentially and to rapidly repeat the sequence.

FIGURE 17.5. Written alternating sequencing tasks. (A) Ramparts. (B) Alternating cursive "m"s and "n"s.

THE FRONTAL EYE FIELDS

The frontal eye fields (FEF, BA8) are the primary cortical area controlling gaze movements. They play a role in intentional (i.e., voluntary) saccadic and smooth pursuit eye movements, maintaining visual fixation, and modulating visual scanning. The FEFs lies within the middle frontal gyrus.

The FEFs receive auditory, tactile, and visual information; thus, they are multimodally responsive. They have afferent and efferent connections with other regions that are involved in eye movements, such as the posterior parietal lobes, superior colliculi, pretectal nuclei, and reticular nuclei eye movement centers, all which in turn influence the brainstem cranial nerve nuclei controlling the extraocular muscles by cranial nerves 3, 4, and 6 (oculomotor, trochlear, and abducens).

Electrical stimulation and irritative lesions of the FEF cause conjugate eye movements (i.e., yoked movements in which the eyes move in the same direction) in the contralateral direction. Conversely, FEF ablation results in an ipsilateral conjugate deviation of the eyes (i.e., both eyes deviate toward the side of the lesion) due to the unopposed action of the intact FEF. This occurs with middle cerebral artery infarcts.

Frontal lobe lesions may produce abnormalities in visual fixation and slowed, inefficient visual scanning. Performance on tasks requiring visual search and attention to visual detail is disrupted, as patients search haphazardly or focus on only one detail while neglecting others. **Hans-Lukas Teuber** (1916–1977) found that patients with frontal lobe lesions were impaired at performing a match-to-sample task that required them to search an array of 48 patterns to find the match to the sample pattern located in the center of the array. **Alexander Luria** (1902–1977), who recorded eye movements as patients examined complex scenes to gain the information necessary to answer a specific question about the stimulus (e.g., how old is the individual in the photo?), found that patients with frontal lesions tended to glance randomly rather than focusing on the elements that provided the needed information, resulting in prolonged response times. A lower percentage of fixations were directed to the eyes, nose, and mouth, and a higher percentage were directed to details such the ears, cheeks, chin, cap, and necktie.

Prefrontal Cortex

Anatomy

Prefrontal cortex consists of association cortex and lies anterior to motor cortex. It consists of granular cortex, having a distinct layer IV (internal granular layer), distinguishing it from the agranular primary motor cortex and dysgranular secondary motor cortex. It also has large layers II (external granular layer) and III (external pyramidal layer), due to its extensive cortico-cortical connections. Prefrontal cortex has widespread afferent and efferent connections, receiving afferents from and projecting to a wide variety of brain regions, making it both accessible to all types of information from the internal and external environment and enabling it to control many other cerebral functions.

This complex connectivity accounts in part for the wide array of behavioral disturbances observed in persons with prefrontal lesions, which are most dramatic with bilateral lesions. Pathology within the prefrontal cortex is associated with changes in personality and loss of cognitive and behavioral control processes known as executive functions. The **executive functions** are conceptualized as high-level top-down processes that control lower-level processes in the service of goal-directed behavior. The core executive functions are inhibitory control, selective attention, working memory, and flexibility. **Inhibitory control** involves the self-control of one's behavior, thoughts, and/or emotions to override strong internal impulses or external lures and instead do what is more appropriate or necessary. **Selective attention** involves the filtering out of irrelevant information by inhibiting or minimizing the impact of internal and external distractions. **Working memory** refers to the temporary holding of information while mentally manipulating it. **Flexibility** is the ability to change perspectives (cognitive flexibility) and adjust behavior to changed demands or priorities (behavioral flexibility). Prefrontal cortex is divided into three anatomical and functional regions: the dorsolateral prefrontal cortex (convexity), orbitofrontal cortex, and medial frontal/anterior cingulate cortex. Injury or disease selectively involving these regions result in distinct syndromes (see "Prefrontal Injury and Disease Syndromes," below).

During phylogenesis, the prefrontal cortex developed disproportionately relative to the rest of the cortex. Based on cytoarchitecture, Brodmann estimated that in relation to the entire cortical mantle, the prefrontal area (i.e., frontal lobe granular cortex) accounts for about 2% in the rabbit, 3% in the cat, 7% in the dog, 8% in the lemur, 9% in the marmoset, 11% in the macaque monkey, 17% in the chimpanzee, and 29% in humans. Thus, the prefrontal cortex has undergone the greatest development in primates, and especially so in humans. In human ontogenesis, the cellular differentiation of the prefrontal cortex is not complete until puberty, and myelination of prefrontal areas continues until early adulthood (roughly age 25). Cyto- and myelo-architectonic development in the orbital areas precedes that of the dorsolateral areas. It is believed that cyto- and myelo-architectonic changes in prefrontal cortex form the basis of cognitive, behavioral, and emotional maturation through adolescence into adulthood.

The Frontal Lobe Controversy

From the early nineteenth century (beginning with Franz Joseph Gall, the founder of phrenology) until the 1930s, the frontal lobes were thought to be the seat of intelligence. This idea originated from the observations that the frontal lobes are dramatically expanded in the human brain, and that patients with frontal lobe tumors have intellectual deterioration.

This notion was challenged in the 1940s and 1950s, beginning with the famous case study of **patient K.M.** published by **Donald Hebb** (1904–1985) and **Wilder Penfield** (1891–1976) in 1940. K.M. underwent bilateral resection of the prefrontal cortex (bilateral frontal lobectomy) for intractable seizures but had no reduction in IQ or any measurable change in mental function. Hans-Lukas Teuber performed a study of over 90 World War II veterans who sustained penetrating missile wounds to the prefrontal cortex, and he did not find consistent cognitive deficits. These observations led to the conclusion that injury to the prefrontal cortex had no impact on cognitive function.

Earlier claims that the frontal lobes were the seat of intelligence were reinterpreted as erroneous, as these findings were based on single case studies of individuals with tumors, whose lesions behave differently from other lesions. Tumors can exert pressure on widespread brain areas and result in symptoms unrelated to the region where the tumor resides. Frontal brain tumors tend to become massive in size because they can grow for longer periods of time and affect both posterior ipsilateral and contralateral brain regions without causing obvious symptoms such as motor, sensory, or language impairments. Thus, the idea that higher functions were localized to the prefrontal cortex was rejected, and the concept of prefrontal cortex as the "silent area" arose.

We now understand that the conclusion that prefrontal cortex injury led to no neurological signs or cognitive impairments was erroneous because it was based on studies using small numbers of subjects, insensitive statistics, and cognitive tasks that are insensitive to the cognitive and behavioral effects of prefrontal cortex pathology.

Phineas Gage

Our understanding of frontal lobe function was revolutionized by the case of **Phineas Gage** (1823–1860). The Gage case is one of the most important in the history of neuroscience because it was the first to show that complex functions related to personality and behavioral regulation are localized within the brain. It remains the most publicized example of personality change following frontal lobe lesions.

Gage worked as the foreman of a railroad track construction gang near the small town of Cavendish, Vermont. Phineas and his coworkers were in the process of blasting granite bedrock into smaller pieces that could be removed. This involved drilling a long hole into the rock, pouring gunpowder into the bottom of the hole, carefully setting a rope-like fuse into the hole using the pointed end of a tamping iron, pouring sand into the hole, and tamping it down with the blunt end of the tamping iron to form a plug. After forming the plug, the crew called a warning, lit the fuse, and ran away. When the lit fuse ignited the gunpowder at the base of the hole, the plug bottled up the force of the explosion and channeled it into the rock, causing it to shatter.

On September 13, 1848, a 13-pound, 3½-foot-long, 2-inch-wide pointed tamping iron shot straight through Phineas's head. He was 26 years old at the time. It is believed that the accident occurred because after the gunpowder was poured into the hole but before the sand was placed, the tamping iron slipped into the hole blunt end first, which sparked the gunpowder and shot the rod directly out of the hole. The rod entered pointed end first through Phineas's left cheek, passed behind his left eye and through his prefrontal cortex, exited his skull at the middle of his forehead just above the hairline, and landed approximately 30 feet away from him.

Miraculously Phineas survived, and he did not even lose consciousness! He was thrown flat on his back, but as his coworkers ran toward him, he sat up and a minute later, he spoke. He was taken by ox cart to Cavendish, approximately ½ mile from the accident. When he arrived, he climbed off the ox cart without help, climbed the stairs of the Cavendish inn where he had been living, and told his landlord the story while he waited for the town physician, **John Martyn Harlow** (1819–1907). Phineas Gage's accident made newspaper headlines.

Dr. Harlow followed the best medical advice of the time, which was to clean the wound, keep it covered, and observe. Several days later, Harlow found Gage, still heavily bandaged, wandering around town in the rain without a coat, planning to walk home to his mother's house in New Hampshire 20 miles away. Ten weeks after the accident, Gage was "fully recovered." He was able to dress and feed himself, speak and understand, and add and subtract; however, Harlow knew that all was not right, and that Gage's personality was altered dramatically. In the spring, Gage returned to Cavendish to reclaim his job as railroad foreman, but his employers quickly came to the same conclusion as Harlow. Eventually, Phineas traveled throughout New England, South America, and California with the tamping iron. He even performed a stint at P. T. Barnum's American "Museum" on Broadway in New York City.

Harlow published an article about the case in the Boston Medical and Surgical Journal (December 1848) titled "Passage of an Iron Rod through the Head." At that time, there was an active debate as to whether functions other than sensory and motor functions were localized in the brain, and a Harvard Medical College surgeon, **Henry Jacob Bigelow** (1818–1890), took an interest in the article. He sent a letter to Harlow encouraging him to document the case by keeping notes and collecting formal statements from witnesses. He invited Harlow and Gage to visit Boston to have the case presented at the medical school. During that visit in 1850, Bigelow made a plaster life mask of Gage's face to record the scars and locations where the tamping iron entered and exited the skull (Figure 17.6).

In 1860, 11 years after the accident, Phineas Gage died after a seizure. No autopsy was obtained. Harlow learned of Gage's death approximately 5 years later and asked Gage's family for permission to have the body exhumed so that the skull could be recovered and kept as a medical record. The family granted permission. Phineas had been buried with the tamping iron, so it was recovered as well. In 1868, Harlow published his second and final article about the case, in which he reported that Gage's

The Frontal Lobes and Associated Disorders | 219

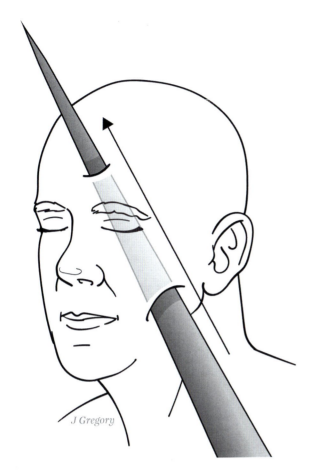

FIGURE 17.6. Path of the tamping iron that shot through Phineas Gage's head.

mind had changed radically since the accident, to the point where friends and acquaintances said that he was "no longer Gage."

The skull and tamping iron are owned by Harvard Medical School in Boston and are on display in the Warren Anatomical Museum in the Francis A. Countway Library of Medicine. Using modern imaging technology and measurements from the skull and tamping iron, **Hanna and Antonio Damasio** and colleagues (1994) modeled the rod's trajectory as a straight line connecting the center of the entry hole and the center of the exit hole. They reported that the brain lesion likely involved the ventromedial portion of both frontal lobes, including the anterior cingulate gyrus and underlying white matter, with more extensive damage in the left hemisphere. The lesion did not involve the motor cortices, Broca's area, or their underlying white matter.

In 1998 a conference on frontal lobe injuries was held in Cavendish, Vermont, commemorating the 150th anniversary of the Gage accident. The conference ended with the unveiling of a bronze plaque memorializing Phineas Gage, explaining what happened to him and his impact on brain science.

Prefrontal Lobotomy

Prefrontal lobotomy (from the Greek *lobos*, "lobe of brain," and *tomos*, "section") involves cutting the white matter connections to and from the prefrontal cortex. In Europe the procedure was known as **prefrontal leucotomy** (from the Greek *leukos*, "white"). The first controlled human lobotomy was performed by the Portuguese neurologist **António Egas Moniz** (1874–1955) in 1936. He believed that psychiatric disorders, like neurologic disorders, had an organic, physical basis, and he viewed the psychological explanations and Freudian theory that dominated psychiatry at the time as unscientific and verging on the metaphysical. He believed that the only hope for treating psychiatric disorders was by using somatic therapies, rather than talk therapy to uncover "intrapsychic conflict"; thus, he was a pioneer of biological psychiatry.

Moniz's argument for prefrontal leucotomy was based on the idea that mental disorders were the result of "fixed thoughts" that interfere with normal mental life, and that the "fixed thoughts" were maintained by nerve pathways in the frontal lobes. He reasoned that destruction of these abnormally stabilized pathways in the frontal lobes would be therapeutic. He did not perform preliminary animal experiments because he believed "in the domain of mental illness, tests with animals are not possible."

Initially, Moniz's method involved drilling burr holes in the patient's skull and injecting a small amount of alcohol (0.2 cc) into the centrum semiovale to destroy the white matter connections of the frontal lobe. By his eighth case, Moniz switched from alcohol to a special wire knife known as a **leucotome**. A burr hole was drilled into each side of the skull, the leucotome was inserted into the brain substance, and with a few sideways movements the white matter fibers were severed. In 1936 he published the results of 20 cases and coined the term **psychosurgery**, a surgical treatment for mental illness. At the First International Conference on Psychosurgery held in Lisbon in 1948, Moniz presented a lecture titled "How I Came to Perform Prefrontal Leucotomy." He won the Nobel Prize in Physiology or Medicine in 1949 for this work.

Walter Freeman (1895–1972) and **James W. Watts** (1904–1994) brought the procedure to the United States and eventually developed the **transorbital lobotomy**, which came to be known as "icepick lobotomy." In this procedure, the leucotome is driven through the thin layer of skull at the top of the eye socket by a mallet, and the pick is wiggled to damage the white matter connections of the frontal lobe (Figure 17.7); it was usually performed bilaterally. This procedure could be performed in only a few minutes in a doctor's office. Freeman himself performed the operation on thousands of people, advocated for it even in those with mild symptoms, and promoted

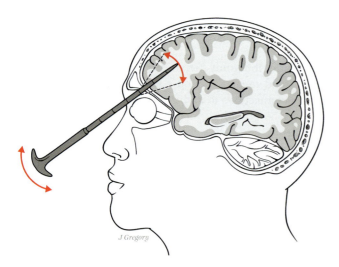

FIGURE 17.7. Transorbital prefrontal lobotomy. The procedure involved inserting a leucotome into the frontal lobe white matter through the patient's eye socket using a hammer and moving the instrument side-to-side; it was usually performed bilaterally. The rationale for the procedure was based on the hypothesis that psychosis was the result of excessive self-reflection in which thoughts circled back on themselves over and over, and that cutting the frontal lobe connections would cut off the endlessly circling thoughts from the rest of the brain.

lobotomy as a casual procedure, claiming that it would become as common as dental work. Lobotomy was most widely practiced in the United States, where approximately 40,000 people were lobotomized.

From the outset, lobotomy was criticized by many in the medical profession, and several widely publicized cases of severe adverse effects on mental function eventually gave the procedure a bad reputation. **Rosemary Kennedy** (1918–2005), sister of President John F. Kennedy, underwent lobotomy by Dr. Freeman, reportedly because her father complained to doctors about her moodiness and interest in boys. The lobotomy had a tragic result, leaving her incontinent, staring blankly at walls for hours, and reducing her language to unintelligible babble. To avoid political scandal, Rosemary's affliction was described as "mental retardation" to the public. Her sister, Eunice Kennedy Shriver, founded the Special Olympics in her honor in 1968.

Howard Dully, one of the youngest recipients of transorbital lobotomy, underwent the procedure in 1960 at age 12 because his stepmother was tired of his "youthful defiance." He was not told about the procedure and had no recollection of it. After years of institutionalization, incarceration, homelessness, and alcoholism, he became sober, received a college degree, and in his 50s began to research what had happened to him as a child. He uncovered the story of his own lobotomy by speaking with his family, relatives of other lobotomy patients, and relatives of Dr. Freeman, and gaining access to Freeman's archives. Dully has told his story in a book titled "My Lobotomy" (2007), in which he reveals that he always felt different, and wondered if something was missing from his soul.

With the advent of **chlorpromazine (Thorazine)** in the 1950s, the use of lobotomies rapidly declined and had ceased by the early 1970s.

Conclusions regarding the role of the prefrontal cortex in human behavior are difficult to draw from the prefrontal lobotomy patient population because while the procedure provided clean, relatively precise prefrontal lesions, it was often performed in individuals with significant psychiatric abnormalities. Thus, post-surgical testing is often confounded by factors other than the prefrontal white matter lesions.

Prefrontal Injury and Disease Syndromes

The most obvious and striking effects of frontal lobe damage in humans are a marked change in social behavior and personality and a deterioration in goal-oriented behavior, without loss of general intelligence, motor, sensory, or memory functions. Individuals with frontal lobe lesions perform fairly well on most standardized structured cognitive tests but have significant difficulty in daily life. There is often a dissociation between the results achieved on cognitive testing and the ability to apply those tested abilities. These behavioral changes are difficult to measure, but family and friends report that the person is no longer the same.

Frontal lobe lesions result from a variety of etiologies, including ruptured anterior communicating artery aneurysms, traumatic brain injury (especially orbitofrontal contusions), focal cortical neurodegenerative disease, tertiary syphilis infection, and frontal lobe meningiomas. Tertiary syphilis infection shows a predilection for the frontal regions and typically produces disinhibition and grandiose manic syndromes. Frontal meningiomas typically occur in either the parasagittal or cribriform plate regions; parasagittal meningiomas affect the medial aspects of the frontal lobes and commonly cause bilateral leg weakness, while cribriform plate meningiomas affect the orbitofrontal lobes and often result in behavioral disinhibition.

As stated above, the prefrontal cortex is divided into three anatomical and functional regions: the dorsolateral prefrontal cortex (convexity), orbitofrontal cortex, and medial frontal/anterior cingulate cortex. Each of these regions has anatomical connections to discrete areas of basal ganglia and thalamic nuclei, and segregated tracts, forming three systems that mediate distinct cognitive, behavioral, and emotional processes. Selective destruction of these systems underlies three distinct prefrontal syndromes: dorsolateral-dysexecutive syndrome, orbitofrontal-disinhibited syndrome, and medial frontal-akinetic/apathetic syndrome.

Several other syndromal models of frontal lobe lesions have been put forth previously. **D. Frank Benson** (1928–1996) and **Dietrich Blumer** (1929–2017) distinguished a "pseudodepressed" syndrome and a "pseudopsychopathic" syndrome. These correspond respectively to two syndromes distinguished by **Marcel Mesulam**, a frontal abulia syndrome characterized by a loss of initiative and diminished motivation, flattened affect, outward display of apathy and indifference, reduced verbal output, and behavioral slowness; and a frontal disinhibition syndrome characterized by disinhibition, childishly foolish behavior, and euphoria.

The Dorsolateral Prefrontal Syndrome (Dysexecutive Syndrome)

The **dorsolateral prefrontal syndrome**, also known as the **dysexecutive syndrome**, is characterized by inability to plan and form strategies in the service of problem-solving, impaired mental flexibility and perseveration, and poor self-monitoring.

PLANNING, STRATEGY FORMATION, AND PROBLEM-SOLVING

Individuals with dorsolateral prefrontal lobe lesions are especially impaired at developing novel cognitive strategies for problem-solving. This includes relatively simple strategies, as well as complex strategies that involve organizing a coherent series of steps or sub-goals to achieve an overall goal in a multi-component problem (i.e., breaking down the goal into sub-goals). A consequence of poor strategy formation is a poorly organized approach to problem-solving. A real-world example is difficulty solving arithmetic problems due to an inability to generate and execute even simple two- or three-step programs due to incomplete analysis of the problem, and an approach characterized by a series of impulsive, fragmentary arithmetic operations. This is not due to a primary disorder of calculation abilities (i.e., primary acalculia; see Chapter 28, Alexia, Agraphia, and Acalculia).

MENTAL FLEXIBILITY AND PERSEVERATION

Mental flexibility is the ability to shift response strategies and engage in divergent thinking. It is therefore also referred to as *set shifting*. Impaired mental flexibility leads to perseveration. **Perseveration** is defined as the continued production of a response when it is no longer appropriate. It is the application of old responses to new situations where such responses are inappropriate. Thus, perseveration results from an inability to inhibit inappropriate responses.

Perseveration is elicited by situations that require cognitive and behavioral flexibility, in which changing task demands require the individual to shift response strategies.

SELF-MONITORING

Individuals with frontal lobe lesions have difficulty using feedback from the environment to regulate or modify their behavior. In other words, they show an inability to profit from experience. Adaptive behavior also requires limiting one's activity in accordance with rules. Individuals with frontal lobe lesions commonly fail to comply with societal rules and task instructions.

The Orbitofrontal Syndrome (Disinhibited Syndrome)

The **orbitofrontal syndrome**, also known as the **disinhibited syndrome** or the "pseudopsychopathic" syndrome, is characterized by impulsivity, disinhibition, inappropriate social behavior, and emotional dysregulation. It is also often associated with **anosmia** (loss of the sense of smell) due to damage of the olfactory nerves, bulbs, or tracts.

Individuals with lesions confined to orbitofrontal cortex may have normal neuropsychological test performance; thus, diagnosis of the neurobehavioral syndrome may rest on behavioral observation, reports from reliable informants, and structured questionnaires about changes in behavior that reflect frontal lobe pathology. Lesions in this region do not produce a memory disorder unless the lesion includes basal forebrain structures (see Chapter 26, "Memory and Amnesia").

IMPULSIVITY AND DISINHIBITION

Impulsivity and disinhibition are due to a loss of normal impulse control. Manifestations include "doing without thinking," saying whatever pops into one's head with little concern or appreciation for the effect on others or personal consequences, behavior governed only by reinforcers, socially inappropriate behavior (including inappropriate sexual behavior), increased risk-taking behavior, and poor social judgment (lack of social tact) due to disregard for social rules and norms.

EMOTIONAL DYSREGULATION

Emotional dysregulation (emotional dyscontrol) manifests in a variety of ways, including inappropriate jocular affect, euphoria, lability, and over-reactivity. Excessive and abnormal joviality and a hollow jocularity characterized by inappropriate puerile (childish) humor (often about others) and inappropriate laughing are known as *Witzelsucht* (German, "joke addiction"). **Pathological**

joking is distinct from the "pathological laughing" of pseudobulbar affect (see Chapter 24, "The Motor System and Motor Disorders"). Signs and symptoms of euphoria include elevated mood, exaggerated self-esteem, fatuousness (foolish, inane comments and behavior), and facetiousness (not meant to be taken seriously or literally). Manifestations of lability and over-reactivity include becoming angry or upset after minor provocation and quickly returning to the previous calm state or crying inappropriately and uncontrollably.

The Medial Frontal/Anterior Cingulate Syndrome (Akinetic/Apathetic Syndrome)

A frequent cause of damage to the medial frontal/anterior cingulate cortex is stroke in the anterior cerebral artery or ruptured anterior communicating artery aneurysm.

The **medial frontal syndrome**, also known as the **akinetic syndrome** and the **apathetic syndrome**, is characterized by lack of initiative (aspontaneity), apathy, memory impairment, leg weakness, and urinary incontinence.

BEHAVIORAL ASPONTANEITY

Behavioral aspontaneity, also known as **frontal adynamia**, refers to reduced behavior and speech output. The core defect underlying behavioral aspontaneity is an inability to generate and sustain one's motivation; this manifests as difficulty initiating daily activities. Individuals who exhibit aspontaneity may stay in bed and watch television the entire day. This condition presents a major obstacle to rehabilitation and creates conflict between the affected individual and family members, as the behavior is erroneously interpreted and attributed to laziness on the individual's part, as opposed to the brain injury.

Behavioral aspontaneity ranges in severity. Bilateral lesions can result in a severe form known as **akinetic mutism** in which the person does not move (**akinesia**) or speak (**mutism**) spontaneously but is not paralyzed or aphasic. Milder lesions result in **abulia** (from the Greek, *aboulia*, "non-will"), a disorder of diminished motivation. It is characterized by a lack of will, drive, or initiative for action, speech, and thought, and manifests as psychomotor slowing, prolonged speech latency, poverty of action, and poverty of speech. Abulic individuals are severely impaired in their ability to initiate and self-regulate purposeful behavior. They respond normally to a direct external stimulus but lapse back into inactivity upon its withdrawal (e.g., eating continuously when in the presence of food regardless of the presence of internal satiety signals, and not eating in response to hunger signals when food is absent from the immediate environment). Since primary abulia is not secondary to depression, antidepressant medications do not lead to improvement as they do in the context of depression. Socioenvironmental factors, such as loss of incentives or lack of perceived control, also are not causal or contributing.

APATHY

Apathy is an absence or lack of feeling, emotion, interest, or concern. Apathetic individuals have a reduced interest in activities and social interaction, reduced motivation to initiate goal-directed activities, and reduced emotional reactivity (emotional blunting).

IMPAIRED MEMORY

Medial frontal damage often involves the basal forebrain, which plays a role in memory (see Chapter 26, "Memory and Amnesia"). The medial frontal syndrome, therefore, often includes episodic memory disorder. It is characterized by anterograde amnesia with a flat learning curve, temporally graded retrograde amnesia, and **confabulation**, the production of false or distorted memories that is not purposeful and without intent to deceive the examiner.

LEG WEAKNESS AND URINARY INCONTINENCE

Leg weakness and urinary incontinence are due to involvement of the medial frontal region; there is a medial frontal **micturition center**.

Signs and Symptoms Due to Diffuse Lesions

Widespread and diffuse damage to prefrontal cortex results in impaired voluntary attentional control, stimulus-bound behavior, and frontal release signs (see Chapter 30, "The Neurological Examination").

VOLUNTARY ATTENTIONAL CONTROL

Voluntary attentional control involves the ability to attend to stimuli that are relevant to the task at hand and to inhibit distraction by (i.e., ignore) incidental stimuli that are irrelevant to the task at hand. It involves selective attention, sustained attention, and rapid alternation of attention. Voluntary attentional control is a prerequisite for performance of many everyday tasks (e.g., having a coherent conversation, reading, preparing a meal). Attentional control (i.e., the ability to concentrate) is compromised with diffuse frontal lesions.

STIMULUS-BOUND BEHAVIOR

Stimulus-bound behavior (**environmental dependency syndrome**) refers to behavior that is not the result of intended or "willed" action on the part of the individual, but rather is involuntary, non-purposeful, outside of the person's control, reflexive in nature, and elicited by environmental stimuli. French neurologist **François Lhermitte** (1921–1998) first described two types of environmental dependency phenomena: imitation behavior and utilization behavior, described below. These behavioral syndromes reflect a deficit in self-guided behavior, with excessive dependence on environmental stimuli and lack of independence from external stimuli in guiding behavior. They are observed in those with frontal lobe stroke, other focal frontal lesions, and frontotemporal dementia due to frontotemporal lobar degeneration. They reflect an imbalance between frontal systems that normally mediate internally generated (motivated), voluntary, reward-driven behavior, and parietal systems that mediate environmentally driven, reflexive/reactive behavior.

Imitation behavior is the tendency to imitate the examiner's gestures and movements. These behaviors occur without volitional intent, and the individual is unable to inhibit these behaviors, even when explicitly instructed not to imitate.

Utilization behavior refers to the automatic elicitation of an instrumentally correct motor response to environmental cues and objects that is inappropriate for the social context. Persons with utilization behaviors will automatically reach for and "utilize" objects within their reach in an "object-appropriate" manner when object use is inappropriate for that particular social context. For example, in a neuropsychological testing session a patient with utilization behavior might automatically pick up a pen and paper on a table and begin writing something without being asked to do so. Utilization behavior has also been referred to as **magnetic apraxia** because tactile and visual presentation of objects compels the affected person to grasp and use them. It has been hypothesized that external stimuli trigger automatic, exploratory behaviors, but that such responses are normally inhibited if they are unnecessary, maladaptive, or inappropriate to the larger external context or internal intentions and goals. In persons with frontal lobe lesions, this inhibition is lost.

There are a variety of more complex forms of environmental dependency syndromes. **Forced hyperphasia** is compulsive, involuntary, environmentally dependent speaking, such as calling out the names of objects in the room, calling out the actions and gestures of people in the room, or calling out every road sign. In **spontaneous television actor participation**, the affected individual speaks to the actors of a movie/television show as if he/she were part of it. In **Zelig syndrome** (**Zelig-like syndrome**, named after a character in a 1983 Woody Allen film) the affected individual assumes a social role that is in keeping with the environmental context, such as speaking in an accent when in the company of persons who speak with such an accent. In **forced person-following** the person forcibly follows any person in his/her sight.

Behavioral Variant Frontotemporal Dementia

Frontotemporal lobar degeneration (FTLD) is a slowly progressive neurodegeneration that most prominently affects the frontal and temporal lobes (Figure 17.8). Frontotemporal lobar degeneration results in a clinical syndrome known as **frontotemporal dementia** (FTD), which is characterized by either slowly developing changes in personality and social behavior or slowly developing impairment of language. FTD encompasses a variety of clinical syndromes: behavioral variant of FTD, language variant FTD (agrammatic and semantic variants of primary progressive aphasia, described in Chapter 27, "Language and the Aphasias"), and variants associated with parkinsonism (progressive supranuclear palsy and corticobasal syndrome, described in Chapter 11, "The Basal Ganglia") or motor neuron disease (FTD-MND, described in Chapter 24, "The Motor System and Motor Disorders"). In FTD the focal atrophy may or may not be apparent on MRI structural imaging. In cases with no apparent brain atrophy, hypometabolism in the frontal lobes and anterior temporal lobes is often apparent on FDG-PET functional imaging.

Neuropathologically, FTLD is associated with abnormal intracellular accumulation of insoluble proteins that are identified with immunohistochemical techniques, and has been found to be heterogeneous, with two main categories based on the specific abnormal protein

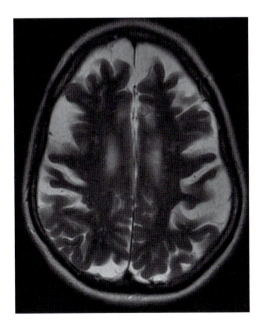

FIGURE 17.8. Axial T2 image of the brain demonstrates bilateral frontal lobe volume loss with prominent sulci and surrounding extra-axial spaces in frontotemporal lobar degeneration.

inclusions: a form with neuronal and glial inclusions of hyperphosphorylated tau protein (FTLD-tau) and a form with TAR DNA-binding protein 43 (TDP-43) inclusions (FTLD-TDP). Frontotemporal lobar degeneration was originally known as **Pick's disease**; **Arnold Pick** (1851–1924) was the first to describe a clinical syndrome associated with selective frontotemporal atrophy and **Pick bodies** on microscopic tissue examination. These are abnormal spherical intracytoplasmic inclusions that are now known to be composed of abnormal clumps of tau protein; thus, Pick's disease is a tauopathy and an FTLD-tau subtype. There is no one-to-one relationship between the specific FTLD neuropathology and FTD clinical phenotype, although there are some correlations.

Early and prominent frontal lobe involvement results in **behavioral variant FTD** (bvFTD), also known as **frontal variant FTD**. Behavioral variant FTD presents with personality and behavior change early in the course of the disease; thus, it is primarily a neuropsychiatric and neurobehavioral disorder. The average age of onset is 54, with some individuals affected as young as their 20s. Onset after 75 is rare.

Behavioral variant FTD has two behavioral syndromes that reflect differences in the topographic distribution of cortical pathology and atrophy: impulsive-disinhibited FTD and apathetic FTD. In **impulsive-disinhibited bvFTD**, the predominant involvement is in the orbitofrontal (ventromedial) and anterior temporal cortices. In **apathetic bvFTD**, there is more widespread frontal involvement extending into the dorsolateral frontal cortex.

The impulsive-disinhibited form of bvFTD is characterized by a range of symptoms, such as losing manners, making inappropriate comments, becoming more extroverted, and impulsive behaviors such as excessive spending, and increased or inappropriate sexual interest. There may be repetitive, ritualized, compulsive behaviors (e.g., hoarding), restlessness and agitation, and a change in food preferences (especially carbohydrate craving). There may be distractibility, pathological joking, or lack of empathy. There is a lack of insight, so individuals with bvFTD typically are unaware of the behavioral changes.

Those with impulsive-disinhibited bvFTD *may* exhibit a dysexecutive syndrome, especially if the atrophy extends into the dorsolateral frontal cortex. If the atrophy is relatively restricted to the orbitofrontal region, social behavior and emotion are affected most prominently, but despite gross behavioral change, there may be no apparent executive dysfunction. Thus, while the presence of executive dysfunction supports a diagnosis of bvFTD, its absence does not rule out a diagnosis of bvFTD.

In apathetic FTD, the behavioral symptoms reflect a disorder of diminished motivation, abulia, and apathy. Such disorders are characterized by reductions in overt behavior, thought content, and emotional responses. Apathetic bvFTD is often associated with executive dysfunction.

Summary

The human frontal lobes have two major functional subdivisions: motor cortex and prefrontal cortex. Primary motor cortex, which resides within the precentral gyrus, gives rise to the descending corticospinal and corticobulbar motor pathways and plays a critical role in the execution of voluntary movements. Unilateral lesions of primary motor cortex, as often occurs with stroke, results in a contralateral hemiplegia or hemiparesis. Secondary motor areas, which lie immediately anterior to primary motor cortex and send afferents to it, include the premotor area, supplementary motor area, and frontal eye fields. These areas play a role in movement programming.

Prefrontal cortex, located anterior to the motor areas of the frontal lobes, has widespread and complex connections. Pathology within the prefrontal cortex is associated with changes in personality and loss of cognitive and behavioral control processes (executive functions) that underlie adaptive, goal-directed behavior. The case of Phineas Gage was the first to demonstrate that complex functions related to personality and behavioral regulation are localized within the brain, and as such his case was pivotal in the history of behavioral neuroscience.

We now appreciate that prefrontal cortex is divided into three anatomical and functional regions: the dorsolateral prefrontal cortex, orbitofrontal cortex, and medial frontal/anterior cingulate cortex. Selective lesions to these areas underlie distinct prefrontal syndromes: a dorsolateral-dysexecutive syndrome, orbitofrontal-disinhibited syndrome, and medial frontal-akinetic/apathetic syndrome.

Additional Reading

1. Archibald SJ, Mateer CA, Kerns KA. Utilization behavior: clinical manifestations and neurological mechanisms. *Neuropsychol Rev.* 2001;11(3):117–130. doi:10.1023/a:1016673807158
2. Boettcher LB, Menacho ST. The early argument for prefrontal leucotomy: the collision of frontal lobe theory and psychosurgery at the 1935 International Neurological Congress in London. *Neurosurg Focus.* 2017;43(3):E4. doi:10.3171/2017.6.FOCUS17249
3. Cerami C, Cappa SF. The behavioral variant of frontotemporal dementia: linking neuropathology to social cognition. *Neurol Sci.* 2013 Aug;34(8):1267–1274. doi:10.1007/s10072-013-1317-9. Epub 2013 Feb 3. PMID: 23377232.
4. Conchiglia G, Della Rocca G, Grossi D. On a peculiar environmental dependency syndrome in a case with frontal-temporal damage: Zelig-like syndrome. *Neurocase.* 2007;13(1):1–5. doi:10.1080/13554790601160558
5. Cummings JL. Frontal-subcortical circuits and human behavior. *Arch Neurol.* 1993;50(8):873–880. doi:10.1001/archneur.1993.00540080076020

6. Damasio H, Grabowski T, Frank R, Galaburda AM, Damasio AR. The return of Phineas Gage: clues about the brain from the skull of a famous patient [published correction appears in Science. 1994 Aug 26;265(5176):1159]. Science. 1994;264(5162):1102–1105. doi:10.1126/science.8178168
7. Funayama M, Takata T. Forced person-following: a new type of stimulus-bound behavior. Neurocase. 2019;25(3-4):75–79. doi:10.1080/13554794.2019.1638944
8. Graff-Radford NR, Woodruff BK. Frontotemporal dementia. Semin Neurol. 2007;27(1):48–57. doi:10.1055/s-2006-956755
9. Harlow JM. Passage of an iron rod through the head [1848]. J Neuropsychiatry Clin Neurosci. 1999;11(2):28–283. doi:10.1176/jnp.11.2.281
10. Lanata SC, Miller BL. The behavioural variant frontotemporal dementia (bvFTD) syndrome in psychiatry. J Neurol Neurosurg Psychiatry. 2016 May;87(5):501–511. doi:10.1136/jnnp-2015-310697. Epub 2015 Jul 27. PMID: 26216940; PMCID: PMC4755931.
11. Leblanc R. Against the current: Wilder Penfield, the frontal lobes and psychosurgery. Can J Neurol Sci. 2019;46(5):585–590. doi:10.1017/cjn.2019.48
12. Lhermitte F. "Utilization behaviour" and its relation to lesions of the frontal lobes. Brain. 1983;106 (Pt 2):237–255. doi:10.1093/brain/106.2.237
13. Lhermitte F, Pillon B, Serdaru M. Human autonomy and the frontal lobes. Part I: Imitation and utilization behavior: a neuropsychological study of 75 patients. Ann Neurol. 1986;19(4):326–334. doi:10.1002/ana.410190404
14. Lhermitte F. Human autonomy and the frontal lobes. Part II: Patient behavior in complex and social situations: the "environmental dependency syndrome." Ann Neurol. 1986;19(4):335–343. doi:10.1002/ana.410190405
15. Moniz E. I succeeded in performing the prefrontal leukotomy. J Clin Exp Psychopathol. 1954;15(4):373–379.
16. Moniz E. Prefrontal leucotomy in the treatment of mental disorders [1937]. Am J Psychiatry. 1994;151(6 Suppl):236–239. doi:10.1176/ajp.151.6.236
17. Miller BL, Cummings JL, eds. The human frontal lobes: functions and disorders. 3rd ed. Guilford; 2020.
18. Neary D, Snowden J. Fronto-temporal dementia: nosology, neuropsychology, and neuropathology. Brain Cogn. 1996;31(2):176–187. doi:10.1006/brcg.1996.0041
19. Neary D, Snowden J, Mann D. Frontotemporal dementia. Lancet Neurol. 2005;4(11):771–780. doi:10.1016/S1474-4422(05)70223-4
20. Prioni S, Redaelli V, Soliveri P, Fetoni V, Barocco F, Caffarra P, et al. Stereotypic behaviours in frontotemporal dementia and progressive supranuclear palsy. Cortex. 2018;109:272–278. doi:10.1016/j.cortex.2018.09.023
21. Rascovsky K, Hodges JR, Knopman D, Mendez MF, Kramer JH, Neuhaus J, et al. Sensitivity of revised diagnostic criteria for the behavioural variant of frontotemporal dementia. Brain. 2011;134(Pt 9):2456–2477. doi:10.1093/brain/awr179
22. Rosness TA, Engedal K, Chemali Z. Frontotemporal dementia: an updated clinician's guide. J Geriatr Psychiatry Neurol. 2016;29(5):271–280. doi:10.1177/0891988716654986
23. Snowden JS, Goulding PJ, Neary D. Semantic dementia: A form of circumscribed cerebral atrophy. Behav Neurol. 1989;2(3):167–182.
24. Snowden JS, Harris JM, Thompson JC, Kobylecki C, Jones M, Richardson AM, et al. Semantic dementia and the left and right temporal lobes. Cortex. 2018;107:188–203. doi:10.1016/j.cortex.2017.08.024
25. Stuss DT, Knight RT, eds. Principles of frontal lobe function. 2nd ed. Oxford University Press; 2012.
26. Teuber HL. The riddle of frontal lobe function in man [1964]. Neuropsychol Rev. 2009;19(1):25–46. doi:10.1007/s11065-009-9088-z
27. Wittenberg D, Possin KL, Rascovsky K, Rankin KP, Miller BL, Kramer JH. The early neuropsychological and behavioral characteristics of frontotemporal dementia. Neuropsychol Rev. 2008;18(1):91–102. doi:10.1007/s11065-008-9056-z

18

Stroke and Vascular Cognitive Impairment

Introduction

Stroke is defined as the acute onset of a non-convulsive, neurologic deficit that is caused by a disturbance of the cerebral circulation. The signs and symptoms of stroke vary widely from person to person; however, in all strokes the neurologic deficit has an abrupt onset (within seconds or minutes). The term *stroke* conveys the idea that the person has "been struck," reflecting the abruptness with which the neurologic deficit develops. Stroke is due to either ischemia (a loss of blood flow to a focal area of neural tissue) or hemorrhage (rupture of a blood vessel and bleeding). This chapter describes the pathophysiology and anatomic classification of ischemic stroke; transient ischemic attack; the pathophysiology of hemorrhagic stroke; stroke diagnosis, treatment, and prevention; and vascular cognitive impairment.

Stroke is the most common neurologic disorder and the most common cause of brain lesions. Ischemic stroke accounts for 88% of all strokes, while hemorrhagic stroke accounts for 12%. Most strokes are due to pathology within the arterial system. Problems involving the cerebral venous system are far less common than those involving the arterial supply because occlusions and hemorrhages occur less often in the venous system, and because there is extensive collateral circulation (**anastomoses**) within the cerebral venous system.

Ischemic Stroke

Pathophysiology

Ischemic stroke is caused by a loss of blood flow to a focal area of neural tissue. This is most often due to **obstruction** of a brain artery by **atheroma** (**atherosclerotic plaque**, a fatty deposit in the inner lining of an artery) and/or a blood clot, but it may also be due to cerebral **hypoperfusion** (insufficient blood flow). Both obstruction of an artery and hypoperfusion to a region of brain tissue result in an **infarction**, a region of tissue death (**necrosis**) caused by a local lack of oxygen. In response to ischemia, brain tissue undergoes **liquefactive necrosis**. The brain tissue is dissolved by enzymes and transformed into a liquid, viscous mass. After some time, white blood cells remove the debris, eventually leaving a cavity (rather than a scar, as occurs with myocardial infarction); thus, there is no filling in of the lost tissue space.

THROMBOEMBOLIC STROKE

Arterial obstruction occurs by one of two mechanisms: thrombosis or embolism. Thus, stroke due to blockage within a blood vessel is also referred to as **thromboembolic stroke**. About two-thirds of ischemic strokes due to arterial occlusion are attributed to thrombosis, and about one-third are attributed to embolism.

A **thrombus** is an obstruction that forms inside an artery and remains in place (**in situ**). Thrombi usually consist of a blood clot that forms around an atherosclerotic plaque. Atheromatous plaques most commonly occur at the bifurcation of vessels, especially at the origin of the internal carotid artery (ICA), middle cerebral artery (MCA), and posterior cerebral artery (PCA). It is rare for the cerebral arteries to develop plaques beyond their first major branching point. Thrombosis also occurs in the small penetrating arteries, cerebral veins, and venous sinuses. Symptoms of thrombotic stroke typically evolve over minutes to hours.

An **embolism** (**thromboembolism**) is a thrombus that dislodges and travels until it can no longer pass through the arterial vasculature. Embolic strokes therefore have an extremely abrupt onset. Cerebral emboli typically originate from the heart or large extracranial arteries (aortic arch, common carotid arteries, internal carotid arteries, and vertebral arteries), and they are accordingly classified as either cardiac or noncardiac. Vascular manipulation during medical procedures such as coronary artery angiography, coronary artery angioplasty and stenting, cardiac surgery, cardiopulmonary bypass, thoracic aortic surgery, carotid angiography, carotid endarterectomy, carotid angioplasty and stenting, and cerebral angiography all carry risk of atheromatous cerebral embolism.

Emboli generally pass as far distally as their size permits along the superficial vascular tree. They rarely follow the sharp angles of the arterial branches, and therefore do not pass to deep brain structures. Approximately 80% of all emboli entering the brain involve the anterior circulation, and 20% involve the posterior circulation. This breakdown is proportional to the distribution of cerebral blood flow. Emboli entering the anterior cerebral circulation most often enter the MCA or its branches rather than the anterior cerebral artery (ACA), since the MCA is a direct extension of the ICA, and it carries the majority (75%–80%) of anterior circulation blood flow.

Embolic strokes often undergo **hemorrhagic transformation** within 12–48 hours. When emboli disintegrate, the damaged vessel is reperfused and then ruptures upon re-exposure to the full force of the arterial pressure. Such hemorrhage is frequently asymptomatic since the tissue involved has already been damaged. Infarcts are therefore classified as **pallid** (pale, i.e., devoid of blood), **hemorrhagic**, or **mixed**.

When blood flow to a region of brain tissue is reduced, the extent of resulting tissue damage depends on the size of the occluded vessel, the duration of the ischemia, and the severity of the ischemia (i.e., reduced vs. absent cerebral blood flow). The severity of the ischemia is related to the availability of compensatory collateral blood flow, and the speed of occlusion. The potential for compensatory collateral blood flow depends on the location of the occlusion in relation to anastomotic channels (alternate routes for blood flow); the presence of anastomoses is highly idiosyncratic among individuals. With obstructions that lie proximal to the circle of Willis, healthy communicating arteries (i.e., the anterior and posterior communicating arteries) will allow for collateral flow and there will be no infarction. If the obstruction occurs in the vicinity of diseased vessels, there will be no collateral flow and the obstruction will result in an infarction. For example, the clinical manifestation of complete occlusion of an ICA can range from silent (i.e., asymptomatic) to lethal due to massive infarction of most of the hemisphere. Speed of occlusion also impacts the severity of ischemia, since gradual narrowing of a vessel due to thrombosis allows time for collateral channels to open, as compared to sudden occlusions due to embolism.

HYPOPERFUSION

The **watershed areas**, which receive dual blood supply from the most distal branches of two large arteries, receive the lowest blood flow and are particularly vulnerable to ischemic injury with hypoperfusion. Cerebral hypoperfusion can result in focal infarctions in the watershed regions.

Anatomic Classification of Ischemic Stroke

There are three general classifications of ischemic stroke based on anatomy: territorial stroke, lacunar stroke, and watershed stroke.

TERRITORIAL STROKE

Territorial strokes are due to occlusion of an artery or arterial branch, which produce infarctions confined to the territory supplied by the artery. The presenting neurological deficit depends on the artery affected, the hemisphere affected, and the extent of the infarct. The fact that the arterial blood supply is generally consistent among individuals forms the basis of **stroke syndromes**, a specific pattern of signs and symptoms. Studies examining the relationship between anatomical lesions and clinical

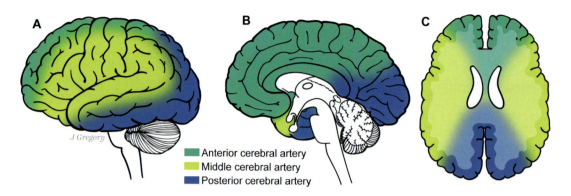

FIGURE 18.1. The vascular territories supplied by the anterior, middle, and posterior cerebral arteries. (A) Lateral surface. (B) Medial surface (midsagittal plane). (C) The axial plane.

signs and symptoms in patients with territorial ischemic strokes have greatly advanced our knowledge of functional neuroanatomy.

Recall that the anterior circulation consists of the ICAs, MCAs, ACAs, and deep branches of the MCAs and ACAs, while the posterior circulation consists of the basilar artery, vertebral arteries, PCAs, pontine arteries, cerebellar arteries (superior, anterior inferior, and posterior inferior cerebellar arteries), and deep branches of the PCAs (see Chapter 5, "Blood Supply of the Brain").

The territories supplied by the MCA, ACA, and PCA are illustrated in Figure 18.1. Regarding the cerebral cortex, the MCA covers a large territory, supplying the lateral aspects of the frontal, temporal, and parietal lobes (see Figure 5.2 in Chapter 5). The ACA supplies the medial and superior aspect of the frontal and parietal lobes (see Figure 5.3 in Chapter 5). The PCA supplies the occipital lobes and inferomedial aspects of the temporal lobes (see Figure 5.5 in Chapter 5). Territorial stroke syndromes due to occlusion of the large vessels supplying the cerebral cortex are described in Table 18.1.

The posterior circulation supplies the cerebellum and brainstem through branches from the vertebrobasilar system. With respect to the deep structures of the cerebrum, the ACA and MCA supply the basal ganglia through the medial and lateral lenticulostriate arteries, respectively. The PCA supplies parts of the thalamus through perforator arteries, and most of the hippocampus. The anterior choroidal artery, which branches off the ICA prior to the MCA-ACA bifurcation, supplies the choroid plexus that lines the anterior part of the temporal horns, as well as several deep brain structures (parts of the internal capsule, parts of the thalamus, optic tract, lateral geniculate nucleus, lateral aspect of the cerebral peduncle, tail of the caudate nucleus, internal segment of the globus pallidus, head of the hippocampus, and amygdala).

There are many **brainstem stroke syndromes**, some common and some rare; many involve ischemia in the distribution of the basilar or vertebral arteries (i.e., the posterior circulation). The cardinal clinical feature of brainstem stroke is crossed sensory findings (ipsilateral face and contralateral body numbness), and/or crossed motor findings (ipsilateral face and contralateral body weakness). The brainstem stroke syndromes can roughly be classified as focal lesions affecting either the medial or lateral aspects of the midbrain, pons, or medulla (see Table 8.1).

LACUNAR STROKE

Lacunar infarcts are small areas of focal ischemic necrosis located in the subcortical structures (basal ganglia, thalamus, and brainstem) and deep brain white matter. They are 3–15 mm in diameter; infarcts less than 3 mm in diameter are referred to as **microinfarcts**. The **lacunes** (lacuna, "hole") consist of small cerebrospinal fluid-filled cavities. They result from **small vessel disease** affecting the deep penetrating arterioles (Figure 18.3). These small vessels have lumen diameters of 40–500 μm that arise directly from and at right angles to the larger vessels at the base of the brain. There is no gradual step-down in size, as occurs in the distal cortical vessels. Their lumens become narrowed or occluded by degeneration of the arterial wall. Because these small vessels arise at right angles to their parent arteries, emboli, which preferentially travel in the main flow stream of the parent artery, usually do not account for lacunar infarcts.

Lacunar infarctions are frequently observed as an incidental finding on brain magnetic resonance imaging (MRI) in older people without a history of discrete neurological symptoms. Those that do not produce any specific neurological signs or symptoms are referred to as **silent lacunar infarctions**. As lacunes accumulate, however, neurological symptoms may develop, depending on the number and distribution of the lesions. **Lacunar strokes**, also known as **microvascular strokes** or **small vessel strokes**, are lacunar infarctions that produce neurological signs and/or symptoms of small subcortical or

TABLE 18.1 Large Vessel Cortical Stroke Syndromes

Left MCA superior division	• Nonfluent (Broca's) aphasia • Right arm and face weakness • Right arm and face sensory loss • Working memory and executive function deficits
Left MCA inferior division	• Fluent (Wernicke's) aphasia • Apraxia • Gerstmann syndrome (agraphia, acalculia, finger agnosia, right-left disorientation)
Left MCA stem	Complete left MCA syndrome is a combination of left MCA superior and inferior division syndromes
Right MCA superior division	• Left arm and face weakness • Left arm and face sensory loss • Working memory and executive function deficits • Left hemineglect
Right MCA inferior division	• Left hemineglect • Anosognosia • Constructional disorder ("constructional apraxia") • Dressing apraxia
Right MCA stem	Complete right MCA syndrome is a combination of right MCA superior and inferior division syndromes
Left ACA	• Right leg weakness and sensory loss • Executive dysfunction
Right ACA	• Left leg weakness and sensory loss • Executive dysfunction
Left ICA	A combination of left MCA and left ACA syndromes
Right ICA	A combination of right MCA and right ACA syndromes
Left PCA	• Right homonymous hemianopia • Pure alexia • Verbal memory deficit
Right PCA	• Left homonymous hemianopia • Spatial memory deficit
Bilateral PCA (top of the basilar syndrome)	• Cortical blindness • Global anterograde amnesia

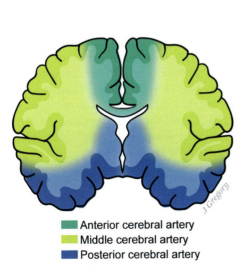

FIGURE 18.2. The vascular territories in the coronal plane.

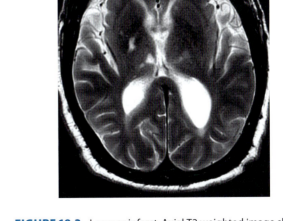

FIGURE 18.3. Lacunar infarct. Axial T2 weighted image shows lacunar infarcts in bilateral basal ganglia and bilateral thalami.

brainstem lesions consisting of motor and/or sensory deficits; they do not result in disturbances of consciousness, behavior, or cognitive function.

Small vessel disease occurs with a variety of pathological processes that affect the small arteries, arterioles, and capillaries of the brain, particularly the penetrating arteries. The deep white matter of the cerebral hemispheres is particularly susceptible to hypoperfusion and ischemia. It receives its blood supply from small caliber penetrating arteries and arterioles. Additionally, white matter receives less blood supply than gray matter. Many types of arteriopathy underlie small vessel ischemic disease.

Arteriosclerosis (microatheroma plaque) and **lipohyalinosis** (lipohyaline deposits within the vessel walls that cause wall thickening and reduced lumen diameter) cause the small penetrating arteries and arterioles to become narrowed, resulting in chronic hypoperfusion and ischemia of the deep cerebral white matter.

Cerebral autosomal dominant arteriopathy with subcortical infarcts and leukoencephalopathy (CADASIL) is a rare, autosomal dominant arteriopathy affecting the smooth muscle cells of the small vessels. CADASIL is caused by a mutation of the NOTCH3 gene on chromosome 19. It results in small lacunar infarcts and diffuse white matter disease in the periventricular white matter and centrum semiovale, basal ganglia, thalamus, and brainstem. Onset of stroke and cognitive decline typically begins in early to middle adulthood in the absence of other vascular risk factors. Average age of onset for CADISIL-related stroke is age 45, and dementia typically develops by the sixth decade of life. CADASIL is one of the most frequent hereditary neurological disorders. Diagnosis of CADASIL is made by genetic testing. The clinical manifestations of CADASIL are: (1) migraines with aura occurring between the ages of 20–40; (2) transient ischemic attacks or subcortical ischemic strokes occurring from age 40; (3) mood disturbances; and/or (4) cognitive impairment, which may occur without other clinical symptoms. By age 65, two-thirds of CADASIL patients have dementia. The onset of cognitive symptoms is usually insidious; early in the course, attention and executive functions are most frequently affected. CADASIL is considered the archetype of **subcortical ischemic vascular dementia**. The dementia often is associated with motor disturbances, including **pseudobulbar palsy**, a syndrome characterized by paresis of the tongue, dysphagia (difficulty swallowing), and dysarthria (unclear articulation of speech), due to lesions of the upper motor neurons of the corticobulbar system that project to cranial nerve motor nuclei within the medulla (i.e., CN9–CN12).

Cerebral autosomal recessive arteriopathy with subcortical infarcts and leukoencephalopathy (CARASIL) is another hereditary vascular dementia syndrome. It is due to mutations in the HTRA1 gene.

Inflammatory and immunologically mediated small vessel disease includes nervous system vasculitis associated with autoimmune connective tissue disorders such as systemic lupus erythematous, Sjögren's syndrome, and scleroderma. Other causes of small vessel disease include post-radiation angiopathy and non-amyloid microvessel degeneration in Alzheimer's disease.

WATERSHED STROKE

Watershed infarcts (**border zone infarcts**, **distal field infarcts**) are ischemic lesions that occur in characteristic locations at the junction between two main arterial territories. They account for approximately 10% of all ischemic strokes. The watershed regions are particularly vulnerable when there is a sharp drop in the systemic blood pressure (hypotension); thus these strokes usually occur with cardiac output failure (e.g., cardiac arrest), severe hypotension during surgery, and significant carotid stenosis in the presence of an incompetent circle of Willis (which causes perfusion failure distal to the site of stenosis). Watershed infarcts may also be caused by **microemboli** (extremely small emboli) that lodge preferentially in the watershed areas because of their small size.

There are two types of border zone infarcts: cortical (external) and subcortical (internal). Cortical border zone infarcts are believed to result from hypoperfusion, microemboli (from the heart or atherosclerotic plaques in major arteries), or a combination of hypoperfusion and microemboli. Severe occlusive disease of the ICA causes both hypoperfusion and embolization, and the two act synergistically as hypoperfusion impedes the clearance (washout) of emboli. The cortical watershed zones are located at the borders of the anterior, middle, and posterior cerebral artery territories. The MCA-ACA territory border lies within the frontal lobe; it is known as the **anterior cortical border zone**. The MCA-PCA territory border lies within the parieto-occipital area; it is known as the **posterior cortical border zone**. Cortical watershed strokes most frequently occur in the anterior cortical border zone. Such infarctions classically present with weakness of the proximal arm and leg muscles with preserved distal strength (**man in a barrel syndrome**) because they affect the trunk and proximal limb representations within the motor cortex (see the motor homunculus in Figure 13.13, Chapter 13). Anterior cortical border zone infarcts are more common than posterior ones because of the high prevalence of ICA disease. Another common pattern of cortical watershed stroke involves the distal territories of the ACA, MCA, and PCA, essentially due to a combination of anterior cortical border zone and posterior cortical border zone infarcts, resulting in a crescent shaped infarction when viewed from the lateral view of the brain (Figure 18.4).

Subcortical (deep) border zone infarcts are caused mainly by compromised hemodynamics (the flow of blood in the circulatory system). The subcortical border zones are located at the MCA-lenticulostriate, ACA-lenticulostriate,

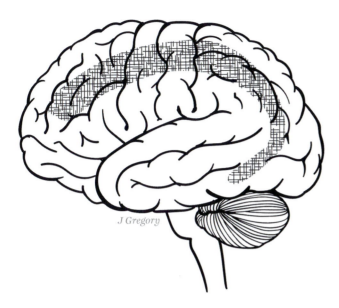

FIGURE 18.4. A common pattern of cortical watershed stroke involves the distal territories of the ACA, MCA, and PCA, essentially due to a combination of anterior cortical border zone (MCA-ACA) and posterior cortical border zone (MCA-PCA) infarcts, resulting in a crescent shaped infarction (hatched area) when viewed from the lateral aspect of the brain.

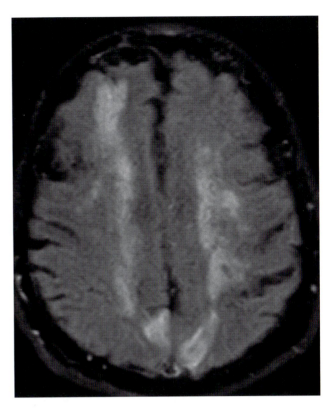

FIGURE 18.5. Watershed infarct. Axial FLAIR image shows deep border zone infarcts involving the bilateral corona radiata.

ACA-medial striate, MCA-anterior choroidal, and PCA-anterior choroidal junctions. A common pattern of deep border zone stroke involves multiple infarctions organized in a linear fashion, parallel to the lateral ventricles in the centrum semiovale or corona radiata, appearing like a string of beads; when confluent, the lesions produce a band-like appearance (Figure 18.5).

Watershed infarcts also occur at the borders between the territories of the major cerebellar arteries, and between the territories of the spinal arteries.

Transient Ischemic Attacks

A **transient ischemic attack** (TIA) is a brief episode of focal neurological signs and/or symptoms with sudden onset due to transient focal ischemia, due to either thrombotic or embolic processes. By conventional clinical definition, if the neurological symptoms resolve within 24 hours the ischemic event is defined as a TIA (although most resolve within 30 minutes), and if the symptoms last longer than 24 hours the event is diagnosed as stroke. With advances in brain imaging technology, however, many patients with symptoms lasting less than 24 hours have been found to have an infarction. Thus, the definition of TIA has been revised as a brief episode of neurological dysfunction caused by focal brain ischemia, typically lasting less than 1 hour and without imaging evidence of infarction.

The symptoms of TIAs are quite varied and depend on the vessels affected. Large vessel TIAs usually are due to embolic events. Small vessel TIAs result from reduced flow through stenosed vessels, or from emboli from microatheroma within the penetrating artery. Ninety percent of TIAs occur in the anterior circulation.

Amaurosis fugax (Greek, "fleeting darkness") is a form of TIA characterized by sudden transitory monocular blindness or blurred vision that appears as a curtain coming down over the visual field. Vision is usually restored within 5–15 minutes, often in a reverse manner from the onset (i.e., curtain coming up). Amaurosis fugax is a sign of ipsilateral carotid artery disease. It often occurs with emboli cast off from an atherosclerotic carotid artery that obstruct the lumen of the ophthalmic artery (the first branch of the ICA) or the retinal artery (a branch of the ophthalmic artery), causing a decrease in blood flow to the retina.

Transient ischemic attacks are a warning sign and risk factor for ischemic stroke. When TIAs precede thrombotic strokes, the TIA and stroke typically present with similar symptoms because they affect the same vascular territory. When TIAs precede embolic strokes, symptoms typically vary between TIAs and between the TIA(s) and the stroke, because different vascular territories are affected. Evaluation and treatment should be initiated soon after the TIA, since severe permanent strokes often follow TIAs, particularly within the first 2 days. Patients with anterior circulation TIAs should be evaluated for carotid stenosis by carotid Doppler ultrasound study and/or angiography.

Hemorrhagic Stroke

Pathophysiology

Hemorrhagic stroke involves rupture of a brain blood vessel. There are two types of cerebral hemorrhage: **intracerebral hemorrhage** (ICH) and **subarachnoid hemorrhage** (SAH). Hemorrhagic stroke is far less common than ischemic stroke; 12% of all strokes are hemorrhagic, with 9% due to ICH and 3% due to SAH. Hemorrhagic stroke has a higher mortality than ischemic stroke, especially SAH.

INTRACEREBRAL HEMORRHAGE

In ICH, a cerebral artery ruptures and blood is released from the vessel directly into the brain parenchyma under high pressure, forming a hematoma (Figure 18.6). Intracerebral hemorrhages are most frequently located deep in the brain because they primarily affect the small penetrating arteries. The bleeding may rupture into the ventricular system or through to the cortical surface into the subarachnoid space. In addition to the area of the brain directly injured by the hemorrhage, hematoma and edema compress and displace brain tissue, and may cause additional damage through herniation of brain tissue, compression of the brainstem, and occlusion of cerebral arteries.

Both ICH and ischemic stroke present with sudden onset of focal neurological signs. Some cases of ICH may be clinically indistinguishable from ischemic stroke when the effects are confined to a specific vascular territory. In most cases, however, the clinical picture between ICH and ischemic stroke differs because (1) ICH usually involves regions supplied by more than one artery, and (2) ICH is more likely to present with symptoms of increased intracranial pressure (i.e., sudden onset severe headache with nausea and/or vomiting), seizures, and altered level of consciousness or coma.

The most common cause of ICH is chronic hypertension. Chronic hypertension leads to degenerative changes such as **microaneurysms** (**Charcot-Bouchard aneurysms**) in the muscle and elastic tissue of the blood vessel walls, which are most pronounced in the small penetrating branches of the middle cerebral, posterior cerebral, and basilar arteries that supply the deep structures of the brain. The most common sites of hypertensive hemorrhage are the basal ganglia (60% are in the putamen), thalamus, pons, and cerebellum. Since the cortex is often relatively spared, ICH symptomatology tends to be neurological (i.e., sensorimotor) rather than behavioral.

SUBARACHNOID HEMORRHAGE

In SAH a cerebral blood vessel (usually an artery) on the surface of the brain ruptures, and bleeding occurs within the subarachnoid space (Figure 18.7). Bleeding into the subarachnoid space may occur spontaneously or may occur secondary to trauma. Primary, nontraumatic SAH is stroke. Secondary, traumatic SAH is not stroke; trauma is the most common cause of SAH.

Subarachnoid hemorrhage differs from other stroke syndromes in that it causes little immediate focal effect on the brain. Symptoms of SAH include severe headache

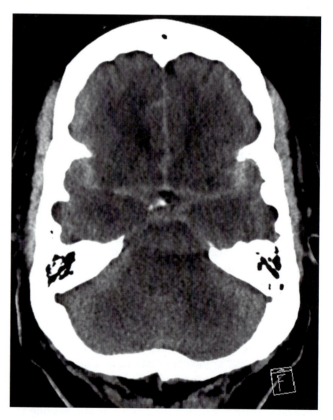

FIGURE 18.7. Subarachnoid hemorrhage. Axial CT shows diffuse subarachnoid hemorrhage in basal cisterns and bilateral frontal and temporal sulci.

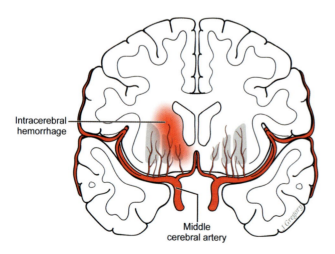

FIGURE 18.6. Intracerebral hemorrhagic stroke.

with a very sudden onset (**thunderclap headache**), nausea and/or vomiting, neck stiffness (**nuchal rigidity**) due to meningeal irritation, seizures, and confusion or lowered level of consciousness. Persons with SAH typically describe the headache as the "worst of their lives" and feeling like they are "being hit in the head with a bat." Prodromal (warning) headaches from minor blood leakage, known as **sentinel headaches**, are common and may occur several hours to several months before the rupture. SAHs are graded according to severity of symptoms.

Most primary SAHs (80%–90%) are caused by the rupture of an intracranial **aneurysm**; approximately 5% of primary SAHs are caused by bleeding from an **arteriovenous malformation** (AVM). Up to 50% of all cases of aneurysmal SAHs are fatal. Untreated ruptured aneurysms have a very high risk (20%–50%) of re-bleeding within the first 2 weeks, and such re-bleeding carries an even higher mortality rate of nearly 85%.

Bleeding from AVMs is often venous rather than arterial; thus the symptoms are typically less severe, and the mortality rate is lower than with arterial bleeding from ruptured aneurysms. Therefore, a history of recurrent episodes of SAH is more common in persons with an AVM than with aneurysm.

Cerebral Amyloid Angiopathy

Cerebral amyloid angiopathy (CAA) is a vasculopathy involving β-amyloid deposition within the walls of small to medium-sized blood vessels, predominantly affecting the cortical vessels. This results in weakened vessel walls, repeated hemorrhage (typically lobar), and superficial **siderosis**, which is the deposition of **hemosiderin** (an iron-rich breakdown product of blood seen in the chronic phase of any type of brain bleed) on the pial surface of the CNS. Neuroimaging markers for CAA include lobar microbleeds, lobar hemorrhages, cortical superficial siderosis, white matter hyperintensity, and convexal SAH. Convexal SAH is an unusual pattern of subarachnoid bleeding in which the bleeding is localized to the convex surface of the brain, without extending into the interhemispheric fissures, basal cisterns, or ventricles. The bleeding is unlikely to be due to aneurysm rupture because most aneurysms arise from vessels on the circle of Willis.

Stroke Diagnosis

The diagnosis of stroke is based on establishing through clinical assessment and neuroimaging that a stroke has occurred, and whether the cause is ischemic or hemorrhagic. Sudden, painless loss of neurological function that localizes to the anatomic vascular territory of a major cerebral artery or its branches is presumed to be acute ischemic stroke in the absence of countervailing evidence.

Computerized tomography (CT) is the preferred neuroimaging modality in the acute phase of suspected stroke because it is excellent for visualizing the acute bleeding of ICH and SAH. It is important to ascertain whether the cause of stroke is ischemic or hemorrhagic, because if the stroke is ischemic in nature the patient may be eligible for thrombolytic therapy (see "Thrombolysis," below), but if the stroke is hemorrhagic then thrombolysis is contraindicated. If CT shows SAH, cerebral angiography is performed to assess vascular anatomy, current bleeding site, and possible presence of other aneurysms, as 10%–30% of all aneurysmal SAH cases have multiple aneurysms. In ischemic stroke, brain imaging also may confirm the presence of infarction and delineate the vascular territory affected.

Stroke Treatment

Ischemic Stroke

THROMBOLYSIS

In ischemic stroke, a zone of mild to moderately ischemic and dysfunctional brain tissue lies between normally perfused brain tissue and the center (core) of the infarction, which is known as the **ischemic penumbra** (Figure 18.8). Hypoperfusion in the penumbra causes ionic and metabolic dysfunction but is not severe enough to result in structural damage. The central goal in treating acute ischemic stroke is to limit brain infarct size by opening the blocked artery and restoring blood flow to the penumbra.

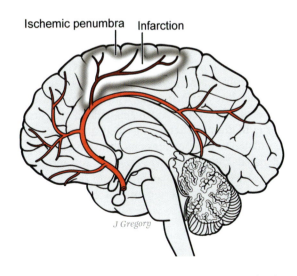

FIGURE 18.8. In ischemic stroke, the ischemic penumbra lies between normally perfused brain tissue and the core of the infarction.

This is accomplished by administering the **thrombolytic** "clot-buster" drug **tissue plasminogen activator** (tPA); this drug is also used to treat other conditions with blood clots, such as myocardial infarction and pulmonary embolism.

Tissue plasminogen activator is a naturally occurring enzyme that is produced by cells in blood vessel walls and is involved in the natural breakdown of blood clots; a genetically engineered version is used for therapeutic purposes. A blood clot consists of a mesh of fine threads of insoluble **fibrin** in which erythrocytes, leukocytes, and platelets are embedded. Tissue plasminogen activator catalyzes the conversion of **plasminogen** (a plasma protein) to **plasmin**, an enzyme that breaks down fibrin and thereby dissolves clots. It may be administered intravenously or intra-arterially directly at the clot site.

The window of opportunity for intravenous tPA is only within 3 hours of symptom onset. The window of treatment opportunity is extended to within 6 hours of symptom onset with intra-arterial delivery of tPA. Intra-arterial thrombolysis involves delivering tPA directly at the site of arterial occlusion by a neurovascular specialist at a hospital with a stroke center. The occluded vessel is visualized by cerebral angiography, a flexible microcatheter is inserted into an artery (usually in the groin area) and navigated to the site of the clot, and tPA is directly applied to the site of obstruction. With intra-arterial administration, blood flow is restored almost immediately.

Since tPA is contraindicated in hemorrhagic stroke, the guidelines for its use include a CT scan prior to administration to exclude the presence of hemorrhage.

CAROTID ENDARTERECTOMY, ANGIOPLASTY, AND STENTING

The incidence of ischemic stroke from carotid artery stenosis can be reduced by carotid endarterectomy, carotid artery angioplasty, or carotid artery stenting in patients with high-grade stenosis or ulcerative lesions, as revealed by carotid Doppler scan. In **carotid endarterectomy**, the atheromatous plaque is removed surgically. **Carotid angioplasty** is a minimally invasive **endovascular** (within the vessel) procedure performed to widen narrowed or obstructed arteries or veins. A deflated balloon attached to a catheter (a **balloon catheter**) is guided into the narrowed vessel, inflated to force expansion of the blood vessel, and then deflated and withdrawn. Stenting is another minimally invasive endovascular procedure in which a tiny mesh tube (a **stent**) is inserted into the carotid artery, deployed, and left in place to widen the vessel and increase blood flow.

Subarachnoid Hemorrhage

Intracranial aneurysms are treated by radiologically guided **endovascular coiling (aneurysm coiling)**. This is a minimally invasive procedure that is performed to block blood flow into an aneurysm. A catheter is maneuvered to the aneurysm site, and platinum coils are released. Over time, a clot forms inside the aneurysm, blocking blood flow into the aneurysm and eliminating the risk of aneurysm rupture. Less commonly, intracranial aneurysms are treated by an older technique known as **aneurysm clipping**. The aneurysm is accessed via **craniotomy** (surgical removal of a portion of bone from the cranium) and a metal clip is placed around the neck of the aneurysm to prevent blood from entering it.

Arteriovenous malformations are treated by three methods: (1) surgical removal of the AVM network, (2) radiosurgery, or (3) intra-arterial embolization. In radiosurgery, ionizing radiation causes the abnormal vessels to shrink, thereby reducing the size of the AVM. **Embolization** involves placing tiny catheters upstream of the AVM and injecting coils or liquid adhesives (glue) under X-ray guidance to close the connection between arteries and veins.

Intracerebral Hemorrhage

Surgical treatment of putamenal, thalamic, or pontine hemorrhage has not been successful because the surgery usually worsens destruction of the deep-seated tissue. Evacuation of cerebellar hemorrhage can be life-saving, as surgery can prevent brainstem compression.

Stroke Risk Factors and Stroke Prevention

Risk factors for adult stroke include age, arterial hypertension, hyperlipidemia (high cholesterol), diabetes mellitus, carotid stenosis, intracranial stenosis, atrial fibrillation, history of TIA, cigarette smoking, obesity, physical inactivity, cocaine use, and excessive alcohol use. Age is the strongest risk factor; risk for ischemic stroke doubles every decade after age 55. Hypertension is the most important modifiable risk factor for stroke. History of TIA increases the risk of ischemic stroke by four times as compared to an age-matched sample.

Primary stroke prevention (i.e., in those with no prior stroke or TIA) and **secondary stroke prevention** (i.e., in those with prior stroke or TIA) strategies focus on controlling vascular risk factors. Anything that promotes cardiovascular health also promotes cerebrovascular health. Interventions include reducing elevated blood pressure with antihypertensive medication and salt intake reduction, controlling blood sugar in those with diabetes, and treating hyperlipidemia with statins and low-fat diet. In patients with a history of ischemic stroke, TIA, emboligenic cardiac or arterial disease, stroke prevention also involves anticoagulants which slow down blood clotting. In patients with carotid artery stenosis, embolic stroke risk can be reduced by endarterectomy, angioplasty, or stenting. Antithrombotic medications (e.g., aspirin) are indicated for secondary stroke prevention in patients with a history of ischemic stroke. Lifestyle modification strategies to reduce stroke risk include smoking cessation, weight loss in overweight individuals, limiting alcohol intake, regular exercise, and a heart-healthy diet. Patients fulfilling all five criteria of low-risk lifestyle (no smoking, regular physical activity ≥ 30 minutes per day, healthy nutrition, moderate alcohol consumption, body mass index < 25 kg/m^2) reduce their stroke risk by 80% compared to patients fulfilling none of the criteria.

Vascular Cognitive Impairment (VCI)

Prior to the discovery that Alzheimer's pathology is the leading cause of dementia, it was believed that arteriosclerosis was the leading cause, and dementia was commonly referred to as "hardening of the arteries." Following the discovery of Alzheimer's pathology, dementia due to cerebrovascular disease was conceptualized as being due to multiple cortical infarcts of different sizes and locations and was referred to as **multi-infarct dementia** (MID). Prior to the advent of CT and MRI, multi-infarct dementia was clinically differentiated from dementia due to Alzheimer's disease based on course and associated neurologic signs. Multi-infarct dementia has an abrupt onset, stepwise progression, and is associated with focal neurological signs, while Alzheimer's disease, which is a cortical neurodegenerative disease, has an insidious onset, gradually progressive course, and absence of sensory and motor findings on neurological examination. With advances in neuroimaging, we now appreciate that there is a broader range of cerebrovascular pathology that results in cognitive impairment.

Definition

In 2011, the American Heart Association/American Stroke Association (AHA-ASA) released a scientific statement titled "Vascular Contributions to Cognitive Impairment and Dementia" that proposed new diagnostic criteria defining cognitive impairment and dementia due to vascular causes. These guidelines were developed in response to major criticisms of previous diagnostic criteria that were modeled after the diagnostic criteria for dementia due to Alzheimer's disease, which required significant memory impairment and therefore excluded dementia syndromes without prominent amnesia.

The AHA-ASA classification system uses the umbrella term **vascular cognitive impairment** (VCI) to refer to all forms of cognitive disorder of vascular origin. The criteria exclude those with active drug or alcohol abuse/dependence within the past 3 months, and those with delirium. The classification differentiates two levels of VCI based on severity: **vascular dementia** (VaD) and **vascular mild cognitive impairment** (VaMCI).

Vascular dementia is defined as a decline in at least two cognitive domains based on cognitive testing (with a minimum of four cognitive domains assessed, including executive/attention, memory, language, and visuospatial functions) that undermine instrumental activities of daily living (IADLs) or activities of daily living (ADLs). Vascular mild cognitive impairment is defined as a decline in at least one cognitive domain that does not undermine IADLs or ADLs. It can be further qualified by the four sub-classifications used to describe mild cognitive impairment of neurodegenerative origin: amnestic, amnestic plus other domains, non-amnestic single domain, and non-amnestic multiple domains.

These diagnoses are further classified as probable or possible. For a diagnosis of probable VCI, there is (1) cognitive impairment and neuroimaging evidence of cerebrovascular disease, and (2) a clear temporal relationship between a vascular event and onset of cognitive deficit(s), or a clear relationship in the severity and pattern of cognitive impairment and diffuse subcortical cerebrovascular disease. For a diagnosis of possible VCI, there is (1) cognitive impairment and neuroimaging evidence of cerebrovascular disease, but (2) there is no clear relationship (temporal, severity, or cognitive pattern) between a vascular disease and the cognitive impairment. Thus, there is cognitive impairment, and the clinical history and symptoms suggest vascular disease, but no confirming imaging studies are available, or there is another neurodegenerative disease or condition that could explain the cognitive decline.

This results in five classifications of VCI: probable VaD, possible VaD, probable VaMCI, possible VaMCI, and unstable VaMCI. This last diagnosis applies to persons with a diagnosis of probable or possible VaMCI whose symptoms revert to normal, as occurs when there is cognitive recovery following a vascular event.

Clinical Presentation of Vascular Cognitive Impairment

Vascular cognitive impairment is a clinical syndrome with heterogenous underlying pathology consisting of large vessel ischemic strokes, strategic infarcts, small vessel disease, hemorrhagic lesions, or any combination of these factors. Consequently, the clinical presentation of VCI also is heterogeneous. The onset and course of VCI varies, depending on the underlying pathology. VCI due to a single infarct has an abrupt onset. VCI due to multiple large vessel infarcts and/or strategic infarcts has a stepwise progression. VCI due to small vessel disease has an insidious onset and gradual progression.

Cognitive impairment following stroke is common, although not all strokes result in cognitive impairment. Single large vessel ischemic strokes or hemorrhagic strokes may result in **post-stroke dementia**, if the cognitive deficits meet criteria for a diagnosis of VaD. **Strategic infarcts**, ischemic lesions in brain regions supplied by an arterial branch that are critical for higher cortical function, result in more circumscribed cognitive deficits. For example, left angular gyrus infarction results in Gerstmann syndrome, the constellation of finger agnosia, right-left disorientation, acalculia, pure agraphia (see Chapter 15, "The Parietal Lobes and Associated Disorders"). Stroke limited to the hippocampus, which is critical for episodic memory formation, results in a persistent anterograde amnesia. Stroke in the PCA distribution may result in damage to the hippocampus and occipitotemporal visual association cortex (the ventral visual stream), resulting in amnesia and visual agnosia (see Chapter 14, "The Occipital Lobes and Visual Processing"). If the infarct involves the primary visual cortex, then the episodic memory disorder will be accompanied by a contralateral hemianopia, and if the infarct affects both PCA territories (since both PCAs originate from the same parent vessel) and both primary visual cortices, then the episodic memory disorder will be accompanied by cerebral blindness. Strategic infarct in the caudate can disrupt connections between the caudate and frontal lobe and can result in symptoms that mimic behavioral variant frontotemporal dementia, including disinhibition, increased craving for sweets, apathy, and loss of empathy. Bilateral medial thalamic infarction may result in **diencephalic amnesia** (see Chapter 10, "The Diencephalon").

The term *multi-infarct dementia* may be applied to cases of dementia due to multiple cerebral infarctions involving cerebral cortex, subcortical white matter, and/or subcortical gray matter structures, and a temporally concurrent stepwise progression of cognitive impairment. Atrial fibrillation, a cardiac arrhythmia that causes blood clots and emboli, is associated with multiple cortical infarcts. Multi-infarct dementia represents a small subset of those with VCI.

Lacunar infarcts and white matter lesions due to small vessel disease are common and may be the most frequent cause of vascular cognitive impairment. White matter lesions are evidenced as diffuse, confluent white matter hyperintensities (areas of increased brightness when visualized by T2-weighted MRI), typically in the periventricular white matter (Figure 18.9). Other terms for these white matter lesions are diffuse white matter disease, subcortical vascular encephalopathy, **leukoencephalopathy**, subcortical leukoencephalopathy, periventricular arteriosclerotic leukoencephalopathy, and **leukoaraiosis**. White matter

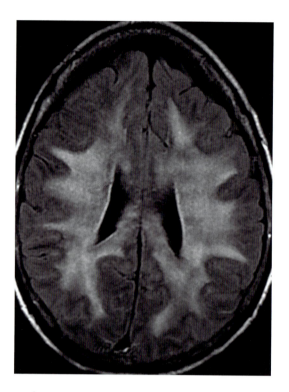

FIGURE 18.9. Leukoencephalopathy. Axial FLAIR image demonstrates diffuse white matter hyperintensity involving bilateral corona radiata, frontal, and parietal white matter.

lesions are associated with slowed cognitive processing, executive dysfunction, depression, and gait impairment. The term **lacunar state** has been used to refer to dementia due to multiple lacunar infarcts, while the term **Binswanger's disease** is used to refer to dementia occurring in the context of extensive periventricular white matter pathology. While the term *small vessel disease* has become synonymous with lacunar infarcts and white matter lesions, it is important to recognize that the parenchymal lesions detected by neuroimaging are a *marker* for underlying small vessel alterations; the small vessels themselves are not, with currently available technology, imageable in vivo. Furthermore, small vessel disease also results in large subcortical intracerebral hemorrhages and microbleeds.

VCI frequently involves mixtures of these cerebrovascular pathologies, which may also coexist with other pathologies such as Alzheimer's disease, complicating the clinical picture. Postmortem studies of persons with dementia show high rates of coexisting Alzheimer's disease plaque and tangle pathology and small vessel disease; the pathologies appear to be additive or synergistic in producing the dementia, underlying the concept of **mixed dementia** (dementia of mixed etiology).

Since VCI is heterogeneous in its underlying pathology and localization of lesions, there is no single neuropsychological profile. The specific presenting symptoms depend on the location and nature of the underlying pathology.

Thus, statements regarding neuropsychological test findings may characterize groups of patients but may not apply at the individual patient level. Studies of patients with VaD and large vessel infarcts have shown concomitant microvascular disease in the subcortical white matter; the combination of cortical or strategic infarcts and small vessel ischemic damage results in a complex clinical picture with mixed signs and symptoms.

Coexisting neurological symptoms are common in VCI, including sensorimotor deficits, gait disturbance, dysarthria, extrapyramidal signs, and urinary incontinence. Depression is also common. The depression may be a psychological response to cognitive deficits and their impact on daily functioning, but the fact that depression occurs with higher frequency in those with VCI compared to those with cognitive impairment due to other causes suggests that the depressive symptoms may be a manifestation of vascular lesions that disrupt frontal-subcortical-limbic networks involved in mood regulation.

Neuropsychological Assessment and Differential Diagnosis

The neuropsychological presentation of VCI is driven by the extent of focal and diffuse vascular brain injury. Psychomotor slowing is ubiquitous in VCI; it is attributed to pathology within the frontal-subcortical circuitry of the deep frontal white matter, basal ganglia, and thalamus. In the early stages of VaD, particularly in those with pathology predominantly affecting the subcortical white matter and nuclei, the neurocognitive profile is typically characterized by deficits in cognitive processing speed, working memory, executive functions (phonemic verbal fluency, cognitive flexibility, response inhibition), and memory (learning and recall, with preserved recognition memory); this pattern typifies subcortical dementia. Multiple studies have shown positive correlations between white matter lesions on MRI and executive and processing speed deficits.

A neurocognitive profile of executive dysfunction that includes deficits in working memory and cognitive processing speed, in the absence of episodic memory impairment and anomia, suggests a vascular etiology underlying the dementia. This suspicion is strengthened if there are significant vascular risk factors, and other conditions that can produce a subcortical dementia profile (e.g., normal pressure hydrocephalus) have been ruled out. In differentiating VaD from Alzheimer's dementia, vascular pathology tends to result in poorer lexical verbal fluency than semantic verbal fluency, while Alzheimer's dementia tends to result in poorer semantic verbal fluency than lexical verbal fluency. The memory disorder of VCI tends to be associated with deficits in learning and retrieval, but no rapid forgetting as in Alzheimer's dementia.

When VCI is suspected, it is critical to obtain a neuroimaging study. However, nearly two-thirds of individuals over the age of 70 exhibit vascular lesions on MRI, many with no cognitive symptoms or impairment. Silent strokes outnumber clinically manifest strokes by more than 11:1, and about 1 in 10 adults harbors a silent stroke by an average age of 63. The question is whether the extent of vascular burden revealed by neuroimaging is sufficient to account for the cognitive signs and symptoms. Clinical radiologic investigations use labels such as "age-related," "mild," "moderate," and "severe" to describe the extent of vascular damage; severe or extensive vascular disease is etiologically relevant, regardless of comorbid conditions.

Treatment Recommendations for Patients with Vascular Cognitive Impairment

Controllable risk factors for cerebrovascular disease and VCI are the same as those for stroke. As some of the risk factors for stroke and VCI are modifiable, early identification of vascular disease is critical. Treatment recommendations for those with VCI are aimed at reducing and/or eliminating risk factors, such as controlling hypertension, controlling hypercholesterolemia, controlling diabetes mellitus, controlling atrial fibrillation, smoking cessation, weight reduction, reducing intake of calories and saturated fats, limiting alcohol consumption, and increasing physical exercise. Vascular cognitive impairment due to small vessel disease is progressive and may benefit from primary prevention. Thus, accurate and early diagnosis is important.

Summary

Stroke is the most common neurologic disorder and the most common cause of brain lesions. Ischemic stroke is caused by loss of blood flow to a focal area of neural tissue due to obstruction (thrombosis or embolism) or hypoperfusion. Ischemic stroke results in a focal area of infarction that may be classified anatomically as territorial infarction, lacunar infarction, or watershed infarction. Hemorrhagic stroke involves rupture of a weakened brain blood vessel. Hemorrhagic strokes also are classified anatomically as ICH or SAH, and most often they originate from aneurysm or arteriovenous malformation.

Vascular cognitive impairment is a form of cognitive disorder due to cerebrovascular pathology, including large vessel ischemic strokes, strategic infarcts, small vessel disease, hemorrhagic lesions, or any combination of these. Since the underlying pathology and lesion location is heterogenous, so too are the signs and symptoms.

Additional Reading

1. Azarpazhooh MR, Hachinski V. Vascular cognitive impairment: a preventable component of dementia. *Handb Clin Neurol*. 2019;167:377–391. doi:10.1016/B978-0-12-804766-8.00020-0
2. Chabriat H, Joutel A, Dichgans M, Tournier-Lasserve E, Bousser MG. Cadasil. *Lancet Neurol*. 2009;8(7):643–653. doi:10.1016/S1474-4422(09)70127-9
3. Damasio H. A computed tomographic guide to the identification of cerebral vascular territories. *Arch Neurol*. 1983 Mar;40(3):138–142. doi: 10.1001/archneur.1983.04050030032005. PMID: 6830451.
4. Dichgans M, Leys D. Vascular cognitive impairment. *Circ Res*. 2017;120(3):573–591. doi:10.1161/CIRCRESAHA.116.308426
5. Ferrer I, Vidal N. Neuropathology of cerebrovascular diseases. *Handb Clin Neurol*. 2017;145:79–114. doi:10.1016/B978-0-12-802395-2.00007-9
6. Gorelick PB, Scuteri A, Black SE, et al. Vascular contributions to cognitive impairment and dementia: a statement for healthcare professionals from the American Heart Association/American Stroke Association. *Stroke*. 2011;42(9):2672–2713. doi:10.1161/STR.0b013e3182299496
7. Gorelick PB, Counts SE, Nyenhuis D. Vascular cognitive impairment and dementia. *Biochim Biophys Acta*. 2016;1862(5):860–868. doi:10.1016/j.bbadis.2015.12.015
8. Graff-Radford J. Vascular cognitive impairment. *Continuum (Minneap Minn)*. 2019;25(1):147–164. doi:10.1212/CON.0000000000000684
9. Iadecola C, Duering M, Hachinski V, et al. Vascular cognitive impairment and dementia: JACC Scientific Expert Panel. *J Am Coll Cardiol*. 2019;73(25):3326–3344. doi:10.1016/j.jacc.2019.04.034
10. Jellinger KA. Pathology and pathogenesis of vascular cognitive impairment: a critical update. *Front Aging Neurosci*. 2013;5:17. Published 2013 Apr 10. doi:10.3389/fnagi.2013.00017
11. Norling AM, Marshall RS, Pavol MA, et al. Is hemispheric hypoperfusion a treatable cause of cognitive impairment? *Curr Cardiol Rep*. 2019;21(1):4. Published 2019 Jan 19. doi:10.1007/s11886-019-1089-9
12. Ogoh S. Relationship between cognitive function and regulation of cerebral blood flow. *J Physiol Sci*. 2017;67(3):345–351. doi:10.1007/s12576-017-0525-0
13. Saver JL. Time is brain—quantified. *Stroke*. 2006;37(1):263–266. doi:10.1161/01.STR.0000196957.55928.ab
14. Skrobot OA, Attems J, Esiri M, et al. Vascular cognitive impairment neuropathology guidelines (VCING): the contribution of cerebrovascular pathology to cognitive impairment. *Brain*. 2016;139(11):2957–2969. doi:10.1093/brain/aww214
15. Skrobot OA, O'Brien J, Black S, et al. The Vascular Impairment of Cognition Classification Consensus Study. *Alzheimers Dement*. 2017;13(6):624–633. doi:10.1016/j.jalz.2016.10.007
16. Skrobot OA, Black SE, Chen C, et al. Progress toward standardized diagnosis of vascular cognitive impairment: guidelines from the Vascular Impairment of Cognition Classification Consensus Study. *Alzheimers Dement*. 2018;14(3):280–292. doi:10.1016/j.jalz.2017.09.007
17. Warsch JR, Wright CB. The aging mind: vascular health in normal cognitive aging. *J Am Geriatr Soc*. 2010;58 Suppl 2:S319–S324. doi:10.1111/j.1532-5415.2010.02983.x

Epilepsy

Introduction

Epilepsy is a group of neurological disorders characterized by recurrent seizures. A seizure is a sudden, uncontrolled electrical disturbance in the brain. There are many different types of seizures, and they are classified based on (1) where they start in the brain (onset); (2) whether awareness is affected; and (3) whether motor symptoms are present at seizure onset. There is no single pathology underlying all seizures, however, they all share a common mechanism that involves disruption in the balance between inhibitory and excitatory neuronal activity that results in hyperexcitability. This chapter discusses electrophysiological methods used in the diagnosis of epilepsy, seizure classification, epilepsy syndromes, medical treatment, seizure focus localization, functional mapping, surgical treatment, and neuromodulation treatments by electrical stimulation.

Definitions

The word *seizure* is derived from the Latin word *sacire*. The word *epilepsy* is derived from the Greek word *epilepsia*. Both words mean "to take possession of, suddenly and forcibly." But the terms *seizure* and *epilepsy* are not synonymous; recognizing this distinction is essential.

A **seizure** is essentially an electrical storm in the brain in which an abnormal, hypersynchronous discharge (i.e., greater than normal simultaneous neuronal firing) from a group of neurons in the brain gives rise to paroxysmal (i.e., sudden, brief, and self-limited) neurological symptoms. Most seizures resolve spontaneously in one to three minutes. **Epilepsy** is a chronic neurological disorder that is characterized by recurrent seizures. A history of at least two seizures is required for a diagnosis of epilepsy, also known as **seizure disorder**. Epilepsy is a very common neurological disorder, affecting approximately 0.5%–1% of the world population. Epilepsy is not a disease per se; it is an umbrella term encompassing many types of clinical syndromes known collectively as the *epilepsies*. Epilepsy is treated with antiepileptic drugs and sometimes surgery.

Many different conditions may cause seizures, thus there is no single pathology underlying all seizures. However, they all disrupt the balance between inhibitory and excitatory neuronal activity, which results in hyperexcitability. Etiologically, seizures may be classified into two broad classes, provoked and unprovoked. This distinction is critical in determining the course of treatment and in determining an epilepsy diagnosis, which requires at least two unprovoked seizures on separate days (i.e., at least 24 hours apart).

Provoked seizures, also referred to as **acute symptomatic seizures**, are isolated seizures that are caused by a systemic toxic/metabolic disturbance or an identifiable acute brain insult. Toxins can cause seizures by altering the balance of excitation and inhibition in the nervous system or interfering with energy metabolism. Toxic etiologies include drugs (e.g., cocaine, amphetamine) and drug withdrawal (e.g., alcohol, benzodiazepine). Metabolic abnormalities cause seizures by interfering with energy metabolism, changing osmolality, or producing endogenous toxins. Metabolic etiologies include electrolyte abnormalities (e.g., low or high blood sodium level), low or high blood sugar (i.e., hypoglycemia, hyperglycemia), and a wide variety of metabolic diseases. Provoked seizures of toxic/metabolic etiology are treated with interventions specific to the underlying cause. Diagnostically, provoked seizures due to identifiable acute brain insult occur within 7 days of the insult. Causes include traumatic brain injury (TBI; including that caused by neurosurgical intervention), stroke (ischemic or hemorrhagic), hypoxic brain injury, and CNS infection.

Unprovoked seizures are classified into two broad categories: (1) seizures of unknown etiology (i.e., seizures that occur without evidence of a provocative cause); or (2) **remote symptomatic seizures** due to a cerebral lesion or as a late effect of an acute injury or insult (i.e., more than 7 days after the onset of the acute brain insult). Causes include TBI, stroke, arteriovenous malformation, tumors or other mass lesions, and brain infections. Recurrent unprovoked seizures define epilepsy, a predisposition to seizures from genetic susceptibility or a chronic pathologic process.

The EEG and Epilepsy Diagnosis

The **electroencephalogram** (EEG) is an electrophysiological monitoring method in which the electrical activity of the cerebral cortex is recorded through electrodes placed on the scalp. Electroencephalography was introduced in 1929 by **Hans Berger** (1873–1941) in a publication titled "On the Electroencephalogram of Man." Berger was searching for the mechanism underlying telepathic communication, which of course he did not find, but he did discover that the EEG of a sleeping brain was very different from that of an awake brain. Based on these observations, he put forth the revolutionary idea that the electrical currents of the brain change based on its functional status (e.g., excited, relaxed, drowsy, asleep).

The routine clinical EEG lasts 20–30 minutes. It is typically used to (1) distinguish seizures from other types of paroxysmal neurological events; (2) characterize seizures for the purpose of treatment; and (3) localize the region of brain from which a seizure originates (i.e., the **seizure focus**) for workup in patients who are candidates for epilepsy surgery. EEG is also used in the evaluation of sleep disorders, evaluation of altered levels of consciousness (e.g., coma), and intraoperative monitoring of depth of anesthesia.

The International 10-20 System

The EEG is recorded using an array of 19 electrodes that are placed on the scalp according to the standard **International 10-20 System** for electrode placement (Figure 19.1). This system is so named because the inter-electrode distance is

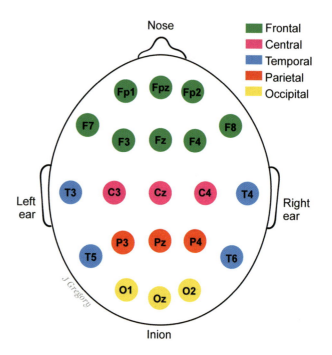

FIGURE 19.1. The International 10-20 System for electrode placement. The location of each electrode is specified by a code that consists of an uppercase letter indicating the underlying lobe (F, T, P, and O) or a central location (C), and either a number or lowercase letter, with even numbers indicating the right hemisphere, odd numbers indicating the left hemisphere, or z (for zero) indicating midline. Fp refers to frontopolar.

10% or 20% of the total distance between two landmarks of the skull, the **nasion** (the depressed area between the two eyes just superior to the bridge of the nose and inferior to the forehead) and the **inion** (the small bony protuberance of the occipital bone at the base of the skull).

The location of each electrode is specified by a code that consists of an uppercase letter combined with either a number or lowercase letter. The uppercase letters F, T, P, and O stand for frontal, temporal, parietal, and occipital, respectively; Fp refers to frontopolar. C refers to electrodes on the central/coronal plane. The even numbers 2, 4, 6, and 8 refer to electrodes placed over the right hemisphere; the odd numbers 1, 3, 5, and 7 refer to electrodes placed over the left hemisphere; and z ("zero") refers to electrodes placed on the midline. The lower the number, the closer the position is to midline. The 19 electrodes are named as follows: Fp1, Fp2, Fz, F3, F4, F7, F8, Cz, C3, C4, T3, T4, T5, T6, Pz, P3, P4, O1, and O2. Additional electrodes may be placed at smaller proportional distances to more precisely assess the electrical activity within a certain brain region.

The electrodes are flat metal discs that are positioned on the head, sometimes using a cap with fixed electrodes. The EEG signal represents a difference between the voltages at two electrodes, recorded as a single **channel** (waveform). The multiple EEG channels are represented in a logical and orderly arrangement known as the **EEG montage**. There are two main types of montages: monopolar and bipolar (Figure 19.2). In **monopolar (referential) montages**, all electrodes are referenced to a single common electrode (e.g., the earlobe) for each hemisphere, or a mathematical combination of signals (e.g., average reference). In **bipolar montages**, each electrode is referenced to an adjacent electrode in series. The **bipolar longitudinal montage** ("double banana") consists of a display in which each channel connects adjacent electrodes from anterior to posterior (e.g., Fp1–F3, F3–C3, C3–P3, P3–01); the **bipolar transverse montage** links adjacent electrodes from left to right. In general, referential montages are useful for judging the magnitude of the abnormality, while bipolar montages are useful for localizing the area of abnormal activity.

Physiological Basis of the EEG

Neurons are constantly producing electrical signals that are generated by the diffusion of ions across cell membranes (see Chapter 2, "Electrical Signaling in Neurons"). The EEG measures postsynaptic potentials of the apical dendrites (i.e., dendritic potentials) of cortical pyramidal neurons, which are oriented perpendicular to the cortical surface (Figure 19.3). When the apical dendrite receives excitatory input by an afferent pathway, it results in an excitatory postsynaptic potential (EPSP). The EPSP is caused by cations (positively charged ions) flowing from the extracellular fluid into the apical dendrite. This influx leaves a region of relative negative charge within the extracellular fluid at the synapse. The positive inward current of the EPSP flows down the dendrite and moves back outward at a site distant from the synapse. This generates a **dipole** in the extracellular fluid surrounding the apical dendrite, with a negative charge distal to the cell body and a positive charge proximal to the cell body. A scalp electrode is closer to the distal end of the apical dendrite and therefore detects the negative potential in the extracellular space that is generated by EPSPs. Thus, the EEG reflects inputs to neurons; it does *not* reflect action potentials. By convention, negative potentials are depicted as upward

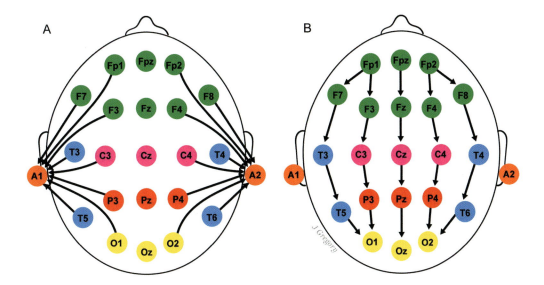

FIGURE 19.2. The EEG signal represents a difference between the voltages at two electrodes recorded as a single channel, and the EEG represents multiple channels. (A) In the monopolar (referential) montage, all electrodes are referenced to a single common electrode (e.g., the earlobe) for each hemisphere. (B) In the bipolar montage, each electrode is referenced to an adjacent electrode in series.

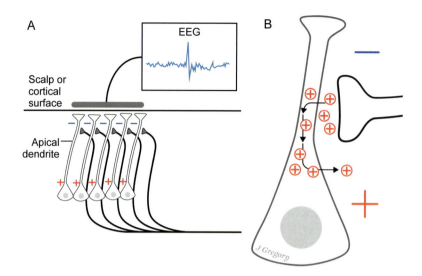

FIGURE 19.3. (A) The EEG measures postsynaptic potentials of the apical dendrites from a population of neurons beneath the electrode. (B) When the apical dendrite receives excitatory input, cations flow from the extracellular fluid into the dendrite, leaving a region of relative negative charge within the extracellular fluid at the synapse, which is detected by the scalp electrode. Thus, the EEG reflects excitatory inputs to neurons; it does *not* reflect action potentials.

deflections because they reflect excitatory input, and positive potentials are depicted as downward deflections because they reflect inhibitory input.

The scalp electrodes cannot detect signals generated by single neurons because the potentials are very small, and the electrodes lie a considerable distance away from the cell with several layers of tissue (CSF, meninges, bone, scalp) interposed between the cortical pyramidal cells and the recording electrode. The cerebral cortex, however, has two properties that enable us to detect summed electrical potentials from a large group of neurons at the gyral surface, using electrodes placed at the scalp surface: (1) pyramidal cells all have the same orientation within the cerebral cortex; and (2) many cortical pyramidal cells are activated synchronously due to modulatory inputs from the thalamus and reticular formation and an intrinsic cortical pacemaker mechanism. When thousands of neurons in a discrete cortical region are activated simultaneously, the summation of the dipoles creates an electrical potential that is detectable at the scalp surface. Each EEG electrode reflects the summed activity of thousands to millions of neurons that lie within about 6 cm² of cortex at the gyral surface beneath the electrode. Cortical tissue that is buried within sulci does not contribute to the EEG signal because the pyramidal cells in these regions are oriented parallel to the electrode surface, and therefore dipoles that lie at the same depth cancel each other out.

EEG Interpretation

The EEG trace is a graphical depiction that shows the brain's activity. It appears wavelike because the summed neural activity is, to varying degrees, synchronous and periodic. The EEG is interpreted with respect to **amplitude** (distance between the midline of the wave and its crest or trough), **frequency** (number of cycles per second, measured in Hz), and **waveform** (shape of the wave).

The amplitude of the potentials at the scalp surface is normally in the 20–100 microvolt range, having been attenuated from the 5 to several hundred millivolt range after traveling through the meninges, skull, and scalp; thus amplification of the signal is necessary (this is not true for EKG). The frequency range is 0.5–30 Hz and is divided into four frequency bands: **delta** (0–3+ Hz), **theta** (4–7+ Hz), **alpha** (8–13+ Hz), and **beta** (> 14 Hz) (Figure 19.4). The EEG rhythms are named by Greek letters (but not in Greek alphabetic order). Berger introduced the concept of alpha and beta waves (rhythms) in 1929. Delta and theta rhythms were discovered later; theta was chosen because it was believed that these waves originated within the thalamus.

The EEG pattern depends on the age of the subject and level of alertness or sleep. In awake and healthy adults, the EEG is desynchronized and symmetrical, with a low-amplitude mixture of alpha and beta waves, no slow brain waves (i.e., delta or theta), and no abnormal bursts of electrical activity. Beta activity is recorded from the center or front of the head, when the patient is alert and thinking with eyes open or closed. Alpha activity is recorded from the posterior head regions, especially during relaxed wakefulness with eyes closed; with eye opening or mental alerting activities, the amplitude of alpha activity attenuates. Theta activity is seen if the patient is awake but drowsy. Delta activity is seen in normal adults only during moderate to deep sleep.

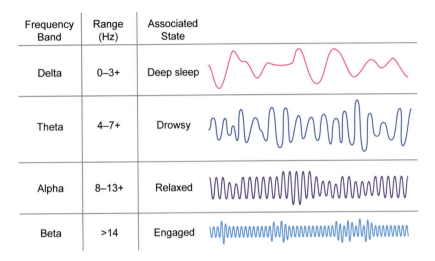

FIGURE 19.4. The EEG frequency range is divided into four frequency bands: delta (0–3+ Hz), theta (4–7+ Hz), alpha (8–13+ Hz), and beta (> 14 Hz).

Abnormal EEG findings indicate significant dysfunction of the underlying cortex. These abnormal findings include slowing of brain waves (delta and theta waves in the EEG of an awake and alert adult), asymmetry (different patterns of electrical activity across the two hemispheres), and sudden bursts of electrical activity (spikes and sharp waves).

EEG abnormalities can be diffuse or focal. Localized pathologic slowing often correlates with focal brain lesions such as brain tumors, cerebral abscesses, and subdural hematomas. Diffuse slow activity may signify an **encephalopathy** (disease, damage, or malfunction of the brain, manifest as an altered mental state) affecting the entire brain. This includes drug intoxication, encephalitis (brain infection), metabolic disorders (e.g., diabetic ketoacidosis) that change the chemical balance in the body, and delirium (which is often due to toxic-metabolic disturbances). An EEG that shows no electrical activity, referred to as a "flat" or "straight-line" EEG, indicates no cortical function. This occurs in coma, severe drug-induced sedation, lack of oxygen or blood flow to the brain, and of course, death.

During a seizure, a large group of cortical neurons fires in synchrony, resulting in abnormally large, paroxysmal (i.e., sudden onset and recurrent) discharges/intermittent high-voltage waves, and disturbances of cortical function. The most common epileptiform discharges are spikes, sharp waves, and spike-wave complexes (Figure 19.5). **Spikes** are very fast waves (< 70 msec duration) that have a pointed shape and clearly stand out from other EEG brain activity. **Polyspikes** are a series of spikes occurring in rapid succession. **Sharp waves** are less fast and sharp than spikes (70–200 msec duration). **Spike-wave complexes** are spike waves followed by a slow wave. **Polyspike-wave complexes** consist of a series of spikes occurring in rapid succession, followed by a slow wave.

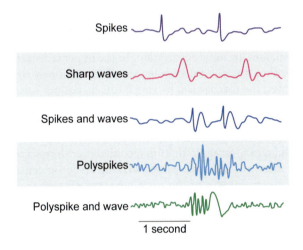

FIGURE 19.5. The most common epileptiform discharges are spikes, sharp waves, spike-wave complexes, polyspikes, and polyspike and wave.

Epileptiform discharges may be apparent only on EEG, in which case they are referred to as **electrographic** or **subclinical seizures**. Hypersynchronous discharges that produce subjective symptoms or objective signs are referred to as **clinical seizures**. Depending on the distribution of discharges, clinical seizures manifest in a wide variety of ways. The resultant effects on behavior depend upon the etiologic locus (where the seizure is originating from within the brain), and whether the abnormal discharge remains localized or spreads throughout the brain.

Seizure activity is divided into stages: pre-ictal, ictal, and post-ictal. Many people who have seizures report that they know in advance that a seizure is imminent by **pre-ictal** phenomena. The pre-ictal period can be divided into a prodromal stage and aural stage. The **prodrome** involves mood and/or behavior changes without specific electrophysiological changes, preceding the seizure by hours or days. The **aura** consists of a sensation that occurs briefly before a seizure; bad odor and strange abdominal sensations are common auras. The **ictus** refers to the seizure itself; the ictal stage is characterized by electrographic and clinical seizure activity. The **post-ictal** stage is a period that follows the active phase of the seizure, in which the individual does not feel normal and may experience confusion, somnolence, irritability, apathy, or depression. They may be unaware of having had a seizure. During the post-ictal period, the EEG may remain abnormal, usually with disorganized, slow background activity. The **interictal period** is the period of time between seizures.

Most standard EEG recordings occur during the interictal period. The recording is made over a period of 20–45 minutes with the patient sitting or lying down, in both awake and drowsy states. In those with seizure disorders, **epileptiform discharges** may be apparent during the interictal period or may be revealed during the transition to sleep. If not, they may become apparent with **activation procedures** such as hyperventilation, photic stimulation, or sleep deprivation. In the **hyperventilation** procedure, the patient breathes deeply and quickly for 3 minutes; the mechanism by which this brings out epileptiform discharges is unknown. In the **photic stimulation** procedure, a strong strobe light is placed directly in front of the patient's eyes. In healthy individuals, this produces no change in the alpha rhythm recorded over the occipital lobe or causes a normal **photic driving** response consisting of entrainment of the occipital alpha rhythm to the stimulus frequency of the strobe light. Spread of the response with abnormal waves is evidence of abnormal cortical excitability.

Ambulatory EEG monitoring allows for extended recordings over 24–72 hours, as the patient can move about and carry on with their daily activities.

If the standard or ambulatory EEG is not sufficient for diagnosing seizure type or localizing seizure focus, it may be necessary to record the EEG while the patient is having a seizure. In **video-EEG monitoring** the patient's behavior and EEG are simultaneously recorded, usually over several days during hospitalization at an epilepsy center. Patients are taken off their antiseizure medications (if they are taking any) to maximize the chances of recording both electrographic and clinical seizures.

Since the cortical electrical signals recorded by scalp electrodes are generated several centimeters below the recording electrodes and must pass through CSF, meninges, skull, and scalp tissue layers that have electrical resistance, the electrical signal has poor spatial resolution. When patients are being considered for neurosurgical intervention and noninvasive studies do not sufficiently localize or precisely define the zone of epileptogenic tissue, it may be necessary to localize the seizure focus with greater resolution than scalp EEG recordings allow. This can be achieved using invasive electroencephalography techniques that record the electrical activity directly from the surface of the cortex or from deep structures. In **electrocorticography** (ECoG), electrical activity is recorded directly from the surface of the cerebral cortex using a grid or strip of electrodes implanted subdurally on the surface of the cortex, generating an **electrocortigram**. In **stereoelectroencephalography** (SEEG), recordings may also be made from deep structures, such as the hippocampus and amygdala, through **depth electrodes** implanted by **stereotactic surgery** using a three-dimensional coordinate system to locate small targets (Figure 19.6).

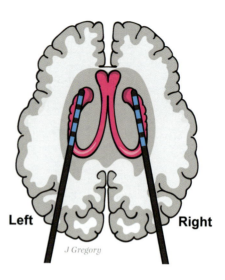

FIGURE 19.6. Depth electrodes placed directly into the hippocampi.

Classification of Seizures

There are many different types of clinical seizures, each with characteristic behavioral changes and electrophysiologic disturbances. **Seizure semiology** is the description of the signs and symptoms of a seizure from onset throughout the course. The earliest signs and symptoms of a seizure may allow for lateralization (i.e., the hemisphere) and localization (i.e., the lobe) of seizure onset. The **International League Against Epilepsy (ILAE) Classification of Seizures**, revised in 2017, is a system that categorizes seizures and epilepsy syndromes (Figure 19.7). Seizures are classified according to three features: (1) site of origin (focal vs. generalized); (2) level of awareness during the seizure (aware vs. impaired awareness); and (3) whether motor symptoms are present at seizure onset (motor vs. non-motor).

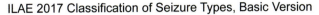
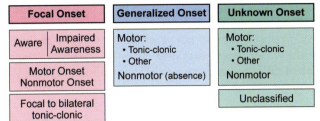

FIGURE 19.7. The basic ILAE 2017 operational classification of seizure types.
Source: Fisher RS, Cross JH, D'Souza C, French JA, Haut SR, Higurashi N, Hirsch E, Jansen FE, Lagae L, Moshé SL, Peltola J, Roulet Perez E, Scheffer IE, Schulze-Bonhage A, Somerville E, Sperling M, Yacubian EM, Zuberi SM. Instruction manual for the ILAE 2017 operational classification of seizure types. *Epilepsia*. 2017 Apr;58(4):531–542. 11/epi.13671. Epub 2017 Mar 8. PMID: 28276064.

Focal vs. Generalized Seizures

The classification system divides seizures based on the onset and spread of abnormal electrical activity. This dimension distinguishes focal onset seizures, generalized seizures, seizures of unknown onset, and focal to bilateral seizures.

Focal onset seizures, also known as **localized seizures** (formerly referred to as *partial seizures*), originate within a specific cortical region in one hemisphere due to a focal cortical lesion or EEG abnormality, and they remain confined to one hemisphere (Figure 19.8). In other words, they have a focal (localized) onset as determined by EEG or clinical symptoms. The location of the seizure focus determines the clinical manifestations. **Generalized onset seizures** begin bilaterally (within both hemispheres) and have diffuse cerebral cortical involvement, as indicated by EEG and clinical manifestations (see Figure 19.8). **Seizures of unknown onset** are unclassified with respect to the nature of the onset due to inadequate information, such as when the seizure onset was not witnessed and/or routine EEG study results are not yet available or have not shown epileptiform activity. With additional information, such seizures may later become classified as either of focal or generalized onset.

In **focal to bilateral seizures**, the seizure has a localized onset but the discharge spreads to both sides of the brain (see Figure 19.8). There is a failure to contain the focal seizure activity by normal inhibitory processes, and the discharge spreads to involve both hemispheres via the corpus callosum and/or the reticular formation of the upper brainstem or thalamus. These seizures were previously referred to as *secondarily generalized seizures*, in contrast to *primary generalized seizures* which have a generalized onset. Thus, while previous classification systems distinguished primary vs. secondarily generalized seizures, the current system uses the term *generalized* only to refer to the nature of the seizure *onset* (i.e., primary generalized seizures are now referred to as "generalized seizures," and secondarily generalized seizures are now referred to as "focal to bilateral seizures").

Seizures with Preserved Awareness vs. Impaired Awareness

Seizures are also classified according to whether awareness is intact or impaired during the seizure, based on whether the person retains awareness of self and the environment throughout the seizure. This dimension only applies to focal seizures because generalized seizures always affect a person's awareness or consciousness due to diffuse and bilateral cerebral cortical involvement. Therefore, focal seizures are further classified as focal aware seizures, focal impaired awareness seizures, and focal seizures awareness unknown. The distinction between focal seizures with vs. without impaired awareness is of practical importance because poorly controlled focal seizures with impaired

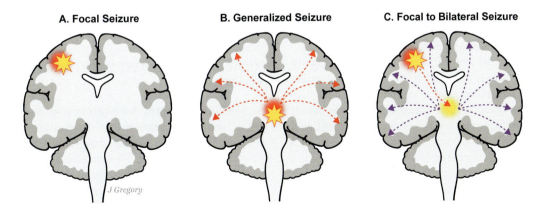

FIGURE 19.8. Seizures are classified based on onset and spread of abnormal electrical activity. (A) Focal onset (localized) seizures originate within a specific cortical region, and they remain confined to one hemisphere. (B) Generalized seizures begin bilaterally and have diffuse cerebral cortical involvement. (C) Focal to bilateral seizures have a focal onset but the discharge spreads to both sides of the brain.

awareness increase patient risk; thus activities such as driving and operating dangerous machinery must be restricted.

During **focal aware seizures**, the person remains awake and alert throughout the seizure, and afterward they have full recollection of the event. They may or may not be able to speak or respond to questions and commands during the seizure. During **focal impaired awareness seizures**, awareness is impaired at some point during the seizure (i.e., all aspects of cognition are affected). Focal impaired awareness seizures often have a seizure focus in the temporal lobe or limbic areas; less frequently, they have a frontal localization. In some cases, it is not known whether the focal seizures are associated with impaired awareness; in these cases, the seizures are classified as **focal seizures awareness unknown**. Further investigation may clarify whether the seizures are associated with impaired awareness or not, allowing for reclassification.

Motor vs. Non-Motor Seizures

Motor phenomena during seizures include **tonic movements** (stiffening), **clonic movements** (rhythmic jerking), **myoclonic movements** (twitching), and automatisms. **Automatisms** (automatic movements) are coordinated, repetitive movements that appear semi-purposeful and occur with impaired awareness. **Oro-alimentary automatisms** include lip-smacking, lip-pursing, chewing, licking, tooth-grinding, and swallowing. **Manual automatisms** include rubbing, fumbling, and picking movements of the hands. **Vocal automatisms** include grunts, shrieks, and incoherent mumbling. **Verbal automatisms** involve utterances consisting of words or phrases (also known as **ictal speech**). Automatisms may also consist of complex acts such as turning the head in an orientation response, walking and running movements, wandering around a room appearing to inspect things with a blank look, or undressing. **Sexual automatisms** consist of repeatedly grabbing or fondling the genitals, and pelvic thrusting movements. Complex acts that were in process before the seizure onset (e.g., walking, driving, turning the pages of a book) may continue during the seizure.

Motor phenomena also include **atonic seizures (drop attacks)** in which there is a sudden loss of muscle tone due to seizure activity while remaining conscious. Such seizures typically last less than 15 seconds and involve the whole body or just part of it, causing the person to suddenly fall to the ground, the head to drop, the eyelids to droop, or a loss of hand grip causing the person to drop anything they are holding. Atonic seizures often begin in childhood and mostly affect children and adolescents. Those who have full body atonic seizures may get injured when they fall, and they often wear a helmet for head protection.

Focal onset seizures are classified as motor or non-motor onset seizures based on whether motor phenomena are the first prominent sign. By contrast, non-motor focal onset seizures involve a wide array of non-motor phenomena at seizure onset, such as autonomic responses (e.g., changes in heart rate, breathing, profuse sweating), sensory seizures, behavioral arrest (i.e., blank stare, halted talking, halted movement or ambulation), cognitive seizures, and emotional seizures. Generalized onset seizures are classified as motor vs. non-motor, based on whether they involve motor phenomena at seizure onset. Generalized onset non-motor seizures consist of absence seizures.

Focal Onset Seizures

Focal onset seizures originate within a specific cortical region in one hemisphere. The **epileptologist** localizes the seizure focus based on a combination of the clinical signs and symptoms, EEG data, and MRI data, which in some cases may show a lesion such as mesial temporal sclerosis, cortical dysplasia, arteriovenous malformation, tumor, or **encephalomalacia** (a focal softening of brain tissue after infarction, infection, trauma, or other injury).

Focal onset seizures are sub-classified along two dimensions: preserved awareness vs. impaired awareness, and motor onset vs. non-motor onset.

Focal Onset Seizures with Preserved Awareness

Focal onset seizures with preserved awareness (formerly referred to as *simple partial seizures*) are brief, lasting seconds to less than 2 minutes. Focal onset aware seizures are further classified as focal motor seizures with preserved awareness (formerly *simple partial motor seizures*), focal sensory seizures with preserved awareness (formerly *simple partial sensory seizures*), focal cognitive seizures (formerly referred to as *psychic seizures*), and focal emotional seizures (also formerly referred to as *psychic seizures*).

FOCAL MOTOR SEIZURES WITH PRESERVED AWARENESS

Focal motor seizures with preserved awareness consist of involuntary movements such as tonic posturing or clonic movements that are restricted to one body part on one side of the body. The involuntary movement may gradually spread ("march") from one muscle group to adjacent muscle groups on the same side of the body, which is referred to as a **Jacksonian motor seizure**. Focal motor seizures with preserved awareness arise from seizure foci in the frontal motor cortex; thus they are also referred to as **frontal lobe seizures**. The abnormal movement usually begins in the hand, face, or toes; the high incidence of focal motor epilepsy originating with movements in these areas reflects the disproportionately large amount of cortical tissue that represents these body parts. A transient post-ictal weakness or paralysis in the affected region that lasts minutes to hours after a focal motor seizure is known as **Todd paralysis**.

FOCAL SENSORY SEIZURES WITH PRESERVED AWARENESS

Focal sensory seizures with preserved awareness arise from sensory cortices, giving rise to **illusions** (distortions of true sensations) and **hallucinations** (sensations of something that is not there). Focal sensory seizures usually consist of somatosensory illusions/hallucinations, but they can also involve the special senses of vision, audition, taste, and smell. The sensory seizure may be followed by a post-ictal sensory deficit due to focal cortical dysfunction following neuronal hyperexcitation in the brain region affected by the seizure; this Todd phenomena is conceptually similar to the Todd's paralysis following focal motor seizure.

Somatosensory seizures typically give rise to numbness, tingling, or "pins-and-needles" sensations. Less commonly, they give rise to sensations of crawling (formication), electricity, or movement of the body part. They rarely give rise to pain or thermal sensations. Somatosensory seizures are nearly always due to seizure activity in the parietal lobe somatosensory cortex, in the cerebral hemisphere contralateral to the side of the sensory phenomenon. The location of the symptoms on the body surface corresponds to the location of the seizure focus within the somatotopic map of the postcentral gyrus. In most cases, the onset of the sensory seizure is in the lips, fingers, or toes, due to the disproportionately large cortical representation of these body parts. The sensation may remain restricted or may spread to adjacent parts of the body, following the pattern of the somatotopic map and resulting in a Jacksonian sensory march.

Olfactory seizures typically involve perception of an exteriorized (i.e., perceived as coming from the environment), disagreeable, or foul odor that is otherwise unidentifiable. Olfactory hallucinations are a common aura that precedes focal seizures with impaired awareness. The aura constitutes a focal seizure with preserved awareness. Olfactory hallucinations are associated with foci in the region of the parahippocampal gyrus or the **uncus** of the temporal lobe. John Hughlings Jackson therefore referred to these seizures as **uncinate fits**. The neuronal discharge begins in the uncus and spreads into the temporal lobe. These seizures have localizing value but do not have lateralizing value.

Gustatory seizures typically consist of an unpleasant taste. They occur with seizure foci in the temporal lobe, insula, or parietal operculum, and while these seizures have localizing value, they do not have lateralizing value.

Visual hallucinations are the hallmark of occipital seizures, a relatively rare form of focal epilepsy. They typically begin in the visual field contralateral to the affected visual cortex and may remain static or move. Unformed visual hallucinations are characterized by fleeting sensations of sparks or flashes of light or color that are simple in shape. They are usually due to a seizure focus in or near primary visual cortex of the occipital lobe. Formed visual hallucinations consist of persons, animals, objects, figures, or scenes. They occur with seizure activity in visual association cortex, in the posterior parietotemporal region. Less commonly, visual seizures may consist of negative phenomena such as scotoma or hemianopia.

Auditory seizures are also rare. Seizure foci in one superior temporal gyrus are associated with buzzing or roaring sensations, while more formed hallucinations such as voices (e.g., repeating unrecognizable words) or the sound of music tend to occur with foci in posterior temporal lobe. These seizures have localizing value (i.e., they originate from a temporal lobe) but they do not have lateralizing value, because the sounds are usually not perceived as coming from one side of space; when they are, the localization is not a reliable sign of seizure lateralization.

Visceral sensations due to seizure activity are often vague and indefinable but may include nausea. Visceral sensations are among the most frequent type of aura, and they typically originate in orbitofrontal, mesial temporal, and insular cortical areas.

FOCAL COGNITIVE SEIZURES WITH PRESERVED AWARENESS

Focal cognitive seizures involve a specific alteration in cognitive function at seizure onset. The specific cognitive alteration is out of proportion to other relatively unimpaired aspects of cognition, and the person remains aware of self and the environment. This stands in contrast to focal impaired awareness seizures in which all aspects of cognition are impaired. Focal cognitive seizures may occur as an isolated phenomenon, or as an aura with progression to a focal impaired awareness seizure or a focal to bilateral seizure.

There are a wide variety of focal cognitive seizures, and the alteration can be either a **negative symptom** (i.e., deficit) or **positive symptom** (i.e., presence of an abnormal behavior, belief, or perceptual experience). Focal cognitive symptoms involving negative symptoms include focal cognitive seizure with expressive aphasia, anomia, receptive aphasia, auditory agnosia, conduction aphasia, alexia, agraphia, acalculia, memory impairment, right-left confusion, and neglect. These clinical ictal features provide clues to the region of ictal onset, just as do focal motor and focal sensory seizures with preserved awareness.

Positive cognitive symptoms at seizure onset include altered perceptions of familiarity, of **déjà vu** (new things or experiences seem familiar) and **jamais vu** (familiar things or experiences seem foreign); **depersonalization** (feeling one is not oneself); **derealization** (the world seems unreal or dream-like); altered perception of time (e.g., time slowing down or speeding up); **autoscopy** (experiencing one's body in extrapersonal space); **flashbacks** (fragments of certain memories or scenes recur with striking clarity); and **forced thinking** (recurrent intrusive thoughts). These phenomena are typically associated with seizures starting in the temporal lobe but have no lateralizing value.

FOCAL EMOTIONAL SEIZURES WITH PRESERVED AWARENESS

Focal emotional seizures are characterized by an alteration in subjective mood or affective behavior at seizure onset. The affective behaviors and/or subjective mood phenomena have a sudden onset and a stereotypic pattern (i.e., similar event-to-event), and they are unrelated to environmental stimuli. Focal emotional seizures may occur as isolated phenomena, or as auras with progression to a focal impaired awareness seizure or a focal to bilateral seizure. If the focal seizure involves an alteration in subjective mood or affective behavior during the seizure but not at the very onset, this feature is used as a seizure descriptor after the seizure is classified according to its onset feature.

Fear (anxiety, panic) is the most common ictal affective experience, and it can involve a very dramatic emotional display. **Focal emotional seizures with fear/anxiety/panic** arise from mesial temporal regions, especially the amygdala.

Focal emotional seizures with laughing are characterized by laughter or giggling at seizure onset, usually unaccompanied by the subjective emotion of happiness. This seizure type often occurs with seizures arising in the hypothalamus due to hamartoma (a noncancerous tumor made of an abnormal mixture of cells), but they may also occur in seizures arising from the frontal or temporal lobes. **Gelastic seizures** feature laughter or giggling at any time during the seizure.

Focal emotional seizures with crying are characterized by stereotyped crying, which may or may not be accompanied by the subjective emotion of sadness. This seizure type is rare. They often arise from the hypothalamus due to hamartoma, but they may also occur in seizures arising from the frontal or temporal lobes. **Dacrystic seizures** feature crying at any time during the seizure. Dacrystic seizures often accompany gelastic seizures.

Focal emotional seizures with pleasure are characterized by the subjective emotional experience of pleasure, bliss, joy, enhanced personal well-being, heightened self-awareness, or ecstasy at seizure onset. This seizure type is rare. They arise from the anterior insular cortex.

Focal emotional seizures with anger are characterized by the subjective emotional experience of anger at seizure onset. They may be accompanied by aggressive behavior that is not purposeful or goal-directed; however, it is more common to see aggressive behavior in the postictal period. This seizure type is rare. They arise from prefrontal or mesial temporal regions.

Rarely, a feeling of extreme embarrassment with reddening of the face occurs as an aura. Embarrassment was first reported as an ictal phenomenon in a man with an astrocytoma in the medial frontal lobe. He experienced extreme embarrassment as an aura, with progression to a focal impaired awareness seizure. This suggests that electric discharge in the prefrontal cortex may produce the opposite effect of prefrontal lesions, in which there is a lack of insight and absence of the capacity for embarrassment.

Focal Onset Seizures with Impaired Awareness

Focal unaware seizures (previously known as *complex partial seizures* and *psychomotor seizures*) involve an alteration or impairment of awareness at some time during the seizure. These usually have a focus in limbic and autonomic areas of the temporal lobe (**temporal lobe seizures**) but may also have a frontal lobe focus. They have many forms. They are sub-classified as having motor or non-motor signs and symptoms at seizure onset. Non-motor signs and symptoms encompass a wide variety of sensory, autonomic, and affective phenomena.

Generalized Seizures

Generalized seizures are a diverse group of seizures, all of which are characterized by generalized spike or polyspike-and-slow-wave discharges that arise within and rapidly engage bilaterally distributed networks, without underlying structural brain abnormalities. These seizures begin at deep midline subcortical structures (upper brainstem and thalamus) and spread to both cerebral hemispheres simultaneously. Due to disrupted electrical activity throughout the cerebral cortex, awareness and consciousness are always affected. If motor involvement is present, it is always manifested bilaterally. Generalized seizures are thought to be caused by metabolic factors, and that a genetic component underlies many of these disorders.

Generalized seizures are further classified as generalized motor seizures and generalized non-motor seizures.

Generalized Motor Seizures

Generalized motor seizures are characterized by motor involvement at the onset that is bilateral, symmetric, and accompanied by impaired awareness. This group includes generalized tonic-clonic seizures, tonic seizures, clonic seizures, atonic seizures, and myoclonic seizures.

Generalized tonic-clonic seizures (GTCs) were previously known as *grand mal seizures* and *convulsive seizures*. A GTC is a violent and frightening spectacle, and this is what most people picture when they think of seizures. The EEG shows generalized polyspikes. GTC seizures have two phases: tonic and clonic. During the tonic phase all muscles contract, producing a rigid extension of arms, legs, and head (in effect, a decerebrate posture; see Chapter 25, "Disorders of Consciousness"). The respiratory muscles also are in a tonic contraction; thus respiration ceases, and the person may become cyanotic (i.e., bluish or purplish in color due to lack of oxygen). There may be urinary and/or fecal incontinence. This phase lasts from several to approximately 30 seconds. As the rate of neuronal discharge slows, the clonic phase begins and the muscles jerk rhythmically. This phase usually lasts approximately 1–3 minutes. Breathing recommences at the end of this phase. Post-ictal lethargy and confusion often last minutes to several hours. Post-ictally, brain electrical activity may remain abnormal, showing background suppression and then diffuse slowing.

Tonic seizures are characterized by sustained muscle contraction in flexor and extensor muscle groups, producing limb and trunk rigidity that lasts seconds to several minutes. Since tonic seizures are generalized, they involve the musculature bilaterally and usually symmetrically (in contrast to focal motor seizures). The EEG usually shows generalized, low-voltage, high-frequency polyspikes. **Clonic seizures** consist of bilateral rhythmic (i.e., regularly repetitive) jerking movements. **Atonic seizures** are characterized by a sudden but brief (several seconds) loss of muscle tone that causes the person to fall to the floor; these seizures are therefore also known as **epileptic drop attacks**. There is a loss of consciousness during the seizure. **Myoclonic seizures** consist of sudden, brief, shock-like involuntary jerking movements due to seizure activity. Myoclonic seizures occur in a variety of epilepsy syndromes. Not all myoclonic movements represent seizures (there are normal forms of myoclonus, and pathological myoclonus due to other causes).

Generalized Non-Motor Seizures

The category of generalized non-motor seizures consists of **absence seizures** (formerly *petit mal seizures*). These are brief, 3–20-second episodes of staring with a vacant expression, during which the person is unaware of their surroundings and unresponsive. Absence seizures are largely a disorder of childhood, not typically associated with abnormalities in development or intelligence, and usually resolve by age 18. A generalized 3 Hz spike-wave discharge is the EEG signature of absence seizures. Absence seizures are often provoked by hyperventilation. As onset and offset of the episodes is abrupt, there are no post-ictal phenomena. Absence seizures are further classified as typical, atypical, myoclonic, or with eyelid myoclonia. Typical absence seizures have a rapid onset and offset, complete loss of awareness, and the 3 Hz spike-wave discharge.

Status Epilepticus

Status epilepticus is broadly defined as a prolonged seizure or multiple seizures with incomplete return to baseline between seizures. It is classified broadly into two types: convulsive and nonconvulsive. Generalized tonic-clonic status epilepticus is the most common form of status epilepticus. Nonconvulsive status epilepticus is continuous or near-continuous generalized electrical seizure activity lasting for at least 30 minutes without physical convulsions; it is characterized by abnormal mental status, unresponsiveness, and oculomotor abnormalities. Status epilepticus is a neurological emergency; it causes neuronal damage and is associated with significant morbidity and mortality (in adults, mortality is about 20%). A loss of endogenous inhibitory mechanisms contributes to the transformation of a single seizure into status epilepticus, and the self-sustaining nature of the condition. Intervention is aimed at rapid termination of seizure activity; intravenous fast-acting lorazepam (a benzodiazepine) is the first-line treatment.

Etiology: Idiopathic (Primary) vs. Symptomatic (Secondary) Epilepsy

Another dimension of classifying seizures is according to presumed etiology. In **idiopathic (primary) epilepsy**, the cause is unknown. In **symptomatic (secondary) epilepsy**, an underlying structural brain lesion has been identified. In **cryptogenic epilepsy**, there likely is a cause but it has not been identified.

A wide variety of brain lesions are epileptogenic. **Post-traumatic seizure disorder** (**post-traumatic epilepsy**) following TBI is the most common cause of secondary epilepsy. Gliosis and scar tissue formation produce the epileptic focus. The incidence of a post-traumatic seizure disorder is greatest with penetrating head injuries where the dura has been breached and there is direct damage of the brain tissue, such as with gunshot wounds and depressed skull fractures. Closed head injury, however, is a more

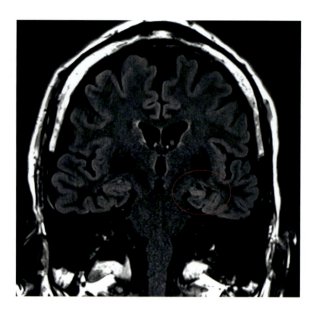

FIGURE 19.9. Mesial temporal sclerosis. Coronal FLAIR MRI at the level of the hippocampus demonstrates left hippocampal volume loss with FLAIR hyperintensity.

Non-Epileptic Seizures

Non-epileptic seizures (NES), also known as **psychogenic seizures** and **pseudoseizures**, are episodes that simulate convulsive or nonconvulsive seizures but are not the result of an abnormal neuronal discharge. They are due to **functional neurological symptom disorder** (conversion disorder, neurological symptoms not accounted for by known neurological mechanisms and incompatible with neurological pathophysiology), or **malingering** (feigned symptoms). Non-epileptic seizures can be mistaken for seizures and are therefore treated with antiepileptic drugs, but they are often treated with high or even intoxicating doses because they do not respond to treatment. Of patients referred for presurgical epilepsy evaluation, 25% have NES. Non-epileptic seizures and epilepsy, however, are not mutually exclusive; 10%–40% of NES patients also have epilepsy.

Pseudoseizures often involve an unconventional motor display such as asynchronous thrashing of the limbs, repeated side-to-side head movements, striking out at persons trying to restrain the person, screaming or talking during the ictus, and eyes kept forcefully closed (whereas the lids open or show clonic movement in genuine convulsions). These events tend to be prolonged for many minutes or hours, unlike seizures. Non-epileptic seizures usually lack features that are common in GTCs, such as tongue biting, incontinence, hurtful falls, and post-ictal confusion. Pseudoseizures tend to be precipitated by emotional factors and tend to occur in the presence of observers. Diagnosis of NES is best established by video-EEG monitoring.

frequent cause of post-traumatic seizures simply because it is much more common than penetrating head injury. The likelihood that epilepsy will develop following TBI is maximal immediately following the trauma, and it declines steadily over the next 2 years.

Neoplasms, particularly gliomas, are another common cause of secondary epilepsy. In persons with brain tumors, 20%–40% have seizures as the onset symptom, and an additional 20%–45% have seizures at some in the course of the disease.

Mesial temporal sclerosis is an atrophic gliotic process involving the hippocampus (most prominently the CA1 and CA3 subfields) and adjacent cortex of the hippocampus (Figure 19.9). It is a very common cause of **temporal lobe epilepsy**. It is not clear to what extent mesial temporal sclerosis may be caused initially by seizures, as it usually develops in those with a history of prolonged febrile seizures during childhood.

Small regions of developmental focal **cortical dysplasia** also tend to be epileptogenic. These are developmental focal abnormalities of the cortical cytoarchitecture with disordered layering and displaced cells.

Over 50% of epilepsy cases are idiopathic (i.e., the cause of seizures is not identifiable). Although electrical activity is periodically abnormal, the brain itself has no macroscopic or microscopic structural abnormalities. It is believed that disorders of metabolism within the neuron and instability of the neuronal membrane are responsible for many cases of idiopathic epilepsy, and there is mounting evidence of a genetic basis. Onset of idiopathic epilepsy occurs most often in childhood and adolescence.

Epilepsy Syndromes

Seizures are symptoms of neurologic dysfunction and are but one manifestation of many neurologic diseases. An epilepsy syndrome is a group of signs and symptoms. Epilepsy syndromes are defined based on the unique combination of seizure type(s), typical age of onset, typical EEG findings, other clinical features, and expected course. Syndromal diagnosis is important because it has implications for pathogenesis, prognosis, and treatment.

Lennox-Gastaut Syndrome

Lennox-Gastaut syndrome is a form of childhood-onset epilepsy with onset from 2 to 6 years of age. It is characterized by daily, multiple seizures of various types; the range of seizure types is greater than in any other epilepsy syndrome. Atonic drop attacks most frequently occur in the setting of Lennox-Gastaut syndrome. In addition to seizures, children with this syndrome frequently have arrested or slowed development and behavior disorders. The syndrome is characterized by a distinctive, slow (1–2 Hz) spike and wave EEG pattern.

Rasmussen's Syndrome

Rasmussen's syndrome is a rare inflammatory disease of unknown cause that affects a single cerebral hemisphere. This syndrome is also known as **Rasmussen's encephalitis** or **chronic focal encephalitis**. It usually affects children under the age of 15, with an average age of onset of 6 years. The acute stage of active inflammation lasts 4–8 months. During this stage there are frequent and severe seizures (usually focal), and a progressive loss of motor, speech, and cognitive function, often with contralateral deficits such as hemiparesis and hemianopia. The inflammation causes permanent damage, leading to atrophy. Frequent seizures also contribute to progressive brain damage in the affected hemisphere during the acute stage. In the chronic (residual) stage, the inflammation is no longer active, but there are stable, neurological deficits due to permanent damage and atrophy caused by inflammation and frequent seizures.

Landau-Kleffner Syndrome

Landau-Kleffner syndrome is a rare disorder with onset at 3–7 years of age. It produces seizures and a progressive loss of language following a period of normal development. It has therefore also been referred to as **acquired aphasia with convulsive disorder**.

Reflex Epilepsy

Reflex epilepsy is an epileptic syndrome in which seizures are triggered by a specific sensory stimulus, activity, or cognitive event. Some people who have reflex epilepsy have multiple seizure triggers.

The most common form of reflex epilepsy is **photosensitive epilepsy**, in which flashing lights, strobe lights, or alternating patterns of light and dark provoke seizures. Other external sensory stimuli may also precipitate reflex seizures. **Audiogenic seizures** are triggered by sudden unexpected noises (startle), voices, or other specific sounds. In **musicogenic epilepsy**, seizures are triggered by specific musical themes. Specific tactile stimuli may also trigger seizures, such as a prolonged tactile stimulus applied to a certain part of the body, immersing a body part in hot water (**hot water epilepsy**), and immersion of the body in warm water or the act of exiting out of water (**bathing epilepsy**). Reflex seizures may also be precipitated by more complex activities, including eating (**eating epilepsy**) and playing card games (**card game–induced reflex epilepsy**), or cognitive events involving higher cerebral functions (e.g., reading, writing, calculation). It is believed that focal reflex seizures are triggered by activation of specific cortical territories by certain stimuli.

Reflex epilepsies are relatively rare, occurring in only 5% of all epilepsies. Most people with reflex epilepsy also suffer from spontaneous seizures, but some have pure reflex epilepsy. Reflex seizures may be generalized or focal, but most (about 85%) are GTCs.

Differential Diagnosis

In epilepsy diagnosis, the first question that needs to be answered is whether the episodes are epileptic in nature. Paroxysmal events are therefore classified as epileptic or non-epileptic. Non-epileptic paroxysmal events are further classified as physiological or psychogenic. Physiological events that may imitate seizures include **syncope** (fainting, a temporary loss of consciousness due to a transient decrease in blood flow to the brain), some migraine headaches, narcolepsy, paroxysmal dyskinesia, and transient ischemic attack (TIA). Psychological disorders that may present like seizures are panic attacks and psychogenic seizures. Most of these disorders can be identified and ruled out based on the clinical history. For example, both focal seizures and TIAs produce transient focal neurological symptoms. These events differ, however, in that TIAs usually last 15 minutes or more, whereas focal seizures last from seconds to 5 minutes. Furthermore, the symptoms of TIAs usually are negative, rather than positive as seen in focal seizures.

Electroencephalography is the mainstay tool for determining whether paroxysmal neurological events are epileptic in nature. EEG abnormalities are often present during the interictal period in patients with seizure disorders, and interictal epileptiform discharges are the electrophysiologic signature of an epileptogenic brain. The interictal EEG, however, may be normal in epileptic patients; thus a normal interictal EEG does not rule out epilepsy. In such cases, various methods that activate epileptiform activity, such as sleep deprivation, hyperventilation, and photic stimulation, may be used to provoke interictal epileptiform discharges. If interictal epileptiform discharges have not been recorded during standard or ambulatory EEG, a recording of the ictal EEG is required for a definitive diagnosis. This is accomplished by video-EEG monitoring. Epileptiform abnormalities recorded during behavioral seizures are diagnostic of epilepsy.

Medical Treatment

The treatment of choice for seizure disorders is **antiepileptic drugs** (AEDs). These medications were previously referred to as **anticonvulsants**, but this is inaccurate because many seizures do not involve convulsive movements. The AEDs are often described as either "older" or "newer" generation, based on whether they were in widespread use before the 1990s. The older generation AEDs include barbiturates (e.g., phenobarbital), benzodiazepines (e.g., clonazapam), carbamazepine, phenytoin, and valproate. The newer generation AEDs were specifically designed to selectively target a mechanism believed to be critical for the epileptic seizures and to have less adverse side effects. These include felbamate, lamotrigine, gabapentin, pregabalin, topiramate, tiagabine, oxcarbazepine, levetiracetam, vigabatrin, and zonisamide.

Different types of seizures are best treated by different medications. Thus, the AED of choice depends on the specific epilepsy syndrome. While the mechanisms of drug action are not understood completely, it is believed that all AEDs generally function to stabilize the neuronal membrane by blocking neuronal excitation during rapid rates of neuronal discharge. Many AEDs are GABA agonists, GABA being the major inhibitory neurotransmitter in the brain.

Because AEDs reduce neuronal excitability, they may have adverse effects on cognition. AEDs usually have fewer cognitive side effects when used in **monopharmacy** (use of a single drug to treat a disorder) in doses within the standard therapeutic range. The goal of epilepsy treatment, however, is complete elimination of seizures; this requires **polypharmacy** (use of multiple drugs to treat a disorder) and/or high AED doses in some patients, increasing the risk of cognitive side effects. The newer AEDs generally have fewer cognitive side effects than older AEDs. Of the older generation AEDs, the barbiturates and benzodiazepines are associated with the greatest risk of cognitive impairment.

Antiepileptic drugs are also commonly used in Neurosciences Intensive Care Units to prevent seizures (**seizure prophylaxis**) in patients with a variety of acute cerebral insults (e.g., TBI, intracranial hemorrhage, ischemic stroke, cerebral venous sinus thrombosis, meningitis) and neurosurgery patients.

Epilepsy Surgery

Surgical treatment is indicated if the seizures arise from a structural lesion such as an abscess, cyst, vascular malformation, or tumor. In such cases the surgery is performed essentially to treat the lesion.

Most cases of idiopathic epilepsy are well controlled with AEDs. AEDs result in complete or nearly complete seizure control in approximately 70% of all epilepsy patients, and significant reduction in the number and severity of seizures in an additional 20%–25%. In some cases, however, the seizures are medically refractory. In cases where the seizure focus is well circumscribed it may be possible to remove it surgically. Patients with focal seizures and a unilateral temporal lobe focus are among the best surgical candidates. Candidacy for surgery is determined by a constellation of tests aimed at precisely defining the localization of the seizure focus and the functional importance of that tissue, as part of a presurgical evaluation.

Seizure Focus Localization

Analysis of the interictal EEG can provide evidence of localized cortical dysfunction, but EEG activity at the time of seizure onset is most informative for localizing the seizure focus. Localization of the seizure focus may be accomplished by careful analysis of the clinical and EEG features revealed by video-EEG monitoring. When noninvasive studies do not sufficiently localize or precisely define the zone of epileptogenic tissue, electrocorticography and depth electrode recordings may provide more precise seizure focus localization.

Functional imaging may be used to supplement these methods of seizure focus localization. During the interictal period, positron emission tomography (PET) reveals the seizure focus as a focal area of hypometabolism, and single-photon emission computerized tomography (SPECT) reveals the seizure focus as a focal area of hypoperfusion (see Chapter 29, "Brain Imaging"). Because SPECT imaging tracers have a relatively long half-life, SPECT imaging can also be used to visualize ictal focal hyperperfusion if tracer is injected during a seizure and the patient is scanned within several hours.

Intracarotid Amobarbital Procedure and Functional Mapping

When seizures originate from a region of cortex that may be critical for functions such as language and memory (i.e., the temporal lobe), it is necessary to map the function around the site of the planned **resection** (surgical excision) to delineate critical brain areas. The surgeon wants to remove the tissue generating the seizures, but also wants to spare tissue necessary for critical cognitive functions in order to minimize postoperative deficits such as aphasia and amnesia. This may be accomplished by examining the effects of temporary inactivation of the entire hemisphere and small areas of cerebral cortex.

The **intracarotid amobarbital procedure** (IAP), also known as **amytal ablation** or the **Wada test** (developed by Juhn Atsushi Wada), involves temporary pharmacological inactivation of one hemisphere. This is accomplished by injecting sodium amobarbital, a short-acting barbiturate that is effective for several minutes, into one internal carotid artery. The Wada test is effectively a temporary simulation of the planned resection that demonstrates whether the contralateral unanesthetized hemisphere can support language and memory. The patient is presented with language and memory tasks to establish the cerebral language representation and predict which patients are at risk for developing a postsurgical aphasia and/or amnesia. Wada testing assesses whether the hemisphere contralateral to a unilateral seizure focus (and contralateral to the injection) can sustain language and memory function following temporal lobectomy. Although the posterior hippocampus is not directly supplied by the distribution of the internal carotid artery, depth electrode EEG recordings show slowing in this region following intracarotid amobarbital administration. If injection on the side planned for lobectomy produces a transient generalized memory loss, this indicates that there is also a lesion affecting the hippocampus on the opposite side, and that lobectomy would result in a severe amnestic syndrome—a very undesirable consequence. The Wada test, however, is invasive and carries a risk of cerebrovascular complications. An emerging alternative for making informed decisions regarding elective surgery based on assessment of cognitive risk is fMRI, which is noninvasive and safe.

To determine the function served by the tissue in the local vicinity of the seizure focus, electrical stimulation is applied directly to the local brain area and its effect on behavior is evaluated, a procedure known as **functional mapping**. Stimulation studies may be conducted intraoperatively (during surgery) or extraoperatively (outside of surgery). **Victor Horsley** (1857–1916) is credited with the first use of intraoperative brain stimulation in humans. In the twentieth century, **Wilder Penfield** (1891–1976) published the most expansive and influential series of intraoperative mapping studies. Electrical stimulation in primary sensory and motor areas produces positive responses (i.e., sensations and movements). Negative responses involving the disruption of function are elicited outside of primary zones. The stimulation produces a temporary lesion by interfering with naturally occurring signals. Clinical functional mapping studies within the temporal lobe focus on determining whether tissue in or near the planned resection is necessary for language and/or memory. Stimulation in regions critical for language produce speech arrest; stimulation in regions critical for memory disrupt memory formation. Few studies have examined localization of functions outside of the temporal lobe or within the right hemisphere.

Some epilepsy centers also use magnetoencephalography, a noninvasive technique that measures the magnetic fields produced by the brain's electrical currents, in the presurgical assessment to localize seizure foci and perform functional mapping.

Surgical Procedures

The goal of epilepsy surgery is to cure epilepsy by resecting an area of epileptogenesis or disrupting the spread of seizure activity.

Temporal lobectomy is the most frequently performed surgical procedure for the treatment of seizures (Figure 19.10). A standard anterolateral temporal lobectomy involves resection of the anterior 4–5 cm of the temporal lobe (lateral cortex, parahippocampal gyrus, and hippocampus). **Topectomy** (**corticectomy**) involves resection of an area of neocortex outside the anterior temporal lobe. The boundaries of these resections are typically determined using electrocorticography. **Lesionectomy** is the surgical resection of structural brain lesions that are epileptogenic, such as low-grade neoplasms, vascular malformations, and cortical dysplasias.

Hemispherectomy involves resection of most or all of one hemisphere. It is used to treat epilepsies in which seizures arise over most or all of one hemisphere. This occurs with diffuse cortical dysplasia, Rasmussen's encephalitis, pan-hemispheric Sturge-Weber syndrome, and large perinatal infarcts. Since most of these conditions involve devastating injury to the hemisphere early in development, new deficits (other than loss of fine motor control in the contralateral hand and some contralateral visual field loss) are rare.

Some surgical procedures are performed for palliative purposes to reduce the severity of the seizures. **Multiple subpial transection** (cutting across tissue) is used to treat seizures arising from extratemporal cortex, by disrupting the horizontal connections within the cortex that are

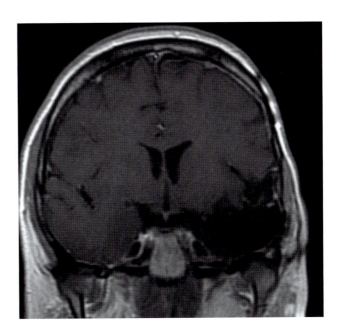

FIGURE 19.10. Coronal T1-MRI post-contrast imaging shows left temporal lobectomy.

necessary for synchronizing cortical activity, in zones that should not be resected such as primary motor, primary sensory, and language cortices.

Corpus callosotomy involves transection of the corpus callosum to disrupt the spread of seizures from one hemisphere to the other. It usually involves sectioning the anterior two-thirds of the callosum to avoid a major disconnection syndrome (see Chapter 13, "The Cerebral Cortex"), although the callosotomy may be completed later if seizure response is less than expected, especially if baseline neurologic function is so impaired that a major disconnection syndrome would not affect the patient's quality of life. This procedure is performed in focal onset epilepsy cases where surgical resection is inappropriate or risky. The primary indication for callosotomy is atonic drop attacks (which are quite potentially injurious because of their sudden onset), although it is effective for other seizure types. The intervention is palliative rather than curative, as the disconnection slows inter-hemispheric seizure propagation and provides patients with a warning or altogether prevents the requisite secondary bilateral synchrony underlying drop attacks.

Neuromodulation

Another approach to epilepsy treatment for patients with medically refractory epilepsy is **neuromodulation** by electrical stimulation. The mechanism of action in treating epilepsy by electrical stimulation of the nervous system is poorly understood, but it is believed that stimulation prevents or aborts seizures by disrupting activity in pathological networks that generate epileptic activity. Three neuromodulation techniques are currently approved for epilepsy: vagal nerve stimulation, deep brain stimulation, and responsive neurostimulation.

Vagal nerve stimulation activates vagal afferents with the goal of modulating the activity of seizure-related circuitry within the brain. The precise therapeutic mechanisms are not well understood. An advantage of this technique is that it influences the activity of pathological brain circuits without the invasiveness of intracranial surgery. **Deep brain stimulation** involves electrical stimulation of subcortical structures by depth electrodes. Targets for treating seizures include the thalamic nuclei, hippocampus, subthalamic nucleus, and cerebellum. Vagal nerve stimulation and deep brain stimulation both provide stimulation according to a regular schedule, but patients may also activate stimulation if they have a seizure warning. **Responsive neurostimulation** monitors and records neural activity from cortical or depth electrodes implanted within the seizure focus and delivers stimulation directly to the seizure focus in response to detected electrographic seizure patterns. This technique is used in patients whose seizure focus or foci can be identified but who are not candidates for surgical resection, such as those with multiple seizure foci (e.g., bitemporal epilepsy) or those with a seizure focus in so-called **eloquent cortex** (areas of cortex necessary for language, motor, or sensory functions). Intracranial electroencephalography is often used to identify the seizure focus, and functional mapping is used to confirm that the patient is not a surgical candidate.

Summary

A seizure is a sudden, abnormal, uncontrolled electrical disturbance in the brain. Epilepsy is a group of neurological disorders characterized by recurrent seizures. There are many different types of seizures, and they are classified according to three features: (1) site of origin (focal vs. generalized); (2) level of awareness during the seizure (aware vs. impaired awareness); and (3) whether motor symptoms are present at seizure onset (motor vs. non-motor).

Electroencephalography, an electrophysiological monitoring method in which the electrical activity of the brain is recorded through scalp electrodes, is an important tool in epilepsy diagnosis and treatment. It allows the epileptologist to distinguish seizures from other types of paroxysmal neurological events, to characterize the seizure type, and to localize the seizure focus. Epilepsy syndromes are defined based on the unique combination of features, including seizure type(s), typical age of onset, typical EEG findings, other clinical features.

Antiepilepsy drugs result in complete seizure control or reduction in the number and severity of seizures in most patients. Surgical treatments are used in cases with medically refractory focal onset seizures originating from a well-circumscribed seizure focus. Epilepsy surgery encompasses curative resective surgeries, and palliative interventions such as corpus callosotomy and implantation of stimulation devices.

Additional Reading

1. Blumenfeld H. Impaired consciousness in epilepsy. *Lancet Neurol.* 2012 Sep;11(9):814–826. doi: 10.1016/S1474-4422(12)70188-6. PMID: 22898735; PMCID: PMC3732214.
2. Devinsky O, Hafler DA, Victor J. Embarrassment as the aura of a complex partial seizure. *Neurology.* 1982;32(11):1284–1285. doi:10.1212/wnl.32.11.1284
3. Fisher RS, Cross JH, French JA, Higurashi N, Hirsch E, Jansen FE, et al. Operational classification of seizure types by the International League Against Epilepsy: position paper of the ILAE Commission for Classification and Terminology. *Epilepsia.* 2017;58(4):522–530. doi:10.1111/epi.13670
4. Gloor P. Experiential phenomena of temporal lobe epilepsy: facts and hypotheses. *Brain.* 1990;113 (Pt 6):1673–1694. doi:10.1093/brain/113.6.1673

5. Gschwind M, Picard F. Ecstatic epileptic seizures: a glimpse into the multiple roles of the insula. *Front Behav Neurosci.* 2016;10:21. Published 2016 Feb 17. doi:10.3389/fnbeh.2016.00021
6. Hamberger MJ. Cortical language mapping in epilepsy: a critical review. *Neuropsychol Rev.* 2007;17(4):477–489. doi:10.1007/s11065-007-9046-6
7. Hogan RE, English EA. Epilepsy and brain function: common ideas of Hughlings-Jackson and Wilder Penfield. *Epilepsy Behav.* 2012;24(3):311–313. doi:10.1016/j.yebeh.2012.04.124
8. Koepp MJ, Caciagli L, Pressler RM, Lehnertz K, Beniczky S. Reflex seizures, traits, and epilepsies: from physiology to pathology. *Lancet Neurol.* 2016;15(1):92–105. doi:10.1016/S1474-4422(15)00219-7
9. Perrine K, Devinsky O, Uysal S, Santschi C, Doyle WK. Cortical mapping of right hemisphere functions. *Epilepsy Behav.* 2000;1(1):7–16. doi:10.1006/ebeh.2000.0026
10. Rahimpour S, Haglund MM, Friedman AH, Duffau H. History of awake mapping and speech and language localization: from modules to networks. *Neurosurg Focus.* 2019;47(3):E4. doi:10.3171/2019.7.FOCUS19347
11. Ritaccio AL, Brunner P, Schalk G. Electrical stimulation mapping of the brain: basic principles and emerging alternatives. *J Clin Neurophysiol.* 2018;35(2):86–97. doi:10.1097/WNP.0000000000000440
12. Rugg-Gunn F, Miserocchi A, McEvoy A. Epilepsy surgery. *Pract Neurol.* 2020 Feb;20(1):4–14. doi: 10.1136/practneurol-2019-002192. Epub 2019 Aug 16. PMID: 31420415.
13. Scheffer IE, Berkovic S, Capovilla G, Connolly MB, French J, Guilhoto L, et al. ILAE classification of the epilepsies: position paper of the ILAE Commission for Classification and Terminology. *Epilepsia.* 2017;58(4):512–521. doi:10.1111/epi.13709
14. Vaddiparti A, Huang R, Blihar D, Du Plessis M, Montalbano MJ, Tubbs RS, Loukas M. The evolution of corpus callosotomy for epilepsy management. *World Neurosurg.* 2020 Sep 2;145:455–461. doi: 10.1016/j.wneu.2020.08.178
15. Wolf P. Reflex epileptic mechanisms in humans: lessons about natural ictogenesis. *Epilepsy Behav.* 2017;71(Pt B):118–123. doi:10.1016/j.yebeh.2015.01.009

20

Traumatic Brain Injury

Introduction

Traumatic brain injury (TBI) is defined as physical trauma to the brain due to impact, penetration, or rapid movement of the brain within the skull, caused by an external mechanical force. Traumatic brain injuries encompass a broad spectrum of severities, from very mild transient changes in brain function resulting in a brief alteration of consciousness with rapid recovery, to catastrophic injuries resulting in severe disability or death. The pathophysiology of TBI is complex, and in many cases involves a combination of pathologies. Primary injuries occur at the time of trauma and are caused by mechanical forces. Secondary injuries are due to delayed processes that evolve after the primary injury and clinically present hours to days after the injury. This chapter describes the epidemiology of TBI, pathophysiology of primary and secondary injuries, classification of injuries, measures of severity and clinical classifications, and chronic traumatic encephalopathy.

Definition

TBI is a form of **acquired brain injury** (i.e., not present at or before birth nor due to genetic abnormalities or neurodegenerative disease). **Non-traumatic acquired brain injuries** have a wide variety of causes that may be due to either internal or external sources, including non-traumatic etiologies such as stroke (ischemic and hemorrhagic), anoxic/hypoxia (e.g., cardiac arrest, respiratory arrest, airway obstruction, acute respiratory distress syndrome, drug overdose, carbon monoxide poisoning), brain neoplasms, brain infection and/or inflammation (e.g., meningitis, encephalitis, brain abscess), and toxins (e.g., lead poisoning, substance abuse).

The terms *head injury*, *head trauma*, and *traumatic brain injury* are sometimes used interchangeably, but they are not synonymous. A head injury is a physical injury to any part of the head; this includes soft tissue wounds of the face or scalp (bruising, swelling, or laceration), skull fractures, and brain injuries. Thus, not all head injuries involve brain injury. On the other hand, some brain injuries are not associated with any *observable* head injury; there is no extracranial tissue injury, no skull fracture, and no brain injury apparent on structural neuroimaging studies.

TBI is evidenced by *any* of the following: (1) loss of consciousness, alteration of consciousness, or post-traumatic amnesia; (2) neurological deficit (e.g., weakness, aphasia) on physical or mental status examination; or (3) positive neuroimaging.

Epidemiology and Etiology

It is estimated that there are over 2,000,000 new cases of TBI annually in the United States. This likely is an underestimate, as many people with mild TBI seek no medical attention after the injury (e.g., concussions in sports) or receive medical treatment outside of facilities that keep incidence data. TBI is responsible for more deaths and disabilities than any other neurologic condition in people under the age of 50. In people over the age of 65, it is second to cancer as the most common cause of death. Approximately 20% of TBIs are classified as moderate-severe, being evenly split between moderate (10%) and severe (10%). The remaining 80% of TBIs are classified as mild.

Peak incidence and risk for TBI is greatest in the late teens and early twenties (15–24 years of age). In this cohort, males outnumber females by 2:1 and TBI is often associated with motor vehicle accidents, contact sports, interpersonal violence, and alcohol abuse. TBI incidence is also high in the elderly and during the first 5 years of life (ages 0–4).

The principal causes of TBI, accounting for almost 85% of all TBI-related hospital visits, are motor vehicle accidents (including pedestrians hit by motor vehicles), falls, and assaults (including child abuse). The relative frequency of each cause varies across the life span. The most common causes of severe TBI across the life span are assaults in the 0–4 age group, motor vehicle accidents in the 15–19 age group, and falls in the 65+ age group. Up to 80% of all instances of fatal child abuse involve direct trauma to the head. TBIs from motor vehicle accidents are most common in young males and frequently involve alcohol. Traumatic spinal cord injury is often accompanied by TBI.

Pathophysiology of TBI

The pathophysiology of TBI is complex, and many TBIs involve a combination of pathologies.

Primary vs. Secondary Injury

Traumatic brain injury processes are classified as primary and secondary. Primary injuries occur at the moment of trauma and are caused by mechanical forces. Secondary injuries are due to delayed processes that evolve after the primary mechanical injury and present clinically hours to days after the injury.

The distinction between primary and secondary injury is an important concept in brain injury management and treatment, which aim to identify potentially preventable complications and prevent secondary brain injury. Even apparently mild brain injuries can result in secondary brain damage that can be fatal or result in severe, persistent disability, giving rise to the clinical phenomenon of patients with brain injury who "talk and deteriorate" or "talk and die." The only way to address primary injury of TBI is through **primary prevention**, measures aimed at preventing the injury altogether by changes in public policy and public culture (e.g., seatbelts, child safety seats, air bags, helmet use, occupational safety standards, speed limits, road engineering and maintenance, alcohol abuse, driver's education, and defensive driving courses).

Primary Injuries

Primary injuries include skull fracture, contusion (bruising), laceration (tearing) of brain tissue, tearing of blood vessels with hemorrhage in or around the brain, and traumatic axonal injury (diffuse axonal injury).

SKULL FRACTURE

A skull fracture is a break in one or more of the eight cranial bones. Skull fracture may or may not be associated with neurological damage. The presence of a skull fracture, however, indicates that the head trauma involved a substantial amount of force and increases the likelihood of brain injury.

There are several types of skull fracture. A **linear skull fracture** is a thin crack in a cranial bone that does not involve bone displacement. This is the most common type of skull fracture. Linear skull fractures usually do not require any intervention. These fractures usually are caused by blunt trauma that results in low-energy transfer over a wide surface area of the skull. A **depressed skull fracture** is a break in a cranial bone in which bone fragments are driven inward toward the brain, sometimes penetrating the dura. These often require surgical intervention and often are due to high-energy transfer, such as a blow from a baseball bat. A **complex depressed skull fracture** is a type of depressed skull fracture in which the dura is torn, increasing the risk of infection. A **basilar skull fracture** (**basal skull fracture, skull base fracture**) is a fracture in the bones at the base of the cranium.

Basal skull fractures may be missed in skull X-rays, but their presence should be suspected with leakage of cerebrospinal fluid (CSF) from the ear (**otorrhea**), blood behind an intact tympanic membrane, **ecchymosis** (bruising) of the skin behind the ear (over the mastoid process of the temporal bone) known as **Battle's sign**, **periorbital ecchymosis** (**racoon eyes**, **panda eyes**), or signs of cranial nerve damage (especially of the olfactory, facial, and auditory nerves).

CONTUSION

Cerebral **contusion** is a focal bruising that typically occurs in the superficial aspect of the brain as a result of the brain hitting against the inner surface of the cranium. A blow to the head can result in the skull bending inward and striking the brain, causing a contusion directly interior to the site of impact. Contusions are most often found in cortical tissue and are most severe on the crests of gyri. The frontal and temporal poles, as well as the inferior surfaces of the frontal and temporal lobes, are particularly susceptible to contusion (Figure 20.1). As the brain moves within the cranium, these regions strike the bony ridges of the inner surface of the skull floor (skull base) such as the sphenoid ridge, anterior clinoid processes, and crista galli of the ethmoid bone (see Chapter 4, "Cranium, Spine, Meninges, Ventricles, and Cerebrospinal Fluid"). By contrast, the inner surface of the skull roof (skullcap) and occipital bone are smooth. Contusions also occur on the brain surfaces adjacent to the tentorium and cerebral falx due to the impact of the soft brain tissue against the rigid surface of the dural reflections, especially the free edge of the tentorium.

Contusions involve **petechial hemorrhage**, tiny punctate bleeds that have a classic "salt and pepper" appearance on CAT scan (Figure 20.2). On autopsy they appear as brownish scars due to **hemosiderin** (a red blood cell breakdown product) deposition. Contusions themselves do not contribute to depressed level of consciousness.

LACERATION

Laceration is a wound produced by the tearing, cutting, or gashing of soft body tissue. **Cerebral laceration** is a tear in brain tissue and brain blood vessels caused by a foreign object or bone fragment from a depressed skull fracture. Cerebral lacerations result in intracranial hemorrhage. Intracranial hemorrhage is classified as **extradural** (epidural) or **intradural**; intradural hemorrhage can be further classified as **subdural**, **subarachnoid**, or **intracerebral** (**intraparenchymal**). Intracerebral hemorrhage can spread into the ventricles and/or subarachnoid space, conditions respectively known as **intraventricular hemorrhage extension** and **subarachnoid hemorrhage extension**.

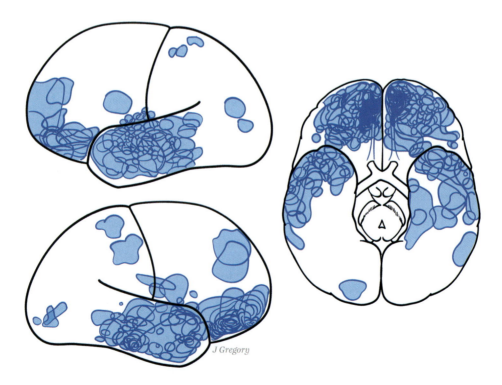

FIGURE 20.1. The frontal and temporal poles, as well as the inferior surfaces of the frontal and temporal lobes, are particularly susceptible to contusion because as the brain moves within the cranium due to head trauma, these regions strike the bony ridges of the inner skull surface.

Source: Courville CB. *Pathology of the central nervous system, part 4*. Mountain View, CA: Pacific; 1937.

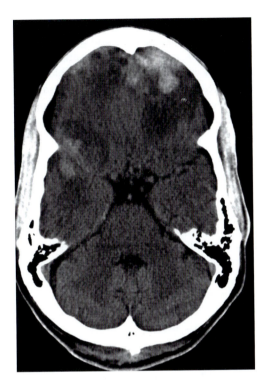

FIGURE 20.2. Axial CT of the head shows bilateral frontal and right anterior temporal pole hemorrhagic contusions.

TRAUMATIC AXONAL INJURY

Traumatic axonal injury (TAI) is a stretching and tearing of axons that occurs with rapid acceleration, deceleration, and/or rotation of the head, during which different parts of the brain (e.g., the cerebral hemispheres and brainstem) move in somewhat different directions and at different speeds. Areas of greatest vulnerability to TAI are large caliber axons, locations where axons change direction, locations where axons enter target nuclei, locations where axons decussate, and the gray-white matter interface of the cerebral cortex where there is a change in tissue density.

Traumatic axonal injury is one of the most common and important pathological features of TBI. It occurs in all severities of TBI and is believed to be the key pathologic substrate of mild TBI. It is particularly associated with motor vehicle accidents. It can occur without head trauma, as in **whiplash** (a **neck sprain** or **strain** injury due to forceful, rapid back-and-forth movement of the neck) and shaken baby syndrome. **Shaken baby syndrome** (**shaken impact syndrome**) occurs in babies and toddlers who are shaken violently (usually out of anger or frustration because the child will not stop crying). Because the child's neck muscles are very weak and cannot fully support the proportionately large head, shaking the child's body causes the head to move violently back and forth, resulting in serious and sometimes fatal brain injury.

Traumatic axonal injury varies in severity. Acute mild TAI produces concussion. Chronic moderate TAI produces mental slowing and deficits of attention and concentration. Severe TAI produces permanent coma or persistent vegetative state (see Chapter 25, "Disorders of Consciousness"). **Coma** is a state of deep unconsciousness in which the patient is unresponsive to and cannot be aroused by *any* stimulation, including painful stimulation such as pinching of the skin on the neck, sternal rubbing, supraorbital pressure, and/or nail bed pressure. In addition to being unarousable, the comatose patient is also unaware of self and environment. **Persistent vegetative state** is a state of unconsciousness accompanied by sleep-wake cycles in which the patient may appear to be awake with eyes opened and involuntary eye movements but is nevertheless unaware of self and environment.

Secondary Injuries

Secondary injury is due to delayed effects of brain injury. Secondary injuries include hematoma, edema (swelling), raised intracranial pressure, brain herniation, impaired cerebral perfusion and cerebral ischemia, infection, scar tissue formation, and cerebral atrophy.

HEMATOMA

Intracranial bleeding from lacerated blood vessels results in a space-occupying **hematoma**, a confined mass of blood (usually clotted) outside of blood vessels that is caused by a break in a blood vessel. Intracranial hematoma is the most common cause of morbidity and mortality in those who have a lucid interval after their injury and then deteriorate. Hematomas occur epidurally (extradurally), subdurally, and intraparenchymally (intracerebrally).

An **epidural hematoma** is a collection of blood on the surface of the brain, between the dura and cranium. Acute epidural hematoma has an elliptical shape (see Figure 4.10 in Chapter 4). These hematomas may develop slowly over hours or days. Epidural hematomas are typically caused by skull fractures associated with bleeding from a branch of the middle meningeal artery, and therefore are usually located in the temporal or temporoparietal region. Occasionally they are caused by rupture of a sagittal or transverse sinus. Treatment involves surgical evacuation of the hematoma.

A **subdural hematoma** is a collection of blood on the surface of the brain, beneath the dura and above the arachnoid membrane. Subdural hematomas are crescent-shaped and typically develop over the entire convexity (see Figure 4.11 in Chapter 4). They are usually due to rupture of bridging veins from the cortical surface to the venous sinuses; however, some are arterial in origin. Subdural hematoma is the most common type of hematoma to develop after TBI, and the majority are due to head trauma.

Subdural hematomas may develop following trivial traumas, such as a minor fall; the trauma may be so trivial that the person may not even recall the event. Subdural hematomas also may develop slowly, over days or weeks after injury due to a slow bleed. They often simulate dementia, with loss of concentration, confusion, and memory loss. Individuals over age 60 are four times more likely to develop chronic subdural hematoma, probably due to widened sulci which result in increased vulnerability of the bridging veins. Risk for subdural hematoma also is increased by anticoagulant medications ("blood thinners") commonly used in patients with high thromboembolic risk to prevent blood clots that may cause stroke, heart attack, deep vein thrombosis, and pulmonary embolism. Subdural hematomas are classified as acute (within first 48 hours after injury), subacute (2–14 days after injury), and chronic (more than 14 days after injury). Treatment of subdural hematoma involves surgical evacuation. Subdural hematomas are associated with a poor prognosis.

An **intraparenchymal hematoma** is a collection of blood within the brain tissue. They occur most commonly in the frontal and temporal lobes; about 25% are subfrontal. They are often multiple. They usually develop 3–5 hours after injury and are a common cause of post-injury deterioration. A "**burst lobe**" occurs when an intraparenchymal hematoma expands to merge with a subdural hematoma.

EDEMA

Edema is swelling from fluid accumulation in body tissues. Cerebral edema may be focal and associated with contusions or hematomas, or it may occur in a generalized diffuse manner. Edema can cause compression of brain structures and herniation.

INCREASED INTRACRANIAL PRESSURE AND BRAIN HERNIATION

Hematoma and edema result in increased **intracranial pressure** (ICP), which may lead to compression of structures (including intracranial blood vessels leading to cerebral ischemia) and brain herniation. Elevated ICP is the most common cause of death in closed head injury, due to compression of lower brainstem structures necessary for vital functions; therefore identifying and treating elevated ICP is imperative in the acute care of patients with head trauma. Signs of increased ICP include certain features on brain imaging, such as effacement of ventricles, sulci, and basal cisterns, and significant midline shift.

IMPAIRED CEREBRAL PERFUSION AND CEREBRAL ISCHEMIA

Cerebral ischemia is caused by impaired cerebral perfusion due to **hypotension** (abnormally low blood pressure), microvascular collapse secondary to raised ICP, or post-traumatic **vasospasm** (a sudden and persistent constriction of the arteries). Traumatic injuries and blood loss are often associated with a drop in systemic blood pressure. Normally, systemic hypotension does not result in a drop in cerebral perfusion pressure because **autoregulation** processes result in a compensatory cerebral vasodilation (see Chapter 5, "Blood Supply of the Brain"). After TBI, however, there may be a loss of autoregulation, and systemic hypotension may have drastic effects on cerebral perfusion.

INFECTION

The presence of a dural tear provides a potential route for infection, resulting in meningitis or a cerebral abscess (see Chapter 22, "Brain Infections"). **Meningitis** is an inflammation of the meninges. A brain **abscess** is an encapsulated pocket containing pus, the organism of infection, and necrotized brain tissue.

POST-TRAUMATIC SEIZURES AND POST-TRAUMATIC EPILEPSY

Post-traumatic seizures are a common complication of TBI and are associated with worse functional outcomes. Those that occur within the first 7 days of TBI are classified as provoked (acute symptomatic) seizures, while those that occur as a late effect (i.e., more than 7 days after TBI) are classified as unprovoked (remote symptomatic) seizures (see Chapter 19, "Epilepsy"). **Post-traumatic epilepsy** is generally defined as one or more unprovoked seizures that occur at least one week after TBI.

Scar tissue formation following TBI frequently leads to spontaneous epileptiform electrical discharges and post-traumatic epilepsy. This occurs because neurons within the scar tissue have altered metabolic function that results in abnormal electrical activity. Risk factors for post-traumatic epilepsy include temporal lobe damage, focal neurological signs, intracranial hematoma, depressed skull fracture, penetrating injury, and severe TBI. The incidence of post-traumatic epilepsy after severe TBI has been found to be as high as 50% in some settings. Risk factors for developing post-traumatic epilepsy include early post-traumatic seizure, dural-penetrating injuries, multiple contusions, and/or subdural hematoma requiring evacuation. Late post-traumatic seizures (i.e., occurring more than a week after injury) may have a long latency; approximately 40% of patients have their first seizure within 6 months of injury, 50%–60% within the first 12 months, and 80% within 2 years of the injury; thus 20% have their first seizure after more than 2 years post-injury. Up to 86% of TBI survivors with a first post-traumatic seizure will have a second seizure within 2 years, providing the rationale for making the diagnosis of post-traumatic epilepsy after a single late post-traumatic seizure rather than requiring recurrent seizures. Occasionally, seizures develop in adults that are explained only by a small scarred cortical contusion that was acquired decades earlier.

Antiepileptic drugs are frequently used for **seizure prophylaxis** in TBI patients, given the high incidence of seizures following TBI and that post-traumatic epilepsy is associated with poorer outcomes following TBI.

CEREBRAL ATROPHY

Traumatic axonal injury induces a neuronal degeneration. Damage to the axonal cytoskeleton, a scaffold within neurons, interferes with axoplasmic transport of cellular products and organelles between the cell body and other regions of the neuron (e.g., axon terminals). These products and organelles accumulate proximal to the lesion and cause axonal swellings, eventually causing the axon to rupture and degenerate. The atrophy particularly affects the long white matter tracts of the cerebral hemispheres and the corpus callosum, especially the genu and splenium.

Classification of Injuries

Traumatic head and brain injuries are classified along multiple dimensions of physical trauma: penetrating vs. non-penetrating head injuries, diffuse vs. focal traumatic brain injuries, and coup vs. contrecoup traumatic brain injuries.

Penetrating vs. Non-Penetrating Head Injuries

Penetrating injury, also known as **open head injury**, involves an open wound caused by an object that penetrates the skull and dura and directly damages the brain parenchyma. Penetrating injuries are caused by bullets, other missiles (i.e., objects entering the body at high speed), knives, and bone fragments from depressed skull fractures. Penetrating injuries typically involve focal brain damage that occurs along the route the object has traveled. The foreign objects may become embedded within the head or may cause "through-and-through" injury with both **entry** and **exit wounds**, as may occur with bullets. Penetrating head injuries tend to be associated with focal neurological signs and symptoms. The larger the lesion, the more severe and general are the deficits. Penetrating head injuries are associated with high risk of infection and post-traumatic seizures; thus they are treated with prophylactic antibiotics and antiseizure medications.

In **non-penetrating injury**, also known as **closed head injury** and **blunt head injury**, the dura is not breached. Damage is due to indirect impact, without entry of a foreign object into the brain. The skull may or may not be fractured. The biomechanics of non-penetrating brain injury include acceleration forces (as when the head is struck by a faster-moving object, such as a bat), deceleration forces (as when a moving head strikes a fixed and solid object, such as a car dashboard or the ground), or a combination of the two. Injury may also occur by motion without impact or external injury, as in whiplash.

Focal vs. Diffuse Injuries

Focal brain injuries are characterized by visible damage that is limited to a well-circumscribed brain region. Focal injuries include contusions to the cortex, laceration, and hematomas.

Diffuse brain injuries involve widespread brain damage and dysfunction. Diffuse pathologies include TAI, ischemic injury, and diffuse cerebral edema. The mildest form of diffuse injury is concussion (see "Concussion," below), whereas severe diffuse brain injury produces coma without mass lesion. Diffuse damage is particularly associated with deficits in attention and concentration, as well as mental slowing.

TBI is often associated with a combination of focal and diffuse injuries, resulting in a mixed clinical profile of focal deficits against a background of deficits due to diffuse injury (e.g., focal frontal deficits against a background of mental slowing and deficits in attention and concentration).

Coup vs. Contrecoup Injuries

Coup injuries are contusions that occur just interior to the site of impact. **Contrecoup injuries** are contusions that occur some distance from the site of impact, as when the brain strikes the inner surface of the skull on the opposite side of the head as the site of impact. The frontal and temporal poles and the inferior surfaces of the frontal and temporal lobes are particularly susceptible to both coup and contrecoup contusion because they strike the rough bony ridges of the inner surface of the skull (Figure 20.3). For example, a fall onto the back of the head may result in contrecoup contusion of the frontal and temporal poles. In contrast, a fall onto the front of the head may not result in contrecoup contusion of the occipital lobe; contrecoup injury of the occipital lobe is rare due to the smooth contour and absence of bony projections on the inner surface of the skull in this area. Impact on the side of the head may also result in contrecoup injury. Contusions that occur on the brain surfaces adjacent to the tentorium and cerebral falx are contrecoup.

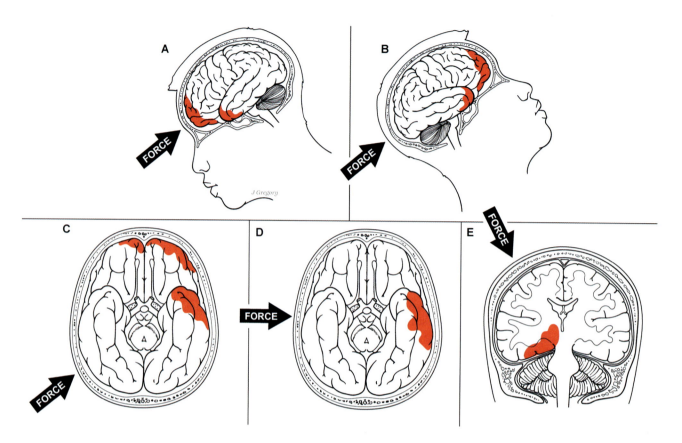

FIGURE 20.3. The frontal and temporal poles and the inferior surfaces of the frontal and temporal lobes are particularly susceptible to coup and contrecoup contusion. (A) Coup injury. (B, C, D, and E) Contrecoup injuries.

Clinical Classification of Acute TBI and Measures of TBI Severity

The severity of acute TBI is often quantified based of three indices of neurological function: the Glasgow Coma Scale, duration of loss of consciousness, and duration of post-traumatic amnesia (Table 20.1).

TABLE 20.1 Common Classification of TBI Severity			
	MILD	MODERATE	SEVERE
GCS score	13–15	9–12	3–8
LOC duration	0–30 min	30 min–24 hr	> 24 hr
PTA duration	< 1 hr	1–24 hr	> 24 hr

GCS: Glasgow Coma Scale; LOC: loss of consciousness; PTA: post-traumatic amnesia.

The Glasgow Coma Scale

The **Glasgow Coma Scale** (GCS) is a 15-point neurological function scale to assess states of altered consciousness, from the mildest confusional state to deep coma. It is widely used by emergency medical technicians (EMTs) and emergency department (ED) personnel to assess severity of TBI in persons with compromised consciousness, but it is also applicable in many other acute medical settings.

The GCS assesses responsiveness by evaluating three behaviors: motor response, verbal response, and eye-opening response (Table 20.2). Subjects are graded according to their best response in each domain. The lowest score in each domain is 1, but the maximum score varies across domains, so that the range of scores for motor responsiveness is 1–6, verbal responsiveness is 1–5, and eye-opening responsiveness is 1–4. The range of total GCS scores is 3–15, with lower scores indicating worse function and higher scores indicating better function (see Chapter 25, "Disorders of Consciousness").

Subscale scores are also often reported, especially when there are factors that preclude assessment of responsiveness in a domain. For example, verbal responding is precluded by intubation and tracheotomy, eye-opening responses may be precluded by facial injuries with ecchymosis and eye swelling, and limb movements may be precluded by other injuries or immobilization for treatment purposes.

The GCS score is used to classify three levels of TBI severity: severe (GCS score ≤ 8), moderate (GCS score 9–12), and mild (GCS score 13–15). Scores are related to the depth of lesions. Lesions in deep central gray matter or the brainstem tend to be associated with lower GCS scores than cortical or subcortical white matter lesions. Initial GCS score is an indicator of injury severity and prognosis, as it correlates with clinical outcome and disability. Lower initial GCS scores are associated with longer duration of coma, higher mortality rates, and worse prospects for rehabilitation.

The GCS is also administered repeatedly over time to chart the course of recovery and/or monitor for deterioration of level of consciousness. The timing and circumstances of the GCS assessment must be taken into account when determining how much weight to give it as a measure of severity and predictor of outcome. The timing of GCS assessment is quite variable. Thus, there are issues relating to which GCS score to use as a measure of severity and predictor of outcome: GCS taken at the scene or in the ambulance by EMT, initial ED GCS, best Day 1 GCS, or worst Day 1 GCS. The World Health Organization (WHO) criteria for classifying TBI severity specify using an initial GCS score assessed at 30 minutes post-injury or at the first opportunity. When the GCS is administered repeatedly, the scores must be interpreted according to the timeline. For example, in patients who deteriorate, early scores may not be representative of the eventual severity of the injury. Similarly, it would be erroneous to diagnose moderate TBI on the basis of an initial low GCS score obtained 10 minutes post-injury if the score recovered to the mild range within 30 minutes post-injury.

Other circumstances must also be considered when interpreting the GCS score as a measure of severity and predictor of outcome. For example, GCS scores may be artificially lowered due to intoxication or sedation for agitation.

Duration of Loss of Consciousness

Duration of loss of consciousness (LOC) is also used as a measure of TBI severity, with LOC < 30 minutes classified as mild, LOC > 30 minutes and < 24 hours classified as moderate, and LOC > 24 hours classified as severe (see Table 20.1). The correlation between duration of LOC and extent of tissue damage, however, is far from perfect. There may be a significant amount of tissue damage without LOC, as may occur with focal lesions such as contusion, laceration, and even severe penetrating head injuries; recall Phineas Gage's missile wound (Chapter 17, "The Frontal Lobes and Associated Disorders").

The accuracy of assessing a brief LOC duration is often limited. A brief LOC might have resolved by the time first responders arrive at the scene. Post-traumatic confusion or amnesia may confound self-report of whether there was a loss of consciousness and its duration; thus, collateral information from a witness is usually more reliable. Some report a period of post-traumatic amnesia as a loss of consciousness because if they have no recall for a period of time, they assume that they were unconscious during that interval. Furthermore, other potential causes of LOC may confound interpretation of presence and/or duration of trauma-related LOC, such as intoxication or syncope.

Post-Traumatic Amnesia

Following resumption of consciousness, there is often a period during which the person is unable to form new memories, known as **post-traumatic amnesia** (PTA). During PTA the person is conscious but has no recall of ongoing events. The interval of PTA is therefore defined as the period of time between return of consciousness and return of continuous memory for day-to-day events. Several studies have found that PTA duration is superior to LOC duration in predicting outcome, particularly future cognitive performance. There is less agreement, however, in injury severity classification systems based on PTA duration. One TBI severity classification system defines injuries with PTA < 1 hour as mild, injuries with PTA 1–24 hours as moderate, and injuries with PTA > 24 hours as severe (see Table 20.1).

Moderate-Severe TBI

Moderate TBI is commonly defined as GCS 9–12, LOC duration > 30 minutes and < 24 hours, and PTA duration 1–24 hours (see Table 20.1). Severe TBI is commonly defined as GCS 3–8, LOC duration > 24 hours, and PTA duration > 24 hours (see Table 20.1).

Cognitive impairment is the core feature of TBI. In general, the greater the severity of acute injury, the greater the severity of long-term cognitive impairment. TBI is also associated with personality and behavioral changes, as well as physical symptoms.

The neurocognitive profile of moderate-severe TBI depends on the extent and pattern of focal and diffuse lesions, but it typically is characterized by deficits in (1) attention, working memory, and speed of information processing; (2) learning and memory; and (3) executive functions. Attention difficulties are very common; these manifest as inability or decreased ability to (1) maintain focused attention in the face of distraction; (2) sustain attention; and (3) divide attention (i.e., multitask). Slowed thinking and

TABLE 20.2 Glasgow Coma Scale (GCS)

EYE OPENING

1	None	Even to supra-orbital pressure
2	To pain	Pain from sternum/limb/supra-orbital pressure
3	To speech	Nonspecific response, not necessarily to command
4	Spontaneous	Eyes open, not necessarily aware
	GCS Eye Opening Score (1–4)	

MOTOR RESPONSE

1	None	To any pain; limbs remain flaccid
2	Extension	Shoulder adducted; shoulder and forearm internally rotated
3	Flexor response	Withdrawal response or assumption of hemiplegic posture
4	Withdrawal	Arm withdraws to pain, shoulder abducts
5	Localizes pain	Arm attempts to remove supra-orbital/chest pressure
6	Obeys commands	Follows simple commands
	GCS Motor Response Score (1–6)	

VERBAL RESPONSE

1	None	No verbalization of any type
2	Incomprehensible	Moans/groans, no speech
3	Inappropriate	Intelligible, no sustained sentences
4	Confused	Converses but confused, disoriented
5	Oriented	Converses and is oriented
	GCS Verbal Response Score (1–5)	
	TOTAL GCS SCORE (3–15)	

reaction time (both simple and choice reaction time) results in low scores on all timed tests, despite the capacity to perform the task accurately. Learning and memory difficulties are characterized by deficits in acquisition, retrieval, and reduced organizational strategies when encoding. Executive dysfunction affects higher-level abilities such as judgment, decision-making, organizational skills, and planning. Executive function deficits often are especially handicapping.

Physical symptoms commonly associated with moderate-severe TBI include headaches, anosmia, balance and dizziness problems, tinnitus and hyperacusis, visual impairments, motor impairments, movement disorders, sleep problems, and fatigue. **Anosmia** is a loss of or reduced olfactory sense; in TBI it is due to damage of the olfactory nerve, bulb, or cortex. Balance and dizziness problems occur with damage of the vestibular branch of the eighth cranial nerve. **Tinnitus** (perception of noise or ringing in the ears) and **hyperacusis** (increased sensitivity to certain frequencies and volume ranges of sound that are perceived as unbearably loud) occur with damage of the cochlear branch of the eighth cranial nerve. Visual impairments include blurred vision, double vision, visual field defects, nystagmus, and abnormal smooth pursuit and saccadic eye movements. Motor impairments include hemiparesis, ataxia, and apraxia. Movement disorders include bradykinesia, tremor, dystonias, and myoclonus. Sleep problems include insomnia (difficulty initiating or maintaining sleep), hypersomnia (excessive sleep or excessive daytime sleepiness), and disturbed sleep-wake (circadian) cycles.

Neuropsychiatric symptoms include depression, anxiety, personality change, anosognosia, and anosodiaphoria. Depression is common following TBI of all severities. The cause is likely multifactorial, arising from both biological consequences of the injury and psychological reaction to the deficits and problems of TBI. Personality change is generally of two types: (1) impulsivity, emotional lability, and socially inappropriate behaviors; or (2) apathy, decreased spontaneity, and emotional blunting. Personality change is usually associated with **anosognosia** (a deficient awareness of deficit) or **anosodiaphoria** (an indifference to deficits without denying their existence). Anosognosia and anosodiaphoria may be misinterpreted as denial; however, denial is a psychological defense mechanism and in most cases of moderate to severe brain injury the underlying cause is neurological rather than psychological.

Diffuse brain damage is particularly associated with attention deficits, slowed processing, and fatigue. Frontal lobe damage is particularly associated with personality change and anosognosia or anosodiaphoria (Chapter 17, "The Frontal Lobes and Associated Disorders").

Adjustment and rehabilitation following moderate-severe TBI are often complicated by obstacles to return to work, obstacles to community integration, family and marital disintegration, and substance abuse. The prevalence rates of pre-injury substance abuse (including alcohol) are much greater in the TBI population than in the general population. Substance abuse following TBI usually reflects resumption of pre-injury substance use patterns.

Mild TBI

Mild TBI (mTBI) is commonly defined as GCS score of 13–15, LOC < 30 minutes, and PTA < 1 hour (see Table 20.1). The diagnosis of mTBI can be challenging because the acute signs and symptoms involving loss or alteration of consciousness and PTA resolve rapidly, and neuroimaging studies are typically negative.

Mild TBI is typically caused by blunt force non-penetrating head trauma. It is heterogeneous, encompassing a broad spectrum of physiological alterations from mild transient neurometabolic changes due to axonal stretching, to structural brain damage. There are two subtypes of mTBI, uncomplicated and complicated. **Uncomplicated mTBI** is defined as a GCS score of 13–15 with negative neuroimaging (e.g., no CT or MRI evidence of edema, hematoma, contusion, or skull fracture). It involves a physiological (metabolic/biochemical) disruption of brain function that results in an alteration or loss of consciousness. **Complicated mTBI** is defined as a GCS score of 13–15 with positive neuroimaging evidence of bleeding, bruising, swelling, and/or skull fracture.

Mild TBI may result in cognitive, physical, and/or behavioral symptoms. Core cognitive deficits associated with mTBI are slowed reaction times and slowed mental processing, attention deficits, forgetfulness, and reduced efficiency of cognitive processing. Attention deficits manifest as poor concentration, heightened distractibility, difficulty doing more than one task at a time, and difficulty conversing with background noise. Activities that were previously performed automatically require more effort and become more error prone.

Physical symptoms of acute mTBI typically consist of nausea, vomiting, dizziness, and headache, but they also include fatigue (e.g., becoming fatigued by afternoon), drowsiness, sleep disturbance, and motor slowing. Behavioral symptoms include irritability and emotional lability.

Concussion

Concussion is the mildest form of TBI, and of course the mildest on the mTBI spectrum. It is a trauma-induced acute alteration in brain function manifesting with neurological signs and symptoms, in the absence of evident structural damage on CT or MRI imaging studies. The diagnosis of concussion is therefore based on clinical assessment. The alteration in brain function is immediate but transient, and manifests as (1) a period of LOC ≤ 30 minutes (GCS score of 13–15 at 30 minutes from onset of LOC); (2) an alteration in mental state (dazed, disoriented, confused); or (3) an interval of PTA < 1 hour. Note that concussion does not necessarily involve any loss of consciousness, and while concussions are usually caused by a blow to the head, in most cases there are no external signs of head trauma (e.g., face or scalp bruising, swelling, or laceration).

The pathophysiology underlying concussion has been elucidated in animal models and corroborated to some extent in human studies. A concussive biomechanical force to the brain triggers a complex cascade of pathophysiologic events involving physical microstructural injury and physiological disturbances. Acceleration and deceleration forces cause axonal stretching, a mild form of TAI that disrupts the integrity of the neuronal membrane. Axonal stretching results in temporary neurophysiological disturbances involving indiscriminate flux of ions into the cell through ion channels that are normally gated, neuronal depolarization and initiation of action potentials, release of excitatory neurotransmitters, massive efflux of potassium, increased activity of membrane ionic pumps to restore homeostasis, glucose metabolism changes, decreased energy (ATP) production, altered cerebral blood flow, inflammatory responses, cytoskeletal damage, and cell death. The physical microstructural injury to axons that is not evident on standard clinical CT or MRI imaging can be detected by advanced imaging techniques such as diffusion tensor imaging (see Chapter 29, "Brain Imaging").

SPORT-RELATED CONCUSSION

Sport-related concussion is caused either directly by a blow to the head, or indirectly by an impulsive force transmitted to the head by a blow elsewhere on the body. In most sports-related concussions there is no LOC, and PTA is brief. The most frequently endorsed subacute symptoms in concussed athletes are headache, a feeling of slowness, drowsiness, difficulty concentrating, and dizziness. Symptoms resolve within 2–28 days. Full recovery is assumed if there are no lingering subjective symptoms, balance testing is normal, and there is no apparent neurocognitive diminishment.

Athletes are at risk for multiple concussions. There is some evidence that previous concussion increases the risk of a future concussion by lowering the threshold for concussion, and that recovery is slower in those who had a previous concussion. With concussion, there may be rapid and complete recovery of cerebral function, yet some brain cells may have been damaged, making the brain more susceptible to the effects of future head trauma. In other words, concussion may reduce **cerebral reserve** (**brain reserve**). The cerebral reserve hypothesis is a heuristic concept that there is redundancy of nervous tissue, that recovery of function is possible if redundant tissue has the capacity to compensate, and that recovery of function is diminished with tissue damage. Within the context of concussion and head trauma, the concept is that multiple mild brain injuries are cumulative, and with each succeeding injury the ability to compensate for lost neurons is diminished.

The standard of care for concussion is observation for 24 hours, and a regimen of rest that includes suspension from contact sports and other activities that put one at risk for head impact and re-injury, any vigorous physical activity, and intensive cognitive activity, followed by gradual resumption of activities. The rationale for avoiding a later impact is that there is a "window of vulnerability" to additional injury, such that the threshold for sustaining a concussion is lowered, and a second concussion sustained prior to resolution of the pathophysiological disturbances following the first concussion might have cumulative effects and prolong recovery. Multiple guidelines recommend that athletes who have sustained a concussion (1) should not return to play the day of injury; (2) should rest (i.e., no vigorous physical activity or heavy mental exertion) until asymptomatic; (3) should not return to play until asymptomatic with exertion; and (4) if previously resolved symptoms return, the athlete should return to the rest phase. The recommendation for rest in concussion management, however, has been criticized for lack of precise definitions of rest and duration of rest. It is also worth noting that the criterion that an athlete diagnosed with concussion should not be allowed to return to the sport on the day of injury cannot be applied in boxing, as one goal of boxing play is to knock the opponent unconscious. The World Medical Association has recommended the general ban of boxing based on the high risk for brain damage.

POST-CONCUSSION SYNDROME

Post-concussion syndrome (PCS) is a diagnosis that has been applied to those with subjective symptoms that persist for more than 3 months after concussion. Headache is the most commonly reported symptom following concussion, and dizziness (a sense of disequilibrium and imbalance) is the second most common. Myriad other symptoms have been reported following concussion, and they are classified as somatic, cognitive, and behavioral. In addition to headache and dizziness, other somatic symptoms include nausea, **photophobia** (light sensitivity), **phonophobia** (noise sensitivity), tinnitus, difficulty focusing vision, postural lightheadedness, and fatigue. Cognitive symptoms include mental "fogginess," difficulty concentrating, and memory problems. Behavioral symptoms include sleep disturbance (hypersomnia, insomnia), irritability, mood lability, anxiety, depression, and personality change.

The validity of PCS is controversial and has been for many years. A big challenge is causally linking subjective, self-reported chronic symptoms and problems that persist long after a concussion, to mTBI that occurred remotely in time. In these circumstances one cannot assume that the symptoms are due to biological effect of the remote injury (i.e., lingering structural damage or physiological dysfunction of the brain). Multiple non-physiological factors may drive symptom reporting long after injury; thus subjective complaints alone should not be used to identify PCS. It may be erroneous to conclude that self-reported symptoms of a remote concussion are causally related to mTBI. The vast majority of people who experience concussion recover quickly and fully. Prospective studies have shown that persistent symptoms are extremely rare. Furthermore, the symptoms of PCS are nonspecific and reported by many other clinical and nonclinical groups. In addition, psychological and social factors may play a role in causing or maintaining the perception and reporting of symptoms long after concussion. If the person has a negative expectation of having symptoms for a long time, this may lead to a **nocebo effect** (expecting a bad outcome leads to a bad outcome). Compensation-seeking patients (i.e., those involved in litigation or applying for disability benefits) have incentives for exaggeration (or over-reporting) of symptoms and problems, and malingering on testing. Persistent post-concussive symptoms are also associated with prior psychiatric history.

TBI Management

Management strategies for patients with TBI focus on preventing secondary injury, as the primary injury cannot be undone. In the acute phase of head trauma, the initial concern is whether the trauma has caused intracranial bleeding (epidural hematoma, subdural hematoma, or a parenchymal hemorrhage) and/or increased intracranial pressure. At the scene of injury and at the hospital ED, the person undergoes neurological assessment to determine if there is a loss or alteration of consciousness. The findings guide medical decisions as to whether neuroimaging is indicated. Computerized tomography (CT) is the imaging modality of choice in acute TBI, as it is fast and accurately detects intracranial hemorrhages and hematomas, and other conditions that are potentially life threatening and treatable by neurocritical care and neurosurgical interventions, such as hematoma and brain edema, which may lead to brain herniation.

Epidural hematomas are treated by draining the pooled blood via a burr hole drilled through the cranium, or ligating the bleeding vessel via **craniotomy** (a procedure in which part of the skull is removed to access the brain and then replaced during the same procedure). Subdural hematomas are treated by **subdural evacuation**. The hematoma is accessed via burr hole(s) in the cranium, or it is accessed by craniotomy if the hematoma is large, or blood has clotted. Asymptomatic small subdural hematomas may simply be monitored by CT; they often are absorbed without additional intervention.

Cerebral edema, an abnormal accumulation of fluid within the brain parenchyma, results in **intracranial hypertension**. It is treated using interventions including hyperosmolar therapy, CSF drainage, and decompressive craniectomy. **Hyperosmolar therapy**, accomplished by intravenous mannitol and hypertonic saline, decreases cerebral blood volume, and creates an osmotic gradient that draws cerebral edema fluid from brain tissue into the circulation. Cerebrospinal fluid drainage is accomplished by an ICP monitor with an external ventricular drain placed in the lateral ventricle. **Decompressive craniectomy** is a neurosurgical procedure in which part of the skull is removed to allow room for brain swelling and is replaced later.

Hyperthermia (≥ 39°C) at the time of injury, as may occur in the context of injuries sustained during sports in the summer, can exacerbate secondary tissue damage; thus, rapid normalization of body temperature may be neuroprotective.

After the acute care period of in-hospital treatment, people with TBI often receive multidisciplinary rehabilitation services aimed at (1) improving the ability to manage activities of daily living independently; (2) addressing specific cognitive, physical, occupational, and emotional challenges; and (3) overall adjustment to a new "self" and new social roles. TBI is often associated with chronic disability, and the impact extends beyond the person with TBI to family, friends, and society at large.

Chronic Traumatic Encephalopathy

In a 1928 *Journal of the American Medical Association* (*JAMA*) article, **Harrison Stanford Martland** (1883–1954) described a clinical syndrome in boxers characterized by dysarthria, ataxia, extrapyramidal signs, and subcortical dementia. He referred to it as **punch drunk syndrome** (aka **punch-drunk encephalopathy**), and it later also came to be referred to as **dementia pugilistica** (from the Latin root *pugil*, "boxer"). The symptoms and signs developed progressively over a long latent period, with an average time of onset 12–16 years after the start of a boxing career. In 1949, **MacDonald Critchley** (1900–1997) introduced the term **chronic traumatic encephalopathy** (CTE) to describe this clinical condition and concluded that it was imperative to perform postmortem studies of the brains of boxers.

In a landmark study published in 1973, **Nick (John Arthur Nicholas) Corsellis** (1915–1994) and colleagues determined the neuropathology of dementia pugilistica by postmortem examination of the brains of 15 former boxers who were institutionalized for dementia prior to their deaths. On macroscopic examination, they found cerebral atrophy, enlargement of the lateral and third ventricles, thinning of the corpus callosum, and **cavum septum pellucidum**, a space between the two layers of the septum pellucidum that is filled with CSF and looks like an additional midline ventricle between the two lateral ventricles (Figure 20.4). Further studies in boxers established an association between dementia pugilistica and exposure to concussions and sub-concussive blows, as measured by total number of fights, number of knockout losses, duration of career, and fight frequency.

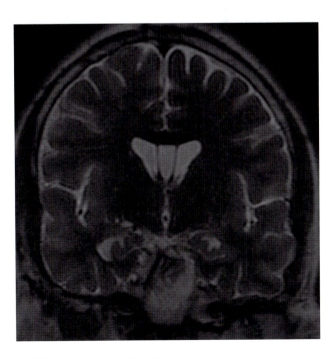

FIGURE 20.4. Coronal T2 MRI of the brain demonstrates cavum septum pellucidum, with CSF space between the septum pellucidum leaflets.

It is important to point out, however, that cavum septum pellucidum may also be congenital, and that in individual cases it may be unclear whether it is congenital or developed due to TBI. Early in development the two layers form with a space between them called the cavum septum pellucidum. This slit-like space is normal in fetuses, but the two layers of the septum pellucidum usually fuse by 3 months of postnatal life. It may, however, persist without producing any symptoms and is considered a normal anatomical variation; incidental (asymptomatic) cava are found in about 1% of brain scans.

In recent years, CTE in contact sports has received a great deal of medical, scientific, and media attention. This began with a journal article published in *Neurosurgery* in 2005 titled "Chronic Traumatic Encephalopathy in a National Football League (NFL) Player" by Bennett Omalu and colleagues. The study reported the autopsy findings of former Pittsburgh Steelers player Mike Webster, who had died suddenly and unexpectedly in 2002. Webster had a history of cognitive and intellectual impairment, psychiatric problems, drug abuse, and suicide attempts prior to his death. Upon autopsy, Webster's gross brain structure looked normal, but Omalu suspected that Webster suffered from CTE, and he had the tissue analyzed. Large accumulations of hyperphosphorylated tau (a microtubule-stabilizing protein) were found. Ann McKee and colleagues have performed postmortem brain studies in over 100 NFL players. They found that the gross structure of the brain in these individuals was normal, but with microscopic neuropathology in 110 of 111 players. The pathognomonic lesion of CTE is described as an irregular, perivascular accumulation of phosphorylated tau aggregates clustered at the depths of cortical sulci in a pattern that differs from other neurodegenerative diseases. McKee and colleagues have also identified CTE in the postmortem brains of professional-level athletes engaged in other contact sports such as ice hockey, rugby, wrestling, and boxing, as well as young amateur football players.

Chronic traumatic encephalopathy is hypothesized to be a progressive neurodegenerative condition that occurs in individuals with a history of repetitive concussive (i.e., symptomatic) and sub-concussive (i.e., nonsymptomatic) blows to the head. It is hypothesized that the neuropathology of CTE gives rise to a clinical syndrome referred to as **traumatic encephalopathy syndrome** (TES). Both CTE and TES are nevertheless controversial entities. One problem is the inconsistency between dementia pugilistica and modern CTE in both clinical features and gross pathology. The clinical syndromes of dementia pugilistica and TES differ; dementia pugilistica is characterized by extrapyramidal motor symptoms, while TES is not. The neuropathologies associated with dementia pugilistica and CTE differ; the brains of boxers with punch drunk syndrome showed gross pathology (as observed by Corsellis) that is not apparent in the brains of NFL players.

Another limitation of this research field involves selection bias. The lesions of CTE cannot be seen on magnetic resonance imaging or any other in vivo imaging technique; they are only revealed on autopsy and by a very specialized, complex staining protocol that is not applied to tissue samples that are not specifically under investigation for CTE. There also is selection bias in terms of the brains that come to pathological examination. The brain donations come from distressed families of individuals with significant histories of dementia, depression, and other neuropsychiatric disturbances. They are not coming from individuals without cognitive and behavioral symptoms who are not concerned that they have CTE. There are also many potential confounding variables in this group, such as years of alcohol abuse, substance abuse, steroid use, and advanced age, all of which may affect pathology burden. Furthermore, the relationship between pathology burden and clinical symptoms is unclear, calling into question the concept of CTE as a disease. Even though many questions about CTE remain, the subject has received a great deal of media attention because of a consolidated class action lawsuit against the NFL charging that the league fraudulently concealed the long-term effects of head trauma.

Summary

TBI is defined as physical trauma to the brain due to impact, penetration, or rapid movement of the brain within the skull caused by an external mechanical force. The pathophysiology of TBI is complex, and in many cases involves a combination of pathologies. Primary injuries, which occur at the time of trauma and are caused by mechanical forces, include skull fracture, contusion, laceration of brain tissue, laceration of blood vessels with bleeding in or around the brain, and traumatic axonal injury. Secondary injuries, which are due to delayed processes that evolve after the primary injury and clinically present hours to days after the injury, include hematoma, edema, raised intracranial pressure, brain herniation, impaired cerebral perfusion and cerebral ischemia, infection, scar tissue formation, and cerebral atrophy. The aim of TBI management and treatment is to prevent secondary injury. Traumatic injuries are also classified as penetrating vs. non-penetrating head injuries, diffuse vs. focal traumatic brain injuries, and coup vs. contrecoup traumatic brain injuries.

The clinical severity of acute TBI is classified as mild, moderate, or severe based on three indices of neurological function: GCS score, duration of LOC, and duration of PTA. Cognitive impairment is the core feature of TBI, and the severity of injury is correlated with the severity of cognitive impairment. TBI is also associated with personality and behavioral changes that are generally more difficult to quantify. Mild TBI encompasses a spectrum of physiological effects, from mild transient neurometabolic changes due to axonal stretching as occur with concussion, to structural brain damage evidenced by neuroimaging, as occurs with complicated mTBI.

Additional Reading

1. Asken BM, Sullan MJ, DeKosky ST, Jaffee MS, Bauer RM. Research gaps and controversies in chronic traumatic encephalopathy: a review. *JAMA Neurol.* 2017;74(10):1255–1262. doi:10.1001/jamaneurol.2017.2396
2. Baldwin GT, Breiding MJ, Dawn Comstock R. Epidemiology of sports concussion in the United States. *Handb Clin Neurol.* 2018;158:63–74. doi:10.1016/B978-0-444-63954-7.00007-0
3. Baxendale S, Heaney D, Rugg-Gunn F, Friedland D. Neuropsychological outcomes following traumatic brain injury. *Pract Neurol.* 2019;19(6):476–482. doi:10.1136/practneurol-2018-002113
4. Corsellis JA, Bruton CJ, Freeman-Browne D. The aftermath of boxing. *Psychol Med.* 1973;3(3):270–303. doi:10.1017/s0033291700049588
5. Corsellis JA. Boxing and the brain [published correction appears in *BMJ.* 1989 Jan 28;298(6668):247]. *BMJ.* 1989;298(6666):105–109. doi:10.1136/bmj.298.6666.105
6. Desai M, Jain A. Neuroprotection in traumatic brain injury. *J Neurosurg Sci.* 2018;62(5):563–573. doi:10.23736/S0390-5616.18.04476-4
7. Filley CM, Kelly JP. White matter and cognition in traumatic brain injury. *J Alzheimers Dis.* 2018;65(2):345–362. doi:10.3233/JAD-180287
8. Giza C, Greco T, Prins ML. Concussion: pathophysiology and clinical translation. *Handb Clin Neurol.* 2018;158:51–61. doi:10.1016/B978-0-444-63954-7.00006-9
9. Iverson GL, Gardner AJ, Shultz SR, Solomon GS, McCrory P, Zafonte R, et al. Chronic traumatic encephalopathy neuropathology might not be inexorably progressive or unique to repetitive neurotrauma. *Brain.* 2019;142(12):3672–3693. doi:10.1093/brain/awz286
10. McAllister T, McCrea M. Long-term cognitive and neuropsychiatric consequences of repetitive concussion and head-impact exposure. *J Athl Train.* 2017;52(3):309–317. doi:10.4085/1062-6050-52.1.14
11. McGinn MJ, Povlishock JT. Pathophysiology of traumatic brain injury. *Neurosurg Clin N Am.* 2016;27(4):397–407. doi:10.1016/j.nec.2016.06.002
12. McKee AC, Abdolmohammadi B, Stein TD. The neuropathology of chronic traumatic encephalopathy. *Handb Clin Neurol.* 2018;158:297–307. doi:10.1016/B978-0-444-63954-7.00028-8
13. McKee AC, Stern RA, Nowinski CJ, Stein TD, Alvarez VE, Daneshvar DH, et al. The spectrum of disease in chronic traumatic encephalopathy [published correction appears in *Brain.* 2013 Oct;136(Pt 10):e255]. *Brain.* 2013;136(Pt 1):43–64. doi:10.1093/brain/aws307
14. McKee AC, Alosco ML, Huber BR. Repetitive head impacts and chronic traumatic encephalopathy. *Neurosurg Clin N Am.* 2016;27(4):529–535. doi:10.1016/j.nec.2016.05.009
15. Omalu BI, DeKosky ST, Minster RL, Kamboh MI, Hamilton RL, Wecht CH. Chronic traumatic encephalopathy in a National Football League player. *Neurosurgery.* 2005;57(1):128–134. doi:10.1227/01.neu.0000163407.92769.ed
16. Omalu B. Chronic traumatic encephalopathy. *Prog Neurol Surg.* 2014;28:38–49. doi:10.1159/000358761
17. Ruff RM, Iverson GL, Barth JT, Bush SS, Broshek DK; NAN Policy and Planning Committee. Recommendations for diagnosing a mild traumatic brain injury: a National Academy of Neuropsychology education paper. *Arch Clin Neuropsychol.* 2009;24(1):3–10. doi:10.1093/arclin/acp006
18. Smith DH, Johnson VE, Trojanowski JQ, Stewart W. Chronic traumatic encephalopathy: confusion and controversies. *Nat Rev Neurol.* 2019;15(3):179–183. doi:10.1038/s41582-018-0114-8
19. Stein DM, Feather CB, Napolitano LM. Traumatic brain injury advances. *Crit Care Clin.* 2017;33(1):1–13. doi:10.1016/j.ccc.2016.08.008
20. Sussman ES, Pendharkar AV, Ho AL, Ghajar J. Mild traumatic brain injury and concussion: terminology and classification. *Handb Clin Neurol.* 2018;158:21–24. doi:10.1016/B978-0-444-63954-7.00003-3
21. Zasler ND, Bigler E. Medicolegal issues in traumatic brain injury. *Phys Med Rehabil Clin N Am.* 2017;28(2):379–391. doi:10.1016/j.pmr.2016.12.012

Brain Neoplasms

Introduction

A **neoplasm**, also known as a **tumor**, is a growth of abnormal cells in some part of the body. A brain tumor, also known as an **intracranial tumor**, is a growth of abnormal cells within the brain or in the tissues immediately surrounding the brain. There are more than 150 types of brain tumors, and they are characterized in multiple ways: intrinsic or extrinsic, invasive or noninvasive, benign or malignant, primary or metastatic, by growth rate, by cell type from which they originate, and by other histological features. The principal categories of intracranial tumors based on cell type of origin are gliomas, meningiomas, pituitary tumors, schwannomas, and metastases. This chapter describes the classification, grading, presenting clinical signs and symptoms, treatment, and neurological effects of intracranial tumors.

Tumor Classification

Tumors grow by expansion, infiltration, or both. **Expansion** refers to enlargement of the tumor around a central core. It tends to occur approximately equivalently in each direction, and thus tumors growing by this mechanism are usually spherical in shape. **Infiltration** (invasion) is the spread of tumor tissue into the interstices of the surrounding tissue.

Intracranial tumors are broadly classified as either intrinsic or extrinsic. **Intrinsic (intracerebral, intra-axial) tumors** arise from within the brain substance. **Extrinsic (extracerebral, extra-axial) tumors** arise from intracranial structures outside the brain substance (e.g., meninges, ependymal cells). Extracerebral tumors produce neurological effects by mechanical compression of the cerebral structures.

Primary brain tumors originate from the tissues of the brain or the brain's immediate surroundings. Primary brain tumors are categorized as glial (originating from glial cells) or non-glial (e.g., originating from meninges, nerves, glands), and benign or malignant. **Benign tumors** are noncancerous and do not spread throughout the body. They are usually encapsulated and segregated from the surrounding tissue, but they may cause neurological dysfunction by compressing adjacent brain tissue. Tumors that do not infiltrate are more likely to be benign. **Malignant tumors** are cancerous; their cells multiply uncontrollably. Tumors that grow by infiltration are far more likely to be malignant and recur after removal. Benign tumors can undergo malignant transformation. Primary brain tumors are far more common in children than adults.

Malignant tumors may **metastasize** (spread) to different parts of the body by detaching from the primary tumor and spreading via the systemic arterial circulation and/or lymphatic channels, forming **secondary tumors** in another part of the body. Secondary tumors are the same type of cancer as the primary tumor. Metastatic spread of primary brain tumors is rare. In adults with malignant brain tumors, metastatic brain tumors originating from outside the brain are far more frequent than primary brain tumors. About 25% of adults with cancer develop metastatic brain tumors.

The **World Health Organization (WHO) Classification of Tumors of the Central Nervous System** classifies central nervous system (CNS) tumors based on two histological features: the cell type from which they arise, and the degree of cellular anaplasia. **Anaplasia** is a cytologic feature of malignant neoplasms that is determined by **biopsy**; the cells appear embryonically immature (i.e., **undifferentiated**) and have an increased capacity for cellular multiplication (i.e., greater **proliferative potential**). This occurs by **dedifferentiation**, a process by which cells develop in reverse, from a more differentiated to a less differentiated state. The gene activity of a differentiated cell type is reprogrammed so that the cell reverts to a less differentiated state, which allows cells to proliferate. Tumor grading is from I to IV based on the degree of histological malignancy. In the context of brain tumors, malignancy is an indication of local growth pattern and capacity to spread within the neuraxis (i.e., how quickly a tumor is likely to grow and spread); it does not refer to the capacity to metastasize to other organs (which is rare).

Grade I tumors are well differentiated; they are benign, non-infiltrative, and grow slowly. **Grade II** tumors are moderately differentiated; they are somewhat infiltrative and relatively slow-growing. **Grade III** tumors are poorly differentiated; they are malignant and infiltrative. **Grade IV** tumors are undifferentiated; they are the most malignant, widely infiltrative, and grow aggressively. Tumor grading forms the basis for treatment planning and prognosis (i.e., likelihood of recovery and recurrence following treatment). Low-grade tumors (Grades I and II) have a better prognosis; they may be observed over time without intervention or may be treated by surgery alone. They are associated with long-term survival. High-grade tumors (Grades III and IV) are the most malignant. They grow rapidly, spread faster than lower grade tumors, and have the highest and fastest mortality rate. They require immediate and aggressive treatment, and they tend to recur rapidly following treatment.

Tumor grade is not the same as the stage of a cancer; **cancer stage** reflects the size and/or extent of the primary tumor and whether there has been lymphatic involvement or metastasis. Cancer stages are from 0 (abnormal cells that are not yet cancer) to IV (the cancer has metastasized).

Gliomas

Gliomas arise from glial cells. The incidence of gliomas is higher in males. There are several types of glioma: astrocytoma, glioblastoma, medulloblastoma, oligodendroglioma, ependymoma, and mixed glioma. Those arising from intraparenchymal glial cells are intracerebral and infiltrative, while those arising from ependymal cells that line the ventricles are extracerebral and non-infiltrative. Gliomas range from relatively benign to highly malignant.

ASTROCYTOMAS

Astrocytomas arise from dedifferentiation of astrocytes. They vary by degree of malignancy. Low-grade astrocytomas occur most frequently along the frontal-temporal convexity but can grow anywhere within the cerebrum.

GLIOBLASTOMAS

Glioblastomas originate from glioblasts (glial precursor cells). They are a grade IV astrocytoma, grow rapidly, and are highly malignant; they are the most aggressive type of nervous system tumor. Glioblastomas occur most commonly in adults over the age of 35. The tumor may be made of a single cell type (**glioblastoma unipolare**) or a variety of cell types (**glioblastoma multiforme**). Glioblastoma multiforme is the most common type of malignant primary brain tumor in adults. Glioblastomas develop their own vascular supply and are prone to hemorrhage.

MEDULLOBLASTOMAS

Medulloblastomas originate from medulloblasts, cerebellar granule cell precursors. They are a highly malignant brain tumor of the cerebellum. They occur almost exclusively in children, and they are the most common type of childhood brain cancer (Figure 21.1).

OLIGODENDROGLIOMAS

Oligodendrogliomas arise from dedifferentiation of oligodendrocytes. They occur most commonly in the deep white matter of the frontal and temporal lobes. They grow slowly, are associated with little or no edema, and tend to calcify.

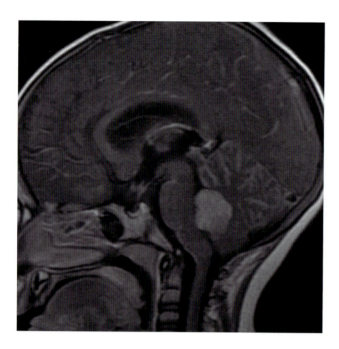

FIGURE 21.1. Sagittal T1 post-contrast image of the brain showing medulloblastoma, a highly malignant brain tumor of the cerebellum that occurs almost exclusively in children. The tumor, centered in the roof of the fourth ventricle, has mass effect on the cerebellum and dorsal brainstem.

EPENDYMOMAS

Ependymomas arise from the ependymal cells lining the ventricles of the brain and spinal cord (Figure 21.2). These tumors are non-infiltrative and have discrete borders with adjacent neural tissue. In the brain, ependymomas most commonly occur at the fourth ventricle. In the spinal cord, ependymomas are the most common type of glioma. Ependymomas are most common in infants and children.

MIXED GLIOMAS

Mixed gliomas are composed of two or more types of glioma tumor cells, most often astrocytes and oligodendrocytes.

Meningiomas

Meningiomas arise from the arachnoid cells of the meninges, and therefore they are extracerebral (Figure 21.3). They are usually very slow-growing and can become very large before giving rise to symptoms. They are often discovered incidentally on brain magnetic resonance imaging (MRI) and computed tomography (CT) studies. Peak incidence is in the sixth and seventh decades of life. Meningiomas occur more than twice as often in women than in men. There is evidence that this difference is due to tumorigenic effects of estrogen and progesterone.

The majority of meningiomas are benign in the sense that they are non-infiltrative and not cancerous, but they may nevertheless be harmful by compressing tissue, producing **mass effect** (compression, distortion, and/or displacement of intracranial contents), and putting pressure on the brain, thereby disturbing brain function. Meningioma symptoms vary by location and size, but headaches and seizures are common initial symptoms. Meningiomas may be multiple. It is not uncommon for these tumors to erode the overlying bone of the skull.

Meningiomas are classified according to their location (Figure 21.4; see Figures 4.1, 4.4, and 4.6 in Chapter 4 for anatomy). Many occur at the sites of the dural folds. Ninety percent of meningiomas are supratentorial. **Convexity meningiomas** are located on the surface of the cerebral cortex, directly under the skull. **Falcine meningiomas** occur in the cerebral falx. **Parasagittal meningiomas** grow on the dural attachment on the external layer of the superior sagittal sinus. **Tentorial meningiomas** grow on the tentorium. **Skull base meningiomas** are on the inferior surface of the brain along the base of the skull; there are several types of skull base meningiomas. **Sphenoid wing meningiomas** grow on the skull base behind the eyes. **Cribriform plate meningiomas**, also known as **olfactory groove meningiomas**, grow along the olfactory nerve within the olfactory grooves of the ethmoid bone cribriform plate. **Suprasellar meningiomas** are located in the **sellar diaphragm** dural fold just above the sella turcica, near the pituitary gland and optic nerve. **Cavernous sinus meningiomas** occur within the cavernous sinus (a dural venous sinus); they are especially challenging to remove and have a high risk for complications due to potential involvement of five cranial nerves (CN2–CN7) and the internal carotid artery.

Posterior fossa (infratentorial) meningiomas are located around the brainstem or cerebellum. They account

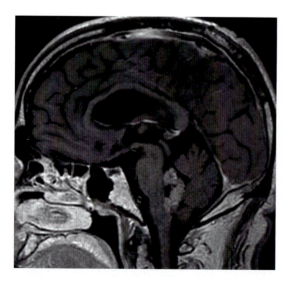

FIGURE 21.2. Ependymoma. Sagittal post-contrast T1 image shows a heterogeneously enhancing mass centered in the floor of the fourth ventricle.

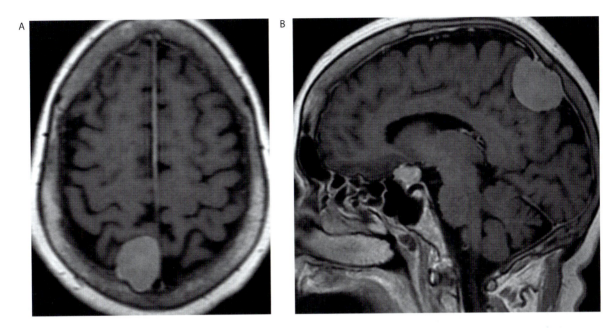

FIGURE 21.3. Meningioma. (A) Axial and (B) sagittal images of a well-circumscribed, enhancing extra-axial mass along the posterior falx. There is mass effect on the subjacent brain parenchyma.

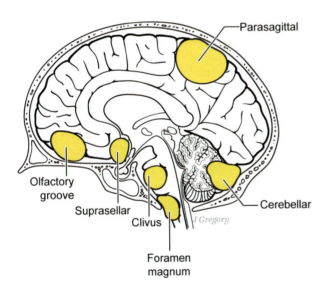

FIGURE 21.4. Meningiomas are classified according to their location.

for approximately 10% of all intracranial meningiomas. Posterior fossa meningiomas often present with cerebellar compression syndrome, brainstem compression syndrome, and features of increased intracranial pressure. The majority of infratentorial meningiomas occur at the **cerebellopontine angle** (the triangle-shaped subarachnoid cistern lying between the cerebellum and pons). **Intraventricular meningiomas** are located within the ventricular system; they may block the flow of cerebrospinal fluid (CSF) through the ventricles and cause hydrocephalus.

Meningiomas are also classified according to cell histology. **Grade I meningiomas** are the slowest growing and considered benign. They are also the most common. If the meningioma is not causing symptoms, its growth may be monitored over time with periodic MRI scans. If the tumor is producing symptoms or there is a high likelihood that it will cause symptoms, then surgical removal is recommended. No additional treatment is required if the tumor is completely removed, but the patient may need radiation treatment after surgery if the excision is incomplete. **Grade II meningiomas (atypical meningiomas)** are slightly more aggressive in growth and have a slightly higher risk of recurrence. **Grade III meningiomas (malignant meningiomas, anaplastic meningiomas)** are the most aggressive but also the rarest, accounting for less than 1% of all meningiomas. Surgery is the first line of treatment, followed by radiation and possibly chemotherapy.

Pituitary Tumors

The pituitary gland is the master gland of the endocrine system. It is a small, pea-sized organ that lies below the hypothalamus and optic chiasm, and within the sella turcica, a saddle-shaped depression at the midline of the sphenoid bone (see Figure 4.1 in Chapter 4 and Figure 10.5 in Chapter 10).

Pituitary tumors grow in or around the pituitary gland. Most are benign as they do not spread to other regions, but they may result in endocrine disturbances by causing hypersecretion or hyposecretion of hormones. They may also cause symptoms due to pressure effects. As the pituitary tumor extends upward and out of the sella turcica, it may compress the overlying optic chiasm and produce bitemporal hemianopsia, loss of vision in the temporal fields of both eyes. Pituitary tumors also are associated with an ice cream cone headache in which the apex of pain points downward toward the center of the head.

There are several types of pituitary tumors, including adenomas and craniopharyngiomas.

PITUITARY ADENOMAS

Pituitary adenomas are benign in that they do not spread to other parts of the body. They do, however, produce significant endocrine abnormalities and are classified based on the hormone secreted by the tumor.

Prolactin-secreting adenomas result in overproduction of prolactin and can cause amenorrhea in women, hypogonadism and low sperm count in men, and infertility. **Growth hormone–secreting adenomas** produce excess growth hormone, resulting in **gigantism** (excessive growth and height due to excessive growth hormone during childhood) and **acromegaly** (abnormal growth of the hands, feet, and face due to excessive growth hormone during adulthood). **Adrenocorticotropic hormone (ACTH)–secreting adenomas** cause the adrenal glands to produce excessive amounts of cortisol (hypercortisolism, Cushing's syndrome). **Thyroid-stimulating hormone (TSH)–secreting adenomas** lead to excessive levels of TSH, resulting in excess production of thyroid hormones T3 and T4 by the thyroid gland, and hyperthyroidism. Hyperthyroidism accelerates metabolism, causing weight loss and other signs and symptoms. Pituitary adenoma, however, is a rare cause of hyperthyroidism. **Oncocytomas** are non-secreting pituitary tumors that make their presence known by pituitary hypofunction.

CRANIOPHARYNGIOMA

Craniopharyngioma is a benign brain tumor of pituitary gland embryonic tissue derived from remnants of a developmental structure, the craniopharyngeal canal, that extends from the floor of the sella turcica to the nasopharynx (Figure 21.5).

Approximately 10%–15% of all pituitary tumors are craniopharyngiomas. These tumors mostly affect children and adolescents. As they grow, they press on the pituitary gland and visual pathways, resulting in hormonal and visual disturbances. They are also often associated with increased intracranial pressure.

Schwannomas

Schwannomas, also known as **neuromas**, are benign cranial nerve tumors that develop from the Schwann cells that provide myelin in the peripheral nervous system. Schwannomas are a type of extracerebral tumor. **Acoustic neuromas**, also referred to as **vestibular schwannomas**, affect the eighth cranial nerve (CN8; Figure 21.6). These are the most common type of neuroma. Common clinical signs are hearing loss, tinnitus, ataxia (due to vestibular dysfunction), and headache. As they grow, they may compress other cranial nerves (CN5, CN7, CN9, and

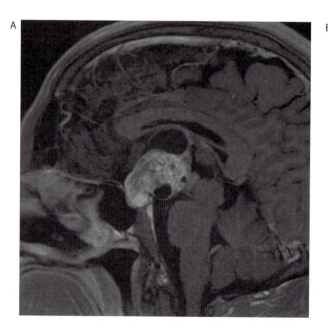
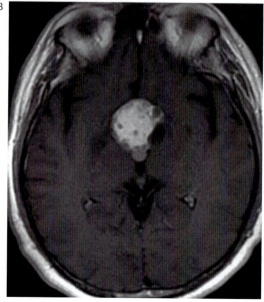

FIGURE 21.5. Craniopharyngioma. (A) Sagittal and (B) axial images of a sellar and suprasellar solid and cystic mass, exerting mass effect on the adjacent structures including the third ventricle, hypothalamus, optic chiasm, and midbrain.

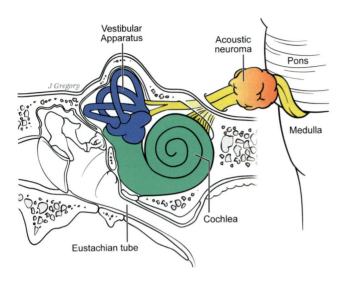

FIGURE 21.6. Acoustic neuromas, also referred to as vestibular schwannomas, affect the vestibulocochlear nerve (CN8).

CN10), and/or the pons and medulla and thereby obstruct CSF flow. Neuroimaging studies and brainstem auditory evoked potentials are most effective in detecting acoustic neuromas. Less commonly, schwannomas arise from CN5, CN9, CN10, and CN7, in descending order of frequency.

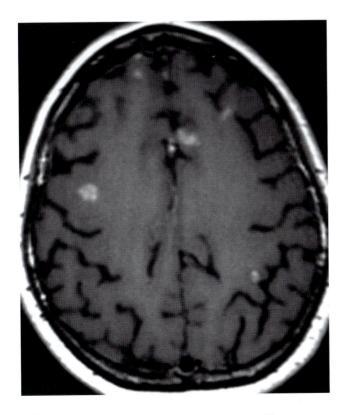

FIGURE 21.7. Brain metastases from breast cancer. Numerous enhancing parenchymal lesions in the bilateral frontal lobes and the left precentral gyrus.

Metastatic Tumors

Metastatic brain tumors are common; approximately 25% of patients who die of cancer have intracranial metastases. Metastatic brain tumors originate from various primary sites. In descending order of frequency, brain metastases develop most often in those with lung cancer, malignant melanoma, renal cell carcinoma, breast carcinoma, and colorectal carcinoma. In adults, metastatic brain tumors are twice as common as primary brain tumors, whereas in children, primary brain tumors are twice as common as metastatic brain tumors. Brain imaging studies have shown that 80% of metastatic brain tumors occur in the cerebral hemispheres, most commonly at the gray-white matter junction. In 70% of cases with brain metastasis there are multiple nodular deposits of tumor throughout the brain, distinguishing metastatic from primary brain tumors (Figure 21.7). The prognosis for brain metastases is dismal.

Clinical Signs and Symptoms

The most common presentations of intracranial tumor are: (1) focal neurological signs and symptoms; (2) seizures; (3) headache and other signs and symptoms of elevated intracranial pressure; and (4) endocrine abnormalities. Some brain tumors, however, are discovered with very mild, vague symptoms such as slow comprehension or reduced ability to sustain mental activity.

Focal Neurological Signs and Symptoms

Focal neurological deficits are the most common presenting sign of intracranial tumors. These deficits, including cognitive and behavioral signs and symptoms, often localize to the tumor site. For example, a frontal lobe tumor may result in personality change and an occipital tumor may result in a visual field defect. As the tumor expands, the deficit progresses. Focal neurologic deficits reflect both the location of the tumor and its rate of growth; slowly growing tumors result in more gradual progression of symptoms than rapidly growing tumors.

The neurological, cognitive, and behavioral signs induced by a tumor may persist after tumor removal due to structural damage caused by the tumor, structural damage caused by neurosurgical intervention, and edema.

Seizures

In adults with brain tumors, 25%–30% experience seizures as their first symptom, while 40%–60% experience seizures at some time during the tumor's course. New-onset seizures are a common first sign of brain tumor in adults. Unexplained seizures with adult onset should be regarded as a possible sign of brain tumor until proven otherwise. Seizures can occur with benign, slowly growing tumors, as well as malignant, rapidly growing tumors.

Headache and Increased Intracranial Pressure

Signs and symptoms of increased intracranial pressure most commonly include headache, nausea, vomiting, papilledema (swelling of the optic disk), bradycardia (slowing of the heart rate), diplopia (double vision), and decreased mental ability.

Headache is the most common symptom of brain tumor. Headaches associated with increased intracranial pressure are characteristically most severe when lying down, due to increased venous pressure associated with the supine position. The headaches often worsen with coughing, sneezing, straining during bowel movement, lifting, bending down, or sudden exertion.

As the volume of intracranial contents increases with neoplasm growth and edema, structures become compressed and dysfunctional. The lumen of affected veins and capillaries collapses, as do the ventricles, sulci, and subarachnoid space. If the intracranial pressure increases sufficiently, the brainstem may become compressed, interfering with vital life functions. CSF flow is usually compromised due to expansion of the space-occupying lesion, which further increases intracranial pressure and may result in herniation and/or hydrocephalus.

Frontal and parietal tumors may result in subfalcial herniation, which compresses the diencephalon. Temporal lobe tumors may induce descending transtentorial (temporal lobe) herniation through the tentorial notch, exerting pressure on the midbrain. This results in hemiparesis due to compression of the cerebral peduncle, as well as compromise of the amygdala and hippocampus. Cerebellar tumors can cause tonsillar herniation (descent of the cerebellar tonsils through the foramen magnum) and result in brainstem compression. Compression of the diencephalon and mesencephalon interferes with function of the reticular activating system, resulting in rapid change in the level of consciousness.

Endocrine Abnormalities

Most endocrine disturbances associated with intracranial tumors result from pituitary-hypothalamic involvement, but increased intracranial pressure may also contribute. As described above, these syndromes may result from either increased or decreased hormonal production.

Diagnosis and Prognosis

Diagnosis of brain tumors rests on the history of signs and symptoms, neurologic examination, and structural imaging by MRI or CT for differential diagnosis (e.g., ruling out hemorrhage) and determination of tumor location/size and extent of edema for treatment planning. In some cases, a lumbar puncture or brain biopsy may be performed in order to determine the tumor type.

Clinical course and outcome are correlated with tumor type, but factors such as intracranial pressure, location and size of the tumor, patient age, and general condition of the patient also play a role. Because intracranial tumors frequently cause increased intracranial pressure due to tumor growth, edema, obstruction of the ventricular system, hemorrhage of the tumor itself, and compression of tissue, brain tumors can never be considered truly "benign." Tumors of the diencephalon or brainstem may produce devastating deficits even though their size may be small. Infiltrative tumors have considerable mortality.

Treatment

Brain tumors are treated by surgical tumor resection, radiation therapy, chemotherapy, and various combinations of these approaches. The goal of treatment for low-grade brain tumors is curative. The goal of treatment for high-grade brain tumors is to decrease neurological symptoms, maximize survival, and minimize morbidity; thus, it is palliative. Depending on the type and stage of cancerous tumors, treatment may cure the cancer (destroy cancer cells so that they are undetectable in the body and do not grow back), control the cancer (slow the growth and prevent its spread), or ease symptoms (shrink tumors to ease pain or reduce pressure). Treatment also often involves steroid medications to reduce swelling and anti-seizure medication to reduce seizures.

Surgical Intervention

Surgical removal is typically the first line of treatment. Tumors on the surface of the brain are accessed by **craniotomy**. For meningiomas, treatment outcome is good if the tumor is in an area that is easily accessible, such as the cerebral convexity or cerebral falx.

Tumors that cannot be reached using a traditional surgical approach are treated by **neuro-endoscopy**, in which the tumor is accessed through small holes (about the size of a dime) in the skull, or through the mouth or nose. The endoscope is navigated to the tumor site to allow visualization of the tumor, and microsurgical instruments are navigated to the site through channels located within the endoscope. This approach also minimizes the amount of tissue damage in the surgical location. In many cases it is not possible to completely remove a tumor; partial resection (**debulking**), however, may prolong survival. Depending on the type of tumor, surgery may be followed by radiation therapy and/or chemotherapy to prevent regrowth.

Intraoperative brain mapping using direct cortical stimulation may be used to guide surgical resection, to allow for extensive tissue resection while minimizing postoperative neurological deficits.

Radiation Therapy

Genes control the rate of cell division and multiplication. Cancer cells divide quickly, and therefore malignant tumors grow rapidly. Radiation damages DNA and causes cells to lose their ability to replicate. Because cancer cells replicate rapidly, they have higher levels of **radiosensitivity** and are more vulnerable and more likely to be damaged by radiation. Thus, radiation causes tumor cells to lose their ability to reproduce; over time, cancer cells die, and the tumor gradually shrinks. Radiation, however, also affects cell division of normal tissues (though to a lesser extent), leading to undesired tissue damage and side effects. Furthermore, radiation does not kill all cells immediately; cells that replicate quickly are affected soon after treatment, but cells that replicate slowly are affected well after treatment. Thus, radiation therapy (RT) rests on a balance between killing cancer cells and minimizing damage to normal cells.

Ionizing radiation is used in cancer treatment; it creates ions in cells by removing electrons from atoms and molecules. There are different types of ionizing radiation (e.g., photon beam, particle), and they differ with respect to amount of energy; the more energy in the radiation, the deeper the tissue penetration. There are also various methods of applying radiation; **external beam radiation** is the most widely used type of RT. The **radiation oncologist** tailors the type and mode of RT to the cancer type and location.

Whole-brain radiation therapy (WBRT) has been the mainstay treatment for brain metastases, which, as indicated earlier, are widely distributed. However, it is toxic to the brain and causes **radiation encephalopathy** (see Chapter 23, "White Matter Disease"). Advances in radiation technology now allow for a more localized delivery of radiation, and consequently the use of WBRT is declining. **Stereotactic radiotherapy** is a form of RT involving image-guided delivery of high-dose radiation to a focal region, targeting tumors that are small and well-defined. It is also known as **stereotactic radiosurgery**, although it does not involve tissue incision. If the radiation is given in multiple doses, it is referred to as **fractionated stereotactic radiotherapy**. Compared to WBRT, stereotactic radiotherapy results in minimal radiation delivery to healthy tissues and a high dose of radiation at the target, and it is associated with better disease control and minimal effect on normal brain tissue. This type of RT can also treat tumors that are too difficult to reach with standard neurosurgery, or if there are contraindications to the more invasive surgical approach.

Chemotherapy

Chemotherapy involves using drugs to slow the growth, stop the growth, or destroy cancer cells. Sometimes it is used as the only treatment, but more often it is used in combination with other treatment approaches. There are many different chemotherapy drugs, and they are administered in a variety of ways. Most often, chemotherapy is delivered by intravenous infusion, but some agents may be administered orally (by pill or capsule) or by injection. They may be delivered directly to the area of the body affected by cancer, such as the abdomen (intraperitoneal chemotherapy), chest cavity (intrapleural chemotherapy), CNS (intrathecal chemotherapy), or through the urethra into the bladder (intravesical chemotherapy). Chemotherapy drugs may also be applied directly to the tumor during surgery by thin disk-shaped wafers or may be injected into a vein or artery that directly feeds a tumor.

Chemotherapy is associated with neurotoxic effects, particularly when administered in high doses, intrathecally (i.e., into the CSF by way of the lumbar cistern), and when combined with RT. The term *chemobrain* has been used to refer to cognitive decline following chemotherapy. In patients with brain tumors, the role of chemotherapy remains unclear given the direct effects of brain tumors on brain function, and the fact that most of those receiving chemotherapy also receive RT. In patients with non-CNS cancers, the literature is mixed, with some studies showing persistent cognitive difficulties with chemotherapy and others showing transient difficulties that resolve.

Neurological Effects

Neurological effects of brain neoplasms depend upon the brain regions affected and the rate of tumor growth. Slow-growing tumors bring on associated signs and symptoms in such a gradual fashion that often neither the patient nor family members can specify onset. Slow-growing frontal lobe tumors tend to be associated with change in personality and behavior, apathy, motor disturbance, and loss of spontaneity and initiative. Rapidly expanding frontal tumors result in confusional states with rather abrupt onset. In general, if psychiatric symptoms are present early in the course, the neoplastic disorder is likely within the frontal regions, temporal regions, or both.

The various behavioral and neurological signs induced by a tumor may persist after successful tumor removal, due to neurosurgical intervention and structural damage caused by the tumor. In addition, post-treatment neurological and cognitive dysfunction may be related to radiation therapy, chemotherapy, and side effects of other medications (e.g., corticosteroids, antiepileptics). In general, the most focal, lateralized signs are related to the site of pathology (i.e., core of the tumor), while nonfocal, non-lateralized findings are related to the secondary diffuse effects.

Patients with brain tumor may undergo neuropsychological assessment for multiple reasons: (1) to determine the nature and severity of impairment, if any (e.g., to determine whether a small, slow-growing meningioma is associated with any cognitive or behavioral signs); (2) to monitor cognitive and behavioral functioning over time or with treatment; and (3) to guide planning for rehabilitation and supportive care.

Summary

There are more than 150 types of brain tumors, and they are characterized along multiple dimensions: intrinsic or extrinsic, invasive or noninvasive, benign or malignant, primary or metastatic, growth rate, cell type from which they originate, and other histological features. The principal categories based on cell type of origin are gliomas, meningiomas, pituitary tumors, schwannomas, and metastases. The most common presentations of brain tumor are focal neurological signs and symptoms, seizures, headache and other signs and symptoms of elevated intracranial pressure, and endocrine abnormalities. Brain tumors are treated by surgical tumor resection, radiation therapy, chemotherapy, and various combinations of these approaches.

Neurological effects of brain neoplasms depend upon the brain regions affected and the rate of tumor growth. The neurological signs induced by a tumor may persist after successful tumor removal, due to structural damage caused by the tumor and neurosurgical intervention. Post-treatment neurological dysfunction may also be related to radiation therapy, chemotherapy, and side effects of other medications.

Additional Reading

1. Jacob J, Durand T, Feuvret L, Mazeron J-J, Delattre J-Y, Hoang-Xuan K, et al. Cognitive impairment and morphological changes after radiation therapy in brain tumors: a review. *Radiother Oncol.* 2018;128(2):221–228. doi:10.1016/j.radonc.2018.05.027
2. Louis DN, Perry A, Wesseling P, Brat DJ, Cree IA, Figarella-Branger D, Hawkins C, Ng HK, Pfister SM, Reifenberger G, Soffietti R, von Deimling A, Ellison DW. The 2021 WHO Classification of Tumors of the Central Nervous System: a summary. *Neuro Oncol.* 2021;23(8):1231–1251. doi: 10.1093/neuonc/noab106
3. Warrington JP, Ashpole N, Csiszar A, Lee YW, Ungvari Z, Sonntag WE. Whole brain radiation-induced vascular cognitive impairment: mechanisms and implications. *J Vasc Res.* 2013;50(6):445–457. doi:10.1159/000354227
4. Wefel JS, Witgert ME, Meyers CA. Neuropsychological sequelae of non-central nervous system cancer and cancer therapy. *Neuropsychol Rev.* 2008;18(2):121–131. doi:10.1007/s11065-008-9058-x

Brain Infections

Introduction

Brain infection is a relatively rare but important cause of mortality and chronic neurological morbidity. Due to the potential for adverse consequences, it is important for clinicians who work with patients with neurological disorders to be familiar with common brain infections. This chapter describes the basics of brain infection and specific examples that are of particular interest.

The suffix -*itis* is used to denote inflammation of an organ (e.g., bronchitis, appendicitis). Inflammation is a biological response of the immune system that can be triggered by a variety of harmful factors, such as pathogens, cell damage (due to trauma, infarction, burns, radiation), foreign bodies, toxins, and autoimmune diseases (in which the body mistakenly recognizes its own cells or tissues as harmful). The causes of inflammation therefore can be infectious or noninfectious; infection involves inflammation, but inflammation does not necessarily indicate infection.

Inflammation can selectively affect specific tissues of the nervous system. **Meningitis** is an inflammation confined to the meninges. **Encephalitis** is inflammation of the brain parenchyma (not to be confused with **encephalopathy**, which describes any medical condition impacting the brain's function). **Pachymeningitis** is inflammation confined to the dura mater. **Leptomeningitis** is inflammation confined to the pia and arachnoid membranes. **Meningoencephalitis** is encephalitis with meningeal involvement. **Ependymitis** (also known as **ventriculitis**) is an inflammation of the ventricular lining. **Myelitis** is a generalized inflammation of the spinal cord parenchyma. **Encephalomyelitis** is a generalized inflammation of the brain and spinal cord. **Radiculitis** is an inflammation of the nerve root(s). **Neuritis** is an inflammation of peripheral nerve(s).

Classification

Central nervous system (CNS) infections are classified by the location of infection and the offending pathogen. The three most common types of CNS infection based on location are meningitis, encephalitis, and brain **abscess**, a localized area of brain infection consisting of an encapsulated pocket containing pus, the organism of infection (i.e., bacteria or fungi), and necrotized brain tissue. The clinical presentation of CNS infection depends on the tissues involved. Meningitis is characterized by fever, headache, and neck stiffness. Encephalitis is characterized by altered mental status. Brain abscess is characterized by focal neurological signs and symptoms.

Infectious diseases are caused by a variety of pathogenic agents classified as viruses, bacteria, fungi, parasites, and prions. Bacteria and viruses are the most common causes of CNS infections. Viruses are composed of small pieces of genetic material (nucleic acid) surrounded by a thin coat of protein; they are acellular (i.e., not composed of cells) and cannot multiply outside of a host cell. Bacteria are unicellular prokaryote organisms; they contain DNA but lack a distinct nucleus. Fungi and parasites are eukaryotes; their cells have a defined nucleus containing DNA. Fungi encompass a large, diverse group of organisms that reproduce by spores and live by decomposing and absorbing the organic matter on which they grow. Parasites are single-celled and multi-cellular organisms that live in or on another species, from which they take nourishment. Prions are the smallest known infectious agents and are composed of a single protein without any nucleic acid.

Many pathogens cause disease by invading a host (i.e., entering the host body), adhering to specific host cells, colonizing, and inflicting damage on those tissues. The infectious agent can be transmitted from person to person, either directly (e.g., via skin contact) or indirectly (e.g., via contaminated food or water). The terms "infection" and "disease" are not synonymous; **infection** is the process whereby a pathogen invades a host, while **disease** is the process by which a pathogen harms a host. For example, one may be infected with the severe acute respiratory syndrome coronavirus 2 (SARS-CoV-2) but not necessarily develop coronavirus disease (COVID-19).

Our bodies have defense mechanisms to prevent infection, as well as defense mechanisms to prevent disease should infection occur. Infectious agents vary in their ease of transmission (degree to which they are **contagious**) and likelihood of causing disease (**virulence**). **True pathogens** are infectious agents that cause disease in virtually any host; **opportunistic pathogens** are potentially infectious agents that cause disease in hosts with compromised immune systems (e.g., due to immunosuppressive treatment, cancer chemotherapy, prolonged corticosteroid use, human immunodeficiency virus, lymphoma), but rarely in individuals with healthy immune systems.

Portal of Entry

Infectious agents enter the body by way of the gastrointestinal tract, respiratory tract, skin inoculation (e.g., animal or insect bite), or exchange of blood or other tissue (e.g., blood transfusion, shared syringe, sexual transmission, tissue transplant). The site at which a pathogen enters the body is the **portal of entry**.

Pathogens rarely reach the brain because of three levels of protective mechanisms: (1) systemic immune responses that destroy organisms at the primary site of infection and within blood; (2) the blood-brain barrier; and (3) the meninges. When infectious agents enter the CNS, they do so by (1) hematogenous spread (i.e., via blood circulation) from a remote tissue infection; (2) local extension of infection from contiguous structures (e.g., ear, sinus, or dental infections); (3) direct invasion of the CNS by trauma (e.g., depressed skull fracture, penetrating traumatic brain injury, neurosurgery (including implanted medical devices such as shunts); and (4) axonal transport through peripheral nerves into the CNS. When pathogens reach the CNS, they may be fatal unless treated because the CNS has weaker immune protection than the rest of the body. Compared to infections involving other organ systems, therefore, CNS infections are more difficult to treat and have greater potential for morbidity and mortality. Early recognition and rapid treatment of acute CNS infections is critical; it can be life-saving and can reduce chronic neurological morbidity.

Meningitis

Meningitis is an inflammation of the meninges, the protective covering surrounding the brain and spinal cord. It is the most common manifestation of CNS infection. The most common cause of meningitis is viral infection, but it may also be caused by bacterial, fungal, or parasitic infection. Meningitis may also have noninfectious causes, such as cancer or brain injury.

The cardinal clinical features of acute meningitis are severe headache, fever, and **nuchal rigidity** (stiff neck), in the context of relatively preserved mental status and absence of focal neurologic signs. Common late clinical features include seizures, cranial nerve palsies, deafness, focal neurologic signs, stupor, and coma.

Viral Meningitis

Viral meningitis is the most common type of meningitis and can be caused by a wide variety of viruses. Headache is the most common presenting symptom and is often severe; it is frequently associated with fever and stiffness of the neck and spine on forward bending due to meningeal irritation.

Viral meningitis has a relatively benign course. Symptoms are generally mild and usually resolve within 1–2 weeks without treatment. It rarely progresses to delirium, stupor, coma, or death. Long-term neurologic sequelae are rare. In most cases there is no specific treatment, but some patients are treated with antiviral medications.

Bacterial Meningitis

Bacterial meningitis is caused by bacterial infection; meningitis caused by infectious agents other than bacteria is referred to as **aseptic** (**nonbacterial**) **meningitis**. The cardinal acute symptoms of bacterial meningitis are sudden fever, severe headache, and nuchal rigidity. There may also be acute cerebral edema, symptoms of increased intracranial pressure (e.g., photophobia, nausea, and vomiting), seizures (due to cortical inflammation), communicating hydrocephalus, focal neurological signs, cognitive decline, and delirium. The incidence of bacterial meningitis is greatest in children and young adults. Immunosuppression, generally used to treat other disorders, is a major risk factor.

Untreated bacterial meningitis has an extremely high death rate, thus prompt diagnosis and treatment are essential. Even with appropriate treatment, the mortality rate is about 10%. In patients suspected of having bacterial meningitis, spinal tap (lumbar puncture) cerebrospinal fluid (CSF) analysis and culture provide definitive diagnosis by findings of white blood cells and identification of the specific offending bacterial microorganism. Neuroimaging does not always aid in diagnosis; there is meningeal enhancement in some patients, but its absence does not rule out the condition. Bacterial meningitis is treated with high-dose intravenous antibiotics that cross the blood-brain barrier, and corticosteroids that treat inflammation and prevent neurological sequelae. Antibiotics are occasionally administered directly into the CSF by intrathecal injection (i.e., into the subarachnoid space of the spinal cord).

Long-term neurologic/neurobehavioral morbidity is seen in 15%–25% of survivors, typically consisting of some combination of sensorineural hearing loss, focal neurological deficits, seizures, or cognitive impairment. The sensorineural hearing loss is due to damage of the cochlea; bacteria reach the cochlea through the cochlear aqueduct. Focal neurological deficits are most commonly caused by infarction but may also be due to other cerebral pathologies such as intracerebral bleeding and brain abscess. Such cerebral pathologies may be associated with seizures. Bacterial toxins and inflammatory response cytotoxins may also produce neuronal damage and contribute to neurologic morbidity. Slowed cognitive processing and mild memory deficit are typical sequelae of bacterial meningitis.

Encephalitis

Encephalitis is inflammation of the brain parenchyma. Acute encephalitis is a febrile illness characterized by the abrupt onset of headache and altered mental state. Other common features include seizures (generalized or focal) and signs of upper motor neuron damage (i.e., hyperreflexia, spasticity, Babinski sign). Some patients develop hemiparesis, aphasia, ataxia, limb tremors, and cortical blindness.

The majority (> 90%) of encephalitis cases are caused by viruses that reach the brain by way of the bloodstream. Once the virus reaches the brain parenchyma, it usually results in widely disseminated infection of neurons and glia, cerebral edema, and an inflammatory response with infiltration of lymphocytes and macrophages. The immune response often terminates the infection, but the person may be left with permanent neurologic sequelae. More than 100 viruses have been implicated in viral encephalitis. The most common are herpes simplex type 1 virus, varicella zoster virus, Epstein-Barr virus, adenoviruses, enteroviruses, arboviruses (**AR**thropod-**BO**rne viruses from ticks, insects, spiders, and mosquitos), and cytomegalovirus. Treatment for viral encephalitis may involve antiviral drugs.

Assessment for acute encephalitis includes blood work and/or lumbar puncture and CSF fluid analysis to identify the pathogen (recall from Chapter 4, "Cranium, Spine, Meninges, Ventricles, and Cerebrospinal Fluid," that the extracellular fluid of the brain is continuous with the CSF). Neuroimaging by CT or MRI is used to identify cerebral edema and mass effect. Electroencephalography (EEG) may provide both diagnostic and prognostic information. In patients with altered level of consciousness, EEG is prognostic beyond the patients' clinical condition; normal EEG is associated with greater likelihood of survival. Treatment of encephalitis often includes anticonvulsants to prevent seizures and corticosteroids to reduce swelling.

Herpes Simplex Virus Encephalitis

Herpes simplex virus encephalitis (HSVE) is the most frequent, sporadic, and fatal encephalitis in humans in the Western world. It is caused by the herpes simplex virus type 1 (HSV-1). This virus is frequently acquired during childhood through oral mucosal infection, producing cold sores around the mouth. Following a primary infection in childhood, HSV-1 remains dormant within the trigeminal ganglia. It can be activated years later and enter the CNS, although the specific mechanism is unknown.

The clinical presentation of HSVE is characterized by an acute or subacute encephalopathy (diffuse brain dysfunction), the hallmark of which is altered mental state (e.g., lethargy, confusion, delirium). There may be seizures and/or coma. Mortality is high (70%) within 7–14 days of onset because the virus multiplies rapidly. Early diagnosis and treatment by antiviral therapy (with acyclovir) are critical to prevent tissue damage; prognosis depends on when treatment is initiated.

HSV-1 has a predilection for the inferior-medial temporal lobes and orbitofrontal brain regions, beginning unilaterally and then spreading to the contralateral hemisphere (see Figure 12.5 in Chapter 12). This anatomical pattern is likely related to the pathway by which the activated virus enters the CNS. Survivors of HSVE often have significant neuropsychiatric sequelae due to involvement of the medial temporal and frontal lobes, consisting of severe memory disorder, extreme alterations of personality, behavioral disorders, language deficits, and olfactory hallucinations or anosmia (loss of olfactory sense). Bilateral lesions are of course associated with worse cognitive and behavioral outcomes than unilateral lesions. Several cases of HSV-1 encephalitis have presented with Klüver-Bucy syndrome due to bilateral damage of the amygdalae (see Chapter 12, "Limbic Structures").

The majority of HSVE cases have an abnormal EEG, with epileptiform discharges superimposed on a disorganized background over one or both temporal lobes.

Rabies Encephalitis

Rabies encephalitis is a rapidly progressive CNS infection that results in rapid neurological deterioration and usually progresses to death. It is caused by rabies virus and is most commonly transmitted to humans via a bite by a rabid animal (i.e., it is transmitted through the animal's saliva). The rabies virus is a neurotropic virus, as it infects nerve cells. It travels from the entry site via peripheral nerve motor neurons to the spinal cord, or to the brainstem when the face is bitten, via retrograde axoplasmic flow, and then to the brain, where it disseminates widely.

In low-resource countries, the most common form of rabies is the dog (canine) variant. The vast majority (95%) of worldwide human deaths due to rabies are in Asia and Africa, with particularly high rates in India. In countries where public health measures have been implemented to reduce the number of rabies-infected dogs (i.e., control of stray dog populations, vaccination of domestic animals), the bat rabies variant accounts for most cases, often without awareness of a bat bite or even exposure to a bat. In the United States, other animals that may transmit rabies include raccoons, skunks, and foxes; however, any mammal may contract and transmit rabies. Non-bite exposure can also occur, such as through inhalation in bat-infested caves.

Initial symptoms of CNS rabies infection are nonspecific and consist of fever, headache, and malaise. Within days, 80% of cases develop classic encephalitic rabies, characterized initially by hyperactivity and progressing to episodes of fluctuating consciousness, agitation, bizarre behavior, and hallucinations. In the later stages of the disease, characteristic neurologic symptoms include hydrophobia due to pain on swallowing and hypersalivation. Twenty percent of cases develop paralytic rabies, characterized by ascending paralysis and quadriplegia.

Rabies is diagnosed based on clinical signs and symptoms and is confirmed by laboratory tests. Early diagnosis of infection, however, has no impact on prognosis. There is no effective treatment once the clinical symptoms appear. The mortality rate is near 100%, and death occurs within 3–10 days after symptom onset. Rabies encephalitis can be prevented after recognized animal bites and potential exposure to the virus by post-exposure prophylaxis, consisting of a course of human rabies immune globulin and rabies vaccine initiated on the day of exposure.

Brain Abscess

An abscess is a localized collection of pus in an enclosed tissue space (i.e., not on a tissue surface) in any part of the body. Abscesses result from a localized infection in any solid tissue; most result from bacterial infection, but they may also occur with fungal infection, parasites, and other foreign substances. Abscesses are a defense mechanism to prevent the spread of infection. When a foreign agent enters the body, the immune system sends infection-fighting white blood cells to the affected area. In the process, some nearby tissue dies and a cavity forms that is filled with pus, which consists of a mixture of living and dead white blood cells, dead tissue, and the foreign substance.

There are two main classes of **intracranial abscess**: (1) **brain abscesses** located within the brain parenchyma; and (2) subdural or epidural **empyemas**, collections of pus within a natural anatomical cavity (Figure 22.1). Most brain abscesses arise by spread through the blood from infection at a distant site (e.g., purulent pulmonary infection, bacterial endocarditis). Metastatic abscesses from hematogenous spread are usually located in the distal territory of the middle cerebral arteries. Brain abscesses may also arise by direct extension (spread) from contiguous structures such as ear, sinus, or dental infections, or by direct introduction via penetrating head trauma, including neurosurgery.

The infection usually begins as a localized encephalitis with focal softening, necrosis, and inflammation. As the process continues, fibroblasts (which are cells that play a critical role in the immune response to tissue injury) proliferate at the edges, forming a firm capsule wall known as the abscess lining. A variable amount of edema surrounds the lesion. The space-occupying lesion expands slowly.

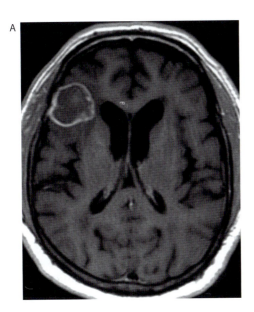
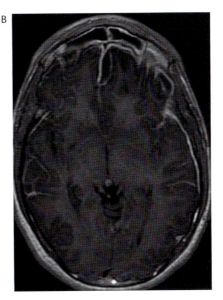

FIGURE 22.1. (A) Brain abscess. Axial post-contrast T1 image demonstrates a thick, rim-enhancing lesion in the right frontal lobe. (B) Subdural empyema. Axial post-contrast T1 image demonstrates peripherally enhancing fluid collection along the anterior falx and anterior to the left frontal lobe. There is additional diffuse reactive dural thickening and enhancement.

Signs and symptoms of brain abscesses are typically subacute in onset and are related to the size and location of the lesion and increased intracranial pressure. Because brain abscesses are slowly expanding space-occupying lesions, their clinical manifestations may be similar to those of some brain tumors. The cardinal symptom of brain abscess is a relentless, progressive headache. This is usually followed by focal neurologic signs and symptoms with insidious onset and gradual progression. The specific focal manifestations depend on the location of the abscess (e.g., lesions in the frontal cortex may produce hemiparesis, lesions in the occipital cortex of one hemisphere may cause homonymous visual defects). Over time, focal neurologic signs and symptoms become more prominent, and psychomotor slowing, lethargy, and confusion emerge. Other signs include intermittent fever, and focal or generalized seizures. As the mass expands, increased intracranial pressure becomes more pronounced. Eventually, the abscess may expand to cause brain herniation (see Chapter 4, "Cranium, Spine, Meninges, Ventricles, and Cerebrospinal Fluid"), or it may rupture into the ventricle and produce ependymitis. If untreated, the brain mass may be lethal.

Diagnostic evaluation includes contrast-enhanced MRI or CT, which shows the abscess as a ring-enhancing lesion, and aids in assessment of potential complications such as mass effect and hydrocephalus. Lumbar puncture is not performed due to potential for transtentorial herniation. Treatment is with antibiotics or antifungal agents, and typically surgical drainage or CT-guided stereotactic aspiration.

Fungal Infections of the CNS

Fungal infections are much less common than bacterial and viral infections, and they occur most frequently as opportunistic infections in those with compromised immune systems. The CNS fungal infections present with various clinical syndromes, including meningitis, encephalitis, and cerebral abscesses. Treatment involves addressing the cause of the suppressed immune system and an extended course of a specific antifungal agent.

Parasitic Infections of the CNS

Parasites are a diverse group of organisms broadly classified as single-celled protozoa or multicellular metazoa. All parasites that affect humans can potentially involve the CNS. Neurocysticercosis is the most common parasitic infection of the CNS; toxoplasmosis, schistosomiasis, and echinococcosis occur less frequently.

Neurocysticercosis

Cysticercosis is a parasitic infection caused by ingestion of the larvae (eggs) of the tapeworm parasite, *Taenia* *solium*. This causes intestinal tapeworm infection, or entrance of larvae into various tissues including muscle and brain and formation of cysts. When the brain and/or spinal cord are affected, the condition is called **neurocysticercosis**. The highest rates of infection with the organism occur in low-resource areas of Latin America, Asia, and Africa.

Cysticercosis is acquired through the fecal–oral route. A tapeworm carrier sheds larvae through feces, which spread through food, water, or contaminated surfaces. The eggs are then ingested when a person eats contaminated food, drinks contaminated water, or places contaminated fingers in their mouth. Once exposed to gastric acid in the human stomach, the eggs turn into larval cysts, cross the gastrointestinal tract, and migrate via the vascular system to the brain, muscle, eyes, and other structures. The larval cysts may remain in the brain as viable cysts for years or degenerate and form calcified granulomas (focal collection of inflammatory cells at the site of infection).

The most common signs and symptoms of neurocysticercosis are seizures and headaches, but signs and symptoms depend on the location and number of cysts in the brain (Figure 22.2).

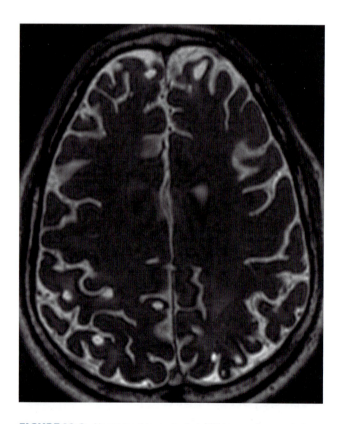

FIGURE 22.2. Neurocysticercosis. Axial T2 image demonstrates numerous rounded hyperintense lesions throughout the brain parenchyma.

Prion Diseases

Prion diseases, also known as **transmissible spongiform encephalopathies** (TSEs), are a group of rapidly progressive, neurodegenerative diseases that can affect animals and humans. They are characterized by spongiform changes in the CNS; microscopic examination of infected neural tissue reveals many small round or oval empty spaces, imparting a spongy appearance. There is no treatment for the TSEs and they are invariably fatal.

The animal prion diseases are **scrapie** (affecting sheep and goats), **bovine spongiform encephalopathy** (affecting cattle), **chronic wasting disease** (affecting deer and elk), **feline spongiform encephalopathy** (affecting domestic and nondomestic cats), **transmissible mink encephalopathy**, **exotic ungulate spongiform encephalopathy** (affecting mammals with hooves), and **spongiform encephalopathy of primates**.

The first recognized TSE was scrapie. One of its clinical signs is that affected sheep compulsively scrape off their fleece against rocks, trees, or fences, apparently due to an itching sensation; hence the name *scrapie*. The transmissible nature of scrapie was established when an outbreak was linked to immunization with a viral vaccine prepared from sheep tissues, including brain and spinal cord, that were later discovered to have been exposed to natural scrapie infection.

The human TSEs are kuru, Creutzfeldt-Jakob disease, fatal familial insomnia, and Gerstmann-Sträussler-Scheinker syndrome. They are further classified as acquired (transmitted from a source outside the body), hereditary (due to a genetic mutation that is inherited from a parent), and sporadic (without known risk factors for the disease). Creutzfeldt-Jakob disease, fatal familial insomnia, and Gerstmann-Sträussler-Scheinker syndrome occur worldwide. These diseases are rare, with an overall incidence of human prion disease of one case per million. Creutzfeldt-Jakob disease accounts for the majority of cases; fatal familial insomnia and Gerstmann-Sträussler-Scheinker syndrome are extremely rare. Kuru, appearing only in a society with cannibalistic practices, is an acquired TSE but has been eradicated since the outlawing of cannibalism (see "Kuru," below).

The TSEs are unique among the infectious diseases in that the causative agent consists of an abnormal protein that replicates without nucleic acid. Indeed, the term *prion* derives from "proteinaceous infectious particle." Proteins are long chains of amino acids that fold into a unique shape; the **protein conformation** (folded shape) is an important determinant of protein function. Prion protein (PrP^C) is normal cell-surface glycoprotein that is commonly found in the cell membrane of neurons. Its role in neuronal function is not fully understood.

Human PrP^C is 253 amino acids long; it is encoded by a single gene known as *PRNP* that is located on chromosome 20.

Normally, PrP^C is stable in shape and does not cause disease. The protein conformation, however, can flip into a "misfolded" shape. The misfolded prion, PrP^{Sc}, is toxic to the body and causes disease. It is also infectious; it binds to PrP^C molecules and induces them to change conformation and convert into PrP^{Sc}, producing a chain reaction that propagates the disease within the brain. The misfolded PrP^{Sc} is insoluble and aggregates (clumps together) into large assemblies, accumulating during disease progression. Thus, the abnormal prions replicate without use of nucleic acids (DNA or RNA), and unlike all other types of infectious agents, they are resistant to sterilization by heat and ultraviolet radiation.

In sporadic forms of prion disease, the transformation of PrP^C into PrP^{Sc} occurs spontaneously due to an error in the cell's machinery that makes proteins and controls their quality. Sporadic cases of prion disease arise in middle or old age, presumably because the cumulative likelihood that PrP^C can spontaneously flip to PrP^{Sc} increases with age. In hereditary forms of prion disease, infectious prions arise from *PRNP* gene mutations. The mutated *PRNP* gene creates PrP^{Sc} molecules, which bind to PrP^C molecules and induce them to convert into PrP^{Sc}. In acquired forms of prion disease, the PrP^{Sc} comes from outside the body, binds to PrP^C molecules of the host, and converts them into PrP^{Sc}.

The prion disorders are characterized by long incubation periods and short clinical duration before death occurs. This implies that the abnormal prions accumulate for many years before symptom onset. Once symptoms begin, however, the disorder rapidly worsens.

Kuru

Kuru was the first identified neurodegenerative disease due to an infectious agent. The term *kuru* means "to shiver" or "tremble." The symptoms of kuru include muscle twitching, other involuntary movements, loss of coordination, behavioral changes, and dementia. It is usually fatal within 1 year of symptom onset.

Kuru reached epidemic proportions in the 1950s and 1960s among the Fore people, an isolated population in the highlands of Papua New Guinea. When researchers came into contact with the Fore in the 1950s, they found that among a tribe of about 11,000, up to 200 people per year (primarily women and children) were dying of an inexplicable illness. Research quickly ruled out toxins and genetic causes. **Shirley Lindenbaum** (1932–2016), a medical anthropologist from the City University of New York, came to suspect that the disease was related to the tribe's practice of funerary endocannibalism, in which the women ate the cooked brains of dead relatives and sometimes fed them to children. The Fore believed that it was better that the body of a relative was eaten by people who loved the deceased, rather than by worms and insects, and that women's bodies were better able to house and tame the spirit of the dead body.

Lindenbaum's hypothesis was tested by a research group at the National Institutes of Health. They injected infected human brain tissue into chimpanzees, and several months later the monkeys developed symptoms of kuru, indicating that the disease is acquired. Kuru has a long incubation period in humans (average 10–13 years, up to several decades), and since the prohibition of endocannibalism in the 1950s, cases of kuru continued to occur for years but eventually decreased. During 2003–2008 only two kuru-related deaths were reported, indicating that the disease has been nearly or completely eradicated.

Creutzfeldt-Jakob Disease

Creutzfeldt-Jakob disease (CJD) is a rapidly progressive neurodegenerative disease characterized by a rapidly progressive dementia, myoclonic jerking, and ataxia. Average age of onset is 60 years. It is invariably fatal; about 70% of affected individuals die within 1 year of symptom onset. Creutzfeldt-Jakob disease is the most common of the human TSEs. It occurs in sporadic, hereditary, and acquired forms, in decreasing order of frequency. **Sporadic CJD** accounts for approximately 85% of cases in the United States, while **genetic CJD** accounts for 10%–15% of cases. Acquired forms of CJD are very rare now.

There are two forms of acquired CJD: iatrogenic CJD and variant CJD. **Iatrogenic CJD** is transmitted by exposure to nervous system tissue or CSF through medical procedures such as dura mater grafts, transplanted corneas, brain implantation of inadequately sterilized electrodes, and injections of contaminated pituitary growth hormone derived from human cadaver pituitary glands. Since 1985, all human growth hormone used in the United States has been synthesized by recombinant DNA procedures to eliminate the risk of transmitting CJD. Both brain biopsy and autopsy pose a small risk of accidental infection for those handling brain tissue. **Variant CJD**, first described in 1996 in the United Kingdom, is acquired by eating meat from cattle affected by bovine spongiform encephalopathy; it is also known as **mad cow disease**. There have been several hundred cases, mostly in the United Kingdom.

Diagnostic testing for CJD includes brain MRI, EEG, and CSF fluid analysis. MRI shows abnormalities in the caudate/putamen or at least two cortical regions. EEG shows disease-typical periodic sharp wave complexes in the majority of patients. CSF testing for the biomarker 14-3-3 protein may help support the diagnosis; it is a normal finding in CSF but may be elevated in prion diseases. Definitive diagnosis of CJD requires neuropathologic and/or immunodiagnostic testing of brain tissue obtained by biopsy or autopsy. Biopsy, however, is usually not performed in suspected cases of CJD unless it is needed to rule out a treatable disorder.

Fatal Familial Insomnia

Fatal familial insomnia (FFI) is a rare prion disease that is primarily genetic, but also may occur sporadically. It is caused by an abnormal variant of the *PRNP* gene and is inherited in an autosomal dominant manner. There are a very small number (less than 100) of sporadic cases without *PRNP* gene mutation, in which case the disorder is referred to as **sporadic fatal insomnia** (SFI).

Fatal familial insomnia is characterized by progressively worsening insomnia and rapidly progressive dementia. Insomnia is usually the initial sign, typically beginning suddenly and worsening rapidly. When sleep is achieved, there may be vivid dreams. Some individuals, however, present with changes in cognition and behavior as the earliest sign of the disease. Disease onset usually occurs in mid-life and progresses to coma and death, usually within 12–18 months of symptom onset. Prior to coma, there is often total inability to sleep.

Additional neurological signs emerge as the disease progresses, such as loss of coordination (ataxia), muscle twitches and jerks (myoclonus), abnormal jerky eye movements (nystagmus), tremor, double vision (diplopia), problems swallowing (dysphagia), slurred speech (dysarthria), hallucinations, and delirium. There may be symptoms of autonomic dysfunction, including disturbances of body temperature regulation (hypothermia, hyperthermia), rapid heart rate (tachycardia), high blood pressure (hypertension), episodes of hyperventilation, excessive sweating (hyperhidrosis), excessive salivation (sialorrhea), excessive tear production (hyperlacrimation), loss of appetite, and weight loss.

In FFI, aggregation of misfolded prion molecules and neurodegeneration primarily affect the thalamus, a diencephalic structure that plays an important role in a wide variety of neurological functions, including the sleep-wake cycle.

Gerstmann-Sträussler-Scheinker Syndrome

Gerstmann-Sträussler-Scheinker syndrome (GSS) is an extremely rare prion disease that is almost always inherited and is found in only a few families around the world. Aggregation of misfolded prion molecules and neurodegeneration primarily affects the cerebellum. Signs and symptoms generally develop between ages 35 and 50. Typical early symptoms include ataxia (resulting in gait change and other problems with coordination) and balance problems. As the disease progresses, there may be dysarthria, nystagmus, dysphagia, hearing problems, and visual disturbance. Most affected individuals develop dementia.

Human Immunodeficiency Virus

Human immunodeficiency virus (HIV) crosses the blood-brain barrier and enters the CNS early after systemic infection. Initially, HIV infection of the brain is usually asymptomatic, although it may result in acute aseptic meningitis or encephalitis. Eventually, however, it is associated with pathologic changes, predominantly within the subcortical white matter and basal ganglia.

HIV infection can lead to acquired immunodeficiency syndrome (AIDS) and development of opportunistic infections. Common opportunistic infections affecting the brain in those with AIDS include cryptococcal meningitis, toxoplasmosis, neurotuberculosis, cytomegalovirus encephalitis, brain lymphomas, and progressive multifocal leukoencephalopathy caused by the JC virus (see Chapter 23, "White Matter Disease").

The HIV virus itself, however, enters the CNS and can produce prominent changes in white matter and subcortical gray matter (basal ganglia, thalamus, brainstem) and result in **HIV-associated neurocognitive disorder**, a classification that excludes neurologic morbidity due to opportunistic CNS infections resulting from immunodeficiency. HIV-associated neurocognitive disorder encompasses a range of disorders from asymptomatic neurocognitive impairment (i.e., substandard performance on neurocognitive testing without subjective symptoms), to mild neurocognitive disorder, to more severe **HIV-associated dementia**, also called **AIDS dementia complex** or **HIV encephalopathy**. HIV-associated dementia is a subcortical dementia, characterized by prominent disturbances in attention and concentration and motor slowing; it is associated with pathologic changes in the brain white matter (leukoencephalopathy).

Prior to the advent of combined antiretroviral therapy, dementia was a common source of morbidity in HIV-infected persons due to CNS opportunistic infections. In the current era in high-resource countries, mild neurocognitive disability accounts for the majority of neurologic morbidity in clinically stable patients. In low-resource countries where combined antiretroviral therapy is not available, opportunistic infections of the CNS continue to account for the majority of AIDS-related neurologic morbidity and mortality.

Lyme Disease

Lyme disease is a tick-borne infection caused by the *Borrelia burgdorferi* spirochete (spiral-shaped bacteria). It is a multisystem disease that was first recognized in 1975 in a cluster of cases occurring in Lyme, Connecticut. Acutely, Lyme disease is characterized by skin rash (a single, enlarging, ring-like erythema) at the site of a tick bite, followed by fatigue and influenza-like symptoms (myalgia, arthralgia, and headache). If untreated, the disease may progress to affect the CNS weeks to months later, producing aseptic meningitis or a fluctuating meningoencephalitis with cranial or peripheral neuritis, well after the skin lesion and systemic symptoms have resolved. In the first stage of Lyme disease, treatment is by oral antibiotic. Once the meninges or nervous system are involved, treatment is by intravenous antibiotic.

Some individuals report persistent symptoms after antibiotic treatment, including vague cognitive symptoms without a prior episode of meningitis or other manifestation of CNS infection. Studies objectively examining neurocognitive function by psychometric testing in those with subjective cognitive symptoms attributed to chronic Lyme disease, however, have yielded variable results and are inconclusive.

Summary

Central nervous system infections are classified by the location of infection and the offending pathogen. The three most common types of CNS infection based on location are meningitis, encephalitis, and brain abscess; CNS infections are caused by viruses, bacteria, fungi, parasites, or prions.

Meningitis, an inflammation of the meninges, is the most common form of CNS infection. Viral meningitis, the most common type, has a relatively benign course, while bacterial meningitis is potentially life-threatening. Encephalitis is inflammation of the brain parenchyma. Examples include herpes simplex virus encephalitis, the most prevalent encephalitis in the Western world, and rabies encephalitis which is far more common in low-resource countries. An abscess is a localized collection of pus in an enclosed tissue space. Intracranial abscesses are classified as either brain abscesses located within the brain parenchyma or subdural or epidural empyemas.

Fungal brain infections are much less common than bacterial or viral infections, and they occur most frequently as opportunistic infections in those with compromised immune systems.

Prion diseases, also known as transmissible spongiform encephalopathies, are a group of rapidly progressive, fatal neurodegenerative disorders characterized by spongiform changes in brain tissue. They have long incubation periods and short clinical duration before death occurs. The human prion diseases are Creutzfeldt-Jakob disease, fatal familial insomnia, Gerstmann-Sträussler-Scheinker syndrome, and kuru.

Additional Reading

1. Antinori A, Arendt G, Becker JT, Brew BJ, Byrd DA, Cherner M, et al. Updated research nosology for HIV-associated neurocognitive disorders. *Neurology*. 2007;69(18):1789–1799. doi:10.1212/01.WNL.0000287431.88658.8b
2. Baldwin KJ, Correll CM. Prion disease. *Semin Neurol*. 2019;39(4):428–439. doi:10.1055/s-0039-1687841
3. Bradshaw MJ, Venkatesan A. Herpes simplex virus-1 encephalitis in adults: pathophysiology, diagnosis, and management. *Neurotherapeutics*. 2016;13(3):493–508. doi:10.1007/s13311-016-0433-7
4. Brandel JP, Knight R. Variant Creutzfeldt-Jakob disease. *Handb Clin Neurol*. 2018;153:191–205. doi:10.1016/B978-0-444-63945-5.00011-8
5. Brouwer MC, Tunkel AR, McKhann GM 2nd, van de Beek D. Brain abscess. *N Engl J Med*. 2014;371(5):447–456. doi:10.1056/NEJMra1301635
6. Collins S, McLean CA, Masters CL. Gerstmann-Sträussler-Scheinker syndrome, fatal familial insomnia, and kuru: a review of these less common human transmissible spongiform encephalopathies. *J Clin Neurosci*. 2001;8(5):387–397.
7. Cracco L, Appleby BS, Gambetti P. Fatal familial insomnia and sporadic fatal insomnia. *Handb Clin Neurol*. 2018;153:271–299. doi:10.1016/B978-0-444-63945-5.00015-5
8. Del Brutto OH, García HH. Taenia solium cysticercosis: the lessons of history. *J Neurol Sci*. 2015;359(1–2):392–395. doi:10.1016/j.jns.2015.08.011
9. Ellis RJ, Calero P, Stockin MD. HIV infection and the central nervous system: a primer. *Neuropsychol Rev*. 2009;19(2):144–151. doi:10.1007/s11065-009-9094-1
10. Geschwind MD. Prion diseases. *Continuum (Minneap Minn)*. 2015;21(6 Neuroinfectious Disease):1612–1638. doi:10.1212/CON.0000000000000251
11. Garcia HH. Neurocysticercosis. *Neurol Clin*. 2018;36(4):851–864. doi:10.1016/j.ncl.2018.07.003
12. Halperin JJ. Neuroborreliosis. *Neurol Clin*. 2018;36(4):821–830. doi:10.1016/j.ncl.2018.06.006
13. Ironside JW, Ritchie DL, Head MW. Prion diseases. *Handb Clin Neurol*. 2017;145:393–403. doi:10.1016/B978-0-12-802395-2.00028-6
14. Jackson AC. Diabolical effects of rabies encephalitis. *J Neurovirol*. 2016;22(1):8–13. doi:10.1007/s13365-015-0351-1
15. Kobayashi A, Kitamoto T, Mizusawa H. Iatrogenic Creutzfeldt-Jakob disease. *Handb Clin Neurol*. 2018;153:207–218. doi:10.1016/B978-0-444-63945-5.00012-X
16. Liberski PP, Gajos A, Sikorska B, Lindenbaum S. Kuru, the first human prion disease. *Viruses*. 2019;11(3):232. Published 2019 Mar 7. doi:10.3390/v11030232
17. Montagna P. Fatal familial insomnia and the role of the thalamus in sleep regulation. *Handb Clin Neurol*. 2011;99:981–996. doi:10.1016/B978-0-444-52007-4.00018-7

18. Rabinstein AA. Herpes virus encephalitis in adults: current knowledge and old myths. *Neurol Clin.* 2017;35(4):695–705. doi:10.1016/j.ncl.2017.06.006
19. Whitley RJ. Herpes simplex virus infections of the central nervous system. *Continuum (Minneap Minn).* 2015;21(6 Neuroinfectious Disease):1704–1713. doi:10.1212/CON.0000000000000243
20. Yoshimura M, Yuan JH, Higashi K, Yoshimura A, Arata H, Okubo R, et al. Correlation between clinical and radiologic features of patients with Gerstmann-Sträussler-Scheinker syndrome (Pro102Leu). *J Neurol Sci.* 2018;391:15–21. doi:10.1016/j.jns.2018.05.012

White Matter Disease

Introduction

Perception, cognition, emotion, and behavior are mediated by neural circuits consisting of multiple gray matter nodes (cortical zones or subcortical nuclei) and white matter tracts connecting those nodes. The gray matter nodes serve local information processing, while white matter serves information transfer. White matter allows for faster communication by enhancing the speed of electrical signaling within neurons, and abnormalities of myelination diminish the efficiency and efficacy of information transfer. White matter, therefore, is critical to optimal brain function. Lesions in white matter connections can undermine the function of distributed neural networks, even when their gray matter components remain intact. This chapter introduces the reader to the basics of white matter and the white matter diseases.

White Matter

Macroscopically, the central nervous system (CNS) is made up two types of tissue, gray matter and white matter. Gray matter consists of aggregates of neuron cell bodies, dendrites, and astrocytes. White matter consists of collections of myelinated axons and is devoid of neuronal cell bodies. The word *myelin* derives from the Greek word for marrow (*myelos*) due to the abundance of white matter within the core of the brain. Myelin has a white appearance due to its high lipid content.

In the brain, gray matter is located on the outer surfaces of the cerebrum and cerebellum, making up the cerebral and cerebellar cortices; white matter is located in the deeper regions of tissue, with aggregates of gray matter nuclei embedded within the white matter. White matter accounts for about 50% of the total brain volume. Because the brain white matter is located beneath the cerebral cortex, it is also referred to as **subcortical white matter**. The white matter immediately adjacent to the lateral ventricles is often referred to as the **periventricular white matter**. In the spinal cord, gray matter is located centrally and is surrounded by white matter.

White matter makes up the projection, commissural, and association fiber tracts of the cerebral cortex and other brain regions. With respect to cerebral cortex, the projection fibers form the afferents and efferents between cerebral cortex and subcortical structures, the commissural fibers (i.e., corpus callosum, anterior commissure) interconnect the two cerebral hemispheres, and the association fibers form both short-range interconnections between adjacent gyri (U-fibers) and long-range interconnections between lobes (see Chapter 13, "The Cerebral Cortex").

The long association fibers generally interconnect cortical regions in different lobes within the same hemisphere (see Figure 13.12 in Chapter 13). These long association fiber bundles include the superior longitudinal fasciculus (which includes the arcuate fasciculus), inferior longitudinal fasciculus, uncinate fasciculus, and cingulum. Each of these long association tracts have a single terminus in the frontal lobe. No other lobe has such rich inter-lobe connectivity. This pattern of connectivity forms the basis for the frontal lobes' function as an executive in the hierarchy of cortical function; other regions of cerebral cortex report to and take direction from the frontal lobes.

White matter is also found in other CNS regions, such as within the cortical mantle (i.e., the bands of Baillarger and line of Gennari; see Chapter 13, "The Cerebral Cortex"), white matter fascicles (bundle of fibers) coursing through subcortical structures such as the thalamus and basal ganglia, and hippocampal afferents and efferents (i.e., alveus, fimbria, fornix, and hippocampal commissure).

Phylogenetically, myelin is a recent development; it is present almost exclusively in vertebrates. Ontogenetically, myelination is crucial to the functional maturation of the nervous system. In species in which birth occurs prior to myelination, the newborn is motorically limited and dependent until myelination is complete. In humans, myelination is a crucial factor in behavioral and cognitive development from infancy to young adulthood. White matter volume continues to gradually increase, peaking at around 50 years of age. In normal aging, white matter volume decreases after age 60, progressing along an anterior-posterior gradient, with the frontal-temporal regions and anterior corpus callosum especially affected. The physiological basis for the age-related decrease in white matter volume and anatomical pattern is unknown. Other age-related white matter changes include disrupted microstructural integrity and white matter lesions; these changes are associated with age-related changes in cognitive function.

Myelin is provided by **Schwann cells** in the peripheral nervous system (PNS) and **oligodendrocytes** in the CNS. The entirety of a single Schwann cell wraps concentrically around an axon, providing a single segment of myelin to a single axon (Figure 23.1). By contrast, each oligodendrocyte has multiple processes extending from its cell body that each form a myelin segment. Each oligodendrocyte provides multiple myelin segments for multiple axons and may provide up to 60 neighboring axons with myelin. Myelin segments are about 1 mm in length. They are also known as **internodes** because they are separated from each other along the axon length by **nodes of Ranvier**. Myelination permits action potentials to propagate down the length of the axon by **saltatory conduction**, which markedly accelerates axonal conduction by up to 100 times (see Chapter 2, "Electrical Signaling in Neurons," and Figure 2.5 in that chapter).

A variety of disorders feature prominent or exclusive white matter pathology, collectively referred to as the **white matter diseases**. These diseases involve defective formation or maintenance of the myelin sheath, thereby disrupting nerve impulse conduction by slowing or blocking signal transmission.

The white matter diseases are classified as either dysmyelinating or demyelinating. **Dysmyelination disorders** involve a failure of normal myelin formation; most are caused by genetic metabolic defects and typically are present during infancy. **Demyelination disorders** involve degeneration of myelin sheath that has formed normally. These disorders typically present in older children and adults following a period of normal development. In demyelination disorders, myelin degradation is primary; demyelinated axons may eventually degenerate, but the neuronal degeneration is secondary to myelin degeneration. By contrast, disorders in which axonal degeneration occurs first (i.e., is primary) and is followed by myelin degeneration are not considered demyelination disorders. Demyelination disorders are further categorized as affecting PNS Schwann cells or CNS oligodendrocytes. The white matter diseases are also classified on the basis of etiology. Primary demyelinating diseases are due to inflammatory autoimmune mechanisms; secondary etiologies of demyelination include toxins, infectious diseases, acquired metabolic disturbances and nutritional deficiencies, and hypoxic-ischemic injury.

White matter diseases affecting the brain are also known as the **leukoencephalopathies**. The nature and severity of clinical manifestations depend on the location and degree of white matter pathology. Focal CNS white matter lesions result in discrete neurological and neurobehavioral signs and symptoms, such as amnesia, specific aphasia syndromes, alexia, hemi-spatial neglect, and specific apraxia syndromes. Diffuse white matter lesions result in more global deficits, including dementia.

Our understanding of white matter diseases was revolutionized by magnetic resonance imaging (MRI). **White matter lesions** are well visualized by T2-weighted, fluid attenuated inversion recovery (FLAIR), and diffusion-weighted imaging MRI sequences (see Chapter 29, "Brain Imaging"). Lesions are revealed as **white matter**

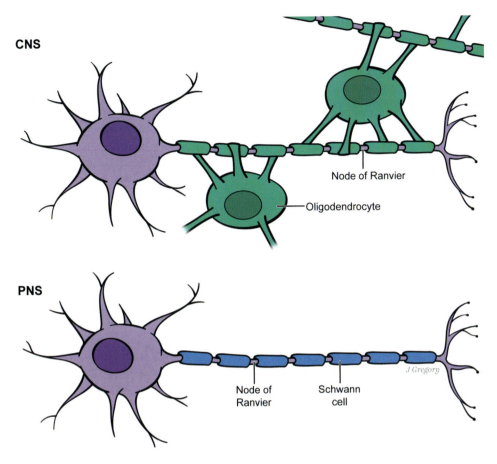

FIGURE 23.1. In the CNS, each oligodendrocyte has multiple processes extending from its cell body that each form a myelin segment. In the PNS, the entirety of a single Schwann cell wraps concentrically around an axon, providing a single segment of myelin to a single axon.

hyperintensities, which are areas of abnormally increased brightness that indicate a change in tissue composition. **Diffusion tensor imaging** (DTI) is used to map the brain white matter tracts (**tractography**).

The term **subcortical dementia** refers to dementia that occurs with a variety of diseases that primarily affect subcortical brain structures (e.g., Huntington's disease, Parkinson's disease, cerebral white matter disease). By contrast to cortical dementias which prominently feature amnesia, aphasia, apraxia, and/or agnosia, subcortical dementias are typified by slowed thinking (**bradyphrenia**), impaired memory characterized by poor recall but preserved recognition, executive dysfunction, and changes of personality and emotion (apathy, inertia, depression). The concept of two broad classes of dementia caused by predominant cortical or subcortical pathology is a useful guide for classifying dementias, although the dichotomy is not strict. Most subcortical dementias show cortical atrophy in later stages, and cortical dementias have subcortical pathology at some point.

Subcortical dementia is also referred to as **fronto-subcortical dementia** and **frontal systems dementia** based on its cognitive and behavioral similarity to frontal lobe disease, and the fact that the lesions producing the syndrome involve subcortical structures (rostral brainstem, thalamus, and/or basal ganglia) and the projections from subcortical structures to the frontal lobes. The term **white matter dementia** has been proposed as a form of subcortical dementia that is specifically due to white matter pathology and distinct from dementias that are due to neurodegeneration of subcortical nuclear structures (gray matter) and that are accompanied by movement disorders.

The Leukodystrophies

The leukodystrophies are a heterogeneous group of **genetic metabolic diseases** (**inborn errors of metabolism**) that usually result in dysmyelination. **Metabolism** refers to the entire range of life-sustaining chemical reactions within living organisms, both **anabolic** (building complex molecules from simpler ones) and **catabolic** (breaking down large molecules into smaller ones). The latter reactions are often **catalyzed** (initiated or sped up) by **enzymes**. Metabolism converts food to cellular energy, provides building blocks for other molecules (e.g., proteins, lipids, nucleic acids), and breaks down chemical reaction waste products. It is the sum of processes necessary for the normal formation, maintenance, and function of the body tissues.

There are more than 50 different leukodystrophies. Each is caused by a specific gene defect that results in a specific enzyme defect, which results in an error in myelin metabolism, which disrupts the establishment or maintenance of myelin. Most leukodystrophies are **autosomal recessive disorders**; two copies of the gene mutation (one from each parent who is a carrier of the gene) are required for disease expression.

The leukodystrophies usually manifest initially in infants and children due to a failure of normal myelination (dysmyelination), but they sometimes manifest initially in teens or young adults, in which case they result in demyelination. They produce a wide variety of neurological disturbances, are progressive, and have a dismal prognosis. This group of rare genetic diseases includes metachromatic leukodystrophy, adrenoleukodystrophy, and vanishing white matter disease. **Metachromatic leukodystrophy** is the most common. MRI shows diffuse, symmetric cerebral white matter abnormality. **Adrenoleukodystrophy** is an X-linked disease of males characterized by both neurologic dysfunction and adrenal insufficiency. MRI shows symmetric white matter abnormality that is most prominent in parieto-occipital regions. **Vanishing white matter disease** is an autosomal recessive disorder that most often affects children, but in some cases, it presents as an early-onset dementia in adults. Longitudinal MRI studies show a pattern of progressive disappearance of white matter.

Autoimmune Inflammatory Demyelinating Diseases

In autoimmune diseases the body's immune system attacks its own tissues. The immune system mistakenly recognizes body cells as foreign and launches an immune response against them. The main autoimmune inflammatory demyelinating diseases are multiple sclerosis (affecting the CNS) and Guillain-Barré syndrome (affecting the PNS).

Multiple Sclerosis

Multiple sclerosis (MS) is the most common CNS demyelinating disease and the most common nontraumatic disabling neurologic disease of young adults. It is an autoimmune disorder in which immune system T-cells attack myelin-forming oligodendrocytes.

The pathologic hallmark of MS is multiple focal areas of myelin loss and gliotic scarring called **sclerotic plaques**. The plaques are irregularly shaped and sharply demarcated (Figure 23.2). Demyelination results in a loss of saltatory conduction along axons. This slows signal transmission but may also block action potential conduction because the internodal axon membrane, which normally is covered in myelin, has few ion channels for propagating action potentials (see Chapter 2, "Electrical Signaling in Neurons").

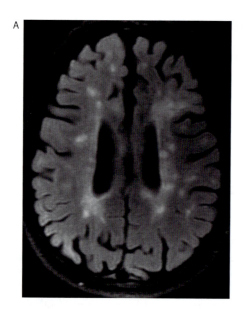
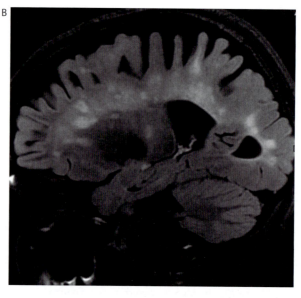

FIGURE 23.2. Multiple sclerosis plaques. (A) Axial and (B) sagittal FLAIR MRI images of the brain showing bilateral hyperintense lesions in a periventricular distribution.

Although myelin is preferentially affected, there may also be significant axonal loss.

Plaques have a predilection for the periventricular white matter (particularly along the lateral aspects of the atria and occipital horns of the lateral ventricles), optic nerves and chiasm, and spinal cord (particularly the corticospinal tracts and dorsal columns). Any CNS structure that contains myelin, however, can harbor MS plaques. Within the cerebrum, MS plaques may affect the projection, commissural, and association fibers of the cerebral hemispheres. Plaques may also be found in CNS gray matter regions such as cerebral cortex, deep nuclei, and brainstem, involving the myelinated axons in these areas without affecting neuronal cell bodies.

Histologically, several processes drive the formation of plaques, including inflammation, oligodendrocyte injury and myelin breakdown, gliosis, axonal degeneration, and remyelination. MS plaques evolve over time from the acute to chronic phases, and are classified as active, smoldering, inactive, and shadow. **Active plaques** occur during the acute phase, causing new or worsening neurological signs and symptoms; they are characterized by immune cells (lymphocytes, microglia, and macrophages) that destroy myelin, and myelin breakdown products. **Smoldering plaques** are characterized by an inactive center surrounded by a rim of activated microglia with little myelin degradation products. With time, glial scars form (**gliosis**), and after a few weeks the plaques reach a burned-out **inactive plaque** stage consisting of demyelinated axons traversing the glial scar tissue without active immune cells. The remaining oligodendrocytes attempt to make new myelin. In some lesions there is partial remyelination resulting in **shadow plaques**, but in other lesions remyelination is ineffective because gliosis creates a barrier between the myelin-producing cells and their axonal targets.

CLINICAL PRESENTATION

A wide variety of neurological deficits may be seen in MS. The clinical signs and symptoms often correspond to the location of the plaques. The plaques have a predilection for the optic nerves, causing **optic neuritis** (inflammation of the optic nerves) and visual loss in many of those with MS. Spinal cord lesions may cause **transverse myelitis**, inflammation across both sides of one spinal segment (level), causing paralysis and somatosensory loss. Other neurological signs include ataxia (incoordination), brainstem signs such as impaired extraocular movements and dysarthria (unclear speech), neuropsychiatric changes, and cognitive decline. Many MS signs and symptoms are attributable to spinal cord and brainstem involvement. Motor and somatosensory signs in the limbs derive primarily from spinal cord lesions. The lower extremities are affected more frequently than the upper extremities, likely because longer axons are more likely to be involved if lesions occur randomly throughout the neuraxis. Physical fatigue is one of the most common and debilitating complaints in those with MS.

Cognitive signs and symptoms are also common and range from very mild changes to severe disabling dementia. The amount of cerebral white matter affected (i.e., lesion area or **lesion burden**) is related to severity of cognitive impairment, although cerebral lesions may be clinically silent. Cognitive symptoms may be present even in the early stages of the disease, and while usually mild, may interfere with day-to-day activities.

The "classic" cognitive profile of MS that is described in the literature is consistent with fronto-subcortical dysfunction, with the most prominent deficits in processing speed and sustained attention, commonly manifesting as difficulty keeping up with conversations and work tasks. Motor slowing may occur early in the course of the disease, but the slowed information processing occurs independent of motor slowing. There may be attention problems, primarily involving working memory; memory deficit characterized by problems with retrieval with spared storage and recognition; and executive dysfunction characterized by reduced abstract thinking, planning, problem-solving, and multitasking abilities. There may be problems with verbal fluency and word-finding, but receptive language skills are generally spared. Focal neurobehavioral syndromes are rare. The literature on MS and cognition, however, was established prior to the introduction of disease-modifying therapies.

Depression and anxiety are common in those with MS. These are understandable emotional reactions to a chronic neurological disease diagnosis that typically is made during early adulthood, especially since the disease is unpredictable, has a fluctuating course, carries a risk of progressing over time to some level of physical disability, and may have a major impact on family and career. Demyelination and neurodegeneration may also result in emotional changes, as may some of the medications used to treat MS, especially corticosteroids.

DIAGNOSIS

Diagnosis of MS requires: (1) dissemination in time (two or more attacks of neurologic signs/symptoms, each lasting a minimum of 24 hours and separated by at least one month); (2) dissemination in space (multiple CNS lesions causing the attacks, as evidenced by neuroimaging); and (3) rule-out of other possible diagnoses.

Diagnosis of MS is based on a combination of clinical history, neurological examination, brain MRI, and lumbar puncture. MRI is the imaging procedure of choice for confirming MS and monitoring disease progression. Gadolinium-enhanced T1-weighted sequences best reveal active lesions, as contrast reveals inflammation. Inactive

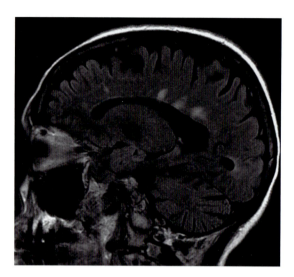

FIGURE 23.3. Dawson's fingers. Sagittal FLAIR image of the brain demonstrates periventricular MS plaques along the axis of the medullary veins, perpendicular to the lateral ventricle.

plaques (i.e., chronic lesions) are best revealed by T2-weighted and FLAIR MRI sequences. **Dawson's fingers**, elongated flame-shaped lesions perpendicular to the lateral ventricle wall, apparent on T2-weighted and FLAIR MRI, represent inflammatory activity surrounding venules (Figure 23.3). They are an important imaging marker in the differential diagnosis of MS, although they are also found in leukoencephalopathy due to cerebral small vessel disease.

Cerebrospinal fluid (CSF) analysis plays an important role in the diagnostic workup in many cases of suspected MS. The hallmark CSF finding in MS is **oligoclonal bands** which are composed of **immunoglobulin G** (IgG) antibodies. In testing, blood and CSF samples are subjected to electrophoresis, which separates proteins. The **IgG index** is a comparison between IgG levels in the CSF and in the blood serum. Elevated CSF IgG is an indicator of chronic immune activation in the CNS. The vast majority of MS patients have an elevated IgG index, however, oligoclonal bands are found in other CNS inflammatory diseases, and are not found in 5%–10% of those with MS. Therefore, CSF analysis by itself cannot confirm or exclude a diagnosis of MS.

The majority of people with MS are diagnosed between the ages of 20–40; the average age of diagnosis is 30. There is usually a delay of several years from time of onset to diagnosis; the transient and variable nature of symptoms makes diagnosis difficult, and other diagnoses need to be ruled out. Late-onset MS, with symptoms starting at or after age 50, is more likely to have a relatively rapid progression with faster accumulation of disability.

DISEASE COURSE

The course of MS varies widely, and four basic MS disease courses (phenotypes) have been defined: clinically isolated syndrome, relapsing remitting, secondary progressive, and primary progressive. **Clinically isolated syndrome** (CIS) is a diagnosis given when there is a single episode of neurologic symptoms (which must last for at least 24 hours) that is caused by inflammation and demyelination of the CNS; those with CIS are at greater risk of developing MS over time (which requires two or more episodes of neurologic signs/symptoms). In **relapsing-remitting MS** (RRMS), the patient experiences partial or total recovery (remission) after attacks (relapses). This is the most common presentation in the early years of disease. In **secondary progressive MS** (SPMS), an initial relapsing-remitting course eventually transitions into a progressive course of neurologic disability. In **primary progressive MS** (PPMS), symptoms and disability worsen continuously from the onset of the disease, without attacks or periods of remission; this is the least common MS subtype.

Radiologically isolated syndrome (RIS) is an incidental imaging finding that is suggestive of MS, without any of the typical physical symptoms of MS. Those with RIS are at greater risk of developing MS over time, and are monitored for potential evolution to MS.

Tumefactive MS is a rare variant of MS that mimics the clinical and radiological features of a neoplasm, presenting with a large intracranial lesion (greater than 2.0 cm in diameter) with mass effect, perilesional edema, and/or ring enhancement with gadolinium contrast on MRI. It often develops into RRMS.

TREATMENT

Treatment of MS has two main components: immunomodulatory therapy for treating acute relapses, and disease-modifying therapies to alter the overall progression of the disease. Specific symptoms such as fatigue, depression, and spasticity also may be treated pharmacologically.

EPIDEMIOLOGY

Multiple sclerosis affects women three times as often as men. Genetic susceptibility plays a role; monozygotic twins are 30% to 50% concordant for MS, while dizygotic twins are only 2.3% concordant. Incidence varies with geographic latitude, particularly for area of residence during childhood, with the highest rates between 45 and 65 degrees north and south of the equator (temperate climates) and the lowest rates near the equator (tropical climates). MS is also most common in northern Europe and the northeastern United States. The reason for these geographic differences is unknown, but the distribution suggests an environmental contribution to the disease. Genetics may partially explain the latitudinal gradient, but the data do not exclude exogenous variables.

Guillain-Barré Syndrome

Guillain-Barré syndrome (GBS) is a rare autoimmune disorder, but the most common peripheral demyelinating disease. It is primarily a demyelinating motor neuropathy that presents as a symmetrical pattern of progressive weakness, beginning in the feet and ascending to the proximal lower extremities and then the upper limbs. It often results in quadriplegia. If the cranial nerves are involved, it leads to facial weakness, dysphagia (trouble swallowing), dysarthria (trouble speaking), and difficulty holding up the head. Autonomic nerve involvement due to demyelination of the vagus and sympathetic nerves results in vital sign lability (i.e., respiratory failure, brady- or tachycardia, hypo- or hypertension, hypo- or hyperthermia). Since GBS does not affect the CNS, cognitive function remains intact.

Guillain-Barré syndrome is due to an autoimmune response triggered by bacterial or viral infection several weeks prior to the onset of weakness. Very rarely, it develops in the days or weeks after a vaccination. It usually lasts several weeks to several months but can last several years. Most recover fully, but some have permanent nerve damage.

Acute Disseminated Encephalomyelitis

Acute disseminated encephalomyelitis (ADEM) is an autoimmune inflammatory demyelinating condition that affects the white matter of the brain and spinal cord. The progression and severity of ADEM vary; some have mild forms of the disorder, while in the most severe cases there may be life-threatening complications such as respiratory failure. It usually presents as an acute-onset, rapidly progressive encephalopathy with polyfocal neurologic deficits that are based upon location of the CNS lesions.

Acute disseminated encephalomyelitis is typically seen in prepubertal children. In many cases a viral infection precedes the development of symptoms by two days to four weeks; however, some cases occur spontaneously without an identified preceding event. The precise etiological mechanism is unknown. While symptoms can be severe, they can be treated with immune-modulating agents such as intravenous immunoglobulin. Outcome is variable and ranges from full recovery to death. Many patients experience persistent cognitive and behavioral deficits, particularly attentional and executive dysfunction.

Toxic Leukoencephalopathy

Toxic leukoencephalopathy is a disorder in which diffuse brain white matter pathology is caused by exposure to agents that produce clinically significant damage to the brain white matter. A wide variety of agents are **leukotoxic** (toxic to white matter), including some drugs of abuse, environmental toxins, cranial irradiation therapy, and therapeutic drugs. The pathophysiology of leukotoxic injury is poorly understood, but a key factor is that leukotoxins are lipophilic, which enables toxins to enter the lipid-rich brain; more specific mechanisms of injury remain unknown.

Toxic leukoencephalopathy was first described in the context of toluene abuse, and some of the principles learned from studying toluene leukoencephalopathy generalize to other leukotoxins. The level of toxin exposure generally predicts the extent of diffuse white matter injury on MRI and severity of neurological dysfunction. Myelin damage may be transient, which portends a good outcome; myelin damage accompanied by axonal damage is associated with worse outcome.

The neurobehavioral consequences of toxic leukoencephalopathy range from mild, transient cognitive dysfunction to stupor, coma, and death. The major clinical focus involves the diagnosis and treatment of acute toxic syndromes and neurologic emergencies with potentially fatal outcomes.

Drugs of Abuse and Environmental Toxins

TOLUENE LEUKOENCEPHALOPATHY

Toluene, also known as **methylbenzene**, is a volatile solvent and a common ingredient in products such as paint, lacquer, glue, gasoline, and other industrial and household products. **Inhalant abuse** is the deliberate inhalation of a volatile substance to achieve an altered mental state; it is also known as volatile substance abuse, solvent abuse, sniffing, huffing, and bagging. Spray paint is particularly popular among inhalant abusers. Compared to other forms of substance abuse, inhalant abuse disproportionately afflicts adolescents, low-income and unemployed adults, people living in isolated rural or reservation settings, and people housed in institutions (psychiatric hospitals, prisons, and residential treatment centers). Low monetary cost and ready availability contribute to these demographic patterns.

Inhalant abuse rapidly produces a state of euphoria, which may be followed by hallucinations, slurred speech and disturbed gait, dizziness, and drowsiness or sleep within seconds to minutes. Chronic abuse leads to irreversible neurological effects such as dementia, ataxia, cranial nerve signs, and corticospinal dysfunction.

Toluene dementia is associated with MRI findings of cerebral and cerebellar white matter pathology. The extent of leukoencephalopathy and degree of cognitive impairment are correlated. The dementia is characterized by prominent deficits in processing speed, sustained attention, memory retrieval, and executive function. Autopsy studies have shown widespread myelin loss without significant gray matter pathology, leading to the conclusion that selective toxic injury to white matter can produce the dementia.

The first description of toxic leukoencephalopathy was in toluene inhalant abusers, and toluene leukoencephalopathy remains the best example of solvent-induced neurobehavioral dysfunction. Other volatile solvents, however, also produce white matter damage and dementia. The impact of chronic, low-level occupational exposure to toluene and other solvents, however, is unclear and remains controversial since the first description of "**chronic painters' syndrome**" in the 1970s.

HEROIN LEUKOENCEPHALOPATHY

Inhalation of heroin vapor, known on the street as "chasing the dragon," has increased since the 1980s because it avoids intravenous administration risks such as HIV and hepatitis infection; however, it can lead to leukoencephalopathy. MRI shows symmetric white matter lesions that especially affect infratentorial structures, the posterior cerebral white matter, and posterior limb of the internal capsule. Autopsy examination shows a **spongiform leukoencephalopathy** (vacuolar degeneration of the white matter). The clinical course is varied, but neurologic complications include dementia, ataxia, and akinetic mutism; approximately 25% of cases have resulted in rapid neurologic deterioration and death. The mechanism of neurologic injury related to heroin inhalation is unknown; an adulterant or the method of drug administration may play a role.

MARCHIAFAVA–BIGNAMI DISEASE

Marchiafava–Bignami disease (MBD) is a rare dementia syndrome that most often occurs with chronic alcoholism. It is characterized by demyelination of the corpus callosum and other white matter tracts. It typically begins in the body of the corpus callosum, followed by the genu, and finally the splenium. It typically affects the central layers of the corpus callosum with relative sparing of the dorsal and ventral extremes of the corpus callosum, which may be seen as the **sandwich sign** on sagittal MRI (with T2 and FLAIR hyperintensities in the central layer with sparing of dorsal and ventral layers). Other white matter regions such as the anterior and posterior commissures, the middle cerebellar peduncle, the corticospinal tracts, and the centrum semiovale are variably involved; the subcortical U-fibers tend to be spared. It is unknown why the corpus callosum is so affected. The initial loss of myelin eventually gives rise to axonal degeneration.

The clinical presentation of MBD is variable. Acute MBD is characterized by mental confusion, disorientation, seizures, and muscle rigidity; most patients progress to coma and eventually die. Subacute MBD is characterized by dementia, dysarthria, and muscle hypertonia; these patients may survive for years. Chronic MBD is characterized by dementia. Diagnosis is based on clinical findings in combination with neuroimaging features.

Most MBD patients are male between 40 and 60 years of age and have a history of chronic alcoholism and malnutrition. The etiology is attributed to a deficiency of all eight B vitamins, which play important roles in nervous system function; vitamin B_{12} is especially important in the synthesis and maintenance of oligodendrocytes and myelin. Administration of vitamin B complex results in improvement in many patients, although some do not recover.

Radiation-Induced Toxic Leukoencephalopathy

Three types of radiation injury can occur in the brain following whole brain radiation therapy for brain tumors, all primarily affecting the cerebral white matter. In temporal order and degree of severity, these are: acute, early delayed, and late delayed radiation encephalopathy.

Acute radiation encephalopathy occurs within the first days to weeks after beginning treatment. It is characterized by headaches, nausea, and worsening of preexisting neurologic deficits. It is a relatively mild syndrome and typically is self-limited. It is believed to result from cerebral white matter edema.

Early delayed radiation encephalopathy develops within several weeks to 6 months after radiation therapy. It is characterized by lethargy, somnolence, and resurgence of neurological signs, and is therefore also referred to as **somnolence syndrome**. White matter hyperintensities are observed on MRI, and the syndrome is ascribed to cerebral demyelination. These white matter changes are transient and there is a slow clinical recovery.

Late delayed radiation encephalopathy emerges months to years after treatment. It is defined radiologically by diffuse deep and periventricular white matter demyelination on MRI that is progressive, without focal lesions. The slow turnover of oligodendrocytes likely accounts for the latency and progressive demyelination. Risk for developing late effects increases with the radiation dose. Clinically, late delayed radiation encephalopathy is characterized by progressive dementia, often with a fatal outcome. The profile is characterized by prominent deficits in processing speed, attention, and executive function.

The use of whole brain radiation therapy has declined since advances in radiation technology now allow for a more localized delivery of radiation. Focal radiation therapy reduces tissue damage, but does produce focal areas of demyelination.

Chemotherapy-Induced Toxic Leukoencephalopathy

Chemotherapy for non-CNS cancers may produce leukoencephalopathy that is neuroradiologically and neuropathologically similar to radiation leukoencephalopathy. **Chemotherapy-induced toxic leukoencephalopathy** is clinically similar to radiation leukoencephalopathy, presenting with lassitude, drowsiness, confusion, memory loss, and dementia. Higher doses are associated with more extensive white matter injury and more severe neurologic complications. The incidence of chemotherapy-induced toxic leukoencephalopathy is increasing as stronger drugs come into use and survival times increase, allowing time for delayed treatment effects to appear.

The first antineoplastic drug recognized to produce leukoencephalopathy was **methotrexate** (administered intravenously or intrathecally), but it is now appreciated that a wide variety of other chemotherapy drugs may have leukotoxic effects as well. Chemotherapy is often administered with radiation, and the toxic effects of combined radiotherapy and chemotherapy are more pronounced than with either modality alone. **Disseminated necrotizing leukoencephalopathy** is a rare syndrome of progressive neurologic deterioration that can occur in patients who had whole brain radiation combined with intrathecal or systemic chemotherapy. It has a sudden and severe onset and is potentially fatal.

The majority of longitudinal cognitive outcome studies have been conducted in breast cancer patients. Criteria for defining the resulting **chemotherapy-related cognitive impairment**, colloquially referred to as "chemo brain" or "chemo fog," are not well standardized in existing studies, so incidence estimates are wide-ranging. In general, the pattern of neurocognitive decline involves the domains of attention (difficulties with working memory load, divided attention, sustained attention), psychomotor speed, executive function, and memory (specifically retrieval deficit). The effects tend to be mild and there is usually full recovery within 1 year after treatment ends, but a subset of patients have longer-term neurocognitive changes that persist for years after chemotherapy ends.

Diffusion tensor imaging, a variant of conventional MRI based on the tissue water diffusion rate that is used for studying white matter architecture and integrity, has revealed subtle forms of toxic leukoencephalopathy, in which normal-appearing white matter on conventional MRI harbors microstructural white matter abnormalities, in patients who have received cancer chemotherapy.

Infectious Demyelinating Disorders

Infectious diseases can selectively affect the brain white matter, with the majority caused by viruses. The pathogenesis of demyelination varies depending on the specific virus and may involve direct effects of the virus on oligodendrocytes or indirect effects of virus-induced immune-mediated reactions. These most notably include human immunodeficiency virus type 1 and progressive multifocal leukoencephalopathy.

Progressive Multifocal Leukoencephalopathy

Progressive multifocal leukoencephalopathy (PML) produces focal demyelination throughout the brain. It occurs almost exclusively among those who are severely immunosuppressed, such as organ transplant patients, persons with leukemia or lymphoma blood cancers, and persons with human immunodeficiency virus infection and acquired immune deficiency syndrome (HIV/AIDS).

Progressive multifocal leukoencephalopathy is caused by the **John Cunningham** (JC) **virus**, named after the first patient in whom it was identified. The JC virus is extremely common, but it is usually harmless in immunologically healthy persons. It lies dormant within certain cells of the body but becomes activated under conditions of suppressed immunity.

Progressive multifocal leukoencephalopathy is characterized by progressive destruction of subcortical oligodendrocytes, particularly within the parieto-occipital lobes. The PML lesions begin as small, patchy, round or oval lesions in the peripheral regions of the white matter, at the gray-white matter interface. The multifocal white matter lesions eventually coalesce to form increasingly larger lesions and may extend to the periventricular areas. This pattern of lesion distribution distinguishes PML from MS, in which lesions are primarily periventricular. Lesions of the corpus callosum, which are common in MS, are relatively rare in PML.

Progressive multifocal leukoencephalopathy presents with cognitive, behavioral, and motor disturbances with insidious onset and rapid progression. The cognitive changes prominently feature inattention and memory loss in the context of relatively preserved language, as well as disturbances of higher visual function such as visual agnosia, pure alexia, and Bálint's syndrome due to parieto-occipital involvement.

Progressive multifocal leukoencephalopathy is a rapidly progressive and fatal disease. In those with HIV/AIDS, the introduction of highly active antiretroviral therapy has reduced opportunistic infections such as PML and has improved outcomes.

Other Viral Infections Associated with Demyelination

Subacute sclerosing panencephalitis (SSPE) is a rare neurological disease of childhood or young adulthood caused by reactivation of the measles virus, 2–10 years after the original viral attack. Initial symptoms usually consist of cognitive and behavioral changes, with progression to dementia, myoclonic jerks, generalized seizures, coma, and in 95% of cases, death. The virus has a predilection for white matter; periventricular and/or subcortical white matter lesions are the most common MRI finding. The incidence in the United States has decreased dramatically with widespread use of the measles vaccine; in low-resource parts of the world the incidence is much higher. There is no effective treatment.

Progressive rubella panencephalitis (PRP) is a very rare progressive neurological disease of adolescence caused by reactivation of the rubella (German measles) virus, 2–10 years after the original viral infection. The disease is characterized by dementia, spasticity, and ataxia. Ongoing neurological deterioration almost invariably leads to coma and subsequent death. Neuropathologically it is characterized by widespread inflammatory demyelination with vasculitis. There is no effective treatment.

Cytomegalovirus is a very common virus that rarely causes illness. **Cytomegalovirus encephalitis** is an opportunistic infection that occurs in severely immunosuppressed individuals, such as persons with untreated AIDS. It typically presents with generalized cognitive decline. MRI shows nonspecific white matter pathology.

Acquired Metabolic Leukoencephalopathy

Central Pontine Myelinolysis

Central pontine myelinolysis (CPM) is characterized by demyelination of the white matter fibers in the central part of the anterior (basal) pons, which houses the

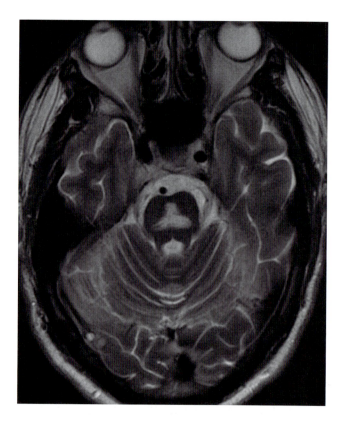

FIGURE 23.4. Central pontine myelinolysis. Axial T2 image of the brain shows hyperintensity within the central pons, with sparing of the periphery and the corticospinal tracts.

corticospinal and corticobulbar tracts (Figure 23.4). The myelinolysis occurs with relative sparing of neuron cell bodies and axons. The volume of demyelination within the pons is variable. Loss of myelin may extend into other brainstem regions, the cerebellum, and the cerebrum, a condition known as **extrapontine myelinolysis** (EPM).

The clinical manifestations of CPM consistently include spastic quadriplegia/quadraparesis and pseudobulbar palsy (head and neck weakness, dysphagia, and dysarthria) due to demyelination of the corticospinal and corticobulbar tracts within the pons. It may result in locked-in syndrome, a rare neurological disorder with complete paralysis of all voluntary muscles except those controlling eye movements (see Chapter 25, "Disorders of Consciousness"). Lesions within the pons may also cause horizontal gaze paralysis/paresis. Because the volume and extent of demyelination are variable in CPM, the spectrum of neurological manifestations and long-term disabilities is diverse. Demyelination extending through the midbrain may result in vertical gaze paralysis/paresis. There may be varying degrees of **encephalopathy** (i.e., altered mental state, possibly including altered level of consciousness reflecting diffuse brain malfunction), and in some cases it may progress to coma or death.

Central pontine myelinolysis occurs as a complication of severe and prolonged sodium depletion (hyponatremia) that is corrected too rapidly; it is therefore also known as **osmotic myelinolysis** or **osmotic demyelination syndrome**. In addition, CPM often occurs in association with alcoholic liver disease or liver transplant. Primary prevention involves identifying individuals at risk and following guidelines for evaluation and correction of hyponatremia. There is no treatment for the underlying cause of CPM; supportive care is aimed at symptom management.

Cobalamin Deficiency

Cobalamin (vitamin B_{12}) is necessary for the maintenance of myelin. Deficiency of this vitamin may result in a variety of neurologic disorders, including peripheral neuropathy or subacute combined degeneration of the spinal cord and brain due to white matter pathology. Less commonly, vitamin B_{12} deficiency may lead to cerebral white matter pathology and dementia. Because B_{12} avitaminosis is easily treated, B_{12} screening is routine in the evaluation of dementia. Cobalamin deficiency is common in older individuals and often is accompanied by cognitive decline and depression. The cerebral white matter lesions are related to the degree of cognitive impairment. There is an association between cobalamin levels and severity of MRI white matter lesions. Vitamin B_{12} replacement results in clinical improvement that parallels neuroradiologic improvement of the leukoencephalopathy.

Hypoxic-Ischemic Leukoencephalopathy

Hypoxic-ischemic encephalopathy is a brain injury caused by inadequate blood oxygenation (e.g., due to respiratory failure, carbon monoxide poisoning) or lack of perfusion to the brain (e.g., due to cardiac arrest, profound hypotension). Both hypoxia/anoxia and ischemia result in inadequate oxygen delivery to brain cells; ischemia also results in inadequate nutrient supply (e.g., glucose) to brain cells. The severity of encephalopathy depends on the extent and duration of hypoxia-ischemia. Particularly affected areas of the brain are the hippocampi, cerebral cortex, watershed regions, and deep white matter.

Hypoxic Leukoencephalopathy

Cerebral hypoxia involves inadequate supply of oxygen carried by circulating blood to the brain. Common causes include carbon monoxide poisoning, respiratory failure due to drug overdose, choking, and strangulation. Carbon monoxide produces hypoxia because it has a greater affinity for hemoglobin than oxygen (i.e., it binds to hemoglobin more efficiently), thereby displacing oxygen and resulting in reduced blood oxygen content and transport to body cells. Drug overdose-related respiratory failure typically involves opiates, alcohol, benzodiazepines, or combinations of these agents.

Acutely, cerebral hypoxia results in selective damage to brain gray matter structures with relative sparing of white matter. Within the first several hours of a hypoxic episode, a consistent injury pattern emerges that is most evident on diffusion-weighted MRI which shows acute tissue injury; certain brain gray matter structures such as hippocampus, cerebral cortex, basal ganglia, and thalami are most vulnerable, while other structures such as the brainstem are relatively spared.

Weeks after the initial hypoxic event, however, there may be widespread demyelination with axonal sparing known as **delayed post-hypoxic leukoencephalopathy**. Those who experience this rare condition have a biphasic course of neurological deterioration. An acute neurologic deterioration immediately after the hypoxic episode is followed by a period of improvement, and then a second phase of neurological deterioration 2–4 weeks following the hypoxic event. Clinically, delayed post-hypoxic leukoencephalopathy has several manifestations characterized on the basis of the leading symptoms, including a form dominated by akinetic-mutism and a form dominated by parkinsonism. The diagnostic hallmark of delayed post-hypoxic leukoencephalopathy is a characteristic leukoencephalopathy evidenced by subcortical white matter hyperintensities on T2-weighted and diffusion-weighted MRI, both of which are used for detection and localization of acute ischemic brain lesions (see Chapter 29, "Brain Imaging"). The lesions are typically extensive, bilateral, and symmetric, with sparing of cortical U-fibers, cerebellum, and brainstem.

The mechanisms underlying delayed post-hypoxic leukoencephalopathy are not well understood. Most of the cases reported have been due to carbon monoxide poisoning, where it occurs in up to 3% of this patient population, but it also occurs with other forms of hypoxia.

Ischemic Leukoencephalopathy

The deep white matter of the cerebral hemispheres is particularly susceptible to ischemia because it receives its blood supply from small-caliber penetrating arteries and arterioles, and white matter has less blood supply than gray matter. With aging and cerebrovascular risk factors, especially hypertension, the small penetrating arteries develop **small vessel disease**, degenerative alterations in the vessel walls of the small arteries and arterioles characterized by vessel wall thickening and vessel lumen diameter narrowing. Consequently, the white matter becomes chronically ischemic. Maintenance of the myelin becomes deficient, resulting in "myelin pallor" on microscopic sections. MRI shows **leukoaraiosis,** white matter damage caused by perfusion disturbances within the arterioles perforating through the deep brain structures. A subcortical dementia may develop, often with superimposed focal neurological deficits that are related to lacunar infarcts.

The most common locations of lesions are, in decreasing order of frequency, the subcortical and periventricular white matter, optic radiations, basal ganglia, and brainstem. The lesions have well defined but irregular margins, and they tend to be multifocal; however, the lesions become confluent as they enlarge. Another pattern of deep white matter ischemic damage is a continuous or nearly continuous band of MRI abnormalities bordering the lateral ventricles due to mild interstitial edema (increased water content).

Binswanger's disease is a progressive vascular dementia that occurs in the context of extensive periventricular white matter pathology due to small vessel disease (see Chapter 18, "Stroke and Vascular Cognitive Impairment").

Summary

The white matter diseases are a group of disorders that feature prominent or exclusive white matter pathology, either due to defective myelin formation (dysmyelination) or myelin degeneration (demyelination), thereby disrupting nerve impulse conduction by slowing or blocking signal transmission.

The leukodystrophies are a group of genetic metabolic diseases that result in an error in myelin metabolism, which disrupts the establishment or maintenance of myelin. Most manifest initially in infants and children due to a failure of normal myelination, but they sometimes manifest initially in teens or young adults, in which case they result in demyelination.

The main autoimmune inflammatory demyelinating diseases are MS which affects the CNS, Guillain-Barré syndrome which affects the PNS, and acute disseminated encephalomyelitis which affects the CNS. Multiple sclerosis is the most common CNS demyelinating disease and most common nontraumatic disabling neurologic disease of young adults.

Secondary etiologies of demyelination include a wide variety of toxins, infectious diseases, acquired metabolic disturbances and nutritional deficiencies, and hypoxic-ischemic injury.

Additional Reading

1. Benedict RHB, DeLuca J, Enzinger C, Geurts JJG, Krupp LB, Rao SM. Neuropsychology of multiple sclerosis: looking back and moving forward. *J Int Neuropsychol Soc.* 2017;23(9–10):832–842. doi:10.1017/S1355617717000959
2. Filley CM. White matter: organization and functional relevance. *Neuropsychol Rev.* 2010;20(2):158–173. doi:10.1007/s11065-010-9127-9
3. Filley CM. White matter: beyond focal disconnection. *Neurol Clin.* 2011;29(1):81–97, viii. doi:10.1016/j.ncl.2010.10.003
4. Filley CM. White matter dementia. *Ther Adv Neurol Disord.* 2012;5(5):267–277. doi:10.1177/1756285612454323
5. Filley CM. *The behavioral neurology of white matter.* 2nd ed. Oxford University Press; 2012.
6. Filley CM. Toluene abuse and white matter: a model of toxic leukoencephalopathy. *Psychiatr Clin North Am.* 2013;36(2):293–302. doi:10.1016/j.psc.2013.02.008
7. Filley CM, Fields RD. White matter and cognition: making the connection. *J Neurophysiol.* 2016;116(5):2093–2104. doi:10.1152/jn.00221.2016
8. Filley CM, McConnell BV, Anderson CA. The expanding prominence of toxic leukoencephalopathy. *J Neuropsychiatry Clin Neurosci.* 2017;29(4):308–318. doi:10.1176/appi.neuropsych.17010006
9. Krieger SC, Cook K, De Nino S, Fletcher M. The topographical model of multiple sclerosis: a dynamic visualization of disease course. *Neurol Neuroimmunol Neuroinflamm.* 2016 Sep 7;3(5):e279. doi: 10.1212/NXI.0000000000000279. PMID: 27648465; PMCID: PMC5015541.
10. Krieger SC, Sumowski J. New insights into multiple sclerosis clinical course from the topographical model and functional reserve. *Neurol Clin.* 2018 Feb;36(1):13–25. doi: 10.1016/j.ncl.2017.08.003. Epub 2017 Oct 18. PMID: 29157394.
11. Marner L, Nyengaard JR, Tang Y, Pakkenberg B. Marked loss of myelinated nerve fibers in the human brain with age. *J Comp Neurol.* 2003;462(2):144–152. doi:10.1002/cne.10714
12. Prins ND, Scheltens P. White matter hyperintensities, cognitive impairment and dementia: an update. *Nat Rev Neurol.* 2015;11(3):157–165. doi:10.1038/nrneurol.2015.10
13. Schmahmann JD, Smith EE, Eichler FS, Filley CM. Cerebral white matter: neuroanatomy, clinical neurology, and neurobehavioral correlates. *Ann NY Acad Sci.* 2008;1142:266–309. doi:10.1196/annals.1444.017
14. Schmahmann JD, Pandya DN. Cerebral white matter: historical evolution of facts and notions concerning the organization of the fiber pathways of the brain. *J Hist Neurosci.* 2007;16(3):237–267. doi:10.1080/09647040500495896
15. Sumowski JF, Benedict R, Enzinger C, Filippi M, Geurts JJ, Hamalainen P, et al. Cognition in multiple sclerosis: state of the field and priorities for the future. *Neurology.* 2018;90(6):278–288. doi:10.1212/WNL.0000000000004977

The Motor System and Motor Disorders

Introduction

Nearly all behavior involves receiving information about the world through the senses and acting on the world through movement. The nervous system as a whole directs behavior adaptively in the interests of survival and reproduction, but ultimately behavior is expressed through a system of nerves and muscles that effect change and motion of the individual in the environment—the motor system.

The basic elements of this system are muscles, primary motor neurons, spinal cord, motor cortex, multiple brainstem motor nuclei, basal ganglia, cerebellum, and the neuronal pathways connecting these structures. **Primary motor neurons** synapse directly with muscle cells. Those neurons supplying the muscles of the body originate from the ventral horn of the spinal cord and travel in the spinal nerves. Those neurons supplying the muscles of the face, head, and neck originate from the brainstem cranial nerve motor nuclei and travel in the cranial nerves. Primary motor neurons are also known as **lower motor neurons** (LMNs). Lower motor neurons receive input from **upper motor neurons** (UMNs). Upper motor neurons originate from motor cortex and subcortical motor nuclei (i.e., the red nucleus, superior colliculus, reticular formation nuclei, and vestibular nuclei), and they give rise to the descending motor pathways of the CNS. The multiple UMN pathways converge on and influence the activity of LMNs. Lower motor neurons are also referred to as "**the final common pathway**" because they transmit signals from a variety of sources to skeletal muscles, which execute the movements constituting behavior (Figure 24.1).

Movements can be involuntary (reflexes), voluntary (purposeful, learned), or a combination of both reflexive and intentional components (e.g., walking), and it occurs on a background of postural tone. Motor cortex is involved with motor planning and the execution of voluntary movements, while subcortical motor structures play an important role in postural tone and control. Subcortical structures also play a role in learned motor skills such as riding a bicycle, swimming, typing, driving, playing piano, and juggling. When we learn these behaviors, we focus attention on voluntary, intentional movements, where the motor cortex is the most dominant. After learning has taken place, subcortical motor structures play a greater role so that motor skills can be executed without significant conscious control (colloquially known as "muscle memory").

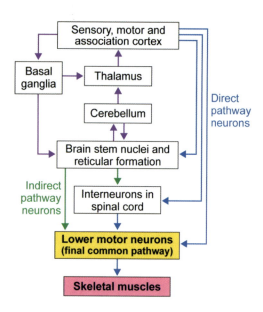

FIGURE 24.1. Lower motor neurons are "the final common pathway" for transmitting neural signals from multiple descending motor pathways to muscle.

The goals of this chapter are to introduce the reader to the basic anatomy of the motor system from muscle to cerebral cortex, and to some motor disorders that arise from pathologies at various levels of the motor system.

Muscle

There are three types of muscle: skeletal, smooth, and cardiac. Skeletal muscle makes up about 40% of the body's mass, while smooth and cardiac muscle together make up about 10%. The motor system is divided into a somatic motor system and a visceral motor system, based on the type of muscle innervated. The **somatic motor system** controls skeletal muscle and is involved in voluntary movement. The **visceral motor system** controls the smooth muscles of the viscera (e.g., gastrointestinal tract and arteries) and cardiac muscle. This system is self-regulating (generally not under voluntary control) and therefore also is referred to as the **autonomic motor system**.

Basic Mechanics of Movement

It is useful to think of the skeleton as the frame of the body and individual bones as the elements of that frame. We have the ability to alter the relative positions of the bones, and hence the overall position of the body. The bones themselves, however, do not receive direction from the nervous system. Rather, it is the skeletal muscles connected to bones via **tendons** that alter the relative positions of the skeletal elements. By contracting, muscles bring two skeletal elements closer together, producing movement about a **joint**. Muscles can only **contract** (shorten) and **relax** (lengthen).

Flexion is movement that decreases the angle between two body parts, while **extension** is movement that increases the angle between two body parts. Muscles are classified as either **flexor muscles** or **extensor muscles**. Muscles are generally organized in **antagonist** (opposing) groups; contraction of one group causes movement in one direction, and contraction of the opposing group produces movement in the opposite direction (Figure 24.2).

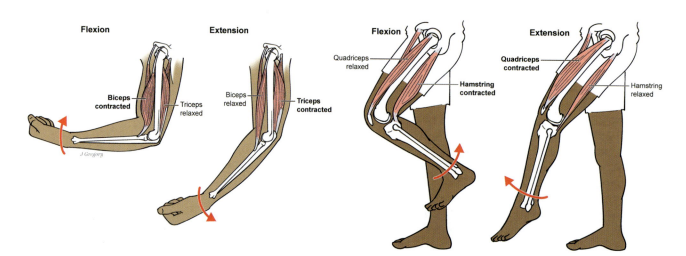

FIGURE 24.2. Muscles are generally organized in antagonist groups. Contraction of flexor muscles causes flexion, a movement that decreases the angle between two body parts. Contraction of extensor muscles causes extension, a movement that increases the angle between two body parts.

Muscle Fibers

Skeletal muscle is composed of elongated cells called **muscle fibers**. Skeletal muscle fibers come in two varieties, extrafusal and intrafusal. **Extrafusal muscle fibers** are the ones that are primarily responsible for muscular contraction and movement; they run the length of the muscle and attach to bones at both ends via tendons. **Intrafusal muscle fibers** are the sensory component of muscle. They monitor the state of muscle contraction and feed that information back to the nervous system. Intrafusal fibers are located within muscle spindles, fusiform (spindle-shaped) sensory (proprioceptor) organs that detect amount and rate of change in the length of a muscle.

MUSCLE FIBER INNERVATION

Skeletal muscle extrafusal fibers are innervated by LMNs. These neurons have large-diameter myelinated axons. In general, nerve fibers are classified according to their axon diameter and myelination state, which determine conduction velocity; in descending order of conduction velocity, the nerve fiber classifications are A-alpha, A-beta, A-gamma, A-delta, B, and C fibers. The neurons that directly innervate skeletal muscle are **alpha motor neurons** and have the most rapid conduction velocity.

The axon terminals of alpha motor neurons make synaptic contact with the extrafusal muscle fiber in an elevated plaque-like region called the **motor end plate**, located on the surface of the muscle fiber at the center of its length. The synapse between the motor neuron axon terminal and the motor end plate of the muscle fiber is called the **neuromuscular junction**. Each muscle fiber receives only one axon terminal and therefore is controlled directly by only one neuron.

Muscle Contraction

The cell membrane of the extrafusal muscle fiber is called the **sarcolemma**. Each muscle fiber is composed of several hundred to several thousand **myofibrils** that run the length of the muscle fiber and muscle. Each myofibril is composed of approximately 1,500 thick **myosin filaments** and 3,000 thin **actin filaments**. The actin and myosin filaments lie side by side; the change in length of the muscle fiber during muscle contraction is due to a chemical interaction between the actin and myosin filaments.

The extrafusal muscle fiber is electrically excitable; it is capable of generating an electrical response. It has a resting potential of about −85 mV. The neuromuscular junction of the somatic motor system uses **acetylcholine** (ACh) neurotransmitter and **nicotinic receptors**. Acetylcholine is deactivated extremely rapidly by enzymatic degradation by **acetylcholinesterase**.

Muscle contraction occurs by the following sequential steps. An action potential (AP) travels down the motor neuron to the neuromuscular junction, which causes the release of a small amount of ACh. Acetylcholine interacts with nicotinic receptors and causes the opening of ACh-gated sodium ion (Na^+) channels in the muscle fiber sarcolemma. When Na^+ enters the interior of the muscle fiber, it causes depolarization and the muscle fiber generates a **muscle action potential**. The AP travels along the sarcolemma and within the muscle fiber along the **transverse tubule system**. The depolarization causes voltage-gated calcium ion (Ca^{2+}) channels to open, which in turn causes the release of calcium ions from the **sarcoplasmic reticulum** into the myofibril. Calcium ions initiate attractive forces between actin and myosin filaments. They act as a cofactor, allowing adenosine triphosphate (ATP) to drive a reaction between actin and myosin, causing the filaments to slide over one another and shorten the length of the muscle fiber. This is the contractile process, and the response of the muscle fiber is called a **muscle twitch**.

The muscle fiber twitch is a discrete response in that it is all-or-none; it does not vary in magnitude. The fiber exists in either the twitch state or not. After a fraction of a second, calcium is removed from the cytoplasm of the muscle cell and is returned to the sarcoplasmic reticulum by calcium pumps, limiting the duration and ending the muscle fiber twitch. Calcium ions remain stored within the sarcoplasmic reticulum until a new muscle AP arrives.

As described above, muscles are composed of many muscle fibers that run the length of the muscle. Unlike muscle fibers, muscles themselves contract along a continuum; the magnitude of muscle contraction is a function of the number of muscle fiber cells that are twitching.

THE MOTOR UNIT

Alpha motor neuron fibers conduct very rapidly (80–120 m/sec). Each alpha motor neuron innervates from 3 to 2,000 muscle fibers (180 on average) in a single muscle. Each muscle fiber is innervated by only one branch from a single motor neuron. A single motor neuron and the set of muscle fibers that it innervates constitutes a **motor unit**; the motor unit is the smallest functional component of the motor system. The collection of motor neurons innervating a single muscle is called a **motor pool**.

The size of the motor unit is determined by the extent of branching of the motor neuron axon. The **innervation** ratio is the number of muscle fibers innervated by a single motor neuron; it is matched to the particular functional demands of the muscle. Small muscles (e.g., in the hand) capable of very fine, exact control of the total force of contraction have a low innervation ratio (i.e., few muscle fibers innervated by a single motor neuron). Large muscles (e.g., the gluteus maximus) that do not have fine control have a high innervation ratio (i.e., many muscle fibers innervated by a single motor neuron). This is analogous to sensory convergence ratios, as in the retina (although in the reverse direction).

The Monosynaptic Stretch Reflex

The **muscle spindle** is a proprioceptive sensory organ in skeletal muscle, consisting of several intrafusal fibers enclosed in a spindle-shaped connective tissue sac. Muscle spindles detect the amount and rate of change in length of a muscle and therefore are also known as **muscle stretch receptors**. They play an important role in the **myotatic reflex** that is crucial for the maintenance of posture. If a person standing upright leans to one side, the postural muscles connected to the vertebral column on the opposite side of the body stretch, and the muscle spindles in those muscles detect the stretching, activating muscle contraction to counteract the stretch and correct the posture. The myotatic reflex is also known as the **monosynaptic stretch reflex** or **deep tendon reflex**.

The monosynaptic stretch reflex circuit is illustrated by the **knee-jerk reflex** (Figure 24.3). When a physician taps the patellar tendon with a reflex hammer, the knee extensor muscle stretches abruptly, causing a reflexive extension of the lower leg (stretch reflex). Muscle stretch is detected by the muscle spindle, a primary afferent nerve fiber that enters the dorsal horn of the spinal cord and synapses in the ventral horn directly on the cell body of an alpha motor neuron, which in turn causes contraction in the same muscle and produces the behavioral reflex. The primary afferent also bifurcates within the ventral horn and innervates an inhibitory interneuron, which in turn innervates an alpha motor neuron that synapses onto the opposing muscle. Because the interneuron is inhibitory, it prevents the opposing alpha motor neuron from firing, thereby reducing the contraction of the opposing muscle. This reciprocal inhibition prevents both groups of muscles from contracting simultaneously. Examination of deep tendon reflexes is an important component of the neurological examination (see Chapter 30, "The Neurological Examination").

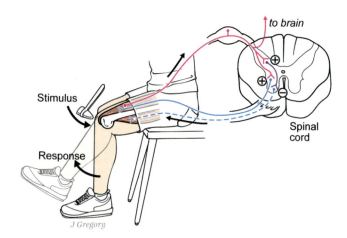

FIGURE 24.3. The patellar tendon (knee jerk) reflex, showing the monosynaptic stretch reflex arc as well as reciprocal inhibition of the antagonist muscle. The sensory neurons make direct excitatory synapses onto the alpha motor neurons within the spinal cord ventral horn, which causes contraction of the quadriceps muscle and sudden leg extension. The sensory neurons also makes excitatory synapses onto inhibitory interneurons within the spinal cord, which inhibit contraction of the antagonist hamstring muscle.

Lower Motor Neurons

Spinal Cord

Within the ventral horn of the spinal cord, the cell bodies of the LMN are arranged somatotopically (i.e., there is a map of the body's musculature). Within a spinal cord segment, the neurons that supply the axial or trunk musculature are located medially, whereas the neurons that supply the limb musculature (arms and legs) are located laterally.

Cranial Nerve Nuclei

Cranial nerve motor nuclei in the brainstem control eye, face, and head movements (see Chapter 6, "The Peripheral Nervous System"). The cranial nerves carrying somatic motor efferents are the oculomotor (CN3), trochlear (CN4), and abducens (CN6) nerves controlling eye movements; the trigeminal nerve (CN5) controlling the muscles of mastication; the facial nerve (CN7) controlling the muscles of facial expression; the glossopharyngeal nerve (CN9) controlling muscles involved in swallowing; the vagus nerve (CN10) providing motor control of soft palate, pharynx, and larynx involved in swallowing and speech; the accessory nerve (CN11) controlling head and shoulder movement; and the hypoglossal nerve (CN12) controlling tongue movements.

Upper Motor Neurons: The Descending Motor Pathways

Motor signals are transmitted *directly* from the cerebral cortex to spinal cord and brainstem cranial nerve motor nuclei through the corticospinal and corticobulbar tracts, respectively (Figure 24.4). Motor signals are also transmitted *indirectly* from the cerebral cortex to spinal cord and brainstem cranial nerve motor nuclei through multiple accessory pathways involving various brainstem motor structures, the cerebellum, and basal ganglia. So, the motor system is organized hierarchically, with the motor cortex at the top of the hierarchy and multiple parallel descending pathways influencing the activity of the primary motor neurons of the spinal cord ventral horn and cranial nerve nuclei.

The motor system also follows another pattern of organization—one of subsystems that serve the differential control of whole-body movements, coordinated limb movements, and independent movement of limbs and fingers. This organization can be seen in the evolution and development of movement. Primitive and infant animals move only with whole body movements (e.g., when fish swim). More advanced and older animals move with coordinated movements (e.g., when amphibians walk). Many mammals use one or two limbs selectively to engage in behaviors such as eating, nest building, and fighting. Relatively independent movements of the arms and fingers are seen only in primates and are most highly developed in humans. The ability to exert fine, conscious control over hand and finger movements was very important in human evolution because the increase in manipulative ability is fundamental to tool use.

Motor Cortex

Motor cortex is located within the frontal lobes. It is divided into primary motor cortex located on the precentral gyrus and secondary motor cortex regions lying anterior to primary motor cortex (see Figure 17.2 in Chapter 17). The motor cortex contains a topographic representation of the body musculature (**motor homunculus**), with the knees to toes represented on the medial surface of the precentral gyrus, the rest of the body represented on the lateral surface of the precentral gyrus, and the face represented most ventrally on the precentral gyrus, closest to the lateral fissure (see Figure 13.13 in Chapter 13). The head and face are represented bilaterally, except for the lower face and tongue. The neurons from the motor cortex representing the body give rise to the **corticospinal tracts** and project to the spinal cord. The neurons from

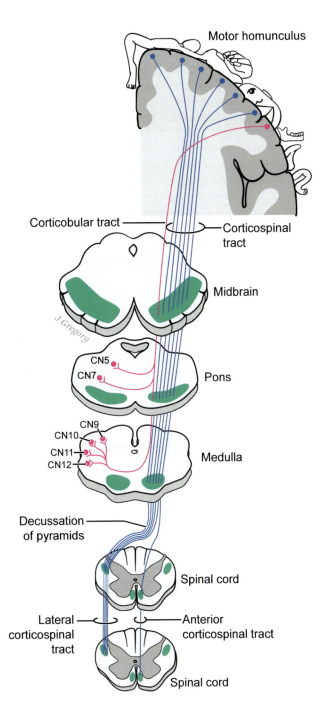

FIGURE 24.4. The corticospinal and corticobulbar tracts control voluntary movements. Corticospinal tract neurons project from primary motor cortex to the spinal cord. Corticobulbar tract neurons project from primary motor cortex to the cranial nerve motor nuclei of the brainstem.

the motor cortex representing the face and head give rise to the **corticobulbar tracts** and project to brainstem cranial nerve motor nuclei (non-oculomotor cranial nerves 5, 7, and 9–12).

Inputs to the primary motor cortex come from premotor areas, somatosensory regions (ventral posterior nuclei of the thalamus, and primary and secondary somesthetic areas), cerebellum (via the ventrolateral nucleus of the thalamus), and basal ganglia (via the ventroanterior nucleus of the thalamus). The premotor areas receive input from the prefrontal and posterior parietal association cortices. The connections between somatosensory and motor areas are organized in a homotopic fashion; the same parts of the body map are interconnected, allowing motor cortex to be informed about the positions of the limbs and speed of movement. Cerebellar and basal ganglia inputs modulate processing within the primary motor cortex, and output from the primary motor cortex.

The primary motor cortex plays a critical role in the execution of voluntary movements. Experimental neurophysiological studies in animals have shown that electrical stimulation of the primary motor area produces isolated movements limited to a single joint or single movement on the opposite side of the body. Single-cell recording studies have shown that action potentials in primary motor cortex occur approximately 60–80 msec before muscle movement. The firing frequency is correlated with the force, direction, extent, and speed of movement.

Pathways from the Cerebral Cortex

CORTICOSPINAL TRACTS

Cortical motor areas representing the body project directly to the spinal cord via the **corticospinal tracts**. Each corticospinal tract contains approximately 1 million fibers that originate from primary and secondary motor cortices, as well as somatosensory areas located posterior to the central sulcus within the parietal lobes. The corticospinal tracts travel from the cerebral cortex of each hemisphere to the ventral horn of the spinal cord, where the cell bodies of the primary motor neurons that directly innervate the body musculature (torso and limbs) reside.

The corticospinal tracts descend through the corona radiata, and then gather to become densely packed and descend within the posterior limb of the internal capsule. Fibers continue their descent through the brainstem, first through midbrain cerebral peduncles, then the pontine tegmentum, and then the **medullary pyramids**, two pyramid-shaped ridges located on the ventral surface of the medulla. The corticospinal tracts are therefore also referred to as the **pyramidal tracts**.

Most corticospinal fibers (85%–90%) cross the midline in the **pyramidal decussation**, a landmark indicating the transition between medulla and spinal cord. The decussating corticospinal fibers then form the **lateral corticospinal tracts** of the spinal cord, so named because they descend in the lateral columns of the spinal cord. The remaining 10%–15% of corticospinal fibers do not cross the midline and instead descend the cord ipsilaterally in the **ventral (anterior) corticospinal tracts,** so named because they descend in the ventral (anterior) columns of the spinal cord (see Figure 7.3 in Chapter 7).

Corticospinal tract neurons terminate within the ventral horn of the spinal cord and control the activity of primary motor neurons, either directly or indirectly via interneurons. The axons of these neurons are among the longest in the CNS; those controlling muscles of the lower body travel from the cerebral cortex to lumbar and sacral levels of the spinal cord. This tract connects the cerebral cortex and spinal cord directly, allowing for very rapid transmission of signals from the cerebral cortex to the periphery.

The cell bodies of the LMNs within the ventral horn have a somatotopic organization; those innervating the axial musculature are located medially, while those innervating the distal musculature (limbs) are located more laterally (see Figure 7.5 in Chapter 7). The lateral corticospinal tract projections terminate at all levels of the spinal cord, especially on motor neurons that innervate the distal limbs (forearms, hands, fingers). This pathway is especially important for manual dexterity. The ventral corticospinal tract projections terminate at the cervical and thoracic levels of the spinal cord, bilaterally within the medial region of the ventral horn, on spinal cord motor neurons that innervate axial (trunk) and proximal limb muscles. This pathway plays a role in movements such as walking, turning the body, and postural movements.

Unilateral damage to primary motor cortex results in a contralateral **hemiplegia** or **hemiparesis**, a paralysis or weakness of the muscles served by the spinal nerves on the opposite side of the body. In milder cases there is a loss of ability to make fine, independent finger movements and a loss of speed and strength in the contralateral limb. Lesions in the descending corticospinal tract rostral to the pyramidal decussation also produce a contralateral hemiplegia. Because lesions of the corticospinal neurons result in axonal degeneration, there is atrophy in the corresponding pyramid that is visibly smaller on gross exam. Lesions of the corticospinal tract caudal to the pyramidal decussation (i.e., lateral corticospinal tract) produce ipsilateral weakness.

CORTICOBULBAR TRACTS

The brainstem can be considered an upward extension of the spinal cord (i.e., into the cranial cavity); it contains motor and sensory nuclei for the face and head regions, in the same way that the dorsal and ventral horns of the spinal cord perform motor and sensory functions for the body.

The **corticobulbar** (**corticonuclear**) **tracts** originate from cortical motor areas representing the face, head, and neck, and they project directly to brainstem cranial nerve motor nuclei (non-oculomotor cranial nerves 5, 7, and 9–12) that control face, head, and neck movements. They descend with the corticospinal tracts through the corona radiata. They continue their descent through the genu (a knee-like part or bend) of the internal capsule, and the midbrain cerebral peduncles, and thereafter travel to the brainstem cranial nerve motor nuclei.

The corticobulbar tracts are analogous to the corticospinal tracts, except that their destination is the brainstem cranial nerve nuclei that direct movement of the face, head, and neck, rather than the spinal cord that directs movements of the trunk and limbs. The corticobulbar tracts are often considered within the designation of the pyramidal tracts, although they do not pass through the medullary pyramids.

Each corticobulbar tract innervates the cranial nerve motor nuclei bilaterally, with two exceptions. The corticobulbar projection to the ventral region of the facial motor nucleus that controls the musculature of the lower face (below the eyes) by the facial nerve (CN7) is contralateral only. Thus, the muscles of the upper face are controlled by both primary motor cortices, while the muscles of the lower face are controlled by the contralateral motor cortex. The corticobulbar projection to the hypoglossal nucleus, which gives rise to the hypoglossal nerve (CN12) and innervates the intrinsic and extrinsic muscles of the tongue, is contralateral only.

Unilateral damage to the cortical motor area representing the face results in paralysis of the contralateral lower quadrant of the face only; the upper face is not paralyzed because it is under the control of the motor cortex of both hemispheres (i.e., each motor cortex projects bilaterally to the facial motor nuclei that in turn control facial movements). In contrast, unilateral damage to a facial motor nucleus or facial nerve results in paralysis of the ipsilateral half of the face.

Indirect Cortical-Brainstem-Spinal Cord Pathways

In addition to influencing motor neurons in the ventral horn of the spinal cord directly by way of the corticospinal tract, motor cortex influences spinal cord motor neurons indirectly through parallel projections to brainstem motor nuclei that in turn project to the spinal cord (**bulbospinal pathways**). These brainstem regions are the **red nucleus, superior colliculus, reticular formation,** and **vestibular nuclei**; they give rise to the **rubrospinal, tectospinal, reticulospinal,** and **vestibulospinal** pathways, respectively. These corticobulbar-bulbospinal (**corticobulbospinal**) pathways are also referred to as the **corticorubrospinal, corticotectospinal, corticoreticulospinal,** and **corticovestibulospinal** pathways. The indirect cortical-brainstem-spinal cord pathways allow cortical motor areas to control complex patterns of muscle activation that are organized within the brainstem.

THE RUBROSPINAL TRACT

The **red nucleus**, located in the midbrain tegmentum, receives large numbers of fibers from the ipsilateral primary motor and premotor cortex (via the **corticorubral tract**), and the contralateral cerebellum. It gives rise to the **rubrospinal tract**, which decussates immediately (at the level of the red nucleus) and descends through the lateral columns of the spinal cord, parallel to the corticospinal tract. Rubrospinal fibers terminate mainly on interneurons within the cervical segments of the spinal cord.

This tract has an inhibitory effect on extensor muscles and an excitatory effect on the flexor muscles of the arms. Lesions rostral to the red nuclei remove cortical inhibition of rubrospinal neurons, resulting in flexion of the upper extremities and extension of the lower extremities. This is known as **decorticate posturing** (see Chapter 25, "Disorders of Consciousness").

RETICULOSPINAL TRACTS

Two descending tracts originating from nuclei within the reticular formation receive projections from most of the sensory systems. The **ventral reticulospinal tract** (**pontine reticulospinal tract**) originates in the pontine reticular formation and descends in the ventral columns of the spinal cord uncrossed. The **lateral reticulospinal tract** (**medullary reticulospinal tract**) originates within the medullary reticular formation and descends in the lateral columns of the spinal cord, both crossed and uncrossed. Neither tract is organized somatotopically.

The lateral and ventral reticulospinal pathways function antagonistically to each other. The ventral reticulospinal pathway terminates on the medial motor neurons and facilitates contraction of the extensor musculature, also known as the **antigravity muscles**. Antigravity muscles such as the back and lower limb extensors act to counterbalance the pull of gravity and maintain an upright posture. The lateral reticulospinal pathway inhibits contraction of these antigravity muscles. Together, the two descending pathways are primarily involved in postural control and organized movements for locomotion (e.g., walking, swimming, and running).

VESTIBULOSPINAL TRACTS

Two descending tracts originate from vestibular nuclei. These nuclei play a role in controlling posture and balance, in response to sensory information from the vestibular apparatus of the inner ear reporting on head position and movement. The cerebellum participates in this control by modulating the activity of the vestibular nuclei.

The **lateral vestibulospinal tract** receives input from the vestibular apparatus otolith organs, which sense direction and speed of linear movement of the head. The tract descends ipsilaterally in the ventral column of the spinal cord and terminates on medial motor neuron pools at all levels of the spinal cord. It facilitates anti-gravity muscles through monosynaptic excitation of extensor motor neurons and di-synaptic inhibition of flexor motor neurons, thereby functioning in the maintenance of balance in response to vestibular signals.

The **medial vestibulospinal tract** receives input from the vestibular apparatus semicircular canals, which sense direction and speed of rotational movement of the head. The pathway is critical to the **vestibulocollic reflex**, which regulates head position by activating neck muscles reflexively in response to head movement and vestibular stimulation, thereby keeping the head on a level plane when walking.

TECTOSPINAL AND TECTOBULBAR TRACTS

The midbrain tectum is composed of two pairs of rounded surface swellings known collectively as the corpora quadrigemina (quadruple bodies); these are the superior and inferior colliculi. The **superior colliculi** receive input from extrageniculate pathways that branch off the optic tracts. The **inferior colliculi** are part of main auditory pathways. The colliculi give rise to the **tectospinal tracts** that descend in the contralateral ventral columns of the spinal cord and terminate on upper cervical spinal cord motor neurons innervating the neck and proximal musculature, and the **tectobulbar tracts** that project to brainstem cranial nerve motor nuclei (CN3, CN4, CN6, CN7, and CN11).

The tectospinal tracts play an important role in orienting body movements in response to visual and auditory stimuli, as well as protective reflex postural movements such as raising the arms in response to sudden visual stimuli and sudden, loud auditory stimuli. The tectobulbar tracts play an important role in orienting eye and head movements in response to visual and auditory stimuli, as well as protective reflex head and eye movements in response to sudden visual and auditory stimuli, such as closing the eyes and turning the head away from the stimulus.

Two Descending Motor Systems

The descending motor pathways can be divided into two major descending spinal cord systems, the lateral motor system and the ventromedial motor system; they control the axial (trunk and head) and distal (arms and legs) musculature, respectively. These systems travel in different columns (white matter regions) and have different terminations within the ventral horn (gray matter regions) of the spinal cord, and therefore they control different muscle groups (see Figures 7.3 and 7.5 in Chapter 7).

THE LATERAL MOTOR SYSTEM

The descending lateral corticospinal and rubrospinal pathways comprise the **lateral motor system** of the spinal cord. These pathways descend in lateral columns of the spinal cord and terminate in the lateral portion of the ventral horn gray matter. The lateral motor system of the spinal cord directs movements of the limbs that are independent of trunk movement.

THE VENTROMEDIAL MOTOR SYSTEM

The **ventromedial motor system** of the spinal cord is composed of the ventral corticospinal, reticulospinal, vestibulospinal, and tectospinal tracts. These pathways descend in the ipsilateral ventral columns of the spinal cord and terminate in the medial portion of the ventral horn gray matter, many terminating bilaterally and at multiple spinal cord levels. These pathways direct movements of the trunk and proximal parts of the limbs (upper legs and arms).

COMPARING THE LATERAL AND VENTROMEDIAL MOTOR SYSTEMS

Both the lateral system and ventromedial system have at least one projection from the cortex and one from the brainstem. The lateral system projections decussate and have unilateral terminations within the spinal cord, while the ventromedial systems generally do not decussate and have bilateral terminations.

The vast majority of synaptic contacts on primary motor neurons are made by interneurons; thus, most of the input from descending motor pathways influences motor neurons indirectly via interneurons. Interneurons integrate input from these various sources and in turn direct the motor neurons.

At most spinal cord levels, the corticospinal and rubrospinal fibers of the lateral system terminate mainly on interneurons. However, in the cervical enlargements of the cord where the hands and fingers are represented, many corticospinal tract fibers terminate directly on motor neurons. This is in line with the fact that primary motor cortex has an extensive representation of the hand and fingers, allowing for fine movement control. Lesions of the lateral motor system disrupt the independent use of limbs for reaching and grasping, without affecting walking.

The ventromedial system pathways are characterized by divergent distribution of their terminals and have numerous collaterals that synapse on interneurons in many spinal cord segments, in line with the fact that these pathways play a role in whole body movements that involve many muscles. These pathways are important in maintaining posture and balance, which rely on the proximal muscles. Lesions of these ventromedial pathways produce severe postural impairments.

Basal Ganglia and Cerebellum

Motor cortex also projects to the cerebellum (via the pontine nuclei in the cortico-ponto-cerebellar pathway) and basal ganglia (via corticostriate pathways). These regions in turn project back up to primary and secondary motor cortex via the thalamus (with the cerebellum projecting via the ventrolateral nucleus, and the basal ganglia projecting via the ventroanterior nucleus). Through this "side loop" circuitry, the cerebellum and basal ganglia provide feedback that modulates the activity of cortical motor areas.

Lower Motor Neuron versus Upper Motor Neuron Lesions

Disruptions of signaling between LMNs and muscle lead to paresis (muscle weakness) or paralysis, **muscle wasting** (atrophy), **hypotonia** (decreased muscle tone), **hyporeflexia** (decreased reflexes), and muscle **fasciculations** (uncontrollable twitching). Disruptions of signaling between UMNs and LMNs also lead to paresis and paralysis, but with **spasticity** (a form of hypertonia in which continuous contraction of muscles produces muscle tightness and stiffness that can interfere with movement) rather than hypotonia, and **hyperreflexia** (exaggerated reflexes) rather than hypotonia. On neurological examination, paresis or paralysis that is accompanied by hypotonia and hyporeflexia indicates LMN pathology, while paresis or paralysis that is accompanied by hypertonia and hyperreflexia indicates UMN pathology.

Bulbar versus Pseudobulbar Palsy

Bulbar palsy is a disorder of the LMNs of the cranial nerves originating from the medulla; these are the glossopharyngeal (CN9), vagus (CN10), accessory (CN11), and hypoglossal (CN12) nerves. Individuals with bulbar palsy have signs and symptoms of the affected cranial nerves: **dysphagia** (difficulty swallowing), **dysarthria** (unclear speech), flaccid paresis, atrophy and fasciculation of the muscles supplied by those cranial nerves, weakness of the palate, dribbling of saliva, nasal speech, and reduced or absent gag reflex. Bulbar palsy can occur with many types of neuropathology, including vascular lesions, degenerative disease, brainstem neoplasm, and inflammatory disorders.

Pseudobulbar palsy is a clinical syndrome that, similar to bulbar palsy, presents with dysphagia and dysarthria. The etiology, however, is **supranuclear** (above the cranial nerve motor nuclei). The lesion bilaterally affects the corticobulbar tracts (i.e., the UMNs that transmit signals to the LMNs) that innervate the cranial nerve motor nuclei of CN9–CN12, but it also affects innervation of the trigeminal (CN5) motor nucleus and facial (CN7) motor nucleus.

Pseudobulbar palsy differs from bulbar palsy in that it includes lack of facial expression and difficulty chewing. Furthermore, pseudobulbar palsy presents with UMN signs (spastic tongue, exaggerated gag and jaw jerk reflexes) and lacks LMN signs (atrophy and fasciculations of the tongue, absent gag reflex). The lesions producing pseudobulbar palsy may occur at any level of the corticobulbar pathways (cerebral cortex, corona radiata, internal capsule, cerebral peduncles, or the brainstem structures rostral to the cranial nerve motor nuclei).

Pseudobulbar palsy can occur with many types of neuropathology; the most common cause is multiple cerebral infarctions, but it also occurs with other processes that cause bilateral corticobulbar tract lesions, including multiple sclerosis, cerebral anoxia, and brainstem tumors.

PSEUDOBULBAR AFFECT

Pseudobulbar palsy may be accompanied by **pseudobulbar affect**, characterized by brief (seconds to minutes) and uncontrollable "emotional" outbursts, usually consisting of laughing or crying. The emotional expression is disproportionate or inappropriate to the social context (external circumstances), and incongruous with the person's mood (internal emotional state). Other terms that have been used to refer to this condition include **emotional incontinence** and **pathological laughter and crying**.

Pseudobulbar affect occurs in the context of a variety of neurologic disorders, including stroke, traumatic brain injury, brain neoplasm, multiple sclerosis, amyotrophic lateral sclerosis, Alzheimer disease, Parkinson's disease, and progressive supranuclear palsy. The mechanism underlying pseudobulbar affect is unclear, but it is believed to be a disinhibition syndrome in which there is a loss or reduction of cortical inhibitory pathways of a brainstem "emotional" center related to laughing or crying behaviors. The cortical inhibitory pathways are believed to involve the frontal lobes, corticobulbar projections, and corticopontine-cerebellar circuits.

Diagnostic Testing

Electromyography (EMG) and nerve conduction studies (NCS) measure the electrical activities of muscles and peripheral nerves, respectively. They are the most important tests in diagnosing **neuromuscular disorders**, which encompass disorders affecting muscle, the nerves that directly control muscle (LMNs), or communication between nerve and muscle.

Electromyography measures muscle electrical activity in response to stimulation of the nerve innervating the muscle. A thin needle electrode is inserted into the muscle to record the electrical activity during a voluntary contraction and at rest. Motor neuron degeneration produces characteristic abnormal electrical signals.

Nerve conduction studies measure the speed and size of signals within peripheral nerves. Two disc electrodes are placed on the skin surface over the nerve that is being evaluated. One electrode delivers a mild electrical pulse (stimulating electrode), and the other records the nerve response (recording electrode).

EMG and NCS are important tools for diagnosing and differentiating lower motor neuron diseases, as well as **radiculopathy** (compression of a spinal nerve root as a result of spinal joint disease, such as pinched nerve due to disc herniation), **peripheral neuropathies** (e.g., due to diabetes), **mononeuropathy** (single nerve damage, such as in carpal tunnel syndrome), neuromuscular transmission disorders (e.g., myasthenia gravis), and primary muscle diseases (e.g., muscular dystrophies).

Detailed brain imaging helps identify structural lesions affecting UMNs that may be due to etiologies other than neurodegeneration, such as stroke, multiple sclerosis, or neoplasm.

Myasthenia Gravis

Myasthenia gravis is a disease that interferes with transmission at the skeletal muscle neuromuscular junction. The hallmark of this disease is muscle weakness, particularly during sustained activity. It commonly affects muscles controlling the eyelids, resulting in eyelid drooping (**ptosis**); muscles controlling eye movements, resulting in double vision (diplopia) and other eye movement anomalies; and muscles controlling facial expression, speaking, chewing, and swallowing. Myasthenia gravis is an autoimmune disease that targets nicotinic acetylcholine receptors, reducing the number of functional receptors at the neuromuscular junction and the efficiency of neuromuscular synaptic transmission. Cholinesterase inhibitors (drugs that interfere with acetylcholinesterase, the enzyme that degrades acetylcholine at the neuromuscular junction) alleviate the signs and symptoms of myasthenia gravis by increasing the concentration of acetylcholine at the neuromuscular junction.

Cerebral Palsy

Cerebral palsy (CP) is a group of nonprogressive developmental neurological motor disorders affecting movement, posture, balance, and/or coordination. It is caused by abnormalities of the fetal or infant brain due to a variety of etiologies. It is usually diagnosed within the first 2 years after birth; signs and symptoms appear during infancy or preschool years. Cerebral palsy is the most common cause of motor disability. It may occur with or without other neurological impairments (e.g., sensory/perceptual, cognitive, communication, behavior, seizure disorders) due to brain injury beyond the motor system.

Cerebral palsy has multiple causes, including fetal/infant stroke, fetal/infant traumatic brain injury, maternal infection during pregnancy, infant infection (e.g., bacterial meningitis, viral encephalitis), and birth-related asphyxia. Risk factors for CP include premature birth (< 28 weeks), breech presentation, and low birth weight. Most cases (85%–90%) are classified as congenital due to brain damage that occurred before or during birth, although the specific cause is often unknown. A small percentage of cases are classified as acquired due to brain injury that occurred after birth, usually infection or traumatic brain injury.

Cerebral palsy is classified along multiple dimensions, including type/nature of the motor disorder, distribution of motor impairment, etiology, structural brain abnormalities on neuroimaging, and severity of impairment. The four main subtypes of CP based on the primary underlying motor abnormality are: spastic, athetoid, ataxic, and mixed. Each type is due to a different neurological abnormality.

Spastic cerebral palsy is the most common type of CP, affecting more than 70% of individuals with CP. It is characterized by impairment of voluntary skilled movements and hypertonia (abnormally high muscle tone, resulting in muscle tightness and stiffness). Spastic CP occurs more frequently in children who are born preterm. Spastic CP is due to damage to the pyramidal motor system (motor cortex and/or corticospinal tract). It is often associated with MRI findings of periventricular leukomalacia (softening of the brain white matter) and/or intraventricular hemorrhage. Spastic CP is classified into three topographical subtypes based on which limbs are affected. In **spastic hemiplegic CP**, movement impairment primarily affects one side of the body. In **spastic diplegic CP**, movement impairment primarily affects the legs. In **spastic quadriplegic CP**, movement impairment prominently affects all four limbs.

Dyskinetic cerebral palsy is the second most common type of CP, affecting 15%–20% of those with CP. It is characterized by abnormal involuntary movements that make it difficult to control body movements and coordination. The abnormal involuntary movements in dyskinetic CP are dystonia (abnormally increased muscle tone and prolonged involuntary muscle contractions affecting only some muscles, resulting in an abnormal posture), athetosis (a dyskinesia characterized by slow, writhing, involuntary movements), and chorea (a dyskinesia characterized by successive, involuntary, brisk movements that may resemble fragments of purposeful voluntary movements). In dyskinetic CP, dystonia, athetosis, and chorea may exist alone or together in different combinations. Dyskinetic CP is typically caused by lesions of the basal ganglia or thalamus that occur during brain development, often due to hypoxic-ischemic brain injury.

Ataxic cerebral palsy is the least common form of CP, affecting 5%–10% of those with CP. It is characterized by prominent deficits in balance and coordination. The main neuroanatomical structure affected by ataxic CP is the cerebellum.

Mixed cerebral palsy is characterized by a mixture of motor signs. The most common mixed form involves a combination of dyskinetic and spastic CP. Mixed CP is due to damage to multiple components of the brain's motor system.

Motor Neuron Disease

The **motor neuron diseases** (MNDs) are a group of neurodegenerative disorders that specifically affect motor neurons. The MNDs are classified according to whether degeneration affects LMNs, UMNs, or both LMNs and UMNs, as well as whether they are inherited or sporadic. Degeneration of LMNs is often apparent on gross postmortem examination as the ventral roots of the spinal cord are atrophic compared to the dorsal roots. The most common MND is amyotrophic lateral sclerosis; other MNDs include primary lateral sclerosis and progressive bulbar palsy.

Amyotrophic Lateral Sclerosis

Amyotrophic lateral sclerosis (ALS), also called **Lou Gehrig's disease** or **classical motor neuron disease**, is a neurodegenerative disorder affecting both upper and lower motor neurons. It is characterized by loss of muscle strength and muscle wasting. Individuals with ALS gradually lose the ability to move the arms and legs, and to hold the body upright; approximately 75% also lose the ability to speak, chew, and swallow. Some have only LMN involvement, a variant known as **progressive muscular atrophy**. A significant proportion of those with ALS also develop frontotemporal dementia, a variant known as **frontotemporal dementia with motor neuron disease** (FTD-MND).

Amyotrophic lateral sclerosis begins most commonly in people between 40 and 60 years of age. Men are affected more often than women. Most cases occur sporadically, but about 10% of cases are familial.

Most individuals with ALS die from respiratory failure, when muscles of the diaphragm and chest wall fail to function properly. This usually occurs within 3 to 5 years from symptom onset; however, about 10% of individuals with ALS survive for 10 or more years. **Stephen Hawking** (1942–2018), the world-famous theoretical physicist, had an early onset, slowly progressive form of ALS and lived with the disease for over 50 years.

Progressive Bulbar Palsy

Progressive bulbar palsy, also called **progressive bulbar atrophy**, is a neurodegenerative disorder affecting the LMNs of CN9–CN12 (glossopharyngeal, vagus, accessory, and hypoglossal nerves) that originate from the medulla. The presenting symptoms are **dysphagia** (difficulty swallowing) and **dysarthria** (unclear speech). Degeneration usually begins in the hypoglossal nucleus (CN12), producing flaccid paresis, atrophy, and fasciculations of the tongue, with bilateral involvement from the outset. The degeneration ascends gradually to affect the cranial nerve motor nuclei that innervate the pharynx, larynx, and soft palate, resulting in dysphagia, dysarthria, drooling saliva, dysphonia (specifically hoarse voice), and loss of palatal and gag reflexes.

The degeneration may ascend to involve the facial motor nucleus (CN7) controlling the muscles of facial expression and the trigeminal motor nucleus (CN5) affecting the muscles of mastication, thereby giving rise to a mask-like face and loss of chewing. Onset of progressive bulbar palsy usually occurs between 50 and 70 years of age; most people diagnosed with this disorder eventually progress to ALS (**bulbar-onset ALS**).

Primary Lateral Sclerosis

Primary lateral sclerosis is a neurodegenerative disorder affecting UMNs originating from the motor cortex. It often affects the legs first, followed by the trunk, arms, and hands, and finally the muscles innervated by the lower cranial nerves. It is characterized by slow, clumsy, effortful movements; stiff legs and arms; and slowed, slurred speech. Affected individuals commonly experience pseudobulbar affect.

Onset of primary lateral sclerosis is very gradual, most commonly beginning in people between 40 and 60 years of age; symptoms progress gradually over years. It differs from ALS in that with primary lateral sclerosis, LMNs are spared, disease progression is slow, and expected life span is normal.

Apraxia

The term *praxis* refers to the execution of skilled movements (also referred to as learned movements). Such movements are required for tool use and communication by gesture or vocalization, and typically require the upper extremities and the orofacial muscles used in speaking. These behaviors evolved over the course of human history, undoubtedly depending on changes in brain circuitry involved in motor control. But skilled movements also may require the torso (e.g., hula hooping, belly dancing), the lower legs and feet (e.g., tap dancing), or the whole body (e.g., gymnastic moves, ball skills such as batting in baseball or serving in tennis).

Apraxia encompasses a spectrum of disorders characterized by the inability to perform learned movements, that is not attributable to more elemental motor deficits such as paresis (weakness), bradykinesia (slowed movements), dystonia (increased or decreased muscle tone), dyskinesia (e.g., tremor, chorea, athetosis), or ataxia (incoordination due to cerebellar pathology or loss of proprioception). It also is not attributable to sensory deficit, language comprehension disorder, or a generalized cognitive impairment (e.g., inattention). Proper execution of skilled movements requires sequencing and spatial organization of motor commands. Apraxia is a disturbance of stored motor knowledge for skilled movements due to motor system dysfunction at the level of the cerebral cortex, without involvement of primary motor cortex. It has been referred to as a **cognitive motor disorder** and **motor agnosia**.

Apraxia occurs in a variety of clinical contexts, but most commonly in stroke and neurodegenerative disease. Within the context of neurodegenerative disease, it is a hallmark sign and diagnostic feature of corticobasal syndrome, which is characterized by parkinsonism plus apraxia (see Chapter 11, "The Basal Ganglia"). In cases where apraxia exists with another motor disturbance, such as the asymmetric akinesia and rigidity of corticobasal syndrome, the disorder of skilled movements cannot be explained by the more elemental motor disturbance.

The first description of apraxia is attributed to John Hughlings Jackson (1835–1911), who in 1861 described a neurological condition in which patients were unable to perform movements to command but had no difficulty performing the same movements spontaneously. The term *apraxia* (meaning *without action*) was introduced in 1871 by **Heymann Steinthal** (1823–1899), a German linguist, to denote the faulty use of everyday objects (e.g., a fork and knife) by patients with aphasia. In 1900, **Hugo Leipmann** (1863–1925) published the first single-case study of a patient with left hemisphere stroke who was unable to imitate simple hand positions or perform pantomimes (gesturing the use of an object) with his right hand. The marked difference in the patient's performance when using the right versus left hand led Leipmann to infer that the disorder could not be attributed to poor comprehension, and that the patient had a specific deficit in the ability to perform skilled movements with the right hand.

In 1905, Leipmann published the first group study that firmly established the clinical validity of apraxia. He studied right-handed patients with either right or left hemiplegia and tested their ability to use their unaffected hand to pantomime to command, imitate gestures, and perform skilled movements using actual tools and objects. Liepmann found that approximately half of the patients with right hemiplegia (left hemisphere lesions) had apraxia, while none of the patients with left hemiplegia (right hemisphere lesions) did. Many of the apraxic patients also had aphasia, but the aphasia did not account for the apraxia as the patients were unable to imitate gestures (a task that does not require language). Leipmann also observed that patients with apraxia typically performed best when using actual tools and objects, less reliably when asked to imitate the use of a tool as demonstrated by the examiner, and worst when asked to pantomime the use of a tool.

Leipmann hypothesized that the left hemisphere stores the "space-time plans" of learned actions, and that these plans were stored in the left parietal lobe. The space-time plans, referred to as "movement representations," "movement formulae," **"praxicons"** and **"motor engrams"** (motor instructions in memory), were believed to be conveyed via association fibers to a "central region" (motor cortex) to command the motor system to adopt the appropriate spatial positions of the relevant body parts over time when performing learned actions.

Leipmann described three forms of apraxia: limb-kinetic apraxia, ideomotor apraxia, and ideational apraxia, as well as unilateral and bilateral forms. He believed that the motor engrams are stored in the inferior parietal lobule of the left hemisphere and transcoded into innervatory patterns at the premotor cortices, which in turn drive the primary motor cortex. Leipmann also believed that the different forms of apraxia were due to disruption at different stages in the cortical processing of voluntary action.

Norman Geschwind (1926–1984) proposed an anatomical model of apraxia based on disconnection between posterior temporal language areas (i.e., Wernicke's area) and left premotor cortex due to a lesion within the superior longitudinal fasciculus that normally connects the two cortical areas. With such lesions, language comprehension is preserved, but the ability to perform actions to verbal command is impaired. Geschwind also hypothesized that such lesions disconnect posterior visual association areas and left premotor cortex, resulting in impaired gesture imitation.

The term *apraxia* has since been applied to a wide variety of neurobehavioral disorders beyond the three forms described by Liepmann (limb-kinetic, ideomotor, and ideational apraxias). These include buccofacial, truncal, unilateral limb, conduction, constructional, dressing, oculomotor, gait, and magnetic apraxias, as well as apraxia of speech (verbal apraxia), various disconnection apraxias, and apraxic agraphia. A variety of classification schemes have been proposed, some distinguishing body-part specific apraxias (e.g., oculomotor, gait, buccofacial, limb apraxias) and others distinguishing task-specific apraxias (e.g., dressing apraxia, constructional apraxia, apraxia of speech, and apraxic agraphia). The term *apraxia* also has been applied to conditions that are not the result of primary disorders of programming learned movements; these misclassified movement abnormalities are better categorized separately from the apraxias but will nevertheless be addressed here following the convention of the field (see "Apraxia-Like Syndromes," below).

Ideomotor Apraxia

Ideomotor apraxia is the most common type of apraxia; thus, it is often referred to simply as *apraxia*. It is characterized by impairment in the spatiotemporal organization of skilled movements. It affects both **transitive movements** (movements that involve using a tool, utensil, or instrument) and **intransitive movements**. Manual dexterity, however, is normal. Movement abnormalities are apparent when the person is asked to pantomime, imitate movements, and/or demonstrate the use of objects. There is a hierarchy of difficulty in performing these tasks, with pantomime the most difficult, imitation of intermediate difficulty, and actual tool use the least difficult.

Ideomotor apraxia is characterized by the inability to generate a spatial plan for movement and to translate the plan into details of angular joint motions. The movements are incorrect in form, but the goal or intent of the act is usually recognizable as correct. Consequently, the errors most characteristic of ideomotor apraxia are spatial errors, of which there are three types: postural, spatial orientation, and spatial movement errors. These errors are most apparent when asked to pantomime transitive movements (e.g., make a movement as if holding and using a utensil or tool). **Postural errors** include failing to position the hand correctly toward the imaginary tool. **Spatial orientation errors** include movements that fail to orient the imaginary tool toward the target. For example, when asked to pretend to cut a slice of bread with a knife, the person may orient the imaginary knife in an arc around their body rather than in a consistent sagittal plane. **Spatial movement errors** involve moving the limb through space incorrectly due to incorrect joint movements, despite the core movement (e.g., cutting, pounding, twisting) being executed correctly. For movements that require the coordination of movement at two or more joints, persons with ideomotor apraxia may use one joint movement primarily (typically the proximal joint), or stabilize a joint that should be moving while moving a joint that should be stable. For example, when asked to pantomime the use of a screwdriver, the person with apraxia may fix the wrist and elbow and rotate the shoulder, rather than fixing the wrist and shoulder and twisting at the elbow.

Ideomotor apraxia may selectively involve the limbs (**limb apraxia**), face and mouth (**buccofacial/oral apraxia**), or trunk (**truncal apraxia**). Limb ideomotor apraxia may be unilateral or bilateral.

ASSESSMENT

Diagnosis of apraxia requires ruling out non-apraxic motor and sensory disorders as the underlying cause of impaired skilled movements. As apraxia undermines the ability to use tools and thereby may affect activities of daily living, patients and their caregivers should be asked about ability to perform activities that involve household tools, such as utensils, a toothbrush, a hammer, and scissors.

The diagnosis of apraxia is complicated when there is a coexisting aphasia or dementia. When left-hand apraxia occurs in the context of right hemiparesis, those affected often attribute their clumsiness to use of their non-dominant hand. Since apraxia is usually mildest when using actual objects and people are rarely asked to pantomime or imitate, right-handed individuals with right hemiparesis and left-hand apraxia rarely complain spontaneously and are often unaware of the apraxic disturbance.

Testing for apraxia should be hierarchical, first administering the most sensitive tasks involving pantomiming actions to verbal commands (e.g., show me how you cut paper with scissors; pretend to brush your teeth holding an imaginary toothbrush; pretend to comb your hair; pretend to flip a coin; pretend to hammer a nail; pretend to pour water from a pitcher into a glass). The movement must be mimed without nonverbal cues from the examiner. Both hands should be tested because apraxia may be unilateral; however, the examiner should not ask the patient to perform the same command sequentially with each hand because visual self-cueing will improve performance on the second trial. In addition to observing a patient's performance for precision, timing, and location of the movements, the clinician should ascertain if they are disturbed by or aware of their own movement errors.

As ideomotor apraxia may be body-part specific, assessment should include examination of skilled movements of the limbs, face, and body. **Limb ideomotor apraxia** is tested by asking the patient to pantomime and imitate motor acts involving the limbs, both transitive (object-related) movements and intransitive gestures.

When asked to pantomime transitive movements, those with limb ideomotor apraxia often make a type of postural error known as **body part as object** in which they use a body part as a tool rather than pretending to hold the tool. For example, when asked to pantomime the use of scissors, the patient will use their fingers as if they were the blades. When asked to pantomime brushing teeth, they may use their fingers as if they were the toothbrush. If the patient uses body part as object, they should be instructed to not use the body part as the object, but rather to pretend that they are really holding the object; those with limb ideomotor apraxia persist in making body part as object errors despite this instruction. If the patient fails pantomiming transitive movements, the examiner assesses the ability to imitate gestures. If the patient cannot imitate the action, they should be provided with the actual object and again asked to follow the command. The presence of the object gives the patient additional visual and proprioceptive cues that may facilitate performance. Ability to perform intransitive communicative gestures (**emblems**) should also be tested, such as waving good-bye, saluting, and making a peace sign. **Buccofacial ideomotor apraxia** is tested by asking the patient to pantomime and imitate motor acts involving the buccofacial musculature, such as blowing out a candle, drinking with a straw, and sticking out the tongue. Patients should be discouraged from cueing themselves by pretending to have the object (e.g., straw, match) in their hands (by gently restraining their hands), as self-cueing facilitates performance. The examiner should look for incomplete, unrelated, or opposite motor acts (e.g., they may inhale while blowing out the imaginary match). **Truncal ideomotor apraxia** is tested by asking the patient to pantomime and imitate movements such as bowing, curtsying, and pretending to kick a ball.

Those with ideomotor apraxia have the greatest difficulty with pantomiming transitive movements to verbal command. They may improve by imitating actions demonstrated to them, but imitation is often defective as well. They may also improve when using the actual object, but again performance often remains defective. Thus, the severity of ideomotor apraxia can be graded as follows: mild if there is only a failure to pantomime; moderate if there is a failure to pantomime and imitate; and severe if there is a failure to pantomime, imitate, and perform the behavior with an actual object.

LOCALIZATION AND PATHOPHYSIOLOGY

Ideomotor apraxia occurs with lesions of the inferior parietal lobule within the language-dominant hemisphere, or projections from this region to left hemisphere motor association cortex carried in the arcuate fasciculus. There is no conclusive evidence for the anatomical correlate of impairments in producing specific body-part apraxias (limbs, mouth/face, and trunk). Because ideomotor apraxia occurs with lesions in the inferior parietal lobule of the language-dominant hemisphere, many of those with ideomotor apraxia are also aphasic. Ideomotor apraxia is most often associated with stroke and degenerative dementia of the Alzheimer type.

Unilateral limb apraxia occurs on the left side of the body and thus is also known as **left-sided apraxia**. It is due to lesions in the anterior corpus callosum, which prevents information from traveling from the left motor association cortex to the right motor association cortex. Because it is a disconnection syndrome, it is also known as **callosal apraxia**.

Ideational Apraxia

Ideational apraxia, also referred to as **conceptual apraxia**, was defined by Leipmann as an inability to sequence the different components of a complex motor act to carry out an "ideational plan." It is a breakdown in performance of tasks that involve a series of steps. The person can perform each individual step but cannot integrate the parts accurately to complete the sequence. The disorder affects spontaneous motor sequences as well as those initiated after verbal command, undermining the ability to manipulate the environment effectively and perform instrumental activities of daily living (e.g., cooking a meal, making a bed).

Those with ideational apraxia make "content errors." They generate well-formed movements, but the movements are incorrect for the tool; thus, they pantomime using a tool as if it were another tool. They are unable to select the actions associated with the use of specific tools or objects (tool-object action knowledge). For example, when asked to pantomime how to use a screwdriver, the patient may pantomime a hammering movement.

Ideational apraxia is seen most often in the context of bilateral diffuse brain disease, particularly in Alzheimer's dementia.

Apraxia of Speech

Apraxia of speech (**verbal apraxia**) is an impairment in the cortical motor programming of speech movements. It is not attributable to a primary motor deficit (e.g., paralysis, weakness, incoordination of the speech musculature) or other language disorder. It is contrasted with buccofacial ideomotor apraxia, which affects volitional movements of the tongue, jaw, and lips during non-speech tasks. Apraxia of speech occurs with left frontal lesions adjacent to Broca's area. It also occurs as a rare neurodegenerative syndrome (**primary progressive apraxia of speech**).

Apraxic Agraphia

Apraxic agraphia is an acquired disorder of writing in which the ability to make the skilled movements needed to form letters is impaired. Writing is characterized by poor letter formation, while oral spelling ability is preserved. Apraxic agraphia is not attributable to disturbance of language function, sensorimotor function, or knowledge of letters and/or words. Apraxic agraphia is a peripheral (non-aphasic) agraphia. Clinico-anatomical studies are limited, but there is evidence that apraxic agraphia occurs with lesions of the superior parietal lobule within the language-dominant hemisphere.

Disconnection Apraxias

Disconnection apraxias, also referred to as **disassociation apraxias**, are caused by white matter lesions that selectively disconnect motor areas from specific sources of input (i.e., verbal, visual, or tactile). The leads to modality-specific apraxias, with errors in movements evoked by stimuli within one modality, with preserved ability to perform movements evoked by stimuli within other modalities. There are several forms of disconnection apraxia. **Verbal-motor disassociation apraxia** is characterized by the inability to perform movements to verbal command despite preserved ability to imitate and use tools. **Visuomotor disassociation apraxia** is characterized by the inability to imitate gestures to visual stimuli, with preserved ability to perform movements to verbal command. **Tactile-motor disassociation apraxia** is characterized by impaired hand movements that involve interacting with an object (i.e., transitive movements), while intransitive movements are preserved.

Conduction Apraxia

Conduction apraxia is like ideomotor apraxia, except that imitation of learned transitive and symbolic movements is impaired, while pantomime of these movements is relatively preserved. This condition received its name because the selective deficit of imitation in conduction apraxia is like the selective deficit of repetition in conduction aphasia. Lesion localization of this rare form of apraxia is unknown.

Apraxia-Like Syndromes

As stated above, the term *apraxia* has been applied to conditions that are not due to primary disorders of programming skilled movements and therefore are not true apraxias. Even though these are not true apraxias, they will be described using the conventional nomenclature. The non-apraxic "apraxias" include limb-kinetic, constructional, dressing, oculomotor, eyelid opening, gait, and magnetic apraxias.

LIMB-KINETIC APRAXIA

Limb-kinetic apraxia was one of the three apraxic subtypes identified by Leipmann. It is defined as an acquired inability to make fine, precise, independent finger movements (i.e., loss of dexterity), and thus it has also been called **clumsy hand**. Tests of limb-kinetic apraxia include buttoning clothes, opening a safety pin, picking up a dime from a flat surface, and rapidly rotating a quarter between the thumb, index, and middle fingers. When asked to pick up a dime from a flat surface, patients with limb-kinetic apraxia cannot make a pinching movement with thumb and index finger, but rather they slide the dime off the table into the palm.

Limb-kinetic apraxia is often unilateral, affecting the hand contralateral to a hemispheric lesion of premotor cortex lying on the convexity. Studies have shown that this disorder is related to dysfunction of the corticospinal pathways controlling hand movements, and that therefore it is a lower-level motor disorder rather than a higher-level disorder of cortical movement programming. The most common cause of limb-kinetic apraxia is stroke.

CONSTRUCTIONAL APRAXIA

The term **constructional apraxia** was coined by **Karl Kleist** (1879–1960) in 1934 to describe impaired ability to construct (draw, assemble, and build) spatial forms in the absence of a lower-level movement disorder or visuoperceptual deficit. Kleist proposed that constructional apraxia was due to an altered connection between visuospatial functions and the kinetic engrams that control manual activity.

Constructional apraxia, also referred to as **visuospatial apraxia**, **visuoconstructive apraxia**, and the preferred terms **constructional impairment** and **visuoconstructive disability**, is not a true apraxia because it is not due to a primary disorder of specific programming movements required to make constructions. The term came to be used for *any* impairment in constructional ability, as manifested on tasks such as spontaneous free drawing, drawing from a model, stick pattern constructions, block designs, three-dimensional constructions, and puzzle assembly. Because inability to perform such tasks is usually due to visuospatial disorders, neglect, or executive dysfunction rather than a disorder of movement programming, the terms **constructional impairment** and **visuoconstructive disability** have replaced the term *constructional apraxia*. Similarly, the term *constructional praxis*, used to refer to the ability to assemble, join, or articulate parts and arrange the component elements in their correct spatial relationships to construct a single unitary structure, has been replaced by the term **constructional ability**. The ability to draw and create other types of constructions requires more than the organization of skilled hand movements.

Constructional disorders occur with parietal lesions of either hemisphere, as well as frontal lesions; thus, there is no single anatomical or functional etiology. Constructional disorders that occur with frontal lobe lesions are due to defective planning and strategy formation; performance improves if planning cues are provided.

Constructional disability is one of the most common behavioral alterations in Alzheimer's disease. Because of the high incidence of constructional impairment in dementia, such tasks are good screening tests in patients of advancing age who present with vague psychiatric or neurological signs and/or symptoms. Construction is a high-level perceptuomotor task that involves the integration of occipital, parietal, and frontal lobe functions. Because of the extensive cortical area necessary to perform constructional tasks, early subtle brain damage frequently disrupts performance. Most patients do not complain spontaneously of such difficulties unless their professions require such abilities (e.g., architect, engineer), and often they are quite surprised to find that they are unable to draw a clock or copy a block design.

Constructional ability is complex and entails many cognitive processes, including visuospatial processing, spatial attention, praxis, and executive functions. Constructional impairment therefore is not indicative of pathology localized to any specific region of cerebral cortex nor affecting any specific domain of function. It is a sign of gross cortical dysfunction, and intact performance on tests of constructional ability can be used to rule out more fundamental deficits in visuospatial processing, spatial attention, and the executive functions of planning and organization. Clinical examination of constructional impairment should include several tests that tap somewhat different aspects of constructional ability, such as reproduction drawing, drawing to command, and reproducing block designs. Reproduction drawings should sample both two- and three-dimensional designs. Separate sheets of paper for each design may be necessary for patients who are highly distractible or perseverative.

DRESSING APRAXIA

Dressing apraxia is the inability to dress oneself correctly due to the inability to orient garments and align them correctly to body parts (e.g., trying to put both arms into one shirt sleeve). It is not attributable to primary motor or sensory deficit or gross cognitive impairment (i.e., delirium or dementia). Dressing apraxia often occurs in the context of a neglect syndrome (e.g., failure to dress the left half of the body) and therefore occurs with right parietal lesions. However, it also may be associated with visuospatial impairment or body schema disorder. Dressing apraxia may occur with constructional impairment.

OCULOMOTOR APRAXIA

Oculomotor apraxia (gaze apraxia) is a visuomotor disorder in which pursuit and visually guided saccades are impaired, but saccades not dependent on vision (i.e., saccades to command, saccades to remembered targets, saccades to sounds, and spontaneous saccades) are preserved. Since the disorder does not involve learned skilled movements, it therefore is not a true apraxia. Oculomotor apraxia usually occurs as a component of Bálint's syndrome. It does, however, rarely occur as an isolated deficit resulting from bilateral posterior parietal lobe lesions affecting the **parietal eye fields** that surround the posterior, medial segment of the intraparietal sulcus.

APRAXIA OF EYELID OPENING

Apraxia of eyelid opening is characterized by the inability to voluntarily open the eyes bilaterally that is not attributable to paralysis. The most common etiology is an idiopathic focal dystonia, but the pathogenesis is poorly understood.

GAIT APRAXIA

Gait apraxia refers to impaired ambulation characterized primarily by **gait ignition failure** (hesitation or inability to initiate gait from a static position) that is not attributable to motor weakness or sensory impairment. However, walking is not an intentionally learned motor skill, but rather a repetitive motor pattern generated by spinal mechanisms and modified by brainstem structures; thus the disturbance is not strictly a true apraxia. This term has most often been applied to the gait disturbance in normal pressure hydrocephalus, but also to the gait disturbance in vascular dementia.

MAGNETIC APRAXIA

Magnetic apraxia was first described by **Derek Denny-Brown** (1901–1981) in 1958, as a disorder in which tactile and visual presentation of objects compels the individual to manually explore and grasp the object, with inability to release the object voluntarily. There is an associated **gegenhalten**, an involuntary resistance to **passive movement** (movement of a body part induced by another person and without voluntary motion on the part of the patient). Magnetic apraxia may extend beyond prominent and persistent instinctive grasping of hand, to affect other body parts such as the mouth or foot when they contact or are close to an object. Denny-Brown observed that the disorder was usually due to frontal lobe lesions, thus he also referred to it as **frontal apraxia**. The term *magnetic apraxia* has been applied to a variety of behavioral disorders that are conceptualized as forms of environmental dependency syndrome, including the frontal release sign of **forced grasping** (see Chapter 30, "The Neurological Examination"), manual groping behavior (see "Alien Hand Syndrome," below), and utilization behavior (see Chapter 17, "The Frontal Lobes and Associated

Disorders"). Magnetic apraxia is associated with frontal lesions and neurodegenerative disease (corticobasal degeneration, Alzheimer's disease, and progressive supranuclear palsy).

Alien Hand Syndrome

The **alien hand syndrome** (AHS), also known as **Dr. Strangelove syndrome**, is a rare disorder of involuntary yet complex, goal-directed hand movement. In other words, the limb performs movements that appear purposeful but occur autonomously without conscious control or intention. The disorder was first described by **Kurt Goldstein** (1878–1965).

Alien hand syndrome has been reported with a variety of lesion locations, including the corpus callosum, prefrontal cortex, supplementary motor area, posterior parietal cortex, anterior cingulate, thalamus, and basal ganglia. It occurs with a variety of brain pathologies, most commonly with callosotomy, anterior cerebral artery stroke, midline tumor, and neurodegenerative disease, particularly corticobasal degeneration. Those with AHS, agnosia, apraxia, and early-onset dementia likely have corticobasal degeneration.

Most cases of AHS are due to disruption of interhemispheric connections and/or connections between the parietal lobes and frontal areas. The underlying neural mechanisms remain unclear, but it has been proposed that the abnormal movements are a release phenomenon, as motor cortex is released from conscious control by neural systems involved in intentional planning. Three anatomical variants of AHS have been described, two anterior (callosal and frontal) and one posterior.

Callosal variant AHS occurs with callosal lesions, particularly those involving the anterior third of the rostrum. This variant often features **intermanual conflict**, a phenomenon in which the alien hand counteracts the intended action of the normally functioning hand. In other words, the two hands perform opposing purposeful movements so that they appear to be fighting against each other. Since callosal variant AHS occurs with callosal lesions, it is accompanied by other interhemispheric disconnection syndromes.

Frontal variant AHS is associated with lesions of the supplementary motor area, dominant medial prefrontal cortex, cingulate cortex, or anterior corpus callosum. This variant, also known as **anarchic hand**, is characterized by **manual groping behavior** (the hand seems to search constantly for nearby objects), grasping, and compulsive manipulation of objects that may be difficult to release. Additional symptoms may include ideomotor apraxia and perseveration. Utilization behavior, a type of environmental dependency syndrome that occurs with frontal lobe lesions, has been considered a bilateral form of AHS. Individuals with this disorder automatically reach for and "utilize" objects within their reach in an "object-appropriate" manner when such object use is inappropriate. However, unlike AHS, utilization behavior is not associated with a sense that the hands behave autonomously and without conscious control, thus it is not considered to be a form of AHS.

Posterior variant AHS is the least common form of AHS. It is associated with posterolateral parietal, occipital, or thalamic lesions, and it may be accompanied by hemianesthesia, hemianopia, visuospatial neglect, or optic ataxia. This variant generally involves the non-dominant hand, and it frequently features avoidance response and involuntary levitation. **Avoidance response** is characterized by the alien hand unintentionally withdrawing from contact with stimuli. **Involuntary levitation** is characterized by the alien hand levitating absent the person's intention.

Summary

The basic elements of the motor system are muscles, primary motor neurons, spinal cord, motor cortex, multiple brainstem motor nuclei, basal ganglia, cerebellum, and the neuronal pathways connecting these structures. Primary motor neurons, also known as lower motor neurons, synapse directly with muscle cells; those neurons supplying the muscles of the body originate from the ventral horn of the spinal cord and travel in the spinal nerves, while those supplying the muscles of the face, head, and neck originate from the brainstem cranial nerve motor nuclei and travel in the cranial nerves. Lower motor neurons receive input from upper motor neurons, which originate from motor cortex and subcortical motor nuclei. The motor neuron diseases are a group of neurodegenerative disorders that specifically affect motor neurons. The most common motor neuron disease is ALS.

Apraxia is a disorder of movement programming that manifests as a selective loss of the ability to perform skilled movements. It is due to motor system dysfunction at the level of the cerebral cortex, exclusive of primary motor cortex. A variety of classification schemes have been proposed, some distinguishing body-part specific apraxias (e.g., oculomotor, gait, buccofacial, limb apraxias) and others distinguishing task-specific apraxias (e.g., dressing apraxia, constructional apraxia, apraxia of speech, and apraxic agraphia). Some of these conditions are not due to primary disorders of programming skilled movements and therefore are not true apraxias.

Alien hand syndrome is a rare disorder of involuntary, complex hand movements that appear purposeful but occur autonomously without conscious control or intention. It occurs with a variety of lesion locations, including the corpus callosum, prefrontal cortex, supplementary motor area, posterior parietal cortex, anterior cingulate, thalamus, and basal ganglia. It also occurs with a variety of brain pathologies, most commonly with callosotomy, anterior cerebral artery stroke, midline tumor, and neurodegenerative disease (particularly corticobasal degeneration). Three anatomical variants of AHS have been described: callosal variant, frontal variant, and posterior variant.

Additional Reading

1. Alfaro A, Bernabeu Á, Badesa FJ, García N, Fernández E. When playing is a problem: an atypical case of alien hand syndrome in a professional pianist. *Front Hum Neurosci*. 2017;11:198. Published 2017 Apr 24. doi:10.3389/fnhum.2017.00198
2. Biran I, Chatterjee A. Alien hand syndrome. *Arch Neurol*. 2004;61(2):292–294. doi:10.1001/archneur.61.2.292
3. Brown RH, Al-Chalabi A. Amyotrophic lateral sclerosis. *N Engl J Med*. 2017;377(2):162–172. doi:10.1056/NEJMra1603471
4. Buxbaum LJ, Randerath J. Limb apraxia and the left parietal lobe. *Handb Clin Neurol*. 2018;151:349–363. doi:10.1016/B978-0-444-63622-5.00017-6
5. Denny-Brown D. The nature of apraxia. *J Nerv Ment Dis*. 1958;126(1):9–32. doi:10.1097/00005053-195801000-00003
6. Foundas AL. Apraxia: neural mechanisms and functional recovery. *Handb Clin Neurol*. 2013;110:335–345. doi:10.1016/B978-0-444-52901-5.00028-9
7. Foundas AL, Duncan ES. Limb apraxia: a disorder of learned skilled movement. *Curr Neurol Neurosci Rep*. 2019;19(10):82. Published 2019 Nov 12. doi:10.1007/s11910-019-0989-9
8. Gainotti G, Trojano L. Constructional apraxia. *Handb Clin Neurol*. 2018;151:331–348. doi:10.1016/B978-0-444-63622-5.00016-4
9. Goldenberg G. Apraxia: the cognitive side of motor control. *Cortex*. 2014;57:270–274. doi:10.1016/j.cortex.2013.07.016
10. Gross RG, Grossman M. Update on apraxia. *Curr Neurol Neurosci Rep*. 2008;8(6):490–496. doi:10.1007/s11910-008-0078-y
11. Hassan A, Josephs KA. Alien hand syndrome. *Curr Neurol Neurosci Rep*. 2016;16(8):73. doi:10.1007/s11910-016-0676-z
12. Heilman KM. Apraxia. *Continuum (Minneap Minn)*. 2010;16(4 Behavioral Neurology):86–98. doi:10.1212/01.CON.0000368262.53662.08
13. Heilman KM. There is more than imitation. *Cortex*. 2014;57:275–308. doi:10.1016/j.cortex.2014.01.022
14. Heilman KM, Watson RT. The disconnection apraxias. *Cortex*. 2008;44(8):975–982. doi:10.1016/j.cortex.2007.10.010
15. Heilman KM. Ideational apraxia: a re-definition. *Brain*. 1973;96(4):861–864. doi:10.1093/brain/96.4.861
16. Kent RM. Cerebral palsy. *Handb Clin Neurol*. 2013;110:443–459. doi:10.1016/B978-0-444-52901-5.00038-1
17. Mendoza G, Merchant H. Motor system evolution and the emergence of high cognitive functions. *Prog Neurobiol*. 2014;122:73–93. doi:10.1016/j.pneurobio.2014.09.001
18. Miller A, Pratt H, Schiffer RB. Pseudobulbar affect: the spectrum of clinical presentations, etiologies and treatments. *Expert Rev Neurother*. 2011;11(7):1077–1088. doi:10.1586/ern.11.68
19. Ochipa C, Rothi LJ, Heilman KM. Conduction apraxia. *J Neurol Neurosurg Psychiatry*. 1994;57(10):1241–1244. doi:10.1136/jnnp.57.10.1241
20. Watson RT, Heilman KM. Callosal apraxia. *Brain*. 1983;106 (Pt 2):391–403. doi:10.1093/brain/106.2.391

Disorders of Consciousness

Introduction

This chapter describes the **disorders of consciousness**, namely coma, vegetative state, minimally conscious state, and delirium, as well as the coma-like states of locked-in syndrome and brain death. **Consciousness** may be defined as a neurological state in which one is fully awake and aware of self and environment and has normal responses to external stimulation and inner needs. **Unconsciousness** is the opposite; it is a neurological state of overtly (i.e., observable) diminished responsiveness to environmental stimuli and unawareness of self and environment.

Consciousness has two basic dimensions: wakefulness and awareness. **Wakefulness** is also referred to as **arousal** or **alertness**. Behavioral signs of wakefulness include eye-opening, gaze, facial expression, and body posture; the contrast between waking and sleeping illustrates a difference between states of arousal. **Awareness** refers to the contents of consciousness. It is the ability to perceive and be cognizant of both the external environment and internal thoughts and emotions, including those relating to physiological states such as hunger and pain.

Normally, consciousness fluctuates over a 24-hour cycle, between the waking state and the sleep state, from which one can be readily awakened. These states and their rhythm are regulated by physiological processes involving a complex interplay of brainstem and diencephalic nuclei, and their interactions with the cerebral cortex. During the normal waking state, one is both awake and aware. During the non-rapid eye movement sleep state, one is neither awake nor aware. During rapid eye movement sleep state, one is not awake but may be aware of dream content.

Wakefulness is generally a prerequisite for awareness, excluding the rapid eye movement sleep state which is unique in this regard. Wakefulness, however, does not depend on awareness; one can be awake but unaware, as occurs in some disorders of consciousness. The disorders of consciousness, therefore, are characterized by either an alteration in awareness, or an alteration in both wakefulness *and* awareness (since disorders that undermine wakefulness also undermine awareness). There is a spectrum of states of altered level of consciousness, the mildest of which reflect a selective disturbance of awareness to varying degrees, and the more severe which reflect disturbances of awareness and wakefulness to varying degrees.

Clouding of consciousness is characterized by reduced awareness of the environment and inattention, without frank disorientation. **Confusional state**, often referred to simply as **confusion**, is an altered awareness characterized by disorientation, bewilderment, and difficulty following commands. **Lethargy**, also referred to as **somnolence**, is a state of reduced awareness and wakefulness in which the patient sleeps most of the day; they can be aroused by moderate stimuli but drift back to sleep when not actively stimulated. When aroused, they are confused and may be incomprehensible. **Obtundation** is a more severe state of reduced awareness and wakefulness in which the patient is difficult to arouse; constant stimulation or painful stimuli are required. Responses to stimulation are slowed and the patient is disoriented. **Stupor** is a deeper unresponsive state from which the patient can be aroused only transiently with vigorous physical painful stimuli and responds with reflex movements, groaning, mumbling, or restless movements; the patient is otherwise unresponsive and makes no meaningful or purposeful movements. **Coma** is a state of unarousable unresponsiveness. Thus, level of consciousness exists on a continuum, from full consciousness in which the individual is awake and aware, through various levels of semiconsciousness, to coma in which the individual is neither awake nor aware and cannot be aroused. Qualitative descriptors used to describe the spectrum of states between the two ends of the consciousness spectrum, however, are imprecise, and it is preferable to use coma grading scales to objectively measure level of consciousness and monitor it over time.

Many different insults affecting the CNS, including primary brain injury, intoxication by alcohol or other drugs, systemic illness, and metabolic abnormalities, can produce loss or alteration of consciousness. Transient episodes of loss of consciousness with sudden onset lasting seconds to minutes occur with syncope (fainting) due to an acute reduction of blood flow to the brain, some generalized seizures due to disruption of brain electrophysiology, and some concussions in which head trauma produces a loss of consciousness lasting less than 30 minutes. Coma is a state of deep unconsciousness in which a person is unresponsive and cannot be awoken that lasts at least 1 hour.

Altered states of consciousness are states in which wakefulness is preserved but awareness is altered in that there is a misrepresentation of reality. These include sleep-related hallucinations that occur when falling asleep (hypnagogic hallucinations) or waking up (hypnopompic hallucinations), hallucinations and delusions that occur within the context of psychotic episodes or psychedelic drug experiences, some epileptic seizures, and hypnotic, trance, and dissociative states.

Neurological Assessment: The Glasgow Coma Scale

Initial neurological examination of the patient with loss or alteration of consciousness briefly assesses response to pain, motor function (including whether there is abnormal posturing), eye-opening, and verbalization. Cranial nerve examination assesses extraocular movements and pupil symmetry, as well as pupillary, corneal, cough, and gag reflexes (see Chapter 30, "The Neurological Examination"). Absence of focal neurological findings suggests a metabolic, infectious, or toxicologic cause of coma.

Results of the neurologic examination are quantified roughly by standardized, quantitative coma grading scales such as the **Glasgow Coma Scale** (GCS), the first standardized tool and the most widely used tool to quantify level of consciousness. Such scales provide a simple way to assess level of consciousness, quantify the severity of coma, monitor recovery or deterioration over time, and predict long-term outcome. They are also used to guide early management of patients with acute brain pathology (e.g., traumatic brain injury, stroke, brain infection, drug overdose, poisoning).

The GCS is a 15-point neurological function scale that assesses responsiveness by evaluating three behaviors: motor response, verbal response, and eye-opening response (see Table 20.2 in Chapter 20). Subjects are graded according to their best response in each domain. The lowest score in each domain is 1; the maximum score varies across domains, so that the range of scores is 1–6 for motor responsiveness, 1–5 for verbal responsiveness, and 1–4 for eye-opening responsiveness. The range of total scores is 3–15, with lower scores indicating worse function and higher scores indicating better function. Scores of 3–8 indicate coma. With coma, the patient has no response (no speech, no eye-opening, no motor response) to any stimulation, including painful stimuli (e.g., supraorbital pressure, pinching of the skin on the neck, sternal pressure, nail-bed pressure). GCS scores are related to the depth of the lesions. Lesions in deep central gray matter or brainstem tend to be associated with lower GCS scores than cortical or subcortical white matter lesions.

Motor Response

Motor responsiveness is graded on a 6-point scale (1–6). A score of 6 indicates that the patient can make spontaneous purposeful movements and follow (obey) simple motor commands (e.g., "hold up your arms," "stick out your tongue"). If the patient is unable to follow commands, the examiner applies a pain stimulus (e.g., earlobe pinch, nail-bed pressure, sternal rub, supraorbital pressure, trapezius

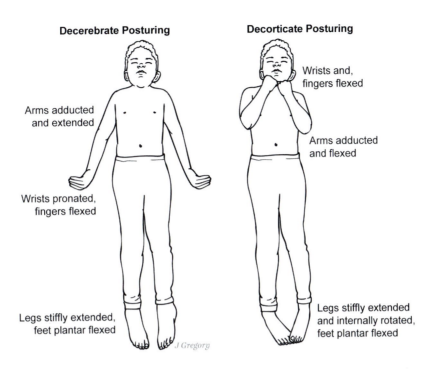

FIGURE 25.1. Decerebrate and decorticate posturing are pathological stereotyped movements that are made in response to noxious stimuli (or passive hyperextension of the head) and indicative of significant brain injury.

squeeze). A score of 5 indicates that the patient makes purposeful movements that locate the pain stimulus and are aimed at removing it (e.g., pushing the stimulus away). A score of 4 indicates that the patient responds to the pain stimulus by limb withdrawal (motor flexion). A GCS score of 3 indicates that the patient responds to the pain stimulus by abnormal posturing with flexion movements (decorticate posturing). A GCS score of 2 indicates that the patient responds to the pain stimulus by abnormal posturing with extension movements (decerebrate posturing). A score of 1 indicates no motor response or movement.

Abnormal posturing in comatose patients involves stereotyped involuntary flexion or extension movements of the extremities that are usually elicited by an external pain stimulus, although they may also occur spontaneously without a stimulus. The abnormal flexion or extension occurs when one set of muscles is incapacitated while the opposing set is not, and the working set of muscles contracts. **Decorticate posturing (decorticate response, decorticate rigidity, flexor posturing)** is characterized by flexion of the arms at the elbow, with flexion of the wrists and fingers and shoulder adduction (Figure 25.1). The legs are extended with internal rotation and plantar flexion. Because the arms are bent in toward the body with clenched fists held on the chest, it is also referred to colloquially as "mummy baby" posture. The lesions producing decorticate posturing interrupt signaling within the corticospinal tracts above the level of the midbrain red nuclei (Figure 25.2). This results in disinhibition of the red nuclei

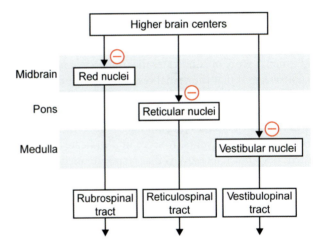

FIGURE 25.2. The mechanisms underlying decerebrate and decorticate posturing involve the vestibulospinal, pontine reticulospinal, and rubrospinal tracts. The vestibulospinal and pontine reticulospinal tracts excite limb extensor motor neurons and inhibit limb flexor motor neurons; the rubrospinal tracts excite upper limb flexor motor neurons. Normally, higher brain centers inhibit these pathways. Extensive lesions at or below the level of the red nuclei result in decerebrate posturing. Extensive lesions above the level of the red nuclei result in decorticate posturing.

and descending rubrospinal tracts, which provide excitatory input to the motor neurons in the cervical spinal cord that innervate the flexor muscles of the upper extremities. Thus, decorticate posturing indicates damage that is localized above the midbrain (cerebral hemispheres, internal capsule, or thalamus). **Decerebrate posturing (decerebrate rigidity, extensor posturing)** is characterized by abnormal extension (adduction and hyperpronation) of the arms. Decerebrate rigidity occurs with lesions localized to the midbrain or lower brainstem regions that involve the red nuclei or rubrospinal tracts. Pontine strokes often result in decerebrate posturing.

Both decorticate and decerebrate posturing are indicative of severe brain damage and are associated with high rates of severe chronic neurological morbidity and mortality. Decerebrate posturing, however, is a more serious condition. The lesions that produce decerebrate posturing are located at a lower level of the nervous system than the lesions that produce decorticate posturing. Progression from decorticate to decerebrate posturing often indicates uncal (transtentorial) or tonsilar (transforaminal) brain herniation.

Verbal Response

Verbal responsiveness is graded on a 5-point scale (1–5). A score of 5 indicates that the patient is oriented to person, place, and time, and makes appropriate verbal responses spontaneously and in response to questions. A score of 4 indicates that the patient is confused; they are capable of speaking in sentences but disoriented (incorrectly answers at least one question regarding orientation to person, place, and time). A score of 3 indicates that the patient is verbally responsive, and that the verbal responses are understandable, but they are not relevant to the question asked. A score of 2 indicates that the patient does not respond with words but instead makes incomprehensible sounds such as moans, groans, or grunts. A score of 1 indicates that the patient does not make any verbal response, including incomprehensible sounds.

Eye-Opening Response

Eye-opening responsiveness is graded on a 4-point scale (1–4). A score of 4 indicates that the patient's eyes open spontaneously (without any prompting). A score of 3 indicates that the patient only opens their eyes in response to a verbal stimulus (i.e., someone speaking to the patient in a normal or louder than normal voice). A score of 2 indicates that the patient only opens their eyes in response to painful stimuli. A score of 1 indicates that the patient makes no eye-opening response to any stimuli, even painful stimuli.

Coma

Coma is a state of unarousable unresponsiveness. The comatose patient is unresponsive to and cannot be aroused by *any* stimulation, including painful stimulation. There is no eye-opening, motor response, or vocal response. As there is a loss of arousal, there is a complete absence of awareness of self and environment.

Coma occurs with a variety of conditions that cause a failure of brain function, including structural brain pathologies such as traumatic brain injury, stroke, anoxic brain injury, brain infection, brain tumor, epidural hematoma, subdural hematoma, and increased intracranial pressure. It also may occur with severe systemic illness (e.g., infection), metabolic abnormalities (e.g., hypoglycemia, hyperglycemia), and high levels of drug or alcohol intoxication. A reversible coma may also be induced pharmacologically for medical purposes. General anesthesia is essentially a pharmacologically induced reversible coma (although physicians often refer to it as "putting the patient to sleep"). Medically induced coma is also used to treat refractory status epilepticus (continuous seizures that fail to respond to antiseizure medications) by suppressing brain activity and aborting seizures, and other neurocritical care conditions by reducing the brain's demands for blood, oxygen, and glucose and reducing brain swelling to protect the brain from secondary injury.

Coma due to severe brain injury of vascular, traumatic, or anoxic origin is usually time-limited, not lasting longer than several weeks. It either progresses to brain death (permanent loss of all brainstem and higher brain functions) or it improves to a vegetative state, minimally conscious state, or to full recovery of consciousness. In cases of incomplete recovery, the patient may remain permanently in a vegetative or minimally conscious state. Advances in emergency and intensive care medicine have increased survival rates following severe brain damage; the existence of patients in brain death, coma, vegetative, and minimally conscious states is largely made possible by advances in life-saving interventions and technologies.

Pathophysiology of Coma

Consciousness is an active neurological process. A conscious state depends on intact cerebral hemispheres and an intact **ascending reticular activating system** (ARAS) interacting with the cerebral hemispheres. The ARAS originates in the upper brainstem reticular formation and projects to the midline and intralaminar nuclei of the thalamus, which in turn project diffusely throughout the cerebral cortex via the diffuse (nonspecific) thalamic projection system. The ARAS maintains a constant, fluctuating stimulation of higher brain centers.

Wakefulness is a prerequisite for awareness. Wakefulness requires normal activity of the ARAS. Awareness requires normal activity of the cerebral cortex, but normal cerebral function is dependent on normal ARAS function. Without the steady input from the ARAS, the cortex does not function efficiently, and the person cannot think clearly, learn effectively, or socially relate meaningfully.

There are many causes of coma, but all involve: (1) bilateral, diffuse cerebral hemisphere dysfunction; (2) brainstem (midbrain and pons) ARAS dysfunction through a focal brainstem lesion or metabolic derangement; or (3) a combination of the two. Coma does not occur with unilateral cerebral hemisphere pathology, unless it involves a large lesion that causes secondary pressure on the contralateral hemisphere or brainstem structures. In contrast, profound coma may result from very small infarctions in the brainstem affecting the ARAS.

The medulla is crucial for vital functions (i.e., functions necessary for survival). It contains autonomic reflex control centers that are crucial to cardiovascular and respiratory functions. Cardiovascular centers regulate heart rate and blood pressure to ensure adequate blood circulation throughout the body. Respiratory centers monitor the acidity of blood by chemoreceptors; when the lungs fail to remove enough carbon dioxide from the blood, the blood becomes acidic (the pH decreases), which triggers reflexive breathing movements by contraction of intercostal and diaphragm muscles. In coma, medulla function is preserved (which is the reason a comatose person survives). There is, however, a global decrease in cerebral perfusion and metabolism that is 50%–70% of the normal range, demonstrated by single photon emission computed tomography (SPECT) and positron emission tomography (PET) functional neuroimaging.

Etiology

Causes of coma encompass a wide range of etiologies. Metabolic/systemic disorders are most common and include exogenous toxins (e.g., drug or alcohol intoxication, medication overdose, CNS depressants, poisons, other environmental toxins), endogenous toxins generated with organ failure (e.g., liver, kidney), hypoxia (e.g., due to pulmonary disease, respiratory failure, or carbon monoxide poisoning), systemic infection (e.g., sepsis), ischemia (e.g., due to decreased cardiac output), metabolic disturbances (hypoglycemia, hyperglycemia, hyponatremia, hypernatremia, hypercalcemia), endocrine abnormalities, seizures (including status epilepticus), and hypo- or hyperthermia. Structural neurological causes of coma may be either focal or diffuse and include traumatic brain injury (including subdural or epidural traumatic hematomas), stroke (ischemic or hemorrhagic), brain tumor, acute hydrocephalus, raised intracranial pressure, encephalitis, primary CNS infections (e.g., meningitis, encephalitis), and brainstem compression occurring directly from a mass lesion or indirectly by herniation (e.g., transtentorial or uncal herniation caused by a supratentorial mass lesion, or tonsillar herniation caused by an infratentorial mass lesion).

The most common causes of coma encountered in the emergency department, in descending order, are drug overdose, hypoxic-ischemic insult secondary to cardiac arrest, stroke (e.g., intracerebral hemorrhage, subarachnoid hemorrhage, cerebellar hemorrhage with brainstem herniation, basilar artery thrombosis), and trauma.

Examination

Acute loss or alteration of consciousness is a sign of life-threatening illness and a medical emergency, requiring prompt diagnosis and intervention in order to preserve life and brain function. Initial physical examination of the patient with loss of consciousness is directed to assessment of the patient's airway, breathing, and circulation (ABCs), and the need for immediate intervention, such as cardiopulmonary resuscitation in cases of cardiac arrest, removing airway obstruction, providing assisted ventilation, and positioning the patient to alleviate effects of shock that may occur with blood or extracellular fluid loss. Neurological examination of the unresponsive patient is quantified roughly by standardized, quantitative coma-grading scales such as the GCS.

The differential diagnosis of coma etiology is extensive, and identification of the underlying etiology often requires laboratory tests and neuroimaging. The underlying cause of the coma must be treated as soon as possible. Initial laboratory tests often include electrolytes, complete blood count, arterial or venous blood gas analysis, and toxicology testing. Neuroimaging may be useful to detect hemorrhage, mass effect, or other structural abnormalities. Additional imaging and/or laboratory testing may be necessary.

Prognosis

Prognosis of coma is related directly to clinical features such as etiology, depth of coma, and length of coma.

Etiology serves as an important factor for assessing potential outcome. In general, those with toxic-metabolic causes of coma (e.g., drug and/or alcohol overdose, poisoning) have better prognosis than those with coma due to other causes. Subarachnoid hemorrhage and stroke carry the worst prognosis. Hypoxic-ischemic coma is most likely to evolve into a vegetative state.

Depth of coma as evaluated by GCS score is another clinical feature predictive of outcome. Patients who display eye-opening have a better chance of recovery compared to patients whose eyes remain closed, and verbally responsive patients have a better chance of recovery than verbally unresponsive patients.

Duration of coma is also a predictor of outcome. As the time that a patient remains in a coma increases, the likelihood that they will progress into a prolonged vegetative state increases and the chance of good recovery decreases.

The Vegetative State

The **vegetative state**, also known as **unresponsive wakefulness syndrome**, is characterized by regaining of the sleep-wake cycle and appearance of wakefulness due to eye-opening, but with no indication of awareness of self or surroundings. There are no volitional responses; spontaneous movements and motor responses to deep pain can occur, but the movements are purposeless.

The primary feature of this disorder is the uncanny appearance of awareness. The dichotomy of this apparent visual alertness differentiates this group of patients from those in coma. Individuals in a vegetative state can be aroused from sleep, with eye-opening and electroencephalographic arousal, but they show no meaningful responses to external stimuli and no signs of experiencing emotions.

The criteria for diagnosis of vegetative state are: (1) no evidence of awareness of self or environment, and an inability to interact with others; (2) no evidence of sustained, reproducible, purposeful, or voluntary behavioral responses to visual, auditory, tactile, or noxious stimuli; (3) no evidence of language comprehension or expression; (4) intermittent wakefulness as manifested by the presence of sleep-wake cycles; (5) sufficiently preserved hypothalamic and brainstem autonomic functions to permit survival with medical and nursing care; (6) bowel and bladder incontinence; and (7) variably preserved cranial nerve and spinal reflexes.

Recovery ultimately depends on the extent of brain injury. Some patients recover slowly and regain awareness, while others with more severe or irreversible brain injury do not improve. A vegetative state lasting more than 1 month is considered a **persistent vegetative state**. Vegetative states of traumatic etiology have a better prognosis than those due to anoxic-ischemic brain injuries. A **permanent vegetative state** is defined as a vegetative state lasting more than 3 months following anoxic-ischemic brain injury and more than 12 months for traumatic brain injury. At this point, there is little chance for meaningful recovery of neurologic functions.

The vegetative state occurs when the brainstem is spared, but there is widespread severe damage to both cerebral hemispheres above the brainstem, that involve the reticular system projections into the thalamus or disconnect the reticulocortical and reticulolimbic pathways. This causes alterations in arousal, but the full picture of coma will not result because the brainstem portion of reticular system innervates the nuclei of the extraocular nerves; therefore such patients can open their eyes and look about. The cortex, however, is not sufficiently stimulated to produce voluntary movement or speech. In vegetative state patients, overall cerebral blood flow and metabolism are 40%–50% of the normal range, as determined by SPECT and PET. In permanent vegetative state patients, cerebral blood flow and metabolism values drop to 30%–40% of the normal range of values; the loss of metabolic function over time is the result of progressive neuronal degeneration.

The Minimally Conscious State

The **minimally conscious state** (MCS) is a condition of severely altered consciousness in which there is behavioral evidence of some awareness of self or environment as demonstrated by some signs of voluntary (nonreflexive) behavior, but the patient is unable to communicate meaningfully. These voluntary behaviors are minimal or inconsistent, but nonetheless they are driven consciously and are a clear improvement over the simple reflex responses of coma and the vegetative state. Such behavioral signs may include purposeful behaviors to environment stimuli (e.g., pursuit eye movement or sustained fixation in response to visual stimuli, reaching for objects, smiling or crying in response to emotional stimuli, vocalizations or gestures in direct response to the linguistic content of questions), following simple commands, gestural or verbal yes-no responses (regardless of accuracy), and intelligible verbalization.

The MCS is usually transient, but some remain in this state permanently for years or decades. Emergence from the MCS is defined by ability to communicate reliably through gestural or verbal yes-no responses, or functional use of objects. Similar to the vegetative state, MCS of traumatic etiology has a better prognosis than MCS due to anoxic-ischemic brain injuries. The prognosis for recovery from MCS is much more promising than that of vegetative state. Functional imaging studies of MCS patients have shown that overall cerebral blood flow and metabolism are decreased relative to normal, but to a lesser degree than in vegetative state patients.

The most sensitive diagnostic tool to distinguish MCS from vegetative state is the **Coma Recovery Scale-Revised** (CRS-r). This measure is designed to detect subtle changes in neurobehavioral status of patients with a disorder of consciousness, using the diagnostic criteria for MCS and vegetative state. The scale consists of 29 items grouped hierarchically into 6 sub-scales: auditory function, visual function, motor function, verbal (oromotor) function, communication, and arousal level.

Delirium

Delirium is a disorder of consciousness in which arousal is intact but awareness is altered, with acute onset and

> **BOX 25.1** The American Psychiatric Association's Fifth Edition of the Diagnostic and Statistical Manual of Mental Disorders (DSM-5) Diagnostic Criteria for Delirium
>
> A. Disturbance in attention (i.e., reduced ability to direct, focus, sustain, and shift attention) and awareness (reduced orientation to the environment).
> B. The disturbance develops over a short period of time (usually hours to a few days), represents an acute change from baseline attention and awareness, and tends to fluctuate in severity during the course of a day.
> C. An additional disturbance in cognition (e.g., memory deficit, disorientation, language, visuospatial ability, or perception).
> D. The disturbances in Criteria A and C are not better explained by a preexisting, established, or evolving neurocognitive disorder and do not occur in the context of a severely reduced level of arousal such as coma.
> E. There is evidence from the history, physical examination, or laboratory findings that the disturbance is a direct physiological consequence of another medical condition, substance intoxication or withdrawal (i.e., due to a drug of abuse or to a medication), or exposure to a toxin, or is due to multiple etiologies.
>
> *Source:* American Psychiatric Association. *Diagnostic and Statistical Manual of Mental Disorders.* 5th ed. Washington, DC: American Psychiatric Association; ©2013:596. All rights reserved.

fluctuating course (Box 25.1). The core cognitive disturbances in delirium are inattention (reduced ability to direct, focus, sustain, and shift attention) and disorientation to time and place. Additional disturbances in cognition include disorganized thought form and disordered thought content (i.e., hallucinations, delusions). Delirious individuals produce incoherent conversations that drift from the central point; they are unable to maintain a coherent stream of thought or action. They are inconsistent and confabulatory in reporting recent events. The cognitive disturbance has an acute (abrupt) onset over several hours to several days; it fluctuates in severity over the course of a day, characterized by waxing and waning of symptoms with intervals of lucidity. Thus, level of attention and orientation to the environment fluctuates. Many additional terms are associated with, or have been used to refer to delirium, including *acute confusional state, altered mental status, toxic metabolic encephalopathy, acute brain syndrome, acute cerebral insufficiency,* and *acute brain failure.*

There are several subtypes of delirium. **Hyperactive delirium** (hyperkinetic delirium) is characterized by increased motor activity and restless agitation. This form of delirium is often referred to as a state of **agitated confusion**. It commonly occurs in a severe form of alcohol withdrawal, known as **alcohol withdrawal delirium** or **delirium tremens**. **Hypoactive delirium** (hypokinetic delirium) is characterized by lethargy, somnolence, lack of initiation, and slow reaction time. **Mixed delirium** is characterized by a mixed presentation that alternates between hyperactive and hypoactive states. In **attenuated delirium syndrome**, some but not all criteria for a diagnosis of delirium are met.

Delirium is a clinical syndrome resulting from a wide variety of physiological insults. Precipitating factors include a wide variety of conditions that originate outside the brain, such as infection with fever, drug or alcohol intoxication or withdrawal, acute metabolic disturbances due to renal or hepatic failure, severe nutritional deficiency, trauma, or surgery, as well as primary neurological causes such as traumatic brain injury, stroke, or brain infection (Table 25.1). In young patients, the most common cause of delirium is toxicity, usually due to medication side effects in children and drug abuse in young adults. In the elderly, infection is the most common cause, but medication side effects are also a common precipitating factor. About 80% of individuals in the stages prior to death experience delirium. Treatment of delirium is directed to correcting all possible causative factors and eliminating medications that promote delirium.

Vulnerability to developing delirium varies with predisposing factors, including age, preexisting cognitive impairment, preexisting brain disease, sleep deprivation, sensory impairment, immobility, dehydration, untreated or poorly managed pain, and poorly controlled medical conditions (hepatic disease, renal disease, diabetes, hypertension). The aged brain is less able to tolerate or adapt to physiologic perturbations; healthy older adults are more prone to developing delirium after contracting a urinary tract infection, but this is very uncommon in young adults. Cognitive impairment (reduced brain reserve) and preexisting brain disease lower the tolerance threshold for developing delirium.

Hospitalized, intensive care unit (ICU), critical care (e.g., on ventilator), and postsurgical patients are particularly prone to develop delirium, especially if they are elderly. Because many ICU patients with delirium have psychotic symptoms (e.g., visual hallucinations), the condition is often inappropriately referred to as "ICU psychosis." Hospital-acquired delirium, however, is most often of the hypoactive type and therefore is likely to go unrecognized. Conditions that promote delirium in hospitalized patients include advanced age, sleep deprivation, unrelieved pain, prolonged bed rest, major surgery, systemic inflammation, and **deliriogenic** drugs. **Postoperative delirium** is very common in elderly patients, especially those with preexisting signs of early dementia.

TABLE 25.1 Delirium Etiologies

Infection	CNS infections, sepsis, systemic infection (e.g., urinary tract infection)
Drug withdrawal	Alcohol, benzodiazepines, sedative-hypnotics, opiates
Acute metabolic disturbances	Acidosis, alkalosis, electrolyte imbalances (sodium, glucose, calcium, magnesium), organ failure (hepatic, renal)
Acute traumatic injuries	TBI, burn injury, polytrauma, exsanguination, pain
CNS pathology	Abscess, tumor, metastases, stroke, inflammatory or autoimmune disorders
Hypoxia	Anemia, CO poisoning, hypotension, pulmonary/respiratory failure, cardiac failure
Nutritional deficiencies	Vitamin B_{12}, folate, thiamine, niacin
Endocrinopathies	Hyper/hypoadrenocorticism, hyper/hypoglycemia, hypothyroidism, hyperparathyroidism, adrenal hyper- or hypofunction
Acute vascular events	Hypertensive encephalopathy, stroke, arrhythmia, shock, vascular operations (microembolism)
Toxins or drugs	Tricyclic antidepressants, anticholinergics, dopaminergics, serotonergics, narcotic analgesics, benzodiazepines, GABA agonists (e.g., propofol), corticosteroids, H2-receptor antagonists, sedative hypnotics, anticonvulsants, antiparkinsonian drugs, sympathomimetics, anti-inflammatory drugs, anti-neoplastic drugs, illicit drugs, toxic chemicals
Heavy metals	Lead, manganese, mercury
Major surgery	Surgical trauma, pain, hyponatremia, infection, iatrogenic medication effects, transplant rejection

The pathogenesis of delirium is not fully understood, but there is evidence that disruption of neurotransmitter systems and systemic inflammatory cytokine response may play a role. The EEG is typically abnormal, with generalized theta or delta slow-wave activity, poor organization of the background rhythm, and loss of reactivity of the EEG to eye-opening and closing.

Neurobehavioral measures to identify delirium include the Confusion Assessment Method (CAM), the Confusion Assessment Method for the Intensive Care Unit (CAM-ICU), the Delirium Symptom Interview (DSI), the Intensive Care Delirium Screening Checklist (ICDSC), the Delirium Detection Score (DDS), and the Delirium Rating Scale (DRS). Once delirium is identified, in-depth cognitive testing should be postponed until the delirium clears. A marked deficit in attention, as occurs in delirium, undermines higher-level cognitive processes. Low neurocognitive test scores in a patient with delirium do not reflect additional neurocognitive disorder; they simply reflect delirium.

Syndromes That Mimic Disorders of Consciousness

Locked-In Syndrome

Locked-in syndrome (LIS) is a rare neurological condition that is typically characterized by complete paralysis of all voluntary muscles except those controlling eye movements and blinking. Individuals with LIS are bedridden and completely reliant on caregivers. There is a loss of voluntary control of breathing and loss of ability to chew and swallow; thus, most patients require tracheostomy, mechanical ventilation, and a feeding tube.

Individuals with LIS are fully conscious. They have preserved sleep-wake cycles with normal wakefulness. They are aware of their environment, as their sensory systems and cognitive function are unaffected. They comprehend spoken and written language. The only problem is in motor output. Since the neural pathways controlling eye blinking and eye movements are intact, those with LIS can communicate with others by blinking and/or eye movements to signal yes-no responses or through eye-controlled, computer-based communication technology.

Locked-in syndrome is often caused by a focal bilateral lesion within the upper (rostral) region of the ventral pons (pontine tegmentum) that interrupts all descending corticospinal and corticobulbar motor pathways (i.e., upper motor neurons originating from the cerebrum). Such a lesion deprives lower motor neurons of the spinal cord and brainstem cranial nerve nuclei (which directly control the body's skeletal muscles) of input from higher brain centers, resulting in loss of all voluntary movement except eyelid blinking and vertical (up and down) eye movements. These movements are spared because they are controlled by the oculomotor nucleus of the oculomotor nerve (cranial nerve 3) and the vertical gaze center (i.e., the rostral interstitial nucleus of the medial longitudinal fasciculus of the midbrain), respectively, which are both located in the midbrain. In LIS there is no damage to the ARAS or brain structures above the level of the lesion; thus consciousness (arousal, awareness, and cognition) is preserved. The EEG shows normal brain activity and sleep-wake cycles. Metabolism in the supratentorial gray matter has been found to be normal in PET functional neuroimaging studies.

The most common cause of LIS is brainstem ischemia due to occlusion at the proximal and middle part of the basilar artery, from which the pontine arteries arise. Other causes of selective focal pontine lesions and LIS include hemorrhagic stroke, pontine demyelination due to central pontine myelinolysis or multiple sclerosis, trauma, tumor, and abscess. Locked-in syndrome, however, may also occur with pathologies outside of the pons that affect motor neuron efferents, such as amyotrophic lateral sclerosis (ALS) and Guillain-Barré syndrome.

In many cases such as stroke, those with LIS were initially comatose but gradually regained consciousness while remaining paralyzed. Because they appear awake but unresponsive, they may be misdiagnosed as in a vegetative state. Locked-in syndrome is also referred to as **pseudocoma**, **de-efferented** state, and **cerebromedullospinal disconnection** because those affected appear to be in a vegetative state, are "disconnected" from the motor system, and often have a focal lesion of the basal pons affecting the corticospinal and corticobulbar descending motor pathways.

There are three variants of LIS based on degree of motor output: (1) **classical LIS**, in which there is a loss of all body movements except blinking and vertical eye movements; (2) **incomplete LIS**, in which some voluntary movements other than eye movements are preserved; and (3) **total LIS**, in which the eyes are paralyzed as well. Since in total LIS there is complete body paralysis and the patient cannot demonstrate consciousness by responding to questions with blinking or eye movements, preservation of cortical function is demonstrated on EEG by normal brain activity, sleep-wake cycles, and attention.

Brain Death

Brain death is defined as the irreversible loss of *all* brain function, including the brainstem. All neurons within the brain (cerebrum, cerebellum, and brainstem) are dead. Brain death differs from coma in that there is a loss of all brainstem function and it is irreversible. The three essential findings of brain death are: (1) coma (i.e., total absence of arousal and awareness); (2) absence of brainstem reflexes; and (3) **apnea** (cessation of breathing). Spinal reflexes may remain intact.

Modern medicine has transformed the course of terminal neurologic disorders, as vital functions can be maintained artificially via mechanical ventilators and other advanced critical care interventions for a long time after the brain has ceased to function. In cases of brain death, however, life support is futile. In many jurisdictions, brain death is a criterion for legal death and the principal prerequisite for donation of organs for transplantation.

The diagnosis of brain death is primarily clinical; it requires total unresponsiveness (i.e., coma), absence of brainstem reflexes, and apnea in the setting of devastating neurological injury. In cases where confounding factors may obscure the clinical examination (e.g., severe hypotension/shock, hypothermia, CNS depressants, neuromuscular blocking agents, metabolic encephalopathies, brainstem encephalitis), it may be necessary to demonstrate absence of brain metabolic activity or blood flow to make the diagnosis.

Neurological examination must show absence of spontaneous movement, decerebrate or decorticate posturing, seizures, shivering, response to verbal stimuli, and response to noxious stimuli administered through a cranial nerve pathway. Complete loss of brainstem function must be evidenced on two clinical assessments of brainstem reflexes. Examination must demonstrate loss of pupillary reactivity (pupillary light reflex) and absence of corneal, oculovestibular, and oculocephalic reflexes. It must also show loss of **bulbar reflexes** related to the function of cranial nerves 9 (glossopharyngeal) and 10 (vagus) and the medulla, as reflected by absent cough and gag reflexes, as well as failure of the heart rate to increase by more than 5 beats per minute after 1–2 milligrams of intravenous atropine (a response that is dependent on the vagus nerve and its nuclei).

Finally, failure of ventilatory drive must be evidenced by a single **apnea test** that is usually performed after the second brainstem reflex test. The apnea test serves to examine whether there is loss or preservation of medullary function through the delivery of a hypercarbic respiratory stimulus (high blood carbon dioxide level). This procedure requires close patient monitoring for respiratory efforts, as ventilator support is removed temporarily until the partial pressure of carbon dioxide in the blood increases to a specific threshold, as indicated by pulse oximetry. Total absence of respiratory efforts in the presence of increased blood carbon dioxide level (hypercarbia) is an indicator of absence of brainstem function.

Functional imaging of cerebral perfusion by SPECT and cerebral metabolism by PET show a "hollow skull phenomenon" in which there is an absence of radiotracer in patients who are brain dead, confirming the absence of neural function in the whole brain.

Summary

Consciousness may be defined as a neurological state in which one is fully awake and aware of self and environment and has normal responses to external stimulation and inner needs. Unconsciousness is the opposite; it is a neurological state of overtly (i.e., observable) diminished responsiveness to environmental stimuli and unawareness of self and environment. Consciousness has two basic dimensions: arousal (wakefulness, alertness), which refers to the level of consciousness; and awareness, which refers to the contents of consciousness. Awareness requires wakefulness; however, wakefulness does not require awareness, as one can be awake but unaware. Level of consciousness exists on a continuum, from full consciousness in which the individual is awake and aware, to coma in which the individual is neither awake nor aware and is in a state of unarousable unresponsiveness. The vegetative state and minimally conscious state lie between the two ends of the consciousness spectrum. The vegetative state (unresponsive wakefulness syndrome) is characterized by regaining of the sleep-wake cycle and appearance of wakefulness due to eye-opening, but without awareness of self or surroundings. The minimally conscious state is a condition of severely altered consciousness with some behavioral signs of some awareness of self or environment, as reflected by signs of voluntary (nonreflexive) behavior. Delirium is a disorder of consciousness in which arousal is intact, but awareness is altered with acute onset and fluctuating course and a core cognitive disturbance of inattention (reduced ability to direct, focus, sustain, and shift attention) and disorientation. Coma-like states include locked-in syndrome, which is characterized by normal consciousness (awake and aware) but complete paralysis of all voluntary muscles except those controlling eye and blinking movements, and brain death, in which there is an irreversible loss of all brain function, manifest as coma plus an absence of brainstem reflexes and apnea.

Additional Reading

1. Bayne T, Hohwy J, Owen AM. Reforming the taxonomy in disorders of consciousness. *Ann Neurol.* 2017;82(6):866–872. doi:10.1002/ana.25088
2. Bayne T, Hohwy J, Owen AM. Are there levels of consciousness? *Trends Cogn Sci.* 2016;20(6):405–413. doi:10.1016/j.tics.2016.03.009
3. Bernat JL. Nosologic considerations in disorders of consciousness. *Ann Neurol.* 2017;82(6):863–865. doi:10.1002/ana.25089
4. Fischer DB, Truog RD. What is a reflex? A guide for understanding disorders of consciousness. *Neurology.* 2015;85(6):543–548. doi:10.1212/WNL.0000000000001748
5. Goldfine AM, Schiff ND. Consciousness: its neurobiology and the major classes of impairment. *Neurol Clin.* 2011;29(4):723–737. doi:10.1016/j.ncl.2011.08.001
6. Gosseries O, Zasler ND, Laureys S. Recent advances in disorders of consciousness: focus on the diagnosis. *Brain Inj.* 2014;28(9):1141–1150. doi: 10.3109/02699052.2014.920522
7. Kondziella D. Functional neuroimaging in disorders of consciousness: raising awareness for those with decreased awareness. *Neuroscience.* 2018;382:125–126. doi:10.1016/j.neuroscience.2018.03.046
8. Laureys S, Owen AM, Schiff ND. Brain function in coma, vegetative state, and related disorders. *Lancet Neurol.* 2004 Sep;3(9):537–546. doi: 10.1016/S1474-4422(04)00852-X. PMID: 15324722.
9. Laureys S, Pellas F, Van Eeckhout P, et al. The locked-in syndrome: what is it like to be conscious but paralyzed and voiceless? *Prog Brain Res.* 2005;150:495–511. doi:10.1016/S0079-6123(05)50034-7
10. Smith E, Delargy M. Locked-in syndrome. *BMJ.* 2005;330(7488):406–409. doi:10.1136/bmj.330.7488.406
11. Teasdale G, Jennett B. Assessment of coma and impaired consciousness: a practical scale. *Lancet.* 1974;2(7872):81–84. doi:10.1016/s0140-6736(74)91639-0
12. Teasdale G, Maas A, Lecky F, Manley G, Stocchetti N, Murray G. The Glasgow Coma Scale at 40 years: standing the test of time [published correction appears in *Lancet Neurol.* 2014 Sep;13(9):863]. *Lancet Neurol.* 2014;13(8):844–854. doi:10.1016/S1474-4422(14)70120-6

Memory and Amnesia

Introduction

Memory is a general term for the mental processes involving the acquisition and retention of information, and the retrieval of stored information. **Amnesia** is a selective disruption of memory function in the context of preserved other cognitive abilities (intelligence, attention, language, visuospatial function, and executive functions); it occurs in a variety of neurological conditions that disrupt brain memory networks. **Anterograde amnesia** is the inability to form new memories following the onset of brain injury or disease. **Retrograde amnesia** involves difficulties recalling both personal autobiographical information from the past and publicly available information such as world news events. Although persons with anterograde amnesia generally have some retrograde amnesia, they do not lose all information learned prior to the brain injury or disease. Previously learned information that is preserved includes semantic knowledge (about objects, facts, and word meanings), as well as motor and cognitive skills (e.g., how to drive a car, play the piano, use appliances, read, perform arithmetic).

This chapter describes the anatomy and clinical features of anterograde amnesia, pathologies that commonly produce amnesia, transient amnesia syndromes, and the various forms of memory that have been elucidated by studying individuals with memory disorders. A great amount of our current understanding of the neurology of memory has come from the study of a single case—H.M.—and this chapter therefore begins with a discussion of this important historical development.

The Case of H.M.

Very little was known about the neurology of memory until 1957, when the field was revolutionized by a report by William Beecher Scoville and Brenda Milner in the *Journal of Neurology, Neurosurgery, and Psychiatry* titled "Loss of Recent Memory after Bilateral Hippocampal Lesions." In this paper, Scoville and Milner described patient H.M. (**Henry Gustav Molaison**, 1926–2008), who at the age of 27 underwent bilateral **temporal lobectomy** as an experimental operation to treat epilepsy. The surgery was performed by Scoville at Hartford Hospital in Hartford, Connecticut, on September 1, 1953. The resections involved the anterior 6 centimeters of both temporal lobes, including the cortex, underlying white matter, and medial temporal structures consisting of the hippocampal complexes and the amygdalae.

The surgery was successful in reducing seizures, but it unexpectedly resulted in a profound, pervasive, and permanent inability to establish new memories involving the conscious recollection of events and facts (anterograde amnesia). H.M. forgot events nearly as quickly as they occurred. He could not learn new people or places that he encountered since the surgery. Even after extensive contact with them, he always behaved as if they were new encounters for him. Indeed, he had no conscious recall of anything that he had done or that had happened since the surgery. By contrast, he had no apparent deficits in intellectual, perceptual, or other cognitive functions. H.M. was studied for five decades, with much of this work conducted by **Brenda Milner** (1918–) and **Suzanne Corkin** (1937–2016). He is the most studied and best-known research subject in the history of neuroscience.

Prior to H.M.'s case, it was believed that memory functions were widely distributed throughout the cerebral cortex and not anatomically distinct from the neural structures serving intellectual and perceptual functions. This notion was based largely on the experimental work of **Karl Lashley** (1890–1958), who pioneered the experimental study of learning and memory in mammals with focal surgical lesions of the cerebral cortex. Lashley sought the location of the **engram**, or memory trace. His findings led him to propose the **principle of mass action** in which memory is distributed throughout the cortex, rather than localized to a specific brain region, and the related **principle of equipotentiality** in which all brain areas are equally able to perform a task (although he allowed for functional specialization of sensory and motor functions). By these principles, the severity of cognitive dysfunction is proportional to the volume of brain tissue damaged, regardless of location.

H.M. revolutionized our understanding of the neurological basis of memory by establishing the fundamental principles that (1) memory is a distinct cerebral function, separable from other perceptual and cognitive abilities; and (2) the medial temporal lobes are crucial for memory.

Detailed studies of H.M.'s impaired and spared functions led to other revelations as well. In addition to anterograde amnesia, impaired ability to recall events since the brain surgery, H.M. had a retrograde amnesia, impaired ability to recall events that had occurred prior to the brain surgery (Figure 26.1). The loss of memory for autobiographical information extended to approximately 3 years before the surgery. The retrograde amnesia was temporally graded, with memory for events that occurred closer to the time of surgery more affected than memory for events that had occurred more distant in time to the date of surgery, within the

FIGURE 26.1. Anterograde amnesia is an impaired ability to recall events occurring after the onset of brain damage. Retrograde amnesia is an impaired ability to recall events that occurred prior to the brain damage; it is temporally graded.

3-year presurgical period. H.M.'s memory for more remote events from his childhood, teens, and early twenties, however, was intact. This phenomenon was demonstrated years later by formal testing assessing H.M.'s ability to recognize faces of people who were famous in the decades from 1920–1970 (Famous Faces Test). Compared to age-matched controls, H.M. had very poor recognition of people who were famous in the 1950s and 1960s (the postmorbid period), but good recognition memory of people who were famous in the 1920s–1940s (the premorbid period).

Although H.M. had a severe anterograde amnesia and forgot recent information very rapidly, he was able to retain some information for a brief period (immediate memory), as evidenced by his ability to participate meaningfully in conversations and to repeat strings of 6 to 7 digits (see "Forms of Memory Based on Time Span," below). He could retain some types of verbal information (e.g., digit strings) for longer periods of time through rehearsal (working memory). If his attention was diverted from rehearsing the information, however, he forgot it. Rehearsal was not possible for other types of information (e.g., nonverbal stimuli such as faces and designs), and he forgot the stimuli within less than a minute.

H.M. was also capable of learning new motor, perceptual, and cognitive skills in which "memory" was evidenced by improved performance with practice. This was initially demonstrated using a **mirror drawing visuomotor coordination task** (Figure 26.2). This task measured H.M.'s ability to learn to trace the outline of a 5-pointed star using reversed visual cues of his hand and the star provided by reflection in the mirror; it took 10 trials for H.M. to acquire this skill. Remarkably, he retained the skill across 3 days, although he had no conscious recollection of performing the task previously. This was the first hint that there was more than one kind of memory, and that this other form of memory does not require medial temporal lobe structures. Support for this hypothesis came from additional studies showing that H.M. was capable of learning other manual coordination tasks. For a time, it was thought that this additional form of memory was specific to motor skills, but it was soon discovered that H.M. also was capable of learning and retaining the perceptual skill of mirror reading; thus the domain of preserved memory was broader. These findings suggested a distinction between two broad classes of memory: declarative and procedural.

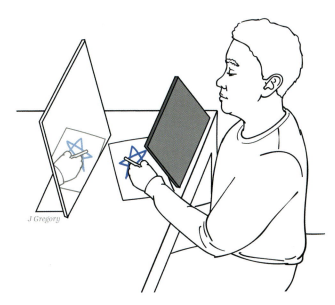

FIGURE 26.2. The mirror drawing visuomotor coordination task. This task was used to measure H.M.'s ability to learn to trace the outline of a 5-pointed star using reversed visual cues of his hand and the star provided by reflection in the mirror. H.M. was able to acquire this skill and retain it over days since the training, yet he had no conscious recollection of having performed the task previously.

The observations of H.M.'s impaired and spared abilities suggested four additional observations about the neural organization of memory: (1) Because H.M.'s memory was selectively impaired (i.e., he had a "circumscribed" deficit, and his intellectual and perceptual functions were unaffected), the damaged structures are specific to memory; (2) Because H.M. was able to acquire new motor and perceptual skills despite his debilitating and pervasive memory impairment, memory is not a unitary process; (3) Because H.M. could retain some information for a brief period of time immediately after first encountering it and could maintain the information for a longer period of time with rehearsal, the damaged structures are not necessary for these abilities (i.e., immediate memory and working memory); and (4) Because H.M. could remember events and facts from time periods remote to his surgery, the damaged structures are not the storage sites for long-term memory.

Anatomy of Medial Temporal Amnesia

H.M.'s bilateral temporal lobectomy involved resection of the anterior 6 centimeters of both temporal lobes, including the cortex, underlying white matter, hippocampal complexes (hippocampal formation and adjacent parahippocampal gyrus; see Figure 16.1 in Chapter 16) and the amygdalae. Despite the title of the 1957 publication ("Loss of Recent Memory after Bilateral Hippocampal Lesions"), it was unknown whether H.M.'s amnesia was attributable entirely to the hippocampal removal, or whether resection of the other temporal lobe structures contributed; this was an acknowledged limitation. It remained to be determined which of the resected structures were responsible for the amnesia.

Experimental studies of temporal lobe memory functions in monkeys with similar and more limited resections addressed these questions. Early studies failed to find that temporal lobe lesions disrupted learning and memory in monkeys, but eventually it became clear that this was because the behavioral tasks used in these studies involved instrumental (operant) conditioning and classical (Pavlovian) conditioning, which entail forms of implicit memory that are not mediated by the medial temporal lobe. **Mortimer Mishkin** (1926–2021) developed a **delayed non-match to sample task** and found that it was extremely sensitive to medial temporal lobe lesions in monkeys. During the acquisition phase of the task, the monkey is presented with a series of objects (sample stimuli) one at a time. After the acquisition phase, recognition memory is tested by presenting a series of two objects, one familiar (a sample stimulus from the acquisition phase) and the other novel, and the monkey is rewarded for choosing the novel stimulus. The number of sample stimuli presented (i.e., length of the list of objects to be remembered) during acquisition varied, as did the time interval between stimulus presentation and visual recognition memory testing. Postoperatively, animals with hippocampal ablations require prolonged training to learn the task or are unable to learn the task at all.

Using this animal model of temporal lobe memory impairment, Mishkin and others eventually showed that the severity of memory impairment depends on both the locus and extent of damage within the medial temporal lobe. Damage limited to the hippocampus is sufficient to produce significant and long-lasting anterograde amnesia. Damage that includes both the hippocampus and adjacent cortices of the parahippocampal gyrus produces more severe memory impairment than damage limited to the hippocampus. The severity of memory impairment is not related simply to extent of damage (as it would be if the mass action principle were correct); the specific structures involved are also critical. When hippocampal region lesions extend anteriorly to include the amygdala, the memory impairment does not increase.

Case studies of other patients with anterograde amnesia due to well-localized temporal lobe lesions from ischemic or anoxic episodes who underwent both detailed neuropsychological testing and postmortem histological analysis have been consistent with the conclusions based on the experimental data.

H.M. underwent brain MRI studies in 1992 and 1993 to identify the precise areas that were removed. He had not undergone MRI previously due to concerns that the clips used to close the dura in 1953 (and prior to the invention of MRI) might be metallic and MRI-incompatible, requiring much research to determine the type of clips used and their material composition. In 1997, Suzanne Corkin and colleagues published H.M.'s MRI findings. They showed that the resection extended approximately 5 cm posteriorly in both hemispheres (less extensively than the original description of the resection extending 6 cm caudally from the temporal pole), and that it was not restricted to the hippocampus and amygdala but also included the parahippocampal gyrus.

Pathologies Causing Medial Temporal Amnesia

Common causes of medial temporal amnesia include anoxic-hypoxic injury, bilateral compromise of the posterior cerebral arteries (PCAs), viral (herpes simplex type I) encephalitis, and early-stage Alzheimer's disease. Hypoxia (e.g., due to respiratory arrest, carbon monoxide poisoning, suffocation, near-drowning) and global cerebral ischemia (e.g., due to cardiac arrest) result in pronounced hippocampal damage that is also often accompanied by more diffuse, less severe cortical damage. The CA1 region of the hippocampus is most vulnerable.

Bilateral PCA infarction results in **amnestic stroke**, as the PCAs supply the inferomedial portion of the temporal lobes, including the hippocampi and parahippocampal areas. Because both PCAs arise from a single parent vessel (the basilar artery), blood flow to this region may be compromised bilaterally.

Herpes simplex type I viral encephalitis produces tissue necrosis in the inferomedial temporal and frontal lobes, often quite asymmetrically (see Figure 12.5 in Chapter 12). Chronic anterograde amnesia is a common clinical sequela; other sequelae include confusion, frontal and temporal seizures, neuropsychiatric signs and symptoms (e.g., emotional lability, depression, anxiety, psychosis), and marked behavioral disturbance (e.g., hyperactivity, impulsivity, disinhibition, aggression).

Early-stage Alzheimer's disease is characterized by a selective impairment in episodic memory, with insidious onset and gradual progression; the earliest neuropathological changes occur in the hippocampus and entorhinal cortex. Later stages are characterized by more diffuse cognitive dysfunction, including deficits in semantic memory and remote autobiographical memory, due to extensive neocortical pathology and atrophy.

Material-Specific Anterograde Amnesia

Studies of patients with unilateral temporal lobectomy, unilateral electroconvulsive therapy (ECT), and intracarotid amobarbital procedure have revealed that hippocampal dysfunction in the language-dominant hemisphere produces **verbal memory** deficits, while hippocampal dysfunction in the non-language-dominant hemisphere produces **nonverbal memory** deficits. These deficits are referred to as **material-specific memory loss**, in contrast to the **general (global) amnestic syndrome** seen with bilateral medial temporal involvement.

Temporal lobectomy in the language-dominant (usually left) hemisphere produces learning and memory impairments for verbal material, whether stimuli are presented visually or aurally, and whether retention is measured by rate of learning, free recall, or recognition. Temporal lobectomy in the non-language-dominant (usually right) hemisphere produces learning and memory impairments for nonverbal material (i.e., information that cannot be encoded verbally), whether presented visually or aurally. This includes visuospatial information such as object locations and nonverbal pictorial material.

Originally, ECT was administered to both hemispheres, but amnesia occurred as a side effect for a period of time after the procedure. Studies of unilateral ECT revealed that it is as effective as bilateral ECT in treating depression, and while bilateral and left-sided ECT disrupt verbal memory, right-sided ECT does not and has better tolerability.

The **intracarotid amobarbital procedure** (**Wada procedure**; see Chapter 19, "Epilepsy") involves temporary pharmacological inactivation of one hemisphere by injecting amobarbital, a short-acting (several minutes) barbiturate into one internal carotid artery. Although the posterior hippocampus is not directly supplied by the distribution of the internal carotid artery, depth electrode EEG recordings show slowing in this region following intracarotid amobarbital administration. In order to predict risk for developing a post-surgical amnestic syndrome, the patient is presented with language and memory tasks while the amobarbital is active to assess whether the hemisphere that is not anesthetized (i.e., the hemisphere contralateral to the unilateral seizure focus) can sustain memory function. If injection on the side planned for lobectomy produces a transient generalized memory loss, this indicates that a lesion is also affecting the hippocampus on the opposite side and lobectomy would result in a pervasive amnestic syndrome, a very undesirable consequence. Similar to findings on the effects of temporal lobectomy and ECT in the language-dominant hemisphere, Wada studies also indicate that pharmacological inactivation of the language-dominant hemisphere results in verbal memory deficits.

Remote Memory and Retrograde Amnesia

In H.M. and other patients with medial temporal lobe damage, a retrograde amnesia accompanies the anterograde amnesia. The retrograde amnesia affects recollection of autobiographical information, news events, and famous people from the recent past. Recollection of such information from the remote past remains intact. Because hippocampal lesions do not disrupt recall of remote memories, the hippocampus is not the site of long-term memory storage, and it is not involved in accessing **long-term memory** (memories that are stored for an extended time).

The retrograde amnesia typically has a temporal gradient, with recent memories more likely lost than old (remote) memories, a principle known as **Ribot's Law** because it was proposed by Théodule-Armand Ribot (1839–1916). The temporal gradient of retrograde amnesia implies that the hippocampus plays a role in the early stage of long-term memory formation and maintenance, but its role diminishes as time passes and a more permanent long-term memory gradually develops (**memory consolidation**). Thus, the hippocampus is believed to convert transient memories into new, stable representations (long-term memories) that depend on neocortex. If the hippocampus is not functional at the time of learning, there is a failure in long-term memory formation.

According to this view, recall of a long-term memory involves reactivation of distributed neocortical regions that were active at the time of initial encoding of the event. When the memory is new and stored as a recent memory, this neocortical reactivation requires hippocampal activity. Once the memory is fully consolidated and stored as a remote memory, neocortical reactivation can occur independently of hippocampal activity.

The extent of retrograde amnesia is variable, depending on the locus and extent of damage. In H.M., it extended approximately 3 years prior to the surgery, while in other cases it has extended several decades back. However, memories from early life are always spared.

Amnesia Due to Lesions in Non-Medial Temporal Lobe Structures

In the 1970s, it became apparent that lesions outside of the medial temporal lobe could also cause amnesia. These structures are the fornices, diencephalon (thalamus and hypothalamus), and basal forebrain, and they form circuits with the hippocampal formation.

Fornix

The **fornix** (Latin, "arch") is a C-shaped bundle of nerve fibers connecting the hippocampus to the diencephalon and basal forebrain (see Figure 16.2 in Chapter 16). It is the main efferent system of the hippocampus, but it also carries some afferents to the hippocampus. Bilateral surgical

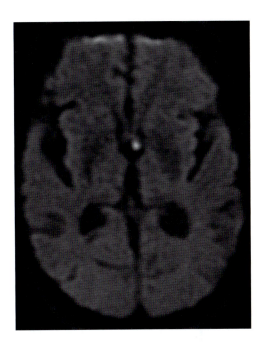

FIGURE 26.3. Axial diffusion-weighted image shows diffusion hyperintensity involving the left fornix due to infarction.

transection of the fornices, as has occurred with removal of tumors near the third ventricle, results in anterograde amnesia similar to that caused by medial temporal lesions. Fornix infarction also results in anterograde amnesia (Figure 26.3).

Diencephalic Amnesia and Korsakoff's Syndrome

Any midline damage that affects diencephalic structures produces amnesia. For example, bilateral thalamic infarcts that affect the dorsomedial nuclei, anterior thalamic nuclei, and mammillothalamic tracts produce severe anterograde amnesia, as well as retrograde amnesia. Thalamic tumors also produce amnesia, but studying their effects is not useful in differentiating the critical brain regions because tumors exert pressure on other brain regions and often produce hydrocephalus due to obstructing the third ventricle.

Unilateral thalamic lesions may produce a material-specific amnesia, with left-sided damage producing anterograde amnesia for verbal material and right-sided damage producing anterograde amnesia for nonverbal visuospatial material. **Patient N.A.** sustained a penetrating brain injury at age 22 when a miniature fencing foil passed through his left nostril into the brain, resulting in a material-specific amnesia for verbal information.

KORSAKOFF'S SYNDROME

A common form of amnesia is **Korsakoff's syndrome** (**Korsakoff's amnesia**). It is a chronic, severe impairment of anterograde memory often associated with alcoholism. Korsakoff's syndrome is similar to the global amnestic syndrome that occurs with medial temporal damage, except that confabulation is more common in Korsakoff's syndrome.

Confabulation is the production of false information without intention to deceive and without awareness that the information is false. The amnestic person believes that the false memory is real, and thus confabulation has been described as "honest lying." It is believed that confabulation is a compensatory mechanism to fill gaps in one's memory; however, not all amnesiacs confabulate, and the mechanism underlying confabulation remains unknown. Confabulated memories can be provoked (occur in response to a memory probe) or spontaneous (occur without an eliciting stimulus).

Another disorder associated closely with alcoholism is **Wernicke's encephalopathy**, consisting of acute confusion, ataxia, nystagmus, and ophthalmoplegia. Wernicke's encephalopathy often precedes Korsakoff's syndrome, and they are often denoted together as **Wernicke-Korsakoff syndrome**.

Korsakoff's syndrome is a form of **diencephalic amnesia**. It is associated with lesions to the mammillary bodies of the hypothalamus and the anterior thalamic nuclei, which are connected by the mammillothalamic tracts. These nuclei and tracts have anatomical connections to the medial temporal lobe (see Figure 12.2 in Chapter 12). This likely explains why medial temporal and diencephalic lesions lead to the same core deficit.

The lesions of Korsakoff's syndrome are caused by thiamine (vitamin B_1) deficiency. Thiamine deficiency occurs most commonly with alcoholism. Because alcohol is highly caloric, food intake is often decreased, leading to a deficit in vitamin intake. Furthermore, alcohol may increase the body's requirement for B vitamins and hinder thiamine absorption. Other causes of thiamine avitaminosis also produce Korsakoff's syndrome, including diseases or surgeries of the digestive tract and dialysis. Thiamine replacement reverses Wernicke's encephalopathy but not Korsakoff's amnesia.

Basal Forebrain Amnesia

The **basal forebrain** is a complex of subcortical nuclei that lie immediately posterior to the orbital prefrontal cortices. These include the septal nuclei, diagonal band of Broca, nucleus accumbens, and nucleus basalis of Meynert. The nucleus basalis is the major source of cholinergic innervation to the neocortex.

Basal forebrain damage produces anterograde and retrograde amnesia, as well as personality changes affecting social judgment and affect. The complex anatomy of the basal forebrain makes it difficult to elucidate the pathophysiology of amnesia due to lesions in this region. The most common cause of basal forebrain amnesia is subarachnoid hemorrhage from anterior communicating artery (ACom) aneurysms. The ACom is part of the circle of Willis; 30%–40% of all ruptured aneurysms are in the ACom. Tumor resection in the basal forebrain region can also cause amnesia.

Frontal Amnesia

Individuals with frontal lobe lesions often have difficulty on recent memory tasks, but it is debated whether this is due to a true memory disorder in which both encoding and retrieval mechanisms are impaired. Rather, the apparent memory deficits result from poor search and retrieval strategies, as well as the inability to inhibit irrelevant associations (i.e., a type of executive dysfunction).

Transient Amnesia Syndromes

Transient Global Amnesia

Transient global amnesia (TGA) is an acute onset of anterograde amnesia that is temporary, lasting 2–24 hours. It typically affects individuals between ages 50 and 70.

A common feature of TGA is repetitive questions in an attempt to reorient, such as "What day is it?" and "What am I doing here?" Information can be retained for several seconds (i.e., immediate memory), but it is lost rapidly when the person is distracted. There is an accompanying temporally graded retrograde amnesia to varying degrees; individuals experiencing a TGA episode are typically unable to report what they were doing when the episode began. They lack awareness of the memory impairment, although they are often perplexed, anxious, and sense that something is wrong. After the episode resolves, there is a dense amnestic gap for events that occurred during the episode; thus they are not troubled by having experienced an episode of TGA, although family and friends who witnessed it may be.

Transient global amnesia is diagnosed on the basis of patient examination and clinical history by the following criteria: (1) an episode of anterograde amnesia witnessed by an observer; (2) no clouding of consciousness or loss of personal identity; (3) cognitive impairment limited to amnesia; (4) no focal neurological or epileptic signs; (5) no recent history of head trauma or seizures; and (6) resolution of symptoms within 24 hours.

The pathophysiology underlying TGA remains unknown, but transient dysfunction of the hippocampus appears to play a role. Diffusion-weighted MRI performed 24–72 hours after symptom onset may show small areas of restricted diffusion of water in the CA1 field of the hippocampus, indicating tissue pathology.

Recurrence is uncommon, with only 6%–10% of patients with TGA experiencing a second episode. There is no evidence of persistent cognitive impairment following an episode of TGA. Patients with TGA do not have an increased risk for stroke.

Other Transient Amnesia Syndromes

Other syndromes of acute onset amnesia that may mimic TGA clinically are transient ischemic attack in the vascular distribution of the PCA, transient epileptic amnesia, and amnestic stroke.

Transient ischemic attack (TIA) is caused by a temporary blockage or insufficiency of blood flow to a focal brain region, resulting in focal neurological symptoms that are transient and that resolve fully without causing infarction. The majority of diagnosed TIAs resolve within 30 minutes. The arterial obstruction is usually embolic, although less commonly it is thrombotic. Persons with TIA have a higher frequency of stroke risk factors as compared to the general population and are at increased risk for future cerebrovascular events and stroke, unlike those with TGA.

Transient epileptic amnesia (TEA) is isolated memory impairment due to seizure activity. **John Hughlings Jackson** (1835–1911) first described a case of TEA in 1888 in a patient referred to as **Dr. Z**. Clinically, episodes of TEA are similar to episodes of TGA, and many cases are initially diagnosed as TGA. However, TEA episodes are distinguished from TGA by their brevity, lasting only a few minutes and less than 60 minutes, compared to episodes of TGA which last 2–24 hours. TEA also recurs because it is due to a seizure disorder, but TGA does not. Evidence supporting a TEA diagnosis includes epileptiform EEG abnormalities and concurrent onset of other clinical features of epilepsy, such as lip smacking and olfactory hallucinations. Unlike TGA, TEA is associated with interictal cognitive impairment.

In **amnestic stroke**, acute onset of anterograde amnesia is the dominant symptom. It occurs with lesions in the territories of the PCA, thalamic arteries, or anterior choroidal artery. The stroke symptoms may persist or resolve after some time. Stroke in the territory of the PCA, which supplies the hippocampus, may also result in an accompanying hemianopia due to infarction in the primary visual cortex and/or its inputs.

Dissociative Amnesia

Neurological amnesia results from brain injury or brain disease, particularly when the medial temporal or diencephalic structures are involved. Amnesia, however, may also occur as a dissociative disorder. **Dissociative disorders** are a group of psychiatric disorders involving disturbance of memory, identity, emotion, perception, behavior, or sense of self as an involuntary escape from reality that usually develops as a reaction to psychological trauma. Dissociative amnesia and neurological amnesia present differently.

Dissociative amnesia is characterized by sudden autobiographical memory loss; it may even affect memory for personal identity. Information about the world (semantic knowledge) is preserved, while information about the person's personal life is affected. Despite the profound retrograde amnesia, there is no anterograde amnesia; ability to learn new information is intact. Amnesia for personal identity is always due to dissociative amnesia. In dissociative amnesia, the memory loss typically begins and ends suddenly and is organized around affective rather than temporal dimensions. There is a preceding physical or emotional trauma in most cases, as well as a prior psychiatric history. Dissociative amnesia is most common in combat veterans and sexual assault victims. In **dissociative amnesia with fugue**, the person either loses one's identity or assumes a new identity, and often unexpectedly wanders or travels.

By contrast, in neurologically based amnesia, memory for personal identity is preserved and there is an anterograde amnesia. The memory loss either begins suddenly after abrupt onset of a brain insult such as trauma or stroke, or insidiously if due to disease. Dissociative amnesia is rarer than neurological amnesia.

Forms of Memory

Studies of H.M. and other amnestic patients revealed that although they are unable to consciously recollect events and experiences that occur since the injury, not all types of learning and memory are affected. These findings showed that memory is not unitary, and there have been several classifications of memory subtypes since this realization.

Larry Squire (1941–) initially proposed that there are two forms of memory, declarative and procedural, subserved by different neurological mechanisms. Declarative memory is the conscious recollection of events and facts, corresponding to what is meant by the term *memory* in common language. It is conceptualized as "knowing what" and encompasses memory of facts and events. Procedural memory, by contrast, is memory dedicated to "knowing how" to perform a skill, without the requirement of ability to describe what has been learned.

It later became clear that there are more than just these two forms of memory, and the classification was revised to distinguish two broad types of memory, declarative and nondeclarative, with each further subdivided (Figure 26.4).

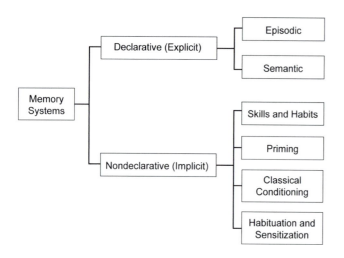

FIGURE 26.4. There are two broad forms of memory, declarative and nondeclarative. Declarative memory (explicit memory) is a form of memory that involves the conscious recollection of events and facts; it encompasses episodic memory and semantic memory. Nondeclarative memory (implicit memory) is a form of memory that does not have a conscious component whereby acquired information is expressed through performance; it encompasses skills and habits, priming, classical conditioning, and simple forms of non-associative learning (habituation and sensitization).

Declarative Memory

Declarative memory (**explicit memory**) consists of conscious recollections for events and facts. **Endel Tulving** (1927–) further distinguished two forms of declarative memory: episodic and semantic.

EPISODIC MEMORY

Episodic memory refers to memories for events or "episodes" that one can travel back to in the mind's eye. It is an autobiographical form of memory for contextually specific events; it is our store of personal experiences. Episodic memory is declarative and is relevant both to recent and remote events. It preserves specific contextual details of events so as to promote conscious recollection of individual experiences. Examples of episodic memory include remembering where you parked the car 2 hours ago, what you did yesterday, and where and how you learned of shocking news events such as the 9/11/2001 attacks, the 01/06/2021 U.S. Capitol attack, or other "where were you when..." moments.

SEMANTIC MEMORY

Semantic memory refers to one's store of knowledge about the world (facts). It includes our knowledge of words, objects, places, and people. This semantic knowledge is retrieved without recalling when and where it was first learned. It is context-free, as its content is generally devoid of information about the time and place in which the information was originally learned. Individuals with amnesia have defective episodic memory but generally retain capacities reflecting semantic knowledge.

Nondeclarative Memory

Nondeclarative memory (**implicit memory**) is an unconscious, non-intentional form of memory whereby acquired information is expressed through performance, and learning is measured with indirect or incidental tasks that make no reference to prior learning episodes at the time of retrieval. *Nondeclarative memory* is an umbrella term that includes a variety of specific forms, including procedural memory, priming, classical (Pavlovian) conditioning (associative learning), and non-associative learning (habituation, sensitization).

The study of nondeclarative memory began with the observation that H.M. was capable of learning new motor and perceptual skills (e.g., mirror drawing, visually guided stylus maze learning, manual coordination tasks, mirror reading), despite having amnesia for the prior episodes of training and testing. As described above, this form of skill learning and memory is known as **procedural memory**. The striatum, cerebellum, and motor cortex play important roles in procedural memory.

Priming is a form of implicit memory by which exposure to a stimulus influences the response to a different subsequent stimulus, without the individual being aware of the connection. Priming effects are demonstrated in the laboratory in several ways, such as: (1) increased accuracy in processing previously exposed stimuli on perceptual identification tasks that require the identification of words or pictures at very brief exposure durations or from degraded representations; and (2) a tendency to generate previously exposed stimuli in response to nominal cues on stem completion. In the real world, words, images, objects, and actions can trigger associations and influence our behavior, a fact that is well known to marketing professionals. The neocortex plays an important role in priming.

Classical conditioning is a form of associative learning in which a neutral stimulus acquires the capacity to evoke a response by being associated with a natural stimulus that automatically evokes a reflexive response (i.e., there is a change in behavior). This form of learning begins with an unconditioned stimulus (US) that naturally (without learning) elicits a reflexive, unconditioned response (UR); US → UR. A neutral stimulus that does not elicit the response is repeatedly paired with the natural US. Over time, the neutral stimulus acquires the capacity to evoke the response originally evoked by the US. Thus, the neutral stimulus becomes a conditioned stimulus (CS), and when presented by itself (i.e., without the US), it elicits a conditioned response (CR); CS → CR. This form of learning was discovered by Russian scientist **Ivan Pavlov** (1849–1936) in his research on the digestive system of dogs. He observed that the dogs, who naturally salivated (UR) when fed (in response to the smell, sight, and taste of food; US), began to salivate (CR) in response to the white lab coats of the laboratory assistants who fed them (CS). In further experimental work, Pavlov paired a neutral bell stimulus with meat, and he conditioned the dogs to salivate in response to the sound of the bell. For simple forms of motor learning, such as the classically conditioned eyeblink response in which an auditory stimulus such as a tone (CS) precedes the presentation of a puff of air directed at the cornea (US), the cerebellum plays an important role. Information about the CS and US enters the CNS by way of modality-specific sensory pathways (in this case, auditory and trigeminal pathways). The sensory pathway carrying the US signal forms the afferent link of a brainstem reflex pathway involving motor neurons that generate eyeblinks; this brainstem reflex circuit operates independently of higher brain areas. The sensory pathway carrying the US signal also projects to the cerebellum, as does the pathway carrying the CS signal. Areas of the cerebellum that receive convergent CS and US inputs are responsible for the acquisition and maintenance of the conditioned response (in this case, eyeblink).

Fear conditioning is another form of classical conditioning, in which organisms learn to predict aversive events by virtue of a neutral stimulus (e.g., a tone) presented in the context of an aversive stimulus (e.g., foot shock). The amygdala plays an important role in fear conditioning (see Chapter 12, "Limbic Structures").

Habituation is a learning process that leads to decreased responsiveness to a stimulus with experience, while **sensitization** is a process that leads to increased responsiveness to a stimulus with experience. **Eric Kandel** (1929–) has pioneered the study of the cellular and molecular basis for these forms of learning and memory using the simple animal model of the Aplysia (sea slug) gill and siphon withdrawal reflex.

Forms of Memory Based on Time Span

Memory has been divided into three forms based on the time span between stimulus presentation and memory retrieval: *immediate, recent,* and *remote*. These terms are descriptive, and the time span implied is not well defined.

Immediate Memory

Immediate memory, also referred to as **immediate recall**, refers to the recall of information immediately after presentation. It has a limited capacity (only about 7 items can be stored at a time) and a limited duration (information is lost with distraction or the passage of time).

The terms *attention span* and *attention capacity* describe this function more accurately. **Attention capacity** refers to the amount of information that can be held "on-line" at a given moment. It is measured by **span tests** (e.g., digit span), which expose the subject to increasingly greater amounts of information and require the individual to reproduce what was seen or heard. The first description of storage capacity limits was the seminal article by **George A. Miller** (1920–2012) titled "The Magical Number Seven, Plus or Minus Two: Some Limits on Our Capacity for Processing Information," published in 1956. The amount of information that can be held in immediate memory can be increased to some extent by chunking the information into larger units, a form of data compression (e.g., 6–7 chunked as 67).

The information held "on-line" has a limited duration; it is lost with the passage of time or distraction. It can be maintained for a longer period of time with rehearsal by working memory. **Working memory** is the ability to hold information in mind *and* mentally manipulate it; it reflects the executive control of attention. A common measure of working memory is the digit span backward task, which measures the number of digits that a person can repeat in reverse order of presentation. Other working memory tasks include spelling words backward and serial sevens (i.e., counting backward from 100 by sevens).

Delirium is a neurocognitive disorder characterized by a core disturbance in attention. Delirious patients perform poorly on immediate memory tasks, while patients with pure memory disorders have an intact attention capacity.

Recent Memory

Recent memory, also referred to as **short-term memory** in clinical contexts, enables recall of information over a period of minutes to days. It is the capacity to remember current, day-to-day events. Recent memory is the most vulnerable form of memory. It is tested by asking patients to recall information after a delay that is filled with distractor tasks to prevent rehearsal. The inability to form new recent memories is known as anterograde amnesia.

Remote Memory

Remote memory refers to the recollection of experiences from the distant past, such as childhood. In clinical contexts as well as in common parlance, remote memory is often referred to as **long-term memory**. Patients with anterograde amnesia usually report that their "short-term memory" is failing but that their "long-term memory" is fine. In patients with a specific defect in new learning (recent memory) since an injury with precise onset, remote memory refers to the recall of events that occurred before the onset of the injury and recent memory defect.

Summary

Memory is a general term for the mental processes involving the acquisition and retention of information. Our understanding of the neurological basis of memory was revolutionized by studies of patient H.M., who underwent an experimental bilateral temporal lobectomy to treat epilepsy in 1953. The surgery was successful in treating the seizures but unexpectedly resulted in a profound anterograde amnesia. Studies of H.M. and other amnestic patients, as well experimental studies performed in animals with similar and more limited resections, have established the following: (1) memory is a distinct cerebral function; (2) the medial temporal lobes are crucial for forming new episodic memories; (3) amnestic patients retain some forms of learning and memory that are demonstrated by changes in behavior rather than the ability to consciously recall events or facts (nondeclarative memory); (4) the severity of memory impairment depends on both the locus and extent of damage within the medial temporal lobe, with damage limited to the hippocampus sufficient to produce significant anterograde amnesia, and damage that includes both the hippocampus and adjacent cortices producing more severe anterograde amnesia.

Common causes of medial temporal amnesia include anoxic-hypoxic injury, bilateral compromise of the PCAs, viral (herpes simplex type I) encephalitis, and early-stage Alzheimer's disease. Lesions outside of the medial temporal lobe, however, may also cause amnesia. These structures are the fornices, diencephalon (thalamus and hypothalamus), and basal forebrain, and they form circuits with the hippocampal formation. Amnesia may also occur transiently, as with transient global amnesia (in which the underlying pathophysiology remains unknown), transient ischemic attack in the vascular distribution of the PCA, or transient epileptic amnesia.

Additional Reading

1. Andrew C, Papanicolaou AC. *The amnesias: a clinical textbook of memory disorders*. Oxford University Press; 2005.
2. Annese J, Schenker-Ahmed NM, Bartsch H, Maechler P, Sheh C, Thomas N, et al. Postmortem examination of patient H.M.'s brain based on histological sectioning and digital 3D reconstruction. *Nat Commun*. 2014;5:3122. doi:10.1038/ncomms4122
3. Augustinack JC, van der Kouwe AJ, Salat DH, Benner T, Stevens AA, Annese J, et al. H.M.'s contributions to neuroscience: a review and autopsy studies. *Hippocampus*. 2014;24(11):1267–1286. doi:10.1002/hipo.22354
4. Bartsch T, Deuschl G. Transient global amnesia: functional anatomy and clinical implications. *Lancet Neurol*. 2010;9(2):205–214. doi:10.1016/S1474-4422(09)70344-8
5. Clark RE, Squire LR. An animal model of recognition memory and medial temporal lobe amnesia: history and current issues. *Neuropsychologia*. 2010;48(8):2234–2244. doi:10.1016/j.neuropsychologia.2010.02.004
6. Clark RE. A history and overview of the behavioral neuroscience of learning and memory. *Curr Top Behav Neurosci*. 2018;37:1–11. doi:10.1007/7854_2017_482. PMID: 29589321.
7. Corkin S. Lasting consequences of bilateral medial temporal lobectomy: clinical course and experimental findings in H.M. *Semin Neurol*. 1984;4(2):249–259. doi:10.1055/s-2008-1041556
8. Corkin S, Amaral DG, González RG, Johnson KA, Hyman BT. H.M.'s medial temporal lobe lesion: findings from magnetic resonance imaging. *J Neurosci*. 1997;17(10):3964–3979. doi:10.1523/JNEUROSCI.17-10-03964.1997
9. Corkin S. What's new with the amnesic patient H.M.? *Nat Rev Neurosci*. 2002;3(2):153–160. doi:10.1038/nrn726
10. Freed DM, Corkin S, Cohen NJ. Forgetting in H.M.: a second look. *Neuropsychologia*. 1987;25(3):461–471. doi:10.1016/0028-3932(87)90071-6
11. Graff-Radford NR, Tranel D, Van Hoesen GW, Brandt JP. Diencephalic amnesia. *Brain*. 1990;113 (Pt 1):1–25. doi:10.1093/brain/113.1.1
12. Jung YC, Chanraud S, Sullivan EV. Neuroimaging of Wernicke's encephalopathy and Korsakoff's syndrome. *Neuropsychol Rev*. 2012;22(2):170–180. doi:10.1007/s11065-012-9203-4
13. Kensinger EA, Ullman MT, Corkin S. Bilateral medial temporal lobe damage does not affect lexical or grammatical processing: evidence from amnesic patient H.M. *Hippocampus*. 2001;11(4):347–360. doi:10.1002/hipo.1049
14. Kril JJ, Harper CG. Neuroanatomy and neuropathology associated with Korsakoff's syndrome. *Neuropsychol Rev*. 2012;22(2):72–80. doi:10.1007/s11065-012-9195-0
15. Kritchevsky M, Chang J, Squire LR. Functional amnesia: clinical description and neuropsychological profile of 10 cases. *Learn Mem*. 2004 Mar–Apr;11(2):213–226. doi: 10.1101/lm.71404. PMID: 15054137; PMCID: PMC379692.
16. Lanzone J, Ricci L, Assenza G, Ulivi M, Di Lazzaro V, Tombini M. Transient epileptic and global amnesia: real-life differential diagnosis. *Epilepsy Behav*. 2018;88:205–211. doi:10.1016/j.yebeh.2018.07.015
17. Miller GA. The magical number seven, plus or minus two: some limits on our capacity for processing information. *Psychol Rev*. 1956 Mar;63(2):81–97. PMID: 13310704.
18. Salat DH, van der Kouwe AJW, Tuch DS, Quinn BT, Fischl B, Dale AM, et al. Neuroimaging H.M.: a 10-year follow-up examination. *Hippocampus*. 2006;16(11):936–945. doi:10.1002/hipo.20222
19. Schmolck H, Kensinger EA, Corkin S, Squire LR. Semantic knowledge in patient H.M. and other patients with bilateral medial and lateral temporal lobe lesions. *Hippocampus*. 2002;12(4):520–533. doi:10.1002/hipo.10039
20. Scoville WB, Milner B. Loss of recent memory after bilateral hippocampal lesions [1957]. *J Neuropsychiatry Clin*

Neurosci. 2000;12(1):103–113. doi:10.1176/jnp.12.1.103

21. Segobin S, Laniepce A, Ritz L, Lannuzel C, Boudehent C, Cabé N, et al. Dissociating thalamic alterations in alcohol use disorder defines specificity of Korsakoff's syndrome. *Brain.* 2019;142(5):1458–1470. doi:10.1093/brain/awz056

22. Squire LR, Amaral DG, Zola-Morgan S, Kritchevsky M, Press G. Description of brain injury in the amnesic patient N.A. based on magnetic resonance imaging. *Exp Neurol.* 1989;105(1):23–35. doi:10.1016/0014-4886(89)90168-4

23. Squire LR. Two forms of human amnesia: an analysis of forgetting. *J Neurosci.* 1981;1(6):635–640. doi:10.1523/JNEUROSCI.01-06-00635.1981

24. Squire LR, Zola SM. Amnesia, memory and brain systems. *Philos Trans R Soc Lond B Biol Sci.* 1997;352(1362):1663–1673. doi:10.1098/rstb.1997.0148

25. Squire LR. The legacy of patient H.M. for neuroscience. *Neuron.* 2009 Jan 15;61(1):6–9. doi:10.1016/j.neuron.2008.12.023. PMID: 19146808; PMCID: PMC2649674.

26. Squire LR, Wixted JT. The cognitive neuroscience of human memory since H.M. *Annu Rev Neurosci.* 2011;34:259–288. doi:10.1146/annurev-neuro-061010-113720

27. Teuber HL, Milner B, Vaughan HG Jr. Persistent anterograde amnesia after stab wound of the basal brain. *Neuropsychologia.* 1968;6(3):267–282. doi.org/10.1016/0028-3932(68)90025-0

28. Thomson AD, Guerrini I, Marshall EJ. The evolution and treatment of Korsakoff's syndrome: out of sight, out of mind? *Neuropsychol Rev.* 2012;22(2):81–92. doi:10.1007/s11065-012-9196-z

Language and the Aphasias

Introduction

Linguistics is the scientific study of language; it is a cognitive science that has the goal of discovering what elements are universal to all languages and what elements are specific to a given language or group of languages. **Neurolinguistics** is a branch of linguistics that is devoted to understanding the brain mechanisms underlying language. Historically, neurolinguistics emerged from **aphasiology**, the study of impaired and spared language abilities resulting from brain damage. With the advent of functional brain imaging which allows us to study the neuroanatomical basis of language in nonaphasics, the study of the neural basis of language has broadened from aphasiology into a melding of linguistics with aphasiology—i.e., neurolinguistics.

Language is a system of symbols and the rules for sequencing those symbols that allows us to communicate our ideas to others. It is the transformation of our thoughts into symbolic representation, often expressed in words through speaking or writing, but we also communicate by gestural symbols (e.g., gestures, sign language). The terms *speech* and *language* are often used interchangeably; however, they are quite different. Speech is a principal route through which we communicate with others, but we also communicate by writing and gesturing. Language provides the common structure that is deployed in speaking, writing, and reading.

Aphasia is a language disorder. The specific deficits affecting spoken language are usually mirrored in written language. Aphasias that are characterized by a prominent deficit in spoken language expression are accompanied by an inability to communicate by writing (agraphia). Aphasias that are characterized by a prominent deficit in spoken language comprehension are accompanied by an inability to comprehend written language by reading (alexia). Aphasias that are characterized by prominent deficits in *both* spoken language expression and comprehension are accompanied by alexia *and* agraphia.

This chapter discusses the basic features of oral language, the clinical assessment of language function, the classic aphasia syndromes, and the primary progressive aphasia syndromes.

Oral Language

Oral language is a system through which we use spoken words to communicate with others, both in terms of expression and comprehension. Three basic features of oral language are phonology, syntax, and semantics. The traditional neurological model of language postulates a division between language production and comprehension processes, but contemporary theories propose that there are primary brain systems for phonology, semantics, and syntax, each contributing to both the comprehension and production of language. A basic understanding of these features is therefore important in the study of aphasia because they may selectively break down.

Phonology refers to the collection of distinctive sounds within a particular language, including their correct pronunciation and rule-based organization of sounds within syllables. **Phonemes** are the smallest units of sound that make up a language. The phonemes include the short and long vowel sounds, the consonant sounds, and sounds made by specific letter combinations (e.g., sounds like /th/ and /sh/). Phonemes distinguish one word from another, and changing a phoneme in a word changes the way that word is pronounced as well as its meaning, although the phoneme itself does not have meaning. Phonology, which deals with assembling abstract sound units that are combined into words, is distinct from **phonetics**, the perception and production of speech sounds, which requires the sensory skills of hearing and motor skills for speaking. Practically, however, it is often difficult to distinguish phonetic from phonological errors in spoken language, since they both result in speech sound errors. **Semantics** refers to the meaning of words and phrases in a language, encompassing both vocabulary (the **lexicon**) and figurative language (e.g., metaphors and idioms). **Syntax** refers to the rules governing sentence structure, the way that words and phrases are arranged in a sentence for it to make sense and convey meaning.

Speech

Speech is a complex motor task involving coordinated activity of approximately 100 respiratory, pharyngeal, laryngeal, and orofacial muscles that control the flow of air from the diaphragm, through the vocal folds of the larynx (voicebox), to the speech articulators (mouth, lips, tongue). **Speech disorders** result in a selective impairment in spoken language output; expressive communication by other modes (i.e., writing and signing) is intact, as is comprehension of spoken, written, and sign language. Speech disorders arise from a wide variety of causes, including cleft lip, cleft palate, vocal cord pathology, hearing loss, and intellectual disability. Speech disorders involve impairment in the motor mechanics of speech (articulation), speech fluency, or voice. Neurological disorders commonly affect speech output, especially those involving extrapyramidal, cerebellar, or subcortical pathology.

ARTICULATION

Articulation refers to the clarity of speech, the process by which sounds, syllables, and words are formed when the tongue, jaw, teeth, lips, and palate alter the air stream coming from the vocal folds. Articulation disorders, also known as **dysarthrias**, are speech disorders resulting from weakness or incoordination of the oral muscles (jaw, lips, tongue, palate/roof of the mouth) or speech motor programming deficits. Articulation errors are motor-based and result in **phonetic errors** in producing the sounds of the language. The most common sound misarticulations are substitutions (e.g., *wabbit* for *rabbit*); sound distortions involving substitution of a nonstandard sound for a standard one, such as in a lisp where the *s and z* sound like *th* (e.g., *yeth* for *yes*); and sound omissions (e.g., *han* for *hand*).

Developmental dysarthria is a speech disorder caused by dysfunction of the immature nervous system that delays speech onset and impairs the strength, speed, accuracy, coordination, and endurance of the muscle groups used to speak. **Acquired dysarthria** is a loss of articulation ability that may be caused by a variety of neurologic insults including cerebellar lesions, stroke, degenerative disease (e.g., multiple sclerosis, progressive supranuclear palsy), and toxins (e.g., lead poisoning, carbon monoxide poisoning, long-term alcohol or drug abuse). Dysarthria may also occur transiently with alcohol or drug intoxication. Cerebellar lesions, particularly those involving the superior paravermal region, result in **ataxic dysarthria**.

FLUENCY

Fluency refers to the smoothness and flow with which sounds, syllables, words, and phrases are joined together in a way that sounds natural and uninterrupted. **Speech dysfluency** is characterized by hesitations (difficulty initiating utterances); false starts (words and sentences that are cut off mid-utterance and restarted); repetition of syllables and words; and prolongations of sounds, syllables, words, or phrases. **Stuttering** (stammering) is a speech dysfluency that is usually developmental. Occasionally it is an acquired adult-onset disorder, usually due to vascular or traumatic insults, but may also occur with extrapyramidal disease.

VOICE

Voice (**phonation**) is the audible sound produced by the passage of air through the larynx. It is characterized by three elements: pitch, volume, and tone (quality).

Variations in these elements allow the speaker to convey additional meaning and emotion to the words uttered. **Voice disorders**, also known as **dysphonias**, involve abnormalities in pitch (too high, too low, never changing), volume (**hyperphonia** or **hypophonia**), or tone/quality (harsh, hoarse, breathy, nasal). Hypophonia commonly occurs with Parkinson's disease.

Prosody refers to the variations in melody, loudness, and rhythm of speech. Prosody plays a role in differentiating statements, questions, and exclamations. It also conveys attitudes and emotions. **Dysprosody** is characterized by alterations in the intensity, rhythm, cadence, and intonation of words that are atypical for the language. The most common causes are stroke and traumatic brain injury. **Aprosodia** is a loss of prosody, so that speech is flat, monotonic, amelodic, and devoid of emotional overtones. Aprosodia occurs with lesions in area 44 of the right frontal operculum (i.e., the right hemisphere homologue of Broca's area). **Foreign accent syndrome** is a rare condition in which mild abnormalities in intonation result in a prosodic distortion that sounds like an unusual accent.

APRAXIA OF SPEECH

Apraxia of speech (AOS) is a motor speech programming disorder which selectively disrupts the finely coordinated movements of articulation. Like all other apraxias, AOS is *not* attributable to a primary motor deficit such as paralysis, weakness, or incoordination of the speech musculature; thus it is distinct from dysarthria. Differentiating AOS and dysarthria may be difficult, and in some cases both disorders are present. Writing is preserved, indicating that the disorder is at the level of speech output and not at the deeper level of language.

Those with AOS are no longer able to convert phonological knowledge into the correct verbal-motor commands, and the speech articulators (mouth, lips, tongue) are not placed in the right position at the right time when attempting to produce a given speech sound or word. This results in hesitancy, effortfulness, slow articulation, inconsistent articulation errors, articulatory groping (multiple attempts at trying to pronounce the word correctly), sound sequencing errors, and a form of dysprosody known as **staccato speech** (**scanning speech**) in which spoken words are broken up into separate syllables, separated by noticeable pauses, with an unusual stress pattern on syllables. The speech errors increase with word complexity and word length.

There are two types of AOS, developmental and acquired AOS. **Developmental AOS**, also known as childhood AOS, is a developmental disorder involving a delay or failure to attain normal motor speech programming skills. The causes of developmental AOS are not well understood. **Acquired AOS** is caused by damage to the parts of the brain that are involved in the motor programming of speech, resulting in a loss of normal speaking ability. It may occur with a variety of pathologies (e.g., stroke, traumatic brain injury, space-occupying lesions, focal neurodegeneration), and is especially associated with lesions of the premotor area within the language dominant hemisphere (although it has also been observed with lesions of the anterior insula or caudate head within the language dominant hemisphere).

Aphasia

Aphasia is the loss or deterioration of one or more aspects of the process of comprehending and/or formulating verbal messages due to an acquired disorder of the brain. The most common causes of aphasia are stroke and head trauma. The definition of aphasia excludes motor speech disorders (although some lesions may cause both a motor speech disorder and aphasia). It also excludes language impairments secondary to other cognitive disorders such as thought disorders or auditory disorders. Aphasia is an acquired disorder of language due to damage to parts of the brain involved in language. It is not a developmental language disorder; the term **dysphasia** is used to describe a selective disorder of language development.

Language is typically **lateralized** to the left hemisphere, and especially so in right-handed individuals. It is for this reason that the left hemisphere is often referred to as the dominant hemisphere, although it is more accurate to refer to it as the **language-dominant hemisphere** (when it is). The majority of people (approximately 90%) are left hemisphere dominant for language. Ninety-six percent of right-handers and 70% of left-handers have language functions localized to the left hemisphere. The remaining 4% of right-handers are right hemisphere dominant for language. Of left-handers, 15% are right hemisphere dominant for language, and 15% have bilateral language representation. Aphasia therefore is usually due to damage to the left hemisphere cerebral cortex, and the below discussion assumes left-hemisphere language dominance.

Within the left hemisphere, lesions in different parts of the cerebral cortex produce different patterns of language disturbance. Occasionally, right-handed persons with right hemisphere lesions develop aphasia, a phenomenon known as **crossed aphasia**; these people are right hemisphere dominant for language. Most cases of aphasia that are due to right hemisphere stroke, whether in left- or right-handed persons, produce a **mirror image syndrome** (i.e., same topographic organization, different hemisphere); the representation of language within the right hemisphere mirrors that of the left hemisphere.

Language Assessment

Aphasia is not a unitary disorder; there are different forms. There are several systems for classifying the aphasias. The Boston classification system divides the aphasias into eight classical aphasia syndromes (see "The Classical Aphasia Syndromes," below). In practice, the symptoms exhibited by any individual person may not fall simply into one category or another, because lesions producing cortical damage are not always coextensive with a functional site. The Boston classification system forms the basis for the Boston Diagnostic Aphasia Exam (BDAE), a structured analysis of six behaviors: spontaneous expressive language, naming, repetition (i.e., the ability to repeat words and sentences spoken by another), auditory comprehension, reading, and writing.

Spontaneous expressive language (propositional speech) is defined as spoken verbal language that conveys thoughts (ideas, knowledge, feelings) generated by the speaker; it encompasses language processing in conversation as well as naming. This stands in contrast to **automatic speech (non-propositional speech)**, defined as over-learned or low-content spoken output that can be produced with little awareness of meaning, such as uttering expletives or common expressions, reciting memorized sequences (e.g., consecutive numbers, the alphabet), repeating words and sentences spoken by another, and reading aloud. These two forms of spoken output are distinguished because non-propositional speech may be spared in severe aphasias. In fact, **speech automatisms**, stereotyped and repetitive utterances, often occur in those with severe aphasia.

Spontaneous expressive language is described with regard to fluency, grammatical structure, evidence of paraphasias, and word-finding ability.

FLUENCY

As described above, fluency refers to the smoothness and flow with which sounds, syllables, words, and phrases are joined together. Fluency may be undermined by either speech or language disorders. The various aphasias are classified as either fluent or nonfluent. Fluent language is flowing and effortless; mean phrase length ranges from 5 to 8 words and sentence length is normal. Reduced fluency can range from fully unrecognizable output, to repeated meaningless utterances, or stereotyped phrases (e.g., "yes," "fine"), to agrammatic language characterized by short, truncated phrases that use only the most meaning-laden words and omit relational words. Nonfluent language is extremely labored; there is considerable hesitation and delay in production. Facial grimacing and hand or body gesturing may accompany the struggle to speak. The impairment is not articulatory; nonfluent aphasia may occur without dysarthria, and dysarthria may occur without aphasia.

SYNTAX

Syntax refers to how sentence structure conveys meaning. This includes meaning that results from word ordering. For example, two sentences that use the same words but have different syntactic structure can have very different meanings (e.g., *Mary loves John* versus *John loves Mary*). It also includes the rules that govern transformations of sentence structure that convey the same meaning. For example, the English language uses a subject-verb-object sentence structure (e.g., *John hit the ball*). It allows for an alternate word order to impart the same meaning (*The ball was hit by John*), but not another word reordering (*John was hit by the ball*) that would impart a different meaning. Syntax also encompasses the rules that govern constructions that signal the functional relationships between constituents in a clause (a group of words that contains a subject and a verb that have a relationship); for example, from the sentence *The lion was killed by the tiger*, knowing which animal did the killing and which animal was killed.

Some aphasias are characterized by a breakdown in the grammatical structure of language (i.e., loss of the ability to place words together in grammatically organized sequences). This results in **agrammatism**, a disjointed simplified sentence structure in which substantives (nouns and verbs) are juxtaposed, omitting relational/function words (e.g., articles, prepositions, conjunctions, auxiliary verbs), and erring on word form variations in which a word is modified to express grammatical variants such as verb tense (e.g., past, present, future) and number (singular, plural). Agrammatism is usually associated with nonfluent aphasias. **Telegraphic speech** is a mild form of agrammatism in which grammatical complexity is reduced to short, truncated phrases, with deletion of many articles, prepositions, conjunctions, and auxiliary verbs (e.g., *want eat*). The main message is understandable, but the sentence is not complete. It is a language disorder, not a speech disorder.

PARAPHASIAS

Paraphasia is a language output error characterized by the production of unintended syllables, words, or phrases when speaking. There are several types of paraphasias.

Literal paraphasias, also known as **phonemic paraphasias**, are speech sound errors that reflect impaired phonological structure. Some phonemic features of the intended word are usually preserved, such as the number of syllables and the vowels. Phonemic paraphasias are most evident with polysyllabic words. The errors consist of **phoneme substitutions** (substitution of sounds within words, e.g., *cactus* → "captus"), **phoneme deletions/omissions** (omission of sounds within words, e.g., *elephant* → "elphant"), **phoneme additions/insertions** (insertion of sounds within words, e.g., *cactus* → "cactusk"), and

phoneme transpositions (misplacement of sounds within words, e.g., *animal* → "aminal").

Verbal paraphasias are word substitution errors. There are two types of verbal paraphasias: semantic and remote. If the substituted word is related in a connotative sphere (e.g., saying "mother" instead of *wife*, or "spoon" instead of *fork*), it is then referred to as a **semantic paraphasia**. Capricious substitutions (e.g., saying "dog" instead of *car*) are referred to as **remote paraphasias**.

Neologisms (neologistic paraphasias) are nonsensical, meaningless words (non-words) that are phonologically legal (e.g., "snopel"). In neologisms, the intended word is not identifiable.

NAMING

Anomia refers to word retrieval difficulties, particularly affecting nouns and verbs. Many persons with anomia describe their symptoms as "memory loss" or "memory failure," as they complain of forgetting the names of people, places, and objects. Anomia is characterized by behavioral signs such as prolonged retrieval latencies (word-finding pauses), incorrect word production (typically semantic paraphasias), **circumlocution** (talking around the word), and gesturing the use of objects that cannot be named. Subjectively, anomia is experienced as the familiar "tip-of-the-tongue" phenomenon in which the person is unable to name an object or person, but knows the name and feels that it is just out of reach and about to become available. Phonemic cues (i.e., providing the first phoneme of the word) often aid word retrieval. The anomia may be so severe that it results in **"empty speech,"** characterized by frequent use of nonspecific words for concrete words (*thing, it, place*) and vague circumlocutions. Despite the struggle to find the appropriate word and resulting word-finding pauses, spoken language is fluent.

Anomia is formally assessed by tests of **visual confrontation naming**, which entail showing the patient objects or line drawings of objects and asking them to provide the names. The interpretation of test performance, however, must take into consideration both premorbid verbal skills and that fact that poor performance may be secondary to nonlinguistic deficits such as visual object agnosia. Qualitatively, however, the confrontation naming test performance of anomic patients is characterized by behavioral signs of word-finding pauses, semantic paraphasias, circumlocution, gesturing the use of objects, and improved naming with phonemic cueing.

Anomia is the most common sign and symptom of neurological language disorder. It may occur in the setting of otherwise normal language, or it may be accompanied by other symptoms of aphasia.

The Classic Aphasia Syndromes

There are eight classic aphasia syndromes: Broca's aphasia, Wernicke's aphasia, global aphasia, conduction aphasia, transcortical motor aphasia, transcortical sensory aphasia, transcortical mixed aphasia, and anomic aphasia. These syndromes are differentiated on the basis of three independent signs: fluency, comprehension, and repetition. Naming is impaired in all eight syndromes (Table 27.1).

Broca's Aphasia

In 1861, **Pierre Paul Broca** (1824–1880), a French physician and anatomist, was called to treat a 51-year-old patient named Louis Victor Leborgne for advanced gangrene of the leg. Leborgne, who had lost his ability to speak 21 years earlier at age 31, could produce virtually no language. However, he often uttered the sound "tan" and therefore became known as **"Tan"** in the hospital

TABLE 27.1	The Eight Classic Aphasia Syndromes				
	APHASIA	**FLUENT**	**REPETITION**	**COMPREHENSION**	**NAMING**
Perisylvian	Broca's	No	Poor	Good	Poor
	Wernicke's	Yes	Poor	Poor	Poor
	Global	No	Poor	Poor	Poor
	Conduction	Yes	Poor	Good	Poor
Extrasylvian	Transcortical motor	No	Good	Good	Poor
	Transcortical sensory	Yes	Good	Poor	Poor
	Transcortical mixed	No	Good	Poor	Poor
	Anomic (nominal)	Yes	Good	Good	Poor

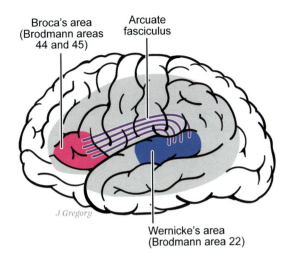

FIGURE 27.1 A lesion of Brodmann areas 44 and 45 of the language-dominant hemisphere (Broca's area) produces an expressive aphasia. A lesion of Brodmann area 22 of the language-dominant hemisphere (Wernicke's area) produces a receptive aphasia. A lesion of the arcuate fasciculus, which connects Wernicke's to Broca's area, produces conduction aphasia. A lesion of the language-dominant hemisphere involving both Broca's and Wernicke's areas produces global aphasia.

where he resided after becoming aphasic. His mouth, tongue, and larynx were intact, and he could comprehend language. Leborgne died one week after Broca examined him. Broca performed an autopsy and found a large lesion centered in the third (inferior) frontal gyrus of the left hemisphere. He identified this region as the seat of the faculty of articulate language, which later became known as **Broca's area**. Subsequently, Broca chronicled 20 additional cases of "aphemia" with autopsy findings of lesions in the left third frontal convolution, through his groundbreaking **lesion-deficit analysis** method for the systematic study of the functional organization of the human brain by examining the relationship between deficits exhibited in clinical cases and postmortem lesion localization.

Lesions in Broca's area produce an **expressive aphasia** characterized by nonfluent language production. The nonfluent output is agrammatic and contains paraphasic errors that typically are of the semantic type. Naming is impaired. Repetition is also impaired. In contrast to language production (expressive language), language comprehension (receptive language) is relatively preserved, although it is rarely normal. Response to word-recognition tasks, simple commands, and routine conversation is generally good, but comprehension of complex syntactical sentences is generally poor. As written language parallels spoken language, there is an associated agraphia (an acquired loss in the ability to communicate through writing due to neurological disorder). People with Broca's aphasia are aware of their expressive language deficit.

Individuals with severe expressive aphasia may be capable of swearing, making statements of self-pity, praying, and singing. It is believed that the ability to produce these types of utterances is mediated by the nondominant (usually right) hemisphere and limbic structures, while the left hemisphere mediates the syntactical, temporal-sequential, motoric, and grammatical aspects of linguistic expression in the majority of people.

Broca's aphasia is produced by lesions in the posterior two-thirds of the inferior frontal gyrus in the pars triangularis (Brodmann area 44) and pars opercularis (Brodmann area 45) of the left hemisphere (Figure 27.1). These lesions are most commonly caused by ischemic stroke in the territory of the superior division of the left middle cerebral artery (MCA, see Chapter 5, "Blood Supply of the Brain"). Broca's aphasia is commonly accompanied by right hemiplegia or hemiparesis due to the proximity of Broca's area to primary motor cortex.

Wernicke's Aphasia

In 1874, German neurologist **Karl Wernicke** (1848–1905) identified a second major aphasia syndrome in 10 patients, with autopsy data on four. The syndrome became known as **Wernicke's aphasia** and was characterized by a core deficit in language comprehension; thus it is also known as a **receptive aphasia**.

The severe comprehension deficit in Wernicke's aphasia is evident at the levels of word recognition, simple commands, and simple conversation. People with Wernicke's aphasia are unable to perform word-picture matching tasks, unable to follow simple commands, and unable to accurately answer simple yes-no questions.

In Wernicke's aphasia, spontaneous expressive language flows effortlessly; mean phrase and sentence lengths are normal, although there is often an increase in number of words per unit time (**logorrhea**). Speech is normal in articulation, rhythm, and melodic tone. Output, however, is devoid of content, consisting of many stock phrases and paraphasias (phonemic, semantic, and neologistic). The copious flow of spoken language with preserved phonology may be so full of paraphasias that its meaning cannot be discerned. The grammatical structure is grossly preserved, but with **paragrammatism**, in which juxtaposed phrases have no meaningful relationships to each other, resulting in logically incoherent language (e.g., *And I want everything to be so talk*). The incomprehensible language output is known as **jargon aphasia** or **word salad**. Lesion-symptom mapping studies have shown that paragrammatism is an independent grammatical disorder resulting from damage to distinct brain systems from those resulting in agrammatism. In right-handed persons with aphasia due to left hemisphere stroke, paragrammatism is associated with damage to the left posterior superior and middle temporal gyri, while agrammatism is associated with damage to Broca's area.

Naming is invariably disturbed and characterized by paraphasias and neologisms. Contextual and phonemic prompting does not improve naming. Repetition is also impaired. People with fluent aphasia are oblivious to their failure in communication.

Since in aphasia, written language deficits usually parallel spoken language deficits, in Wernicke's aphasia there is **alexia with agraphia**. Wernicke's aphasics cannot write spontaneously or to dictation. Those with alexia with agraphia may, however, be able to write their names because it is routine, no longer requires precise acoustic analysis, and has become a motor stereotypy. Wernicke's aphasics can copy verbal material that is presented visually.

The lesions causing Wernicke's aphasia are in the posterior portion of the left superior temporal gyrus (see Figure 27.1). Therefore, this region is also known as **Wernicke's area**. It is part of Brodmann area 22; some authors also include the angular gyrus (Brodmann area 39) of the inferior parietal lobule.

Direct cortical stimulation studies performed for the purpose of mapping areas of eloquent cortex in the clinical preoperative evaluation of neurosurgery patients have confirmed that electrical stimulation in Wernicke's area disrupts language comprehension (recall that outside of primary motor and sensory areas, electrical stimulation produces a temporary lesion by interfering with naturally occurring signals and disrupts function). Lesions in Wernicke's area are most commonly caused by ischemic stroke in the territory of the inferior division of the left MCA.

The defective expressive language of Wernicke's aphasia has been attributed to an inability to monitor language output. The phenomenon of pure word deafness, a selective deficit in comprehending spoken language, however, is inconsistent with this interpretation (see Chapter 16, "The Temporal Lobes and Associated Disorders"). Although Wernicke's area has traditionally been held to play a critical role in language comprehension, modern imaging and neuropsychological studies have revealed that this region plays its major role in the phonologic retrieval stage of speech production, rather than language comprehension. The combination of paraphasic speech production and comprehension impairment that occurs in Wernicke's aphasia, then, results from damage to the phonologic retrieval system and inability to map perceived phoneme sequences to word concepts within the semantic system.

Global Aphasia

Global aphasia, also known as **mixed aphasia**, is characterized by both impaired expression and comprehension. It is essentially a combination of Broca's and Wernicke's aphasias. Spontaneous expressive language and language comprehension are absent or reduced to only a few select words. Repetition is impaired. Reading and writing are

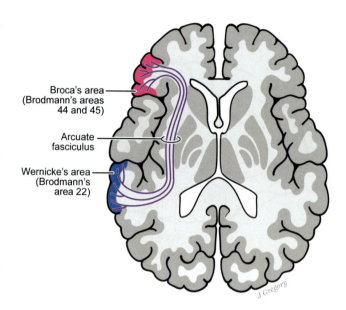

FIGURE 27.2. The arcuate fasciculus connects Wernicke's and Broca's areas.

likewise severely impaired. Global aphasia is the most severe form of aphasia, as well as the most common form.

Global aphasia is caused by large lesions in the perisylvian region that damage most or all of the anterior and posterior language areas (see Figure 27.1). The most common cause is embolic occlusion of the internal carotid artery or the origin of the MCA in the language-dominant hemisphere. It is almost always associated with right hemiplegia and right hemisensory loss.

Conduction Aphasia

Conduction aphasia is a rare form of aphasia characterized by a selective deficit in the ability to repeat words and sentences spoken by another person, in the context of preserved expressive and receptive language. Most commonly the repetition errors are paraphasias, word substitutions, or word omissions. Function words, as opposed to nouns, are most commonly affected. Persons with this disorder are aware of their errors. The critical lesion is in the **arcuate fasciculus**, resulting in a disconnection between Wernicke's language comprehension area and Broca's language expression area (Figure 27.2; also see Figure 27.1). There also is an associated impairment in carrying out verbal commands, as the commands are prevented from reaching the motor cortex of either hemisphere.

The Extrasylvian Aphasias

Broca's, Wernicke's, global, and conduction aphasias are all examples of **perisylvian aphasias**; they are due to damage in the vicinity of the Sylvian fissure. Perisylvian aphasias all involve the inability to repeat words and sentences spoken by another. **Extrasylvian aphasias** occur with damage outside the area surrounding the Sylvian fissure in the left hemisphere. These aphasias are characterized by intact repetition, with disruption of other language functions. The extrasylvian aphasias are also referred to as the **transcortical aphasias**. **Ludwig Lichtheim** (1845–1928) introduced the term *transcortical*; he believed that the lesions were outside of primary language cortex and in adjacent areas of cortex that he referred to as "areas of concepts."

There are three forms of transcortical aphasia: transcortical motor aphasia, transcortical sensory aphasia, and mixed transcortical aphasia. The transcortical aphasias are usually due to infarcts within the cortical watershed zones (see Figure 18.4 in Chapter 18). The most common pathology is watershed infarction due to decreased cerebral circulation (as occurs with cardiac arrest or significant stenosis of the carotid artery) or hypoxia (as occurs with carbon monoxide poisoning).

Transcortical motor aphasia is similar to Broca's aphasia, except that repetition is preserved. It results from lesions in the frontal lobe of the language-dominant left hemisphere, anterior or superior to Broca's area at the MCA-ACA interface. These lesions may disrupt pathways that connect the supplementary motor area and Broca's area. **Transcortical sensory aphasia** is similar to Wernicke's aphasia, except that repetition is preserved. It results from lesions in the parietal and temporal areas, posterior to Wernicke's area at the MCA-PCA interface. **Mixed transcortical aphasia** is similar to global aphasia, except that repetition is preserved. In fact, repetition is the only spared language ability. Repetition is easy to initiate in these patients because they have a tendency to be **echolalic** (i.e., repeat what others say). The lesions that produce mixed transcortical aphasia are extensive, crescent-shaped infarcts within the cortical watershed zones at both the MCA-ACA and MCA-PCA interfaces, with intact perisylvian structures allowing repetition to be preserved (see Figure 18.4 in Chapter 18).

Nominal (Anomic) Aphasia

Anomia occurs in all of the classic aphasias. It can also occur in isolation without other speech or language impairment, a condition referred to as *anomic aphasia* or *nominal aphasia*. **Anomic aphasia** is characterized by word-finding pauses, circumlocution, semantic paraphasias, and gesturing the use of objects that cannot be named. The naming deficit is present in spontaneous conversation as well as in naming objects, regardless of the sensory modality by which the object is presented (e.g., visual, auditory, somatosensory).

Isolated anomia is typically associated with lesions in the inferotemporal or polar temporal cortex of the language-dominant hemisphere, outside the classic language regions.

Other Aphasias

Subcortical Aphasia

Aphasia can occur with lesions to subcortical structures, namely the basal ganglia and thalamus (especially the ventrolateral and anteroventral nuclei). **Subcortical aphasia** results in both speech and language defects. In the acute phase, patients usually are mute, recovering gradually to a hypophonic, slow, dysarthric output. The language defects are variable between patients, and the profile does not conform to any of the classical cortical-related aphasia syndromes. Frequently, right hemiparesis or hemisensory impairments accompany the aphasia. There is a strong tendency for subcortical aphasia to resolve, especially when caused by hemorrhagic lesions. If the causative lesion also involves the language cortex, recovery will be incomplete.

Thalamic aphasia occurs with thalamic lesions in the left hemisphere that involve the ventrolateral nucleus. It is believed to result from a disconnection between thalamic nuclei and cortical language centers. It may present as a fluent aphasia with impaired comprehension and paraphasic errors, or a nonfluent aphasia with hypophonic speech. Repetition is generally intact. Thalamic aphasias tend to resolve relatively rapidly.

Dynamic Aphasia

Dynamic aphasia, also referred to as **verbal adynamia** and **frontal dynamic aphasia**, is characterized by a profound reduction in spontaneous verbal output in the context of well-preserved naming, repetition, and comprehension. The poverty of output manifests as a near-total failure to initiate spoken output, and responses to questions that are sparse, laconic, and initiated only after long pauses. Spontaneous expressive language rests on the ability to generate a message or formulate a verbal thought; impairment in this ability is the hallmark of dynamic aphasia. Those with dynamic aphasia have markedly reduced spontaneous verbal output because they have "nothing to say"; they have lost the capacity for self-generated thought. However, the sense and structure of their spoken language is normal on highly structured tasks that do not require verbal thought generation, such as naming, repetition, and reading. On formal testing, those with dynamic aphasia perform extremely poorly on verbal fluency tasks which require the patient to produce as many words as possible from a category (category verbal fluency) or beginning with a specific letter (letter verbal fluency) in one minute.

Dynamic aphasia occurs with lesions in the left dorsolateral frontal lobe, including the deep white matter. It occurs as a primary language disorder, but also occurs secondary to more generalized behavioral inertia, abulia (a disorder of diminished motivation that is characterized by a lack of initiative), and executive function deficits in those with frontal and fronto-subcortical lesions. It occurs in some patients with frontal lobe dementia or progressive supranuclear palsy, which is often accompanied by a frontal lobe dementia.

The Primary Progressive Aphasias

In 1982, **M. Marsel Mesulam** (1945–) published a report describing six patients with isolated aphasia but no stroke, with a progressive course but no amnesia, in the *Annals of Neurology*. The report was titled "Slowly Progressive Aphasia Without Generalized Dementia," differentiating this disorder from the aphasias that were then well known to the field of neurology. This title conveyed the slowly progressive nature of the course, differentiating this form of aphasia from that due to stroke and the rapidly progressive aphasias seen with brain tumors. The title also conveyed that the language deficit did not occur in the context of generalized dementia, highlighting the difference between these patients and those with Alzheimer's disease dementia. Mesulam coined the term **primary progressive aphasia** (PPA) to reflect a focal clinical presentation of aphasia, in which aphasia is the primary (i.e., initially most salient) defect and is progressive (i.e., due to neurodegenerative disease). Thus, PPA is a clinical syndrome characterized by language disturbance that is relatively pure or isolated (i.e., largely unaccompanied by other cognitive deficits) in the initial phase (at least 2 years) and worsens gradually over time. In the later stages, a global dementia evolves.

A diagnosis of PPA requires that the following be met: (1) the most prominent clinical feature (at onset and for initial stages of disease) consists of communication difficulty, which is the primary contributor to impaired activities of daily living; (2) symptoms should not be attributable to other neurological, psychiatric, or medical disorders; and (3) prominent nonlanguage cognitive or behavioral impairments are not be present initially.

There are three subtypes of PPA: semantic variant, nonfluent/agrammatic variant, and logopenic variant PPA, described in detail below. The focal symptoms reflect disease topography (Figure 27.3). The diagnosis is of PPA is clinical, based on the signs and symptoms described in the medical history and observed on examination. It involves a two-step process of first establishing that the basic criteria of PPA are met, and then establishing the specific PPA subtype. Determining the subtype of PPA can help to elucidate the specific linguistic deficits (semantic, grammatical, phonological) and strengths, which may guide treatment planning. Not all cases of PPA, however, can be classified into one of the established clinical variants; these cases are labeled either **mixed PPA** or **unclassifiable PPA**.

Autopsy examination of the brains of PPA patients have revealed asymmetric focal cortical atrophy of the frontal and temporal lobes, usually most severe in the left hemisphere. As stated above, the symptoms of the different PPA variants reflect disease topography; they do not reflect a specific underlying neuropathology. There is, however, a loose association between PPA phenotype and underlying pathology. The semantic and nonfluent/agrammatic variants are most commonly associated with the frontotemporal lobar degeneration (FTLD) spectrum of pathologies. The semantic variant is most often associated with trans-activation-response DNA binding protein 43, known as FTLD-TDP. The nonfluent/agrammatic variant is most often associated with FTLD-tauopathy. The logopenic variant is associated with Alzheimer's pathology in most cases (although in most cases, Alzheimer's pathology is associated with Alzheimer's type dementia).

Semantic Variant PPA

Semantic variant PPA, also referred to as **semantic dementia**, is characterized by a gradually progressive deterioration in naming ability *and* word comprehension. Persons with semantic dementia usually present with complaints of losing their "memory for words." They speak fluently with preserved phonology (word pronunciation) and syntax but are anomic. They make semantic paraphasias, and there is a characteristic progression of word substitutions, starting with semantically similar category coordinates (e.g., *zebra* → "giraffe"), then higher-familiarity members of the same category (e.g., *zebra* → "horse"), then superordinate category names (e.g., *zebra* → "animal"). Naming does not improve substantially with phonemic

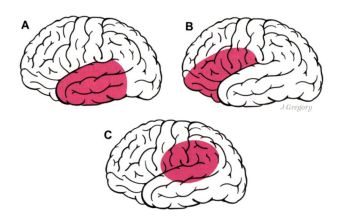

FIGURE 27.3. The three subtypes of primary progressive aphasia (PPA); the focal symptoms reflect disease topography. (A) Semantic variant PPA. (B) Nonfluent/agrammatic variant PPA. (C) Logopenic variant PPA.

cueing. Category fluency is markedly impaired, while letter fluency is relatively spared, at least in the early stage.

There is an associated impairment in comprehension (knowledge of the meaning) of content words (nouns, verbs, adjectives) that usually manifests as patients asking the meaning of words they used to (or likely used to) know. Comprehension deficits are apparent on word-picture matching tests, picture-word matching tests, and tests requiring the patient to generate word definitions. The deficit is strongly dependent on word familiarity. Comprehension of sentence syntax, however, remains intact. As long as the patient knows the meaning of the content words, they can understand even syntactically complex sentences.

Semantic dementia is a disorder of **semantic memory**, our database of knowledge about words *and* facts. The deficit in word production and word comprehension is not confined to word meaning, but extends to more general semantic deficits affecting knowledge about people and objects, although these deficits may be less prominent early in the course. At first the deficit in person knowledge affects the ability to name people, but as it progresses it affects the ability to generate information about people from either their names, faces, or other cues. Eventually the person with semantic dementia is unable to distinguish whether someone is familiar or not, or famous or not. Object knowledge deficits present as difficulty knowing what things are, including the superordinate categories they belong to and knowledge of their physical features and functions. This manifests as inability to identify objects (or pictures of objects) by sorting them into semantically related categories (e.g., tools, fruit, clothing) or distinguish real from imaginary objects, and consequently an inability to use objects correctly. Semantic dementia, therefore, is much more than merely a focal language disorder, as the deficits extend beyond the domain of language.

Elizabeth Warrington (1931–) provided the first modern description of semantic dementia in 1975. She described three patients who were initially diagnosed with visual object agnosia. None had stroke, traumatic brain injury, or other known brain lesion, and in all cases the course was gradually progressive. Warrington demonstrated that they all had deficits in both object and word recognition. She ruled out generalized intellectual impairment, selective perceptual impairment, and selective language impairment as the core defect, and hypothesized that the core deficit was in semantic memory.

The deficits in object and people knowledge in semantic dementia are distinguished from visual object agnosia and prosopagnosia. Those disorders are modality specific, and if information about the object or person is gained through a sensory modality other than vision, recognition is intact, evidencing intact semantic knowledge. Comparison of auditory responsive naming versus visual confrontation naming also discriminates between visual object agnosia and semantic dementia. With visual object agnosia, the core deficit is an inability of the visual system to access the intact semantic knowledge base; thus visual confrontation naming is impaired but auditory responsive naming is preserved. With semantic dementia, the neural structures underlying the semantic knowledge base are compromised; thus both auditory responsive naming and visual confrontation naming are impaired.

Like semantic dementia, early-stage Alzheimer's dementia is often associated with anomia, as evidenced by prolonged word-finding pauses in both spontaneous discourse and naming. Early-stage Alzheimer's dementia and semantic dementia, however, differ in multiple ways. In early-stage Alzheimer's dementia, naming is aided by phonemic cueing and word comprehension is preserved, but not so in semantic dementia. In early-stage semantic dementia, memory for day-to-day events is preserved, but not so in Alzheimer's dementia. This is best demonstrated with nonverbal tests of memory, but also becomes apparent to the examiner who sees the patient over several test days and the patient shows excellent recall for the earlier encounter(s). In early-stage semantic dementia, temporal orientation and drawing skills are also preserved, but not so in Alzheimer's dementia. Advanced stages of semantic dementia and Alzheimer's dementia may be clinically indistinguishable as they both progress to late-stage dementia with global cognitive impairment.

For a diagnosis of semantic variant PPA, both anomia and single-word comprehension impairment must be present, as well as three of the following: impaired object knowledge; surface alexia or surface agraphia; spared repetition; and spared motor speech and grammar. **Surface alexia** is an acquired reading disorder characterized by reading errors for words with irregular spelling-to-sound (grapheme-to-phoneme) correspondence (e.g., *enough*, *debt*, *castle*, and *aisle*). **Surface agraphia** is an acquired writing disorder characterized by spelling errors for words with irregular sound-to-spelling (phoneme-to-grapheme) correspondence (see Chapter 28, "Alexia, Agraphia, and Acalculia").

Neuroimaging studies of semantic dementia patients typically reveal marked atrophy (50%–80% gray matter loss) of the anterior temporal lobes with a rostral-caudal gradient, most prominently affecting the temporal pole and inferior/lateral aspects of the lobe (perirhinal cortex and anterior fusiform gyrus). The atrophy is usually bilateral but quite asymmetric, with the left hemisphere more affected than the right hemisphere in the vast majority of cases (since 90% of people have left hemisphere language dominance). Mesial temporal lobe structures are preserved. The degree of anterior temporal lobe atrophy and the degree of semantic memory impairment are correlated. An imaging-supported diagnosis of semantic dementia requires: (1) a clinical diagnosis of semantic dementia; and (2) imaging evidence of one or more of the following: (a) predominant anterior temporal lobe atrophy, and/or (b) predominant anterior temporal hypoperfusion on SPECT or hypometabolism on PET.

Predominant left temporal atrophy (**left temporal lobe variant semantic dementia**) is associated with the classic clinical syndrome of semantic variant PPA, characterized by progressive naming and word comprehension deficits. Predominant right temporal atrophy (**right temporal lobe variant semantic dementia**) is more associated with visual object and face recognition deficits, as well behavioral disturbances (personality changes, loss of empathy, compulsions); language deficits are less evident. This, of course, assumes left hemisphere language dominance. As the disease progresses, degeneration progresses to the less affected temporal lobe, so that within 3 years of clinical onset, the clinical syndromes become more difficult to distinguish. With further progression, the degeneration extends rostrally into the posterior inferior frontal lobes and atypical behavioral symptoms may develop.

Nonfluent/Agrammatic Variant PPA

Nonfluent/agrammatic variant PPA, also referred to as **nonfluent PPA** or **progressive nonfluent aphasia**, is a syndrome characterized by nonfluent output that is is hesitant, effortful, and reduced in rate, with speech sound errors and/or agrammatism.

In nonfluent/agrammatic variant PPA, the speech sound errors are characterized by groping to articulate words by successive approximations, with editing breaks and retakes, which disrupts the flow of speaking. These difficulties affect both spontaneous expressive language and automatic speech (recitation of memorized sequences, repetition, and reading aloud). Some of the speech sound errors are phonetic motor-based articulation errors; however, most are phonological language-based errors. The disturbance in the phonological construction of words results in phoneme deletions (e.g., rhinocerus → *rhinorus*), additions (e.g., rhinocerus → *rhinoceronus*), substitutions (errors in phoneme selection, e.g., rhinocerus → *rhinonerus*), and transpositions (errors in phoneme ordering, e.g., rhinocerus → *rhinorecus*). These errors occur mostly with polysyllabic words. The phonemic paraphasias are well-formed, undistorted speech sounds of the language. Misplacement of sounds within words indicates that the errors occur at the level of abstract word representation, rather than at the level of articulation. By contrast, phonetic errors which are due to motor speech (articulation) deficit result in sounds or combinations of sounds that may not occur in the language. An additional distinction between phonologic and phonetic speech sound errors is that phonological errors are inconsistent, as the speaker makes different mistakes even when pronouncing the same word repeatedly, whereas phonetic errors are usually consistent as the speaker makes the same mistake when pronouncing the same word several times. The speech sound errors of nonfluent/agrammatic variant PPA are often described as apraxia of speech, which is considered a disturbance at the level of motor programming; however, most involve processes that are related to language processing rather than motor programming of the articulators.

Agrammatism is a breakdown in the ability to use grammatical rules to construct and comprehend sentences. Comprehension of grammatically simple sentences is preserved, but comprehension of grammatically complex sentences that signal logico-grammatical relationships is impaired, as demonstrated by tests of syntactic processing.

For a diagnosis of nonfluent/agrammatic PPA, at least one of the core features of agrammatism or effortful, halting speech with inconsistent speech sound errors and distortions ("apraxia of speech") must be present, and two of the following associated features must also be present: impaired sentence comprehension, spared single word comprehension, and spared object knowledge.

Neuroimaging studies of nonfluent/agrammatic PPA patients typically reveal marked left anterior insular and posterior frontal atrophy. As the disease progresses, it extends rostrally into the posterior inferior frontal lobes, and motor symptoms (e.g., limb apraxia, parkinsonism, dysphagia) may develop. An imaging-supported diagnosis of nonfluent/agrammatic variant PPA diagnosis requires: (1) a clinical diagnosis of nonfluent/agrammatic variant PPA; and (2) imaging evidence of one or more of the following: (a) predominant left posterior fronto-insular atrophy on MRI, and/or (b) predominant left posterior fronto-insular hypoperfusion on SPECT or hypometabolism on PET, of course assuming left hemisphere language dominance.

Logopenic Variant PPA

Logopenic variant PPA, also referred to as **logopenic progressive aphasia**, is a clinical syndrome characterized by two core features: (1) impaired single-word retrieval in spontaneous spoken language and naming, and (2) impaired sentence and phrase repetition. Phonemic paraphasias in spontaneous speaking and naming that are well articulated and without distortions are also common (they are phonologic errors, not phonetic). Persons with logopenic progressive aphasia speak in a slow and halting manner, with frequent pauses due to anomia. Their anomia is usually less severe than in persons with semantic dementia, and they usually make phonemic paraphasias rather than semantic paraphasias. For a diagnosis of logopenic progressive aphasia, both core features of word-finding difficulty and impaired repetition must be present, as well as three of the following: phonological errors in speech; spared single word comprehension/object knowledge; spared motor speech; and an absence of agrammatism.

The fundamental deficit in logopenic PPA is in auditory working memory. This underlies the difficulty with sentence and phrase repetition. Persons with logopenic progressive aphasia also have a deficit in sentence comprehension due to problems holding lengthy information in auditory working memory. The deficit in sentence comprehension is related to the length and predictability of the sentence, rather than its grammatical complexity.

In logopenic progressive aphasia, the area of greatest atrophy involves the left posterior temporal cortex and inferior parietal lobule (supramarginal and angular gyri). Imaging studies show predominant left posterior perisylvian or parietal atrophy on MRI, and predominant left posterior perisylvian (parietal) hypoperfusion on SPECT imaging and hypometabolism on FDG-PET imaging. An imaging-supported diagnosis of logopenic variant PPA diagnosis requires the following: (1) a clinical diagnosis of logopenic variant PPA; and (2) imaging evidence of at least one of the following results: (a) predominant left posterior perisylvian or parietal atrophy on MRI, or (b) predominant left posterior perisylvian or parietal hypoperfusion on SPECT or hypometabolism on PET, of course assuming left hemisphere language dominance.

Logopenic variant PPA is usually associated with Alzheimer's disease pathology, and of the PPA syndromes, LPA is the most common aphasia phenotype of Alzheimer's disease.

Summary

Language is a system of symbols and the rules for structuring those symbols that allows us to communicate with others through speaking, reading/writing, and signing, in both expression and comprehension. Aphasia is loss or deterioration of one or more aspects of the process of comprehending and/or formulating verbal messages due to an acquired brain disorder. Language is lateralized to the left hemisphere in 90% of people (96% of right-handers and 70% of left-handers). Consquently, aphasia is usually due to damage to the left hemisphere cerebral cortex. Within the left hemisphere, lesions in different parts of the cerebral cortex produce different patterns of language disturbance.

There are eight classical aphasia syndromes, differentiated on the basis of three independent signs: fluency, comprehension, and repetition. Anomia is present in all. Broca's, Wernicke's, global, and conduction aphasias are classified as perisylvian aphasias; they are due to damage in the vicinity of the Sylvian fissure and all involve impaired repetition. Broca's aphasia is characterized by pronounced deficit in expressive langage, Wernicke's aphasia is characterized by pronounced deficit in language comprehension, and global aphasia is characterized by pronounced deficit in both expression and comprehension. Conduction aphasia is characterized by a relatively isolated deficit in repetition. The extrasylvian aphasias, also known as transcortical aphasias, occur with damage outside the area surrounding the Sylvian fissure, usually due to infarcts within the cortical watershed zones. They feature intact repetition. They are transcortical motor aphasia (similar to Broca's aphasia, except that repetition is preserved), transcortical sensory aphasia (similar to Wernicke's aphasia, except that repetition is preserved), and mixed transcortical aphasia (similar to global aphasia, except that repetition is preserved). Anomic aphasia is characterized by isolated but marked anomia.

The primary progressive aphasias are due to focal cortical neurodegeneration. There are three subtypes: semantic variant, nonfluent/agrammatic variant, and logopenic variant PPA. The focal symptoms reflect disease topography. Semantic variant PPA is characterized by a gradually progressive deterioration in naming ability *and* word comprehension; however, it is more broadly a disorder of semantic memory. Nonfluent/agrammatic variant PPA is characterized by nonfluent speech, speech sound errors, and agrammatism. Logopenic variant PPA is characterized by impaired single-word retrieval in spontaneous spoken language and naming, and impaired sentence and phrase repetition. Not all cases of PPA, however, can be classified into one of the established clinical variants; these are labeled either mixed PPA or unclassifiable PPA.

Additional Reading

1. Binder JR. The Wernicke area: modern evidence and a reinterpretation. *Neurology.* 2015 Dec 15;85(24):2170–2175. doi:10.1212/WNL.0000000000002219. Epub 2015 Nov 13. PMID: 26567270; PMCID: PMC4691684.
2. Brambati SM, Rankin KP, Narvid J, Seeley WW, Dean D, Rosen HJ, Miller BL, Ashburner J, Gorno-Tempini ML. Atrophy progression in semantic dementia with asymmetric temporal involvement: a tensor-based morphometry study. *Neurobiol Aging.* 2009 Jan;30(1):103–111. doi:10.1016/j.neurobiolaging.2007.05.014. Epub 2007 Jul 2. PMID: 17604879; PMCID: PMC2643844.
3. Broca P. On the site of the faculty of articulated speech [1865]. *Neuropsychol Rev.* 2011;21:230–235. doi:10.1007/s11065-011-9173-y
4. Broca P. Remarks on the seat of spoken language, followed by a case of aphasia [1861]. *Neuropsychol Rev.* 2011;21:227–229. doi:10.1007/s11065-011-9174-x
5. Damasio AR. Aphasia. *N Engl J Med.* 1992;326(8):531–539. doi:10.1056/NEJM199202203260806
6. Damasio H, Damasio AR. The anatomical basis of conduction aphasia. *Brain.* 1980;103(2):337–350. doi:10.1093/brain/103.2.337
7. Damasio AR, Geschwind N. The neural basis of language. *Annu Rev Neurosci.* 1984;7:127–147. doi:10.1146/annurev.

ne.07.030184.001015
8. Gainotti G, Barbier A, Marra C. Slowly progressive defect in recognition of familiar people in a patient with right anterior temporal atrophy. *Brain*. 2003;126(Pt 4):792–803. doi:10.1093/brain/awg092
9. Gainotti G. A metanalysis of impaired and spared naming for different categories of knowledge in patients with a visuo-verbal disconnection. *Neuropsychologia*. 2004;42(3):299–319. doi:10.1016/j.neuropsychologia.2003.08.006
10. Gainotti G. Anatomical functional and cognitive determinants of semantic memory disorders. *Neurosci Biobehav Rev*. 2006;30(5):577–594. doi:10.1016/j.neubiorev.2005.11.001
11. Gainotti G. Is the right anterior temporal variant of prosopagnosia a form of "associative prosopagnosia" or a form of "multimodal person recognition disorder"? *Neuropsychol Rev*. 2013;23(2):99–110. doi:10.1007/s11065-013-9232-7
12. Geschwind N, Levitsky W. Human brain: left-right asymmetries in temporal speech region. *Science*. 1968;161(3837):186–187. doi:10.1126/science.161.3837.186
13. Geschwind N. The organization of language and the brain. *Science*. 1970 Nov 27;170(3961):940–944. doi:10.1126/science.170.3961.940. PMID: 5475022.
14. Gorno-Tempini ML, Dronkers NF, Rankin KP, Ogar JM, Phengrasamy L, Rosen HJ, et al. Cognition and anatomy in three variants of primary progressive aphasia. *Ann Neurol*. 2004;55(3):335–346. doi:10.1002/ana.10825
15. Gorno-Tempini ML, Hillis AE, Weintraub S, Kertesz A, Mendez M, Cappa SF, et al. Classification of primary progressive aphasia and its variants. *Neurology*. 2011;76(11):1006–1014. doi:10.1212/WNL.0b013e31821103e6
16. Harciarek M, Kertesz A. Primary progressive aphasias and their contribution to the contemporary knowledge about the brain-language relationship. *Neuropsychol Rev*. 2011;21(3):271–287. doi:10.1007/s11065-011-9175-9
17. Hart J Jr, Berndt RS, Caramazza A. Category-specific naming deficit following cerebral infarction. *Nature*. 1985;316(6027):439–440. doi:10.1038/316439a0
18. Henry ML, Wilson SM, Ogar JM, Sidhu MS, Rankin KP, Cattaruzza T, Miller BL, Gorno-Tempini ML, Seeley WW. Neuropsychological, behavioral, and anatomical evolution in right temporal variant frontotemporal dementia: a longitudinal and post-mortem single case analysis. *Neurocase*. 2014;20(1):100–109. doi:10.1080/13554794.2012.732089
19. Hurley RS, Mesulam MM, Sridhar J, Rogalski EJ, Thompson CK. A nonverbal route to conceptual knowledge involving the right anterior temporal lobe. *Neuropsychologia*. 2018 Aug;117:92–101. doi:10.1016/j.neuropsychologia.2018.05.019. Epub 2018 May 23. PMID: 29802865; PMCID: PMC6344946.
20. Lazar RM, Mohr JP. Revisiting the contributions of Paul Broca to the study of aphasia. *Neuropsychol Rev*. 2011;21(3):236–239. doi:10.1007/s11065-011-9176-8
21. Montembeault M, Brambati SM, Gorno-Tempini ML, Migliaccio R. Clinical, anatomical, and pathological features in the three variants of primary progressive aphasia: a review. *Front Neurol*. 2018 Aug 21;9:692. doi:10.3389/fneur.2018.00692. PMID: 30186225; PMCID: PMC6110931.
22. Mesulam MM. Slowly progressive aphasia without generalized dementia. *Ann Neurol*. 1982;11(6):592–598. doi:10.1002/ana.410110607
23. Mesulam MM, Rogalski EJ, Wieneke C, Hurley RS, Geula C, Bigio EH, et al. Primary progressive aphasia and the evolving neurology of the language network. *Nat Rev Neurol*. 2014;10(10):554–569. doi:10.1038/nrneurol.2014.159
24. Ojemann GA. Individual variability in cortical localization of language. *J Neurosurg*. 1979;50(2):164–169. doi:10.3171/jns.1979.50.2.0164
25. Ojemann G, Ojemann J, Lettich E, Berger M. Cortical language localization in left, dominant hemisphere: an electrical stimulation mapping investigation in 117 patients. 1989. *J Neurosurg*. 2008;108(2):411–421. doi:10.3171/JNS/2008/108/2/0411
26. Poeppel D, Hickok G. Towards a new functional anatomy of language. *Cognition*. 2004 May-Jun;92(1-2):1–12. doi:10.1016/j.cognition.2003.11.001. PMID: 15037124.
27. Rogalski E, Cobia D, Harrison TM, Wieneke C, Thompson CK, Weintraub S, et al. Anatomy of language impairments in primary progressive aphasia. *J Neurosci*. 2011;31(9):3344–3350. doi:10.1523/JNEUROSCI.5544-10.2011
28. Seeley WW, Bauer AM, Miller BL, Gorno-Tempini ML, Kramer JH, Weiner M, et al. The natural history of temporal variant frontotemporal dementia. *Neurology*. 2005 Apr 26;64(8):1384–1390. doi:10.1212/01
29. Signoret JL, Castaigne P, Lhermitte F, Abelanet R, Lavorel P. Rediscovery of Leborgne's brain: anatomical description with CT scan. *Brain Lang*. 1984;22(2):303–319. doi:10.1016/0093-934x(84)90096-8
30. Snowden JS, Thompson JC, Neary D. Famous people knowledge and the right and left temporal lobes. *Behav Neurol*. 2012;25(1):35–44. doi:10.3233/BEN-2012-0347
31. Snowden JS, Harris JM, Thompson JC, Kobylecki C, Jones M, Richardson AM, et al. Semantic dementia and the left and right temporal lobes. *Cortex*. 2018;107:188–203. doi:10.1016/j.cortex.2017.08.024
32. Warrington EK. The selective impairment of semantic memory. *Q J Exp Psychol*. 1975;27(4):635–657. doi:10.1080/14640747508400525
33. Warrington EK, Shallice T. Category specific semantic impairments. *Brain*. 1984;107 (Pt 3):829–854. doi:10.1093/brain/107.3.829

28

Alexia, Agraphia, and Acalculia

Introduction

Reading, writing, and math are cultural inventions of comparatively recent development in human history. These skills are acquired mostly through formal education and form the basis of literacy and numeracy, which are fundamental and critical to functioning in modern society. These abilities, however, can be disrupted by brain injury and disease. This chapter describes: (1) the acquired loss of reading, writing, and calculation skills, namely alexia, agraphia, and acalculia, respectively; (2) the various forms and classification of each disorder; and (3) their neurological bases.

The term *acquired* indicates that the disorder is one in which there was a loss of a skill due to brain injury or disease after a normal developmental course. By contrast, **dyslexia**, **dysgraphia**, and **dyscalculia** are developmental learning disabilities that are characterized by a failure of normal skill acquisition and development. It is presumed that these specific learning disabilities are due to abnormalities in brain development with probable genetic causes. According to diagnostic classifications, these are (1) specific; (2) persistent; and (3) characterized by performance that is greater than one standard deviation below the expected level for a given age and intelligence; (4) cause significant impediment to academic achievement and/or activities of daily life; and (5) cannot be explained by sensory, neurological, psychiatric, motivational, or other cause, or by inadequate education. Confusingly, the terms *dyslexia*, *dysgraphia*, and *dyscalculia* are occasionally loosely applied to the acquired forms of these disorders (e.g., *acquired dyslexia*).

Alexia, agraphia, and acalculia are disorders that result from disruption of neurological processes. Each disorder has several distinct subtypes, each associated with a distinct set of characteristics and a distinct lesion location, implicating that they each involve disruption of a distinct underlying neurocognitive process.

Some of the disorder subtypes, however, do not represent a distinct syndrome, but rather are *secondary* to a more general syndrome that extends beyond the realm of reading, writing, or number processing/calculation. For example, all these skills may be undermined by visuospatial processing disturbances or hemi-spatial neglect, and the terms *spatial alexia*, *spatial agraphia*, and *spatial acalculia* have been used to refer to the resulting functional impairments. Reading, writing, and written calculations of course require vision and visuospatial processes, and *any* disorder affecting vision, visual exploration, visual recognition, or visuospatial processing will undermine the performance of these skills. Classifying these secondary functional impairments as subtypes of alexia, agraphia, and acalculia may seem nonsensical if you take this to an extreme, like referring to the inability to read visual material, write, or perform written calculations that results from complete vision loss as "blindness alexia," "blindness agraphia," and "blindness acalculia." The inclusion of secondary forms within the classifications of alexia, agraphia, and acalculia is rooted in history, but remains useful to clinicians for conceptualizing and describing functional limitations that arise due to a more general disorder. Furthermore, secondary effects on reading, writing, or calculation may be the first sign(s) of disturbance observed by a clinician. For example, spatial alexia, spatial agraphia, and spatial acalculia are often the earliest signs of posterior cortical atrophy (see Chapter 15, "The Parietal Lobes and Associated Disorders").

Alexia

Alexia is an acquired disorder of reading that is due to brain injury or disease; the person with alexia had normal reading skills prior to the neurologic insult. Alexia usually occurs within the context of aphasia, which is most often caused by stroke affecting the language-dominant hemisphere. In the late nineteenth century, French neurologist **Joseph Jules Déjerine** (1849–1917) showed that alexia may occur independent of the more general language impairment of aphasia. He described two such syndromes: pure alexia (also known as alexia without agraphia), an isolated disorder of written language comprehension; and alexia with agraphia, a disorder of written language comprehension and expression. Additional forms of alexia have been identified since then, including surface alexia and deep alexia.

The reading errors made by persons with alexia are referred to as **paralexias**. Alexia syndromes are typically classified according to the pattern of reading errors revealed during oral reading of words, **pseudowords** (nonwords) which are letter strings that are pronounceable as though they were words using phonological decoding skills (e.g., *drit*), and **pseudo-homophones** which are pseudowords that sound like words when phonologically decoded (e.g., *brane*).

Pure Alexia

Pure alexia is so named because, unlike most cases of alexia, it occurs in pure form without agraphia or other aphasic symptoms. It is also known as **alexia without agraphia** and **occipital alexia**. It occurs with lesions in the left ventromedial occipital region that include the splenium of the corpus callosum (see Figure 14.12 in Chapter 14). Such lesions disconnect the transfer of visual information from posterior visual cortices to the language region located more anteriorly, usually within the left hemisphere. More specifically, such lesions prevent incoming visual information from reaching the left angular gyrus for linguistic interpretation; the angular gyrus itself, however, remains intact. Pure alexia is a **disconnection syndrome** and often is accompanied by other signs of visual-verbal disconnection, such as color anomia (see Chapter 14, "The Occipital Lobes and Visual Processing"). It typically occurs with infarction in the PCA territory of the language-dominant hemisphere affecting occipitotemporal cortex, leaving regions concerned with the nonvisual aspects of language intact. Letter reading is significantly superior to word reading. A region of the left fusiform gyrus (BA37) has been named the **visual word form area** because it is hypothesized to represent the neural substrate of the **orthographic lexicon** (one's inventory of written words); functional imaging in healthy adults during reading tasks shows activity. This region, however, also is active when processing other types of complex visual stimuli, and while it plays an important role in reading, it is not exclusively dedicated to processing words.

Alexia with Agraphia

Alexia with agraphia, also known as **central alexia** and **parietal-temporal alexia**, is a disorder of both reading and writing; thus it may be regarded as a syndrome of acquired illiteracy. The lesion involves the inferior parietal and posterolateral temporal regions of the left hemisphere, most critically the **angular gyrus**, an area of cross-modal association cortex (in contrast to alexia *without* agraphia in which the angular gyrus is intact). Alexia with agraphia may be associated with other deficits depending on the extent of the lesion, including right homonymous visual field defects and components of angular gyrus syndrome such as finger agnosia, right-left disorientation, acalculia (acquired loss of arithmetic skills), and/or anomia.

Surface Alexia

A **phoneme** is the smallest sound unit of a spoken language. Phonemes are combined to make words. A **grapheme** is a letter or group of letters that represents a phoneme. The **orthography** (writing system) of different languages varies with respect to consistency in the correspondence between phonemes and graphemes. For example, Spanish has a highly phonemic orthography in which almost all words have a regular correspondence between phonemes and graphemes; all words are pronounced as written. English has a relatively nonphonemic orthography in which there many **exception words** (**irregular words**) which have an irregular correspondence between phonemes and graphemes and are *not* pronounced as written (e.g., *enough, castle, aisle*).

Observations regarding the nature of paralexias in acquired dyslexia have led to information-processing models of reading. A direct-phonologic processing route relies on grapheme-to-phoneme (letter-to-sound) phonological decoding; it transforms orthographic representations (written words) into phonological representations (read words) directly. An indirect-semantic processing route performs whole word reading; it transforms orthographic representations into semantic representations, which in turn are transformed into phonological representations. Skilled readers utilize both processes when converting written language to spoken language (i.e., reading). The indirect-semantic route allows us to read words with which we are familiar, including both regular and exception words. By contrast, the direct-phonologic route, which relies on the serial decoding of individual graphemes to corresponding phonemes, allows us to generate plausible pronunciations when reading words that we are not familiar with, including pseudowords.

Surface alexia is characterized by errors reading only exception (irregular) words. Persons with surface alexia can read regular words that rely on standard grapheme-phoneme decoding (e.g., *bet, get, jet, let, met, net, pet, set, vet, wet*), but err when presented with exception words that violate standard rules and have an irregular pronunciation relative to their spelling (e.g., *debt*). In surface alexia the indirect-semantic route is dysfunctional, but the direct-phonologic route is intact, allowing for reading via grapheme-to-phoneme decoding rules. Since individuals with surface alexia are unable to perform whole word reading and rely on grapheme-to-phoneme decoding, they are well able to read regular words (e.g., *ship*) and pseudowords (e.g., *yatchet*), both of which rely entirely on grapheme-to-phoneme (letter-to-sound) decoding, but often err reading exception words (e.g., *yacht*). Furthermore, when reading exception words, they make **regularization errors** (**surface errors**) which are consistent with phonologic decoding (e.g., *aisle* → "*azle*"). There is a significant word frequency effect of the paralexias, such that regularization errors occur far more commonly with low-frequency words than high-frequency words.

Those with surface alexia perform poorly on reading tasks that require phonologic analysis, such as homophone matching tasks and pseudo-homophone matching tasks. Homophone matching involves reading an irregular word and identifying the homophone (i.e., word with the same pronunciation) from an array of four choices (e.g., *sight* = *site*). Pseudo-homophone matching involves reading an irregular (exception) word and matching its pronunciation to a pseudoword from an array of four choices (e.g., *ocean* = *oshen*).

Surface alexia is often seen in the setting of semantic dementia, in which case it is accompanied by other signs of semantic knowledge deficit, including anomia, reduced word comprehension, and poor category fluency, the quickness with which one can generate words from a semantic category of nouns (e.g., animals). In this clinical context the surface alexia is a manifestation of the degradation of semantic knowledge.

Interestingly, in the case of semantic dementia, surface alexia is readily observed in English readers since English has a relatively high percentage of words with irregular spellings, but is not observed in Spanish readers since Spanish lacks exception words. In those with semantic dementia who are bilingual for English and Spanish, the surface alexia is apparent in their reading of English but not Spanish.

Deep Alexia

The defining feature of **deep alexia** is the production of **semantic paralexias**, a type of reading error in which the word read aloud, despite being erroneous, is related in meaning to the written target word. The semantic relationship may take many forms, including synonyms (e.g., *lawyer* → "*attorney*"), antonyms (e.g., *hot* → "*cold*"), semantic subordinates (e.g., *bird* → "*robin*"), semantic superordinates (e.g., *celery* → "*vegetable*"), and semantic associates (e.g., *house* → "*garden*"). In addition to the semantic errors, there also is marked difficulty reading unfamiliar words and pseudowords, which relies on grapheme-to-phoneme decoding. Attempts to read unfamiliar words and pseudowords often result in **lexicalization errors** in which a non-word is read as a real word (e.g., *nace* → "*name*"). Thus, those with deep alexia are unable to use phonologic decoding. Furthermore, there is particular difficulty reading abstract (non-imageable) words compared to concrete (highly imageable) words, and difficulty reading function words (words that express grammatical relationships among other words within a sentence, such as *and, the, by, or*). There also are errors reading bound morphemes/affixes (i.e., word elements that cannot stand alone as a word), and the paralexias involve bound morpheme additions, deletions, or substitutions (e.g., reading the word *running* as *runner*).

In deep alexia the direct-phonologic route is completely abolished by neural damage, and the indirect-semantic route is partially impaired. Most cases of deep dyslexia occur within the context of severe aphasia with extensive left hemisphere damage, including much of the left frontal lobe and extending posteriorly.

Spatial Alexia

Spatial alexia (**visuospatial alexia**) is a disturbance in reading ability that arises secondary to a more general disorder of visuospatial processing or disorders of spatial attention such as hemi-spatial neglect and simultanagnosia. Spatial alexia therefore is not a true alexia due to specific disturbance in the neural circuitry underlying reading. For example, in hemispatial neglect due to right parietal lesions, the left half of space is neglected. When reading, the left half of the page is neglected and consequently the reader is unable to comprehend written paragraphs. Ability to read text on the right half of the page, however, is normal. This form of spatial alexia has also been referred to as **neglect alexia**.

Agraphia

Agraphia is an acquired disorder of spelling and/or writing due to neurological damage in individuals with normal premorbid literacy skills. Most often, agraphia occurs in the context of aphasia following left hemisphere lesions. Typically, the writing deficit is at least as severe as the spoken language deficit, and the features of the agraphia parallel those of the aphasia. Thus, the agraphia that accompanies nonfluent expressive aphasia (i.e., Broca's aphasia or transcortical motor aphasia) is characterized by effortful, sparse output with agrammatism, while the agraphia that accompanies fluent receptive aphasia (i.e., Wernicke's aphasia or transcortical sensory aphasia) is characterized by easily written output with indecipherable meaning and paragraphic errors. A **paragraphia** is the unintentional omission, transposition, or insertion of letters, syllables, or words in writing; the written errors (paragraphias) are similar to the spoken errors (paraphasias).

In 1867, **John William Ogle** (1824–1905) reported that agraphia occasionally occurs independent of aphasia. He distinguished two types: a **linguistic agraphia** characterized by spelling errors, and a **motor agraphia** characterized by poor letter formation. It is now recognized that there are many different types of agraphia associated with a wide variety of lesions. Common neurological classifications are **aphasic agraphia** (as described above), **agraphia with alexia** (also known as **parietal agraphia** or alexia with agraphia, as described above), pure agraphia, surface agraphia, spatial agraphia, and apraxic agraphia. In practice, however, it may be difficult to apply these classifications to individual patients due to the presence of mixed features.

Pure Agraphia

Pure agraphia is an acquired disorder of misspelling in the absence of alexia, aphasia, or apraxia. It is associated with focal left parietal lobe pathology. It is an element of **Gerstmann's syndrome**, defined by the tetrad of agraphia, acalculia, finger agnosia, and right-left disorientation. This syndrome has been reported to be pathognomonic for a left angular gyrus lesion, although this is controversial. Agraphia rarely occurs as an isolated disorder.

Surface Agraphia

Surface agraphia, also known as **lexical agraphia**, is characterized by marked impairment in spelling exception words that cannot be accomplished by phoneme-to-grapheme (sound-to-letter) decoding, with significantly better accuracy in spelling regular words and preserved ability to spell pseudowords, in a manner analogous to surface alexia.

In literate persons, the spellings of familiar words are easily recalled as whole words from one's spelling vocabulary (orthographic lexicon). In contrast to this semantic lexical approach, spellings can be assembled by phoneme-to-grapheme decoding using a phonologic processing strategy, which is employed when one is unsure about the spelling of a word, or when spelling an unfamiliar word or pseudoword. Spelling via phoneme-to-grapheme decoding yields correct responses for regularly spelled words, such as *fire*, but phonologically plausible errors for irregularly spelled words, such as *kwire* for *choir*. The spelling disorder of surface agraphia is attributed to dysfunction of the indirect-semantic spelling route, forcing reliance on a phoneme-to-grapheme decoding strategy that produces phonologically plausible regularization errors when reading irregular words, a finding that is most pronounced on low-frequency items (e.g., *yot* for *yacht*). Surface agraphia is often observed in individuals with semantic dementia; the frequent co-occurrence of surface alexia and surface agraphia in this clinical context suggests that deterioration in semantic representations is a key factor.

Spatial Agraphia

Spatial agraphia encompasses a variety of spatial deficits that affect nonlinguistic aspects of writing, such as ignoring the left side of the page, inability to maintain a horizontal writing direction, writing over other words, and incorrect grouping of letters, words, and blank spaces. There is no aphasic deficit, and words are spelled properly. Spatial agraphia characterized by ignoring the left side of the page is a manifestation of a more general unilateral spatial neglect due to right parietal lesions. It is associated with spatial alexia and signs of hemineglect on tests of constructional ability. Spatial agraphia is also observed in posterior cortical atrophy.

Spatial agraphia, just like spatial alexia, does not represent a specific neurocognitive syndrome; it is a historical term that is now used to describe a functional limitation that accompanies a more general visuospatial processing disorder.

Apraxic Agraphia

Apraxic agraphia is a disorder of writing in which the individual cannot form letters legibly but can spell words correctly. The individual letters written are often difficult to recognize and may appear as meaningless scrawls. The poor letter formation is due to impaired motor programming of skilled movements of the hand, so that the spatiotemporal aspects of writing are disturbed.

Apraxic agraphia is not attributable to more elementary sensorimotor, cerebellar, or basal ganglia dysfunction. It is often, but not always, associated with ideomotor apraxia. Apraxic agraphia occurs most commonly with parietal stroke in the hemisphere contralateral to the dominant hand, but it may also occur with neurodegenerative disease. In right-handed individuals, the damage typically involves the left inferior parietal lobule. Lesions in this area may also produce a combination of alexia with agraphia and apraxic agraphia.

The left inferior parietal lobule contains spatial representations of the movements required to write; a specialized region of premotor cortex known as **Exner's area** then converts these spatial representations into innervatory patterns, which in turn drive primary motor cortex. Apraxic agraphia therefore can also be caused by left hemisphere lesions affecting the connections from the inferior parietal lobe to frontal premotor cortex. Unilateral left-hand apraxic agraphia may occur with a lesion in the corpus callosum that disconnects Exner's area in the left hemisphere from the right hemisphere's premotor and motor areas.

Agraphia Due to Non-Apraxic Motor Disturbances

Non-apraxic disturbances of motor function that affect the regulation of movement force, speed, amplitude, and coordination also affect the ability to form legible letters. **Micrographia** is the production of abnormally small letters. It is common in Parkinson's disease and is due to defective control of the force, speed, and amplitude of handwriting movements resulting from the basal ganglia pathology. Cerebellar pathology produces incoordination and tremor, and it may result in poor handwriting due to irregular and disjointed hand movements. Damage to primary motor cortex or the corticospinal tract producing hemiparesis of the dominant hand will also affect handwriting; when the hemiparesis is marked, individuals typically shift to writing with the nondominant hand. These elemental disturbances of motor function are revealed by the neurological examination and do not reflect a cognitive disorder.

Acalculia

Number processing and calculation abilities are complex cognitive processes that rely on language, spatial, and executive function abilities. Brain injury and disease are frequently associated with a decline in ability to perform calculations, and while neuropsychological evaluations may include screening items, they rarely include in-depth examination of number processing and calculation abilities. Nevertheless, neuropsychologists and other healthcare professionals working with patients with brain injuries and diseases should be familiar with the various presentations of number processing and calculation disorders.

Acalculia is an acquired impairment in the ability to perform mathematical calculations (i.e., in persons with previously normal calculation abilities) due to brain injury or disease. The term (in German, *akalkulia*) was coined by **Salomon Eberhard Henschen** (1847–1930) to refer to a selective disturbance in calculation ability due to brain damage. In 1925, he published a review of 305 cases in the literature and 67 of his own patients; he concluded that calculation disturbances may occur independent of language impairments and proposed an anatomical substrate. **Hans Berger** (1873–1941) distinguished two types of acalculia in 1926: **primary acalculia** in which there is a specific disturbance in computational abilities; and **secondary acalculia**, the more common of the two forms, in which problems performing calculations arise from more general defects in other cognitive disturbances (e.g., aphasia, neglect, executive dysfunction, alexia). This distinction remains useful clinically. Since then, several classifications of the acalculias have been proposed. In 1961, **Henry Hécaen** (1912–1983) proposed three variants of acalculia based on the presumed mechanism of the deficit: anarithmetia (primary acalculia), spatial acalculia, and acalculia associated with alexia and/or agraphia for numbers. **Alexander Luria** (1902–1977) also distinguished three variants: primary acalculia, optical (visuoperceptual) acalculia, and frontal acalculia, emphasizing that calculation disturbances can result from diverse brain pathologies. Current neurological classifications distinguish primary acalculia and five types of secondary acalculia: aphasic, alexic, agraphic, spatial, and dysexecutive (frontal) acalculias.

Primary Acalculia

Primary acalculia is also known as **anarithmetia, anarithmia,** and **anarithmetria**. Deficits in primary acalculia include difficulty understanding numerical symbols (digits, Arabic numerals, Roman numerals, written number words), arithmetical symbols (+, −, ×, ÷), and procedural rules (such as "carry over," "borrow"); poor estimation abilities; and poor number comparison abilities. Since primary acalculia is a fundamental defect, the difficulties are apparent regardless of whether tasks are presented or responded to in an oral or written format. In addition, the inability to understand numerical concepts and perform basic arithmetic operations cannot be explained by aphasia, apraxia, alexia, or spatial disorder. Individuals with this disorder may be able to count aloud and perform other rote numerical learning, such as the multiplication tables. Primary acalculia is most often associated with focal lesions affecting the left parietal lobe, especially when the angular gyrus is involved. Cases of pure primary acalculia (i.e., without additional deficits) caused by focal lesions, however, are extremely rare. **Progressive primary acalculia** occurs as a biparietal variant of Alzheimer's disease (non-visual variant posterior cortical atrophy), with greater pathology in the left inferior parietal lobule.

Acalculia is an element of Gerstmann's syndrome, which is defined by the tetrad of acalculia, finger agnosia, agraphia, and right-left disorientation. While the four Gerstmann symptoms can occur in isolation, their association has prompted hypotheses that a unitary, elementary cognitive process links calculation, finger gnosis, knowledge of left and right, and writing, based on anthropological and developmental observations. For example, children begin to learn numbers and calculation by using their fingers. Number systems are grounded in counting by fingers. From the Latin *digitus*, the word *digit* means both "number" and "finger." Base-10 numeral systems are predominant simply because we have 10 fingers. The first forms of writing were based on carved signs encoding numerosity (as in tally sticks). Children learn left-right discrimination using their hands.

Aphasic, Alexic, and Agraphic Acalculias

Aphasic acalculia is a form of acalculia observed in aphasia that results from a primary language deficit; the associated alexia and/or agraphia for written verbal language also results in an alexia and/or agraphia for numbers. The overall error rate in various calculation tasks correlates with the severity of the language deficit; thus, global aphasia is associated with the greatest deficits in language and calculations. In addition, the qualitative error pattern varies between different types of aphasia.

Alexic acalculia is due to alexia for numbers without aphasia; ability to perform mental calculations is preserved. In cases of pure alexia (alexia without agraphia) in which letter reading is significantly superior to word reading, there are greater difficulties reading numbers composed of several digits (compound numbers) than reading single digits; when reading compound numbers, the person exhibits decomposition and digit-by-digit reading (e.g., 59 becomes 5, 9). In cases of alexia with agraphia, the alexia includes an inability to read written numbers and mathematical symbols. Because of lesion location, there is often an associated primary alcalculia in which reading and writing difficulties plus computational disturbances result in severe acalculia.

Agraphic acalculia is due to agraphia for numbers without aphasia. This may be due to an apraxic agraphia or other motor agraphias.

Spatial Acalculia

Spatial acalculia is a secondary disorder of calculation in which errors arise because of neglect, misalignment of numbers, reversal of digits (e.g., 12 for 21), or inversions (e.g., 9 for 6). This form of acalculia is caused primarily by right hemisphere lesions, particularly when the parietal lobe is involved. It is frequently associated with hemispatial neglect, spatial alexia, constructional apraxia, and general spatial disorders. Usually there is no difficulty in counting, simple mental operations, or successive operations (i.e., adding or subtracting a certain quantity successively, as in serial sevens).

Dysexecutive Acalculia

Dysexecutive acalculia, also known as **frontal acalculia**, refers to cases in which dysexecutive syndrome interferes secondarily with the ability to perform calculations. Persons with frontal lobe damage have particular difficulty with mental operations as compared to written operations, as well as performing successive operations and mathematical problem-solving. This may be due to attention difficulties, perseveration, impulsivity, or loss of complex mathematical concepts.

Summary

Alexia is an acquired disorder of reading due to brain injury or disease. It is commonly observed in right-handed individuals following damage to the language-dominant left hemisphere, and usually occurs within the context of aphasia, but it may also occur independent of aphasia, and with a variety of subtypes. Pure alexia (alexia without agraphia) is an isolated disorder of written language comprehension. Alexia with agraphia is a selective disorder of written language comprehension and expression. Surface alexia is characterized by errors reading exception words with regularization errors. Deep alexia is characterized by semantic paralexias, a type of reading error in which the word read aloud is related in meaning to the written target word. Spatial alexia (visuospatial alexia, neglect alexia) is a disturbance in reading ability that is secondary to hemispatial neglect or other visuospatial disorders.

Agraphia is an acquired disorder of spelling and/or writing due to brain injury or disease. Like alexia, agraphia most often occurs in the context of aphasia following left hemisphere lesions, but it may also occur independent of aphasia, and there are a variety of subtypes. Pure agraphia is a rare, acquired disorder of misspelling in the absence of alexia, aphasia, or apraxia; it is associated with focal left parietal lobe pathology. Surface agraphia is characterized by marked impairment in spelling exception words with preserved spelling of words with regular correspondences between phonemes and graphemes; it is analogous to surface alexia. Spatial agraphia encompasses a variety of spatial deficits that affect nonlinguistic aspects of writing, such as ignoring the left side of the page, inability to maintain a horizontal writing direction, writing over other words, and incorrect grouping of letters, words, and blank spaces. Apraxic agraphia is a disorder of writing in which the individual cannot form letters legibly due to impaired motor programming of skilled movements of the hand but can spell words correctly. It is a manifestation of a more general ideomotor apraxia and occurs with lesions in the inferior parietal lobule of the hemisphere that is contralateral to the dominant hand, or the connections from the inferior parietal lobe to frontal premotor cortex. Agraphia may also be secondary to non-apraxic disturbances of motor function.

Acalculia is an acquired disorder of number processing and/or calculation skills due to brain injury or disease affecting the left inferior parietal lobe. In primary acalculia there is a specific disturbance in computational abilities that is due to a loss of numerical concepts, including numerical symbols, arithmetical symbols, procedural rules, estimation abilities, and number comparison abilities. It is a fundamental defect that cannot be explained by another cognitive disorder; thus the inability to understand numerical concepts and/or perform basic arithmetic operations is apparent regardless of whether tasks are presented or responded to in an oral or written format. By contrast, secondary acalculia refers to problems performing calculations that arise from other more general cognitive disturbances, such as aphasia, alexia, agraphia, visuospatial disorders, or executive function disorders. The secondary acalculias are therefore classified as aphasic acalculia, alexic acalculia, agraphic acalculia, spatial acalculia, and dysexecutive (frontal) acalculia.

Additional Reading

1. Ardila A, Rosselli M. Spatial agraphia. *Brain Cogn.* 1993;22(2):137–147. doi:10.1006/brcg.1993.1029
2. Ardila A. On the origins of calculation abilities. *Behav Neurol.* 1993;6(2):89–97. doi:10.3233/BEN-1993-6204
3. Ardila A, Rosselli M. Spatial acalculia. *Int J Neurosci.* 1994;78(3-4):177–184. doi:10.3109/00207459408986056
4. Ardila A, Rosselli M. Acalculia and dyscalculia. *Neuropsychol Rev.* 2002;12(4):179–231. doi:10.1023/a:1021343508573
5. Benson DF. The third alexia. *Arch Neurol.* 1977;34(6):327–331. doi:10.1001/archneur.1977.00500180021004
6. Bub D, Kertesz A. Deep agraphia. *Brain Lang.* 1982;17(1):146–165. doi:10.1016/0093-934x(82)90011-6
7. Bub DN, Arguin M, Lecours AR. Jules Dejerine and his interpretation of pure alexia. *Brain Lang.* 1993;45(4):531–559. doi:10.1006/brln.1993.1059
8. Bub D. Alexia and related reading disorders. *Neurol Clin.* 2003;21(2):549–568. doi:10.1016/s0733-8619(02)00099-3
9. Coslett HB. Acquired dyslexia. *Semin Neurol.* 2000;20(4):419–426. doi:10.1055/s-2000-13174
10. Delazer M, Karner E, Zamarian L, Donnemiller E, Benke T. Number processing in posterior cortical atrophy: a neuropsycholgical case study. *Neuropsychologia.* 2006;44(1):36–51. doi:10.1016/j.neuropsychologia.2005.04.013
11. Grafman J, Passafiume D, Faglioni P, Boller F. Calculation disturbances in adults with focal hemispheric damage. *Cortex.* 1982;18(1):37–49. doi:10.1016/s0010-9452(82)80017-8
12. Grossman M, Libon DJ, Ding XS, Cloud B, Jaggi J, Morrison D, et al. Progressive peripheral agraphia. *Neurocase.* 2001;7(4):339–349. doi:10.1093/neucas/7.4.339
13. Heilman KM, Coenen A, Kluger B. Progressive asymmetric apraxic agraphia. *Cogn Behav Neurol.* 2008;21(1):14–17. doi:10.1097/WNN.0b013e318165b133
14. Henderson VW. Jules Dejerine and the third alexia. *Arch Neurol.* 1984;41(4):430–432. doi:10.1001/archneur.1984.04050160096022
15. Henderson VW. Alexia and agraphia: contrasting perspectives of J.-M. Charcot and J. Hughlings Jackson. *Neurology.* 2008;70(5):391–400. doi:10.1212/01.wnl.0000298680.47382.61
16. Henderson VW. Alexia and agraphia from 1861 to 1965. *Front Neurol Neurosci.* 2019;44:39–52. doi:10.1159/000494951
17. Judd T, Gardner H, Geschwind N. Alexia without agraphia in a composer. *Brain.* 1983;106 (Pt 2):435–457. doi:10.1093/brain/106.2.435
18. Keller C, Meister IG. Agraphia caused by an infarction in Exner's area. *J Clin Neurosci.* 2014;21(1):172–173. doi:10.1016/j.jocn.2013.01.014
19. Kim HJ, Chu K, Lee KM, Kim DW, Park SH. Phonological agraphia after superior temporal gyrus infarction. *Arch Neurol.* 2002;59(8):1314–1316. doi:10.1001/archneur.59.8.1314
20. Laine M, Niemi P, Niemi J, Koivuselkä-Sallinen P. Semantic errors in a deep dyslexic. *Brain Lang.* 1990;38(2):207–214. doi:10.1016/0093-934x(90)90111-s
21. Magrassi L, Bongetta D, Bianchini S, Berardesca M, Arienta C. Central and peripheral components of writing critically depend on a defined area of the dominant superior parietal gyrus. *Brain Res.* 2010;1346:145–154. doi:10.1016/j.brainres.2010.05.046
22. McCloskey M, Caramazza A, Basili A. Cognitive mechanisms in number processing and calculation: evidence from dyscalculia. *Brain Cogn.* 1985 Apr;4(2):171–96. doi:10.1016/0278-2626(85)90069-7. PMID: 2409994.
23. McCloskey M, Aliminosa D, Macaruso P. Theory-based assessment of acquired dyscalculia. *Brain Cogn.* 1991 Nov;17(2):285–308. doi:10.1016/0278-2626(91)90078-m. PMID: 1799455.
24. Mendez MF, Moheb N, Desarzant RE, Teng EH. The progressive acalculia presentation of parietal variant

Alzheimer's disease. *J Alzheimers Dis*. 2018;63(3):941–948. doi:10.3233/JAD-180024
25. Nolan KA, Caramazza A. Modality-independent impairments in word processing in a deep dyslexic patient. *Brain Lang*. 1982;16(2):237–264. doi:10.1016/0093-934x(82)90085-2
26. Ripamonti E, Aggujaro S, Molteni F, Zonca G, Frustaci M, Luzzatti C. The anatomical foundations of acquired reading disorders: a neuropsychological verification of the dual-route model of reading. *Brain Lang*. 2014;134:44–67. doi:10.1016/j.bandl.2014.04.001
27. Roeltgen DP, Heilman KM. Lexical agraphia: further support for the two-system hypothesis of linguistic agraphia. *Brain*. 1984;107 (Pt 3):811–827. doi:10.1093/brain/107.3.811
28. Roux FE, Dufor O, Giussani C, Wamain Y, Draper L, Longcamp M, et al. The graphemic/motor frontal area Exner's area revisited. *Ann Neurol*. 2009;66(4):537–545. doi:10.1002/ana.21804
29. Scarone P, Gatignol P, Guillaume S, Denvil D, Capelle L, Duffau H. Agraphia after awake surgery for brain tumor: new insights into the anatomo-functional network of writing. *Surg Neurol*. 2009;72(3):223–241. doi:10.1016/j.surneu.2008.10.074
30. Turkeltaub PE, Goldberg EM, Postman-Caucheteux WA, Palovcak M, Quinn C, Cantor C, et al. Alexia due to ischemic stroke of the visual word form area. *Neurocase*. 2014;20(2):230–235. doi:10.1080/13554794.2013.770873
31. Utianski RL, Duffy JR, Savica R, Whitwell JL, Machulda MM, Josephs KA. Molecular neuroimaging in primary progressive aphasia with predominant agraphia. *Neurocase*. 2018;24(2):121–123. doi:10.1080/13554794.2018.1454963
32. Varley RA, Klessinger NJ, Romanowski CA, Siegal M. Agrammatic but numerate. *Proc Natl Acad Sci U.S.A.* 2005;102(9):3519–3524. doi: 10.1073/pnas.0407470102.

Brain Imaging

Jacqueline C. Junn and Suzan Uysal

Introduction

Neuroimaging is a branch of medical imaging that involves the in vivo depiction of central nervous system (CNS) anatomy and function. Structural imaging, such as computed tomography (CT) and magnetic resonance imaging (MRI), provides anatomic detail of the brain and structural abnormalities (e.g., tumors, herniation, hemorrhage, developmental abnormalities). Functional brain imaging, such as single-photon emission tomography (SPECT) and proton emission tomography (PET), measures and visualizes the physiological activities of the brain. This chapter provides a basic overview of the structural imaging modalities of plain film X-ray, angiography, CT, MRI, and cranial ultrasound, as well as the functional imaging modalities of SPECT, PET, and functional magnetic resonance imaging (fMRI).

The introduction of **computed tomography** in 1972 by **Sir Godfrey Newbold Hounsfield** (1919–2004) revolutionized the field of neuroradiology and the clinical neurosciences because it allowed for the direct visualization of intracranial structures and pathologic changes in the cerebral parenchyma by noninvasive means. Prior to the advent of CT, imaging of the head structures was limited to X-ray imaging of the skull by plain film radiography, blood vessels by catheter angiography, and ventricles by pneumoencephalography. **Pneumoencephalography** is an obsolete procedure for imaging the ventricular system and subarachnoid space that involved replacing cerebrospinal fluid (CSF) with air (which served as a contrast medium and was introduced into the spinal subarachnoid space through lumbar puncture) and taking an X-ray of the head; it was used to reveal conditions such as hydrocephalus and brain atrophy. Prior to CT, catheter angiography and pneumoencephalography also were used to detect masses that displace and/or deform the cerebrovasculature or ventricles, respectively. The anatomy of the CNS was inferred from indirect observations of mass effect on these structures, suggesting adjacent intracranial pathology. In many cases, diagnostic confirmation was only possible by neuropathologic examination at autopsy.

CT was also revolutionary in that it introduced the technique of cross-sectional imaging, **tomography** (Greek, *tomo*, "slice"). By contrast, conventional plain radiography produces two-dimensional **projection images** based on the way the X-ray beam passes through the body, akin to shadows that reflect the degree of absorbed radiation. With projection imaging, the structures in the path of the X-ray beam are superimposed in the resulting image, as if looking *through* the patient, whereas cross-sectional imaging eliminates superimposition of structures, as if looking *inside* the patient. Cross-sectional imaging also allows for digital data processing to improve tissue contrast to view the anatomy in various planes (axial, coronal, or sagittal).

Plain Film X-Ray

Plain film X-ray is an imaging technique that uses X-rays, high-energy electromagnetic radiation that can pass through the body. The electromagnetic radiation is emitted from an X-ray tube directed at a subject, and an X-ray detector (such as radiographic film, or a digital X-ray detector) on the other side of the subject that registers the amount of radiation that passed through the subject. The resulting static two-dimensional image represents the degree to which the X-ray photons have passed freely through or have been absorbed (attenuated) by anatomical structures. **Attenuation** is the reduction of the X-ray beam intensity as it traverses matter. The denser the tissue, the greater the X-ray attenuation. High-density tissue such as bone appears white, and low-density tissue such as lung (due to air content) appears black. It is often necessary to obtain multiple images of the same body part from different viewpoints.

X-ray imaging is particularly useful for assessing bony structures and foreign bodies. With respect to the head, it can show skull and facial fractures and skull remodeling or erosion from underlying tumors or cysts (Figure 29.1). It does not allow evaluation of intracranial structures.

Angiography

Angiography produces images of the cerebral vasculature (arteries and veins) and allows for detection of vascular abnormalities. **Conventional arterial catheter angiography** involves injecting a radio-opaque contrast into the vascular system to make blood vessels opaque to X-rays and stand out against surrounding tissues (Figure 29.2). **Fluoroscopic images** are created by passing a continuous X-ray beam through the head and displaying the resulting images on a monitor to visualize movement of the contrast agent, as an X-ray movie. The flow of blood as the contrast circulates through the vasculature is observed, first through arteries, then into capillaries, and finally into veins.

Arterial catheter angiography is an important tool in the workup for neurovascular disorders such as vascular occlusions, vascular injury, and vascular abnormalities such as aneurysms, arteriovenous malformations, and moyamoya disease (see Figures 5.10, 5.11, and 5.12 in Chapter 5). Prior to CT and MRI, conventional angiography was also used to detect the presence of space-occupying lesions inferred from distortion in the shape and/or position of vessels. Today, noninvasive CT angiography (CTA) and magnetic resonance angiography (MRA) are more routinely used than conventional angiography, which is invasive and carries greater risks. However, it does produce the most detailed images of the cerebral vasculature; therefore it is used when CTA and MRA do not provide a definitive diagnosis, and for further characterization of and therapy for vascular lesions.

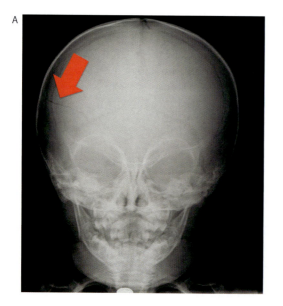
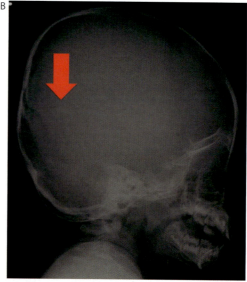

FIGURE 29.1. (A) Anteroposterior and (B) lateral skull X-rays of a child demonstrate a right parietal fracture.

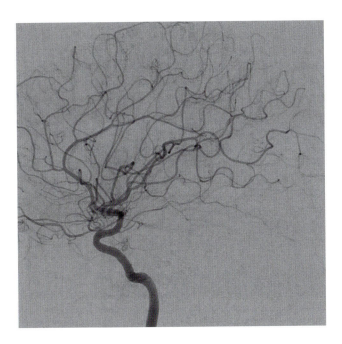

FIGURE 29.2. A lateral angiogram of the internal carotid artery.

Structural Brain Imaging

Structural neuroimaging provides information about the shape, size, and integrity of structures in the brain. The main imaging modalities for studying the structural anatomy of the brain are CT and MRI, and a variety of protocols for these have been developed for specific imaging purposes. A third structural imaging modality is ultrasound, but that is used primarily for infants. Selection of the most effective imaging modality and whether to use contrast is based on the clinical indication, medical history, the differential diagnoses regarding the underlying etiology of neurological signs and symptoms, spatial and temporal resolution of the images, and risks to the patient.

Imaging Planes

The three basic planes used in medical imaging are axial, coronal, and sagittal planes (see Figure 1.5 in Chapter 1).

Axial sections divide the structure into superior and inferior portions. Axial plane images are oriented as if the patient is lying supine and their feet are directed toward you, the viewer. The left side of the body is represented on the right side of the image, the right side of the body is represented on the left side of the image, the anterior aspect (front) of the head is represented at the top of the image, and the posterior aspect (back) of the head is represented at the bottom of the image.

Coronal sections divide the structure into anterior and posterior portions. Coronal plane images are oriented as if the patient is standing upright and facing the viewer. The left side of the body is represented on the right side of the image, the right side of the body is represented on the left side of the image, the top (superior aspect) of the head is represented at the top of the image, and the bottom (inferior aspect) of the head is represented at the bottom of the image.

Sagittal sections divide the structure into left and right portions. Sagittal plane images are oriented as if the patient is standing upright, and you are viewing them from their left side. The anterior aspect of the head is represented at the left side of the image, and the posterior aspect of the head is represented at the right side of the image.

Contrast Enhancement

The ability to visualize pathology on CT or MRI scans is heightened by **contrast enhancement**. **Contrast media** are chemical agents that improve the **contrast** (differences in image intensity of a structure against its surrounding tissue) and aid in the characterization of pathology. **Iodinated contrast media**, contrast agents that contain iodine, are used for CTs; gadolinium-based contrast agents are commonly used for MRIs.

Contrast is administered by injection into a large peripheral vein, either rapidly as a single large dose (bolus injection) or more slowly by infusion. Normally, diffusion of the contrast agent from the bloodstream to CNS is limited by the blood-brain barrier (see Chapter 5, "Blood Supply of the Brain"), but in many pathologies, the blood-brain barrier is weakened and contrast leaks from the intravascular compartment into the extravascular interstitial fluid compartment, leading to post-contrast enhancement of the lesion (Figure 29.3). In the CNS, contrast enhancement involves the intravascular and extravascular compartments independently.

Intravascular enhancement (**vascular enhancement**) forms the basis of angiography. It is related to blood volume and blood flow and allows detection of cerebrovascular abnormalities (e.g., stenosis, vasodilatation, aneurysm, vascular malformation, neovascularity, and shunting). **Extravascular enhancement** (also known as **interstitial enhancement**) is related to alterations in blood-brain barrier permeability, allowing for visualization of a variety of pathologies, including neoplasms, inflammation, demyelination, and infection. In some pathologies such as neoplasms, new abnormal blood vessels may develop, which tend to be leakier than normal capillaries, resulting in extravascular enhancement.

Interpretation Basics

MASS EFFECT AND HERNIATION

Under normal conditions, structural neuroimaging shows that the intracranial contents have a clearly

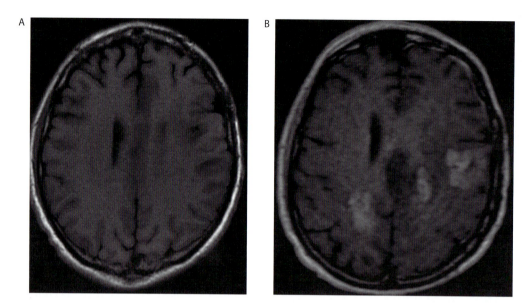

FIGURE 29.3. MRI with and without contrast. (A) Without contrast. (B) With contrast, showing enhancing metastatic lesions in the right parietal and left frontal lobes.

defined midline with bilateral symmetry. **Mass effect** is the compression, distortion, and/or displacement of the intracranial contents. It may be caused by (1) focal intracranial **space-occupying** lesions such as neoplasms, non-neoplastic masses (e.g., arachnoid cyst, abscess), or localized hemorrhage; or (2) brain edema (swelling). Mass effect manifests in many ways (see Figure 4.9 in Chapter 4), including **effacement** (elimination/disappearance of a space) of sulci, partial or complete effacement of ventricles, **midline shift** (displacement of midline structures), and **herniation** (shift of tissue from its normal location into an adjacent space). Mass effect and brain herniations can result in compression of normal brain tissue and blood vessels, leading to further neurological dysfunction and ischemia.

Intracranial pressure (ICP) is the pressure within the cranium. Under normal circumstances, ICP ranges from 5 to 15 mmHg, with brain parenchyma accounting for 80% of the intracranial volume, and CSF and blood each accounting for 10%. The **Monroe-Kellie doctrine** stipulates that the total volume of brain, CSF, and intracranial blood remains constant. An increase in the volume of any one of these three components must be accompanied by a reduction in the volume of the other components, otherwise ICP will increase. In a compensated state, ICP remains normal when there is an increase in volume of any of the three intracranial contents, primarily through reducing venous blood volume by displacement out of the intracranial space, reducing intracranial CSF volume by diversion into the spinal subarachnoid space, or a combination of the two. For example, an increase in brain volume due to edema results in decreased volume of the CSF space, as evidenced by effacement of local sulci, adjacent subarachnoid cisterns, and/or portions of the ventricles. When compensatory mechanisms fail, however, ICP will increase rapidly, and brain parenchyma will herniate from areas of high pressure to areas of low pressure. This may occur with edema, hydrocephalus, or space-occupying lesions (e.g., hematoma, neoplasm, abscess).

The first step in evaluating a structural brain imaging study is to determine whether there are signs of impending herniation that would indicate emergent intervention. The ventricles and basal subarachnoid cisterns (see Figure 4.6 in Chapter 4) are examined for effacement. Herniation of the cerebellar tonsils produces crowding at the foramen magnum and may compress the brainstem, compromising the functions of the pons and medulla, which are responsible for respiration and cardiac rhythm control. Effacement of the **quadrigeminal cistern** (located dorsal to the superior and inferior colliculi of the midbrain tectum) is a marker for ascending **transtentorial herniation**. Effacement of the **suprasellar cistern** (located above the sella) is a marker for **uncal herniation**. **Subfalcine herniation** is a displacement of brain parenchyma under the cerebral falx. With greater degrees of midline shift, the foramen of Monro of one hemisphere becomes effaced, leading to lateral **ventricle entrapment** in which the exit path for CSF is obstructed, and the isolated ventricle becomes dilated due to continued CSF production.

INTRA-AXIAL VERSUS EXTRA-AXIAL LOCATION

Accurate locating of a lesion is an important factor in developing an appropriate differential diagnosis.

Extra-axial lesions are located outside the brain parenchyma but within the cranium; the intraventricular space is also considered a subdivision of the extra-axial compartment. Extra-axial lesions may displace brain parenchyma. Extra-axial lesions may be further characterized by their relationship to the meninges and sutures; extra-axial intracranial hemorrhage may be epidural, subdural, subarachnoid, or intraventricular. Extra-axial intracranial masses may be extradural, intradural, or intraventricular.

Intra-axial lesions are located within the brain parenchyma. The differential diagnosis of intra-axial lesions is much more extensive than extra-axial lesions and may include primary or secondary neoplastic, infectious/inflammatory, congenital, metabolic, posttraumatic, and vascular processes.

INTRACRANIAL HEMORRHAGE

One of the most common indications of neuroimaging is evaluating for the presence of **intracranial hemorrhage**, which can be seen in epidural, subdural, subarachnoid, intraparenchymal (intracerebral), or intraventricular compartments, or in multiple compartments.

Epidural and subdural hemorrhages are typically due to trauma. The hematomas compress the adjacent brain and often require emergency surgical evacuation. Both can be associated with parenchymal brain damage.

Epidural hematomas have a **lentiform** (lens-shaped, biconvex, elliptical, lemon-shaped) appearance on imaging (see Figure 4.10 in Chapter 4). They have a convex inner margin against the brain because the periosteal (outer) layer of the dura adheres tightly to the overlying **cranium** (**brain case**); thus blood cannot extend easily along the potential epidural space. They can cross the midline because they are above the dura, but they do not cross the cranial suture lines because of the tight attachment of the dura at these locations. Epidural hematomas tend to cause local mass effect due to their shape and resistance to diffuse spread and produce focal neurological signs and symptoms. They are often associated with calvarial fractures.

Subdural hematomas have a **crescentic** appearance (crescent-shaped, concave, banana-shaped), with a concave inner margin paralleling the cortical margin of the adjacent brain (see Figure 4.11 in Chapter 4). This is because the meningeal (inner) layer of the dura adheres loosely to the underlying arachnoid membrane, allowing blood to extend easily along the potential subdural space. For this reason, subdural hematomas are more extensive than epidural hematomas. Because subdural hematomas lie beneath the dura, they can cross the cranial sutures. However, they cannot cross the dural reflections, and therefore they do not cross the midline; instead, they extend along the dura of the falx into the interhemispheric fissure and onto the tentorium. Subdural hematomas tend to present with disturbances of mentation and consciousness, rather than focal or lateralizing signs.

Subarachnoid hemorrhage (SAH) is most often caused by ruptured aneurysm or trauma (traumatic SAH). Most aneurysms are located near vascular branch points; thus, most aneurysmal subarachnoid hemorrhages are observed within the suprasellar cistern, surrounding the circle of Willis, and the Sylvian cisterns surrounding the middle cerebral arteries.

Intracerebral hemorrhage (ICH), also known as **intraparenchymal hemorrhage**, is hemorrhage within the brain substance (Figure 29.4). The etiology for ICH is more extensive than for other types of intracranial hemorrhage; different causes include trauma/contusion, hypertensive hemorrhage, hemorrhagic conversion of arterial or venous infarcts, underlying vascular lesion (including AVM, cavernous angioma, and aneurysm), underlying neoplasm, vasculopathies (e.g., amyloid angiopathy), underlying coagulopathy, and encephalitis. If there is a history of trauma, contusion is suspected; these hemorrhages are particularly located within the anterior/inferior frontal lobes or anterior or posterior temporal lobes. If the hemorrhage originates within the basal ganglia or thalamus, a hypertensive etiology is strongly suspected, although an underlying vascular or neoplastic lesion is also a possibility. If the patient is young or without significant medical history, an underlying arteriovenous malformation should be considered. If the patient is elderly and has evidence of chronic hemorrhages elsewhere, amyloid angiopathy is a possible etiology.

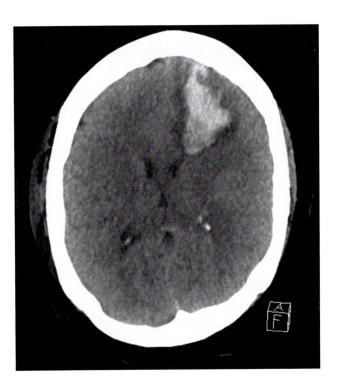

FIGURE 29.4 CT scan showing intracerebral hemorrhage in the left frontal lobe.

HYDROCEPHALUS VERSUS VOLUME LOSS

Evaluation of the ventricles includes determining shape, location, and size. Ventricular dilation can be due to parenchymal volume loss or hydrocephalus. Ventricular dilation that is accompanied by widening of the cortical sulci and gyral thinning suggests that the ventricular expansion is due to parenchymal atrophy (or lack of brain development when observed in pediatric patients). Ventricular dilation that is *not* accompanied by widened sulci suggests hydrocephalus. Ventricular dilation that is accompanied by effaced sulci suggests communicating hydrocephalus.

Once hydrocephalus is diagnosed, the next step is to determine whether it is noncommunicating (obstructive) or communicating (see Figure 4.12 in Chapter 4). In noncommunicating hydrocephalus, there is an obstruction of CSF ventricular outflow. In communicating hydrocephalus, CSF can flow into the subarachnoid space, but there is an overproduction of CSF or obstruction at the arachnoid villi where CSF drains into the dural venous sinuses; thus all the ventricles are enlarged.

Normal pressure hydrocephalus (NPH) is a type of communicating hydrocephalus. Mild cases may be difficult to diagnose by imaging, but key imaging features for diagnosis and selection of shunt-responsive patients are: (1) ventricular enlargement not entirely attributable to cerebral atrophy or congenital enlargement (Evans index > 0.3); (2) no macroscopic obstruction to CSF flow; and (3) callosal angle less than 90 degrees. The **Evans Index** is the ratio of maximum width of the frontal horns and the maximal internal diameter of the skull at the same level (Figure 29.5). The **callosal angle** is measured between the lateral ventricles on a coronal MR image at the level of the posterior commissure and perpendicular to the anteroposterior commissure plane (Figure 29.6).

CEREBRAL EDEMA

Cerebral edema, swelling due to increased water content in the brain, occurs in many neurological conditions and can be potentially life-threatening. Signs and symptoms are related to secondary mass effect, which can lead to structural compression, vascular compromise resulting in cerebral ischemia, and herniation. Clinical and radiologic manifestations of cerebral edema are usually reversible in the early stages; in severe cases, emergency **decompressive craniectomy** may be required.

The major categories of cerebral edema are cytotoxic, vasogenic, and combined, but there are other types, such as interstitial edema. Neuroimaging often yields clues that allow for characterization of edema, which may be critical for early and accurate diagnosis of the underlying neurological cause and guiding medical intervention (e.g., osmotherapy, diuretics, hypothermia, corticosteroids).

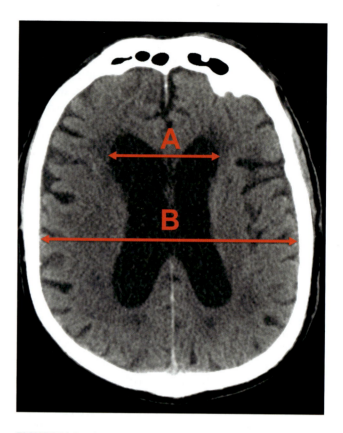

FIGURE 29.5. The Evans Index is the ratio of maximum width of the frontal horns and the maximal internal diameter of the skull at the same level (A/B). A ratio below 0.3 is normal; a ratio greater than 0.3 indicates ventriculomegaly. The Evans Index in this case was 0.41.

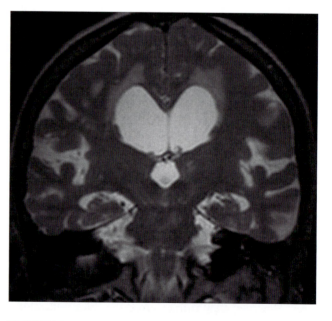

FIGURE 29.6. A coronal T2 MR image of the brain in a patient with normal pressure hydrocephalus. The callosal angle is the angle between the lateral ventricles on a coronal MR image at the level of the posterior commissure. A callosal angle less than 90 degrees is typical for normal pressure hydrocephalus.

In **cytotoxic cerebral edema**, there is an abnormal diffusion of water from the extracellular to the intracellular space, causing neuronal swelling. The trigger for this is a shutdown of sodium-potassium pumps, an energy-dependent neuronal membrane protein responsible for the high extracellular and low intracellular sodium ion concentration in the CNS (see Chapter 2, "Electrical Signaling in Neurons"). When energy delivery to cells fails, as occurs with cerebral infarction, cells are unable to maintain the sodium-potassium pump function. Consequently, sodium ions flow down the osmotic gradient and accumulate within neurons, drawing chloride and water along. This results in cellular swelling due to the redistribution of water from extracellular to intracellular compartments. Normally on CT, gray matter is brighter (hyperdense) compared to the white matter. However, with cytotoxic edema (e.g., infarct), the gray matter's density decreases, leading to a loss of the **gray-white matter differentiation** on neuroimaging (Figure 29.7). Once the presence of possible cerebral infarction is established, the vasculature is evaluated for underlying stenosis, thrombus, or occlusion by computed tomography angiography (CTA), magnetic resonance angiography (MRA), and/or conventional angiography.

In **vasogenic cerebral edema**, there is breakdown of the blood-brain barrier due to vascular injury that disrupts tight endothelial junctions of the vessel walls (see Chapter 5, "Blood Supply of the Brain"). As a result, intravascular fluid leaks from the capillaries into the extracellular space; this extracellular edema mainly affects white matter. Vasogenic edema can be seen with a variety of conditions including neoplasm, infection, encephalitis, hemorrhage, venous thrombosis, and arteriovenous malformation. It also occurs with **posterior reversible encephalopathy syndrome** (PRES), a clinico-radiological syndrome characterized by headache, seizures, altered mental status (confusion, decreased level of consciousness), and white matter vasogenic edema typically affecting the posterior occipital and parietal lobes of the brain. It is due to an acute hypertensive crisis with inability to autoregulate in the setting of acute blood pressure change. Although the exact mechanism is not well understood, it is believed to be secondary to disruption of the blood-brain barrier.

In **combined cerebral edema**, there is a global loss of gray-white matter differentiation and effacement of the sulci, ventricles, and basal cisterns. Combined cerebral edema is associated with trauma, hypoxic-ischemic encephalopathy, infection, inflammation, and metabolic or toxic conditions.

Interstitial cerebral edema is associated with hydrocephalus and therefore is also known as **hydrocephalic edema**. Increased intraventricular pressure causes CSF to diffuse across the ventricular ependymal lining into the extracellular space of the periventricular white matter. The combination of **ventriculomegaly** (enlarged ventricles) and increased periventricular water content on imaging is suggestive of interstitial edema in acute obstructive hydrocephalus.

ENCEPHALOMALACIA

Encephalomalacia is the softening of brain tissue that occurs as a late manifestation of injury. It is the result of

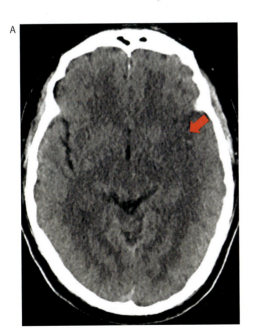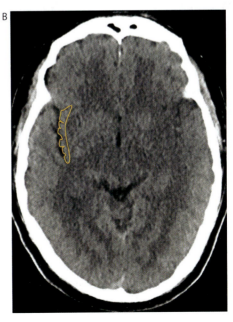

FIGURE 29.7. An axial CT image of the brain demonstrates loss of insular ribbon (loss of gray-white matter differentiation) due to cytotoxic edema from a left MCA infarct. (A) The red arrow points to the infarct. (B) The yellow outline indicates the intact insular ribbon in the right hemisphere.

liquefactive necrosis, in which injured brain tissue is dissolved by enzymes. Areas of encephalomalacia are often surrounded by a rim of **gliosis** (proliferation glial cells). The most common causes of encephalomalacia are infarction, hemorrhage, and trauma (including surgery). Areas of encephalomalacia may result in focal neurological deficit, may serve as a seizure focus, or may even be asymptomatic. On gross pathologic inspection, encephalomalacia is apparent as decreased firmness of brain tissue and blurred cortical margins.

Computed Tomography

CT uses X-rays and enables the distinction between different anatomical structures based on the differential absorption of X-rays by different tissues. The CT machine consists of a large ring-shaped scanner and a platform that moves the patient, usually lying supine, through the scanner. The ring has an X-ray emitter on one side that passes a narrow beam of X-rays through the body, and X-ray detectors directly opposite the emitter that measure the amount of radiation that has passed through. The ring rotates around the body in the imaging plane, and the information from all angles for each slice is combined by computer to generate a cross-sectional image. As the patient is moved through the scanner, a series of successive axial (horizontal) image slices of the patient's body are obtained. For this reason, the scans are acquired in the axial plane (and hence the previous name, "computed axial tomography"), but computer algorithms may be used to reconstruct the original digital image data into other anatomic planes (i.e., sagittal and coronal).

In CT scans, structures are distinguished and characterized based on tissue density; the denser the tissue, the greater the X-ray attenuation. CT measures the local X-ray attenuation coefficients of three-dimensional cube-shaped tissue volume elements, known as **voxels** (volume pixels). Density is measured in **Hounsfield Units** (HU), with water assigned a value of zero HU and appearing dark. For each slice, the voxel data are translated into **pixel data**, the picture elements that make up the two-dimensional digital image of each slice. The resulting grayscale images consist of a matrix arrangement of pixels, and each pixel has a grayscale value that represents the X-ray attenuation of the tissue voxel that it represents.

Regions that attenuate X-rays more than water (e.g., bone, blood, soft tissue) have positive HU values, appear bright, and are **hyperdense**. Regions that attenuate X-rays less than water (e.g., fat, air) have negative HU values, appear darker than water, and are **hypodense**. Thus, in CT, bone appears white, and CSF appears dark (see Figures 4.10 and 4.11 in Chapter 4). Bone, calcification, and metal are the densest (brightest), acute blood is denser than soft tissue, water is denser than fat, and air is the least dense (darkest).

Compared to MRI, CT scans have a faster acquisition time and lower cost. CT is also particularly helpful for identifying blood, calcification, and bone pathologies. A non-contrast head CT is the mainstay imaging modality in emergency and critical care settings, as it can be used to assess quickly for hemorrhage, herniation, and hydrocephalus (the "three Hs"), all of which may require emergency neurosurgical intervention. It is the first-line imaging modality for (1) head trauma, as it can quickly assess for skull fractures and acute hemorrhage; (2) acute stroke; and (3) identifying focal brain injuries with mass effect and potential for midline shift and herniation. Non-contrast CT can reveal calcifications within the brain parenchyma or vasculature. Calcification may occur with a wide spectrum of pathologies, including genetic developmental disorders (e.g., Fahr's disease), infectious diseases (e.g., neurocysticercosis), and neoplasms. Calcification also occurs with normal aging without associated pathology (physiologic/age-related calcifications), within the pineal gland, habenula, choroid plexus, basal ganglia, and dura mater.

INTERPRETATION

CT scans are used to evaluate for the presence of bleeding and structural abnormalities of the cisterns, brain, ventricles, and bone. It is important to remember that multiple pathologies can occur concurrently.

The radiologist examines for evidence of blood, such as extradural hematoma, subdural hematoma, SAH, ICH, or intraventricular hemorrhage (IVH). The appearance of blood varies over time; a more acute hematoma appears hyperdense as compared to a chronic bleed. Epidural hematomas, subdural hematomas, and subarachnoid hemorrhages are extra-axial fluid collections (within the skull but outside the brain parenchyma). Because epidural hematomas cannot cross skull sutures, they may compress brain structures rapidly and produce brainstem herniation. Thus, CT imaging is extremely important and may prompt neurosurgical evacuation.

The subarachnoid cisterns, as well as the cerebral sulci and cerebellar folia, are examined for effacement, asymmetry, and blood due to SAH (see Figure 18.7 in Chapter 18). Four key cisterns are the **ambient cistern** surrounding the midbrain, the **suprasellar cistern** superior to the sella turcica, the **quadrigeminal cistern** posterior to the midbrain tectum, and the **Sylvian cistern** across the insular surface within the Sylvian fissure (see Figure 4.6 in Chapter 4).

The brain is examined for sulcal effacement, abnormal shifts of brain tissue, herniation, and changes in tissue composition and gray-white matter differentiation. With sulcal effacement, there is a loss of the normal dark CSF within the sulci. Abnormal shifts of brain tissue and herniation include subfalcine, uncal, transtentorial (superior or inferior), and tonsillar herniation. Gray-white matter differentiation refers to the appearance of the interface between parenchymal white matter and gray matter, which can be used to differentiate cytotoxic and vasogenic edema. Loss of the **insular ribbon sign** of normal gray-white differentiation may be an early sign of middle cerebral artery infarction (see Figure 29.7).

The ventricles are examined for their size, morphology, and symmetry. The radiologist may also examine for IVH and choroid plexus anomalies. Ventriculomegaly occurs with hydrocephalus and brain atrophy. Ventricular effacement may result from cerebral edema, a mass, or intracranial hemorrhage. Hyperdensity within the ventricular system may be due to IVH, infection, hypercellular mass, or calcified choroid plexus.

Bone is examined for fractures of the calvarium or other osseous pathologies.

LIMITATIONS

In comparison to MRI, CT has lower contrast resolution; therefore evaluation of the brain parenchyma is better with MRI. As CT is susceptible to several types of image **artifact**, distortions in the image that are unrelated to the anatomical structure may occur. Image artifacts can degrade image quality, sometimes to the point of making images uninterpretable and diagnostically unusable; they may even simulate certain types of pathology. Common CT artifacts include patient motion and metallic object streak artifacts. **Motion artifact** is created by patient motion during scanning, which results in blurring and misregistration of images. Metallic objects such as aneurysm clips can create **metallic streak artifacts** that may entirely obscure adjacent structures. Another limitation of CT is the use of ionizing radiation.

COMPUTED TOMOGRAPHY ANGIOGRAPHY

Computed tomography angiography (CTA) is a type of **contrast-enhanced CT** that uses **radiopaque contrast** and X-ray beams to visualize the vasculature (Figure 29.8). Computer post-processing software can be used to remove nonvascular structures and create three-dimensional reconstruction of the vessels.

This technique is used in the evaluation of SAH, ICH, and ischemic stroke. In the context of SAH and ICH, it allows for detection of aneurysms, vascular malformations, and ongoing bleeding. In the context of ischemic stroke, it allows for detection of occlusion, stenosis, or dissection within the intracranial and extracranial carotid arteries and vertebrobasilar system.

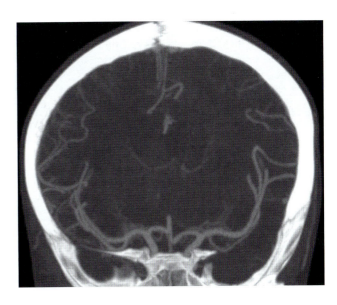

FIGURE 29.8. A coronal CT angiogram image of the head.

Arterial catheter angiography and CTA both require use of ionizing radiation and intravenous contrast. Limitations of CTA are: (1) it provides a static image from one time point during a single vascular phase (e.g., the arterial phase) and does not permit evaluation of flow mechanics; and (2) it has lower resolution than catheter angiography and does not visualize very small arteries such as the perforating arteries that supply the deep brain structures.

COMPUTED TOMOGRAPHY PERFUSION IMAGING

CT perfusion imaging quantifies blood flow through the brain parenchyma and is an important adjunct to brain CT and CTA in the context of ischemic stroke (Figure 29.9). It enables differentiation of salvageable ischemic brain tissue within the penumbra from irreversible tissue necrosis within the infarct core. CT perfusion imaging parameters can detect areas of focal cerebral ischemia at an early stage and can aid in planning neurovascular interventions such as intra-arterial thrombolysis and mechanical clot retrieval. It is also useful for the diagnosis and treatment of vasospasm in the setting of SAH.

Magnetic Resonance Imaging

MRI provides higher contrast resolution than CT, allowing for improved delineation of soft tissue and more detailed information about structure. Therefore, it is more sensitive in evaluating for soft tissue pathologies such as infection, neoplasm, other parenchymal infiltrative processes.

MRI involves placing the body in a strong magnetic field; the strength of the magnet is measured in **Tesla** (T)

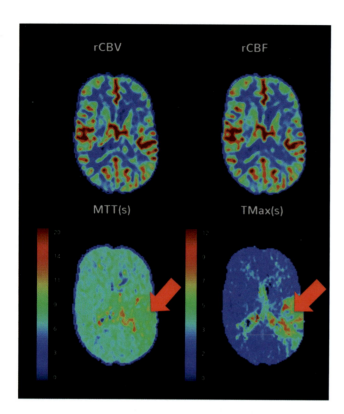

FIGURE 29.9. CT perfusion imaging showing ischemic penumbra within the left MCA territory. Regional cerebral blood flow (rCBF), regional cerebral blood volume (rCBV), mean transit time (MTT), and time-to-maximum (Tmax) are used to identify the infarct core and the ischemic penumbra, as patients with a small core and a large penumbra are most likely to benefit from reperfusion therapies. The infarct core area has prolonged MTT or Tmax, with decreased rCBF and rCBV. The ischemic penumbra has prolonged MTT or Tmax, with reduced rCBF and near normal or increased rCBV.

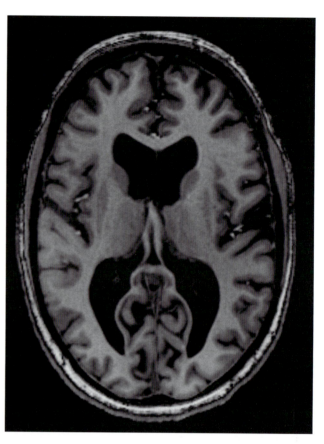

FIGURE 29.10. 7T MRI axial T1 image of the brain.

units. Most clinical scanners are 1.5T or 3.0T, although there are varying strengths below 1.5T and up to 10.5T. Higher Tesla scanners use stronger magnets and produce more detailed images (Figure 29.10).

MRI is based on the facts that (1) hydrogen atoms are abundant in the body since the human body is 70% water and there are two hydrogen protons in every water molecule, and (2) the hydrogen nucleus consists of a single proton which behaves as a tiny bar magnet. The protons are normally oriented in different directions, but when placed in an external magnetic field such as the MRI, the protons align. This alignment is disrupted by adding radiofrequency (RF) pulses. During each pulse, the protons all tilt on an angle and **resonate** (wobble) at the same frequency. When the RF pulse is turned off, the protons return to their normal aligned state in the magnetic field in a process called **relaxation**, and in so doing emit RF energy. The rate at which the emitted RF signals dissipate over time varies across different tissue types, depending on the proportion of water in the tissue. The emitted signals are measured by the receiver coil, and the data are converted into intensity levels to create grayscale images consisting of a matrix arrangement of pixels.

Within arteries, however, the situation is different because, unlike solid tissue, blood flows. An RF pulse causes the hydrogen nuclei protons to resonate, but when the RF pulse is turned off, the protons have moved and there is no relaxation signal; thus the arteries look like there is nothing inside them. The term **flow void** refers to a loss of signal that occurs within vessels that contain vigorously flowing blood; it is generally synonymous with vascular patency; lack of flow void occurs with altered blood flow dynamics (e.g., cerebral venous thrombosis).

There are two ways in which the proton relaxes, T1 and T2 relaxations. Tissue contrast/intensity is determined by its T1 and T2 relaxation times. The most common MRI sequences are T1-weighted and T2-weighted scans. Each sequence has some T1 and T2 weighting; however, it is the overall predominant weighting that determines what the sequence is. In general, T1-weighted images are superior in visualizing anatomic detail. T1 post-contrast images make abnormalities brighter, especially when there is breakdown in the blood-brain barrier. T2-weighted images are helpful when visualizing pathology such as edema.

Prior to undergoing an MRI scan, patients must be screened for any potential magnetic metal devices or foreign bodies. Those patients with implanted electronic devices such as cardiac pacemakers, cochlear implants, nerve stimulators, and drug infusion pumps may not be able to get an MRI study, as some of these devices are made from ferromagnetic materials which are unsafe for MRI. The MRI machine is always on, and the strong magnetic field pulls the ferromagnetic objects toward the MRI machine and turns them into dangerous projectiles. Therefore, ferromagnetic objects should never be brought into the MRI environment. Additionally, non-ferromagnetic implants may also be problematic as they can potentially cause tissue heating. Therefore, to ensure everyone's safety, every MRI facility has a comprehensive screening procedure and safety protocols.

There are several limitations to MRI beyond the need to screen patients for MRI safety. Some patients experience claustrophobia while in the scanner, and the noise generated by the machine can cause patients to become anxious. MRI images take longer to obtain than CT, and motion artifacts are common, compromising image quality and image interpretation.

T1-WEIGHTED MRI

In T1-weighted images, CSF appears hypointense ("black"), gray matter appears isointense ("gray"), and white matter appears hyperintense ("brighter gray"; Figure 29.11). The contrast of T1-weighted images can be enhanced by injecting gadolinium to help visualize pathologies involving breakdown of the blood-brain barrier, such as inflammation, tumors, and abscesses.

T2-WEIGHTED MRI

In T2-weighted images, CSF appears white, gray matter appears light gray, and white matter appears dark gray (hypointense compared to gray matter); thus, T2 images have nearly the opposite contrast of T1 images (see Figure 29.11). T2-weighted images provide good discrimination between gray and white matter areas and are helpful when examining for white matter pathologies.

FLUID ATTENUATED INVERSION RECOVERY MRI

Fluid Attenuated Inversion Recovery (FLAIR) MRI is similar to T2-weighted MRI (Figure 29.12), but it suppresses the fluid signal (i.e., makes it dark). This sequence increases the distinction between dark CSF and bright areas of abnormality, allowing for improved imaging of white matter abnormalities by suppressing the CSF signal. This sequence, therefore, provides excellent visualization of periventricular white matter pathology and the interstitial edema that occurs with acute obstructive hydrocephalus, in which the ventricular ependymal lining is disrupted and CSF migrates into the brain parenchyma around the ventricles (usually involving the lateral ventricles).

DIFFUSION-WEIGHTED MRI

Diffusion-weighted MRI (DW-MRI), also known as diffusion-weighted imaging (DWI), measures

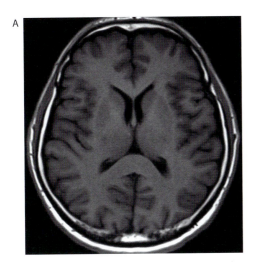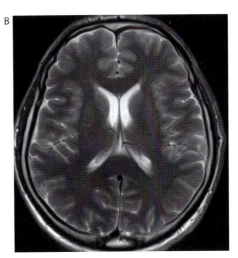

FIGURE 29.11. Axial MRI image of the head in (A) T1-weighted sequence and (B) T2-weighted sequence.

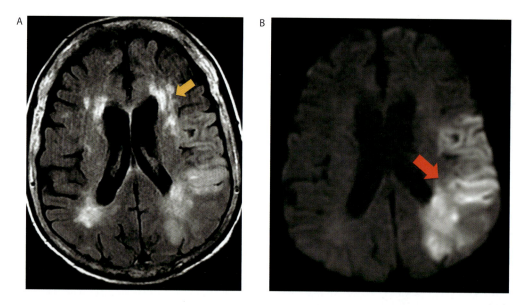

FIGURE 29.12. Acute infarct. (A) FLAIR MRI. (B) Diffusion-weighted MRI. The diffusion image (B) shows hyperintensity in the left MCA territory (red arrow), with matching FLAIR image (A) hyperintensity, and additional areas of background microvascular ischemic disease in the bilateral corona radiata (yellow arrow).

diffusional motion of water molecules, quantified by a parameter known as the **apparent diffusion coefficient** (ADC). DWI represents the combination of T2 signal and diffusion values; thus the images are described in terms of both signal intensity and whether water can move around less (reduced diffusion). Reduced diffusion implies cell injury and appears as bright signal on the diffusion-weighted images and dark on the ADC map. This sequence is highly sensitive to acute infarction (see Figure 29.12). Reduced diffusion also occurs with other pathologies, such as highly cellular tumors and abscesses.

DIFFUSION TENSOR IMAGING TRACTOGRAPHY

Diffusion tensor imaging (DTI) is an MRI technique that measures the directionality and magnitude of water diffusion, thereby allowing evaluation of white matter microstructural integrity. Normally, diffusion of water molecules in white matter is **anisotropic** (directional), occurring parallel to the direction of the neural tract. This contrasts with **isotropic** diffusion, which is unrestricted (or equally restricted in all directions) and random. **Fractional anisotropy** (FA) is the most common DTI index measure, expressed as a value between zero and one. A value of zero indicates that diffusion is isotropic, as occurs in CSF. The FA approaches 1.0 in healthy white matter, signifying normal water diffusion along the direction of a tract where fibers are highly organized and parallel to each another. Reduced anisotropy is characteristic of structural white matter lesions, as occurs in a variety of disorders including stroke, leukoaraiosis, traumatic brain injury, neoplasms, and multiple sclerosis.

DTI is insensitive to detecting: (1) injury in small white matter tracts due to its low spatial resolution, and (2) axonal injury in crossing white matter regions with complex anatomy where there is no single predominant direction of the axons.

Fiber tractography is a 3D reconstruction technique to assess neural tracts using data collected by DTI. The predominant direction of diffusion indicates how the fibers are oriented in a 3D coordinate system, represented using a color code in which red indicates transverse fibers (i.e., right-to-left or left-to-right), green indicates anteroposterior fibers, and blue indicates craniocaudal fibers (head-to-foot or foot-to-head).

SUSCEPTIBILITY WEIGHTED IMAGING

Susceptibility weighted imaging (SWI) is particularly sensitive to compounds that distort the local magnetic field, and therefore can detect and differentiate blood products and calcium. It is particularly useful to look for cerebral microbleeds (microhemorrhages), which are evidenced by small deposits of iron-rich **hemosiderin**, the end product of blood degradation in brain; common causes include trauma (hemorrhagic diffuse axonal injury), cerebral amyloid angiopathy, and chronic hypertensive encephalopathy. SWI shows the iron deposits within the basal

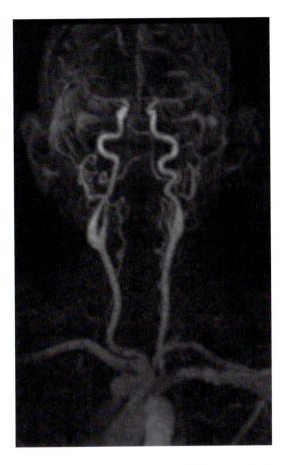

FIGURE 29.13. A coronal post-contrast MRA image of the head and neck.

ganglia that occur in **neurodegeneration with brain iron accumulation** (NBIA), a group of rare genetic neurodegenerative diseases with parkinsonism. SWI also shows abnormal calcium deposits that may occur with a variety of pathologies. **Fahr's disease** is a rare inherited neurological disorder characterized by abnormal deposits of calcium in the basal ganglia and dentate nuclei of the cerebellum, with movement disorder and neuropsychiatric features.

MAGNETIC RESONANCE ANGIOGRAPHY

Magnetic resonance angiography (MRA) is an alternative to conventional angiography and CT angiography (Figure 29.13). It can be performed without contrast in a procedure known as **non-contrast enhanced MRA** (time-of-flight MRA), which eliminates the need for radiation and contrast.

MRA is performed to screen patients for possible ischemia or possible underlying vascular pathology; however, CTA has higher spatial resolution and typically provides a more detailed anatomic evaluation of the vascular system.

MR PERFUSION-WEIGHTED IMAGING

MRI perfusion quantifies the relative amount of blood flowing into tissue. This information is useful in several clinical scenarios, including defining the ischemic penumbra in ischemic stroke and for the evaluation of tumors due to tumor angiogenesis and reduced integrity of the blood-brain barrier.

INTRAOPERATIVE NEUROIMAGING

Neuroimaging is also used during brain surgery. Intraoperative MRI allows for accurate navigation and stereotactic surgical guidance (especially for biopsy and when lesions are small or deep seated). Intraoperative MRI requires specialized operating room suites and specialized scanners, with the MRI magnet stored in an adjacent room. The patient is either moved to the magnet or the MRI magnet is moved to the patient to obtain images. Only MRI-compatible equipment and surgical instruments are allowed in the MRI operating room.

Ultrasound

A third structural imaging modality is **ultrasound**, which uses high-frequency sound waves that are inaudible (i.e., above the frequency range of human hearing). It is based on the principle that when sound waves strike an object, they are reflected as an echo. Different tissues reflect the sound waves to different degrees; ultrasound waves travel through soft tissue and fluids but bounce back off denser tissues. By measuring the echo waves, it is possible to evaluate the appearance of organs, tissues, and vessels in size, shape, and consistency, and to detect abnormal masses. The ultrasound transducer (probe) both emits ultrasound waves and detects the ultrasound echoes reflected by the structure under investigation. The reflected waves are transformed by software that produces an image known as a **sonogram**.

Cranial ultrasound is most often performed in infants. The fontanelles provide an **acoustic window**, allowing the ultrasound beam to freely pass into and back from the brain. Cranial ultrasound is used to evaluate for brain malformations, hydrocephalus, and intraventricular hemorrhage and/or hypoxic-ischemic injury in preterm neonates. It is safe because it does not use radiation and does not require sedation. Cranial ultrasound is also used in adult neurosurgical patients for intra-operative imaging once the skull has been opened, to evaluate tumors and facilitate their safe removal.

Doppler ultrasound measures blood flow to and within the brain through relatively thin transtemporal bone acoustic windows. It is used to detect conditions that affect blood flow such as vascular stenosis, occlusion, or vasospasm, and raised intracranial pressure. It is also used as an intraoperative monitor during surgical procedures that place the brain at risk for cerebral hyper- or hypoperfusion, or gaseous or particulate embolization, such as carotid endarterectomy, aortic arch procedures, and coronary artery bypass surgery.

Functional Neuroimaging

Functional neuroimaging techniques measure aspects of brain function and metabolism. They can be divided into those that use molecular imaging probes and those that use MRI principles.

A molecular imaging probe is a naturally occurring compound or drug that has been labeled with a **radioisotope** (radioactive isotope) without altering its bioactivity. These labeled compounds are called **tracers**. The tracer is injected into the bloodstream and distributed throughout the body. As the radioisotope undergoes radioactive decay, the tracer emits a signal that can be detected by an external scanner.

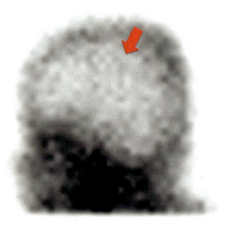

FIGURE 29.14. Brain SPECT scan showing lack of intracranial tracer activity seen in a patient suspected of brain death.

Single Photon Emission Computed Tomography

Single photon emission computed tomography (SPECT) is a functional imaging modality that shows the three-dimensional distribution of radiotracers. The radioactive tracers used in SPECT emit single gamma-ray photons (those with the highest energy in the electromagnetic radiation spectrum) and have a relatively long **half-life** (amount of time for one-half of a substance's radioactivity to decay). The emitted gamma rays are detected by a **gamma camera** consisting of a ring of gamma-ray photon detectors, and tomographic computation constructs the tomographic images.

Regional cerebral flow imaging (rCBF) can be measured using SPECT, which is helpful in assessing neuronal activity by looking for areas of hypoperfusion. Clinical uses for SPECT rCBF mapping include assessment for suspected early-stage cortical neurodegenerative disease, epileptic focus localization, traumatic brain injury, and brain death (Figure 29.14).

In addition to visualizing brain perfusion, SPECT is used for visualizing molecular activity. The **DaTscan** visualizes a tracer that binds to the **dopamine transporter** molecule by SPECT imaging (see Figure 11.3 in Chapter 11). The dopamine transporter molecule is located on the axon terminals of dopaminergic neurons and is responsible for dopamine reuptake from the synaptic cleft. In Parkinson's disease, there is a loss of nigrostriatal dopaminergic neurons and reduced DaT tracer binding in the striatum. This test is used in the diagnostic workup for Parkinson's disease and differential diagnosis of other disorders with parkinsonism.

SPECT is widely available and relatively inexpensive, but it lacks anatomical resolution. Functional and structural imaging data, however, can be combined to create images with both functional activity and structural detail. In **SPECT-CT**, the images from structural CT and functional SPECT scans are combined by **co-registration**.

Positron Emission Tomography

Positron emission tomography (PET) is another functional imaging modality. It has higher spatial resolution than SPECT, but it is expensive and technologically complex.

A common tracer utilized in PET brain imaging uses deoxyglucose labeled with the positron emitter ^{18}F-flouride ion (FDG-PET). Active neurons take up the radiolabeled deoxyglucose as readily as glucose but metabolize it much more slowly. Consequently, the radiotracer remains in the active neurons long enough for computer-generated tomographic images to be formed. This technique can be used to assess metabolic activity in a variety of functional

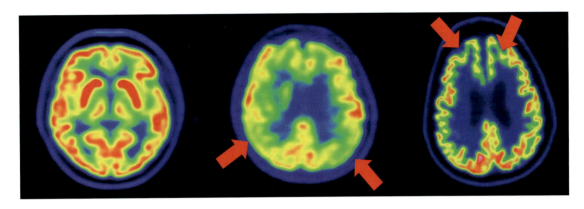

FIGURE 29.15. PET CT demonstrating (A) normal brain metabolism, (B) bilateral parietal hypometabolism in Alzheimer's disease, and (C) frontal metabolism in frontotemporal dementia.

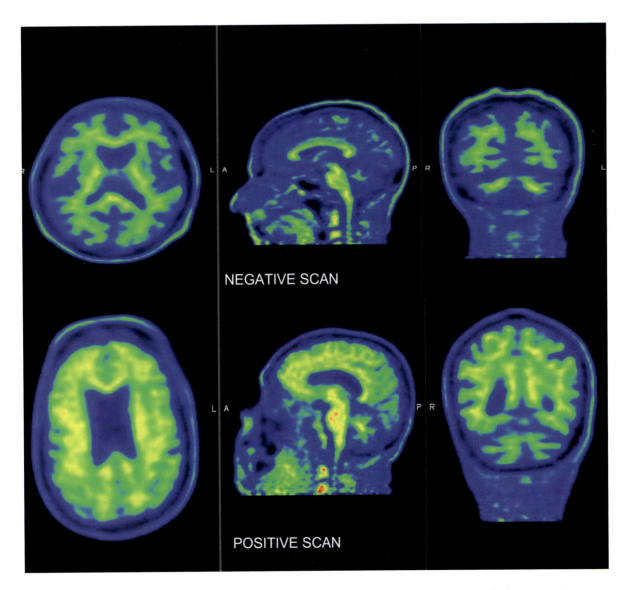

FIGURE 29.16. Amyloid scan. Amyloid PET imaging demonstrating amyloid deposition in the patient on the lower row of images.

states. Epileptic foci and areas of infarction may show up as regions of hypometabolism. In the early stages of Alzheimer's disease, there is bilateral hypometabolism in parietal cortex when no structural damage is apparent on CT or MRI, and as the disease progresses the hypometabolism spreads to more regions and larger regions of the cortex (Figure 29.15).

Many different compounds can be labeled with positron-emitting isotopes for molecular imaging, including drugs and hormones, making it possible to map out the location and density of receptors and other molecules. Amyloid imaging uses PET and amyloid tracers to detect beta amyloid plaques, a pathologic hallmark of Alzheimer's disease (Figure 29.16).

Functional MRI

Functional MRI (fMRI) is an MRI protocol that portrays areas of brain activity as changes in blood flow based on the **blood oxygen level dependent** (BOLD) technique. This technique takes advantage of the differences between the magnetic properties of oxygenated hemoglobin (oxyhemoglobin) and deoxygenated hemoglobin (deoxyhemoglobin); oxyhemoglobin is repelled away from a magnetic field, while deoxyhemoglobin is weakly attracted to magnetic fields. In metabolically active regions of the brain, neuronal activity is associated with a greater supply of oxygenated blood than is required, resulting in a higher-than-normal ratio of oxygenated to deoxygenated blood. MRI signals that reflect hemoglobin-deoxyhemoglobin ratios can provide a measure of regional changes in blood flow and reveal regional brain activation that occurs during the performance of specific perceptual, cognitive, or behavioral tasks. Because fMRI measures changes in blood oxygenation, it indirectly reflects cerebral metabolism, which indirectly reflects neural activity.

This imaging technique is used extensively in research because it utilizes readily available equipment, is completely noninvasive, and does not require exposure to radioisotopes. It also provides superior spatial resolution as compared to SPECT and PET. In clinical practice, fMRI is used primarily in presurgical patients for localizing eloquent cortex (e.g., speech, motor function) prior to brain tumor resection.

Summary

The ability to directly image structures and pathologies of the brain has evolved rapidly since the introduction of CT in the 1970s and MRI in the 1980s, with many advanced imaging techniques continuing to be developed. The technologies fall into two broad classes: structural imaging, which involves the in vivo depiction of CNS anatomy and structural abnormalities; and functional brain imaging, which measures and visualizes the physiological activities of the brain such as metabolism and perfusion, as well as imaging of specific molecules. Advances in imaging technologies have revolutionized the clinical neurosciences, allowing for more rapid and accurate diagnosis of brain injuries and diseases, precision interventions, and improved patient outcomes. A basic understanding of brain imaging techniques and interpretation is necessary for clinicians who encounter neuroimaging study reports in their work with patients with brain injuries or diseases.

Additional Reading

1. Annavarapu RN, Kathi S, Vadla VK. Non-invasive imaging modalities to study neurodegenerative diseases of aging brain. *J Chem Neuroanat.* 2019;95:54–69. doi:10.1016/j.jchemneu.2018.02.006
2. Arenaza-Urquijo EM, Przybelski SA, Lesnick TL, Graff-Radford J, Machulda MM, Knopman DS, et al. The metabolic brain signature of cognitive resilience in the 80+: beyond Alzheimer pathologies. *Brain.* 2019;142(4):1134–1147. doi:10.1093/brain/awz037
3. Bigler ED. Structural image analysis of the brain in neuropsychology using magnetic resonance imaging (MRI) techniques. *Neuropsychol Rev.* 2015;25(3):224–249. doi:10.1007/s11065-015-9290-0
4. Blumenfeld H. Introduction to clinical neuroradiology. In: Blumenfeld H, *Neuroanatomy through clinical cases.* 3rd ed., pp. 85–123. Sinauer Associates; 2021.
5. Chanraud S, Zahr N, Sullivan EV, Pfefferbaum A. MR diffusion tensor imaging: a window into white matter integrity of the working brain. *Neuropsychol Rev.* 2010;20(2):209–225. doi:10.1007/s11065-010-9129-7
6. Chen JE, Glover GH. Functional magnetic resonance imaging methods [published correction appears in *Neuropsychol Rev.* 2015 Sep;25(3):314]. *Neuropsychol Rev.* 2015;25(3):289–313. doi:10.1007/s11065-015-9294-9
7. Evans AC, Janke AL, Collins DL, Baillet S. Brain templates and atlases. *Neuroimage.* 2012;62(2):911–922. doi:10.1016/j.neuroimage.2012.01.024
8. Lebby PC. *Brain imaging: a guide for clinicians.* Oxford University Press; 2013.
9. Lockhart SN, DeCarli C. Structural imaging measures of brain aging. *Neuropsychol Rev.* 2014;24(3):271–289. doi:10.1007/s11065-014-9268-3

10. Madden DJ, Bennett IJ, Song AW. Cerebral white matter integrity and cognitive aging: contributions from diffusion tensor imaging. *Neuropsychol Rev.* 2009;19(4):415–435. doi:10.1007/s11065-009-9113-2
11. Mueller BA, Lim KO, Hemmy L, Camchong J. Diffusion MRI and its role in neuropsychology. *Neuropsychol Rev.* 2015;25(3):250–271. doi:10.1007/s11065-015-9291-z
12. Roalf DR, Gur RC. Functional brain imaging in neuropsychology over the past 25 years. *Neuropsychology.* 2017;31(8):954–971. doi:10.1037/neu0000426

30
The Neurological Examination

Suzan Uysal and Stephen Krieger

Introduction

The neurological examination is a medical evaluation of nervous system function that seeks to determine: (1) if there is evidence of a neurologic problem; (2) the possible neuroanatomical site of pathology (i.e., localize the lesion); (3) the underlying etiology; and (4) an informed treatment plan. It is organized into seven components: (1) mental status; (2) cranial nerve function; (3) motor function; (4) sensory function; (5) reflexes; (6) coordination; and (7) station and gait. The mental status component assesses cognitive and behavioral functions related to the cerebrum. The physical components primarily evaluate the sensory and motor functions of the nervous system. Several instruments are used to perform the physical neurological exam: a reflex hammer, two tuning forks (256 Hz and 128 Hz), cotton applicators with wooden shafts, an ophthalmoscope, a penlight, and a visual acuity card. Some tests use no instruments at all. The overall goal of the exam is to create a compact profile of neurologic function in a short examination period.

While the core of the exam is standard, each exam is tailored to address specific clinical questions raised by the clinical history, such as: (1) Is there evidence of a neurological disorder? (2) Is the neurological dysfunction diffuse or focal, and if focal what is the localization? (3) What pathology underlies the neurological disorder? Neurological localization involves determining which level of the nervous system is dysfunctional (e.g., cerebral cortex, subcortical white matter, deep gray matter structures, cerebellum, brainstem, spinal cord, spinal nerves, peripheral nerves, neuromuscular junction, muscles). Knowledge of the functional anatomy of the nervous system makes it possible to interpret symptoms and identify the localization of a lesion to make a correct anatomic diagnosis. Determining etiology involves generating a differential diagnosis of types of pathology that could account for the signs and symptoms. Note that the patient may be referred for additional investigative studies for the purposes of localization and/or differential diagnosis, such as neuroimaging, neurophysiological studies, laboratory studies, or neuropsychological examination. The neurologist integrates the results of the neurological examination and other studies to arrive at a diagnosis and determine a treatment plan. If the neurologist finds that the problem is not neurological in nature, they will refer the patient to an appropriate specialist.

The Clinical History

The first critical step in determining the etiology involves taking an accurate **clinical history** of the specific (observable) **signs** and (subjective) **symptoms** from the patient and/or other informant, and/or the medical record, including the nature of onset (e.g., sudden, subacute, or insidious), date/time of onset, duration, and course (i.e., static, progressive, resolving, fluctuating). The neurologist also makes behavioral observations while listening to the patient recount the history of the presenting problem, and throughout the examination. The clinical presentation of the patient therefore encompasses both the symptoms reported by the patient and the signs observed by the neurologist. Questions raised by the clinical history and behavioral observations together guide the conduct of the exam. The clinical diagnostic process involves generating a list of all conditions consistent with the signs and symptoms. The exam seeks to narrow this list as much as possible.

The Mental Status Exam

The mental status assessment is focused on cortical function. It evaluates the patient's level of alertness, orientation (to person, place, and time), and higher-level aspects of brain function, such as attention, language, memory, and thought process. Mental status is always assessed first, as it is an essential part of the clinical history that guides the conduct of rest of the examination.

Mental status is assessed by observing the patient in conversation and often is supplemented by standardized **mental status screening tests** such as the Mini-Mental Status Examination (MMSE) or the Montreal Cognitive Assessment (MoCA). This component of the neurological exam is aimed at detecting delirium, dementia, and other cognitive abnormalities.

It is necessary to establish that the patient is alert and oriented before assessing higher cognitive functions. **Alertness** is the degree to which the patient is awake and responsive. **Orientation** reflects the patient's knowledge of person (*What is your name?*), place (*Where are we?*), and time (*What is today's date?*). Normal alertness and orientation on examination is typically abbreviated as **AOX3**. **Level of consciousness** encompasses arousability and responsiveness to environmental stimuli (see Chapter 25, "Disorders of Consciousness"). Is the patient conscious and fully responsive? Can they remain focused in conversation? There is a spectrum of states of altered level of consciousness, the mildest reflecting a disturbance of awareness (clouding of consciousness and confusional state), and the more severe reflecting disturbances of wakefulness and awareness based on the degree of response to stimuli (somnolence, obtundation, stupor, and coma). A comatose person demonstrates no volitional response, even to painful or noxious stimulation such as pressure applied to the nail bed.

Delirium, also referred to as acute confusional state, is characterized by a core disturbance in attention (i.e., the ability to direct, focus, sustain, and shift attention) and orientation, with abrupt onset and fluctuation from agitation to lethargy over the course of a day. It is a clinical syndrome that reflects disturbance in the physiological function of the brain. It can be caused by a wide variety of toxic and metabolic factors; thus, when delirium is identified it should trigger a search for the underlying physiological condition.

Other localizable impairments of mental status include aphasias, apraxias, and neglect syndromes. Abnormal findings on the mental status component of the neurological exam often prompt neurologists to seek further delineation of the problem by referring the patient for **neuropsychological assessment** to elucidate the nature and severity of cognitive deficits for the purposes of diagnostic clarification and/or treatment planning. Neuropsychological assessment is an objective, quantitative evaluation of cognition and behavior.

The Cranial Nerve Exam

The cranial nerves emanate from the cranium carrying sensory afferents from the periphery to the cranial nerve nuclei, and motor efferents from the cranial nerve nuclei to the effector organs (i.e., muscles and glands) of the periphery (discussed in greater detail in Chapter 6, "The Peripheral Nervous System," and Chapter 8, "The Brainstem"). The cranial nerves serve essential functions for human life, and impairments in their function suggests a disease process affecting the brainstem or skull base, and therefore are almost always clinically significant.

To summarize, there are 12 pairs of cranial nerves: olfactory (CN1), optic (CN2), oculomotor (CN3), trochlear (CN4), trigeminal (CN5), abducens (CN6), facial (CN7), vestibulocochlear (CN8), glossopharyngeal (CN9), vagus (CN10), accessory (CN11), and hypoglossal (CN12). They serve three general functions: (1) motor and sensory innervation of the head and neck; (2) innervation of the special sense organs (olfaction, vision, audition, vestibular, gustation); and (3) parasympathetic innervation. Some cranial nerves are composed entirely of sensory afferent fibers (CN1, CN2, CN8); some are composed entirely of motor efferent fibers (CN3, CN4, CN6, CN11, CN12); and some are mixed, possessing both afferent and efferent fibers (CN5, CN7, CN9, CN10). Some cranial nerves carry parasympathetic fibers of the autonomic system (CN3, CN7, CN9, CN10).

Lesions of the cranial nerves that carry motor efferents or the cranial nerve nuclei that give rise to those efferents result in cranial nerve **palsies**, weakness or paralysis of the muscles supplied by the nerve. These are facial nerve palsy, the gaze palsies, and bulbar palsy. **Facial nerve palsy** results from dysfunction of the facial nerve (CN7), which innervates the muscles of facial expression. The **gaze palsies** result from dysfunction of the oculomotor (CN3), trochlear (CN4), or abducens (CN6) nerves that innervate the extraocular muscles that control eye movement and position. Impaired function of the lower cranial nerves that carry efferents originating from within the medulla, namely the glossopharyngeal (CN9), vagus (CN10), and hypoglossal (CN12) nerves that provide motor control of the pharynx, larynx (voice box), and tongue, lead to dysphagia, dysarthria, and dysphonia, respectively. **Dysphagia** is a swallowing disorder characterized by difficulty moving and manipulating food, liquids, medications, or secretions through the mouth, throat, and/or esophagus into the stomach. Manifestations include choking on liquids, difficulty handling secretions, and difficulty chewing (which can lead to aspiration pneumonia and be life-threatening). **Dysarthria** is a speech disorder due to weakness or incoordination of the oral muscles (jaw, lips, tongue, palate/roof of the mouth) that results in slurred speech. **Dysphonia** is a voice disorder in which there are abnormalities in pitch (too high, too low, never changing), volume (hyperphonia or hypophonia), or tone/quality (harsh, hoarse, breathy, nasal).

The cranial nerves constitute the afferent and efferent links of reflex arcs in the brainstem. Reflexes involving cranial nerves that are often assessed in the cranial nerve exam are the pupillary light reflex (afferent CN2, efferent CN3), accommodation reflex (afferent CN2, efferent CN3), corneal reflex (afferent CN5, efferent CN7), vestibulo-ocular reflex (afferent CN8, efferents CN3, CN4 and CN6), and gag reflex (afferent CN9, efferent CN10).

The cranial nerve exam is a systematic examination of cranial nerve function. It is conducted in sequential fashion, providing "top to bottom" information about the integrity of the brainstem at the levels of the midbrain, pons, and medulla by the **4-4-4 rule**: (1) the nuclei of first four cranial nerve pairs (CN1–CN4) are located the midbrain or higher subcortical structures; (2) the nuclei of the second four cranial nerve pairs (CN5–CN8) are located in the pons; and (3) the nuclei of the last four cranial nerve pairs (CN9–CN12) are located in the medulla or upper cervical cord (see Figure 6.6 in Chapter 6). The exam does not simply assess cranial nerve function in numerical order from 1–12, however; it also incorporates functional groups (e.g., those controlling voluntary eye movements, those serving as the afferent and efferent links of a reflex arc). Since the cranial nerves are paired, the examiner compares the function of each. Asymmetric function is indicative of unilateral pathology.

For the purposes of neurological examination, functions served by the cranial nerves may be summarized as follows:

- Smell: olfactory nerve (CN1)
- Visual acuity, visual fields, ocular fundi: optic nerve (CN2)
- Pupillary reactions: optic and oculomotor nerves (CN2 and CN3)
- Extraocular movements: oculomotor, trochlear, and abducens nerves (CN3, CN4, and CN6)
- Facial sensation, movement of the jaw, corneal reflex (sensory): trigeminal nerve (CN5)
- Facial movements, gustation, and corneal reflex (motor): facial nerve (CN7)
- Hearing and balance: vestibulocochlear nerve (CN8)
- Swallowing, elevation of the palate, gag reflex, and gustation: glossopharyngeal and vagus nerves (CN9 and CN10)
- Shrugging of the shoulders and turning the head: accessory nerve (CN11)
- Movement and protrusion of the tongue: hypoglossal nerve (CN12).

CN1 Testing

CN1 function is not tested routinely during the screening neurologic examination, but it should be assessed when patients complain of loss of smell, in cases of closed head injury when shearing forces can sever olfactory nerve fibers as they traverse the cribriform plate of the skull base, and in other cases where there is any suspicion of subfrontal pathology that may affect the olfactory bulbs or olfactory cortex (e.g., from toxin inhalation).

The olfactory nerve is tested using aromatic nonirritant materials such as coffee, clove, tobacco, or soap; irritating substances such as alcohol or ammonia are not used because they stimulate CN5 (the trigeminal nerve, maxillary division). One nostril is tested at a time with the eyes closed.

Trauma of the olfactory nerves at the point of their traversing the cribriform plate and/or orbitofrontal lesions affecting the olfactory bulbs, tracts, and/or olfactory cortex can result in **anosmia**, a partial or total loss of the sense of smell. This nerve does not cross the midline; therefore, a unilateral nerve lesion results in ipsilateral anosmia. Loss of smell may also be a symptom of neurodegenerative and dementing illnesses.

CN2 Testing

The optic nerve is responsible for input of visual information to the brain. A complete lesion of an optic nerve results in monocular blindness of the affected eye. Evaluation of the optic nerves and higher visual pathways includes fundoscopic examination of the interior surface of the eyes, visual field testing, and visual acuity testing.

The **ophthalmoscope (funduscope)** is used to examine the **fundus**, the interior surface of the eye (retina, optic disc, macula, fovea, and posterior pole), to detect pathology such as papilledema (optic disc swelling caused by increased intracranial pressure), optic atrophy (optic neuropathy), vascular changes of the retina, and other abnormalities. Monocular visual field testing using static visual perimetry, which measures thresholds to spots of light, sampling the full extent of the visual field of each eye, may further elucidate issues discovered through ophthalmoscopic examination; this and some disorders of the optic nerve may be referred to the ophthalmologist.

Preliminary testing of the visual fields is done by quadrant using finger movements. This is called **confrontation visual field testing**. The examiner stands directly across from the patient and, with their arms outstretched, gradually moves their fingers, in a plane exactly halfway between the examiner and the patient, from beyond the examiner's visual field (i.e., where they are not seen) into their peripheral vision, and then toward their central vision. The patient is instructed to indicate when they first detect the fingers within their visual field. The objective is to determine whether the patient has full visual fields, and the location and extent of visual field defects if present; the examiner uses their own visual field as a point of reference and "maps out" the patient's visual field defect. A visual field defect present only in one eye suggests a lesion within that eye or the optic nerve. A visual field defect present in the same location in both eyes (e.g., a homonymous hemianopsia) suggests a unilateral lesion in the visual pathway after the optic chiasm, at the optic tract, optic radiation, or primary visual cortex (see Chapter 14, "The Occipital Lobes and Visual Processing"). If there is suspicion or evidence of a field defect, the patient will be referred for more sensitive in-depth visual field testing. Normal visual fields are noted as **VFFTC** (*Visual Fields Full to Confrontation*).

If there is no field defect, unilateral spatial neglect is tested by **double simultaneous stimulation**, the simultaneous presentation of two stimuli, each to the opposite visual field. In patients with extinction, the patient reports single stimuli throughout the visual field, but when presented with bilateral stimuli, the patient detects stimuli in only one-half of the visual field and extinguishes stimuli in the opposite visual field. Unilateral spatial neglect and extinction most commonly occur with lesions of the right parietal lobe association cortex (see Chapter 15, "The Parietal Lobes and Associated Disorders").

Mild deficits of visual acuity are tested using the common **Snellen chart**, which displays letters of decreasing size to determine the smallest letters that can be read, or by determining if the patient can read newsprint at a normal reading distance. With severe deficits, light perception, movement perception, and finger counting may be used. In most cases, the neurologist will refer a patient with reduced visual acuity to an ophthalmologist.

CN2 and CN3 Testing

The optic (CN2) and oculomotor (CN3) nerves are tested together by examining pupil size, pupil shape, pupil reactivity to light, and accommodation. Normal pupils are equal in size, round, and reactive to light and accommodation; this is typically abbreviated as **PERRLA** (*Pupils Equal, Round, Reactive to Light and Accommodation*). The notation **PERRL** indicates that accommodation was not tested.

PUPIL SIZE

Pupil size is the product of the balance between sympathetic innervation of the iris dilator muscles and parasympathetic innervation of the iris constrictor muscles. Under dim light conditions or when the amount of light falling on the retina decreases, sympathetic activity dilates the pupil. Under bright light conditions or when the amount of light falling on the retina increases, parasympathetic activity constricts the pupil.

Normal pupil size is approximately 2–4 mm under normal room light conditions. Pupil size is examined as the patient fixates on a distant point. Asymmetry between the two pupils (**anisocoria**) reflects dysfunction of either parasympathetic (CN3) innervation of the iris sphincter muscle that causes pupil constriction, or sympathetic innervation of the iris dilator muscle that causes pupil dilation. If, for example, the *right pupil is larger* than the left under room light conditions, this raises the question of whether there is a lesion producing sympathetic denervation of the left eye or a lesion producing parasympathetic denervation of the right eye. This is assessed by comparing the asymmetry under normal room light and dim light conditions. If the asymmetry is greatest in dim light, then sympathetic innervation of the left eye is disrupted; the lesion prevents the left eye from dilating in dim light. If the asymmetry is greatest in normal room light, then parasympathetic innervation of the right eye is disrupted; the lesion prevents pupillary constriction that normally is elicited by bright light and the pupil is abnormally dilated (**mydriasis**). Such unilateral pupil dilation occurs with ipsilateral lesions of CN3 (oculomotor nerve). Unilateral compression of CN3 can be caused by an aneurysm or uncal herniation due to a mass (space-occupying) lesion. These are archetypal examples of how a cranial nerve deficit can herald a life-threatening condition.

PUPILLARY LIGHT REFLEX

The pupillary light reflex is tested by shining a penlight into one eye and observing the reactions of both eyes. The response of the ipsilateral eye is the direct pupillary light reflex; the response of the contralateral eye is the consensual pupillary reflex. Normally the direct and consensual responses are equal.

The neural circuit mediating the pupillary light reflex consists of a three-neuron chain (Figure 30.1). The afferent link of the reflex pathway consists of optic tract fibers that project from each eye to the midbrain pretectal area bilaterally. Second-order neurons from each pretectal area in turn project to the midbrain **Edinger-Westphal nuclei** (the origin of preganglionic parasympathetic axons in the oculomotor nerve) bilaterally. The efferent link of the reflex pathway consists of oculomotor nerve fibers that provide parasympathetic innervation driving the pupillary constrictor muscles ipsilaterally. If all links in the pathways are intact, shining a light in one eye results in constriction of both pupils at an equal rate and to a similar degree.

The degree of dilation or constriction of the pupils is determined by the level of illumination registered in the brain from input to both eyes. In the case of unilateral lesions of the afferent arc (CN2), when light is shown in the affected eye, there is no pupillary response to light in either eye. This is referred to as an **afferent pupillary defect** or a **Marcus-Gunn pupil**. This defect is most apparent when rapidly "swinging" a penlight back and forth between the two eyes. If both optic nerves are normal, there is no change in pupil diameter as the light swings from one eye to the other because the total registered illumination is constant. If one optic nerve has a lesion, pupil diameter changes as the light moves from eye to eye. When the light moves from the normal eye to the abnormal eye (i.e., from more to less registered illumination), the pupils dilate; when the light moves from the abnormal eye to the normal eye (i.e., from less to more illumination), the pupils constrict.

Unilateral lesions of the efferent arc (CN3) result in failure of the pupillary light reflex in the affected eye when light is shone in either eye due to disruption of the parasympathetic innervation to the iris. It also results in anisocoria that is greatest in bright light, and **ptosis** (drooping) of the upper eyelid.

Since the neural circuitry mediating the pupillary light reflex does not involve the lateral geniculate nucleus (LGN), optic radiations, or visual cortex, patients with cerebral blindness have intact pupillary light reflexes.

In patients who are unconscious after head trauma, examination of the size of the pupils and their reaction to light has great importance. The intracranial course of the oculomotor nerve makes it especially vulnerable in cases of temporal herniation caused by increased intracranial pressure. Because the parasympathetic fibers are superficial in the nerve, slowness of the pupillary light reflex is an early sign of third nerve compression.

ACCOMMODATION REFLEX

When an observer visually fixates on a near object or an object that is moving toward them, the eyes converge, the lenses become more convex, and the pupils constrict (**miosis**). This coordinated response of the convergence eye movements, lenses, and pupils is known as the **accommodation reflex**, also known as the **near triad**. This reflex is tested by observing the pupillary response as the eyes converge when the patient visually follows an object (typically the examiner's finger) as it moves from a distant point toward the nose to a point about 6 inches from the eyes. As gaze shifts from the far point to the near point, the eyes converge and there should be an associated pupillary constriction that is equal in both eyes.

The afferent link of the accommodation reflex consists of optic nerve (CN2) fibers. The efferent link originates from the Edinger-Westphal nucleus of the oculomotor nerve (CN3), which initiates ciliary muscle contraction, causing the lens to become more convex.

The pupillary light and lens accommodation reflexes both utilize the optic nerve (CN2) as the afferent link and the Edinger-Westphal nucleus and oculomotor nerve

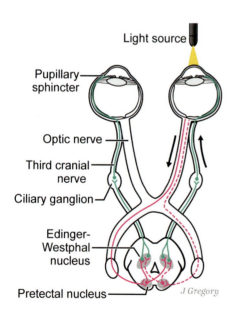

FIGURE 30.1. The neural circuit mediating the pupillary light reflex consists of a three-neuron chain. The afferent link of the reflex pathway consists of optic tract fibers that project from each eye to the midbrain pretectal nuclei bilaterally. Second-order neurons project to the midbrain Edinger-Westphal nuclei bilaterally. The efferent link of the reflex pathway consists of oculomotor nerve fibers that provide parasympathetic innervation driving the pupillary constrictor muscles ipsilaterally. If all links in the pathways are intact, shining a light in one eye results in constriction of both pupils at an equal rate and to a similar degree.

(CN3) fibers as the efferent link, but the pathways differ. The pathway of the accommodation reflex is more complex and extensive than that of the light reflex and includes the occipital cortex. Additionally, within the Edinger-Westphal nucleus the pathways mediating the near triad and the light reaction are segregated. Consequently, there may be dissociation between light reaction and accommodation reflexes with lesions in the pathway beyond CN2.

CN3, CN4, and CN6 (Eye Movement) Testing

Six **extrinsic (extraocular) eye muscles** control eye movements: four rectus muscles (superior, inferior, medial, and lateral) and two oblique muscles (superior and inferior). These six muscles function as three antagonistic pairs: the lateral and medial rectus muscles, the superior and inferior rectus muscles, and the superior and inferior oblique muscles. The position and movement of each eye is determined by the pattern of contraction of all six extrinsic muscles. Furthermore, movement of the two eyes is coordinated so that the eyes move together (**yoked** or **conjugate eye movements**).

The eyes move along three axes: horizontal, vertical, and anterior-posterior. Rotation of the eye around the horizontal axis produces elevation (upward eye movements) and depression (downward eye movements). Rotation of the eye around the vertical axis produces **abduction** (movement laterally/temporally, away from midline) and **adduction** (movement medially/nasally, toward the midline). Rotation of the eye around the anterior-posterior axis produces torsional movements of **intorsion** (bringing the top of the eye toward the nose) and **extorsion** (bringing the top of the eye away from the nose). It is less obvious to us that the eyes can move in a torsional direction, as compared to vertical and horizontal eye movements, but this ability is demonstrated by looking in a mirror and keeping your line of sight fixed on your eyes as you tilt your head left and right. As the head tilts, the position of the eyes relative to the world remains stable (i.e., the top of the eye stays on top) because the eyes make torsional movements within the head.

The six extraocular muscles are controlled by three cranial nerves: CN3 (oculomotor), CN4 (trochlear), and CN6 (abducens). The trochlear nerve (CN4) controls the superior oblique, the abducens (CN6) controls the lateral rectus (the abducens abducts), and the oculomotor nerve (CN3) controls the remaining four extraocular muscles (medial rectus, superior rectus, inferior rectus, inferior oblique). The function of these cranial nerves is assessed by observing for abnormalities in eye position at rest and testing eye movements (smooth pursuit, convergence, and saccades). Abnormalities consist of **dysconjugate gaze** (strabismus/misalignment of the eyes) and **nystagmus**, a rhythmic oscillation of the eyes that is normal only with extreme lateral gaze and the optokinetic reflex (see "Eye Movement Reflexes," below).

CRANIAL NERVE GAZE PALSIES

Lesions of the cranial nerve nuclei or nerves controlling the extraocular muscles cause weakness in the specific extraocular muscle(s) innervated. Since there is resting tone in all the eye muscles, isolated weakness in one muscle results in dysconugate gaze, with inability to direct both eyes in the same direction and deviation of the affected eye due to the unopposed action of the remaining extraocular muscles (Figure 30.2). Dysconjugate

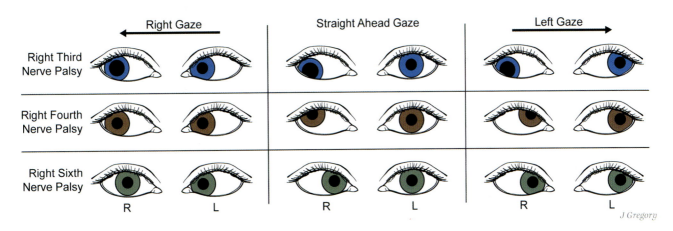

FIGURE 30.2. Cranial nerve gaze palsies viewed from the vantage point of an examiner. An oculomotor (CN3) palsy is characterized by an eye that is abducted and depressed ("down and out") when looking straight ahead. On lateral gaze in the direction opposite to the affected eye (e.g., right eye looking left), the affected eye cannot adduct. A trochlear (CN4) palsy is characterized by upward drift of the eye when looking straight ahead. On lateral gaze in the direction opposite to the affected eye (e.g., right eye looking left), the eye can adduct but the upward deviation increases. An abducens (CN6) palsy is characterized by an eye that is adducted when looking straight ahead. On lateral gaze in the same direction as the affected eye (e.g., right eye looking right), the affected eye cannot abduct.

gaze also results in **diplopia** (double vision) when both eyes are opened; the person may attempt to ameliorate the double vision by adjusting their head position so that their "good eye" lines up with the affected eye. Diplopia may be characterized as horizontal (images separated horizontally), vertical (images separated vertically), or diagonal (one image slightly tilted). The image separation may be maximal in a particular direction of gaze. The range of eye movements is examined by asking the patient to follow a target (usually the examiner's finger or a penlight) with the eyes only, through six cardinal fields of gaze (upper right, right, lower right, left, upper left, lower left).

Unilateral lesions of the oculomotor nerve (CN3) or nucleus result in **oculomotor palsy**, which is characterized by an eye that is abducted and depressed ("down and out") on gaze straight ahead, due to the unopposed action of the lateral rectus (controlled by the abducens) and superior oblique (controlled by the trochlear). On lateral gaze in the direction opposite to the affected eye, the affected eye cannot adduct. This produces a predominantly horizontal diplopia. Lesions of the oculomotor nucleus or nerve also result in ptosis (drooping of the upper eyelid), a "blown" (dilated) pupil, and loss of pupillary light and accommodation reflexes.

Unilateral lesions of the trochlear nerve (CN4) or nucleus, which innervates the superior oblique, result in **trochlear palsy** in which the affected eye drifts upward relative to the normal eye on gaze straight ahead. On lateral gaze in the direction opposite to the affected eye, the eye can adduct but the upward deviation increases. This results in a diagonal diplopia (skew diplopia); the patient sees two visual fields separated vertically, one from each eye. The diplopia is most pronounced when the person attempts to use convergent down-gaze, such as when reading or walking downstairs. To compensate, rather than making down-gaze movements, patients will tilt the head forward by tucking the chin, so that the visual fields from the eyes are aligned and the images from both eyes can align properly.

Unilateral lesions of the abducens nerve (CN6) or nucleus, which innervates the lateral rectus muscle, result in **abducens palsy**, which is characterized by an eye that is adducted on gaze straight ahead. On lateral gaze in the direction of the affected eye, the affected eye cannot move beyond the midline (abduct). This produces a horizontal diplopia. A person with abducens palsy usually keeps the head turned somewhat to the side of the lesion to compensate for the loss of lateral motion of the eye. Because the abducens runs a long course on the ventral surface of the pons, it is particularly vulnerable to increased intracranial pressure, but dysfunction, especially when bilateral, is not localizing (i.e., it does not necessarily suggest a pontine lesion).

CENTRAL CONTROL OF EYE MOVEMENTS

Eye movements direct the fovea to visual stimuli and maintain foveal fixation during target movement and head movement. This requires precise coordination among the motor neurons controlled by CN3, CN4, and CN6.

There are two general types of eye movements, conjugate and vergence (non-conjugate). With **conjugate movements**, the two eyes move the same amount in the same direction. There are two main types of conjugate eye movements, smooth pursuit and saccadic. **Smooth pursuit eye movements** are smooth movements used when visually tracking an object that is moving at a fixed distance from the observer. **Saccadic eye movements** are fast, step-like movements in which the eyes move quickly from one point of fixation to another. These are the eye movements used to explore the world visually, such as when scanning scenes, pictures, and reading. With **vergence movements**, the two eyes move in opposite directions when the point of fixation moves toward or away from the observer. Conjugate and vergence movements are smoothly coordinated so that images of the outside world fall on the two retinas in proper binocular registration.

Smooth pursuit eye movements are tested by asking the patient to follow the examiner's finger (which is held at least 2 feet from the patient's eyes to minimize convergence) without moving their head, as it moves through the full range of six cardinal directions of gaze (i.e., the examiner moves their finger in an H pattern), representing maximal individual muscle strength: up and out (superior rectus), up and in (inferior oblique), lateral movement/abduction (lateral rectus), medial movement/adduction (medial rectus), down and in (superior oblique), and down and out (inferior rectus).

Convergence movements are tested by asking the patient to visually fixate on an object (e.g., the examiner's finger) as it moves slowly toward a point between the patient's eyes. Convergence is produced by the medial recti, and divergence is produced by the lateral recti.

Saccadic eye movements are tested by having the examiner hold their hands 1 foot apart, about 1.5–2 feet in front of the patient, and asking the patient to look from one hand to the other.

The neural circuitry underlying the coordination of movement between the two eyes is complex and depends on multiple structures and pathways that influence the activity of the oculomotor, trochlear, and abducens nuclei. This is accomplished by (1) a network of internuclear neurons, (2) brainstem premotor networks, (3) cortical eye movement centers, and (4) inputs from cerebellum, basal ganglia, and vestibular structures. **Internuclear neurons** make connections between oculomotor, trochlear, and abducens nuclei, and are carried in the **medial longitudinal fasciculus**. The **premotor networks** are functional units of neurons at the brainstem level, consisting of the **horizontal gaze center** (including the paramedian pontine reticular formation and the abducens nucleus of the pons) controlling horizontal eye movements, and the **vertical gaze center** (i.e., the rostral interstitial nucleus of the medial longitudinal fasciculus of the midbrain) controlling vertical eye movements. The **cortical eye movement centers** consist of the **parietal eye fields** controlling visual fixation, and the **frontal eye fields** controlling voluntary horizontal saccadic eye movements. The ocular motor system therefore may be divided by anatomic location into **infranuclear, nuclear, internuclear,** and **supranuclear** components. Eye movement testing allows the neurologist to distinguish the level of the disturbance.

Conjugate gaze palsies affect the ability to move both eyes in the same direction. These palsies can affect gaze in a horizontal or vertical direction. **Vertical gaze palsy** can selectively affect gaze in an upward or downward direction; it is a prominent feature of progressive supranuclear palsy.

EYE MOVEMENT REFLEXES

Eye movement reflexes may also be assessed, particularly in comatose patients and infants who cannot follow commands. These reflexes function to stabilize the retinal image in response to movement of the head (the vestibulo-ocular reflex) and movement of the whole visual field relative to the head (the optokinetic reflex).

The **vestibulo-ocular reflex**, also known as the **oculocephalic reflex** and the **doll's eye reflex**, occurs in response to head movement (i.e., changes in head acceleration), which stimulates the vestibular system. Head movements cause the image of the world to sweep across the retina. Vestibulo-ocular reflexive eye movements are in the opposite direction of and compensate for the head movement, in the service of stabilizing the retinal image and allowing fixation on an object while moving. This reflex is tested in patients with lowered states of consciousness by holding the patient's eyes open and passively moving their head from side to side rapidly (**passive movements** are movements of a patient's body part that are performed by the examiner, without voluntary movement on the part of the patient). The reflex is intact (**positive doll's eye sign**) if the eyes move in the direction opposite to that in which the head is moved, indicating preserved function of the brainstem (Figure 30.3). The reflex is absent (**negative doll's eye sign**) if the eyes remain fixed mid-orbit, indicating brainstem injury affecting the gaze centers at the level of the midbrain and pons. In conscious patients the reflex is suppressed by visual fixation.

Optokinetic reflex eye movements occur in response to movements of the whole visual field relative to the

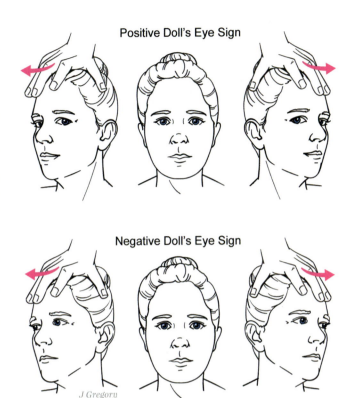

FIGURE 30.3. The vestibulo-ocular reflex. The reflex is intact (positive doll's eye sign) if the eyes move in the direction opposite to that in which the head is passively moved, indicating preserved function of the brainstem gaze centers. The reflex is absent (negative doll's eye sign) if the eyes remain fixed mid-orbit, indicating brainstem injury affecting the gaze centers at the level of the midbrain and pons.

head, as when looking out of the side window of a moving vehicle such as a car or train. These eye movements also are intended to stabilize the retinal image (i.e., they are an expression of the fixation tendency). Smooth pursuit eye movements make up the slow tracking phase, and saccades make up the fast phase. This reflex is examined by testing for **optokinetic nystagmus**, a normal form of nystagmus that is elicited by presenting a rotating striped drum within the visual field of the patient. The gaze is fixed automatically on one of the stripes and follows it by smooth pursuit until the stripe leaves the visual field, at which point a saccade is made to fix the gaze on the next target; the cycle of smooth pursuit and saccade is repeated.

CN5 Testing

The trigeminal nerve (CN5) has three branches that carry sensory information from the face: **ophthalmic** (V1), **maxillary** (V2), and **mandibular** (V3). The motor root of the trigeminal nerve travels exclusively in the mandibular division; it innervates the **muscles of mastication** that are used for chewing.

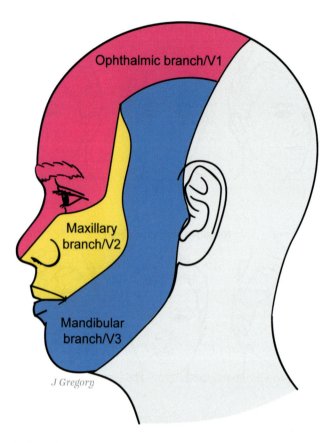

FIGURE 30.4. Dermatomes of the trigeminal nerve: forehead (ophthalmic branch/V1), cheek (maxillary branch/V2), and chin (mandibular branch/V3).

The sensory component of the trigeminal nerve is tested by light touch with a cotton wisp and light pinprick applied to the face with the patient's eyes closed. The three dermatomes are tested: forehead (V1), cheek (V2), and chin (V3) (Figure 30.4). The patient is asked to indicate when they detect a stimulus, and with pinprick, whether the sensation is sharp or dull. The neurologist compares the strength of sensation between the two sides of the face. Diminished sensation occurs with lesions anywhere along the pathway, from the nerve to the primary somatosensory cortex. This includes the ipsilateral nerve or main sensory nucleus of the trigeminal, or the contralateral ascending sensory trigeminothalamic tract, ventroposterior medial nucleus of the thalamus, thalamic radiation, or somatosensory cortex. A nerve lesion results in a loss of sensation in one of the ipsilateral dermatomes. A cerebral lesion (i.e., above the level of the trigeminothalamic tract decussation) results in loss of sensation on the opposite side of the face.

The motor component of this nerve is assessed by asking the patient to clench the teeth while the examiner palpates the **masticatory (masseter) muscles**. If both sides of the face are innervated normally, the masseters have equal bulk and are both capable of contraction. The motor component of this nerve is also assessed by asking the patient to open their mouth against resistance applied by the examiner at the base of the patient's chin. Normally the mandible travels vertically, but with a nuclear or nerve lesion, the jaw will deviate to the same side as the lesion due to weakness on that side.

CN7 Testing

The facial nerve (CN7) provides innervation to the muscles of facial expression and carries taste afferents from the anterior two-thirds of the tongue.

Motor function of the facial nerve is initially assessed by observation of the face at rest, during spontaneous emotional expression (smiling, laughing, frowning), and during talking. Facial asymmetries such as downturned mouth, sagging cheek (flattened **nasolabial fold**), or sagging lower eyelid (widened **palpebral fissure**) are indicators of facial muscle weakness.

Facial muscle strength is assessed by having the patient wrinkle their forehead by looking upward, raise their eyebrows, smile showing their teeth, frown, and puff out both cheeks. The examiner observes for asymmetry and difficulty performing these movements. Subtle unilateral weakness results in a lag in the movement on one side of the face, while more severe weakness results in both lag and diminished amplitude of the movement. Strength against a resistance is tested by having the patient close their eyes while the examiner attempts to open them, and keep their cheeks puffed out against pressure. Responses are normally symmetric.

When one entire side of the face is paralyzed, the lesion involves either the ipsilateral facial motor nucleus or facial nerve (i.e., it is "peripheral"). This is known as **facial nerve palsy**. Idiopathic facial nerve palsy, also known as **Bell's palsy**, is an acute inflammatory reaction of CN7 that is characterized by pain and paralysis of the muscles controlling facial expression. The etiology is unknown; most cases recover spontaneously.

When movement of the forehead is spared on the paretic side of the face, the lesion is above the level of the facial motor nucleus (i.e., central). This nucleus has a dorsal region which innervates the muscles of the upper face and a ventral region which innervates the muscles of the lower face. The dorsal region receives cortical input from primary motor cortex bilaterally, while the ventral region receives only contralateral input. Thus, a supranuclear (upper motor neuron) corticobulbar tract lesion affecting fibers that control facial movements produces clear-cut paresis only in the muscles of the contralateral lower quadrant of the face. By contrast, a lower motor neuron lesion to the facial motor nucleus or nerve results in paralysis of all facial muscles on the same side of the injury, as the nerve is the "final common pathway" to the muscles of facial movement on that side.

Taste on the anterior two-thirds of the tongue is tested by having the patient identify familiar compounds such as a sugar and salt solution placed on each side of the tip of the tongue using a cotton applicator.

CN5 and CN7 Testing

The trigeminal (CN5, V1 division) and facial (CN7) nerves are tested together by the **corneal (blink) reflex**. The trigeminal provides the afferent link and the facial nerve provides the efferent link of this reflex.

The corneal reflex is an involuntary blink response elicited by touching the cornea. Its evolutionary purpose is to protect the eyes from foreign bodies. This reflex is tested by having the patient look up and away from the examiner while the examiner lightly touches the **sclera** of the eye (i.e., the white) with a sterile cotton swab or wisp of cotton (avoiding contact with the pupil and iris). Stimulation of the cornea of one eye normally elicits a blink response in both eyes. The response of the eye touched is the **direct blink response**, and the response of the opposite eye is the **consensual blink response**. Damage to the afferent (CN5) link results in an absent corneal reflex in both eyes when the affected side is stimulated. Damage to the efferent link (CN7) leads to a decreased blink on one side, regardless of the side of corneal stimulation.

CN8 Testing

The vestibulocochlear nerve (CN8) mediates hearing and balance.

Auditory function is assessed grossly by the patient's ability to hear fingertips rubbing. The patient sits with their eyes closed and the examiner stands in front of them, with their hands near the patient's ears (one on each side) and out of the patient's sight. The examiner rubs their fingers together on one side, at first very gently, and then with increasing force until the patient detects the sound. Both ears are tested.

If a deficit is detected, the neurologist must determine whether it is due to conduction deafness or sensorineural deafness. **Conduction deafness** is an inability of sound waves to pass through the external ear or middle ear. This occurs with obstruction of the external ear (typically due to wax buildup) or damage to the eardrum (tympanic membrane) or middle ear ossicles (e.g., otosclerosis). **Sensorineural deafness** results from lesions of the cochlea, cochlear nerve, or auditory pathways up to the brainstem. The Weber test and Rinne test are performed to differentiate conduction deafness from sensorineural deafness.

In the **Weber test**, the base of a vibrating tuning fork (typically 256 or 512 Hz) is held against the vertex of the head (i.e., midline top of the head, near the midpoint of the sagittal suture). In normal hearing (or bilateral, symmetrical hearing loss of either type), the sound is equal in both ears. If the sound is unequal between the two ears, this indicates the presence of a unilateral hearing loss, either conduction or sensorineural. With unilateral conduction deafness, bone conduction is better than air conduction and the sound is louder in the affected ear because distraction from external sounds (air conduction) is reduced in that ear. With unilateral nerve deafness, both bone and air conduction are impaired, and the sound is louder in the normal ear. This test also can be accomplished without a tuning fork by simply asking the patient to hum. In conduction deafness, the affected ear hears the hum louder. (This effect can be demonstrated by humming while sticking your finger in one ear, simulating conduction deafness by obstructing the external ear; the ear with the finger in it hears the hum louder). In sensorineural deafness, the normal ear hears the hum louder.

A positive Weber test is followed up with a Rinne test to compare air conduction with bone conduction. In the **Rinne test**, a vibrating tuning fork (again typically 256 or 512 Hz) is placed on the mastoid process behind each ear for 2–3 seconds, and then rapidly moved to just outside the ear and kept there for an additional 2–3 seconds. The patient reports in which position they hear the tuning fork louder. If the patient hears the sound louder when the tuning fork is next to the external auditory meatus, air conduction is better than bone conduction, which is normal (i.e., there is no significant conductive hearing loss). If the patient hears the sound louder when the tuning fork is held against the mastoid process, bone conduction is better than air conduction, and there is a significant conductive hearing loss in that ear.

Vestibular function is assessed indirectly by observing for nystagmus and gait instability. Vestibular function can also be tested with the **turning test**, in which the patient is asked to march in place with their arms outstretched and eyes closed. Normally, the patient will remain in the same position, but if there is a vestibular lesion the patient will gradually and unintentionally turn toward the side of the lesion.

Patients with reduced hearing or vestibular function may be referred to an otolaryngologist for in-depth audiological testing or vestibular testing. One important cause of sensorineural hearing loss and/or vestibular dysfunction is an **acoustic neuroma** (Schwann cell tumor of CN8).

CN9 and CN10 Testing

The glossopharyngeal (CN9) and vagus (CN10) nerves are tested by examining the soft palate and uvula, as well as assessing the gag reflex.

The patient is asked to open their mouth and say "ahhh." Normally, the soft palate and uvula elevate symmetrically. The neurologist determines whether palatal elevation is symmetric or asymmetric by observing whether the uvula is at midline or deviates. A lesion of the vagus nerve (CN10) affects the motor branches of the pharynx and soft palate and produces a 10th nerve palsy, characterized by deviation of the uvula and the posterior pharyngeal wall away from the lesion.

The **gag reflex** (**pharyngeal reflex**) is tested on the left and the right sides by touching the palate or posterior pharyngeal wall with a tongue depressor or the wooden end of a long Q-tip. Normally, a gag response is elicited by stimulation of either side. This elicits a consensual, symmetric, brisk, and brief elevation of the soft palate and contraction of pharyngeal muscles. The afferent link of the reflex arc is CN9, and the efferent link is CN10. With CN9 lesions, the afferent link is disrupted; stimulation on one side produces less response than stimulation of the other side, but the response is symmetric. With CN10 lesions, the efferent link is affected. With stimulation of either side, palate elevation is asymmetric, with retraction of the palate toward the normal side due to the unopposed pull of the muscle.

CN11 Testing

The accessory nerve (CN11) carries efferents innervating the ipsilateral **sternocleidomastoid** and **trapezius muscles**, which mediate head and shoulder movements, respectively.

This cranial nerve is assessed by examining for wasting of the trapezius muscle and drooping of the shoulder while observing the patient from behind. The strength of the trapezius muscle is tested by asking the patient to shrug their shoulders while the examiner applies resistance against this motion (by pressing down on the patient's shoulders with their hands). The strength of the two sides is compared. Contraction of the sternocleidomastoid rotates the head to the opposite side and tilts the chin upward. The strength of the sternocleidomastoid muscle is assessed by having the patient rotate their head against a resistance. Strength is compared between the two sides.

Peripheral (lower motor neuron) lesions (i.e., the nerve or cranial nerve nucleus) produce weakness of the ipsilateral trapezius and sternocleidomastoid muscles. This results in reduced shoulder shrug on the side of the lesion and reduced strength to rotate the head in the direction opposite to the lesion. The upper motor neuron supply to the lower motor neurons is mostly crossed for the trapezius, but mostly uncrossed for the sternocleidomastoid. Supranuclear lesions therefore produce contralateral trapezius weakness and ipsilateral sternocleidomastoid weakness, resulting in reduced shoulder shrug on the side opposite to the lesion and reduced strength to rotate the head in the direction opposite to the lesion.

CN12 Testing

The hypoglossal nerve (CN12) mediates tongue movements used during speech and eating. It carries efferents that originate from the hypoglossal nucleus in the medulla and innervate the intrinsic and extrinsic muscles of the tongue. The hypoglossal nucleus is under the influence of descending corticobulbar fibers that originate from the face region of the motor cortex of the opposite hemisphere and decussate within the medulla just above the hypoglossal nucleus. Nerve, nuclear, or supranuclear (corticobulbar) lesions that affect CN12 function cause dysarthria and dysphagia. Signs of dysphagia include drooling and poor control of tongue movements.

CN12 function is tested by asking the patient to protrude ("stick out") their tongue and rapidly move it from side to side. Tongue strength is also assessed by asking the patient to protrude their tongue into each cheek while the examiner pushes against it.

A lower motor neuron lesion affecting CN12 causes weakness on the ipsilateral side of the tongue. Consequently, the tongue deviates from midline toward the side of the lesion due to the unopposed action of the intact muscle contralaterally. A lower motor neuron lesion also results in **fasciculations** (muscle twitches) and denervation atrophy of the intrinsic muscles on the affected side, making the surface of the tongue look wrinkled. An intramedullary lesion at the nuclear level commonly has bilateral effects because the hypoglossal nuclei are close to midline and each other, and are usually both affected. An upper motor neuron lesion of the corticobulbar projection to the contralateral hypoglossal nucleus results in contralateral weakness and deviation of the tongue toward the strong side, without atrophy or fasciculations.

The Motor Exam

The objective of the motor examination is to detect abnormalities of the motor system by evaluating: appearance of the muscles, muscle tone, strength, presence of abnormal movements, reflexes, coordination, posture, and gait. Motor functions are tested in a head-to-foot order, comparing one side to the other to observe for asymmetries. Asymmetries are more apparent if the examiner proceeds from side to side rather than testing one entire side before proceeding to the other. Knowledge about handedness is necessary in assessing strength and coordination because these are expected to differ between the two sides of the body. Additionally, proximal muscles are compared to distal muscles.

Appearance

Appearance of the musculature is observed for: (1) bulk of the muscles, (2) level of resting activity, (3) fasciculations, and (4) posture.

Severe muscle atrophy associated with weakness usually signals damage to lower motor neurons.

Abnormalities in level of resting activity are hypokinesia (reduced amplitude of movement) and hyperkinesia (excessive and abnormal movements).

Fasciculations are visible twitches (spontaneous contractions) in a resting muscle that look like worms moving below the dermis. Fasciculations occur in a group of muscle fibers that are innervated by a single motor neuron, due to spontaneous firing of the neuron. Fasciculations may be a symptom of a lower motor neuron disorder, or they may be caused by hypersensitivity to acetylcholine, the neurotransmitter used at the neuromuscular junction (e.g., due to up-regulation of nicotinic acetylcholine receptors in response to denervation caused by motor neuron disease).

Muscle Tone

Muscle tone refers to residual muscle tension (**tonus**) when muscles are at rest. It reflects a continuous, passive partial contraction of the muscles, and is the basis for the slight resistance that normal relaxed muscle has to passive movement (stretch) in the resting state. Muscle tone helps maintain posture. Tone can be abnormally increased or decreased.

Muscle tone is evaluated by asking the patient to relax as the examiner passively moves the neck and limbs at several joints using smooth, gentle movements, feeling for resistance or rigidity. Passive movements that are commonly tested in the neurological exam include rotating the neck from side to side, flexing and extending the neck, alternately flexing and extending the elbow and wrist, and alternately flexing and extending the knee and ankle. Normally, there is a very slight resistance to passive movement.

Decreased resistance to movement is referred to as **hypotonia**, **atonia**, or **flaccidity**. Hypotonia is characteristic of cerebellar dysfunction, lower motor neuron injury/disease (**neuropathy**), or muscle disease (**myopathy**).

Increased resistance to movement is referred to as **hypertonia** or **rigidity**. Rigidity is characteristic of upper motoneuron and basal ganglia dysfunction. There are several types of rigidity. **Spasticity**, also known as **clasp-knife rigidity**, is characterized by a sudden increase in tone followed by sudden relaxation (the "clasp-knife" or "jack-knife" response) in response to a very fast movement. Resistance to slow movements may be normal. Spasticity is characteristic of upper motor neuron lesions. **Lead pipe rigidity** is characterized by a steady increase in resistance throughout the movement. **Cogwheel rigidity** is characterized by a rachet-like increase in resistance. Both lead pipe and cogwheel rigidity are associated with Parkinson's disease. **Paratonia** is failure to relax muscles during muscle tone assessment, despite multiple solicited attempts at relaxation. In **oppositional paratonia** ("gegenhalten") the patient seems to be voluntarily pushing against the passive movements. In **facilitatory paratonia** ("mitgehen") the patient seems to be voluntarily assisting the passive movements. These responses, however, are involuntary. Paratonia is often associated with diffuse frontal lobe disease.

Abnormal tone may also result in postural abnormalities. In Parkinson's disease, hypertonia predominanlty affects the flexor muscles of the neck, trunk, and limbs, resulting in a flexed (stooped) posture. **Focal dystonia** results in a sustained abnormal posture of one part of the body (e.g., hand, neck).

Strength

Abnormalities of strength (power) are associated with weakness or paralysis. The examiner's expectations for a patient's muscle strength take into account the patient's age, sex, muscle bulk, and physical condition. Abnormalities of strength occur with disorders at the levels of upper motor neuron, lower motor neuron, neuromuscular junction, or muscle.

Strength testing samples proximal and distal muscles of the upper and lower extremities. Body movements routinely examined include shoulder abduction, elbow flexion, wrist extension, hip flexion, knee extension, and ankle dorsiflexion. Strength is tested by asking the patient to move the limb to a position where the muscle being tested is maximally active, and then to maintain the position against resistance provided by the examiner.

Muscle strength is usually graded on a scale ranging from 0 to 5, and documented as a fraction where the numerator denotes the patient's score out of a normal (maximal) score of 5: 0/5 = no movement; 1/5 = trace of contraction (i.e., visible muscle movement, but no movement at the joint); 2/5 = active movement when gravity is eliminated; 3/5 = active movement against gravity but not against resistance; 4/5 = active movement against gravity and resistance, but not full strength; 5/5 = normal strength.

Subtle signs of decreased strength are apparent on tests of **drift**. Arm drift is tested by asking the patient to hold their arms extended forward, with palms up, for 15–30 seconds, with their eyes closed (otherwise visual compensation occurs). The examiner watches for gradual drift downward and pronation (inward rotation) of the hand. Leg drift is typically tested with the patient lying supine (on the back) with the legs bent at the knee and held at a 30-degree angle. If there is weakness, the leg will waver and drop within 30 seconds.

Lower motor neuron lesions occur with disease or damage to the anterior horn cells or their axons. Such lesions produce paralysis or paresis. They also produce flaccidity, fasciculations (due to spontaneous motor unit discharges), pronounced atrophy (i.e., **denervation atrophy**), reduced or absent deep tendon reflexes (see "Deep Tendon Reflexes," below), and a normal plantar reflex in which a noxious stimulus to the sole of the foot causes plantar flexion (see "The Plantar Reflex and Babinski Sign," below).

Upper motor neuron lesions occur with disease or damage to many levels of the CNS (cerebral cortex, corona radiata, internal capsule, cerebral peduncles, brainstem, and spinal cord). They are characterized by paralysis or paresis, but without fasciculations or significant atrophy. They also are associated with **spasticity** (increased muscle tone with heightened deep tendon reflexes), **clonus** (deep tendon reflex responses characterized by rhythmic muscular contractions and relaxations), and Babinski sign (abnormal plantar reflex in which a noxious stimulus to the sole of the foot causes plantar extension; see "The Plantar Reflex and Babinski Sign," below).

It is worth considering the pattern of weakness upon completion of the motor exam. The patient may have one-sided weakness (a hemiparesis); weakness of both legs (paraparesis); weakness of all four limbs (quadriparesis); a more diffuse pattern such as proximal muscle weakness (which suggests muscle disease); or focal weakness of a specific set of muscles (which suggests a lesion affecting a single nerve supplying those muscles).

Abnormal Movements

After considering bulk, tone, and strength, it is important to also observe and characterize any abnormal or involuntary movements. Such movements include tremor, chorea, and athetosis. **Tremor** is a rhythmic, involuntary movement that is symmetric about a midpoint within the movement, with both portions of the movement occurring at the same speed. There are various tremor types, broadly classified as resting tremor and action tremor. **Resting tremor** occurs at rest and diminishes during voluntary movement; it is characteristic of Parkinson's disease and parkinsonism. **Action tremor** (**volitional tremor**) occurs during voluntary movement. There are several sub-classifications of action tremor. **Postural tremor** occurs when a person maintains a position against gravity, such as holding the arms outstretched. **Kinetic** tremor occurs with any voluntary movement. **Intention tremor** occurs with purposeful movements directed at a target (e.g., reaching); it is characteristic of cerebellar pathology.

Chorea is abrupt, brief, irregular (non-stereotyped), jerky movements. **Athetosis** is slow, writhing, involuntary movement that is most pronounced in the hands and fingers. Both chorea and athetosis may occur with basal ganglia lesions affecting the striatum.

The Sensory Exam

The sensory exam component of the neurological exam assesses the general body sense of somatosensation. The special senses of smell, taste, vision, hearing, and equilibrium are tested separately in the cranial nerve exam component of the neurological exam.

Somatosensation consists of multiple submodalities: **fine touch** (also known as **discriminative touch**), **crude touch** (also known as **nondiscriminative touch**), **proprioception** (body position and movement sense), **thermoception** (temperature sensation), and **nociception** (pain sensation). Fine touch enables us to sense objects on the skin surface with detailed information about shape, size, texture, amount of pressure, location, and direction of movement, allowing for object identification. Crude touch enables one to sense the presence of an object but not to localize or identify it (see Chapter 15, "The Parietal Lobes and Associated Disorders"). The reason why these two forms of touch are differentiated clinically is that they are carried by two different somatosensory pathways, and lesions in somatosensory cortex may abolish fine touch but not crude touch.

The sensory exam component of the neurological exam assesses fine touch, proprioception, vibration, temperature, and pain. **Vibratory sensation**, also known as **pallesthesia**, is elicited by oscillation of objects on the skin and the simultaneous activation of multiple types of mechanoreceptors within the superficial and deeper layers of the skin that signal touch and deep pressure, respectively. Fine touch, proprioception, and vibration sensations rely on the dorsal column system, while crude touch, temperature, and pain sensation rely on the spinothalamic tract. Thus, lesions in the dorsal column system usually result in loss of fine touch, proprioception, and vibration sense, while crude touch, temperature and pain sensation modalities remain intact. Lesions in the spinothalamic tract result in loss of temperature and pain sensations, while fine touch, proprioception, and vibration sense modalities remain intact.

Since loss of one submodality in a pathway is often associated with the loss of the other modalities conducted by the same tract in the affected area, it is not necessary to test all submodalities carried by the two pathways. Temperature sensation, also known as **thermoception**, is generally not assessed, as it seldom produces any additional information beyond pain sensation testing. Crude touch is not assessed in the neurological examination of alert patients; however, if a patient has a loss of fine touch sensation the neurologist will determine whether crude touch is also affected. In patients with altered level of consciousness in emergency department and critical care settings, crude touch sensation is assessed in determining the degree of responsiveness to stimuli.

Sensory functions are tested in a head-to-foot order, with one side compared to the other to observe for asymmetries, as with the motor exam. Lesions at different levels of the CNS (e.g., peripheral nerve, nerve root, spinal cord, brainstem, cerebrum) will produce sensory symptoms that follow different distributions. If a patient presents with complaints of loss of sensation or abnormal sensation over an area, or if the history otherwise suggests that there may be a sensory deficit, a more detailed sensory examination is conducted to outline of the affected region. The nature and pattern of the sensory findings will allow the neurologist to identify the anatomical basis.

Discriminative Touch

Light touch is tested with a wisp of cotton. After the patient identifies any locations on the skin where they feel that their sensation is abnormal, the examiner tests by touching the patient's skin lightly with the wisp while the patient's eyes are closed and asks them to say "yes" when they feel the light touch on the skin. The examiner maps the limits of the deficit. Different regions of the skin may be tested based upon the examiner's hypotheses.

In cases that require more precise testing of thresholds, **von Frey hairs** are used. This is a set of 20 nylon monofilaments of ascending diameters and stiffness. To determine sensory thresholds, the filament is placed perpendicularly to the skin with slowly increasing force until it bends.

Proprioception

Proprioception is the sense of joint position and movement. It is tested by determining the patient's ability to perceive passive movement when their eyes are closed (i.e., without visual cues). For example, the examiner may hold the most distal joint of a finger or toe by its side, move it slightly up or down, and ask the patient to say whether the finger is moving and in which direction ("up" or "down"). If position sense is normal distally, it need not be tested proximally; however, if it is impaired distally, then testing proceeds proximally until position sense is found to be normal.

Proprioception is also tested by asking the patient to perform movements that normally can be accomplished under proprioceptive guidance only. With their eyes closed, the patient is asked to touch their nose, to bring their forefingers together with the arms outstretched, and to perform alternating movements such as touching the thumb with each finger. The examiner may also pose one of the patient's hands in a particular position in space and ask them either to imitate that position with the other hand or find the thumb of the posed hand with the other hand. Both hands are tested (with tasks for the hands reversed). A true proprioceptive loss will prevent the patient from imitating the position of the posed hand and finding it in space.

Vibration

Vibratory sensation is tested using a 128 Hz tuning fork that is placed over the bony prominences along the body surface (bone is an effective resonator) such as the knuckles, wrists, toes, ankles, spine, and collarbone. The vibrating fork is allowed to run down until the moment that vibration is no longer perceived, and the fork is then transferred quickly to the corresponding joint on the opposite limb for comparison. **Pallhypesthesia** is a diminished sense of vibration. Vibration sense testing is commonly used in the diagnosis of dorsal horn dysfunction and polyneuropathy.

Pain

Pain sensation (nociception) is tested by pinprick. The patient is asked whether the stimulus feels sharp, and whether it feels the same as compared to other areas chosen, based on the hypothesis being tested (e.g., comparing right vs. left can distinguish brain from cord lesion).

Cortical Somatosensory Function

Cortical somatosensory analysis is necessary for perception of shape and texture, as well as object recognition. Damage to the sensory cortex or the thalamocortical projections to somatosensory cortex results in contralateral asterognosis, agraphesthesia, and extinction. Cortical somatosensory function is tested while the patient's eyes are closed. Accurate evaluation of integrative cortical sensory processes requires that there is sufficient primary sensation.

Stereognosis is the ability to identify solid objects by touch, based on their shape, texture, and weight, in the absence of cues from other sensory modalities (e.g., visual, auditory). Manual stereognosis requires intact dorsal column–medial lemniscus pathways conveying discriminative touch and proprioceptive information from the periphery to primary somatosensory cortex, and intact cortical somatosensory areas within the parietal lobe, as well as fine motor control allowing for tactile exploration. To test stereognosis, the examiner places an object in one of the patient's hands and asks them to palpate the item and identify it. Objects that are commonly tested include a paper clip, quarter, penny, dime, and key. Both hands are tested separately. **Astereognosis** is a loss of tactile object recognition due to a deficit in the cortical somatosensory processing affecting shape, texture, and/or weight perception, which undermines the ability to form a mental tactile image (see Chapter 15, "The Parietal Lobes and Associated Disorders"). Unilateral astereognosis with intact primary sensation occurs with lesions in the contralateral parietal cortex. By contrast, **tactile agnosia** is a selective impairment of tactile object recognition in which somatosensory processing is intact, but there is an inability to associate accurate tactile perceptions of objects with stored knowledge. This occurs when somatosensory association cortex is disconnected from the semantic memory store located in the inferior temporal lobe.

Graphesthesia is the ability to identify form patterns drawn on the skin with a stylus (see Chapter 15, "The Parietal Lobes and Associated Disorders"). The examiner tests this cortical somatosensory function by drawing numbers and/or letters on the fingertips or palm of the patient's hand using the wooden end of a cotton applicator and asking the patient to identify the drawn forms without visual cues. The patient is asked to close their eyes and hold up their hand with the palm oriented toward their face, and the examiner writes the numbers and letters oriented so that they are upright relative to the patient. Both hands are tested. Unilateral **agraphesthesia** with intact primary sensation occurs with lesions in the contralateral parietal cortex. This may occur due to a deficit in cortical somatosensory processing determining directionality of lines written on the skin, which is necessary for form recognition. In some cases, however, agraphesthesia is due to a disconnection between the somatosensory association cortex and language-related areas necessary for number and letter recognition more generally. Agraphesthesia will of course also be observed in patients with alexia for numbers and letters (see Chapter 28, "Alexia, Agraphia, and Acalculia").

Double simultaneous stimulation tests for **extinction,** a mild form of sensory hemi-inattention (neglect) in which there is a failure to detect stimuli from one half of space. Within the somatosensory realm, the ability to detect stimuli is evaluated while the patient's eyes are closed using pinprick or touch applied to the hand, cheeks, and feet, tested unilaterally and bilaterally. The patient is asked to identify which side of the body was touched and whether both sides were touched. The patient with mild neglect detects unilateral stimuli whether presented to the left or right side, but with double simultaneous stimuli, the patient extinguishes the stimulus within the neglected hemi-space (i.e., only detects stimuli in the non-neglected hemi-space). In other words, detection of unilateral stimuli is normal on both sides of the body, but with bilateral stimuli, the subject fails to detect the stimulus within the affected hemi-space. Most often, the left hemispace is affected, and right parietal lobe dysfunction is the underlying cause. Thus, extinction is characterized by neglect of contralesional stimuli only in the presence of ipsilesional stimuli. Extinction can also be tested within visual and auditory modalities. Within the visual modality, it may be necessary to distinguish hemispatial neglect from hemianopia. Those with hemianopia are aware of their field cut and make compensatory head and eye movements, while those with neglect appear to lack awareness of the continuity of visually presented objects and do not make head and eye movements to explore the neglected half of space.

The Reflex Exam

The deep tendon reflexes and plantar response are routinely examined. In special situations, other reflexes may also be tested.

Deep Tendon Reflexes

Deep tendon reflexes (DTRs), also known as **muscle stretch reflexes** or **myotatic reflexes**, are monosynaptic (single synapse) reflexes. The reflex circuit consists of a two-neuron reflex arc (see Figure 24.3 in Chapter 24). The sensory afferents originate from muscle stretch (muscle spindle) receptors and enter the spinal cord via the dorsal roots. These afferents make excitatory synaptic contacts directly on the dendrites and cell bodies of alpha motor neurons within the ventral horn of the spinal cord. The alpha motor neuron axons form the efferent limb of the reflex arc. They exit the spinal cord via the ventral roots and synapse on skeletal muscle. The DTRs are tested by striking the tendon with a reflex hammer. This causes rapid stretch of the muscle, which stimulates muscle stretch receptors and generates an afferent signal that is conducted to the spinal cord, which in turn generates an efferent signal that is conducted back to the muscle, eliciting a reflexive muscle contraction.

Five stretch reflexes are typically tested, each corresponding to a particular spinal nerve root and muscle: the **brachioradialis** tendon located at the wrist (C5–C6); the **biceps** tendon located at the elbow crease (C5–C6); the **triceps** tendon located at the back of the upper arm just above the elbow (C6–C7); the **patellar** tendon located just below the kneecap (L3–L4); and the **Achilles** tendon located at the ankle (S1). The results of the DTR exam are often recorded by drawing a stick figure and indicating the strength of the five major reflex sites.

The reflex response is evaluated with respect to threshold (the amount of hammer force necessary to obtain contraction), velocity, strength, and duration of contraction. It is also evaluated for "overflow" of the reflex response in which nearby muscles that were not tested contract. The DTRs are graded on the following scale: 0 = absent reflex; 1+ = hypoactive reflex; 2+ = normal reflex; 3+ = brisk (hyperactive without clonus) reflex; 4+ = unsustained clonus; 5+ = sustained clonus (Figure 30.5). The "+" after the number simply reflects that the number refers to a DTR grade and not a motor strength grade. Deep tendon reflexes rated as 0, 4+, or 5+ are abnormal. Deep tendon reflexes rated as 2+ are normal. Deep tendon reflexes rated as 1+ or 3+ are normal, unless they are asymmetric, there is a large difference between the arms and legs, or there are associated abnormalities in muscle tone or muscle strength.

Hyporeflexia, a depressed or absent reflex response, results from dysfunction of some component of the reflex arc (sensory receptors, sensory afferents, spinal cord, motor efferents, or muscle). **Hyperreflexia** (reduced threshold, short latency, prolonged response, widened reflexogenic zone) results from increased excitability of the lower motor neuron pool due to a suprasegmental lesion (above the level of the spinal reflex arc) that interrupts any or all the upper motor neuron pathways that descend from the cortical, subcortical, midbrain, and brainstem levels and normally modulate the reflex arc activity.

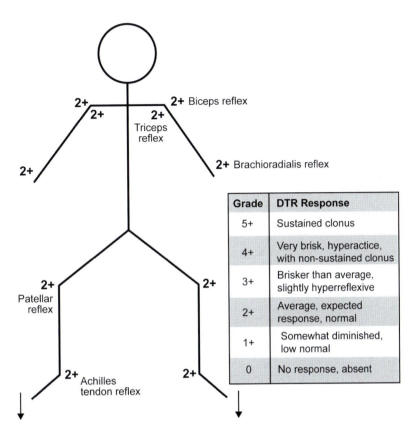

FIGURE 30.5. Deep tendon reflex (DTR) testing findings are often recorded using a stick figure. Typically, five reflexes are tested: biceps, triceps, brachioradialis, patellar (knee), and Achilles (ankle). The DTR reflexes are graded on a scale from 0 to 5, followed by a "+" after the number. Ratings of 2+ are normal. The results of the plantar response test are often recorded on the DTR stick figure drawing, with up- or down-going arrows at the feet representing the direction of the plantar response.

Absent or reduced DTRs are indicative of lesions that directly disrupt the reflex arc (i.e., the peripheral sensory input or the lower motor neuron output), either due to peripheral neuropathy or muscle disease. Hyperactive (brisk) DTRs are indicative of suprasegmental lesions affecting the corticospinal (upper motor neuron) tracts, due to a loss of descending inhibitory signals that modulate the activity of the local spinal reflex circuit.

Primitive Reflexes

Primitive reflexes are reflexes that are normally present during early development but are abnormal when observed in adults. The primitive reflexes are mediated by subcortical structures, and as the descending cortical motor pathways gain functionality with progressive myelination, these reflexes come under inhibitory control by 1–2 years of age.

THE PLANTAR REFLEX AND BABINSKI SIGN

The **plantar reflex** is a **superficial reflex** (elicited by sensory afferents from skin rather than muscle). It is elicited by a noxious stimulus to the sole of the foot (the **reflex receptive field**) which causes immediate flexion of the toes as a normal defensive withdrawal response. It is a **spinal reflex** mediated by circuitry within the spinal cord, but it is influenced by higher centers of the CNS.

On neurological examination, the plantar reflex is elicited by stroking of the sole of the foot with a pointed object (e.g., wooden end of a cotton applicator, tongue depressor, pen, pointed end of a key), moving up the lateral plantar side of the foot from the heel toward the small toe, and then across the metatarsal pads to the base of the big toe (Figure 30.6). The applied stimulus should be uncomfortable but not painful. Normally the big toe moves in a plantar (relating to the sole of the foot) fashion due to flexion (downward contraction), often with flexion and adduction of the other toes. Thus, the toes are said to be *down-going*. This **flexor plantar response** is a normal withdrawal reflex to noxious stimulation.

An **extensor plantar response** is characterized by extension (dorsiflexion) of the big toe, often with extension and abduction ("fanning") of the other toes (*up-going* toes). An extensor plantar response is also known as a **Babinski reflex** or **Babinski sign**. It is a primitive reflex that occurs normally in children up to 2 years of age. When present in children older than age 2 or in adults, it indicates CNS pathology with loss of upper motor neuron control over

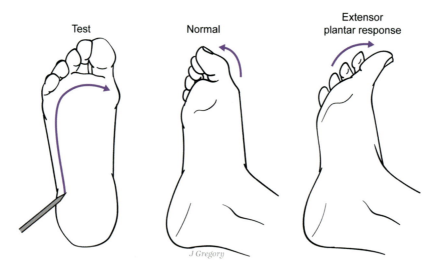

FIGURE 30.6. The plantar reflex is elicited by stroking of the sole of the foot with a pointed object, moving up the lateral plantar side of the foot from the heel toward the small toe and then across the metatarsal pads to the base of the big toe. Normally this results in toe flexion and adduction (*down-going toes*). The Babinski sign is an abnormal response characterized by toe extension and abduction (*up-going toes*).

the reflex circuit. A more severe form of the abnormal response includes dorsiflexion at the ankle and flexion at the knee and hip.

The results of the plantar response test are often recorded on the stick figure drawing used to record the results of the deep tendon reflex exam, with up- or down-going arrows at the feet representing the direction of the plantar response (i.e., up-going arrows indicate a positive Babinski sign; see Figure 30.5).

OTHER FRONTAL RELEASE REFLEXES

Other primitive reflexes may reappear with CNS lesions in which there is a loss of cortical inhibition. They are predominantly associated with frontal lobe pathology (especially of vascular, neurodegenerative, or traumatic etiology) and are therefore also known as **frontal release signs**. These primitive reflexes are not routinely tested; they are tested in patients suspected of having Parkinson's disease or other brain disease or injury affecting the frontal lobes. The frontal release reflexes involve the palms (grasp and palmomental reflexes) and face (glabellar, snout, suck, and rooting reflexes). Of these reflexes, the grasp reflex is the most informative, as it is specific to frontal lobe injury and localizes to the contralateral supplementary motor area in the medial frontal lobe.

The **grasp reflex** consists of flexion of the fingers that is elicited by stimulation of the palm. Attempts to withdraw the stimulus often result in strengthening the grasp. The reflex is present at birth and usually disappears by 6 months of age, but reappears later in life with frontal lobe disease. The adult patient with a positive grasp reflex may even be unable to release the grasp when requested to do so, a phenomenon referred to as **magnetic apraxia**.

The **glabellar reflex** is a primitive reflex elicited by tapping on the forehead between the eyebrows (the glabella area). Normally, adults stop blinking after a few taps as they habituate to the stimulus. It is abnormal for the blink response to persist with repetitive tapping (**Myerson's sign**); this is often seen in Parkinson's disease.

Coordination/Cerebellar Exam

Ataxia is an abnormality of coordination. Loss of proprioception and vestibular pathology can affect performance on coordination tests, but such deficits will manifest on other portions of the neurologic examination. Ataxia that results from loss of proprioception is known as **sensory ataxia** or **proprioceptive ataxia**, while ataxia that results from loss of vestibular function is known as **vestibular ataxia**.

In patients with normal strength and intact sensation, deficits on tests of coordination reflect cerebellar dysfunction. **Cerebellar ataxia** due to lesions of the cerebellar hemispheres is characterized by decomposition of movement, dysmetria, intention tremor, and dysdiadochokinesis.

Decomposition of movement occurs with complex movements involving more than one joint and is due to a defect in the timing of different components of the movement. This results in movements that are performed in jerky stages and as a series of successive single simple movements, rather than as a continuous smooth movement.

Dysmetria refers to a lack of accuracy in voluntary movements. The most common type of error is hypermetria, an overshooting of directed movements of the limbs (or eyes) toward a target. This inability to stop movement at a chosen spot is due to a delay in the timing of the movement termination of antagonistic muscles that check the movement. Patients with cerebellar lesions can also exhibit hypometria, undershooting directed movements of the limbs (or eyes) toward a target due to premature arrest of the movement before the target is reached.

Intention tremor is a tremor that arises when performing voluntary movements directed at a target. The rhythmic oscillations worsen with targeted movements such as grasping an object. The tremor originates at the fulcrum at the joint (e.g., shoulder or hip) and increases in amplitude as the target is approached (i.e., end-point tremor). This symptom is due to defective feedback control from the cerebellum on a cortically initiated movement.

Dysdiadochokinesis is an impaired ability to perform rapidly alternating movements using antagonistic muscle groups, such as repetitive pronation and supination of the hand. It is due to impaired timing of initiation of each successive movement.

Coordination tests include finger-to-nose, heel-to-shin, and rapid alternating movements. In the **finger-to-nose test**, the patient rapidly and alternately touches their own nose and the examiner's finger, which is held out at arm's length from the patient and moves between touches. The examiner observes for whether the movements are smooth and accurate, or if there is dysmetria (e.g., past pointing) and/or intention tremor. In the **heel-to-shin test**, the patient lies supine, the foot is dorsiflexed, and the heel is placed on the opposite knee and slid smoothly down the tibia (shinbone) to the ankle. In patients with cerebellar ataxia, the heel wobbles from side to side. Tests of **rapid alternating movements** of the hand include repetitively touching the thumb to the forefinger, touching the thumb to each finger in rapid succession, alternately pronating and supinating the hand, rotating the hand as if screwing in a light bulb, tapping the thigh with extended fingers, and tapping the floor with the foot. The patient is asked to repeat the sequences as rapidly as possible. Ataxic patients exhibit dysdiadochokinesis; they are unable to maintain a rhythmic alternation, and the movements are unequal in both amplitude and speed.

Station and Gait

Abnormalities of **station** (stance) and **gait** (walking) have a wide variety of potential causes, including abnormalities in upper motor neurons, lower motor neurons, cerebellum, basal ganglia, vestibular sense, proprioceptive sense, and muscle. Assessment of station includes observation of stance and the Romberg test. Assessment of gait typically involves observation of gait and **tandem gait** (heel-toe walking); other tests include walking on heels and walking on toes.

Three senses aid in maintaining balance: vision, vestibular sense, and proprioception. All three of these sensory modalities provide input into the cerebellum, either directly or indirectly. The patient must have two out of three of these senses intact to maintain balance while standing. It is optimal to test stability of station before assessing the more complex task of gait.

The **Romberg test** provides information related to the function of the cerebellum and vestibular system. The patient stands with their feet together, first with eyes open and then with eyes closed. Most people will sway slightly as they change from eyes open to closed but are able to make appropriate postural adjustments. Closing the eyes eliminates one sense, but the other two senses still function, and the person does not sway significantly. If either proprioception or the vestibular sense is impaired, the patient is able to maintain balance with eyes open, but sways excessively nearly to the point of falling when the eyes are closed because under this condition only one of the three senses is operating. Thus, a positive Romberg sign, defined as excessive postural swaying only when the eyes are closed, indicates either proprioceptive or vestibular dysfunction. A vestibular deficit will be accompanied by a patient report of vertigo (the sensation that the environment is spinning), while a proprioceptive deficit will be confirmed by an abnormality on proprioceptive sensory testing. Patients with cerebellar ataxia have excessive postural swaying when the eyes are open and the swaying may increase when the eyes are closed, but this is not a positive Romberg sign.

To examine gait, the patient is asked to walk approximately 10 steps, turn around, and return to the examiner with the eyes open. The examiner observes for gait symmetry, base of gait (i.e., horizontal stride width when both feet are in contact with the ground), gait evenness, gait rhythm, step length, height of lifted knees (e.g., a normal or high-stepping gait), arm swing (normal or reduced), and efficiency of turning. If gait appears normal, tandem gait is observed because it will exaggerate any instability.

Damage to many parts of the nervous system may produce gait disturbance. The **cerebellar gait** is characterized by a wide base, unsteadiness, irregularity of steps, and lateral veering. The patient may compensate by shuffling, keeping both feet on the ground simultaneously. The **gait of sensory ataxia** is characterized by brusque walking movements. The legs are flung abruptly forward and outward as the patient steps forward, often lifted higher than necessary, with stamping of the feet so that the entire sole strikes the ground at once. As these patients are more reliant on visual cues to regulate their motion, they carefully watch the ground and their legs as they walk. Their gait disturbance is greatly exaggerated when they are deprived of visual cues, such as when walking in the dark. Sensory ataxia is due to impaired proprioception and occurs with damage of the peripheral nerves, posterior roots, posterior columns, medial lemnisci, or spinocerebellar degeneration syndromes (e.g., Freidrich ataxia). Normal pressure hydrocephalus is associated with **magnetic gait**, a hypokinetic, apractic gait disorder in which the feet look as though they are stuck to the floor. Patients with unilateral weakness have an asymmetrical or lopsided gait; with spastic hemiplegia (i.e., due to upper motor neuron damage), increased muscle tone in the leg results in a stiff leg that is hyperextended (and therefore longer than the other) without flexion. This leads to contralateral trunk lean, which raises the ipsilateral pelvis to clear the paralyzed leg. It also leads to circumduction in which the leg is dragged around the body with the toe tracing a semicircle on the floor, moving outward and then inward as it advances, rather than the normal straightforward movement. In addition, the forearm is flexed. Gait disturbance may also result from non-neurologic causes. For example, an **antalgic** (limping) **gait** may develop to avoid pain while walking.

Summary

The neurological examination is the neurologist's formal evaluation of nervous system function. It is organized into seven components: (1) mental status, (2) cranial nerves, (3) motor system, (4) reflexes, (5) sensory system, (6) coordination, and (7) station and gait. The examiner tailors the exam to clinical questions; however, each domain of function is addressed. Neurological localization involves determining the level of the nervous system that is dysfunctional (e.g., cerebral cortex, subcortical white matter, deep gray matter structures, cerebellum, brainstem, spinal cord, spinal nerves, peripheral nerves, neuromuscular junction, and/or muscles). Knowledge of the functional anatomy of the nervous system makes it possible to interpret symptoms and localize lesions to make a correct anatomic diagnosis. Determination of etiology involves generating a differential diagnosis of types of pathology that could account for the signs and symptoms, and takes into account the clinical history (e.g., onset, course, and duration). Other investigative studies, such as neuroimaging studies, neurophysiological studies, laboratory studies, and neuropsychological examination, may also contribute to localization and differential diagnosis. The neurologist integrates the results of the neurological examination and other studies to come to a diagnosis and determine an appropriate treatment plan, such as medical intervention, surgery, or rehabilitation.

Additional Reading

1. Blumenfeld H. The neurologic exam as a lesson in neuroanatomy. In: Blumenfeld H, *Neuroanatomy through clinical cases*. 3rd ed., pp. 49–82. Sinauer Associates; 2021.
2. Boes CJ. The history of examination of reflexes. *J Neurol*. 2014;261(12):2264–2274. doi:10.1007/s00415-014-7326-7
3. Hillis JM, Milligan TA. Teaching the neurological examination in a rapidly evolving clinical climate. *Semin Neurol*. 2018;38(4):428–440. doi:10.1055/s-0038-1667135
4. Rohkamm R. *Color atlas of neurology*. 2nd ed. Thieme Stuttgart; 2014.
5. Weiner WJ, Goets CG, Shin RK, Lewis SL, eds. *Neurology for the non-neurologist*. 6th ed. Lippincott Williams & Wilkins; 2010.

Neuropsychological Assessment

Introduction

Neuropsychological assessment is an in-depth, performance-based evaluation of cognitive and behavioral functions (also referred to as higher brain functions). It is performed by **neuropsychologists** who have specialty training and expertise in brain-behavior relationships; the effects of brain injuries and disease on brain function; psychometric measurement of mental processes relating to brain function; and the diagnosis of cognitive, behavioral, and emotional disorders associated with brain injury or disease. The neuropsychological assessment can be thought of as an expansion of the mental status component of the neurologic examination. Both examinations evaluate nervous system functions, and they are complementary; some patients may require both. This chapter describes the process of clinical neuropsychological assessment and the structure of the resulting report.

The neuropsychologist tailors the examination to the clinical question(s) and the patient. It begins with an interview, in which the neuropsychologist elicits the history of signs and symptoms from the patient and/or other informant such as a family member, and a review of the medical record. This information forms the basis of hypotheses regarding the underlying cause of signs and symptoms or about the nature of the deficits resulting from a known neurological diagnosis. After the interview, the neuropsychologist performs formal testing using instruments that are designed and validated to measure specific aspects of brain function (e.g., memory, language), as well as symptom inventories to assess for mood, behavior, and psychiatric changes. The testing component of the evaluation is akin to a single-subject experiment in which the neuropsychologist challenges the patient with structured tasks that involve presenting stimuli and recording the patient's responses. Throughout the entire interaction with the patient, the neuropsychologist makes **behavioral observations** that can inform the diagnostic or descriptive conclusions.

The neuropsychologist communicates the findings in a structured written **neuropsychological assessment report** that is typically organized into the following sections: history, behavioral observations, testing procedures and test results, conclusions, and recommendations. The report typically emphasizes cognitive function, but may address behavioral, affective, and psychiatric changes. Changes that are attributable to brain injury or brain disease are referred to as **neurocognitive disorders**, **neurobehavioral** disorders, **neuroaffective disorders**, and **neuropsychiatric disorders**, respectively.

The Referral Question

Physicians refer patients for neuropsychological assessment to answer specific questions stated as the purpose of referral or **referral question**, or because of symptoms stated as the **chief complaint**. The referral question or chief complaint is a brief statement of the suspected or known problem and referral source's reason for requesting the evaluation. The entire neuropsychological report aims to answer the referral question, which is either diagnostic or descriptive of the nature and extent of cognitive deficits due to a known brain condition.

Diagnostic neuropsychological assessments focus on determining the etiology of signs and symptoms. Some assessments involve differentiating normal from pathological states in patients with concerns about cognitive changes; for example, identifying typical age-related memory changes versus early signs of Alzheimer's disease. Other assessments may involve differentiating neurological disorders; for example, semantic dementia from logopenic progressive aphasia in a patient presenting with complaints of anomia (see Chapter 27, "Language and the Aphasias"). Because brain disease often manifests initially as emotional and/or behavioral change, patients with an acute onset of psychiatric symptoms are sometimes referred for neuropsychological assessment. A third category of assessments therefore involves differentiating neurological from primary psychiatric disorders, for example, dementia versus **depressive pseudo-dementia**, or frontotemporal dementia syndrome with diminished motivation, abulia, and apathy versus depression (see Chapter 17, "The Frontal Lobes and Associated Disorders"). Finally, some assessments involve differentiating neurological disorders from neurological signs and symptoms that are inconsistent or incongruent with neurological disease or injury, such as **functional neurological disorder** in which neurological signs and symptoms are psychogenic in origin, or feigned signs and symptoms as occurs in factitious disorder or malingering.

Descriptive neuropsychological assessments describe the nature and extent of cognitive deficits from a known brain condition. Such assessments are performed for the purposes of treatment planning, such as the need for medical intervention (e.g., medication) or rehabilitation services, and guiding decisions about functional status, such as the patient's ability or readiness to return to work or school, or ability to live independently and safely. They may also be repeated after some interval to monitor clinical course (e.g., recovery, progression) or treatment efficacy.

Note that some neuropsychological assessments are conducted for forensic rather than clinical purposes. Forensic neuropsychological assessments (not covered in this chapter) address legal questions that can assist in determinations relating to disability claims, civil litigation, or criminal litigation.

History

The History section of the report typically summarizes the following background information in subsections: history of the presenting problem or present illness; previous examinations and studies; medical history (including psychiatric history and medications); family medical history; developmental history; and education, work, and social history.

History of the Presenting Problem/Present Illness

The richest source of information for developing hypotheses to be tested by the neuropsychological assessment usually is a carefully elicited clinical history of the presenting problem or present illness. People with brain injury or disease, however, may not be able to provide a complete and accurate history; therefore it is often necessary to obtain background information from someone close to the patient (e.g., a family member) and the medical record, if available. Even if the patent can provide a history, information from a corroborating source may help to ensure accuracy. Patients with memory loss may confabulate, and this may not be immediately apparent to the neuropsychologist. Additionally, patients with cognitive deficits from brain injury or disease may be unaware of their deficits (anosognosia).

This clinical history presents a chronological description of the illness or symptoms, including the nature and date of onset, duration, course, and associated behavioral, psychiatric, and/or neurological changes, including information about prior care and treatment related to the present illness. It also includes information about the impact of the illness on **activities of daily living**, encompassing **basic activities of daily living** (i.e., skills required to manage one's basic physical needs, including personal hygiene or grooming, dressing, toileting, transferring or ambulating, and eating) and **instrumental activities of daily living** (i.e., the ability to care for oneself and one's home).

Previous Examinations and Studies

Results of other exams and studies performed to investigate the current illness or condition must also be taken into consideration and included in the report. This includes a description of the findings from neurological examination, neuroimaging studies, and blood tests, but may also include results from other neurodiagnostic tests such as electroencephalography, angiography, functional neuroimaging, cerebrospinal fluid study, and prior neuropsychological examinations.

Medical History

The medical history presents information about neurologic disorder or disease, systemic disease that may be associated with changes in brain function, surgical history, psychiatric history including drug and alcohol use, medications, and toxin exposure.

Family Medical History

In certain cases, family medical history is potentially relevant, including neurologic and major psychiatric disease in other family members, and history of heritable neurological diseases.

Developmental History

History of birth complications, developmental delays, and neurodevelopmental disorders (e.g., attention deficit hyperactivity disorder, learning disabilities) must be taken into consideration, as they may impact test performance and interpretation.

Because handedness and cerebral dominance for language are related, handedness is always reported. Observation of the hand used for writing may not be an accurate reflection of natural handedness in older individuals because in the past, many natural left-handers were taught to write with the right hand; thus the neuropsychologist also asks the patient about natural handedness.

Education, Work, and Social History

Educational and occupational background of the patient helps the neuropsychologist to gauge expectations for test performance. Information about family structure and social supports is often relevant, particularly in cases where there has been a decline in the fundamental skills required to independently care for oneself (i.e., activities of daily living).

Behavioral Observations

Behavioral observations provide crucial information that may aid in diagnosis. Careful observation of the patient's behavior is important because several syndromes are diagnosed primarily by their behavioral manifestations, such as delirium (acute confusional state), frontal lobe syndrome, anosognosia, neglect, and apathy.

Observation begins upon first encounter with the patient and continues throughout the evaluation. The neuropsychologist makes a series of observations about general appearance, sensory and motor function, attention, speech and language, memory, thought form and content, behavior, affect, and interpersonal relatedness. The Behavioral Observations section of the neuropsychological assessment report is organized into subsections reflecting these categories.

Appearance

Observations regarding general appearance address hygiene, grooming, and dress (e.g., whether the patient is dressed and groomed appropriately, unkempt, or malodorous). This subsection may also include whether the patient's appearance matches their chronological age, and other notable observations, such as signs of hemispatial neglect (e.g., shaving or applying makeup to only half of the face).

Sensory and Motor Function

These observations address sensory and motor limitations, whether the patient uses assistive devices, and whether such limitations may confound the test data. They also include whether activity level is normal, hypoactive/hypokinetic (e.g., psychomotor slowing, lack of spontaneous movements, delayed initiation of movements), or hyperactive/hyperkinetic (e.g., psychomotor agitation). This subsection may also include observations of abnormal movements such as tics, tremors, dyskinesias, facial grimaces, bizarre gestures, or parkinsonian symptoms such as stooped posture, bradykinesia, rigidity, resting tremor, festinating gait, hypomimia (masked facies), micrographia, and hypophonia.

Speech and Language

Spontaneous speech is observed for rate, volume, articulation, and prosody. Speech rate is observed as normal, pressured, unusually fast, or unusually slow. Speech volume is observed as normal, hyperphonic, or hypophonic. Speech articulation is observed as clear versus dysarthric. Prosody is observed as normal versus aprosodic (monotone).

Observations regarding language address both receptive and expressive language. Spoken language comprehension is observed for ability to understand questions and test instructions. Expressive language is observed for fluency, grammatical structure (normal versus agrammatic or telegraphic), paraphasias, and signs of anomia such as word-finding pauses, circumlocution, and gesturing object use.

Cognitive Process

Cognitive process observations address awareness, attention, memory, thought form, and thought content.

Observations regarding awareness address whether the patient has insight into their current situation/condition and the reason for undergoing evaluation, or whether they are unaware of or rationalize cognitive losses.

Observations regarding attention address whether the patient can sustain attention to the task at hand, or is distracted easily (e.g., by noises in the environment that most people can filter out) and needs to be called back repeatedly to the task at hand. They may also pertain to observing for disorders of spatial attention (i.e., unilateral spatial neglect or simultanagnosia).

Observations regarding memory address whether the patient presents a coherent, logical, and chronological account of their history, or if they present their history with a disorganized chronology, omit major events (e.g., death of a spouse or recent hospitalization, injury, surgery), confabulate, or are unsure and continuously refer to their companion for confirmation or provision of information. Observations regarding memory also address whether the patient can retain information such as questions and test instructions.

Thought form is observed for whether the patient exhibits orderly thinking and maintains a coherent stream of thought that is linear, logical, and goal-directed, or whether there is a **formal thought disorder** in which there is an apparent breakdown in the logical organization of thoughts (i.e., incoherent thinking). There are a variety of forms of formal thought disorder. **Circumstantial** thinking is circuitous, non-direct thinking with excessive, unnecessary detail that digresses but eventually returns to the main point of conversation or answers the question. **Tangential** thinking also involves excessive and irrelevant detail, but the patient never reaches the main point of conversation or answers the question. **Derailment** (**loosening of associations**) involves shifting thoughts that are loosely associated with each other. **Flight of ideas** involves rapidly shifting loosely associated thoughts, with the patient jumping from one point to another in an incoherent manner.

Thought content is observed for signs of paranoid ideation (e.g., suspiciousness), delusions, illusions, hallucinations, phobias, or obsessions.

Behavioral Regulation

Behavioral regulation is observed for signs of disinhibition, impulsivity, perseveration, impersistence, stimulus-bound behavior, and abulia (e.g., offering no information except in response to direct questioning, answering with single words).

Affect

Observations regarding **affect**, the outward expression of mood, describe the quality of the patient's emotional expression as **euthymic** (normal), dysphoric (depressed, tearful, sad, hopeless), anxious (tense, sweating, tremor), apathetic/unconcerned, irritable, angry, hostile, or elated/euphoric (inappropriate optimism, boasting). Affect is also described with respect to its appropriateness, consistency, intensity (amplitude), and range. Appropriateness refers to whether the affect displayed is appropriate to the context and the patient's mood and thoughts, or inappropriate to context (e.g., laughing when talking about a disabling injury or a death) and incongruent with the patient's mood or ideas (e.g., crying without the accompanying feelings of sadness). Consistency pertains to whether affect is stable and even, or labile with rapid changes unrelated to external events. Comment is made about whether the intensity of expression of feelings is normal, or whether affect is flat (i.e., lacking) or blunted (i.e., diminished), especially when talking about issues that normally would be expected to engage the emotions. Comment is also made about whether the patient displays a broad range of affect, or whether the range was constricted, with little variability.

Observations regarding affective behavior also consider whether there are signs of pseudobulbar affect, such as pathological crying or laughing that may occur with brain injury or disease.

Comportment, Tact, and Interpersonal Relatedness

The neuropsychologist also observes the patient's social behavior, evaluating whether comportment and tact are appropriate in the context of a doctor's visit or whether there are socially inappropriate behaviors, such as crossing interpersonal boundaries, interrupting during conversation, loss of manners and social graces, facetiousness, or offensive comments. Interpersonal style is evaluated for whether the patient is congenial, open (candid), engaged, cooperative, and makes eye contact; withdrawn, guarded, evasive, and avoids eye contact; or annoyed, irritable, hostile, cautious, defensive, resistant, easily frustrated, or uncooperative.

Testing Procedures and Results

Neuropsychological assessments are based largely on quantitative cognitive test performance. Multiple domains of cognitive function are mediated by specific and distinct neural circuits and may be compromised selectively by brain injury or disease. These domains include attention, visuoperceptual function, constructional ability, language, learning and memory, abstract thinking and reasoning, and executive functions. Some examinations also include assessment of mood and/or psychiatric signs and symptoms (including behavior change) using symptom inventories.

Neuropsychologists plan the examination guided by the referral question or chief complaint. The test instruments used vary based on the neuropsychologist's clinical judgment, the patient's background and presentation, and hypotheses about the underlying neurological condition and/or lesion localization. Although most neuropsychologists use a core battery of tests, they often tailor the exam with supplemental tests, since no single battery will suit the purposes of all examinations. The specific test instruments used are indicated in the neuropsychological assessment report. Test selection prioritizes tests that have documented validation for use in the condition of interest. They attempt to control for factors that may impact test performance independent of neurological injury or disease, such as sensory limitations, motor limitations, specific learning difficulties, and linguistic factors that may confound the test data. They may use simpler tasks with fewer potential confounds to test hypotheses about various possible reasons for test failure, to eliminate possible contributing factors. They also consider and include in the report other potential confounds to the validity of test data, such as lack of full cooperation with the testing process (e.g., if a patient resents being compelled to undergo testing or is not motivated), or intentionally fabricating or amplifying symptoms for some benefit.

A goal is to approach the examination in a hierarchical manner, beginning with assessment of basic, foundational cognitive processes such as attention and language, then intermediate-level cognitive functions such as spatial cognition and memory, and finally high-level cognitive functions such as abstract thinking, reasoning, and executive functions. Erroneous conclusions can usually be avoided by approaching the assessment with this hierarchy in mind since higher functions depend upon the integrity of more basic functions. To give an example, an inattentive, distractible patient does not efficiently assimilate information presented during the testing, leading to deficient performance across a wide variety of tests, including memory tests. Language is also a foundational cognitive ability. Deficits in language can undermine performance on many tests that require verbal responses, such as tests of verbal memory and verbal abstract thinking. Impaired constructional abilities will confound performance on tests of nonverbal memory that require the patient to draw figures from memory.

Raw test scores are interpreted with respect to normative data, so they are converted to standardized scores that reflect the patient's performance relative to a normative sample that is roughly matched to the patient on relevant demographic variables (e.g., age, education).

The test findings are reported in a structured format organized by domains. These typically include attention, cognitive processing speed, visuoperceptual function, constructional ability, language, learning and memory, abstract thinking and reasoning, and executive functions. Some evaluations may require detailed examination and analysis within a domain; for example, assessments for possible primary progressive aphasia include more extensive evaluation of language functions.

Attention

Attention refers to the cognitive process of selecting specific external or internal stimuli for active processing and filtering out or ignoring extraneous information. The attentive patient can focus on a single stimulus or task and screen out irrelevant stimuli. Attention may be fractionated into attention capacity, executive control of attention, and cognitive processing speed.

Attention requires alertness; however, alertness is not sufficient for attentiveness (see Chapter 25, "Disorders of Consciousness"). Alertness is a more basic arousal process in which the awake patient can respond to *any* stimulus in the environment. The alert but inattentive patient is unable to focus on a single stimulus and is distracted by sounds, movements, and events. The neural system responsible for attention is distributed widely in the brain and includes the ascending reticular activating system (ARAS), thalamus, limbic structures, and neocortex, particularly the parietal and frontal association areas. Attention results from a balance between ascending reticulocortical activation and descending corticoreticular modulation. Limbic structures may add emotional intensity to the object of attention. Conscious voluntary effort and control of attention is provided by the frontal lobes. Because attention represents a complex interaction among these structures, damage in many different areas of the brain can disrupt attention. Damage to the ARAS causes decreased level of consciousness, and this basic deficit in alertness underlies the attentional disturbance. Inattention can occur with lesions of the thalamus, posterior internal capsule, and other subcortical structures. Extensive bilateral cortical damage due to any etiology also causes disturbances in the ability to focus and maintain attention, resulting in distractibility and inability to ignore irrelevant stimuli.

Delirium (confusional state) is a neurocognitive syndrome characterized by core disturbances in awareness of self and environment and attention. The patient is alert but disoriented and has difficulty following commands. Delirium has an abrupt onset over several hours to several days, and fluctuating severity over the course of a day (see

Chapter 25, "Disorders of Consciousness"). Delirium reflects diffuse brain dysfunction and is most often caused by toxic-metabolic disturbances. The attentional disturbance is characterized by gross inability to direct, focus, sustain, and shift attention. Because attention is a necessary precondition for optimal performance on all cognitive tests and inattention is a hallmark symptom of delirium, patients with delirium perform poorly on cognitive tests. Thus, once a diagnosis of delirium is established, there is no benefit of additional cognitive testing until symptoms have resolved.

ATTENTION CAPACITY

The amount of information that can be held in mind at a given moment in time is referred to as **attention capacity** in the parlance of clinical neuropsychologists. Attention capacity is the most basic measure of attention. It is assessed with span tests that present the patient with increasingly greater amounts of information and require them to reproduce what was seen or heard immediately after presentation. For this reason, attention capacity is also referred to as **immediate memory** or **immediate recall**.

Attention capacity is commonly assessed using digit span forward tasks in which the neuropsychologist reads aloud a string of digits at a rate of one digit per second without inflection or breaks. The task begins with a short string consisting of two or three digits, and the length of the strings increases gradually by one digit across trials. Non-aphasic patients without dementia and of average intelligence can accurately repeat a sequence of 5–7 digits.

EXECUTIVE CONTROL OF ATTENTION

The voluntary and flexible control of attention encompasses working memory, selective attention, sustained attention, divided attention, and alternating attention. In the parlance of clinical neuropsychologists, **working memory** is the process of holding information in mind, performing a mental operation on the information, and producing a result. Working memory facilitates complex tasks that involve multiple steps requiring active monitoring or manipulation of information. **Selective attention** is the ability to focus on a task or stream of information in the presence of distracting stimuli. **Sustained attention** (**vigilance**) is the ability to maintain focused attention for a continuous period. **Divided attention** is the ability to attend to more than one relevant stimulus or process at one time. **Alternating attention** involves the ability to switch attention between tasks.

Working memory tasks include recalling digit spans backward, mental sequencing, spelling words backward, and mental calculations. A common working memory task is serial sevens, which involves counting backward from 100 by sevens. Selective attention is often evaluated with cancellation tasks. These typically are paper-and-pencil tests in which the patient visually searches for and marks ("cancels") target items within an array of stimuli (e.g., instances of the letter "A" within an array that contains all letters of the alphabet). Performance measures may include the number of **omission errors** (misses), number of **commission errors** (false alarms), and time to complete the task. Sustained attention is often evaluated by computer-administered continuous performance tasks that present a continuously changing stream of stimuli over a long duration of time, in which a rarely occurring target stimulus is embedded. The patient is required to respond only to the target stimulus.

PROCESSING SPEED

Processing speed refers to the speed and efficiency with which cognitive tasks are performed. It has been conceptualized as reflecting the speed with which the attentional system can process information. Processing speed is evaluated by various timed tasks. Slow cognitive processing, referred to as **bradyphrenia**, is often found in the context of subcortical dementias due to a variety of etiologies.

Visuoperceptual Function and Spatial Cognition

This domain includes assessment for visuoperceptual disorders such as apperceptive visual agnosia, achromatopsia, visual object agnosia, and prosopagnosia (see Chapter 14, "The Occipital Lobes and Visual Processing"). Disorders of spatial cognition include visuospatial processing disorders, topographic disorientation, body schema disorders, and disorders of spatial attention (see Chapter 15, "The Parietal Lobes and Associated Disorders").

In most cases referred for neuropsychological assessment where there is no specific concern about disorders of visual agnosia or spatial cognition, specific specialized testing is not necessary. Visual confrontation naming test performance, in which the patient identifies objects in line drawings, can be used to rule out apperceptive agnosia and visual object agnosia. Disorders of spatial attention can be ruled out by visual scanning, visual search, and construction tasks.

VISUOPERCEPTUAL FUNCTION

Apperceptive agnosia will manifest on most tasks involving presentation of visual stimuli, including filling out administrative forms. Formal assessment for this disorder includes **visual form discrimination** testing.

Visual object agnosia is a more specific disorder of object recognition in which visuoperceptual abilities are preserved. Visual confrontation naming tasks can be used to evaluate for this disorder. Patients with visual object agnosia cannot identify objects through the visual modality, but they can provide detailed descriptions of shape, size, contour, position, and number, and their object identification errors usually consist of objects that are similar in shape to the target, demonstrating that they can make use of shape information. Semantic knowledge about the object is intact, as evidenced by the fact that object identification by other sensory modalities (e.g., touch, sound) is preserved.

SPATIAL COGNITION

The brain processes spatial information by forming mental representations of space, including maps of the environment within the immediate visual space, the external environment beyond the immediate visual space, and the body. These mental maps form the basis of a broad range of abilities collectively referred to as **spatial cognition**. Disorders of spatial cognition are particularly associated with parietal lobe function (see Chapter 15, "The Parietal Lobes and Associated Disorders").

Mental representations of visual space provide information about objects' locations and relations among objects. Visuospatial processing disorders include impaired ability to localize objects in space, judge depth, and judge angular orientation (i.e., judgment of line orientation).

Mental representations of the external environment provide information about where locations are in three-dimensional space and their spatial relationships relative to other locations (topographic knowledge); this information forms the basis for our ability to develop and process navigational routes. **Topographic disorientation** results from a deficit in forming and/or retrieving abstract maps of the spatial layout of familiar environments. This manifests as an inability to find one's way in familiar environments and learn new routes; this commonly occurs with Alzheimer's disease. It may be formally assessed by testing the ability to locate items such as countries or cities on a map.

When questions arise around spatial cognition and/or posterior parietal function, the neuropsychologist also assesses for disorders of body schema, namely autotopagnosia, finger agnosia, and right-left disorientation, by testing body part localization and right-left discrimination.

SPATIAL ATTENTION

Unilateral spatial neglect and simultanagnosia are conceptualized as disorders of spatial attention (see Chapter 15, "The Parietal Lobes and Associated Disorders"). These syndromes are particularly associated with lesions of parietal cortex. Performance on tests of constructional ability and tests requiring rapid visual search can rule out disorders of spatial attention. If there is a question of possible unilateral spatial neglect, further assessment may involve horizontal line bisection tasks and cancellation tasks. Simultanagnosia may be assessed by asking the patient to describe a complex visual scene.

Praxis and Constructional Ability

Praxis is the ability to perform skilled (learned) movements such as those required for tool use and communicative gestures. **Apraxia** is a disorder of cortical programming of voluntary, purposeful, skilled movements (see Chapter 24, "The Motor System and Motor Disorders"). There are many forms of apraxia, and apraxia occurs with various neurological disorders. It is especially prevalent in left hemisphere stroke, but is also common in neurodegenerative disorders such as Alzheimer's disease, corticobasal syndrome, and progressive supranuclear palsy. Several structured tests can aid in the identification of apraxia. Both transitive (involving tools) and intransitive (communicative or gestural) movements are assessed. Movements involving the upper limbs, lower limbs, buccofacial muscles, and trunk muscles are assessed.

Constructional ability refers to the ability to assemble, join, or articulate parts and arrange component elements in their correct spatial relationships to construct a single unitary structure. Constructional ability is complex and entails many cognitive processes, including visuospatial processing, spatial attention, praxis, and executive functions. Intact performance on tests of constructional ability rules out more fundamental deficits in visuospatial processing and spatial attention.

Constructional ability is assessed with tasks requiring the patient to copy and draw geometric shapes or pictures of objects, and to reproduce other stimuli such as block constructions and stick constructions that do not require a graphomotor response. Block design tasks require the patient to rearrange blocks with various color patterns on different sides to match a target design. Assessment of constructional ability relies on adequate vision and sufficient motor ability to use paper, pencil, and blocks.

Several specific types of drawing errors are usually accepted as pathognomonic of brain disorders. These include rotation by more than 45 degrees, perseveration or repetition of the entire figure or part of it, fragmentation of the figure or omission of major elements, and significant difficulty in either the integration or the placement of individual parts at the correct angles or locations. "**Closing in**" is a form of stimulus-bound behavior that occurs on copying and block design reproduction tasks; the work is executed directly on the model by drawing over the

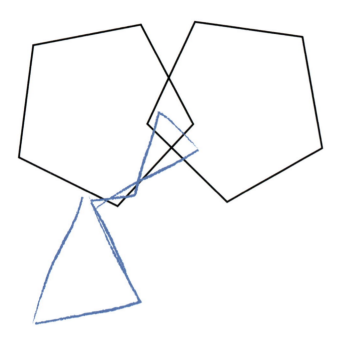

FIGURE 31.1. "Closing in" is a form of stimulus-bound behavior that is observed on copying tasks and is characterized by drawing over the model or tracing over the lines of the model.

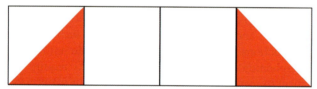

FIGURE 31.2. A block design error that is considered indicative of brain injury or disease is "stringing out" in which blocks are arranged in a row, violating the square matrix configuration of the target design.

model, tracing over lines of the model, or placing blocks directly on top of the stimulus cards (Figure 31.1). Another block design error that indicates brain injury or disease is a broken configuration, where blocks are arranged in a configuration that violates the 2 × 2 or 3 × 3 square matrix of the target design, such as "stringing out" the blocks by arranging them in a single row (Figure 31.2).

Language

Language is evaluated formally with specific attention to spontaneous expressive language, fluency, comprehension, repetition, naming, reading, and writing (see Chapter 27, "Language and the Aphasias"). Behavioral observations establish whether there are overt signs of basic expressive or receptive language difficulties. Spontaneous expressive language is evaluated for fluency, grammatical structure, and paraphasias. Language comprehension is evaluated for the ability to readily understand questions and test instructions.

Patients who do not present with complaints of decline in language function, or who do not show signs of language difficulties during the clinical interview, do not require in-depth assessment of language. A combination of confrontation naming and verbal fluency tasks with screening test items for repetition, reading, and writing usually suffices. Verbal fluency tasks require the patient to produce as many words as possible from a category (**category verbal fluency**) or beginning with a specific letter (**letter verbal fluency**) in one minute. More in-depth assessment of language is required for patients presenting with complaints of decline in language function or signs of language difficulties during the clinical interview.

CONFRONTATION NAMING

Anomia is the most common language deficit; therefore, most neuropsychological test batteries include **confrontation naming** tests for objective and quantitative assessment of this disorder. **Visual confrontation naming** tests typically involve presenting line drawings of objects to the patient and asking for the name of the object. Because severity of anomia varies widely, it is important to assess uncommon, low-frequency words as well as common, high-frequency words. Many anomic patients can name common objects quickly and accurately, but manifest great hesitancy, paraphasias, and circumlocutions when asked to name uncommon objects. It is important to distinguish whether poor performance is due to difficulty retrieving the word versus other factors such as visual agnosia or potential non-neurological confounds (e.g., low level of education, testing in the patient's non-preferred language). Patients with anomia demonstrate their recognition of the object by circumlocution and gesturing its use.

COMPREHENSION

Language comprehension may be evaluated formally independent of language expression with tasks that require minimal or no verbal response, such as answering yes/no questions or pointing to command (e.g., "point to the clock"). Asking the patient to answer general or open-ended questions that require constructing complex answers does not specifically assess language comprehension because it depends upon the integrity of the entire language system. Single-word comprehension may be assessed by word-picture matching tests.

REPETITION

Repetition of spoken language (i.e., repeating words and sentences back to the examiner) is a complex task that can be undermined by disturbances in auditory processing or speech production, or a disconnection between receptive and expressive language functions. Repetition testing is performed using items of ascending difficulty, beginning with single monosyllabic words, and proceeding to complex sentences. Errors include paraphasias, grammatical errors, omissions, and additions. In examinations performed focusing on language functions, evaluation of repetition is especially important since poor performance is a discriminating feature of aphasia syndromes.

READING AND WRITING

For some examinations, formal assessment of reading and writing is useful (see Chapter 28, "Alexia, Agraphia, and Acalculia"). It is important to consider educational background and history of learning disabilities, which are established prior to testing. In the literate, non-aphasic patient, assessment can begin with sentence reading and sentence writing. If a problem is identified with sentences, then single word and letter reading and writing are assessed for greater specificity. Comprehensive language assessment instruments include subtests pertaining to different aspects of reading and writing.

Number Processing and Calculation

Number processing and calculation abilities rely on language, spatial, and executive function abilities (see Chapter 28, "Alexia, Agraphia, and Acalculia"). Brain injury and disease frequently result in a decline in ability to perform calculations, and while neuropsychological evaluations usually include items that screen for this, some cases may require in-depth examination of number processing and calculation abilities. Neurological classifications of number processing and calculation disorders distinguish primary acalculia and five types of secondary acalculia: aphasic, alexic, agraphic, spatial, and dysexecutive (frontal) acalculias.

Learning and Memory

Memory disturbances are the most common complaint in people with brain disorders, but they also are common in those with other neurocognitive disorders (e.g., attentional disorders, anomia), depression, or anxiety.

Episodic memory refers to the ability to consciously recollect events ("episodes") that one can travel back to in the mental experience (see Chapter 26, "Memory and Amnesia"). It enables us to recall information over a period of minutes to days, and to remember current, day-to-day events; it also forms the basis for longer-term memories. Anterograde amnesia is the hallmark sign of hippocampal pathology, but it also occurs with damage to the fornices, diencephalon (thalamus and hypothalamus), and basal forebrain, which form circuits with the hippocampal formation.

The ability to form and maintain new episodic memories is the result of the processes of encoding, storage, and retrieval. **Encoding** is the process by which memories form. Ineffective encoding may occur because of insufficient attention and/or ineffective strategic processes during the acquisition phase. Ineffective encoding results in deficient learning, the process of acquiring new information and improving recall with successive presentations of the same information. **Storage** is the ability to retain information over time. A deficit in storage results in both deficient free recall and deficient recognition memory after a delay. Defects in storage result in rapid forgetting. **Retrieval** is the process by which stored memories are accessed or recalled. Selective retrieval deficits manifest as deficient delayed free recall in the context of intact delayed recognition memory.

In clinical practice, learning refers to the improvement of recall after successive presentations of the same information, while memory refers to the ability to retain the information over a delay. Episodic memory disorder (anterograde amnesia) is the inability to learn and remember new material after a brain insult.

The most sensitive tests of episodic memory are word-list learning and memory tests, which require the patient to learn a list of words over several trials and recall it after a delay filled with distractor tasks to prevent rehearsal. To distinguish whether poor delayed recall is due to impaired retrieval versus impaired encoding and/or storage, the neuropsychologist also tests **recognition memory** by asking the patient to identify targets and foils from a larger list of words. Patients with defective encoding and/or storage fail to recall and recognize stimuli, while those with defective retrieval fail to recall but succeed in recognizing stimuli. Other tests used to assess episodic memory utilize stimuli such as short stories, paired associate word lists, and geometric shapes. Both verbal and nonverbal memory are assessed in many cases.

When evaluating memory, the neuropsychologist must be aware that deficits in more basic cognitive processes can result in poor memory test performance. Performance on memory tests requires sustained attention; inattentive, distractible patients have difficulty performing optimally on memory tests. Disturbances of language function may result in low scores on verbal memory tests, such as word-list learning and memory tasks, and paragraph recall tasks. Valid memory testing presumes that the patient is reasonably attentive, and that no other impairment is primary in undermining task performance.

Abstract Thinking and Reasoning

Abstract thinking, also known as **concept formation**, is the most advanced of the cognitive abilities. It allows us to use concepts to form and understand generalizations and theoretical relationships that go beyond obvious, concrete, salient characteristics. By contrast, concrete thinking is based on direct (but not necessarily current) experience with the world. Problem-solving may or may not involve abstract thinking, but formation of an abstract concept often facilitates problem-solving. **Reasoning** is the use of logic in the service of problem-solving. Abstract thinking and reasoning relate to intelligence; therefore performance on tests of these abilities is interpreted within the context of both educational background and premorbid intelligence. These higher cognitive functions are disrupted by injury or disease affecting the frontal lobes, but deficits may also be particularly prominent in the context of widespread bilateral cortical neurodegenerative disease.

Some tests of abstract thinking and reasoning involve bottom-up inferences from the specific (or concrete) to the general (or abstract). For example, the ability to identify higher-level properties or patterns shared by several specific items or events may be assessed by asking the patient to articulate the shared attributes of dissimilar objects (i.e., finding similarities). Another example is the ability to use and understand metaphors, in which concrete concepts convey generalities. This is often assessed with proverb interpretation (e.g., *Don't judge a book by its cover*; *Still waters run deep*).

Other tests of abstract thinking involve top-down inferences from the general to the specific. This type of reasoning may be assessed with tests based on the game of "20 Questions": the patient is presented with an array of pictures of objects, each of which can be categorized in multiple ways, and asked to deduce which one the examiner has in mind by asking the fewest number of yes/no questions. Successful performance requires the patient to categorize, formulate abstract yes/no questions, and incorporate the examiner's feedback to formulate more efficient questions. Deficits are reflected in the patient's use of ineffective categorization strategies, such as relying on questions that refer to single items; this reflects **concrete thinking**.

Executive Functions

Executive function is an umbrella term for high-level cognitive processes involving the top-down control and direction of lower-level cognitive abilities that enable us to plan and carry out goal-oriented behavior. An individual with preserved general intelligence, no domain-specific cognitive deficits (e.g., amnesia, aphasia, agnosia), and no severe psychiatric illness, but who nonetheless is unable to carry out self-directed, adaptive behavior, is often described clinically as having impaired executive functioning. Impairments in executive function profoundly impact daily function and quality of life, as these cognitive skills are implicated in job performance, social relationships, and instrumental activities of daily living.

The concept of executive functions grew out of early observations of patients with frontal lobe damage. Despite the historical link between executive and frontal lobe function, there is increasing awareness that these terms are not synonymous, since non-frontal brain regions contribute to executive control functions. Executive function tasks activate distributed neural networks that prominently involve the prefrontal cortex, but also include the parietal cortex, basal ganglia, thalamus, and cerebellum. Furthermore, prefrontal cortices have functions outside of the domain of cognitive control processes, particularly with respect to the personality, affect, and behavioral regulation. Executive dysfunction is often associated with frontal lobe pathology, but it is also often associated with cerebral white matter pathology.

The executive functions encompass a wide array of abilities, including initiation, perseverance, executive control of attention, inhibition of automatic responses, planning, organizing, sequencing, cognitive flexibility, and self-monitoring (see Chapter 17, "The Frontal Lobes and Associated Disorders"). Because the executive functions are multifaceted, no single test assesses all aspects of executive function. Furthermore, many executive function tests evaluate more than one aspect of executive function. Prefrontal cortex is divided into three anatomical and functional regions that mediate distinct cognitive, behavioral, and emotional processes; dorsolateral, orbitofrontal, and medial frontal. Dorsolateral prefrontal pathologies are most associated with the dysexecutive syndrome, while orbitofrontal pathologies are more associated with a disinhibited syndrome, and medial frontal-akinetic pathologies are more associated with an apathetic syndrome.

Executive dysfunction may also be reflected in qualitative test performance. Poor planning and organization may be apparent on tests of constructional ability, as the patient takes a piecemeal, disorganized approach. Tests of executive functioning require top-down cognitive control to direct responses according to task goals. **Set loss errors**, also called **rule violation errors**, involve a breakdown in the adherence to task-specific rules. Such errors may reflect poor executive control of attention, a tangential cognitive process, or impulsivity.

INITIATION AND PERSEVERANCE

Behavioral spontaneity deficits are tapped by **letter verbal fluency** tasks in which subjects are asked to generate as many words as they can that begin with a specific letter, within one-minute trials. Poor performance in non-aphasic patients often indicates left frontal dysfunction, assuming no other non-neurological factors confound the test data (e.g., limited facility with the English language, reduced cooperation, feigning symptoms).

Behavioral spontaneity deficits are also associated with unusually slow responses on tasks requiring idea generation, for example, when defining words or identifying the similarity between two items. Patients with abulia perform normally on highly structured tasks that require the patient to respond in specific ways to specific stimuli; therefore, they may present well within the structured exchange of the neurological examination but perform poorly and/or slowly on neuropsychological examination for any task requiring idea generation.

EXECUTIVE CONTROL OF ATTENTION AND INHIBITION OF AUTOMATIC RESPONSES

Executive control of attention may be assessed by **Stroop tasks**, which are used to assess the ability to inhibit cognitive interference. The test involves three conditions: (1) reading color-words (red, green, or blue); (2) naming patches of color (red, green, or blue); and (3) an incongruent (mismatch) condition in which color-words are printed in an incongruent color ink (e.g., the word *red* printed in blue ink) and the patient is required to name the printed color of the words rather than read the color-words (Figure 31.3). The patient is required to perform the task as quickly as possible within a fixed time interval (e.g., 45 seconds) in all three conditions. The incongruent condition requires the selective processing of a less salient stimulus (i.e., ink color) while continuously blocking out the processing of a conflicting prepotent stimulus. This is cognitively challenging and results in slow, error-prone responding. This delay in response time between automatic and controlled processing of information is known as the **Stroop effect** or **Stroop interference effect**; it was originally described by American psychologist J. Ridley Stroop (1897–1973) in 1935 and is one of the best-known phenomena in cognitive science. The Stroop interference effect is particularly pronounced in patients with poor executive control of attention and disinhibition due to frontal lobe dysfunction, and manifests as an inability to suppress the automatic response of reading words.

PLANNING AND STRATEGY FORMATION

Planning is evidenced by the ability to accomplish a desired goal through a series of intermediate steps; it entails organization and sequencing. Planning ability is often assessed by **Tower tests** that are based on the "Tower of Hanoi" puzzle (Figure 31.4). The apparatus consists of three rods and several disks of different sizes that can slide onto any

Trial 1	Trial 2	Trial 3
RED	XXXXX	GREEN
GREEN	XXXXX	BLUE
BLUE	XXXXX	RED
BLUE	XXXXX	BLUE
RED	XXXXX	RED
GREEN	XXXXX	RED
RED	XXXXX	GREEN

FIGURE 31.3. Stroop tasks involves three conditions: (1) reading color-words; (2) naming patches of color; and (3) an incongruent condition in which color-words are printed in an inconsistent color ink and the patient is required to name the color of the ink instead of reading the color-word. In the incongruent condition, responding is slow and error-prone, reflecting the difference between automatic and controlled information processing. The Stroop effect is greater than normal in patients with poor executive control of attention and disinhibition due to frontal lobe dysfunction.

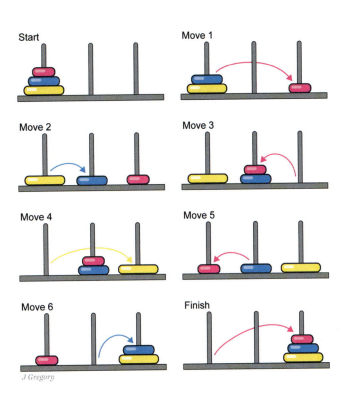

FIGURE 31.4. Tower test. The patient is presented with a starting configuration of disks placed on the rods. The objective is to rearrange the disks on the rods to match a target configuration with the fewest number of moves, while obeying the following rules: (1) only one disk can be moved at a time, and (2) no disk may be placed on top of a smaller disk. Efficient performance requires planning and strategy formation.

rod. For each trial, the patient is presented with a starting configuration of disks placed on the rods and asked to rearrange the disks on the rods to achieve a target configuration. Difficulty increases across trials with the number of disks and the number of moves required. The objective is to rearrange the disks on the rods to match a target configuration with the fewest number of moves, while obeying the following rules: (1) only one disk can be moved at a time, and (2) no disk may be placed on top of a smaller disk. This task requires patients to break the problem into subgoals, mentally plan a sequence of moves to achieve each subgoal, and monitor one's success in achieving each subgoal. Patients with poor planning and problem-solving skills have difficulty formulating an effective strategy and have a trial-and-error approach rather than developing a hierarchy of subgoals to achieve the goal. Impulsive responding without a plan of action also undermines task performance.

Inability to plan and generate strategies may also be apparent on visuoconstruction tasks, as patients fail to mentally break target designs into their component parts and often use a disorganized, piecemeal approach. Surprisingly, despite a piecemeal approach to copying relatively complex geometric designs, their constructions may ultimately be relatively accurate, as the spatial relationships of many elements are preserved. With such patients, constructional ability may be facilitated if they are provided with a partial structure and strategy, such as being asked to draw the components of a complex design sequentially or by providing a grid structure for block design tasks. A consequence of using a poorly organized approach to copying tasks is poor delayed recall, as organization and structure facilitate memory; such apparent "memory" failure is secondary to executive dysfunction and does not reflect primary memory disorder.

COGNITIVE FLEXIBILITY AND PERSEVERATION

Cognitive flexibility (set shifting) is a broad term that refers to the ability to adapt flexibly to changing conditions and to shift between different mental sets, responses, or strategies. Impaired cognitive flexibility can lead to **perseveration**, the application of old responses to new situations where such responses are inappropriate. Cognitive flexibility is frequently assessed by card-sorting tasks in which the patient is required first to deduce a sorting principle (e.g., color, shape, or number) based solely on correct/incorrect feedback, and subsequently, after an undisclosed change in reward contingency, to relinquish the old sorting principle and adopt a new one. Impaired cognitive flexibility results in perseverative responding on such tasks, where the patient fails to change their sorting strategy in response to feedback indicating that the old sorting principle is no longer in effect, and continuously responds incorrectly.

Motor perseveration involves repeating a single movement continuously. For example, when asked to draw a circle, the patient may perseverate by drawing circles until the pen is removed from their hand. Perseveration may also be conceptual and manifest as blending responses. For example, if a patient is asked to write to dictation and then to draw, letters or words may be incorporated into the drawing. Perseveration may be most prominent when tasks are difficult, or the patient is fatigued.

SELF-MONITORING

Self-monitoring is a quality control system for behavior. It involves the ability to "observe" one's performance, evaluate its effectiveness, detect errors, and make corrections to improve performance. It also allows the ability to use feedback from the environment to regulate or modify one's behavior and limit activity according to rules. Deficits in self-monitoring and error detection manifest across a wide variety of tasks, including tests of constructional ability, Tower tests, and card sorting tasks. Patients behave as though they are unaware of their errors and simply continue responding.

MOTOR PROGRAMMING

Motor programming tasks may also evidence executive dysfunction, and they are often used as a component of a behavioral neurological examination. These include manual sequencing tasks, tests of conflicting instructions, and go/no-go tasks. **Manual sequencing tasks** include **Luria's three-step (fist-edge-palm) test**, in which the patient is asked to perform a sequence of three movements (i.e., tap the table with a fist, side of palm, and open palm) repeatedly and as quickly as possible (see Figure 17.4 in Chapter 17). The neuropsychologist demonstrates three sequences of these movements, and then the patient is asked to perform six sequences alone and without errors. **Written alternating sequencing tasks** include drawing an alternating sequence of triangles and squares ("ramparts" design) and writing a series of alternating script "m"s and "n"s (see Figure 17.5 in Chapter 17). **Tests of conflicting instructions**, which require the patient to respond to one movement by performing a different movement, are often used to assess for excessive sensitivity to interference. For example, in the **alternate tapping task**, the patient is asked to tap twice in response to a single tap given by the examiner, and once in response to two taps. Another variant is to ask the patient to put up two fingers (index and middle fingers in the form of a V) when the examiner puts up only one index finger and put up one finger when the neuropsychologist puts up two fingers. **Go/no-go** tests assess inhibitory control. In these tests, one class of trials requires a response ("go" trial), while another class of trials requires withholding a response ("no-go" trial). For example, the patient may be instructed to tap once when the examiner taps once, but to not tap when the examiner taps twice.

Performance Validity Testing

Some patients intentionally fabricate or amplify cognitive symptoms, and this may be evaluated psychometrically by **performance validity testing**. **Malingering** is intentional feigning or exaggeration of symptoms that is motivated by external incentives, such as avoiding work or other duty, evading responsibility, evading incarceration, or obtaining financial gain such as disability pension, insurance settlement, or personal injury damages. **Factitious disorder** is intentional feigning or exaggeration of symptoms that is motivated by internal incentives, such as seeking sympathy for being ill (i.e., a need to "assume the sick role"). Performance validity testing can determine *whether* a patient's performance is an invalid reflection of their actual ability, but it cannot determine *why*; therefore, it cannot differentiate between malingered and factitious presentations per se.

Mood and Personality

Depression and anxiety may both be associated with subjective cognitive symptoms. Therefore, neuropsychological assessments often include screening for depression and anxiety by self-report symptom inventories.

Changes in personality and behavior that are associated with known or suspected frontal lobe pathology may be assessed by symptom questionnaires. Collateral information from family members or caregivers is often sought, as patients with personality and behavior change due to frontal lobe injury or disease typically have poor insight into their symptoms.

Diagnostic/Descriptive Conclusions

The Conclusions section of the report provides an answer the referral question, either to provide diagnostic clarification, or to describe the nature and extent of cognitive deficits from a known brain condition for the purposes of treatment planning and monitoring, or guiding decisions. A critical step in this process involves interpretation of the neuropsychological test data. It is important to recognize that successful performance on any specific test relies on multiple cognitive processes and not just the one that it was designed to tap, as this can lead to incorrect inferences. It is also important to be aware that scores do not communicate responses in full. Low scores on complex tasks may be due to a disturbance in any of the basic functions involved or any combination of them. For example, patients who copy a complex geometric design accurately using a piecemeal fragmented approach, rather than a well-structured organized approach, likely will have poor recall of the figure, not because of a primary disorder of memory but rather because they failed to exploit good perceptual organization which ordinarily would facilitate recall and accurate reproduction of the design. The examiner therefore must use several strategies to elucidate determinants of deficient performance and avoid collecting confounded test data. One strategy involves making qualitative observations about the component processes apparently involved in the patient's obtaining a particular test score. Another strategy involves observing the shared variance among tests by looking at patterns of deficient performance across a wide variety of tests and inferring the common underlying task demands. Multiple failures on seemingly disparate tests may reflect a common disorder. The magnitude of failure on the test will reflect the degree to which the disturbed function is represented in the composition. In many cases, a small number of functional disturbances will parsimoniously account for the many observed deficits in performance.

Diagnostic neuropsychological assessments integrate neuropsychological test findings with behavioral observations and clinical history, including findings from other studies such as neurological exams, neuroimaging studies, and electroencephalography. Each component of the neuropsychological examination is viewed within the context of the whole, and the diagnostic process involves a search for the presence or absence of corroborative features. Diagnostic conclusions summarize the clinical findings, address localization (if relevant), and identify the most likely diagnosis. They may also involve discussion of alternative possible diagnoses, or the rationale upon which differential diagnoses were ruled out and the logic underlying the diagnostic conclusion.

In cases where the underlying etiology is known, the conclusions drawn from the neuropsychological evaluation specify the nature and extent of cognitive deficits and spared abilities. Such assessments are performed for the purpose of guiding treatment and decisions about functional status, usually specified in a Recommendations section. Repeat neuropsychological evaluation is useful to monitor treatment efficacy or to monitor the course of injury recovery or disease progression.

Recommendations

Many neuropsychological assessment reports conclude with a section providing recommendations aimed at improving care, function, adjustment, quality of life, and safety. Further medical diagnostic workup (e.g., neuroimaging study, laboratory studies) is often recommended in cases with no history of neurological injury or disease when the neuropsychological examination reveals cognitive and/or behavioral deficits suggestive of a neurological disorder. Interventions and strategies for compensating for specific cognitive and behavioral disorders are often recommended in cases with a known brain injury, where the purpose of the neuropsychological assessment is to elucidate specific cognitive and behavioral strengths and weaknesses. Recommendations may address the need for assistance in functional activities of daily living such as managing medications or finances, or the ability to return to work, drive, or live independently. Additional services may be recommended, such as cognitive remediation or psychotherapy aimed at facilitating acceptance of and adjustment to losses and a new sense of self that often accompanies brain injury or disease. Specific accommodations at school or work may be recommended to enhance performance and optimize academic or occupational success. Recommendations may also include evidence-based activities and lifestyle changes for maintaining brain health and cognitive function during aging, such as physical exercise, dietary recommendations, social connectedness, and mental stimulation.

Summary

Neuropsychological assessment is an in-depth evaluation of cerebral functional integrity, and cognitive and behavioral function. It aids in establishing a diagnosis or describes the profile of the patient's cognitive deficits and abilities as they relate to a known brain condition, through a detailed examination that is centered on the presenting problem and the patient's known characteristics. The examination involves a clinical interview by which the neuropsychologist elicits the clinical history, behavioral observations, and formal neuropsychological testing. The domains of function evaluated include attention, visuoperceptual function, constructional ability, language, learning and memory, abstract thinking and reasoning, and executive functions. The neuropsychologist interprets and integrates information from the clinical history, behavioral observations, and test results, and communicates the findings in a structured written neuropsychological assessment report that culminates in a presentation of diagnostic conclusions and treatment recommendations.

Additional Reading

1. Gasquoine PG. Historical perspectives on ancient Greek derived "a" prefixed nomenclature for acquired neurocognitive impairment. *Neuropsychol Rev.* 2017;27(2):147–157. doi:10.1007/s11065-017-9346-4
2. Lezak MD, Howieson DB, Bigler ED, Tranel D. *Neuropsychological assessment.* 5th ed. Oxford University Press; 2012.
3. Luria AR. *Higher cortical functions in man.* Basic Books; 1966.
4. Morgan JE, Ricker JH, eds. *Textbook of clinical neuropsychology.* Psychology Press; 2008.
5. Morgan JE, , Baron IS, Ricker JH, eds. *Casebook of clinical neuropsychology.* Oxford University Press; 2010.
6. Parsons MW, Hammeke TA, eds. *Clinical neuropsychology: a pocket handbook for assessment.* 3rd ed. American Psychological Association; 2014.
7. Ravdin LD, Katzen HL. *Handbook on the neuropsychology of aging and dementia.* 2nd ed. Springer; 2019.
8. Scott JG. *The little black book of neuropsychology: a syndrome-based approach.* Springer; 2011.
9. Strub RL, Black FW. *The mental status examination in neurology.* 4th ed. F. A. Davis; 2000.
10. Vanderploeg RD, ed. *Clinician's guide to neuropsychological assessment.* Lawrence Erlbaum Associates; 2000.

Index

For the benefit of digital users, indexed terms that span two pages (e.g., 52–53) may, on occasion, appear on only one of those pages.
Tables, figures, and boxes are indicated by *t*, *f*, and *b* following the page number

4-4-4 rule, 380
5-hydroxytryptamine, 29, 30

A– (organic anions), 18*f*, 18
abducens nerve palsy (sixth nerve palsy), 71, 384
abducens nerves, 67*f*, 68*t*, 304
    overview, 71
    testing, 383–85
abducens nucleus, 71
abduction, 383
ablative surgery, 130
abnormalities, 58–60
abnormal movements, 390
abnormal posturing, 321–22
abscesses, 260, 279–80, 282–83
absence seizures, 249
absolute refractory period, 19*f*, 20
abstract thinking, 406
abulia, 222
acalculia, 194, 207–8, 353–54, 357–58
ACAs (anterior cerebral arteries), 52, 140, 229*f*
accessory nerves, 67*f*, 68*t*, 304
    overview, 73–74
    testing, 388
accessory nucleus, 73–74
accommodation reflex, 382–83
acetylcholine (ACh), 28–29, 303
acetylcholinesterase (AChE), 28, 303
acetylcholinesterase inhibitors (AChEIs), 29, 210
Achilles tendon, 392, 393*f*
Acom (anterior communicating artery), 54*f*, 334
acoustic agnosia, 205–6
acoustic neuroma, 72, 206, 274–75, 387
acoustic window, 373–74
acquired achromatopsia, 178–79
acquired AOS, 342
acquired aphasia with convulsive disorder, 251
acquired brain injury, 256
acquired dysarthria, 341
acromegaly, 117, 274
ACTH (adrenocorticotropic hormone)- secreting adenomas, 274
actin filaments, 303
action potential (AP), 2, 14, 19–21, 25*f*
    muscle, 303
    propagation, 20
    sodium-potassium pump (Na+/K+ pump), 20–21
action tremor (volitional tremor), 390
activation procedures, 244
active plaques, 293
active transport, 16*f*, 17
activities of daily living, 398
acute alcohol intoxication, 105
acute disseminated encephalomyelitis (ADEM), 295
acute radiation encephalopathy, 296
acute symptomatic seizures (provoked seizures), 240
Adams triad (Hakim's triad), 47
ADC (apparent diffusion coefficient), 371–72
addiction, 147
adduction, 383
ADEM (acute disseminated encephalomyelitis), 295
adenosine triphosphate (ATP), 17
ADH (antidiuretic hormone), 116
ADHD (attention-deficit hyperactivity disorder), 33
adipsia, 118
adipsic diabetes insipidus, 118
adrenal medulla, 119
adrenocorticotropic hormone (ACTH)–secreting adenomas, 274
adrenoleukodystrophy, 292
AEDs (antiepileptic drugs), 251–52
affect, 400
afferent paresis, 187
afferent pupillary defect, 382
afferents, 5
afterbrain, 10–11, 11*f*
after-hyperpolarization, 19*f*, 20
agenesis of corpus callosum, 165
aggression, 119, 141
agitated confusion, 325
agnosias, 160, 176
agonist, 27
agrammatic variant PPA, 350
agrammatism, 343
agranular cortex, 161, 213–14
agraphesthesia, 188, 391–92
agraphia, 180, 194, 353–54, 356–57
agraphia with alexia, 356
agraphic acalculia, 358
AHS (alien hand syndrome), 317
ahylognosia, 187
AICA (anterior inferior cerebellar arteries), 53, 98
AIDS dementia complex, 286
Akelaitis, Andrew, 163
akinesia, 126, 215, 222
akinetic mutism, 222
akinetic syndrome, 222
akinetopsia, 179
alcohol withdrawal delirium, 325
alertness, 319, 379
alexia, 180, 353–56
alexia with agraphia, 346, 354
alexia without agraphia. *See* pure alexia
alexic acalculia, 358
alien hand syndrome (AHS), 317
alien limb syndrome, 134
allocentric frame of reference, 192
allocortex, 162
allodynia, 188–89
all-or-none law, 20
alpha frequency wave, 242, 243*f*
alpha motor neurons, 303
alpha-synuclein, 129
ALS (amyotrophic lateral sclerosis), 311
altered states of consciousness, 320
alternate tapping task, 408–9
alternating attention, 402
alveus (alvear pathway), 201
Alzheimer, Aloysius ("Alöis"), 207
Alzheimer's disease, 29, 207–10, 332, 375*f*
    clinical presentation, 207–8
    diagnosis, 210
    epidemiology, 208
    etiology, 209
    pathology, 208–9
    treatment, 210
amaurosis fugax, 231
ambient cistern, 39–40, 368
ambulatory EEG, 244
amino acid neurotransmitters, 30
    GABA (gamma-aminobutyric acid), 30
    glutamate, 30
    glycine, 30

Ammon's horn, 200
amnesia, 29, 207, 329–38
   dissociative amnesia, 335
   forms of memory, 335–37
   forms of memory based on time span, 337–38
   lesions in non-medial temporal lobe structures, 333–34
   medial temporal amnesia, 331–32
   patient H.M., 329–31
   remote memory and retrograde amnesia, 332–33
   transient amnesia syndromes, 334–35
amnestic stroke, 332, 335
amorphognosia, 187
AMPA receptors, 30
amplitude, 202, 242
amygdala, 10, 140–43
   aggression, 141
   anatomy, 140
   fear and anxiety, 142–43
   Klüver-Bucy Syndrome, 141
amygdalae, 140
amygdalectomy, 141
amygdaloid complex, 200
amyloid hypothesis, 209
amyloidosis, 209
amyloid PET imaging, 375f
amyotrophic lateral sclerosis (ALS), 311
amytal ablation, 252–53
anabolic physiological processes, 62, 291–92
anaerobic metabolism, 49–50
analgesia, 22
anaplasia, 271
anarchic hand, 317
anarithmetia, 357–58
anastomosis, 59–60, 226
anesthesia, 22, 71
aneurysm, 233
aneurysm clipping, 234
aneurysm coiling, 59, 234
angiography, 57, 362
angioplasty, 234
angular gyrus, 152, 184–85, 354
anions, 15
anisocoria, 381
anisotropic diffusion, 372
anomia, 194, 207–8, 344
anomic aphasia, 347
anosmia, 70, 207, 221, 264, 380
anosodiaphoria, 194, 264–65
anosognosia, 194–95, 264–65
   anosognosia for hemiplegia, 194–95
   visual anosognosia, 195
ANS. See autonomic nervous system (ANS)
antagonist, 27, 302
antalgic gait, 395–96
anterior cerebral arteries (ACAs), 52, 140, 229f

anterior choroidal arteries, 41–42, 51
anterior cingulate cortex, 139–40
anterior cingulotomy, 140
anterior circulation, 51
anterior clinoid processes, 36
anterior column, 76f, 76–77
anterior commissure, 140, 156
anterior communicating artery (Acom), 54f, 334
anterior cord syndrome, 81–82, 82f
anterior cortical border zone, 230
anterior corticospinal tracts, 306
anterior fontanelle, 37
anterior forceps (forceps minor), 155–56
anterior fossa, 36
anterior horn, 41, 76f, 76
anterior inferior cerebellar arteries (AICA), 53, 98
anterior lobe, 94–95, 95f, 116–17
anterior lobule, 152
anterior median fissure, 76
anterior medullary velum, 41, 84–85
anterior nuclei, 111–12
anterior orbital gyrus, 212–13
anterior perforated substance, 53, 124
anterior-posterior axis, 6
anterior rami, 66
anterior (supraoptic) region, hypothalamus, 114
anterior root, 64, 65f
anterior spinal arteries (ASA), 53, 77
anterior spinothalamic tract, 80, 186
anterior territory infarcts, 113
anterior thalamic perforating artery, 110–11
anterior thalamic territory, 110–11
anterograde amnesia, 329
anterolateral sulcus, 76
anterolateral system, 80, 186
anteroposterior plane, 7–8, 363
anteropulsion, 127
antianxiety drugs, 33
antibodies, 22
anticonvulsants, 251–52
antidepressants, 32–33
antidiuretic hormone (ADH), 116
antiepileptic drugs (AEDs), 251–52
antigravity muscles, 307
antimanics (mood stabilizers), 33
antipsychotics, 31–32
Anton's syndrome (visual anosognosia), 195
anxiety, 142–43
   focal emotional seizures with, 248
   ictal fear, 143
   patient S.M., 142–43
anxiolytics, 33
aorta, 50
aortic arch, 50, 51f
AOS (apraxia of speech), 314

AP. See action potential (AP)
apathetic bvFTD, 224
apathetic syndrome, 222
apathy, 112, 222
aphagia, 118
aphasia, 340, 342–44, 347–48. See also language
   dynamic aphasia, 347–48
   language assessment, 342–44
   subcortical aphasia, 347
aphasic acalculia, 358
aphasic agraphia, 356
aphasiology, 340
aphemia, 158
apnea, 327
apparent diffusion coefficient (ADC), 371–72
appendicular ataxia, 102–3
appendicular muscles, 79
apperceptive agnosia, 176
apperceptive visual agnosia, 176–77
apraxia, 102, 163, 312–17, 403
   apraxia-like syndromes, 315–17
   apraxia of eyelid opening, 316
   apraxia of speech, 314
   apraxic agraphia, 314–15
   conduction apraxia, 315
   disconnection apraxias, 315
   ideational apraxia, 314
   ideomotor apraxia, 313–14
   verbal apraxia, 314–15, 342, 357
aprosodia, 342
aqueduct of Sylvius, 40f, 40–41
arachnoid cysts, 43
arachnoid granulations, 38f, 42
arachnoid membrane, 39–40
arachnoid trabeculae, 39
arachnoid villi, 42
ARAS (ascending reticular activating system), 112, 322
arbor vitae, 95f, 95
archicortex, 162
arcuate fasciculus, 345f, 346f, 346
arcuate fibers (U-fibers), 156
area postrema (chemoreceptor trigger zone), 55–56
Arnold-Chiari malformation, 106
arousal, 319
arterial stenosis, 58
arteries, 51–55
   blood supply of deep structures, 53–54
   cerebral arterial circle (circle of Willis), 54
   defined, 50
   extracranial origin of arteries supplying brain, 50–51
   internal carotid arterial system/anterior circulation, 51–52
   segmentation of major cerebral arteries, 54–55

vertebrobasilar arterial system/posterior circulation, 52–53
watershed zones, 54
arteriograms, 57
arteriopathy, 228
arteriosclerosis, 230
arteriovenous malformation (AVM), 59, 233
arteriovenous shunts, 59
articulation, 341
artifact, 369
ASA (anterior spinal arteries), 53, 77
ascending pathways of spinal cord, 79–80
ascending reticular activating system (ARAS), 112, 322
ascending somatosensory tracts, 85
aseptic meningitis, 43, 281
asomatognosia, 189
aspinous cells, 3
association cortex, 160
association fibers, 6, 156–57
association neurons, 3
association nuclei, 111–12
associative agnosia, 176
associative basal ganglia circuit, 125–26
associative visual agnosia, 163, 177–78
astereognosis, 187, 391
astereopsis, 179
asterixis, 136
astrocytes (astroglia), 4
astrocytomas, 4, 271
asynergy, 103, 394–95
ataxias, 101–4, 394
appendicular ataxia, 102–3
ataxic dysarthria, 103
truncal ataxia, 102
ataxic cerebral palsy, 311
ataxic dysarthria, 103, 341
atheroma, 226–27
atherosclerosis, 54
athetosis, 126, 390
atonia, 389
atonic seizures, 246, 249
ATP (adenosine triphosphate), 17
atrium, 41, 50
atrophy, 309
attention, 401–2
attention capacity, 337, 402
executive control of, 402, 407
processing speed, 402
attentional field, 193
attention-deficit hyperactivity disorder (ADHD), 33
attenuated delirium syndrome, 325
attenuation, 362
atypical antidepressants, 33
atypical antipsychotics, 32
audiogenic seizures, 251
audition, 202–6
cortical auditory disorders, 205–6

deficits, 204–5
ear, 202–4
illusions and hallucinations, 206
pathways, 204
stimulus, 202
auditory affective agnosia, 205–6
auditory association cortex, 160
auditory evoked potentials, 22, 205
auditory labyrinth, 72, 203, 204f
auditory meatus, 202
auditory radiations, 204
auditory receptor hair cells, 203
auditory sound agnosia, 205–6
auditory verbal agnosia, 205
aura, 243–44
auricle (pinna), 202
autoantibodies, 22
autoimmune inflammatory demyelinating diseases, 292–95
acute disseminated encephalomyelitis (ADEM), 295
Guillain-Barré syndrome (GBS), 295
multiple sclerosis (MS), 292–94
autoimmune limbic encephalitis, 146
automatic speech (non-propositional speech), 343
automatisms, 246
autonomic functions, 90
autonomic ganglia, 62–63
autonomic nervous system (ANS), 62–64
dysautonomia, 63–64
hypothalamus, 115
sympathetic and parasympathetic divisions of, 62–63
autoreceptors, 27
autoregulation, 260
autoscopy, 181, 248
autosomal recessive disorders, 292
autosomes, 129
autotopagnosia, 190–91
AVM (arteriovenous malformation), 59, 233
avoidance response, 317
awareness, 319
axial muscles, 79
axial sectioning plane, 7–8, 363
axoaxonic synapses, 3
axodendritic synapse, 3
axolemma, 17–18
axon, 2
axon hillock, 2, 20, 21f
axon terminals, 2–3, 25
axoplasm, 17–18
axosomatic synapses, 3

Babinski, Joseph, 103, 194
Babinski sign, 393–94
bacterial meningitis, 281
balance, 102
Bálint, Rezso, 193

Bálint's syndrome, 194
balloon angioplasty, 58, 234
bands of Baillarger, 161
barbiturates, 30
Bard, Phillip, 119
baroreceptors, 50
basal cistern, 39f, 39–40
basal forebrain, 143, 334
basal ganglia, 10, 123–37
anatomy, 123–26
calcification, 136
Huntington's disease, 135–36
neurodegeneration with brain iron accumulation (NBIA), 137
Parkinson plus syndromes, 131–35
Parkinson's disease, 126–31
Sydenham's chorea, 136
syndromes from pathology, 126
upper motor neurons, 309
Wilson's disease, 136
basal parietotemporal line, 153
basic activities of daily living, 398
basilar artery, 52–53, 54f
basilar membrane, 203
basilar pons, 87–88
basilar skull fracture, 257
basilar sulcus, 87
basis pontis, 87–88
basket cells, 3, 161
basolateral complex, amygdala, 140
bathing epilepsy, 251
Battle's sign, 257–58
BBB (blood-brain barrier), 4, 55–56
BDAE (Boston Diagnostic Aphasia Exam), 343
behavioral aspontaneity, 222
behavioral neuropharmacology. See neuropsychopharmacology
behavioral observations, 397, 399–400
behavioral regulation, 400
behavioral variant FTD (bvFTD), 223–24
Bell-Magendie law, 64
Bell's palsy, 72, 386
benign tumors, 270–71
Benson, D. Frank, 195, 221
Benson's syndrome (posterior cortical atrophy), 195–97
benzodiazepines, 30, 33
Berger, Hans, 240, 357
beta frequency wave, 242, 243f
Betz cells (giant pyramidal cells), 161, 213–14
biceps tendon, 392, 393f
Bigelow, Henry Jacob, 218
bilateral, 7
bilateral visual inattention, 193–94
binge/intoxication stage, 147
binocular disparity, 179
Binswanger's disease, 236, 300
biogenic amines. See monoamines
biopsy, 271

biparietal Alzheimer's disease, 196–97
bipolar cells, 170f
bipolar montages, 241
bipolar longitudinal montage, 241
bipolar transverse montage, 241
bipolar neurons, 3
Bisiach, Edoardo, 192
bitemporal hemianopia, 174
blind spot, 176
blink (corneal) reflex, 387
blood-brain barrier (BBB), 4, 55–56
blood oxygen level dependent (BOLD) technique, 376
Blumer, Dietrich, 221
blunt head injury, 261
Bodamer, Joachim, 178
bodily agnosia (autotopagnosia), 190–91
body, lateral ventricle, 41
body part as object, 313–14
body schema disorders, 190–91
Bogen, Joseph, 164
BOLD (blood oxygen level dependent) technique, 376
border zone infarcts, 230–31, 231f
Boston classification system, 343
Boston Diagnostic Aphasia Exam (BDAE), 343
botulinus toxin, 28–29
Bouillard, Jean-Baptiste, 158
bouquet cells, 161
bovine spongiform encephalopathy, 284
brachial plexus, 66
brachiocephalic (innominate) artery, 50, 51f
brachioradialis tendon, 392, 393f
brachium conjunctivum, 89, 96–97
brachium pontis, 87–88, 96–97
bradykinesia, 126, 127
bradyphrenia, 47, 128, 402
brain, 8–10
    blood supply of, 49–60
    brainstem, 8
    cerebellum, 9
    cerebrum, 9–10
    developmental basis of major subdivisions, 10–11
    limbic system, 10
brain aneurysm, 58f, 58–59
braincase. See cranium
brain death, 50, 327
brain-gut peptides, 30
brain herniation, 42–43, 260
brain infections, 279–87
    abscesses, 282–83
    classification, 279–80
    encephalitis, 281–82
    fungal infections of CNS, 283
    human immunodeficiency virus (HIV), 286
    Lyme disease, 287
    meningitis, 280–81
    parasitic infections of CNS, 283–84
    portal of entry, 280
    prion diseases, 284–86
brain ischemia, 50, 260
brainstem, 5f, 8, 67f, 84–91
    anatomy, 84–86, 92t
    clinical considerations, 91
    medulla, 86–87
    midbrain, 88–89
    pons, 87–88
    reticular formation, 89–91
brainstem auditory evoked potential, 205
brainstem stroke syndromes, 228
brainstem syndromes, 91
brain stimulation reward, 143–44
bregma, 37f, 37
bridging veins, 44
Broca, Paul, 138, 158, 344–45
Broca's aphasia, 158, 344–45
Broca's area, 344–45, 346f
Brodmann, Korbinian, 162
Brodmann areas, 162, 169f, 345f
Brownian motion, 15
Brown-Séquard syndrome, 82f, 82
buccal branch, 72
buccofacial apraxia, 313
buccofacial ideomotor apraxia, 313–14
Bucy, Paul, 141
bulbar, 87
bulbar-onset ALS, 311
bulbar palsy, 87, 309
bulbar reflexes, 327
bulbospinal pathways, 307
burst lobe, 260
bvFTD (behavioral variant FTD), 223–24
bypass, 59–60

CA (catecholamines), 29
CADASIL (cerebral autosomal dominant arteriopathy with subcortical infarcts and leukoencephalopathy), 230
calcarine fissure, 153, 168–69
calcification, 136
calculation, 405
callomarginal artery, 140
callosal angle, 366
callosal apraxia, 314
callosal disconnection, 163
callosal sulcus, 153
calvarium. See cranium
cancer stage, 271
Cannon, Walter B., 119
Capgras syndrome, 178, 208
CARASIL (cerebral autosomal recessive arteriopathy with subcortical infarcts and leukoencephalopathy), 230
carbidopa, 130
card game–induced reflex epilepsy, 251
cardiac muscle, 302
carotid bifurcation, 50
carotid canals, 50
carotid endarterectomy, 56–57, 234
carotid sinus (carotid bulb), 50
carriers, protein, 17
catabolic physiological processes, 62, 291–92
catalyzed reactions, 291–92
catecholamines (CA), 29
category verbal fluency, 404
cations, 15
cauda equina, 75–76
caudal, 6, 7f
caudal solitary nucleus, 73
caudate nucleus, 124, 135f
cavernous sinus meningiomas, 272
cavum septum pellucidum, 267
CBF (cerebral blood flow), 56–57
cell body, 2
center-surround antagonism, 170
central alexia, 346, 354
central canal, 41, 76, 86
central control of eye movements, 384–85
central cord syndrome, 82f, 82
central diabetes insipidus, 118
central fissure, 151
central herniation, 42f, 42
central nervous system (CNS), 35–48
    cerebrospinal fluid (CSF), 41–42
    clinical considerations, 42–48
    cranium, 35–37
    fungal infections of, 283
    meninges, 37–40
    parasitic infections of, 283–84
    peripheral nervous system vs., 5
    spine, 37
    subdivisions of, 9f
    ventricular system, 40–41
central nucleus, amygdala, 140
central pontine myelinolysis (CPM), 91, 298–99
central poststroke pain syndrome, 111
central sulcus, 151
cephalic flexure, 6
cerebellar ataxia, 394
cerebellar circuitry, 97
cerebellar cognitive-affective syndrome, 104
cerebellar cortex, 95, 101f
cerebellar falx, 38f, 38
cerebellar gait, 395–96
cerebellar granule cells, 97
cerebellar infarction, 104–5
cerebellar motor syndrome, 102
cerebellar peduncles, 96–97, 99t
cerebellar tonsils, 95f, 95, 106
cerebellar tremor, 103, 126, 390, 395
cerebellomedullary cistern (cisterna magna), 39f, 39–40
cerebellopontine angle, 87, 105, 272–73
cerebellorubral pathways, 97

cerebellothalamic pathways, 97
cerebellothalamocortical pathway, 97
cerebellum, 5f, 9, 94–106
    anatomy, 94–98
    cerebellar pathology, 104–6
    exam, 394–95
    functional divisions of, 98–101
    signs and symptoms of damage, 101–4
cerebral amyloid angiopathy, 233
cerebral aqueduct, 40f, 40–41
cerebral arterial circle, 54f, 54
cerebral atrophy, 261
cerebral autoregulation, 56, 260
cerebral autosomal dominant arteriopathy with subcortical infarcts and leukoencephalopathy (CADASIL), 230
cerebral autosomal recessive arteriopathy with subcortical infarcts and leukoencephalopathy (CARASIL), 230
cerebral blindness, 174–76
cerebral blood flow (CBF), 56–57
cerebral cortex, 10, 101f, 150–66
    agenesis of corpus callosum, 165
    anatomy, 151–54
    comparative neuroanatomy, 150–51
    cortical connections, 154–57
    cortical disconnections, 163–65
    cortical localization, 157–58
    cytoarchitecture of, 161–63
    functional maps of, 158–59
    pathways from, 306–7
    primary, secondary, and tertiary cortical zones, 159–60
cerebral edema, 42, 366–67
cerebral falx, 38f, 38
cerebral hemispheres, 10, 11
cerebral ischemia, 50, 260
cerebral laceration, 258
cerebral localization, 157
cerebral palsy, 310–11
cerebral peduncles, 88f, 89, 155
cerebral perfusion pressure, 56, 260
cerebral reserve, 266
cerebrocerebellum, 100–1
cerebromedullospinal disconnection, 327
cerebrospinal fluid (CSF), 4, 41–42
cerebrovascular anatomy, 49–60
    abnormalities, 58–60
    arterial supply of brain, 51–55
    blood-brain barrier (BBB), 55–56
    blood supply, basal ganglia, 124–25
    blood supply, brainstem, 86
    blood supply, cerebellum, 98
    blood supply, cingulate cortex, 140
    blood supply, spinal cord, 77
    blood supply, temporal lobes, 201
    blood supply, thalamus, 110–11
    brain venous blood outflow, 55
    cerebral blood flow and metabolism, 56–57
    circulation, 49–51
    imaging of cerebrovasculature, 57–58
cerebrum, 5f, 9–10
    cerebral hemispheres, 10, 11
    diencephalon, 10
cervical branch, 72
cervical enlargement, 78
cervical plexus, 66
chandelier cells, 161
channels, protein, 17
Charcot-Bouchard aneurysms, 232
Charcot-Marie-Tooth disease, 66
Charles Bonnet syndrome, 182
chemical neurotransmission, 24–34
    neuropsychopharmacology, 31–33
    neurotransmitter systems, 28–31
    synaptic transmission, 24–28
chemical senses, 206
chemistry, 15–16
    balance of concentration and electrostatic gradients, 15–16
    concentration gradients, 15
    electrostatic gradients, 15
chemoarchitecture, 25
chemobrain, 277
chemoreceptors, 206
chemoreceptor trigger zone, 55–56
chemosenses, 62
chemotherapy, 277, 297
chemotherapy-related cognitive impairment, 297
Chiari malformation, 105–6
Chiari malformation type I, 106
Chiari malformation type II, 106
Chiari malformation type III, 106
Chiari malformation type IV, 106
chiasmatic cistern, 39f, 39–40, 364, 368
chief complaint, 398
chief trigeminal sensory nucleus, 71
chloride (Cl–) ions, 18f
chlorpromazine (Thorazine), 31
chorea, 126, 390
choreoathetosis, 126
choroid plexus, 41–42
chronic alcoholism, 105
chronic focal encephalitis, 251
chronic painters' syndrome, 296
chronic traumatic encephalopathy, 267
chronic wasting disease, 284
cilia, 203
ciliary ganglion, 382f
cingulate cortex, 139–40, 152, 153
cingulate gyrus, 139f, 139, 153
cingulate herniation, 42f, 42, 364
cingulate motor area, 139–40
cingulate sulcus, 139, 153
cingulotomy, 140
cingulum, 139
circadian rhythms, 108, 121
circle of Willis, 54f, 54
circular sulcus, 152
circulation, 49–51
    extracranial origin of arteries supplying brain, 50–51
    systemic and pulmonary circulations, 50
circumlocution, 344
circumstantial thinking, 400
circumventricular organs, 55–56
CIS (clinically isolated syndrome), 294
cisterna magna, 39f, 39–40
CJD (Creutzfeldt-Jakob disease), 285–86
Cl– (chloride) ions, 18f
clasp-knife rigidity, 309, 389–90
classical conditioning, 336–37
classical LIS, 327
classical motor neuron disease, 311
classic aphasia syndromes, 344–47
classic Chiari malformation, 106
clinical considerations, 42–48
    arachnoid cysts, 43
    brainstem, 91
    electrical signaling in neurons, 22
    epidural and subdural hematomas, 43–45
    hydrocephalus, 45–48
    intracranial pressure, mass effect, and brain herniation, 42–43
    lumbar puncture, 45
    meningeal headaches, 43
    meningiomas, 43
    meningitis, 43
    spinal nerves, 66
    subarachnoid hemorrhage, 43
clinically isolated syndrome (CIS), 294
clinical neurophysiology, 22
clinical seizures, 243
clonic movements, 246
clonic seizures, 249
clonus, 389–90
closed head injury, 261
closing-in, 403–4, 404f
clouding of consciousness, 320
clumsy hand, 315
CN1. *See* olfactory nerves (CN1)
CN2. *See* optic nerves (CN2)
CN3. *See* oculomotor nerves (CN3)
CN4. *See* trochlear nerves (CN4)
CN5. *See* trigeminal nerves (CN5)
CN6. *See* abducens nerves
CN7. *See* facial nerves (CN7)
CN8. *See* vestibulocochlear nerves (CN8)
CN9. *See* glossopharyngeal nerves (CN9)
CN10. *See* vagus nerves (CN10)
CN11. *See* accessory nerves
CN12. *See* hypoglossal nerves (CN12)
CNS. *See* central nervous system (CNS)
cobalamin deficiency, 299
cochlea, 72, 203, 204f
cochlear branch, 203, 204f
cochlear duct, 203

cochlear implants, 204–5
cochlear nerve, 72
cochlear nerve ganglion, 203
cochlear nuclei, 72
cochleotopy, 204
cognitive flexibility, 221, 408
cognitive impairment, 128
cognitive motor disorder. *See* apraxia
cognitive process, 400
cogwheel rigidity, 127, 389
collateral branches, 90–91
collateral sulcus, 153, 200
collateral trigone (atrium), 41, 50
colloid cysts, 46
color anomia, 180–81
color coding, 171
columns, 162–63
coma, 259, 320, 322–23
    etiology, 323
    examination, 323
    pathophysiology, 322–23
    prognosis, 323
Coma Recovery Scale-Revised (CRS-r), 324
combined cerebral edema, 367
commission errors, 402
commissural fibers, 155–56
commissure of the fornix, 201
commissures, 6
commissurotomy, 163
communicating hydrocephalus, 45–46
comparative neuroanatomy, 150–51
complete SCI, 81
complex depressed skull fracture, 257
complex visual hallucinations, 181
complicated mTBI, 265
comprehension, 404–5
computed tomography (CT), 361, 368–47
    interpretation, 368–46
    limitations, 346–47
computed tomography angiography (CTA), 57, 369–47
    CT perfusion imaging, 347
concentration gradients, 15–16
concept formation, 406
conceptual apraxia, 314
conceptual errors, 191
conclusions section, neuropsychological assessment, 409
concrete thinking, 406
conductance decrease synapse, 21
conduction, 120
conduction aphasia, 163, 346
conduction apraxia, 315
conduction deafness, 204, 387
cones, 169
confabulation, 195, 222, 334
confabulatory pseudo-recognition, 195
confluence of sinuses, 38–39, 39f
confrontation naming, 404
confrontation visual field testing, 381

confusional state, 320, 322
congenital malformations, 105–6
conjugate eye movements, 383, 384
conjugate gaze palsies, 385
consciousness, 319–28
    coma, 322–23
    delirium, 324–26
    Glasgow Coma Scale (GCS), 320–22
    level of, 379
    minimally conscious state (MCS), 324
    reticular formation, 90
    syndromes mimicking disorders of, 326–27
    vegetative state, 324
consensual blink response, 387
constructional ability, 315, 403–4
constructional apraxia, 315–16
constructional impairment, 315
contagiousness, 280
contiguity errors, 191
contraction, muscle, 302–4
contralateral, 7
contralesional, 7
contrast enhancement, 363, 364f, 369
contrecoup injuries, 261
contusion, 258
conus medullaris (conus terminalis), 75
convection, 120
conventional arterial catheter angiography, 362
convexity, 152
convexity meningiomas, 272
coordination exam, 394–95
copper chelating agents, 136
co-registration, 374
Corkin, Suzanne, 330
cornea, 172f
corneal reflex, 387
coronal sectioning plane, 7–8, 363
coronal suture, 37f, 37
corona radiata, 155
corpora quadrigemina, 88–89
corpus callosum, 153, 155–56
    agenesis of, 165
    corpus callosotomy, 163–65, 254
Corsellis, Nick (John Arthur Nicholas), 267
cortex, 6
cortical auditory disorders, 205–6
cortical-brainstem-spinal cord pathways, 307–8
cortical deafness, 205
cortical dysplasia, 250
cortical equipotentiality (holism), 157
cortical eye movement centers, 384–85
cortical Lewy body disease, 131–32
cortical sensations, 185
cortical somatosensory function, 391–92
corticectomy, 253
corticobasal syndrome, 134
corticobulbar tracts, 85, 214, 305, 306–7

corticobulbospinal pathways, 307
corticofugal fibers, 154–55
corticomedial region, amygdala, 140
corticopetal fibers, 154–55
corticopontine projections, 87–88
cortico-ponto-cerebellar pathway, 87–88, 97, 100
cortico-ponto-cerebellar-thalamocortical circuitry, 100
corticoreticulospinal pathways, 90, 307
corticorubral tract, 307
corticorubrospinal pathways, 307
corticospinal tracts, 85, 214, 305, 306
cortico-striato-thalamo-cortical loop, 125
corticotectospinal pathways, 307
corticovestibulospinal pathways, 307
cortisol, 117
coup injuries, 261
CPM (central pontine myelinolysis), 91, 298–99
cranial foramina, 36
cranial fossae, 36
cranial nerve nuclei, 87
cranial nerves, 5, 66–74
    abducens nerves (CN6), 71
    accessory nerves (CN11), 73–74
    cranial nerve gaze palsies, 383–84
    exam, 379–88
    facial nerves (CN7), 71–72
    glossopharyngeal nerves (CN9), 72–73
    hypoglossal nerves (CN12), 74
    motor nuclei, 304
    oculomotor nerves (CN3), 70
    olfactory nerves (CN1), 69–70
    optic nerves (CN2), 70
    trigeminal nerves (CN5), 71
    trochlear nerves (CN4), 71
    vagus nerves (CN10), 73
    vestibulocochlear nerves (CN8), 72
cranial ultrasound, 373–74
craniopharyngioma, 274
craniosacral division, 63
craniotomy, 234, 267, 276
cranium, 5, 35–37, 365
crescentic, 365
Creutzfeldt-Jakob disease (CJD), 285–86
cribriform plate, 36
cribriform plate meningiomas, 272
crista galli, 36
Critchley, MacDonald, 267
crossed aphasia, 342
CRS-r (Coma Recovery Scale-Revised), 324
crude touch, 79, 185, 186, 390
crying
    focal emotional seizures with, 248
    pathological, 309
cryptogenic epilepsy, 249
CSF (cerebrospinal fluid), 4, 41–42
CSF tap test (lumbar tap test), 48
CT. *See* computed tomography (CT)

CTA (computed tomography angiography), 57, 369–47
CT perfusion imaging, 347
cuneate fasciculus, 77, 85, 185
cuneate nucleus, 86
cuneus, 153, 168–69
curare, 28–29
cutaneous mechanoreceptors, 185
cutaneous nerve distribution, 66
cutaneous thermoreceptors, 120
cysticercosis, 283–84
cysts, 43
cytoarchitecture of cerebral cortex, 161–63
    columnar organization, 162–63
        cortical cells, 161
        cortical layers, 161
        cytoarchitectural maps, 162
        types of cerebral cortex, 162
cytomegalovirus encephalitis, 298
cytotoxic cerebral edema, 367

DA (dopamine), 29
dacrystic seizures, 248
Dale's principle, 25
Damasio, Hanna and Antonio, 219
Dandy-Walker syndrome, 106
Darwin, Charles, 115
DA transporter, 129
DaTscan, 374
DAT-SPECT scan, 129
Dawson's fingers, 293–94
Dax, Marc, 158
dB (decibel) scale, 202
DBS. *See* deep brain stimulation (DBS)
deafferentation, 111
debulking, 276–77
decerebrate posturing, 89, 321–22
decibel (dB) scale, 202
declarative memory, 336
decomposition of movement, 103, 394–95
decompressive craniectomy, 104–5, 267, 366
decortication, 208
decussation, 7
dedifferentiation, 271
de-efferented state, 327
deep (intrinsic) cerebellar nuclei, 95–96
deep alexia, 355
deep brain stimulation (DBS), 254
    addiction, 147
    major depressive disorder, 147
    Parkinson's disease, 130
deep pontine nuclei, 87–88
deep tendon reflexes (DTRs), 304, 392–93, 393f
deep veins, 55
deinstitutionalization movement, 31
déjà vu, 248
Déjerine, Joseph Jules, 180, 354
Déjerine-Roussy syndrome, 111, 189

delayed non-match to sample task, 331
delayed post-hypoxic leukoencephalopathy, 299
deliriogenic drugs, 325
delirium, 324–26, 379
delirium tremens, 325
delta frequency wave, 242, 243f
delusions, 208
delusions of misidentification, 208
dementia, 12
dementia pugilistica, 267
dementia with Lewy bodies, 131–32, 182
demyelinating neuropathies, 66
demyelination disorders, 290, 297–98
dendrites, 2, 3
dendritic spines, 3
dendropsia, 181
denervating neuropathies, 66
denervation atrophy, 389
denervation supersensitivity, 27
Denny-Brown, Derek, 316–17
de novo, 11
dentate nucleus, 95–96, 101f
depersonalization, 248
depolarization, 19f, 19, 21f
depressed skull fracture, 257
depression
    deep brain stimulation (DBS), 147
    monoamine theory of, 32
depressive pseudo-dementia, 398
depth electrodes, 244
derailment, 400
derealization, 248
dermatomal map, 66
dermatomes, 64–66
descending motor tracts, 85–86, 305–9
descending pathways spinal cord white matter, 80
descriptive neuropsychological assessments, 398
detection threshold, 187
developmental AOS, 342
developmental dysarthria, 341
developmental history, 399
diabetes insipidus, 118
diagnosis
    Alzheimer's disease, 210
    epilepsy, 240–44
    multiple sclerosis (MS), 293–94
    testing motor system, 309–10
    tumors, 276
Diagnostic and Statistical Manual of Mental Disorders (DSM-5), 325b
diagnostic neuropsychological assessments, 398
diaschisis, 111
diencephalic amnesia, 111–12, 236, 334
diencephalon, 10–11, 11f, 108–22
    hypothalamus, 113–21
    thalamus, 108–13

differential diagnosis, 251
diffuse brain injuries, 261
diffuse Lewy body disease, 131–32
diffusion, 15–16, 16f
    balance of concentration and electrostatic gradients, 15–16
    concentration gradients, 15
    electrostatic gradients, 15
diffusion tensor imaging (DTI), 290–91, 372
diffusion-weighted MRI (DW-MRI), 371–72
diplopia, 70, 310, 383–84
dipole, 241–42
direct blink response, 387
directional terms, 6–7, 7f
dirty-tie sign, 132–33
disconnection apraxias (disassociation apraxias), 315
disconnection syndrome, 163, 354
discriminative touch, 79, 185, 390, 391
disease, 280
disease-modifying therapy, 130
disinhibited syndrome, 221–22
disinhibition, 221
disorders of multiple system degeneration. *See* Parkinson plus syndromes
disseminated necrotizing leukoencephalopathy, 297
dissociation in sensory loss, 187
dissociative amnesia, 335
dissociative amnesia with fugue, 335
dissociative disorders, 335
distal, 7
distal field infarcts, 230–31, 231f
divided attention, 402
DM (dorsomedial nucleus), thalamus, 112
doll's eye reflex, 99–100, 180, 193, 385
doll's eye test, 132–33
donepezil, 210
dopamine (DA), 29
dopamine transporter, 374
dorsal, 6, 7f
dorsal-caudal axis, 6
dorsal cochlear nucleus, 204
dorsal column, 76f, 76–77
dorsal column–medial lemniscal system, 79–80, 185–86
dorsal column nuclei, 86, 186
dorsal horn, 76f, 76
dorsal intersegmental tract, 77
dorsal lateral geniculate nucleus, 173
dorsal median sulcus, 76
dorsal motor nucleus of vagus, 73
dorsal pons, 88
dorsal root, 64, 65f
dorsal root ganglia, 64, 65f
dorsal rootlets, 76
dorsal stem (superior division), middle cerebral artery, 52

dorsal stream lesions, 179–80
dorsolateral prefrontal syndrome, 221
dorsomedial nucleus (DM), thalamus 112, 126
double simultaneous stimulation, 192–93, 381, 392
dressing apraxia, 316
drift, 389
drinking, 118–19
drop attacks, 246, 249
drugs
　classification, 31
　defined, 31
　drug reward, 144–45
　toxic leukoencephalopathy and, 295–96
DTI (diffusion tensor imaging), 290–91, 372
DTRs (deep tendon reflexes), 304, 392–93, 393f
Dully, Howard, 220
dural folds, 38
dural reflections, 38f, 38
dural septa, 38
dural venous sinuses, 38–39
dura mater, 37–39
DW-MRI (diffusion-weighted MRI), 371–72
dynamic aphasia, 347–48
dynorphins, 31
dysarthria, 87, 309, 311, 341, 380
dysautonomia, 63–64, 128, 134
dyscalculia, 353
dyschromatopsia, 181
dysconjugate gaze, 383
dysdiadochokinesia (dysrhythmokinesis), 103, 395
dysesthesia, 188–89
dysexecutive acalculia, 358
dysexecutive syndrome, 221
dysgranular cortex, 213–14
dysgraphia, 353
dyskinesias, 126
dyskinetic cerebral palsy, 310–11
dyslexia, 353
dysmetria, 103, 395
dysmorphopsia, 181
dysmyelination disorders, 290
dysphagia, 73, 87, 309, 311, 380
dysphasia, 342
dysphonias, 73, 341–42, 380
dysprosody, 342
dysrhythmokinesis (dysdiadochokinesia), 103, 395
dystonia, 127

ear, 202–4
early delayed radiation encephalopathy, 296
early-stage Alzheimer's dementia, 349
eating epilepsy, 251

ecchymosis, 257–58
echolalia, 347
ECoG (electrocorticography), 244
Economo, Constantin von, 112
edema, 260
Edinger-Westphal nuclei, 70, 382
education history, 399
EEG. See electroencephalogram (EEG)
EEG montage, 241
effacement, 42, 363–64
efferents, 5
egocentric frame of reference, 192
Ehrlich, Paul, 55
electrical potential, 17–18
electrical signaling in neurons
　chemistry concepts governing diffusion, 15–16
　clinical considerations, 22
　membrane potentials, 17–22
　membrane structure and permeability, 16–17
electrocorticography (ECoG), 244
electroencephalogram (EEG), 22, 240–44
　International 10–20 System, 240–41
　interpretation, 242–44
　physiological basis of, 241–42
electrogenic pump, 20
electrographic seizures, 243
electrolytes, 15
electromyography (EMG), 22, 310
electrostatic force, 15
electrostatic gradients, 15–16
eloquent cortex, 254
emblems, 313–14
emboliform nucleus, 95–96
embolism, 227
embolization, 234
emesis, 55–56, 73
EMG (electromyography), 22, 310
emotional dysregulation, 221–22
emotional expression, 115
emotional incontinence, 133, 309
empty speech, 344
empyemas, 282
en bloc turning, 127
encephalitides, 146
encephalitis, 146, 281–82
　herpes simplex virus encephalitis (HSVE), 281–82
　rabies encephalitis, 281–82
encephalitis lethargica (sleeping sickness), 112
encephalocele, 106
encephalo-duro-arterio-synangiosis, 59–60
encephalomalacia, 246, 367–68
encephalomyelitis, 279
encephalo-myo-synangiosis, 59–60
encephalopathy, 243, 298
encoding, 405

end arteries, 124–25
endarterectomy, 58
endbrain, 10–11, 11f
endocrine function
　abnormalities, 276
　hypothalamus, 115–17
endocrine glands, 115–16
endogenous opioids (opioid peptides), 30–31
endorphins, 31
endoscopic third ventriculostomy, 46
endosteal layer, dura, 37–38
endovascular coiling (aneurysm coiling), 59, 234
engram, 330
enkephalins, 31
entorhinal cortex, 201
entrapment, 66
entry wounds, 261
environmental dependency syndrome, 222–23
environmental toxins, 295–96
enzymes, 291–92
EP (evoked potentials), 22, 205
ependymal cells, 4, 41–42
ependymitis, 279
ependymomas, 4, 272
Epi (epinephrine), 29
epidemiology
　Alzheimer's disease, 208
　multiple sclerosis (MS), 294
epidural anesthesia, 82
epidural hematomas, 43–45, 259
epilepsy, 239–54
　classification of seizures, 244–46
　defined, 239–40
　differential diagnosis, 251
　electroencephalogram (EEG) and diagnosis, 240–44
　etiology, 249–50
　focal onset seizures, 246–48
　generalized seizures, 248–49
　non-epileptic seizures, 250
　surgery, 252–54
　syndromes, 250–51
　treatment, 251–52
epileptic drop attacks, 249
epileptic kinetopsia, 179
epileptiform discharges, 244
epileptologist, 246
epinephrine (Epi), 29
episodic memory, 336, 405
epithalamus, 108
EPM (extrapontine myelinolysis), 298
EPSPs (excitatory postsynaptic potentials), 21, 25–26
equilibrium, 98
equilibrium potential (Nernst potential), 18
equipotentiality, principle of, 330
ethmoid bone, 35, 36f, 36

etiology, 158
    Alzheimer's disease, 209
        coma, 323
        delirium, 326t
Evans Index, 366f
evaporation, 120
evoked potentials (EP), 22, 205
exception words (irregular words), 354–55
excitatory postsynaptic potentials (EPSPs), 21, 25–26
executive functions, 126, 217, 406–9
exit wounds, 261
Exner's area, 357
exocrine glands, 115
exocytosis, 25
exotic ungulate spongiform encephalopathy, 284
expansion, tumor 270
explicit memory, 336
expressive aphasia, 345
extension, 302
extensor muscles, 79, 302
extensor plantar response, 393–94, 394f
extensor posturing, 89, 321–22
external (outer) ear, 202
external beam radiation, 277
external carotid artery, 50, 51f
external granular layer, 161
external gross anatomy, 75–76
external lumbar drainage, 48
external medullary lamina, 110
external pyramidal layer, 161
external ventricular drain, 104–5
exteroceptive sense, 79
extinction, 192–93, 392
extorsion, 383
extra-axial, 364–65
extracellular electrode, 17–18
extradural hemorrhage, 258
extrafusal muscle fibers, 303
extrageniculate pathways, 174
extraneuritic plaques, 209
extraocular muscles, 70, 383
extrapersonal space, 189–90
extrapontine myelinolysis (EPM), 298
extrapyramidal syndrome, 123–24
extrapyramidal system, 123–24
extrastriate cortex, 169
extrasylvian aphasias, 344t, 346–47
extravascular enhancement, 363
extraventricular drain, 104–5
extrinsic muscles, 70, 383
extrinsic tumors, 270
eye movements
    central control of, 384–85
    reflexes, 385
eye-opening response, 322

FA (fractional anisotropy), 372
facial expression muscles, 72

facial motor nucleus, 72
facial nerve palsy, 72, 380, 386
facial nerves (CN7), 67f, 68t, 304
    overview, 71–72
    testing, 386–87
facilitated diffusion, 17
facilitatory paratonia, 389
factitious disorder, 409
Fahr's disease, 136, 372–73
falcine meningiomas, 272
family medical history, 399
far neglect, 192
fasciculations, 303, 309, 388, 389
fasciculi, 6, 77f, 77, 156
fastigial nucleus, 95–96
fatal familial insomnia (FFI), 286
fear, 142–43
    focal emotional seizures with, 248
    ictal fear, 143
    patient S.M., 142–43
fear conditioning, 337
feeding, 118
feline spongiform encephalopathy, 284
fenestrations, 55
Ferrier, David, 158–59, 213
festination, 127
FFI (fatal familial insomnia), 286
fibers, muscle, 303
fiber tractography, 372
fibrin, 234
fight or flight response, 62, 119
final common pathway. See lower motor neurons (LMNs)
fine touch, 79, 185, 390, 391
finger agnosia, 191
finger-to-nose test, 103, 395
fissures, 151
fist-edge-palm (Luria manual sequencing task), 215, 216f, 408–9
5-hydroxytryptamine, 29, 30
flaccidity, 389
FLAIR (fluid attenuated inversion recovery) MRI, 371, 372f
flapping tremor, 136
flashbacks, 248
flexibility, 217
flexion, 302
flexor muscles, 79, 302
flexor plantar response, 393
flexor posturing, 89, 307, 321–22
flight of ideas, 400
flocculi, 95
flocculonodular lobe, 94–95, 95f, 96f
Flourens, Jean-Pierre, 157
flow void, 370
fluency, 341, 343
fluid attenuated inversion recovery (FLAIR) MRI, 371, 372f
fluoroscopic images, 362
focal aware seizures, 246

focal brain injuries, 261
focal cerebral ischemia, 50
focal cognitive seizures, 247–48
focal emotional seizures, 248
focal impaired awareness seizures, 246
focal motor seizures with preserved awareness, 246–47
focal neurological deficits, 275
focal onset seizures, 245, 246–48
    with impaired awareness, 248
    with preserved awareness, 246–48
focal seizures awareness unknown, 246
focal unaware seizures, 248
folium, 95
fontanelles, 37
foramen magnum, 36
foramen magnum herniation, 42f, 42, 95
foramen of Magendie, 41
foramen of Monro, 40f, 40–41
foramina, 36
foramina of Luschka, 41
forced grasping, 316–17
forced hyperphasia, 223
forced person-following, 223
forced thinking, 248
forceps major, 155–56
forceps minor, 155–56
forebrain. See cerebrum
foreign accent syndrome, 342
formal thought disorder, 400
fornix, 41, 201, 333
fortification, 181
4-4-4 rule, 380
fourth nerve palsy, 71, 384
fourth ventricle, 40f, 40–41
fovea, 169–70
foveal vision, 170
fractional anisotropy (FA), 372
fractionated stereotactic radiotherapy, 277
Freeman, Walter, 219–20
frequency, 202, 204f, 242
frequency selectivity, 203
frequency tuning, 203
Fritsch, Gustav, 158, 213
frontal acalculia (dysexecutive acalculia), 358
frontal adynamia, 222
frontal amnesia, 334
frontal apraxia, 316–17
frontal bone, 35, 36f, 37f
frontal dynamic aphasia, 347–48
frontal eye fields, 180, 214f, 216–17, 384–85
frontal horn, 41
frontal lobes, 151, 212–24
    anatomy, 212–13
    behavioral variant FTD (bvFTD), 223–24
    motor cortex, 213–17
    prefrontal cortex, 217–19
    prefrontal injury and disease syndromes, 220–23
    prefrontal lobotomy, 219–20

frontal plane (coronal sectioning plane), 7–8, 363
frontal release reflexes, 394
frontal-subcortical dementia syndrome, 47
frontal systems dementia, 128, 291
frontosubcortical dementia, 128, 291
frontotemporal dementia (FTD), 223–24, 375f
frontotemporal dementia with motor neuron disease (FTD-MND), 311
frontotemporal lobar degeneration (FTLD), 223–24
full agonists, 27–28
full consciousness, 319
functional divisions, cerebellum, 98–101
    cerebrocerebellum, 100–1
    spinocerebellum, 100
    vestibulocerebellum, 98–100
functional imaging modality, 57–58
functional mapping, 252–53
functional neuroimaging, 374–76
    functional MRI (fMRI), 376
    positron emission tomography (PET), 374–76
    single photon emission computed tomography (SPECT), 374
functional neurological disorder, 250, 398
fundus, 381
funduscope (ophthalmoscope), 381
fungal infections of CNS, 283
funiculus, 6, 77
fusiform cells, 161
fusiform face area, 178
fusiform gyrus, 168–69

GABA (gamma-aminobutyric acid), 30
Gage, Phineas, 218–19
gag reflex, 388
gait
    exam, 395–96
    in NPH, 47
    Parkinson's disease, 127
    truncal ataxia, 102
gait apraxia, 316
gait ignition failure, 316
galantamine, 210
Gall, Franz Josef, 157
gamma-aminobutyric acid (GABA), 30
gamma camera, 374
ganglia, 6, 170f
ganglionic arteries, 53–54
ganglionic layer, 161
gated ion channels, 17
gating, 17
gaze apraxia, 180, 190, 316
gaze-evoked nystagmus, 103–4
gaze palsies, 380
Gazzaniga, Michael, 164
GBS (Guillain-Barré syndrome), 295
GCS. See Glasgow Coma Scale (GCS)

GCS Eye Opening Score, 264t
GCS Motor Response Score, 264t
GCS Verbal Response Score, 264t
GDP (guanyl nucleotide diphosphate), 26–27
gegenhalten, 316–17, 389
general amnestic syndrome, 332
generalized seizures, 245, 248–49
    generalized motor seizures, 249
    generalized non-motor seizures, 249
    status epilepticus, 249
generalized tonic-clonic seizures, 249
general somatic afferents, 62
general somatic efferents, 62
general visceral afferents, 62
general visceral efferents, 62
genetic CJD, 285
genetic metabolic diseases, 291–92
genetics, Parkinson's disease, 129
geniculocalcarine fibers, 173
genu, 155, 156
Gerstmann, Josef, 194
Gerstmann's syndrome, 194, 356
Gerstmann-Sträussler-Scheinker syndrome (GSS), 286
Geschwind, Norman, 163, 312–13
GH (growth hormone), 117
giant pyramidal cells, 161, 213–14
gigantism, 117, 274
gigantocellular reticular nuclei, 90
glabellar reflex, 127, 394
glabrous skin, 187
Glasgow Coma Scale (GCS), 262–63, 320–22
    eye-opening response, 322
    motor response, 320–22
    verbal response, 322
glia, 3–4
    astrocytes (astroglia), 4
    ependymal cells, 4
    gliomas, 4
    microglia, 4
    oligodendrocytes (oligodendroglia), 4
    Schwann cells, 4
glioblastomas, 271
gliomas, 4, 271–72
gliosis, 4, 293, 367–68
global amnestic syndrome, 332
global aphasia (mixed aphasia), 346
global cerebral ischemia, 50
globose nucleus, 95–96
glossopharyngeal nerves (CN9), 67f, 68t, 304
    overview, 72–73
    testing, 387–88
glutamate, 30
glycine, 30
gnosis, 160
GnRF (gonadotropin-releasing factor), 117
Goldstein, Kurt, 158, 317

Golgi, Camillo, 2
Golgi stain, 2, 161
Golgi Type II neurons, 3
Golgi Type I neurons, 3
gonadotropin-releasing factor (GnRF), 117
gonadotropins, 117
go/no-go tests, 408–9
G-protein (guanine nucleotide binding protein), 26f, 26–27
G protein-coupled receptors, 26–27
gracile fasciculus, 77, 85, 185
gracile nucleus, 86, 186
graded potentials, 25–26
Grades I-III meningiomas, 273
Grades I-IV tumors, 271
granular layer, 97
granule cell, 161
graphemes, 354–55
graphesthesia, 188, 391–92
grasp reflex, 394
gray commissure (lamina X), 76
gray matter, 5f, 5–6
gray-white matter differentiation, 367
growth hormone (GH), 117
growth hormone–secreting adenomas, 274
GSS (Gerstmann-Sträussler-Scheinker syndrome), 286
guanine nucleotide binding protein (G-protein), 26f, 26–27
guanyl nucleotide diphosphate (GDP), 26–27
Guillain-Barré syndrome (GBS), 295
gustation, 206
gyrencephaly, 150–51
gyri, 151, 154f
gyrus rectus, 154, 212–13

habituation, 337
Hakim's triad (Adams triad), 47
half-life, 374
Hallervorden-Spatz disease, 136
hallucinations
    auditory, 206
    focal sensory seizures with preserved awareness, 247
    olfactory, 207
    somatosensory, 188–89
    visual, 181–82
Harlow, John Martyn, 218
Hawking, Stephen, 311
Head, Henry, 190
headache, 276
hearing aids, 204
heat gain center, 120
heat loss center, 120
Hebb, Donald, 217
Hécaen, Henry, 357
hedonic value, 118
heel-to-shin test, 395
Heilman, Kenneth, 193

hematoma, 44, 259–60
hemiachromatopsia, 178
hemi-alexia, 164–65, 180
hemianesthesia, 187
hemianopia (hemianopsia), 174
hemiballismus, 126
hemi-inattention, 192–93
hemiparaplegia, 82
hemiparesis, 54, 214–15, 306
hemiplegia, 54, 214–15, 306
hemispatial neglect, 192–93
hemispherectomy, 253
hemispheres
 cerebellar, 95f, 95
 cerebral cortex, 151
hemorrhagic stroke, 232–33
 cerebral amyloid angiopathy, 233
 pathophysiology, 232–33
hemorrhagic transformation, 227
hemosiderin, 233, 258, 372–73
Henschen, Salomon Eberhard, 357
hepatolenticular degeneration, 136
hereditary motor and sensory neuropathies, 66
herniation, 363–64
heroin leukoencephalopathy, 296
herpes simplex virus encephalitis (HSVE), 146, 281–82, 332
Heschl's gyrus, 159
Hess, Walter, 119
heterotypical cortex, 162
higher-order functions, 10
hindbrain (rhombencephalon), 10, 11f
hippocampal commissure, 201
hippocampal complex, 200
hippocampal-dentate complex, 200
hippocampus, 10, 200–1
history, neuropsychological assessment, 398–99
 developmental history, 399
 education, work, and social history, 399
 family medical history, 399
 medical history, 399
 of presenting problem or illness, 398
 previous examinations and studies, 398–99
Hitzig, Eduard, 158, 213
HIV (human immunodeficiency virus), 286
HIV-associated dementia, 286
HIV-associated neurocognitive disorder, 286
HIV encephalopathy, 286
hobbyism, 130
Hodgkin, Alan, 17–18
holism, 157
Holmes, Gordon, 190
homeostasis, 115
homogenetic cortex, 162
homonymous hemianopia, 174–75

homonymous quadrantanopia, 175
homotypical cortex, 162
homunculus, 158–59
horizontal cells, 161
horizontal fissure, 94–95
horizontal gaze center, 384–85
horizontal plane, 7–8, 363
hormones, 115–16
Horsley, Victor, 253
hot-cross bun sign, 135
hot water epilepsy, 251
Hounsfield, Godfrey Newbold, 361
Hounsfield Units (HU), 368
HPA (hypothalamic-pituitary-adrenal) axis, 117
HPG (hypothalamic-pituitary-gonadal) axis, 117
HPGr (hypothalamic-pituitary-growth) axis, 117
HPP (hypothalamic-pituitary-prolactin) axis, 117
HPT (hypothalamic-pituitary-thyroid) axis, 117
HSVE (herpes simplex virus encephalitis), 146, 281–82, 332
HU (Hounsfield Units), 368
Hubel, David, 162–63
human immunodeficiency virus (HIV), 286
hummingbird sign (penguin sign), 93f, 133
hunger, 118
huntingtin gene, 135
Huntington's disease, 135–36
Huxley, Andrew, 17–18
hydrocephalic edema, 367
hydrocephalus, 45–48
 normal pressure hydrocephalus (NPH), 47–48
 volume loss vs., 366
hydrocephalus ex vacuo, 45
hydrophilic phosphate heads, 16
hydrophobic tails, 16
hyperactive delirium, 325
hyperacusis, 264
hyperalgesia, 188–89
hyperdense, 368
hyperesthesia, 188–89
hyperkinesia, 125
hypermetamorphosis, 141
hypermetria, 103
hypernatremia, 118
hyperosmolar therapy, 267
hyperperfusion, 56
hyperphagia, 118
hyperphonia, 341–42
hyperreflexia, 309, 392
hyperschematica, 189
hyperthermia, 120, 267
hyperthyroidism, 117
hypertonia (rigidity), 127, 389

hyperventilation, 244
hypesthesia, 71
hypoactive delirium, 325
hypodense, 368
hypoesthesia, 187
hypoglossal nerves (CN12), 67f, 68t, 304
 overview, 74
 testing, 388
hypoglossal nucleus, 74
hypokinesia, 125, 126, 127
hypometria, 103
hypomimia, 127
hyponatremia, 299
hypoperfusion, 56, 226–27
hypophonia, 127, 341–42
hypophyseal portal system, 116–17
hyporeflexia, 309, 392
hyposchematica, 189
hyposmia, 70
hypotension, 260
hypothalamic-hypophysial tract, 116
hypothalamic-pituitary-adrenal (HPA) axis, 117
hypothalamic-pituitary-gonadal (HPG) axis, 117
hypothalamic-pituitary-growth (HPGr) axis, 117
hypothalamic-pituitary-prolactin (HPP) axis, 117
hypothalamic-pituitary-thyroid (HPT) axis, 117
hypothalamus, 10, 113–21
 anatomy, 113–15
 autonomic nervous system (ANS), 115
 circadian rhythms, 121
 endocrine function and pituitary gland, 115–17
 function, 115
 hypothalamic syndromes, 121
 motivated behaviors, 117–20
 thermoregulation, 120–21
hypothermia, 120
hypothyroidism, 117
hypotonia, 309, 389
hypoxia, 54
hypoxic-ischemic injury, 56–57
hypoxic-ischemic leukoencephalopathy, 299–300

iatrogenic CJD, 285
ICA (internal carotid artery), 50, 51f
ICH (intracerebral hemorrhage), 232, 234, 365
ICP (intracranial pressure), 42–43, 260, 276, 364
ictal fear, 143
ictal speech, 246
ictus, 243–44
ideational apraxia, 314
ideomotor apraxia, 313–14

idiopathic (primary) epilepsy, 249–50
idiopathic NPH (normal pressure hydrocephalus), 47
IgG (immunoglobulin G), 294
IgG index, 294
ILAE (International League Against Epilepsy), 244
illusions
  auditory, 206
  focal sensory seizures with preserved awareness, 247
  somatosensory, 188–89
  visual, 181–82
imaging, 361–76
  angiography, 362
  of cerebrovasculature, 57–58
  functional neuroimaging, 374–76
  plain film X-ray, 362
  structural brain imaging, 363–74
imaging planes, 363
imitation behavior, 223
immediate memory, 337, 402
immunoglobulin G (IgG), 294
impaired awareness seizures, 245–46, 248
implicit memory, 336–37
impulse control disorders, 130
impulsive aggression, 141
impulsive-disinhibited bvFTD, 224
impulsivity, 221
inactive plaque, 293
inborn errors of metabolism, 291–92
incidental imaging finding, 58–59
incomplete LIS, 327
incomplete SCI, 81
incus, 203
indirect cortical-brainstem-spinal cord pathways, 307–8
infarction, 50
infection, 260, 280
inferior, 6, 7f
inferior cerebellar peduncles, 96–97
inferior cerebellar veins, 98
inferior colliculi, 88–89, 204, 308
inferior division, middle cerebral artery, 52
inferior frontal gyrus, 152, 212
inferior frontal sulcus, 152, 212
inferior hemiretina, 170
inferior horn, 41
inferior medullary velum, 41, 84–85
inferior oblique muscle, 70
inferior parietal lobule, 152, 184–85
inferior rectus muscle, 70
inferior sagittal sinus, 38–39, 39f
inferior salivatory nucleus, 73
inferior surface of cerebral cortex, 154
inferior temporal gyrus, 152, 200
inferior temporal sulcus, 152, 200
inferolateral arteries, 53–54, 110–11
inferolateral territory infarcts, 113
inferolateral thalamic territory, 110–11

infiltration, tumor, 270
infragranular layers, 161
infranuclear components, ocular motor system, 384–85
infratentorial compartment, 38
infundibulum, 38, 113–14, 115
inhalant abuse, 295
inhibition of automatic responses, 407
inhibitory control, 217
inhibitory postsynaptic potentials (IPSPs), 21, 25–26
inion, 37, 240–41
initiation, 406–7
inner ear, 203
innervation ratio, 303–4
innominate artery, 50, 51f
instrumental activities of daily living, 398
insular cistern, 39–40, 368
insular cortex (island of Reil), 152f, 152
insular ribbon sign, 368–69
intention tremor, 103, 126, 390, 395
interbrain. See diencephalon
interhemispheric disconnections, 163
interhemispheric fissure, 151
inter-ictal period, 243–44
intermanual conflict, 317
intermediate zones, 76f, 76, 78–79, 96
internal arcuate fibers, 186
internal capsule, 54, 124, 155
  anterior limb, 124, 155
  genu, 124, 155
  posterior limb, 124, 155
internal carotid arterial system, 51
internal carotid artery (ICA), 50, 51f
internal granular layer, 161
internal gross anatomy, 76–77
  spinal cord gray matter, 76
  spinal cord white matter, 76–77
internal jugular veins, 55
internal medullary lamina, 110
internal pyramidal layer, 161
International 10-20 System, 240–41
International League Against Epilepsy (ILAE), 244
interneurons, 3
internodes, 20, 290
internuclear components, ocular motor system, 384–85
internuclear neurons, 384–85
interoceptive sense, 79
interpeduncular cistern (basal cistern), 39f, 39–40
interpeduncular fossa, 53, 88, 89, 113–14
interpersonal relatedness, 400
interposed nucleus, 95–96
intersegmental spinal reflexes, 81
intersegmental tracts, 77
interstitial cerebral edema, 367
interstitial enhancement, 363
interstitial fluid, 55

interthalamic adhesion, 41, 108–9
intertransverse foramina, 50
interventricular foramen, 40f, 40–41
intervertebral disk, 37
intervertebral foramens, 37, 64
intervertebral space, 37
intorsion, 383
intoxication stage, 147
intra-axial, 364–65
intracarotid amobarbital procedure, 252–53, 332
intracellular microelectrode, 17–18
intracellular recording, 17–18
intracerebral hemorrhage (ICH), 232, 234, 365
intracranial abscess, 282
intracranial atherosclerotic disease, 58
intracranial hemorrhage, 47, 365
intracranial hypertension, 267
intracranial pressure (ICP), 42–43, 260, 276, 364
intracranial self-stimulation, 144
intracranial tumor. See tumors
intradural hemorrhage, 258
intrafusal muscle fibers, 303
intrahemispheric disconnections, 163
intralaminar nuclei, 110
intramural ganglia, 63
intransitive movements, 313
intraoperative neuroimaging, 373
intraparenchymal hematoma, 260
intraparenchymal hemorrhage, 365
intraparietal sulcus, 152, 184–85
intravascular enhancement, 363
intraventricular hemorrhage extension, 258
intraventricular meningiomas, 272–73
intrinsic cerebellar nuclei, 95–96
intrinsic tumors, 270
inverse agonists, 27–28
in vivo, 17–18
involuntary levitation, 317
iodinated contrast media, 363
ion channels, 17
ionizing radiation, 277
ionotropic receptors, 17, 26f, 26
ion pumps (ion transporters), 17
ions, 15, 18f
ipsilateral, 7
ipsilesional, 7
IPSPs (inhibitory postsynaptic potentials), 21, 25–26
iris, 172f
irregular words, 354–55
ischemia, 50, 54
ischemic leukoencephalopathy, 299–300
ischemic penumbra, 233–34
ischemic stroke, 50, 226–31, 372f
  anatomic classification of, 227–31
  pathophysiology, 226–27

transient ischemic attacks, 231
treatment, 233–34
island of Reil, 152f, 152
isocortex, 162
isotropic diffusion, 372
isthmus, 153, 200

Jackson, John Hughlings, 158, 192, 213, 335
Jacksonian march, 189
Jacksonian motor seizures, 213, 246–47
jamais vu, 248
jargon aphasia, 345
John Cunningham (JC) virus, 297
joints, 302
jugular foramen, 55
juxtarestiform body, 96–97

K+ (potassium ions), 18f
Kandel, Eric, 337
Kayser-Fleischer rings, 136
Kennedy, Rosemary, 220
kianate receptors, 30
kinesthesia, 79
kinetic tremor, 390
kinetopsia, 181
Kleist, Karl, 315
Klüver, Heinrich, 141
Klüver-Bucy syndrome, 141
knee-jerk reflex, 80, 304
Korsakoff's syndrome, 112, 333–34
Kraepelin, Emil, 207
kuru, 285

laceration, 258
lacrimal nucleus, 72
lacunar infarcts, 228–30, 229f
lacunar state, 236
lacunes, 228
lambda, 37f, 37
lambdoid suture, 37f, 37
lamina X (gray commissure), 76
Landau-Kleffner syndrome, 251
language, 340–51
  aphasia, 342–44, 347–48
  classic aphasia syndromes, 344–47
  evaluating, 404–5
  language-dominant hemisphere, 342
  observations, 399
  oral language, 341–42
  primary progressive aphasia (PPA), 348–51
large vessel cortical stroke syndromes, 229t
Lashley, Karl, 330
late delayed radiation encephalopathy, 296
lateral, 6, 7f
lateral apertures, 40f, 41
lateral cerebellar nucleus, 95–96
lateral column, 76f, 76–77
lateral corticospinal tracts, 80, 306
lateral fissure, 151

lateral geniculate nuclei (LGN), 70, 111
  primary visual cortex from, 173
  from retina to, 172–73
lateral hemispheres, cerebellum, 96f, 96
lateral horn, 76
lateral intersegmental tract, 77
lateral lenticulostriate arteries, 124
lateral medullary syndrome, 91
lateral motor system, 308–9
lateral occipital gyri, 152–53
lateral occipital sulcus, 152–53
lateral occipitotemporal gyrus, 153, 200
lateral olfactory striae, 206–7
lateral orbital gyrus, 212–13
lateral parietotemporal line, 152
lateral rectus muscle, 70
lateral reticulospinal tract. See medullary reticulospinal tracts
lateral spinothalamic tract, 80, 186
lateral striate arteries, 53–54
lateral sulcus, 151
lateral surface of cerebral cortex, 152–53
lateral ventricles, 40f, 40–41
lateral vestibulospinal tract, 99, 308
lateral zones, 96f, 96
laughter
  focal emotional seizures with, 248
  pathological, 309
L-DOPA (levodopa), 29, 31, 129–30
lead pipe rigidity, 389
learning, 405–6
Leborgne, Louis Victor, 158
left angular gyrus syndrome, 194
left common carotid artery, 50, 51f
left-handed agraphia, 164–65
left-sided apraxia, 164–65, 314
left subclavian artery, 50, 51f
leg weakness, 222
Leipmann, Hugo, 312
Lelong, Lazare, 158
lemniscus, 6
Lennox-Gastaut syndrome, 250–51
lens, 172f
lenticulostriate arteries, 53f, 53–54, 124
lentiform, 365
lentiform nucleus (lenticular nucleus), 124
leptomeningeal cysts, 43
leptomeninges, 37
leptomeningitis, 279
lesion burden, 293
lesion-deficit analysis, 158, 344–45
lesionectomy, 253
lethargy, 320
letter verbal fluency, 404, 406–7
leucotome, 219
leukoaraiosis, 236, 299–300
leukodystrophies, 291–92
leukoencephalopathies. See white matter disease
level of consciousness, 379

levodopa (L-DOPA), 29, 31, 129–30
levodopa-carbidopa, 130
Lewy bodies, 129
lexical agraphia, 349, 356
lexicalization errors, 355
lexicon, 341
LGN. See lateral geniculate nuclei (LGN)
Lhermitte, François, 222–23
Lichtheim, Ludwig, 346–47
Liepmann, Hugo, 205
ligand-gated ion channels, 17, 26f, 26
ligands, 25
Lilliputian hallucinations, 181
limb apraxia, 134, 313
limb ataxia, 102–3
limbic association cortex, 160
limbic cortex, 138, 139, 153
limbic encephalitis, 146
  autoimmune limbic encephalitis, 146
  herpes simplex virus encephalitis (HSVE), 146
limbic lobe, 153
limbic system, 10, 138–48
  amygdala, 140–43
  cingulate cortex, 139–40
  deep brain stimulation (DBS) for psychiatric disorders, 147
  history, 138–39
  limbic encephalitis, 146
  nucleus accumbens, 143–45
  septal region, 145
limb ideomotor apraxia, 313–14
limb-kinetic apraxia, 315
limiting sulcus, 152
Lindenbaum, Shirley, 285
linear skull fracture, 257
lingual gyrus, 153, 168–69, 200
linguistic agraphia, 356
linguistics, 340
lipid solubility, 16f, 16–17
lipohyalinosis, 230
lipophilic lipids, 295
liquefactive necrosis, 226–27, 367–68
LIS (locked-in syndrome), 91, 326–27
Lissauer, Heinrich, 176
lissencephaly, 150–51
literal paraphasias, 343–44
lithium, 33
"Little Albert Experiment, The" (Watson and Rayner), 142
LMNs. See lower motor neurons (LMNs)
lobules, 94–95
LOC (loss of consciousness), 262t, 263
local anesthetics, 22
local circuit neurons, 3
localized seizures. See focal onset seizures
lock and key metaphor, 25
locked-in syndrome (LIS), 91, 326–27
locus coeruleus, 30, 88
Loewi, Otto, 24–25

logopenic variant PPA, 350–51
logorrhea, 345
longitudinal callosal fascicles, 165
longitudinal fissure, 151
longitudinal plane, 7–8, 363
long-term memory, 332–33, 337–38
loosening of associations, 400
loss of consciousness (LOC), 262t, 263
loudness, perceived, 202
Lou Gehrig's disease, 311
lower motor neurons (LMNs), 78, 301
  overview, 304
  upper motor neurons lesions vs., 309
lumbar cistern, 40f, 40
lumbar enlargement, 78
lumbar plexus, 66
lumbar puncture, 45, 48
lumboperitoneal shunting, 46
Luria, Alexander Romanovich, 158, 217, 357
Luria manual sequencing task, 215, 216f, 408–9
Luzzatti, Claudio, 192
Lyme disease, 287

macropsia, 181
macrosomatognosia, 189
macula, 169–70
mad cow disease, 285
magnetic apraxia, 223, 316–17, 394
magnetic gait, 395–96
magnetic resonance angiography (MRA), 57, 373
magnetic resonance imaging (MRI), 369–50
  diffusion tensor imaging (DTI), 372
  diffusion-weighted MRI (DW-MRI), 371–72
  fluid attenuated inversion recovery (FLAIR) MRI, 371
  intraoperative neuroimaging, 373
  magnetic resonance angiography (MRA), 373
  MR perfusion-weighted imaging, 373
  susceptibility weighted imaging (SWI), 372–73
  T1-weighted MRI, 371
  T2-weighted MRI, 371
main trigeminal sensory nucleus, 71
major depressive disorder, 147
malignant tumors, 270–71
malingering, 250, 409
malleus, 203
mammillary bodies, 113–14
mammillary region, 114
mandibular branch, 71, 72, 385, 386f
man in a barrel syndrome, 230
manual automatisms, 246
manual groping behavior, 317
manual sequencing tasks, 408–9

MAO (monoamine oxidase), 29
MAOIs (monoamine oxidase inhibitors), 32
Marchiafava–Bignami disease (MBD), 296
Marcus-Gunn pupil, 382
marginal sulcus, 153, 185
Martinotti cells, 161
Martland, Harrison Stanford, 267
mass action, principle of, 330
massa intermedia, 41, 108–9
mass effect, 42–43, 272, 363–64
mass lesions, 42, 363–64
masticatory (masseter) muscles, 385, 386
material-specific anterograde amnesia, 332
material-specific memory loss, 332
maxillary branch, 71, 385, 386f
MBD (Marchiafava–Bignami disease), 296
MCAs (middle cerebral arteries), 52, 53f, 229f
MCS (minimally conscious state), 324
medial, 6, 7f
medial cerebellar nucleus, 95–96
medial forebrain bundle, 144
medial frontal syndrome, 222
medial geniculate nucleus (MGN), 111, 204
medial-lateral axis, 6
medial lemnisci, 85, 186
medial lenticulostriate arteries, 124
medial longitudinal fasciculus, 384–85
medial occipitotemporal gyrus, 153, 200
medial olfactory striae, 206–7
medial orbital gyrus, 212–13
medial preoptic nucleus, 119–20
medial rectus muscle, 70
medial striate arteries, 53–54, 124
medial surface of cerebral cortex, 153–54
medial temporal amnesia, 331–32
  material-specific anterograde amnesia, 332
  pathologies causing, 332
medial vestibulospinal tract, 99, 308
median aperture, 41
median group nuclei, 90
median plane, 7–8, 363
medical history, 399
medulla, 8, 85f, 86–87
  cranial nerve nuclei, 87
  nuclei regulating vital life functions, 87
  somatosensory nuclei, 86
medulla oblongata, 86
medullary pyramids, 86f, 86, 157, 306
medullary reticular formation, 90
medullary reticulospinal tracts, 90, 307
medulloblastomas, 102, 271
melanin, 124
melatonin, 108
membrane, 16–17
membrane potentials, 17–22
  action potential (AP), 19–21
  postsynaptic potential (PSP), 21–22

resting membrane potential, 18–19
membrane transport proteins, 17
  ion channels, 17
  ion pumps (ion transporters), 17
memory, 329–38
  amnesia due to lesions in non-medial temporal lobe structures, 333–34
  consolidation, 333
  dissociative amnesia, 335
  evaluating, 405–6
  forms based on time span, 337–38
  forms of, 335–37
  impaired, 222
  medial temporal amnesia, 331–32
  patient H.M., 329–31
  remote memory and retrograde amnesia, 332–33
  transient amnesia syndromes, 334–35
menace reflex, 180
meningeal headaches, 43
meningeal layer, dura, 37–38
meninges, 37–40
  arachnoid membrane, 39–40
  dura mater, 37–39
  pia mater, 40
meningiomas, 43, 272–73
meningitis, 43, 260, 280–81
  bacterial meningitis, 281
  viral meningitis, 280
meningoencephalitis, 279
mental flexibility, 221
mesencephalic locomotor region, 90
mesencephalic reticular formation, 90
mesencephalic trigeminal nucleus, 71
mesencephalon. See midbrain
mesial temporal sclerosis, 250
mesocortex, 162
mesocortical pathway, 29f, 29
mesocorticolimbic system, 89
mesolimbic pathway, 29f, 29, 144
Mesulam, M. Marsel, 221, 348
metabolism, 56–57, 291–92
metabotropic receptors, 26f, 26–27
metachromatic leukodystrophy, 292
metallic streak artifacts, 369
metamorphopsia, 181
metastasize, 271
metastatic tumors, 275
metathalamus, 108
metencephalon (afterbrain), 10–11, 11f
methotrexate, 297
methylbenzene, 295
Meyer's loop, 173
Meynert, Theodor, 156–57
MGN (medial geniculate nucleus), 111, 204
Mickey Mouse sign, 93f, 133
microaneurysms, 232
microdialysis, 144
microemboli, 230
microglia, 4

micrographia, 127, 357
microinfarcts, 228
micropsia, 181
microsomatognosia, 189
microvascular strokes, 228–30, 229f
micturition center, 222
midbrain, 8, 10–11, 11f, 88–89
    midbrain tectum, 85f, 88–89
    midbrain tegmentum, 85f, 89
midbrain reticular formation, 90
midcingulate cortex, 139–40
middle cerebellar peduncles, 87–88, 96–97
middle cerebral arteries (MCAs), 52, 53f, 229f
middle ear, 203
middle fossa, 36
middle frontal gyrus, 152, 212
middle (tuberal) region, hypothalamus, 114
middle temporal gyrus, 152, 200
midline shift, 42f, 42, 363–64
midsagittal plane, 7–8, 9f
mild TBI, 265
milk ejection reflex, 116
Miller, George A., 337
Miller, Neal, 143–44
Milner, Brenda, 330
Milner, Peter, 143–44
minimally conscious state (MCS), 324
Mini-Mental Status Examination (MMSE), 379
miosis, 382
mirror drawing visuomotor coordination task, 330, 331f
mirror image syndrome, 342
Mishkin, Mortimer, 173–74, 331
mitgehen, 389
mixed aphasia, 346
mixed cerebral palsy, 311
mixed delirium, 325
mixed dementia, 236
mixed gliomas, 272
mixed nerves, 64
mixed PPA, 348
mixed transcortical aphasia, 347
MMSE (Mini-Mental Status Examination), 379
MNDs. See motor neuron diseases (MNDs)
MoCA (Montreal Cognitive Assessment), 379
modal specificity, 159
moderate-severe TBI, 263–65
modules, 162–63
Molaison, Henry Gustav, 329–30
molecular layer, 97, 161
Moniz, António Egas, 57, 219
monoamine oxidase (MAO), 29
monoamine oxidase inhibitors (MAOIs), 32

monoamines, 29–30
    dopamine (DA), 29
    norepinephrine (NE), 30
    serotonin (5-HT), 30
monoamine theory of depression, 32
monocular blindness, 174
monogenic Parkinson's disease, 129
mononeuropathy, 66, 310
monopharmacy, 252
monopolar montages, 241
monosynaptic stretch reflex, 80, 304
Monro-Kellie doctrine, 45, 364
Montreal Cognitive Assessment (MoCA), 379
mood, 409
mood stabilizers, 33
motion artifact, 369
motion blindness, 179
motivated behaviors, 117–20
    aggression and sham rage, 119
    fight or flight response, 119
    hunger and feeding, 118
    sexual behavior, 119–20
    thirst and drinking, 118–19
motivational state or drive, 117–18
motoneurons, 3, 301
motor agnosia. See apraxia
motor agraphia, 356
motor cortex, 213–17
    history, 213–14
    primary motor cortex, 214–15
    secondary motor cortex, 215–17
    upper motor neurons, 305–6
motor end plate, 303
motor engrams, 312
motor function, 90
motor homunculus, 214, 305
motor map, 213
motor neuron diseases (MNDs), 311–12
    amyotrophic lateral sclerosis (ALS), 311
    primary lateral sclerosis, 311–12
    progressive bulbar palsy, 311
motor perseveration, 408
motor programming, 408–9
motor seizures, 246
motor system, 301–18
    alien hand syndrome (AHS), 317
    apraxia, 312–17
    cerebral palsy, 310–11
    diagnostic testing, 309–10
    exam, 388–90
    grading responsiveness, 320–22
    lower motor neurons, 304
    lower motor neuron vs. upper motor neuron lesions, 309
    motor neuron diseases (MNDs), 311–12
    muscle, 302–4
    myasthenia gravis, 310
    observations, 399
    upper motor neurons, 305–9

motor unit, 303–4
Mountcastle, Vernon, 162–63
movement, 302
movement agnosia, 179
moyamoya disease, 59–60
MRA (magnetic resonance angiography), 57, 373
MRI. See magnetic resonance imaging (MRI)
MR perfusion-weighted imaging, 373
multiform layer, 161
multi-infarct dementia, 235
multimodal association areas (tertiary zones), 160
multiple mononeuropathy, 66
multiple sclerosis (MS), 292–94
    clinical presentation, 293
    diagnosis, 293–94
    disease course, 294
    epidemiology, 294
    treatment, 294
multiple subpial transection, 253–54
multipolar neurons, 3
multisystem atrophy, 134–35
Munk, Hermann, 176
muscarinic receptors, 28
muscle, 302–4
    mechanics of movement, 302
    monosynaptic stretch reflex, 304
    muscle contraction, 303–4
    muscle fibers, 303
    muscle tone, 389
muscle spindle, 304
muscle stretch receptors, 304
muscle twitch. See fasciculations
muscle wasting, 309
musicogenic epilepsy, 251
mutism, 222
myasthenia gravis, 310
mydriasis, 70, 381
myelencephalon, 10, 11f
myelin, 4
myelinated axons, 20, 21f
myelitis, 279
Myers, Ronald, 164
Myerson's sign, 127, 394
myoclonic movements, 246
myoclonic seizures, 249
myofibrils, 303
myopathy, 102, 389
myosin filaments, 303
myotactic stretch reflex, 80, 304
myotatic reflex, 304
myotomes, 66

Na+ (sodium) ions, 18f
Na+/K+ pump (sodium-potassium pump), 20–21
naming, 344
narcolepsy, 33

nasal hemiretina, 170, 173f
nasion, 37, 240–41
nasolabial fold, 386
natural reward, 145
NBIA (neurodegeneration with brain iron accumulation), 137
NCS (nerve conduction studies), 22, 310
NE (norepinephrine), 24–25, 30
near neglect, 192
near triad, 382
necrosis, 226–27
negative antagonists, 27–28
negative doll's eye sign, 385
negative symptom, 247
neglect alexia, 356
neglect syndrome, 112
    neuroanatomical basis of, 193
    variants of, 192–93
neocerebellar syndrome, 102
neocerebellum, 100–1
neocortex, 162
neologisms, 344
neoplasms. See tumors
neostriatum, 124
nephrogenic diabetes insipidus, 118
Nernst potential, 18
nerve cells. See neurons
nerve conduction studies (NCS), 22, 310
nerve deafness, 204, 387
nerve impulse, 19
nerves, 6
nerve sheath, 37
nervous system, 1–12
    glia, 3–4
    neuroanatomy, 5–6
    neurons, 1–3
neural tube, 10
neuraxis (neuroaxis), 6
neurites, 3
neuritis, 279
neuroaffective disorders, 397–98
neuroanatomy, 5–6
    afferents vs. efferents, 5
    brain anatomy, 8–10
    central vs. peripheral nervous system, 5
    developmental basis of major brain subdivisions, 10–11
    directional terms, 6–7
    gray matter vs. white matter, 5–6
    planes of section, 7–8
neuroaxis (neuraxis), 6
neurobehavioral disorders, 397–98
neurocognitive disorders, 397–98
neurodegeneration with brain iron accumulation (NBIA), 137
neurodegenerative disease, 11–12
neuroeffector junction, 3
neuro-endoscopy, 276–77
neurofibrillary tangles, 209
neuroglia. See glia

neurohormones, 115–16
neurolepsis, 31
neuroleptic malignant syndrome, 132
neuroleptics, 31
neurolinguistics, 340
neurological channelopathies, 22
neurological examination, 378–96
    clinical history, 379
    coordination/cerebellar exam, 394–95
    cranial nerve exam, 379–88
    mental status exam, 379
    motor exam, 388–90
    reflex exam, 392–94
    sensory exam, 390–92
    station and gait, 395–96
neuromas. See Schwannomas
neuromodulation, 30, 254
neuromuscular disease, 22, 102, 309–10
neuromuscular junction, 3, 303
neuronal inclusions, 129
neuron doctrine, 2
neurons, 1–3
    classification, 3
    electrical signaling in, 14–22
    muscle fiber innervation, 303
    structural and functional components of, 2–3
    synapse, 3
neuron theory, 2
neuropathy, 66, 389
neuropeptides, 30–31
neuropharmacology, 31
neuroprosthetic device, 204–5
neuroprotective therapy, 130
neuropsychiatric disorders, 128, 397–98
neuropsychological assessment, 379, 397–410
neuropsychological assessment report, 397–98
    behavioral observations, 399–400
    conclusions section, 409
    history section, 398–99
    recommendations, 409–10
    referral question, 398
    testing procedures and results, 400–9
neuropsychologists, 397
neuropsychopharmacology, 31–33
    antidepressants, 32–33
    anti-manic agents (mood stabilizers), 33
    antipsychotics, 31–32
    anxiolytics (antianxiety drugs), 33
    drug classification, 31
    psychostimulants, 33
    sedative-hypnotics, 33
neurosecretion, 115–16
neurosecretory cells, 115–16
neurosurgical revascularization, 59–60
neurotransmitter receptors, 25–27
    autoreceptors, 27
    ionotropic receptors, 26

    metabotropic receptors, 26–27
    receptor regulation, 27
neurotransmitters, 2–3
    mechanisms of drug action, 27–28
    overview, 24–25
    removal of neurotransmitter from synapse, 27
    systems, 28–31
    transmitter release, 25
neutral receptor antagonists, 27–28
nicotinic receptors, 28, 303
nigrostriatal pathway, 29f, 29, 89, 124, 126–27
Nissl method, tissue stain, 161
NMDA receptors, 30
nocebo effect, 266
nociception, 79, 185, 390, 391
nociceptors, 80–81, 185
nodes of Ranvier, 4, 20, 21f, 290
nodulus, 95
nominal aphasia, 347
non-apraxic motor disturbances, 357
noncommunicating hydrocephalus, 45–46
non-contrast enhanced MRA, 373
nondeclarative memory (implicit memory), 336–37
nondiscriminative touch, 79, 185, 186, 390
non-epileptic seizures, 250
nonfluent variant PPA, 350
non-gated channels, 17
non-motor features, Parkinson's disease, 128
non-motor seizures, 246
non-paraneoplastic encephalitis, 146
non-penetrating injury, 261
nonpolar molecules, 17
non-propositional speech, 343
nonspecific thalamic nuclei, 112
non-traumatic acquired brain injuries, 256
nonverbal memory, 332
norepinephrine (NE), 24–25, 30
normal pressure hydrocephalus (NPH), 47–48, 366
noxious stimulus, 80–81
nuchal rigidity, 232–33, 280
nuclear components, ocular motor system, 384–85
nuclear groups, thalamus, 110
nuclei, 6, 114t
nucleus accumbens, 143–45
    anatomy, 143
    reward, 143–45
nucleus ambiguus, 73
nucleus interpositus, 95–96
number processing, 405
nystagmus, 72, 103–4, 383

obex, 41, 86
object-centered frame of reference, 192
oblique plane, 7–8

obstruction, 226–27
obstructive hydrocephalus, 45–46
obtundation, 320
occipital alexia. *See* pure alexia
occipital bone, 35, 36*f*, 37*f*
occipital horn, 41
occipital lobes, 151, 168–82
　anatomy, 168–69
　dorsal stream lesions, 179–80
　ventral stream lesions, 177–79
　retina, 169–71
　visual agnosia, 176–77
　visual disconnection syndromes, 180–81
　visual field, 171–72
　visual field defects and cerebral blindness, 174–76
　visual hallucinations and illusions, 181–82
　visual pathways, 172–74
occipitotemporal line, 152
occipitotemporal sulcus, 153, 200
oculocephalic reflex, 99–100, 180, 193, 385
oculomotor apraxia, 180, 190, 316
oculomotor nerve palsy, 70
oculomotor nerves (CN3), 67*f*, 68*t*, 304
　overview, 70
　testing, 381–85
oculomotor palsy, 384
Ogle, John William, 356
Olds, James, 143–44
olfaction, 206–7
olfactory bulbs, 36, 69–70, 154, 206–7, 207*f*
olfactory filaments/fila, 69–70, 206
olfactory foramina, 36, 69–70
olfactory fossae, 36
olfactory groove meningiomas, 272
olfactory mucosa (olfactory epithelium), 206
olfactory nerves (CN1), 36, 67*f*, 68*t*, 206, 207*f*
　overview, 69–70
　testing, 380
olfactory sulcus, 154, 206–7, 212–13
olfactory tract, 69–70, 154, 206–7
oligoclonal bands, 294
oligodendrocytes (oligodendroglia), 4, 290
oligodendrogliomas, 4, 271
olivopontocerebellar atrophy, 135
omission errors, 402
oncocytomas, 274
on-off fluctuations, 130
open head injury, 261
operculum, 152
ophthalmic artery, 51
ophthalmic branch, 71, 385, 386*f*
ophthalmoscope (funduscope), 381
opiate receptors, 30–31
opioid peptides, 30–31
opportunistic pathogens, 280
oppositional paratonia, 316–17, 389

optic ataxia, 179–80, 190
optic chiasm, 113–14, 114*f*, 172
optic disc, 169
optic foramen, 70
optic nerve fibers, 170*f*
optic nerves (CN2), 67*f*, 68*t*, 169, 172
　overview, 70
　testing, 380–83
optic neuritis, 70, 293
optic neuropathy, 70
optic radiations, 173
optic tracts, 172
optokinetic nystagmus, 103–4, 180, 385
optokinetic reflex, 385
oral apraxia, 313
oral language, 341–42
orbital gyri, 154
orbital sulcus, 212–13
orbitofrontal syndrome, 221–22
organic anions (A–), 18*f*, 18
organ of Corti, 203
organophosphates, 28–29
orientation, 379
oro-alimentary automatisms, 246
orthogonal, 7–8
orthographic lexicon, 354
orthography, 354–55
orthostatic hypotension, 63–64, 134
oscillopsia, 181
oscilloscope, 17–18
osmolality, 116
osmoreceptor cells, 116
osmotic myelinolysis (osmotic demyelination syndrome), 299
ossicles, 203
otorrhea, 257–58
outer (external) ear, 202
oval window, 203–4
oxytocin, 116
Ozeretski alternating motor sequence task, 215, 216*f*

pachymeningitis, 279
pachymeninx, 37
PAG (periaqueductal gray), 88*f*, 89
pain, 391. *See also* nociception
paleocortex, 162
palinopsia, 181
pallesthesia, 79, 390, 391
pallhypesthesia, 391
pallid infarct, 227
pallidotomy, 130
pallium, 150
palpebral fissure, 386
panda eyes, 257–58
Papez, James, 139
Papez circuit, 139*f*, 139
paracentral lobule, 153, 185
paracentral sulcus, 153
paracusia, 206

paragrammatism, 345
paragraphia, 356
parahippocampal gyrus, 139*f*, 153, 200
paralexias, 354
paralinguistic agnosia, 205–6
parallel fibers, 97
paralysis, 309
paramedial zones, 96
paramedian arteries, 110–11
paramedian groups, 90
paramedian territory infarcts, 113
paramedian thalamic territory, 110–11
paraneoplastic encephalitis, 146
paraneoplastic limbic encephalitis, 146
paraphasias, 343–44
paraplegia, 81
parasagittal meningiomas, 272
parasagittal planes, 7–8
parasitic infections of CNS, 283–84
parasomnias, 128
parastriate cortex, 169
parasympathetic system, 62–63, 63*f*
paratonia, 389
paravermal zones, 96
paravertebral ganglia, 63
parenchyma, 1
paresis, 86, 309
paresthesia, 188–89
parietal agraphia, 356
parietal bones, 35, 36*f*, 37*f*
parietal eye fields, 180, 316, 384–85
parietal lobe epilepsy (PLE), 189
parietal lobes, 151, 184–97
　anatomy, 184–85
　anosognosia, 194–95
　Bálint's syndrome, 194
　disorders of cortical somatosensory processing, 187–88
　Gerstmann's syndrome, 194
　left angular gyrus syndrome, 194
　posterior cortical atrophy (Benson's syndrome), 195–97
　somatosensation, 185–87
　somatosensory hallucinations and illusions, 188–89
　spatial attention disorders, 191–94
　spatial cognition disorders, 189–91
parietal operculum, 184
parietal-temporal alexia, 346, 354
parietal-temporal-occipital association cortex, 160
parieto-occipital sulcus, 151, 185
parieto-temporal basal line, 168–69
parieto-temporal lateral line, 168, 199–200
Parkinson plus syndromes, 131–35
　corticobasal syndrome, 134
　diffuse Lewy body disease, 131–32
　multisystem atrophy, 134–35
　progressive supranuclear palsy, 132–34

Parkinson's disease, 29, 126–31, 129f, 357
    genetics of, 129
    hypokinesia and bradykinesia, 127
    Myerson's sign (glabellar reflex), 127
    non-motor features, 128
    other causes of parkinsonism, 130–31
    pathology, 128–29
    postural instability and gait
        impairment, 127
    resting tremor, 127
    rigidity, 127
    treatment, 129–30
Parkinson's disease dementia, 128
pars compacta, 124
pars opercularis, 152, 212
pars orbitalis, 152, 212
pars reticulata, 124
pars triangularis, 152, 212
partial agonists, 27–28
parvocellular reticular nuclei, 90
passive movements, 187, 316–17, 385
passive transport, 17
past-pointing, 103
patellar reflex, 80, 304
patellar tendon, 392, 393f
Paterson, Andrew, 192
pathognomonic cortical sign, 158
pathological joking, 221–22
pathology
    Alzheimer's disease, 208–9
    basal ganglia, 126
    cerebellum, 104–6
    medial temporal amnesia, 332
    Parkinson's disease, 128–29
    thalamus, 112–13
pathophysiology
    coma, 322–23
    hemorrhagic stroke, 232–33
    ideomotor apraxia, 314
    ischemic stroke, 226–27
    of traumatic brain injury (TBI), 257–61
patient K.M., 217
patient N.A., 333
patient S.M., 142–43
pattern processing, 170–71
Pavlov, Ivan, 336–37
PCAs (posterior cerebral arteries), 53, 229f
Pcom (posterior communicating arteries), 51, 54f, 54
pedigree analysis, 129
peduncle, 6
pelopsia, 181
penetrating injuries, 261
Penfield, Wilder, 158–59, 213, 217, 253
penguin sign (hummingbird sign), 93f, 133
peptidases, 31
perforant path, 201
perforating arteries, 53–54
performance validity testing, 409

perfusion, 56
periaqueductal gray (PAG), 88f, 89
pericallosal artery, 140
perikaryon, 2
periorbital ecchymosis, 257–58
periosteal (endosteal) layer, dura, 37–38
peripersonal space, 189–90
peripheral nerve field, 66
peripheral nerves, 66
peripheral nervous system (PNS), 61–74
    central nervous system vs., 5
    cranial nerves, 66–74
    functional subdivisions of, 61–64
    spinal nerves, 64–66
peripheral neuropathy, 66, 310
peristriate cortex, 169
perisylvian aphasias, 344–47, 344t
periventricular white matter, 289
permanent vegetative state, 324
permeability, 15
perpendicular fasciculus, 156–57
PERRLA (pupils equal, round, reactive to light and accommodation), 381
perseverance, 221, 406–7, 408
persistent vegetative state, 259, 324
personality, 409
personal space, 189–90
PET (positron emission tomography), 374–76
petechial hemorrhage, 258
petrous ridge, 36
phantom boarder syndrome, 208
phantom limb, 189
phantosmia, 207
pharmacology, 31
pharmacotherapy, 129–30
pharyngeal reflex, 388
phenothiazines, 31
phenotype, 12
phonagnosia, 205–6
phonation, 341–42
phonemes, 341, 343–44, 354–55
phonemic paraphasias, 343–44
phonetic errors, 341
phonetics, 341
phonology, 341
phonophobia, 43, 266
phospholipid molecules, 16
photic driving, 244
photic stimulation, 244
photophobia, 43, 266
photopsia, 181
photoreceptors, 169, 170f
photosensitive epilepsy, 251
phrenology, 157
physiological nystagmus, 103–4
pia mater, 38f, 40
PICA (posterior inferior cerebellar arteries), 53, 98
Pick, Arnold, 223–24

Pick bodies, 223–24
Pick's disease, 223–24
piloerection, 119
pineal gland, 108
pinna (auricle), 202
pitch, 202
pituitary adenomas, 274
pituitary dwarfism, 117
pituitary fossa, 36
pituitary gland, 115–17
    anterior lobe, 116–17
    posterior lobe, 116
pituitary stalk, 38, 113–14, 115
pituitary tumors, 273–74
pixel data, 368
plain film X-ray, 362
planes of section, 7–8
planning, 221, 407–8
plantar reflex, 393–94, 394f
planum temporale, 200
plasmin, 234
plasminogen, 234
PLE (parietal lobe epilepsy), 189
pleasure, focal emotional seizures with, 248
plexiform layer, 161
plexopathy, 66
plexuses, 66
PML (progressive multifocal leukoencephalopathy), 297–98
pneumoencephalography, 361
PNS. See peripheral nervous system (PNS)
POA (preoptic area), 120
polar artery, 110–11
polar molecules, 17
poles, 152
polydipsia, 118
polymorphic layer, 161
polyneuropathy, 66
polyopia, 181
polypharmacy, 252
polysomnography, 128
polyspikes, 243
polyspike-wave complexes, 243
polysynaptic reflex arc, 80–81
polyuria, 118
pons, 8, 85f, 86f, 87–88
    dorsal pons, 88
    ventral pons, 87–88
pontine arteries, 87
pontine auditory hallucinosis, 206
pontine cistern, 39–40
pontine nuclei, 101f
pontine reticular formation, 90
pontine reticulospinal tracts, 90, 307
pontine tegmentum, 88
pontocerebellar projections, 87–88
pontocerebellar tracts, 97
pontocerebellum, 100–1
portal of entry, 280

positive doll's eye sign, 385
positive somesthetic symptoms, 188–89
positive symptom, 247
positron emission tomography (PET), 374–76
postcentral gyrus, 152, 184
postcentral sulcus, 152, 184
posterior, 6, 7f
posterior cerebral arteries (PCAs), 53, 229f
posterior choroidal arteries, 41–42, 110–11
posterior cingulate cortex, 139–40
posterior circulation, 51, 52–53
posterior clinoid processes, 36
posterior column, 76f, 76–77
posterior column–medial lemniscal system, 185
posterior communicating arteries (Pcom), 51, 54f, 54
posterior cord syndrome, 82f, 82
posterior cortical atrophy, 195–97
posterior cortical border zone, 230
posterior fissure, 94–95
posterior fontanelle, 37
posterior forceps (forceps major), 155–56
posterior fossa, 36
posterior fossa decompression surgery, 106
posterior fossa meningiomas, 272–73
posterior horn, 41, 76f, 76
posterior inferior cerebellar arteries (PICA), 53, 98
posterior lobe, 94–95, 95f, 116
posterior lobule, 152
posterior median sulcus, 76
posterior medullary velum, 41, 84–85
posterior orbital gyrus, 212–13
posterior perforated substance, 53
posterior (mammillary) region, hypothalamus, 114
posterior reversible encephalopathy syndrome (PRES), 367
posterior root, 64, 65f
posterior spinal arteries (PSA), 53, 77
posterior territory infarcts, 113
posterior thalamic territory, 110–11
posterolateral sulcus, 76
postganglionic neurons, 62–63
post-ictal, 243–44
postoperative delirium, 325
post-stroke dementia, 236
postsynaptic, 3
postsynaptic potential (PSP), 14, 21–22, 24
post-traumatic amnesia, 262t, 263
post-traumatic epilepsy, 249–50, 260–61
post-traumatic seizures, 260–61
postural errors, 313
postural hypotension, 63–64, 134
postural tremor, 390
posture, 102, 127
potassium ions (K+), 18f
PPA. See primary progressive aphasia (PPA)

PPMS (primary progressive MS), 294
praxicons, 312
praxis, 403–4
precentral gyrus, 152, 212, 214
precentral sulcus, 152, 212
preclinical stage of disease, 209
precuneus, 185
precursor loading, 129–30
prefrontal association cortex, 160
prefrontal cortex, 214f, 217–19
    anatomy, 217
        frontal lobe controversy, 217–18
        Gage, Phineas, 218–19
prefrontal lobotomy, 219–20
preganglionic neurons, 62–63
pre-ictal period, 243–44
premotor area, 214f, 215
premotor centers, 90
premotor networks, 384–85
preoccipital notch, 152
preoptic area (POA), 120
PRES (posterior reversible encephalopathy syndrome), 367
presbycusis, 204
preserved awareness seizures, 245–48
presynaptic, 3
pretectal nucleus, 382f
prevertebral ganglia, 63
primary acalculia, 357–58
primary auditory cortex, 204
primary brain tumors, 270–71
primary epilepsy, 249–50
primary fissure, 94–95, 95f, 96f
primary injuries, 257–59
primary lateral sclerosis, 311–12
primary motor cortex, 213–14, 214f
primary motor neurons, 3, 301
primary olfactory cortex, 206–7
primary progressive aphasia (PPA), 348–51
    logopenic variant PPA, 350–51
    nonfluent/agrammatic variant PPA, 350
    semantic variant PPA, 348–50
primary progressive MS (PPMS), 294
primary sensory areas, 159
primary sensory neurons, 3
primary somatosensory cortex, 186
primary visual cortex, 169, 173
priming, 336
primitive reflexes, 393–94
prion diseases, 284–86
    Creutzfeldt-Jakob disease (CJD), 285–86
    fatal familial insomnia (FFI), 286
    Gerstmann-Sträussler-Scheinker syndrome (GSS), 286
    kuru, 285
problem-solving, 221
Probst bundles, 165
procedural memory, 336
processing speed, 402
prodrome stage, 243–44

progressive bulbar palsy, 311
progressive multifocal leukoencephalopathy (PML), 297–98
progressive muscular atrophy, 311
progressive nonfluent aphasia, 350
progressive rubella panencephalitis (PRP), 298
progressive supranuclear palsy, 91, 93f, 132–34
projection fibers, 6, 154–55
projection images, 361–62
projection maps, 158–59
projection neurons, 3
prolactin, 117
prolactin-secreting adenomas, 274
proliferative potential, 271
propositional speech, 343
proprioception, 79, 185, 390, 391
proprioceptive ataxia, 394
proprioceptive signals, 100
proprioceptors, 185
prosencephalon, 10, 11f
prosody, 342
prosopagnosia, 178
prosopometamorphopsia, 178, 181
protein conformation, 284–85
provoked seizures, 240
proximal, 7
PRP (progressive rubella panencephalitis), 298
PSA (posterior spinal arteries), 53, 77
psalterium, 201
pseudobulbar affect, 133, 309
pseudobulbar palsy, 133, 230, 309
pseudocoma, 327
pseudo-homophones, 354
pseudoseizures, 250
pseudowords, 354
PSP (postsynaptic potential), 14, 21–22, 24
psychogenic seizures, 250
psychostimulants, 33
psychosurgery, 140, 219
pterygoid processes, 36
ptosis, 70, 310, 382
pulley nerves. See trochlear nerves (CN4)
pull-test, 127
pulmonary arteries, 50
pulmonary circulation, 50
pulmonary veins, 50
pulvinar, 112
pumps, protein, 17
punch drunk syndrome, 267
punding, 130
pupil, 172f
pupillary light reflex, 382
pupillary sphincter, 382f
pupils equal, round, reactive to light and accommodation (PERRLA), 381
pupil size, 381
pure agraphia, 356

pure alexia, 163, 180, 194, 354
pure word deafness, 163, 205
Purkinje cells, 97
Purkinje layer, 97
putamen, 124
pyramidal cells, 3, 161
pyramidal decussation, 86f, 86, 306
pyramidal tracts, 306
pyriform cortex, 200

quadrant achromatopsia, 178
quadrigeminal cistern, 39f, 39–40, 364, 368
quadriplegia, 81

rabies encephalitis, 281–82
"racing car" configuration, 165, 166f
racoon eyes, 257–58
radiation, 120
    therapy, 277
    toxic leukoencephalopathy induced by, 296–97
radiation encephalopathy, 277
radiculitis, 279
radiculopathies, 66, 187, 310
radioisotope, 374
radiologically isolated syndrome (RIS), 294
radiopaque contrast, 369
radiosensitivity, 277
Ramón y Cajal, Santiago, 2
ramus, 66
random errors, 191
raphe nuclei, 30, 90
rapid eye movement (REM) sleep behavior disorder, 128, 131–32
Rasmussen's syndrome, 251
Rayner, Rosalie, 142
rCBF (regional cerebral flow imaging), 374
reactive fear, 143
reading, 405
reasoning, 406
recent memory, 337–38
receptive amusia, 205–6
receptive aphasia, 345
receptive aprosodia, 205–6
receptive field, 170, 171f
receptor agonists, 27
receptor antagonists, 27
receptor down-regulation, 27
receptor subtypes, 26
receptor up-regulation, 27
recesses, 41
reciprocal coordination task, 215, 216f
recognition memory, 405
recommendations, neuropsychological assessment, 409–10
red nucleus, 88f, 89, 307
reduplicative paramnesia, 208
referential montages, 241
referral question, 398
reflex arcs, 80, 81f

reflex epilepsy, 251
reflexes
    brainstem, 91
    exam, 392–94
    eye movements, 385
    spinal cord, 80, 81f
reflex receptive field, 393
refractory period, 19f, 20
regional cerebral blood flow, 56
regional cerebral flow imaging (rCBF), 374
regularization errors (surface errors), 355
relapse stage, 147
relapsing-remitting MS (RRMS), 294
relative refractory period, 19f, 20
relaxation, muscle, 302
relaxation process, 370
relay neurons, 3
release hallucinations, 182
REM (rapid eye movement) sleep behavior disorder, 128, 131–32
remote memory (long-term memory), 332–33, 337–38
remote paraphasias, 344
remote symptomatic seizures, 240
repetition, 405
repolarization, 19f, 19–20
representational neglect, 192
reptilian brain, 8
resection, 252
resonate, 370
respiration, 120
responsive neurostimulation, 254
restiform body, 96–97
resting membrane potential, 18–19
resting tremor, 126, 127, 390
reticular activating system, 90
reticular formation, 89–91, 307
    autonomic functions, 90
    consciousness and sleep-wake cycle, 90
    motor function, 90
    sensory modulation, 90–91
reticular theory, 2
reticulospinal tracts, 307
reticulothalamocortical pathway, 90
retina, 169–71
    color coding, 171
    lateral geniculate nucleus from, 172–73
    pattern processing within, 170–71
    receptive field, 170
retinohypothalamic fibers, 121
retinotopic map, 173
retinotopy, 159–60
retrieval, 405
retrocollis, 132
retrograde amnesia, 329, 332–33
retropulsion, 127
retrosplenial cortex, 139–40
reuptake inhibitors, 32–33
reuptake transporter, 27
reward, 143–45

brain stimulation reward, 143–44
drug reward, 144–45
natural reward, 145
reward pathway, 29
Rexed laminae, 76
rhinal sulcus, 153
rhinencephalon, 138–39, 206–7
rhombencephalitis, 146
rhombencephalon, 10, 11f
rhomboid fossa, 41, 86, 87f
Ribot's Law, 333
right common carotid artery, 50, 51f
right-left disorientation, 191
right subclavian artery, 50, 51f
rigidity (hypertonia), 127, 389
Rinne test, 387
RIS (radiologically isolated syndrome), 294
risus sardonicus, 136
rivastigmine, 210
rods, 169
Rolandic fissure, 151
Romberg position, 102, 395
roof (skullcap), 36, 37f
roof (tectum), 41
rostral, 6, 7f
rostral-caudal axis, 6
rostral solitary nucleus, 72, 73
rostrum, 156
RRMS (relapsing-remitting MS), 294
rubrospinal tract, 89, 307
rule of 4, 91, 92t
rule violation errors, 406

saccadic eye movements, 384
sacral plexus, 66
sagittal fissure, 151
sagittal sectioning plane, 7–8, 363
sagittal suture, 37f, 37
saltatory conduction, 20, 290
sandwich sign, 296
sarcolemma, 303
sarcoplasmic reticulum, 303
sarin nerve gas, 28–29
satiety center, 118
SCA (superior cerebellar arteries), 53, 98
scanning speech, 103, 342
SCAs (spinocerebellar ataxias), 105
schizophrenia, dopamine hypothesis of, 31
Schmahmann's syndrome, 104
Schwann cells, 4, 290, 291f
Schwannomas, 4, 274–75
SCI (spinal cord injury), 81–82
sclera, 387
sclerotic plaques, 292–93
scopolamine, 29, 210
scotomas, 176
scrapie, 284
secondary acalculia, 357
secondary auditory cortex, 204
secondary epilepsy, 249–50

secondary injuries, 259–61
secondary motor cortex, 213–14, 215–17
    frontal eye fields, 216–17
        premotor area, 215
        supplementary motor area, 215
secondary progressive MS (SPMS), 294
secondary somatosensory cortex, 187
secondary tumors, 271
secondary visual cortex, 173–74
secondary zones, 160
second messenger, 26f, 26–27
sedative-hypnotics, 33
SEEG (stereoelectroencephalography), 244
segmental organization, 77–78
segmental spinal reflexes, 81
segmentation of major cerebral arteries, 54–55
seizure focus, 240
seizure prophylaxis, 252, 261
seizures, 244–46
    focal vs. generalized, 245
    motor vs. non-motor, 246
    with preserved awareness vs. impaired awareness, 245–46
    tumors and, 275–76
seizure semiology, 244
selective attention, 217, 402
selective serotonin reuptake inhibitors (SSRIs), 32–33
selectivity, ion channel, 17
self-monitoring, 221, 408
sellar diaphragm, 38, 272
sella turcica, 36, 115
semantic dementia, 348–49
semantic errors (conceptual errors), 191
semantic memory, 336, 349
semantic paralexias, 355
semantic paraphasia, 344
semantics, 341
semantic variant PPA, 348–50
sensitization, 337
sensorimotor basal ganglia circuit, 125
sensorineural deafness, 204, 387
sensory ataxia, 187, 394, 395–96
sensory modulation, 90–91
sensory relay nuclei, 159
sentinel headaches, 232–33
SEPs (somatosensory EPs), 22
septal nuclei, 145
septal region, 145
septohippocampal pathway, 145
septum pellucidum, 41, 145, 153
septum verum, 145
serotonin, 29, 30
serotonin-norepinephrine reuptake inhibitors (SNRIs), 33
set loss errors, 406
set shifting, 221, 408
sexual automatisms, 246

sexual behavior, 119–20
sexually dimorphic nucleus, 119–20
SFI (sporadic fatal insomnia), 286
shadow plaques, 293
shaken baby syndrome (shaken impact syndrome), 259
sham rage, 119
sharp waves, 243
Sherrington, Charles, 115, 213
shivering, 121
short association fibers, 156
short-term memory, 337–38
shuffling gait, 127
shunt, 46
shunt-responsive patients, 48
siderosis, 233
sigmoid (S-shaped) sinuses, 38–39, 39f
signs, 11, 158, 379
silent lacunar infarctions, 228–30
silver staining, 2
simple diffusion, 16f, 17
simultanagnosia, 193–94
single-photon emission computed tomography (SPECT), 57–58, 374
single-unit recording, 17–18
sixth nerve palsy, 71
skeletal muscle, 61–62, 302
skull base, 36
skull base meningiomas, 272
skullcap, 36, 37f
skull fracture, 257–58
skull sutures, 37
sleeping sickness, 112
sleep-related disorders, 128
sleep-wake cycle, 90
small molecule neurotransmitters, 28
small vessel disease, 236, 299–300
small vessel strokes, 228–30, 229f
smoldering plaques, 293
smooth muscle, 302
smooth pursuit eye movements, 384
Snellen chart, 381
SNRIs (serotonin-norepinephrine reuptake inhibitors), 33
social history, 399
sodium (Na+) ions, 18f
sodium-potassium pump (Na+/K+ pump), 20–21
soma, 2
somatic system, 61–62, 302
somatosensation, 62, 185–87
    exam, 390–92
    somatosensory cortex, 186–87
    somatosensory pathways, 185–86
somatosensory association cortex, 160
somatosensory cortex, 186–87
    lesions of, 187–88
somatosensory EPs (SEPs), 22
somatosensory homunculus, 187

somatosensory nuclei, 86
somatosensory radiations, 186, 187
somatostatin, 117
somatotopy, 79, 80, 158, 159–60
somatotropin, 117
Sommer's sector, 200
somnolence, 320
somnolence syndrome, 296
sonogram, 373
sound wave, 202
space-occupying lesions, 42, 363–64
span tests, 337
spasticity, 309, 389–90
spatial acalculia, 358
spatial agraphia, 356–57
spatial alexia, 192, 356
spatial attention, 403
spatial attention disorders, 191–94
spatial cognition, 402–3
spatial cognition disorders, 189–91
spatial integration, 21–22
spatial movement errors, 313
spatial orientation errors, 313
special senses, 62
special somatic afferents, 62
special visceral afferents, 62
specific relay nuclei, 111
SPECT (single-photon emission computed tomography), 57–58, 374
SPECT-CT, 374
speech, 341–42
    apraxia of speech (AOS), 342
    articulation, 341
    fluency, 341
    observations, 399
    voice, 341–42
speech automatisms, 343
speech disorders, 341
speech dysfluency, 341
Sperry, Roger, 164
sphenoid bone, 35, 36f, 36
sphenoid ridge, 36
sphenoid wing meningiomas, 272
spikes, 243
spike-wave complexes, 243
spina bifida, 106
spinal accessory nerves. See accessory nerves
spinal anesthesia, 82
spinal column, 5, 37
spinal cord, 5f, 5, 75–83, 304
    anatomy, 75–83
    epidural and spinal anesthesia, 82
    functional organization of, 78–80
    spinal cord injury syndromes, 81–82
    spinal reflexes, 80–81
spinal cord gray matter, 76f
    functional organization of, 78–79
    overview, 76
spinal cord injury (SCI), 81–82

spinal cord white matter, 76f, 79–80
　ascending pathways, 79–80
　descending pathways, 80
　functional organization of, 79–80
　overview, 76–77
　somatotopy, 80
spinal foramen, 37
spinal nerves, 5, 64–66
　clinical considerations, 66
　dermatomes, 64–66
　myotomes, 66
　peripheral nerves, 66
spinal reflexes, 80–81, 393
spinal tap. See lumbar puncture
spinal trigeminal nucleus, 71, 73, 186
spine, 5, 37
spinocerebellar ataxias (SCAs), 105
spinocerebellar tracts, 85, 100
spinocerebellum, 100
spinoreticular tract, 186
spino-spinal connections, 77
spinotectal tract, 186
spinothalamic system, 186
spinothalamic tracts, 6, 85
spiny cells, 3
spiral ganglion, 203
splenium, 139–40, 156
split-brain surgery, 163
SPMS (secondary progressive MS), 294
spongiform encephalopathy of primates, 284
spongiform leukoencephalopathy, 296
spontaneous expressive language, 343
spontaneous ictal fear, 143
spontaneous television actor participation, 223
sporadic CJD, 285
sporadic fatal insomnia (SFI), 286
sport-related concussion, 265
Spurzheim, Johann, 157
squid giant axon, 17–18
Squire, Larry, 337
ß-amyloid, 209
S-shaped sinuses, 38–39, 39f
SSRIs (selective serotonin reuptake inhibitors), 32–33
staccato speech, 103. See also scanning speech
standard anatomic position, 6
Standardized Field Sobriety Test, 105
stapes, 203
station
　exam, 395–96
　truncal ataxia, 102
status epilepticus, 249
Steele-Richardson-Olszewski syndrome, 91, 93f, 132–34
Steinthal, Heymann, 312
stellate cells, 3, 161
stenting, 58, 234

stereoblindness, 179
stereoelectroencephalography (SEEG), 244
stereognosis, 187, 391
stereotactic radiotherapy, 277
stereotactic surgery, 244
sternocleidomastoid, 73, 388
sticky visual attention, 193–94
stimulus-bound behavior, 222–23
storage, memory, 405
straight sinus, 38–39, 39f
Strangelove syndrome, 317
strategic infarct dementia, 113
strategic infarcts, 236
strategy formation, 221, 407–8
strength, motor, 389–90
stria of Gennari, 169
striate arteries, 124
striate cortex, 169
stria terminalis, 41, 140
striatonigral degeneration, 135
striatum, 124
stroke, 226–37
　diagnosis, 233
　hemorrhagic stroke, 232–33
　ischemic stroke, 226–31
　risk factors and prevention, 234–35
　treatment, 233–34
　vascular cognitive impairment (VCI), 235–37
stroke syndromes, 227–28
Stroop effect, 407
Stroop tasks, 407
structural brain imaging, 363–74
　computed tomography (CT), 368–47
　contrast enhancement, 363
　imaging planes, 363
　interpretation, 363–68
　magnetic resonance imaging (MRI), 369–50
　ultrasound, 373–50
stupor, 320
stuttering, 341
subacute combined degeneration, 299
subarachnoid cisterns, 39–40
subarachnoid hemorrhage, 43, 232–33, 234
subarachnoid hemorrhage extension, 258
subarachnoid space, 39
subcallosal area, 154
subclinical seizures, 243
subcortical aphasia, 347
subcortical dementia, 128, 291
subcortical ischemic vascular dementia, 230
subcortical white matter. See white matter
subdural evacuation, 267
subdural hematomas, 43–45, 259–60
subdural space, 43–44
subfalcine herniation, 42f, 42, 364
substantia nigra, 88f, 89, 123
subthalamic nucleus, 108, 123

subthalamotomy, 130
subthalamus, 108
sulci, 151, 154f
sundowning, 208
superficial reflex, 393
superficial veins, 55
superior, 6, 7f
superior cerebellar arteries (SCA), 53, 98
superior cerebellar peduncles, 89, 96–97
superior cerebellar veins, 98
superior cistern, 39f, 39–40
superior colliculi, 88f, 88–89, 307, 308
superior division, middle cerebral artery 52
superior frontal gyrus, 152, 212
superior frontal sulcus, 152, 212
superior hemiretina, 170
superior-inferior axis, 6
superior medullary velum, 41, 84–85
superior oblique muscle, 70
superior olivary nucleus, 204
superior orbital fissure, 70
superior parietal lobule, 152, 184–85
superior rectus muscle, 70
superior sagittal sinus, 38–39, 39f
superior salivatory nucleus, 72
superior temporal gyrus, 152, 200
superior temporal sulcus, 152, 200
supplementary motor area, 215
suprachiasmatic nucleus, 121
supragranular layers, 161
supramarginal gyrus, 152, 184–85
supranuclear components, ocular motor system, 384–85
supranuclear ophthalmoplegia, 132–33
supranuclear vertical gaze palsy, 132–33
supraoptic nuclei, 116
supraoptic region, 114
suprasellar cistern, 39f, 39–40, 364, 368
suprasellar meningiomas, 272
supratentorial compartment, 38
surface agraphia, 349, 356
surface alexia, 349, 354–55
surface errors, 355
surgery
　epilepsy, 252–54
　tumors, 276–77
susceptibility weighted imaging (SWI), 372–73
sustained attention, 402
Sydenham's chorea, 136
Sylvian cistern (insular cistern), 39–40, 368
Sylvian fissure, 151
sympathetic system, 62–63, 63f
sympathomedullary pathway, 119
symptomatic epilepsy, 249–50
symptomatic therapy, 130
symptoms, 11, 158, 379
synapse (synaptic cleft), 2–3
synaptic transmission, 24–28, 25f
　mechanisms of drug action, 27–28

neurotransmitter receptors, 25–27
neurotransmitters, 24–25
removal of neurotransmitter from synapse, 27
transmitter release, 25
syncope, 63–64, 251, 320
syndromes, 12
syntax, 341, 343
system atrophies, 12
systemic circulation, 50

T (Tesla) units, 369–70
T1-weighted MRI, 371
T2-weighted MRI, 371
tactile agnosia, 188, 391
tactile anomia (tactile aphasia), 163, 188
tactile-motor disassociation apraxia, 315
tactile sensation, 79
tandem gait, 102, 395
tangential thinking, 400
tanycytes, 55–56
tardive dyskinesia, 31–32, 126
targeted temperature management, 56–57
tau, 209
tauopathies, 131, 133, 209
TBI. *See* traumatic brain injury (TBI)
TEA (transient epileptic amnesia), 335
tectobulbar tracts, 308
tectoreticulospinal pathway, 90
tectorial membrane, 203
tectospinal tracts, 307, 308
tectum, 41
tegmentum, 41
teichopsia, 181
telegraphic speech, 343
telencephalon, 10–11, 11*f*
teleopsia, 181
temporal bones, 35, 36*f*
temporal branch, 72
temporal hemiretina, 170
temporal horn, 41
temporal integration, 21–22
temporal lobectomy, 253, 329–30
temporal lobe epilepsy, 250
temporal lobes, 151, 199–211
    Alzheimer's disease, 207–10
    anatomy, 199–201
    audition, 202–6
    olfaction, 206–7
temporal lobe seizures, 248
temporal planum, 200
temporal pole, 200
tendons, 302
tentorial meningiomas, 272
tentorial notch (tentorial incisure), 38
tentorium, 38*f*, 38, 94
terminal buttons, 2–3, 25
territorial stroke, 227–28
tertiary zones, 160
Tesla (T) units, 369–70

tessellopsia, 181
tests of conflicting instructions, 408–9
tetraplegia, 81
Teuber, Hans-Lukas, 176, 217
TGA (transient global amnesia), 334–35
thalamic amnesia, 111–12
thalamic aphasia, 112, 347
thalamic dementia, 113
thalamic pain syndrome, 111, 189
thalamic radiations, 110, 155
thalamic stroke syndromes, 112–13
thalamocortical pathways, 154–55
thalamogeniculate arteries, 53–54, 110–11
thalamotomy, 111, 130
thalamus, 10, 101*f*, 108–13
    anatomy, 108–10
    blood supply, 110–11
    thalamic nuclei and thalamic function, 111–12
    thalamic pathology, 112–13
therapeutic hypothermia, 56–57
thermogenesis, 120–21
thermoreception (thermoception), 120, 390
thermoreceptors, 120, 185
thermoregulation, 120–21
    afferent sensing, 120
    central control, 120
    efferent responses, 120–21
thermoregulatory behaviors, 121
thermosensation (thermesthesia), 79, 185
theta frequency wave, 242, 243*f*
third-nerve palsy, 70
third ventricle, 40*f*, 40–41, 109*f*
thirst, 118–19
thoracolumbar division, 63
Thorazine, 220
thought content, 400
thought form, 400
threshold of excitation, 19*f*, 19
thromboembolic stroke, 227
thrombolysis, 233–34
thunderclap headache, 232–33
thyroid hormones, 117
thyroid-stimulating hormone (TSH)–secreting adenomas, 274
TIA (transient ischemic attack), 231, 335
tic douloureux, 71
tight junctions, 55
tinnitus, 72, 206, 264
tissue plasminogen activator, 233–34
titubation, 102
toluene dementia, 295–96
toluene leukoencephalopathy, 295–96
tomography, 361–62
tonic movements, 246
tonic seizures, 249
tonotopic map, 204
tonsillar herniation, 42*f*, 42, 95
tonus, 389
topectomy, 253

topographical knowledge, 190
topographic disorientation, 190, 403
total LIS, 327
Tower tests, 407–8
toxic leukoencephalopathy, 295–97
    chemotherapy-induced, 297
    drugs of abuse and environmental toxins, 295–96
    radiation-induced, 296–97
tracers, 374*f*, 374
tractography, 290–91
tracts, 6
transcortical aphasias, 346–47
transcortical motor aphasia, 347
transcortical sensory aphasia, 347
transforaminal herniation, 42*f*, 42, 95
transient epileptic amnesia (TEA), 335
transient global amnesia (TGA), 334–35
transient ischemic attack (TIA), 231, 335
transitive movements, 313
transmissible mink encephalopathy, 284
transmissible spongiform encephalopathies (TSEs). *See* prion diseases
transorbital lobotomy, 219–20
transtentorial herniations, 42*f*, 42, 364
transverse myelitis, 293
transverse plane, 7–8, 363
transverse pontine fibers, 87–88
transverse sinuses, 38–39, 39*f*
transverse temporal gyri of Heschl, 200
transverse tubule system, 303
trapezius muscles, 73, 388
traumatic axonal injury, 259
traumatic brain injury (TBI), 256–69
    classification of injuries, 261
    defined, 256–57
    epidemiology and etiology, 257
    mild TBI, 265
    moderate-severe TBI, 263–65
    pathophysiology of, 257–61
    severity of acute TBI, 262–63
traumatic encephalopathy syndrome, 268
trazodone, 33
treatment
    Alzheimer's disease, 210
    epilepsy, 251–52
    multiple sclerosis (MS), 294
    Parkinson's disease, 129–30
    stroke, 233–34
    tumors, 276–77
    vascular cognitive impairment (VCI), 237
tremor, 390
triceps tendon, 392, 393*f*
tricyclic antidepressants, 32
trigeminal motor nucleus, 71
trigeminal nerves (CN5), 67*f*, 68*t*, 304
    overview, 71
    testing, 385–86, 387
trigeminal neuralgia, 71

trochlear nerve palsy, 71, 384
trochlear nerves (CN4), 67f, 68t, 87f, 304
    overview, 71
    testing, 383–85
true pathogens, 280
truncal apraxia, 313
truncal ataxia, 102
    gait, 102
    station, 102
truncal ideomotor apraxia, 313–14
TSEs (transmissible spongiform encephalopathies). See prion diseases
TSH (thyroid-stimulating hormone)–secreting adenomas, 274
tuberal region, hypothalamus, 114
tuber cinereum, 113–14
tuberoinfundibular pathway, 116–17
tuberothalamic artery, 110–11
Tulving, Endel, 336
tumefactive MS, 294
tumors, 4, 270–78
    cerebellum, 105
    classification, 270–75
    clinical signs and symptoms, 275–76
    diagnosis and prognosis, 276
    neurological effects, 277–78
    treatment, 276–77
turning test, 387
two-point discrimination threshold, 187
tympanic membrane, 202
typical antipsychotics, 31–32

U-fibers, 156
ultrasound, 373–50
UMNs. See upper motor neurons (UMNs)
uncal herniation, 42, 364
uncinate fits, 247
uncinate gyrus, 200
uncinate seizures, 207
unclassifiable PPA, 348
uncomplicated mTBI, 265
unconsciousness, 319
uncus, 154, 200, 247
Ungerleider, Leslie, 173–74
unilateral, 7
unilateral limb apraxia, 314
unilateral spatial neglect, 192–93
unilateral tactile anomia, 164–65
unimodal association areas, 160
unipolar neurons, 3
universal cerebellar transform, 97
unprovoked seizures, 240
unresponsive wakefulness syndrome, 324
upper motor neurons (UMNs), 78, 301, 305–9
    basal ganglia and cerebellum, 309
    descending motor systems, 308–9
    indirect cortical-brainstem-spinal cord pathways, 307–8
    lower motor neuron lesions vs., 309

motor cortex, 305–6
    pathways from cerebral cortex, 306–7
Urbach-Wiethe disease, 142–43
urinary incontinence, 47, 222
utilization behavior, 223

VA (ventral anterior nucleus), thalamus, 111
vagal nerve stimulation, 254
vagotomy, 73
vagus nerves (CN10), 67f, 68t, 304
    overview, 73
    testing, 387–88
vagus nerve stimulation, 73
vanishing white matter disease, 292
variant CJD, 285
vascular cognitive impairment (VCI), 235–37
    clinical presentation of, 235–37
    defined, 235
    neuropsychological assessment and differential diagnosis, 237
    treatment, 237
vascular dementia, 235
vascular enhancement, 363
vascular mild cognitive impairment, 235
vascular steal, 59
vasogenic cerebral edema, 367
vasospasm, 260
vegetative processes, 62
vegetative state, 324
veins
    brain venous blood outflow, 55
    defined, 50
vela, 84–85
vena cava, 50
venograms, 57
venous sinuses, 55
ventral, 6, 7f
ventral amygdalofugal pathway, 140
ventral anterior nucleus (VA), 111
ventral cochlear nucleus, 204
ventral column, 76f, 76–77
ventral corticospinal tracts, 306
ventral horn, 76f, 76
ventral intersegmental tract, 77
ventral lateral nuclei of thalamus, 97
ventral lateral nucleus (VL), 111
ventral median fissure, 76
ventral pons, 87–88
ventral posterior nucleus, thalamus, 111, 185
ventral posterolateral nucleus (VPL), thalamus, 111, 186
ventral posteromedial nucleus (VPM), thalamus, 111, 186
ventral reticulospinal tracts. See pontine reticulospinal tracts
ventral root, 64, 65f

ventral rootlets, 76
ventral stem (inferior division), middle cerebral artery, 52
ventral stream lesions, 177–79
ventral striatum, 124, 143
ventral tegmental area, 89, 143
ventricle entrapment, 364
ventricular dilation, 45
ventricular system, 40–41
ventriculitis, 279
ventriculomegaly, 47, 369
ventriculoperitoneal (VP) shunt, 46
ventriculostomy, 104–5
ventromedial motor system, 308–9
VEPs (visual EPs), 22
verbal adynamia, 347–48
verbal apraxia, 314, 342
verbal automatisms, 246
verbal memory, 332
verbal-motor disassociation apraxia, 315
verbal paraphasias, 344
verbal response, 322
vergence movements, 384
vermis, 95f, 95, 96f, 96
vertebrae, 37
vertebral arteries, 50, 51f
vertebral canal, 37
vertebral column, 5, 37
vertebral foramen, 37
vertebral segments, 37
vertebrobasilar system, 51, 52–53
vertical gaze center, 384–85
vertical gaze paresis, 132–33
vertical occipital fasciculus, 156–57
vertigo, 72
vestibular apparatus, 72, 203
vestibular ataxia, 394
vestibular branch, 203
vestibular nerve, 72
vestibular nuclei, 72, 98–99, 307
vestibular nystagmus, 103–4
vestibular schwannoma, 72, 206, 274–75, 387
vestibulocerebellum, 98–100
vestibulocochlear nerves (CN8), 67f, 68t, 203
    overview, 72
    testing, 387
vestibulocollic reflex, 308
vestibulo-ocular reflex, 99–100, 180, 193, 385
vestibulospinal tracts, 99, 307–8
VFFTC (visual fields full to confrontation), 381
vibratory sensation, 79, 390, 391
video-EEG monitoring, 244
viewer-centered frame of reference, 192
vigilance, 402
viral infections, 298

viral meningitis, 280
virulence, 280
visceral hallucinations, 189
visceral system, 61–62, 302
visceral thermoreceptors, 120
visual agnosia, 176–77
visual anomia, 164–65, 180
visual anosognosia (Anton's Syndrome), 195
visual association cortex, 160
visual confrontation naming, 344, 404
visual disconnection syndromes, 180–81
visual EPs (VEPs), 22
visual field, 171f
    defects, 174–76
    overview, 171–72
visual fields full to confrontation (VFFTC), 381
visual form discrimination, 402
visual object agnosia, 163, 177–78
visual pathways, 172–74
visual perimetry, 171
visual release hallucinations, 182
visual variant Alzheimer's disease, 196–97
visual word form area, 354
visuoconstructive apraxia, 315
visuoconstructive disability, 315
visuomotor ataxia, 179–80, 190
visuomotor disassociation apraxia, 315
visuomotor disorders, 190
visuoperceptual function, 402–3
visuospatial alexia, 192, 356
visuospatial apraxia, 315
visuospatial dementia, 195
visuospatial disorders, 190
vital signs, 115
VL (ventral lateral nucleus), thalamus, 111
vocal automatisms, 246
Vogel, Philip, 164
voice, 341–42

volitional tremor, 390
voltage-gated ion channels, 17
voltmeter, 17–18
volume loss, 366
voluntary attentional control, 222
von Frey hairs, 391
voxels, 368
VP (ventriculoperitoneal) shunt, 46
VPM (ventral posteromedial nucleus), thalamus, 111
VPL (ventral posterolateral nucleus), thalamus, 111, 186

Wada test, 252–53, 332
Wagenen, William Van, 163
wakefulness, 319, 322–23
Wallenberg syndrome, 91
Warrington, Elizabeth, 349
watershed areas, 54, 227
watershed infarcts, 230–31, 231f
water-soluble molecules, 16
Watson, John, 142
Watts, James W., 219–20
waveform, 242
WBRT (whole-brain radiation therapy), 277
Weber test, 387
Weigert method, tissue stain, 161
Weisel, Torsten, 162–63
Wernicke, Karl, 156–57, 345
Wernicke-Korsakoff syndrome, 334
Wernicke's aphasia, 345–46
Wernicke's area, 184–85, 345f, 346f, 346
Wernicke's encephalopathy, 334
whiplash, 259
white matter, 5f, 6, 65f, 289–91
white matter dementia, 291
white matter disease, 236, 289–300
    acquired metabolic leukoencephalopathy, 298–99
    autoimmune inflammatory demyelinating diseases, 292–95

    hypoxic-ischemic leukoencephalopathy, 299–300
    infectious demyelinating disorders, 297–98
    leukodystrophies, 291–92
    toxic leukoencephalopathy, 295–97
    white matter, 289–91
white matter hyperintensities, 290–91
WHO Classification of Tumors of the Central Nervous System, 271
whole-brain radiation therapy (WBRT), 277
Willis, Thomas, 123
Wilson's disease, 136
wing-beating tremor, 136
withdrawal reflex, 80–81
withdrawal stage, 147
Witzelsucht, 221–22
Wolpert, Ilja, 193
word salad, 345
work history, 399
working memory, 217, 330, 337, 402
World Health Organization (WHO) Classification of Tumors of the Central Nervous System, 271
writing, 405
written alternating sequencing tasks, 215, 216f, 408–9

X-ray imaging, 362, 368

yoked eye movements, 383

Zangwill, Oliver, 192
Z-drugs, 33
Zelig syndrome, 223
Zihl, Josef, 179
zipper hypothesis of callosal function, 156
zonal organization, cerebellum, 96
zoonopsia, 181
Z, patient, 335
zygomatic branch, 72